Taxonomy and Ecology of Climbers: Climbing Plants of India

Vivek Pandi

Taxonomy and Ecology of Climbers: Climbing Plants of India

Springer

Vivek Pandi
Manipal Centre for Natural Sciences
Manipal Academy of Higher Education
Karnataka, India

ISBN 978-981-19-8644-4 ISBN 978-981-19-8645-1 (eBook)
https://doi.org/10.1007/978-981-19-8645-1

This Springer imprint is published by the registered company Springer Nature Singapore Pte Ltd.
The registered company address is: 152 Beach Road, #21-01/04 Gateway East, Singapore 189721, Singapore

Preface

Climbers, particularly woody climbers, have emerged as an important component of mainstream ecological research in recent decades, owing to a growing consensus on their ecological importance not only in the tropics but also in subtropical and temperate ecosystems. While our understanding of climber ecology has grown tremendously, certain fundamental but critical research topics remain largely unknown. For example, the taxonomic diversity of climbers and the number of species still remain unknown. Climbers have traditionally been overlooked in many forest inventories due to a lack of standard definitions, measurement protocols, and difficulties in assigning climbing mechanisms, among other complexities. As a result, when compared to trees, the diversity of climbing plants is underrepresented. The lack of such precise data limits our understanding of their actual ecological and evolutionary significance. The lack of precise data on climber diversity is truly a global issue. Given resource constraints and a lack of transparency in data sharing, I attempted for the first time in this book to present the taxonomic diversity of climbing plants in India, one of the world's most biodiverse countries. Such research would facilitate our goal of creating a global climber database.

I give a broad overview of climbers in Chapter 1, including their types, morphological characteristics, taxonomic diversity, and positive and negative impacts on forest ecology. The second chapter conducts a systematic review of past and current climber research across multiple research themes. This chapter also provides a thorough examination of some of the focused research themes, such as phytosociology and liana-tree interaction studies. Climber ecology is covered in Chapter 3, specifically the environmental factors that influence climber diversity and distribution. In Chapter 4, I discuss the current state of climber research and its future prospects. In addition, I discuss some of the critical scientific gaps in climber research moving forward. Chapter 5 includes a comprehensive checklist of Indian climbing flora. This also provides in-depth information about each species, such as its distribution, leaf type, flower type, fruit type, and IUCN status. In Chapter 6, I discuss climber radiations in angiosperms at the family level. Based on a thorough review of available literature, the final chapter also provides an updated list of global climbing plant families.

I am confident that this book will draw more attention to these fascinating life forms and stimulate future research.

Karnataka, India
Vivek Pandi

Acknowledgments

I would like to express my sincere gratitude to Prof. N. Parthasarathy for introducing me to the fascinating field of climber (liana) ecology. I would like to show my admiration for the trailblazing scientists whose research on climber ecology and taxonomy has always served as a source of inspiration for me as I wrote this book. My heartfelt gratitude goes to Dr. M. Prithviraj for his unwavering support and encouragement throughout this book project. Prof. M. K. Janarthanam and Prof. D. Narasimhan deserve special recognition for their constant support. My sincere gratitude goes to Dr. Sudhakar Reddy, Dr. Anbarashan, and Mr. Naveen Babu for their invaluable support and outstanding contribution. Special thanks to Mr. Rakshith, Mr. Vishal Nandha Kennedy, Ms. Tanuja Bannur, Mr. Veeresh H Ari, Mr. Sachin Shet, and Mr. Pillutla Manjunaath for their excellent assistance. I would like to express my gratitude to Dr. Karuppusamy and Dr. Dhatchanamoorthy for sharing climber photographs. I thank the MAHE authorities for their help throughout this project. I appreciate the constant motivation and support from my colleagues Dr. Jessy, Dr. Chaitra, Dr. Suranjan, Dr. Navya, and Dr. Naimisha for their constant encouragement and support. For my family, I am grateful. I could not have completed this book without their untiring support. I would like to thank the editors at Springer-Nature for their support during the publication process.

I'm grateful for the amazing world of plants, particularly climbers, which never cease to amaze me.

Contents

About the Author

Vivek Pandi is a plant ecologist who has spent close to a decade studying the ecology and biodiversity of climbers and has contributed significantly to research on Indian climbers. He holds a Ph.D. in Ecology of woody climbers and has a longstanding research interest in understanding the evolution of climbing behavior in plants. He is currently studying the taxonomy, systematics, and phylogenetics of Indian climbing flora. He has published over 15 research materials on various themes of climber ecology. In addition, he has established a one-hectare permanent monitoring plot in India's Western Ghats biodiversity hotspot to understand better the dynamics of liana communities and their impact on forest structure. He is currently employed as a Professor (Assistant) of Evolutionary Biology at the Manipal Centre for Natural Sciences, Manipal Academy of Higher Education (Institute of Eminence) in India.

Climbers—A General Overview

1

Abstract

Climbers, which are essential components of tropical forests, are one of the fascinating groups of plants. Their taxonomic and ecological relevance to forest structure, diversity, and function has received considerable attention in recent decades. This chapter focuses on the general characteristics and distinguishing features of climbers and their role in the forest ecosystem. This chapter also provides an overview of the book.

Keywords

Climbers · Lianas · Tropical forest · Circumnutation · Climbing mechanism · Scramblers · Species diversity

We have always been amazed by the variety of plant growth forms. Whether they are canopy trees, intermediate shrubs, or ground-dwelling herbs, each has evolved a distinct structural and functional strategy to ensure its survival. The plants with extraordinarily long and thin stems that cling to tall trees, binding the canopy together and bearing dense foliage, are common sights in tropical forests worldwide. These plants lack mechanical architecture for self-supporting vertical growth and are classified as climbers. Have we ever wondered why and how plants climb? Why cannot climbers grow erect just like other plants? Such questions have also long piqued the interest of many biologists and naturalists, including Charles Darwin. He discussed the structure and behavior of climbing plants in his classic study, *On the movements and habits of climbing plants* in 1875. In his publication, he drew his inspiration from Gray (1858), who studied tendril movements in selected species of Cucurbitaceae. Furthermore, the peculiarities associated with climbers have been the focus of several ecological studies. For example, the mechanisms of their climbing, circumnutation, anatomical properties (Ayensu & Stern, 1964; Carlquist, 1981;

V. Pandi, *Taxonomy and Ecology of Climbers: Climbing Plants of India*,
https://doi.org/10.1007/978-981-19-8645-1_1

Davis & Evert, 1968; Dobbins, 1969, 1971; Obee, 1934; Schenck, 1892), unusual hydraulic efficiencies (Scholander, Love, & Kanwisher, 1955; Scholander, Ruud, & Leivestad, 1957), and greater ecological amplitude have been continuously studied for the past two centuries.

The climbers can either be nonwoody (vines) or woody (lianas), and they can attach to or ascend host trees using any of the following climbing techniques, viz. Twining, tendril climbing, hook climbing, root climbing, and simple scrambling (Plates 1.1, 1.2, and 1.3). The evolution of these climbing techniques is regarded as one of the key innovations in the evolutionary history of plants (Gianoli, 2004, 2015). Climbers are frequently mistaken for parasites, which derive their nutrients from the host. However, climbers germinate and root in the soil independently but lack the mechanical strength needed for vertical alignment. Therefore, to climb to the forest canopy and be exposed to the environment with plenty of light, they must rely on other external support, primarily trees. The term "structural parasites" thus frequently appears in the literature. It can be challenging to differentiate between tree and climber seedlings during the juvenile stage because many climbers begin their life as self-supporting plants. However, climbers subsequently engage in a rhythmic motion in a circular pattern known as circumnutation to locate a suitable host. The length of circumnutating stems may lead up to 2 m. Some species like *Combretum albidum* (Combretaceae) continue to grow in length along the ground, producing searcher shoots until they find a suitable host (Plate 1.4). This shoot elongation and increased internodal distance are typical characteristics of climbers often used to identify climbers in fossil records (Burnham, 2009, 2015). The presence of climbing mechanisms and a higher leaf-to-stem ratio are other significant features that differentiate climbers from other plant growth forms (Burnham, 2009, 2015).

Climbers are so diverse that they make up one-third of the globally known plant families (Pandi et al., 2022). Gentry (1991) estimated that at least one climber species existed in 131 plant families. According to recent estimates, at least one climber species exists in 163 angiosperm families or 172 Tracheophytes families (Gianoli, 2015). Despite their phylogenetic diversity, climbers repeatedly radiated only within a few taxa. Gentry (1991), for example, found that most Neotropical climbers belonged to just 25% of the plant families. Therefore, diversification among climber clades is not consistent, and, more importantly, climbing habits have evolved many times independently during evolution. Gentry proposed specialized climbing mechanisms to promote diversification among the major climbing clades. Gianoli tested the key innovation hypothesis by comparing several pairs of sister clades with and without climber representations and found that those clades with climbers diversified more than their non-climber counterparts. Nevertheless, the characteristics that led to the success of climbing habits only within a few taxa are still unknown. Some pantropical plant families, for example, Convolvulaceae, Cucurbitaceae, and Menispermaceae are only climbers.

Climbers are one of the most visible aspects of tropical forest ecosystems. Indeed, their presence and abundance frequently differentiate tropical forests from temperate ecosystems. The lesser dominance of climbers in temperate forests is associated with

Plate 1.1 A mustache-like hooks of *Hugonia mystacina*, a hook climber (**a**). Stems of the armed-scrambler *Dalbergia horrida* leaning on to the host (**b**); *Celastrus paniculatus* twining on the host tree's trunk (**c**) Picture credit: Vivek Pandi (**a** and **b**); Subbaiah Karuppusamy (**c**)

Plate 1.2 *Pothos scandens*, a root climber spreading adhesive roots over the host tree's trunk (**a**); *solena amplexicaulis* with spirally coiled tendril (**b**). Picture credit: Vivek Pandi

Plate 1.3 A scrambler liana with multiple ramets forming a cluster in the dry deciduous forests of the northern Western Ghats (**a**); A host tree colonized by multiple liana stems in the dense understory of moist deciduous forests in the northern Western Ghats (**b**). Picture credit: Vivek Pandi

Plate 1.4 *Combretum albidum*, a twining climber forming searcher shoots in the exposed understory to find a suitable host (**a**); The process of circumnutation still continuing after overtopping the tree crown (**b**). Picture credit: Vivek Pandi

the dysfunction of the vascular systems under colder temperatures. Various environmental factors and anthropogenic activities can affect the diversity and abundance of lianas (Schnitzer & Bongers, 2002). Several key abiotic factors, including total rainfall, rainfall seasonality, and soil-edaphic factors, impact liana abundance (Londré & Schnitzer, 2006; Schnitzer & Bongers, 2002). Low rainfall areas, for example, favor liana proliferation (Swaine & Grace, 2007), as liana abundance is known to correlate negatively with rainfall (Dewalt, Schnitzer, & Denslow, 2000; Schnitzer, 2005). Furthermore, the prolonged dry seasons (Restom & Nepstad, 2004) and the presence of suitable trellises can also contribute to liana abundance (Putz, 1984). Lianas are known to benefit greatly from forest disturbances, which may provide them with a selective advantage over trees (Hegarty & Caballé, 1991). However, the impact of anthropogenic disturbances on liana communities is unclear because varying degrees of disturbance produced contrasting results across studies (Laurance et al., 2001; Addo-Fordjour, Anning, Atakora, & Agyei, 2008). With increased light availability, liana abundance has been observed to increase in canopy gaps and forest edges (Laurance et al., 2001). Elevational and latitudinal gradients have also influenced liana structure and diversity (Balfour & Bond, 1993; Gentry, 1991).

The importance of lianas in contributing to total floral diversity and maintaining forest dynamics cannot be overstated (Pandi & Parthasarathy, 2015). In tropical forests, lianas can account for up to 45 percent of the woody stem density and 35 percent of the woody species diversity (Babu & Parthasarathy, 2019; DeWalt et al., 2015; Senbeta et al., 2005). They account for approximately 43 percent and 37 percent of the overall species density and richness of Indian tropical dry evergreen forests (TDEFs) (Babu & Parthasarathy, 2019). In most tropical forests, lianas maintain an average of 5–10% aboveground biomass (AGB). However, tropical forests with dense liana populations may contribute up to 30% of the AGB (DeWalt & Chave, 2004). Despite contributing less to standing biomass, lianas account for roughly 40% of leaf biomass and canopy cover (Rodríguez-Ronderos, Bohrer, Sanchez-Azofeifa, Powers, & Schnitzer, 2016). Furthermore, climbers typically have faster leaf turnover rates (Hegarty, 1991). Nonetheless, lianas' thin stems are insufficient to compensate for the tree biomass they displace (Schnitzer & Bongers, 2011; Van Der Heijden & Phillips, 2008).

Climbers, particularly lianas, are expected to play a significant role in controlling tropical forests' structure, function, and dynamics, given their diversity and abundance. Climbers are well-known for their clumped distribution, which creates a dense entanglement that can be difficult to penetrate (Plate 1.5). Furthermore, they compete aggressively with trees for above- and below-ground resources, resulting in decreased tree growth and function. In many cases, liana colonization increases host tree mortality. It is no surprise that many forest management practices aimed at timber extraction regard them as a nuisance. Lianas reproduce vegetatively by producing ramets (asexual reproduction) and can also reproduce sexually through pollination to maintain their sheer dominance and diversity in a tropical forest (Putz, 1984; Schnitzer & Bongers, 2002). Lianas have a wide range of pollination, dispersal, and phenological systems, which are critical for maintaining biological diversity (Burnham, 2002; Nabe-Nielsen, 2001). Disturbances and resource partitioning also significantly favor or maintain their diversity (Schnitzer & Bongers, 2002). Lianas exhibit thigmonastic movement, which allows them to better sense

Plate 1.5 Dense liana entanglement in the forest understory is often hard to penetrate (**a** and **b**). Photos are taken from the tropical dry evergreen forests of peninsular India. Picture credit: Vivek Pandi

their surroundings and adapt for survival and growth (Putz, 1984). Lianas (both native and exotic) are very successful in living up to their reputation as aggressive invaders, making them one of the most noxious weeds (Forseth & Innis, 2004;

Horvitz & Koop, 2001). Overall, lianas can reduce the growth rate of trees by up to 84 percent and increase their mortality rate by threefold (Van der Heijden, Phillips, & Schnitzer, 2015). They have the potential to significantly alter soil nutrients and water dynamics. For example, lianas in and across forest landscapes redistribute nutrients horizontally and reduce available soil moisture during seasonal droughts (Schnitzer et al., 2012).

Climbers' presence can also be beneficial because it connects the forest canopy, creating a pathway for arboreal animals. They also provide a valuable food source for a diverse range of faunal assemblages. The climbers perform important functions in a forest ecosystem, such as preserving overall habitat heterogeneity and species diversity (Schnitzer & Bongers, 2002). Climbers have traditionally been used for their medicinal value in treating various ailments (Parthasarathy, Pandi, & Anil, 2015). In Cote d'Ivoire, West Africa, the local communities reportedly used 114 medicinally important liana species (Tra Bi, Kouame, & Traore, 2002). Extracts derived from lianas are frequently used to treat diseases such as diabetes, anemia, respiratory disorders, malaria, dental problems, urinary tract infection (UTI), and even cancer (Parthasarathy et al., 2015; Tra Bi, Kouamé, & Traoré, 2005). Common Indian climbers such as *Capparis sepiaria*, *Cayratia pedata*, and *Hugonia mystax* are three liana species that have antiseptic, astringent, and anti-inflammatory properties. In addition to the economic services, they provide ecological services such as food and habitat to different faunal communities (Bongers, Parren, Swaine, & Traoré, 2005; Bongers, Parren, & Traoré, 2005; Martins, 2009; Schnitzer & Carson, 2001). Unlike trees, lianas do not reach their peak in flowering and fruiting, making them an important part of the animal diet during times of scarcity (Bongers, Parren, Swaine, & Traoré, 2005; Bongers, Parren, & Traoré, 2005; Heideman, 1989). Closing the canopy after tree fall also helps to stabilize the microclimate beneath for the survival of shade-tolerant species (Schnitzer & Bongers, 2002).

Despite their obvious presence, Climbers have been largely ignored in inventory studies compared to trees for various reasons (Gentry, 1991; Jacobs, 1976; Schnitzer & Bongers, 2002). However, research on lianas has gained significant momentum in the last two or three decades owing to their potential in forest functioning and dynamics (Schnitzer, Putz, Bongers, & Kroening, 2015; Tang, Kitching, & Cao, 2012). Lianas are abundant in tropical forests worldwide, but their diversity and abundance vary greatly within the tropics. This summarizes the fact that no universal trend is associated with their dynamism and is more indicative that local conditions may be the inherent driving factors for their changes (Addo-Fordjour et al., 2021; Nath, Babu, Dilshad, Dar, & Parthasarathy, 2022). Various studies on changes in liana diversity and abundance have been conducted around the world, and many have reported temporal increases in liana abundance, particularly in Neotropical forests (Ingwell, Joseph Wright, Becklund, Hubbell, & Schnitzer, 2010; Phillips et al., 2002; Schnitzer & Bongers, 2011; Tymen et al., 2016; Yorke, Schnitzer, Mascaro, Letcher, & Carson, 2013). Similarly, Hogan et al. (2017) reported that liana abundance increased in high-intensity land-used areas of north-eastern Puerto Rico, demonstrating how lianas appear to increase in disturbed conditions. Re-inventories in African tropical forests yielded contradictory results, with a decreasing and an increasing trend in liana abundance observed (Addo-Fordjour et al., 2021; Bongers, Ewango, van der Sande, & Poorter, 2020). Globally, only a few studies have considered lianas for long-term monitoring, which hinders our understanding of their adaptations to the changing environment. Similarly, climbers are frequently overlooked regarding their importance to global floral diversity (Gentry, 1991). While climber research has gained momentum in the last two decades (Da Cunha Vargas, Grombone-Guaratini, & Morellato, 2021; Schnitzer et al., 2012; Tang et al., 2012), robust data on the taxonomic diversity of climbers at the regional or global scale remain elusive. This lack of precise data impedes understanding the ecology and evolution of these fascinating life forms. Climbers' dominance and basal area are reportedly increasing worldwide, and such data are critical for understanding the underlying mechanisms and efficient planning, management, and conservation of tropical forests.

This book discusses past and present climber research, identifying potential scientific gaps that impede our understanding of climber diversity, ecology, and evolution. The ecology of climber distribution and the major environmental factors that influence climber diversity and distribution are also discussed. A dedicated chapter discusses the recommendations and potential future of climber research and opportunities. Following a detailed approach, the taxonomic estimate for climbing plants in India is presented for the first time, along with an updated list of global climbing plant families.

References

Addo-Fordjour, P., Agyei, L. A., Ofosu-Bamfo, B., Issifu, I. N., Osei, G. O., Appiah-Kubi, R., & Kroduah, P. O. (2021). Temporal dynamics of liana communities in moist semi-deciduous forest stands with different management histories in Ghana. *Forest Ecology and Management, 489*, 119042.

Addo-Fordjour, P., Anning, A. K., Atakora, E. A., & Agyei, P. S. (2008). Diversity and distribution of climbing plants in a semi-deciduous rain forest, KNUST botanic garden, Ghana. *International Journal of Botany, 4*(2), 186–195.

Ayensu, E. S., & Stern, W. L. (1964). Systematic anatomy and ontogeny of the stem in Passifloraceae. In *Systematic plant studies*. Smithsonian Institution Press.

Babu, K. N., & Parthasarathy, N. (2019). Assessment of liana diversity and carbon stock in differently disturbed tropical dry evergreen forests of southern India. *Tropical Plant Research, 6*(1), 74–89.

Balfour, D. A., & Bond, W. J. (1993). Factors limiting climber distribution and abundance in a southern African forest. *Journal of Ecology, 81*, 93–100.

Bongers, F., Ewango, C. E., van der Sande, M. T., & Poorter, L. (2020). Liana species decline in Congo basin contrasts with global patterns. *Ecology, 101*(5), e03004.

Bongers, F., Parren, M. P., & Traoré, D. (Eds.). (2005). *Forest climbing plants of West Africa: Diversity, ecology and management*. CABI.

Bongers, F. J. J. M., Parren, M. P. E., Swaine, M. D., & Traoré, D. (2005). Forest climbing plants of West Africa: Introduction. In *Forest climbing plants of West Africa: Diversity, ecology and management* (pp. 5–18). CABI.

Burnham, R. J. (2002). Dominance, diversity and distribution of lianas in Yasuní, Ecuador: Who is on top? *Journal of Tropical Ecology, 18*(6), 845–864.

Burnham, R. J. (2009). An overview of the fossil record of climbers: Bejucos, sogas, trepadoras, lianas, cipós, and vines. *Revista Brasileira de Paleontologia, 12*(2), 149–160.

Burnham, R. J. (2015). Climbing plants in the fossil record: Paleozoic to present. In *Ecology of lianas* (pp. 203–220). Wiley Blackwell.

Carlquist, S. (1981). Wood anatomy of Nepenthaceae. *Bulletin of the Torrey Botanical Club, 108*, 324–330.

Da Cunha Vargas, B., Grombone-Guaratini, M. T., & Morellato, L. P. C. (2021). Lianas research in the Neotropics: Overview, interaction with trees, and future perspectives. *Trees, 35*(2), 333–345.

Darwin, C. (1875). *The movements and habits of climbing plants*. John Murray.

Davis, J. D., & Evert, R. F. (1968). Seasonal development of the secondary phloem in Populus tremuloides. *Botanical Gazette, 129*(1), 1–8.

DeWalt, S. J., & Chave, J. (2004). Structure and biomass of four lowland Neotropical forests. *Biotropica, 36*(1), 7–19.

Dewalt, S. J., Schnitzer, S. A., Alves, L. F., Bongers, F., Burnham, R. J., Cai, Z., & Van Melis, J. (2015). Biogeographical patterns of liana abundance and diversity. In *Ecology of lianas* (pp. 131–146). Wiley.

Dewalt, S. J., Schnitzer, S. A., & Denslow, J. S. (2000). Density and diversity of lianas along a chronosequence in a central Panamanian lowland forest. *Journal of Tropical Ecology, 16*(1), 1–19.

Dobbins, D. R. (1969). Studies on the anomalous cambial activity in Doxantha unguis-cati (Bignoniaceae). I. Development of the vascular pattern. *Canadian Journal of Botany, 47*(12), 2101–2106.

Dobbins, D. R. (1971). Studies on the anomalous cambial activity in Doxantha unguis-cati (Bignoniaceae). II. A case of differential production of secondary tissues. *American Journal of Botany, 58*(8), 697–705.

Forseth, I. N., & Innis, A. F. (2004). Kudzu (Pueraria montana): History, physiology, and ecology combine to make a major ecosystem threat. *Critical Reviews in Plant Sciences, 23*(5), 401–413.

Gentry, A. H. (1991). The distribution and evolution of climbing plants. In F. E. Putz & H. A. Mooney (Eds.), *The biology of vines* (pp. 3–49). Cambridge University Press.

Gianoli, E. (2004). Evolution of a climbing habit promotes diversification in flowering plants. *Proceedings of the Royal Society B: Biological Sciences, 271*(1552), 2011–2015.

Gianoli, E. (2015). Evolutionary implications of the climbing habit in plants. In *Ecology of lianas* (pp. 239–250). Wiley Blackwell.

Gray, A. (1858). *Manual of the botany of the northern United States: Including Virginia, Kentucky, and all east of the Mississippi*. Ivison, Phinney.

Hegarty, E. E. (1991). Leaf litter production by lianas and trees in a sub-tropical Australian rain forest. *Journal of Tropical Ecology, 7*(2), 201–214.

Hegarty, E. E., & Caballé, G. (1991). Distribution and abundance of vines in forest communities. In *The biology of vines* (pp. 313–335). Cambridge University Press.

Heideman, P. D. (1989). Temporal and spatial variation in the phenology of flowering and fruiting in a tropical rainforest. *Journal of Ecology, 77*, 1059–1079.

Hogan, J. A., Mayorquin, S., Rice, K., Thompson, J., Zimmerman, J. K., & Brokaw, N. (2017). Liana dynamics reflect land-use history and hurricane response in a Puerto Rican Forest. *Journal of Tropical Ecology, 33*(2), 155–164.

Horvitz, C. C., & Koop, A. (2001). Removal of non-native vines and post-hurricane recruitment in tropical hardwood forests of Florida 1. *Biotropica, 33*(2), 268–281.

Ingwell, L. L., Joseph Wright, S., Becklund, K. K., Hubbell, S. P., & Schnitzer, S. A. (2010). The impact of lianas on 10 years of tree growth and mortality on Barro Colorado Island, Panama. *Journal of Ecology, 98*(4), 879–887.

Jacobs, M. (1976). The study of lianas. *Flora Malesiana Bulletin, 29*(1), 2610–2618.

Laurance, W. F., Pérez-Salicrup, D., Delamônica, P., Fearnside, P. M., D'Angelo, S., Jerozolinski, A., Pohal, L., & Lovejoy, T. E. (2001). Rain forest fragmentation and the structure of Amazonian liana communities. *Ecology, 82*(1), 105–116.

Londré, R. A., & Schnitzer, S. A. (2006). The distribution of lianas and their change in abundance in temperate forests over the past 45 years. *Ecology, 87*(12), 2973–2978.

Martins, M. M. (2009). Lianas as a food resource for brown howlers (Alouatta guariba) and southern muriquis (Brachyteles arachnoides) in a forest fragment. *Animal Biodiversity and Conservation, 32*(1), 51–58.

Nabe-Nielsen, J. (2001). Diversity and distribution of lianas in a neotropical rain forest, Yasuní National Park, Ecuador. *Journal of Tropical Ecology, 17*(1), 1–19.

Nath, S., Babu, K. N., Dilshad, K., Dar, A. A., & Parthasarathy, N. (2022). Impact of anthropogenic disturbances on the liana diversity and need for conservation of resource valued species-evidences from Indian tropical dry evergreen forests. *Taiwania, 67*(1), 129–139.

Obee, D. J. (1934). Noteworthy anatomical features from the wood of some indigenous Kansas shrubs and lianas. *Transactions of the Kansas Academy of Science (1903-), 37*, 83–89.

Pandi, V., Babu, N. K., Anbarashan, M., Reddy, S. C., Borgohain, J., Shynyan, K., Mathew, A. A., Rakshith, H., Joseph, J., Kennedy, V. N., & Parthasarathy, N. (2022). Taxonomic estimates of climbing plants in India: How many species are out there? *Écoscience, 29*, 1–19.

Pandi, V., & Parthasarathy, N. (2015). Diversity and carbon stock assessment of trees and lianas in tropical dry evergreen forest on the Coromandel Coast of India. *Tropical Plant Research, 2*(3), 230–239.

Parthasarathy, N., Pandi, V., & Anil, K. (2015). Liana diversity and their ecosystem services in tropical dry evergreen forest on the Coromandel Coast of India. In *Biodiversity of lianas* (pp. 161–178). Springer.

Phillips, O. L., Martínez, V. R., Arroyo, L., Baker, T. R., Killeen, T., Lewis, S. L., Malhi, Y., Mendoza, A. M., Neill, D., Vargas, P. N., Alexiades, M., Cerón, C., Fiore, A. D., Erwin, T., Jardim, A., Palacios, W., Saldias, M., & Vinceti, B. (2002). Increasing dominance of large lianas in Amazonian forests. *Nature, 418*(6899), 770–774.

Putz, F. E. (1984). The natural history of lianas on Barro Colorado Island, Panama. *Ecology, 65*(6), 1713–1724.

Restom, T. G., & Nepstad, D. C. (2004). Seedling growth dynamics of a deeply rooting liana in a secondary forest in eastern Amazonia. *Forest Ecology and Management, 190*(1), 109–118.

Rodríguez-Ronderos, M. E., Bohrer, G., Sanchez-Azofeifa, A., Powers, J. S., & Schnitzer, S. A. (2016). Contribution of lianas to plant area index and canopy structure in a Panamanian forest. *Ecology, 97*(12), 3271–3277.

Schenck, H. (1892). Contributions to the biology and anatomy of the lianas, in particular the species native to Brazil (No. 4-5). G Fisher.

Schnitzer, S. A. (2005). A mechanistic explanation for global patterns of liana abundance and distribution. *The American Naturalist, 166*(2), 262–276.

Schnitzer, S. A., & Bongers, F. (2002). The ecology of lianas and their role in forests. *Trends in Ecology & Evolution, 17*(5), 223–230.

Schnitzer, S. A., & Bongers, F. (2011). Increasing liana abundance and biomass in tropical forests: Emerging patterns and putative mechanisms. *Ecology Letters, 14*(4), 397–406.

Schnitzer, S. A., & Carson, W. P. (2001). Treefall gaps and the maintenance of species diversity in a tropical forest. *Ecology, 82*(4), 913–919.

Schnitzer, S. A., Mangan, S. A., Dalling, J. W., Baldeck, C. A., Hubbell, S. P., Ledo, A., et al. (2012). Liana abundance, diversity, and distribution on Barro Colorado Island, Panama. *PLoS One, 7*(12), e52114.

Schnitzer, S. A., Putz, F. E., Bongers, F., & Kroening, K. (2015). The past, present, and potential future of liana ecology. In *Ecology of lianas* (pp. 1–10). Wiley Blackwell.

Scholander, P., Love, W., & Kanwisher, J. (1955). The rise of sap in tall grapevines. *Plant Physiology, 30*, 93–104.

Scholander, P., Ruud, B., & Leivestad, H. (1957). The rise of sap in a tropical liana. *Plant Physiology, 32*, 1–6.

Senbeta, F., Schmitt, C., Denich, M., Demissew, S., Velk, P. L., Preisinger, H., Tadesse, W., & Teketay, D. (2005). The diversity and distribution of lianas in the Afromontane rain forests of Ethiopia. *Diversity and Distributions, 11*(5), 443–452.

Swaine, M. D., & Grace, J. (2007). Lianas may be favoured by low rainfall: Evidence from Ghana. *Plant Ecology, 192*(2), 271–276.

Tang, Y., Kitching, R. L., & Cao, M. (2012). Lianas as structural parasites: A re-evaluation. *Chinese Science Bulletin, 57*(4), 307–312.

Tra Bi, F. H., Kouame, F. N., & Traore, D. (2002). Utilisation des lianes dans deux Forêts Classées de l'Ouest de la Côte d'Ivoire. In *Chapitre dans le livre Bongers & Traoré*. CABI.

Tra Bi, F. H., Kouamé, F. N., & Traoré, D. (2005). Utilisation of climbers in two forest reserves in West Côte d'Ivoire. In *Forest climbing plants of West Africa: Diversity, ecology and management* (pp. 167–181). CABI.

Tymen, B., Réjou-Méchain, M., Dalling, J. W., Fauset, S., Feldpausch, T. R., Norden, N., Phillips, O. L., Turner, B. L., Viers, J., & Chave, J. (2016). Evidence for arrested succession in a liana-infested Amazonian Forest. *Journal of Ecology, 104*(1), 149–159.

Van Der Heijden, G. M., & Phillips, O. L. (2008). What controls liana success in Neotropical forests? *Global Ecology and Biogeography, 17*(3), 372–383.

Van der Heijden, G. M., Phillips, O. L., & Schnitzer, S. A. (2015). Impacts of lianas on forest-level carbon storage and sequestration. In *Ecology of lianas* (pp. 164–174). Wiley Blackwell.

Yorke, S. R., Schnitzer, S. A., Mascaro, J., Letcher, S. G., & Carson, W. P. (2013). Increasing liana abundance and basal area in a tropical forest: The contribution of long-distance clonal colonization. *Biotropica, 45*(3), 317–324.

2 Overview of the Research on Climbers from the Past to the Present: A Global Analysis

Abstract

Climbers were among the least studied plant growth forms until recently for a variety of reasons. However, research on climbers, particularly the ecology of woody climbers, has flourished over the past four decades. This exponential rise is linked to a growing recognition of climbers' importance to the dynamics, structure, and function of tropical forests. Since then several studies have been conducted on various research themes covering the broad spectrum of climber biology. In this chapter, I examine the trends in climber research from the past to the present and provide a thorough analysis of climber studies that have advanced our understanding of their biology.

Keywords

Phytosociology · Tropical forests · Stem anatomy · Forest management · Plant–animal interactions · Carbon stock · Ethnobotany

2.1 Introduction

Scientists have long been intrigued by climbers (such as Plumier, 1693; Description des plantes de l'Amérique), but most significant and spectacular research only emerged in the mid-to-late 19th century (for example, Gray, 1858; Darwin, 1875; Schenck, 1892). Charles Darwin explored the structure and behavior of climbing plants through experiments in his seminal work, *On the movements and habits of climbing plants* (1875). He was primarily interested in plant movements; thus, he mainly focused on twiners and tendril-bearers to study. Researchers continued to find the circumnutation process fascinating, particularly the physiology of plant movements (Bremekamp, 1912; Koningsberger, 1922; Koning, 1933). Schenck (1892) was the first to characterize the habit-related traits of climbers at the

V. Pandi, *Taxonomy and Ecology of Climbers: Climbing Plants of India*, https://doi.org/10.1007/978-981-19-8645-1_2

community level, such as the presence of long internodes and climbing processes. In addition to categorizing the climbers, viz. twiners, tendril climbers, root climbers, and scramblers, he also noted the diversity and distribution of the climber families. Schenck also studied the anomalous secondary thickening in many climber species, which served as a precursor for several other studies (Pfeiffer, 1926; Obee, 1934; Dobbins, 1969, 1971; Ayensu & Stern, 1964; Davis & Evert, 1970; Carlquist, 1981).

A greater understanding of the physiology of climbers, particularly the rise of sap in lianas, was made possible by Scholander et al. (1955, 1957). In his book *The Natural History of Palms*, corner (1966) outlined the general traits of rattans, the climbing palms. Hegnauer (1975) conducted pioneer research on the secondary metabolites of specific climbers, demonstrating the promising potential of alkaloids derived from climbers for treating various illnesses (e.g., the anticancer property of *Cocculus* sp.). Until the late twentieth century, many, perhaps most, sparing few studies focused only on climbers' anatomical and physiological characteristics, leaving a substantial knowledge gap regarding their ecology. Few studies—such as those by Lebrun (1937), Steenis (1939), Fox (1968, 1969), and Janzen (1969, 1974)—included climbers in the mainstream ecological studies. The ecology of lianas remains essentially unexplored, according to Jacobs (1976). According to Gentry (1991), lianas are probably the most under-collected of any major habit group of plants. It is about the pioneering works of Gentry (1974, 1985, 1991, 1992, 1995) and Putz (1980, 1984a, b) and the seminal book on the biology of vines edited by Putz and Mooney (1991) that climbers, which were previously overlooked, have garnered attention. Climber research has gained significant traction since then. For example, Schnitzer et al. (2015) found in a systematic review that the publication of lianas (ecology) increased 22-fold between 1980 and 2009.

2.2 Literature Survey and Classification of Research Themes

I referred to 1277 research materials in this systematic review, including books, technical reports, dissertations, research papers, reviews, and book chapters on climbers. I searched for information using the terms climber, liana, vines, and scandent plant in several databases, including Web of Science, liana ecology project, JSTOR, and Google Scholar, without placing any restrictions on the time frame (through December 2021) or the location of the study. Based on the theme of the research, I divided the research materials into 12 major and several minor categories: general (climber-specific articles, including reviews, ecological theories, and concepts, protocols), habit-morphology-climbing behavior and growth form, phyto-sociology (including diversity and distribution, long-term monitoring plots), bio-mass and carbon, anatomy (including biomechanics), physiology (ecophysiology), functional traits (morphology, anatomy, physiology, and biochemistry), liana–tree interactions (association and competition), lianas–carbon–forest management (including invasive characteristics), positive impacts (Ethnobotany, Phytochemistry) phylogeny, and fossil studies. The number of research publications under each broad and sub-category is given in Table 2.1.

Table 2.1 Theme-based categorization (sub-categories included) of climber research publications from the past to the present ($n = 1277$)

Theme of the research	Number of publications
Phytosociology	207
Forest management and Conservation	120
Systematics, Taxonomy, and Floristics	92
Liana–tree interaction	83
Ecophysiology	77
Plant–animal interactions	71
Anatomy	68
Functional traits	65
Invasiveness	46
Habit-morphology and growth form	37
Reproductive biology	35
Biomass and carbon stock	34
Growth and regeneration	30
Ethnobotany	29
Long-term monitoring	27
Genetics and Molecular studies	25
Climbing behavior	18
Technology/application	18
Evolution and phylogeny	17
Biomechanics	15
Phytochemistry	14
Fossil studies	8
General themes/protocol studies/Books	141

2.3 Geographical Distribution of Research on Climbers

Historically, Neotropical regions have dominated climber research (45%), followed by the Paleotropics (27%) (Fig. 2.1). More than 70% of climber research has been conducted in the tropics. Extra-tropical studies on climbers accounted for 4.6% of all research published on climbers. Australia and Oceania accounted for 5% of total climber research output, while intercontinental research involving multiple study sites accounted for 2%. Most climbing studies at the continental level were conducted in South America, followed by Asia and North America (Fig. 2.2). The top 10 countries with the most output of climbing research are included in Table 2.2.

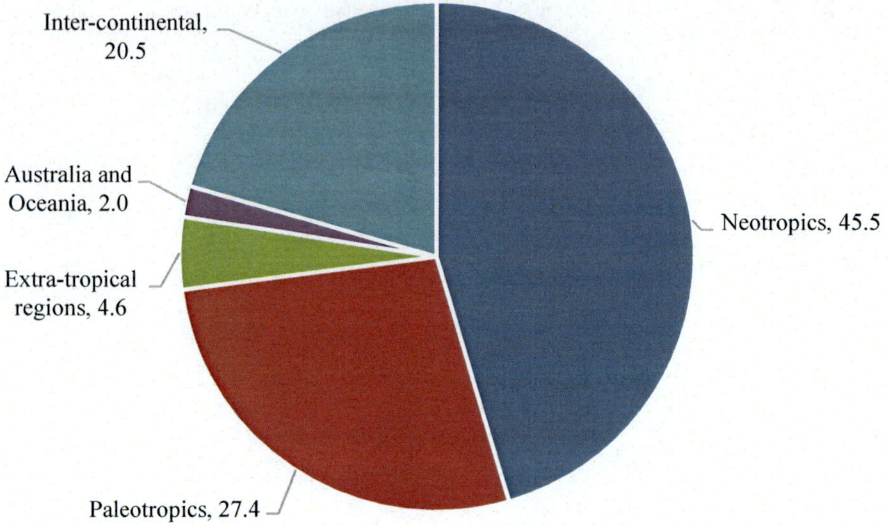

Fig. 2.1 Geographical distribution of climber research—the proportion of climber research output from five broad geographical regions ($n = 1277$)

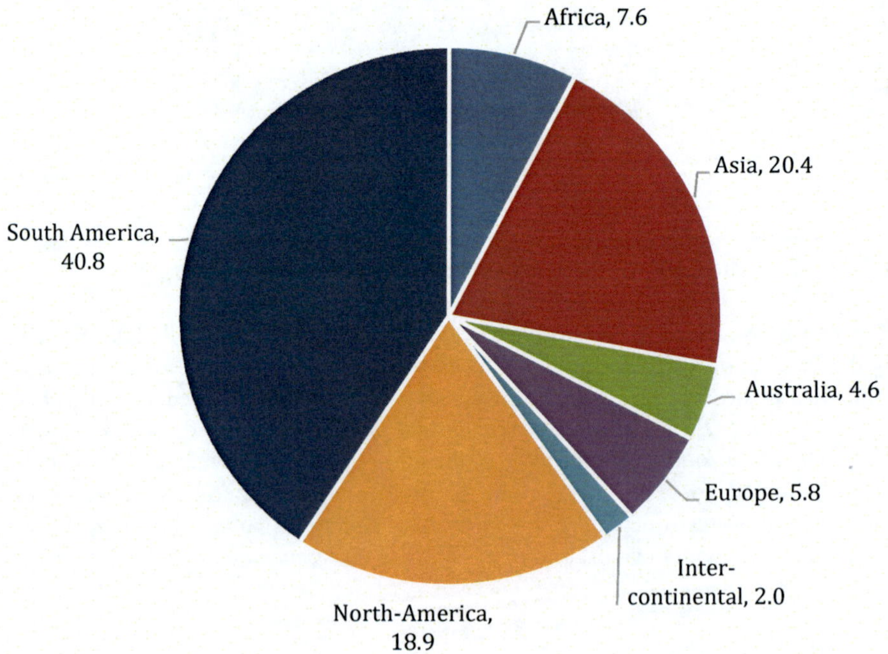

Fig. 2.2 Breakdown of climber research output by percentage on a continental scale ($n = 1277$)

Table 2.2 The top ten countries with the most climber research output across all research themes

Country	Number of studies
Brazil	182
United states of America	116
Panama	101
India	82
China	53
Mexico	47
Malaysia	33
Costa Rica	30
Australia	22
Ghana	17
Bolivia	17
Ecuador	17
Chile	17

2.4 Phytosociology

Phytosociology refers to the functional characterization of the vegetation derived from the floristic composition of a given area (Dengler et al., 2008). Most of the earlier studies included only the trees in the phytosociological studies. Unlike trees or shrubs, climbers were rarely treated as a separate group. Penfound and Howard (1940) were probably the first to have included vines in their inventory carried out in an Evergreen Oak forest of New Orleans. They reported ten climber species in the summary table, which were neither presented nor discussed in the text. Later, Duncan (1967) and Fox (1969) studied the climbers exclusively from the Southeastern states, the USA, and the lowland Dipterocarp forest of Malaysia. While the former described the floristic richness and taxonomic keys to the climbers, the latter studied the diversity and density of climbers in ten 1-acre plots. Fox (1969) also designed a climber-cutting experiment to reduce the damage caused by climbers on mid-sized trees. Gentry (1982) included woody climbers in his attempt to understand the patterns of species diversity in the Neotropics. However, it is Putz (1984a) carried out one of the pioneering studies on estimating the diversity, abundance, and climbing mechanisms of lianas in the tropical moist forest stands of Barro Colorado Island, Panama. Since then, climbers have been noted in numerous ecological inventories (Table 2.3).

Climbers with wider xylem vessels are prone to cold-induced vascular dysfunction (freeze-based embolism) that limits climber diversity and abundance to a greater extent in temperate ecosystems, whereas tropical climbers have benefited from it. But the prevalence of temperate climbers, such as *Vitis* sp., can be exceedingly high or comparable to that of any important tropical climber species (Robertson et al., 1978). Schnitzler (1995) studied the diversity and distribution of lianas in the gallery forests of Rhine valley, France; one of the few earlier studies focused on the community assemblage of temperate lianas. Barring others, only a few studies have

Table 2.3 A comprehensive list of liana quantitative ecological inventories carried out across the world

Forest type	Site location	Area inventoried	DBH threshold	Total number of			Dominant species	Dominant family	References
				Individuals	Species	Family			
Subtropical broadleaved forest	Jomotshangkha Wildlife Sanctuary, Nepal	0.3 ha (30 nos. 10 × 10 m)	≥2 cm DBH	–	38	20	*Mikania micrantha, Piper mipigua, Paederia* sp.	Asteraceae; Piperaceae; Vitaceae	Nepal and Dorji (2020)
Tropical Montane Evergreen Forest	Western Ghats, India	1 ha (10 nos. of 10 × 100 m)	≥1 cm DBH	202	22	14	*Tetrastigma leucostaphylum*	–	Mohandass et al. (2016)
Tropical Forest	Western Ghats, India	5 ha (500 nos. of 10 × 10 m)	≥1 cm DBH	–	81	30	*Jasminum flexile; Salacia oblonga; Ziziphus oenopolia*	Apocynaceae, Menispermaceae; Leguminosae	Pandian and Ravichandran (2019)
lowland dipterocarp, hill dipterocarp, and lower montane forests	Royal Belum State Park, Perak, Malaysia	1 ha (25 nos. of 20 × 20 m)	≥1 cm DBH	–	92	23	–	Annonaceae, Connaraceae, Fabaceae	Rahmad and Akomolafe (2018)
Tropical Montane Evergreen Forest	Western Ghats, India	13.86 ha (154 nos. 30 × 30 m)	≥1 cm DBH	1980	45	21	–	–	Mohandass et al. (2017)
Dry Afromontane-mixed broadleaf-coniferous forests	Afromontane forest, Ethiopia	2.8 ha (72 nos. 20 × 20 m)	≥1 cm DBH	3912	24	15	*Carissa spinarum, Rubus steudneri, Rosa abyssinica*	–	Birhane et al. (2020)
Secondary rain forest	Biological Gardens of the ObafemiAwolowo University, Ile-Ife, Nigeria	1.5 ha (6 nos. of 50 × 50 m)	≥1 cm DBH	–	41	19	*Acacia ataxacantha, Baissea campanulata, Chasmanthera dependens*	Celastraceae	Uwalaka et al. (2020)

Humid Tropical Forests	Yangambi Man and Biosphere Reserve, Democratic Republic of Congo	12 ha	≥2 cm DBH	2368	105	22	–	Celastraceae, Fabaceae, Apocynaceae	Mumbanza et al. (2020)
Tropical rain forest	The Azagny National Park, Côte d'Ivoire	14 ha (280 nos. of 20 × 25 m)	≥1 cm DBH	5346	63	28	*Adenia lobata, Tiliacora dinklagei; Strophanthus gratus*	Fabaceae, Apocynaceae, Icacinaceae	Koffi et al. (2016)
Primary and secondary tropical forest	Penang National park, Malaysia	4.8 ha (30 nos. of 40 × 40 m)	≥2 cm DBH	–	46	15	*Artabotrys oblongus; Strychnos curtisii*	Fabaceae; Apocynaceae; Loganiaceae	Addo-Fordjour et al. (2016)
Mixed rain forest	Congo	20 ha (2 nos. of 200 × 500 m)	≥2 cm DBH	15008	195	34	*Manniophyton fulvum; Rourea thomsonii; Dichapetalum staudfii*	Connaraceae; Euphorbiaceae; Dichapetalaceae	Ewango et al. (2015)
Tropical lowland forest	Bukit panchor forest reserve, Malaysia	4.8 ha (30 nos. of 40 × 40 m)	≥2 cm DBH	–	45	15	*Uncaria screlorophylla; Artabotrys crassifolius; Bauhinia bidentate*	–	Addo-Fordjour et al. (2014)
Tropical montane forest	Nilgiri biosphere reserve, India	1.7 ha (19 nos. of 30 × 30 m)	≥1 cm DBH	1276	15	10	–	–	Mohandass et al. (2014)
Tropical dry evergreen forest	Coromandel Coast, India	8 ha (40 nos. of 0.2 ha)	≥1 cm DBH	5689	77	32	*Strychnos lenticellata; Combretum albidum; Lantana camara*	Combretaceae; Asclepiadaceae; Vitaceae	Anbarashan and Parthasarathy (2013)
Tropical hill forest	Penang hill forest reserve, Malaysia	4.8 ha (30 nos. of 40 × 40 m)	≥2 cm DBH	524	68	22	*Coptosapelta parviflora; Tetracera macrophylla*	Annonaceae; Fabaceae; Apocynaceae	Addo-Fordjour et al. (2013)

(continued)

Table 2.3 (continued)

Forest type	Site location	Area inventoried	DBH threshold	Total number of Individuals	Species	Family	Dominant species	Dominant family	References
Tropical forest	Southern scrap forest reserve, Ghana	0.9 ha (18 nos. of 25 × 20)	≥2 cm DBH	365	57	19	*Salacia elegans; Calycobolus africanus; Landolphia dulcis*	Apocynaceae; Fabaceae; Celastraceae	Addo-Fordjour et al. (2013)
Lower and upper montane forest	Khangchendzonga biosphere reserve, Eastern Himalayas	3 ha (30 nos. of 50 × 20 m)	≥0.2 cm DBH	194	43	28	*Hedra nepalensis; Holboellia latifolia; Actinidia callosa*	Vitaceae; Caprifoliaceae; Ranunculaceae	Chhetri and Bhattarai (2013)
Tropical rain forest	Penang National park, Malaysia	2.4 ha (15 nos. of 40 × 40 m)	≥2 cm DBH	454	46	15	*Artabotrys oblongus; Willughbeia angustifolia; Strychnos curtisii*	Fabaceae; Annonaceae; Apocynaceae	Addo-Fordjour et al. (2012)
Tropical rain forest	Atewa range forest reserve, Ghana	2.5 ha (10 nos. of 0.25 ha)	≥1 cm DBH	429	40	17	*Acacia kamerunsis; Acacia pentagona; Combretum mucronatum*	Fabaceae; Apocynaceae; Rubiaceae	Addo-Fordjour et al. (2012)
Tropical lowland forest	Pasoh forest reserve, Malaysia	2.5 ha (16 nos. of 40 × 40 m)	≥1 cm DBH	1628	167	31	*Byttneria maingayi; Combretum nigrescens; Caesalpinia parviflora*	Annonaceae; Connaraceae; Fabaceae	Nurfazliza et al. (2012)
Lowland tropical forest	Barro Colorado Island, Panama	50 ha (20,000 nos. of 5 × 5 m)	≥1 cm DBH	47183	162	36	*Coccoloba excelsa; Hiraea reclinata; Maripa panamensis*	Bignoniaceae; Sapindaceae; Fabaceae	Schnitzer et al. (2012)

Subtropical semi-deciduous forest	North Argentina	6 ha	≥1 cm DBH	–	51	–	*Chamissoa altisima; Acacia velutina; Adenocalymna marginatum*	–	Campanello et al. (2012)
Tropical rain forest	Yasuni national park, Ecuador	2.5 ha (25 nos. of 50 × 20 m)	≥2.5 cm DBH	751	180	–	*Machaerium cuspidatum; Combretum laxum; Uncaria guianensis*	Fabaceae; Bignoniaceae; Malpighiaceae	Macía (2011)
Tropical lowland forest	Sumaco biosphere reserve, Ecuador	1.6 ha (40 nos. of 20 × 20 m)	≥1 cm DBH	2243	–	–	–	–	Homeier et al. (2010)
Tropical forest	Eastern Ghats, India	55 ha (110 nos. of 5 × 1000 m)	≥1.5 cm DBH	32033	143	37	*Lantana camara; Pterolobium hexapetalum; Ziziphus oenoplia*	Verbenaceae; Rhamnaceae; Mimosaceae	Muthumperumal and Parthasarathy (2010)
Lowland Dipterocarp forest	Imbak canyon, Malaysia	0.2 ha (2 × 0.1 ha)	All lianas	142	23	12	*Spathobolus* sp1; *Spathobolus* sp2; *Bauhinia* sp.	Fabaceae; Annonaceae; Icacinaceae	Kammesheidt et al. (2009)
Seasonal rain forest	Xishuangbanna, China	3 ha	≥2 cm DBH	458	95	32	*Strychnos cathayensis; Byttneria grandifolia; Bousigonia mekongensis*	Loganiaceae; Annonaceae; Papilionaceae	Lü et al. (2009)
Subtropical forest	Xujiaba, China	3.2 ha (8 × 0.4 ha)	≥0.5 DBH	1610	33	17	*Parthenocissus himalayana; Rosa longicuspis; Rubus panioculatus*	Rosaceae; Smilacaceae	Yuan et al. (2009)
Tropical forest	Xishuangbanna, China	1.5 ha (15 nos. of 20 × 50 m)	≥0.2 cm DBH	1022	147	48	*Celastrus monospermus; Craspedolobium schochii*	Fabaceae; Apocynaceae	Cai et al. (2009)

(continued)

Table 2.3 (continued)

Forest type	Site location	Area inventoried	DBH threshold	Total number of Individuals	Species	Family	Dominant species	Dominant family	References
Tropical lowland forest	Hainan Island, China	3 ha (300 nos. of 10 × 10 m)	≥1 cm DBH	1594	70	38	*Mussaenda crosa; Bowringia callicarpa; Byttneria aspera*	Rubiaceae; Papilionaceae	Ding and Zang (2009)
Moist semi-deciduous forest	Bobiri forest reserve, Ghana	1 ha (4 nos. of 50 × 50 m)	≥2 cm DBH	640	27	13	*Griffonia simplicifolia; Montandra guineensis; Calycobolus africanus*	Apocynaceae; Fabaceae	Addo-Fordjour et al. (2009)
Semi-deciduous	Tinte Bepo forest reserve	1.5 ha (6 nos. of 50 × 50 m)	≥2 cm DBH	380	27	12	*Montana guineensis; Griffonia simplicifolia*	Apocynaceae; Fabaceae	Addo-Fordjour et al. (2009)
Semi-deciduous rain forest	KNUST botanical garden, Ghana	2 ha (16 nos. of 25 × 25 m)	≥1.5 cm DBH	951	72	26	*Strophantus bateri; Griffonia simplicifolia*	Apocynaceae; Fabaceae; Liliaceae	Addo-Fordjour et al. (2008)
Subtropical forest	Argentina	0.68 ha (17 nos. of 20 × 20 m)	≥1 cm DBH	841	47	19	*Adenocalymna marginatum; Acacia velutina; Arrabidaea mutabilis*	Bignoniaceae; Fabaceae	Campanello et al. (2007)
Montane forest	Argentina	6 ha	≥2 cm DBH	2346	12	11	–	–	Malizia and Grau (2006)
Tropical evergreen forest	Valparai, Western Ghats, India	4 ha (5 × 0.8 ha)	≥1 cm DBH	2250	60	35	*Connarus sclerocarpus; Calamus gamblei; Strychnos vanprukii*	–	Muthuramkumar et al. (2006)

Tropical dry evergreen forest	Pudukottai, Tamil Nadu, India	4 ha (400 nos. of 10 × 10 m)	≥1 cm DBH	2678	35	22	*Combretum albidum; Derris scandens; Strychnos lenticellata*	Combretaceae; Papilionaceae; Loganiaceae	Reddy and Parthasarathy (2006)
Tropical rain forest	Sapilok forest reserve, Malaysia	2.25 ha (9 nos. of 50 × 50 m)	≥0.5 cm DBH	3026	107	32	*Spathobolous oblongifolius; Tetracera korthalsii*	Fabaceae; Annonaceae; Apocynaceae	DeWalt et al. (2006)
Tropical montane forest	Ethiopia	8.12 ha (203 nos. of 20 × 20 m)	≥2 cm DBH	30917	123	40	*Landolphia buchananii; Paullinia pinnata; Hippocratea Africana*	Asclepiadaceae; Fabaceae; Annonaceae	Senbeta et al. (2005)
Tropical forest (secondary)	Nigeria	0.25 ha (4 nos. of 25 × 25 m)	≥3 cm DBH	448–1152	49	28	*Acacia ataxacantha; Adenia lobata; Dioscorea preussii*	Ampelidaceae; Apocynaceae; Connaraceae	Muoghalu and Okeesan (2005)
Tropical wet forest	Costa Rica	0.78 ha (9 nos. of 24 × 36 m)	≥0.2 cm DBH	1161	60	28	*Moutabea aculeata; Davilla nitida; Paragonia pyramidata*	Bignoniaceae; Dilleniaceae; Polygonaceae	Mascaro et al. (2004)
Subtropical wet forest	Puerto Rica	0.8 ha (20 nos. of 20 × 20 m)	≥1 cm DBH	419	6	6	*Marcgravia* spp.; *Rourea surinamensis; Schlegelia brachyantha*	Macgraviaceae; Connaraceae	Rice et al. (2004)
Tropical rain forest	Yasuni national park, Ecuador	2.4 ha	≥1 cm DBH	4348	311	–	*Machaerium cuspidatum; Paragonia pyramidata; Sciadotenia toxifera*	–	Burnham (2004)

(continued)

Table 2.3 (continued)

Forest type	Site location	Area inventoried	DBH threshold	Total number of Individuals	Species	Family	Dominant species	Dominant family	References
Tropical dry evergreen forest	Cuddalore, Tamil Nadu, India	4 ha (400 nos. of 10 × 10 m)	≥1 cm DBH	3307	39	24	*Strychnos lenticellata; Combretum albidum; Derris ovalifolia*	Combretaceae; Papilionaceae; Loganiaceae	Reddy and Parthasarathy (2003)
Tropical rain forest	Lacandon, Mexico	1.2 ha (4 × 0.3 ha)	≥1 cm DBH	2510	90	34	*Combretum argentinum; Hiraea fagifolia; Machaerium floribundum*	Bignoniaceae; Sapindaceae; Fabaceae	Ibarra-Manríquez and Martínez-Ramos (2002)
Tropical rain forest	Yasuni national park, Ecuador	0.4 ha (2 nos. of 20 × 100 m)	≥1 cm DBH	606	138	43	*Machaerium cuspidatum*	Fabaceae; Celastraceae; Sapindaceae	Nabe-Nielsen (2001)
Tropical evergreen forest	Eastern Ghats, India	8 ha (4 nos. of 100 × 200 m)	≥1.6 cm DBH	384	26	18	*Hiptage benghalensis; Elaeagnus indica; Gnetum ula*	–	Chittibabu and Parthasarathy (2001)
Semi-deciduous forest	Bolivia	2.16 ha (24 nos. of 30 × 30 m)	≥2 cm DBH	2471	51	–	–	–	Pérez-Salicrup et al. (2001)
Tropical moist forest	Barro Colorado Island, Panama	0.04 ha (8 nos. of 5 × 10 m)	≥0.5 cm DBH	1896	89	22	*Maripa panamensis; Prionostema aspera; Petrea aspera*	Bignoniaceae; Sapindaceae	DeWalt et al. (2000)
Tropical evergreen forest	Varagalaiar, Western Ghats, India	30 ha (3000 nos. of 10 × 10 m)	≥1 cm DBH	11200	75	37	*Chilocarpus atrovirens; Olax scandens; Piper nigrum*	Piperaceae; Olacaceae; Apocynaceae	Muthuramkumar and Parthasarathy (2000)

Tropical rain forest	Albertine rift, Uganda	6 ha (26 nos. of 20 × 500 m)	≥1 cm DBH	2783	35–62	41	*Efulensia montana; Landolphia foretiana; Loeseneriella apiculate*	Celastraceae; Apocynaceae; Rubiaceae	Eilu (2000)
Tropical semi-evergreen forest	Eastern Ghats, India	2 ha (200 nos. of 10 × 10 m)	≥1.6 cm DBH	232	–	–	*Ventilago madraspatana; Aganosma cymosa; Sageretia filiformis*	Rhamnaceae; Asclepiadaceae; Verbenaceae	Kadavul and Parthasarathy (1999)
Old-growth evergreen forest	Brazil	0.56 ha (2 nos. of 2 × 1400 m)	≥1 cm DBH	1872	63	24	*Memora schemburgkii; Memora flavida; Connarus* sp.	–	Vidal et al. (1997)
Rain forest	Lower Hunter valley, Australia	1 ha	≥1 cm DBH	520	27	17	–	–	Chalmers (1994)
Lowland Dipterocarp forest	Danum valley, Borneo	8 ha	≥0.6 cm DBH	7054	40	17	–	–	Campbell and Newbery (1993)
Evergreen forest	Australia		≥0.1 m tall	5771	42	–	–	–	Hegarty and Caballé (1991)
Tropical rain forest	Peru and Venezuela	7 ha	≥10 cm DBH	14–26	10-017	–	–	–	Gentry (1988)
Dipterocarp forest	Lambir national park, Malaysia	1 ha	≥1 cm DBH	765	–	–	*Spathobolus multiflorus; Erycine* sp.; *Lodes* sp.	Fabaceae; Icacinaceae	Putz and Chai (1987)
Dipterocarp forest	Sungei tekam forest reserve, Malaysia	1 ha	≥2 cm DBH	376	–	–	–	–	Appanah and Putz (1984)

(continued)

Table 2.3 (continued)

Forest type	Site location	Area inventoried	DBH threshold	Total number of			Dominant species	Dominant family	References
				Individuals	Species	Family			
Semi-deciduous forest	Barro Colorado Island, Panama	1 ha (10 nos. of 0.1 ha)	≥1 cm DBH	3165	65	–	*Hiraea reclinata; Paullinia turbacensis; Petrea aspera*	–	Putz (1984a, b)
Dipterocarp forest	Gunung Mulu National park, Malaysia	2 ha	≥1 cm DBH	2400	–	–	–	–	Proctor et al. (1983)

DBH is the diameter at the breast height or 1.37 m

been conducted on the diversity and distribution of climbers in temperate ecosystems owing to their poor species diversity (e.g., Quigley & Plantt, 1996; Jiménez-Castillo et al., 2007; Carrasco-Urra & Gianoli, 2009; Gianoli et al., 2010; Valladares et al., 2011; Blick & Burns, 2011; Hu, 2011; Schnitzler et al., 2016; Rossell & Eggleston, 2017). The increased interest in lianas paralleled the number of methodologies adopted in different studies. For instance, there were considerable differences in the plot dimension, dbh threshold, and life-form consideration, which made it difficult to compare the liana assemblage between the study locations. To avoid these inconsistencies, Gerwing et al. (2006) and Schnitzer et al. (2008) developed standardized liana censusing techniques.

The diversity and abundance of lianas vary significantly across different geographical regions. These variations have been thoroughly investigated at both the continental (Gentry, 1993) and global scales (Schnitzer et al., 2015, Ecology of lianas). Most earlier studies provided only descriptive and quantitative measurements of the liana abundance across different sites. Subsequently, the later studies addressed various factors including Latitude (Molina-Freaner et al., 2004; Weiser et al., 2007; Lobos-Catalán & Jiménez-Castillo, 2019) rainfall (Longino, 1986; Clinebell et al., 1995; Molina-Freaner et al., 2004; Swaine & Grace, 2007; DeWalt et al., 2010; Guo et al., 2012; Durigon et al., 2013; Lobos-Catalán & Jiménez-Castillo, 2019), temperature (Lobos-Catalán & Jiménez-Castillo, 2019), seasonality (Van der Heijden & Phillips, 2009; DeWalt et al., 2010; Durigon et al., 2013; Carvalho et al., 2016), altitude (Balfour & Bond, 1993; Lieberman et al., 1996; Vazquez & Givnish, 1998; Padaki & Parthasarathy, 2000; Jiménez-Castillo et al., 2007; Chettri et al., 2010; Alves et al., 2012; Addo-Fordjour et al., 2013; Addo-Fordjour et al., 2014; Rossell & Eggleston, 2017; Bruy et al., 2017), edaphic factors (Balfour & Bond, 1993; Collins & Wein, 1993; Clinebell et al., 1995; Ibarra-Manríquez & Martínez-Ramos, 2002; Malizia et al., 2010; Nurfazliza et al., 2012; Addo-Fordjour et al., 2013; Addo-Fordjour et al., 2014; Rodríguez-Quintero & Gianoli, 2016; Yang et al., 2018), trellis availability (Balfour & Bond, 1993; Allen et al., 1997; Baars et al., 1998; DeWalt et al., 2000; Ibarra-Manríquez & Martínez-Ramos, 2002; Malizia et al., 2010), and light availability (Gianoli et al., 2010; Valladares et al., 2011) that are likely to control the diversity, distribution, growth, and abundance of lianas. The effect of forest disturbance can also explain the patterns of liana community assemblage (Addo-Fordjour et al., 2009; Malizia et al., 2010; Nesheim et al., 2010; Addo-Fordjour et al., 2012; Mohandass et al., 2015; Mori et al., 2016; Mohandass et al., 2017), particularly the effects of fragmentation and edge creation (Addo-Fordjour & Owusu-Boadi, 2016; Campbell et al., 2018; Ofosu-Bamfo et al., 2019). The post-fire/hurricane effects on liana communities are broadly reported in few studies (e.g., Allen et al., 2005; Hogan et al., 2017; Uwalaka & Muoghalu, 2017).

The seminal study of Phillips et al. (2002) presented crucial evidence of increasing liana abundance and biomass in tropical forests. The liana dominance hypothesis gained significant momentum as numerous studies supported the findings of Phillips et al. (2002) (e.g., Allen et al., 2005; Londré & Schnitzer, 2006; Ingwell et al., 2010; Laurance et al., 2014; Khadanga et al., 2015; Parthasarathy, 2015; Jones et al., 2017;

Pandian et al., 2017). This increase in liana abundance and biomass can significantly impact the tropical forest structure and function (Schnitzer & Bongers, 2011). Schnitzer and Bongers (2011) also discussed the possible putative mechanisms that explain the dominance of lianas. Contradictory results from a few research refuted the liana dominance hypothesis (Caballé & Martin, 2001; Ewango, 2010). The findings of Ceballos and Malizia (2017), which showed a decrease in liana density but an increase in the basal area in a subtropical montane forest, partially supported the idea. More long-term monitoring investigations must therefore determine the universality of the liana dominance hypothesis.

2.5 Tree–Liana Interaction

Lianas, the structural parasites, mainly rely on trees for their vertical support. Climbers that encountered suitable supports fared better than free-standing climbers. The patterns of resource allocation and the morphology of vines can be influenced by the presence of the host (Den Dubbelden & Oosterbeek, 1995). Finding a suitable host is therefore crucial to liana's life history. However, this intricate relationship has been shown to be harmful to host trees in several ways (Plate 2.1). For instance, liana colonization reduces host tree growth, recruitment, and fertility and raises mortality rates (Putz, 1978, 1984a, b) (Plates 2.2, 2.3, 2.4). Featherly (1941) was probably the first to report the negative impact of climbers on the host trees. Using dendrochronological studies, he proved the impact of grapevines on the growth and mortality of host trees. Lutz (1943) investigated the mechanical damage climbers inflict while twisting or twining around the host trees' branches and stems. Liana infestations can facilitate folivore and arboreal animals' access to tree crowns (Montgomery & Sunquist, 1975; Charles-Dominique et al., 1981). Numerous studies have looked at liana–tree interactions (Table 2.4) and their networks (Sfair et al., 2010, 2016), but they have primarily focused on comparing trees with lianas to trees without them. The variation in infestation rates among different host species was an exciting feature.

How do other species fend off lianas, and why do certain species host more of them than others? Do trees have special defense mechanisms to ward against or shed lianas? Several theories have been put forth to explain these liana–tree interaction patterns. For instance, host trees with buttressed stems may efficiently keep lianas away from the trunk, according to Black and Harper's (1979) theory. However, Boom and Mori (1982) disproved the buttress–liana theory since they could not identify a connection between the exclusion of lianas and the presence or absence of buttresses in the host trees. According to Putz (1984a, b), climbers are absent from trees with flexible stems and large leaves. According to Rich et al. (1987), palm trees are resistant to vine infestation because they lack branches, have robust stems, and occasionally lose their big leaves.

The symbiotic relationships (Janzen, 1973), spines/thorns (Rich et al., 1987), bark characteristics (Reddy & Parthasarathy, 2006), height, bole girth, branching pattern, growth rate, and leaf phenology (Vivek & Parthasarathy, 2017) of the host trees are

Plate 2.1 Dense liana infestations on host trees in a moist deciduous forest (**a**); dry evergreen forest (**b**, **c**); semi-evergreen forest (**d**). Picture credit: Vivek Pandi

Plate 2.2 *Cassytha filiformis*, a parasitic climber formed a dense blanket over the host tree's crown (**a**); *Combretum albidum* completely engulfed the host tree, leaving limited to no light for photosynthesis (**b**); A single host tree carrying laterally infested liana loads are often susceptible for mechanical damage (**c**). Picture credit: Vivek Pandi

Plate 2.3 Dense liana colonization in a degraded dry evergreen forest (**a**); A tree-fall gap aggressively invaded by climbers (**b**). Picture credit: Vivek Pandi

just a few examples of the factors that other studies have linked to the colonization of lianas (Vivek & Parthasarathy, 2016, 2017). However, most studies found conflicting results with a single feature, which made it difficult to comprehend how lianas and their hosts coevolved. Therefore, the combined characteristics may

Plate 2.4 A giant liana species, *Derris scandens* without suitable support on the forest floor of a dry evergreen forest (**a**); *Entada* sp. fallen off the ground along with the host tree in a moist deciduous forest (**b**). Picture credit: Vivek Pandi

improve host trees' effectiveness in discouraging lianas (Sfair et al., 2016); nevertheless, it may differ across lianas with varied climbing tactics (Vivek & Parthasarathy, 2016). To answer if host specificity exists between lianas and trees, Hegarty and Caballé (1991) suggested that lianas and trees do not engage in species-

Table 2.4 A list of selected liana–tree interaction studies conducted across different study sites. The percentage infestation refers to the proportion of trees that had at least one liana

Forest type and location of the study site	Percentage of infested trees	Tree measurement threshold (GBH in cm)	Methodology—definition of infestation	References
Tropical Rainforest, Ghana	56–63	15.7	All contacts (not differentiated into the primary or secondary infestation)	Ofosu-Bamfo et al. (2019)
Tropical dry evergreen forest, south-east India	57.9	10	All primary contacts (not differentiated into trunk or crown infestation)	Vivek and Parthasarathy, (2017)
Tambopata nature reserve, Peru	49.8	31.4	All primary contacts (Trunk and crown infestations differentiated)	van der Heijden et al. (2008)
Subtropical semi-deciduous forest, Argentina	82.4	31.4	NA	Campanello et al. (2007)
Tropical dry evergreen forest, south-east India	63.8	10	Primary and secondary infestations differentiated	Reddy and Parthasarathy, (2006)
Tropical rain forest, Mexico	63.3	31.4	All contacts (not differentiated into the primary or secondary infestation)	Pérez-Salicrup and De Meijere (2005)
Subtropical humid forest, Bolivia	73	31.4	Primary and secondary infestations differentiated	Alvira et al. (2004)
Tropical evergreen forest, Eastern Ghats, India	10	30	All primary contacts (not differentiated into trunk or crown infestation)	Chittibabu and Parthasarathy, (2001)
Tropical evergreen forest, Western Ghats, India	28	30	All primary contacts (not differentiated into trunk or crown infestation)	Muthuramkumar and Parthasarathy, (2001)
Seasonal evergreen forest, Brazil	90	15.7	NA	Gerwing (2001)
Seasonal moist forest, Bolivia	86.3	31.4	All contacts (not differentiated into the primary or secondary infestation)	Pérez-Salicrup et al. (2001)

(continued)

Table 2.4 (continued)

Forest type and location of the study site	Percentage of infested trees	Tree measurement threshold (GBH in cm)	Methodology—definition of infestation	References
Tropical dry forest, Bolivia	77	62.8	All primary contacts (Trunk and crown infestations differentiated)	Carse et al. (2000)
Seasonal dry forest, Bolivia	75	31.4	All primary contacts (not differentiated into trunk or crown infestation)	Killeen et al. (1998)
Wet evergreen forest, Lambir, Malaysia	52.3	31.4	Primary and secondary infestations differentiated	Putz and Chai (1987)
Moist forest, Similajau, Malaysia	53.8	31.4	NA	Putz (1984a, b)

The term "primary infestation" refers to lianas that infested the host directly through the trunk or crown. Secondary infestation is the colonization of lianas on hosts laterally from other sources. GBH is the girth of the tree at breast height

specific interactions (host specificity), confirmed by Garrido-Perez and Burnham (2010). Therefore, it is still unknown whether lianas locate their hosts in a deterministic or opportunistic manner. The adaption of various techniques limited our ability to compare findings across studies. Some research only looked at crown infestation, others only looked at trunk infection, and a few studies looked at both. Therefore, a standardized approach to researching liana–tree interactions is required.

2.6 Lianas in Forest Management Practices

The competition for shared resources like light, water, and soil nutrients is crucial for the growth and survival of plants. When Featherly (1941) described the negative impact of grapevines on the host trees, he probably would not have realized the nature of competition between climbers and trees. Still, he observed a repeated decline in growth and increased mortality among liana–laden trees. However, at present, we have several direct pieces of evidence to show that lianas compete aggressively with trees for aboveground and belowground resources (Schnitzer et al., 2005; Chen et al., 2008; Toledo-Aceves & Swaine, 2008; De Deurwaerder et al., 2016) and outperform trees (Tobin et al., 2012; Pasquini et al., 2015; Wright et al., 2015; César et al., 2018). This resulted in decreased tree growth (Nabe-Nielsen et al., 2009; Ladwig & Meiners, 2009), fecundity (García León et al., 2018), regeneration (Schnitzer & Carson, 2010; Martínez-Izquierdo et al., 2016), and survival. There is ample evidence that liana research will impact forest management, and no wonder climbers are considered a nuisance or threat in forest management practices (Fox, 1968; Smith & McCay, 1979; Chaplin, 1985; McNab & Meeker, 1987).

While ascending through the host to reach the forest canopy, lianas often damage the trees, particularly the trunk deformations. Forest managers are concerned about this since the malformed trees produce low returns. Furthermore, disease and breakage are more likely to affect trees with defects (e.g., Shigo, 1986). Removing climbers has become a standard technique in silvicultural methods to prevent logging damage. Lianas should be removed before logging to minimize damage and liana proliferation after logging to fill in gaps since their flexible stems typically survive falling and proliferate abundantly to colonize the other trees (e.g., Putz, 1985; Vidal et al., 1997; Parren & Bongers, 2001; Gerwing & Uhl, 2002; Pinard & Putz, 1994; Schnitzer & Bongers, 2005; Cesar et al., 2016; Cleary, 2017). However, cutting or removing lianas could be costly and time consuming (Smith & McCay, 1979; Vidal et al., 1997; Pérez-Salicrup et al., 2001). Additionally, lianas may grow back quickly after being removed (Pérez-Salicrup et al., 2001; Parren & Bongers, 2001; Alvira et al., 2004).

The effects of post-logging on liana diversity and abundance are also fairly discussed in the literature (e.g., Putz, 1984a, b; Gardette, 1996; Ding & Zang, 2009; Mesh et al., 2017; Guo et al., 2019). Although comprehensive studies reported lianas' impact on logging, only a few studies have looked at the long-term effects of selective logging on faunal communities (Jamhuri et al., 2018; Cleary et al., 2005) and tree–liana interaction networks (Magrach et al., 2016). While liana cutting may provide silvicultural benefits, it may jeopardize the conservation value of the managed forest. In the future, liana cutting protocols that balance silvicultural goals of minimizing logging damage and post-logging liana proliferation with conservation goals of preserving their beneficial effects on forest ecosystems may be developed (e.g., Sfair et al., 2015).

Apart from their harmful effects on host trees, the invasive climbers can be noxious to their environment. According to Sundarapandian et al. (2016), 55 climbers were listed as one of the worst invasive species on the planet (Global invasive species database, 2013). *Pueraria montana*, recognized for its allelopathic effects (Rashid et al., 2010) and nitric oxide emissions (Hickman et al., 2010), and *Celastrus orbiculatus* are some of the notorious climber species (Pooler et al., 2002; Leicht & Silander, 2006). *Clematis vitalba* (Ogle et al., 2000), *Vincetoxicum rossicum* (Cappuccino, 2004), and *Lantana camara* are the other invasive climbers that pose a severe threat to the natural vegetation.

2.7 Positive Impacts: Animal Plant Interactions; Ethnobotany; Phytochemistry; Carbon Stock

In addition to their detrimental effects, lianas can be beneficial by boosting ecosystem processes, structural diversity, and habitat heterogeneity. According to standard estimates, climbers contribute 25% to the overall diversity of woody species (Gentry, 1991; Pérez-Salicrup et al., 2001) or even more (up to 50%) of woody species in liana-dense forests (Muthumperumal & Parthasarathy, 2009; Vivek & Parthasarathy, 2015). Lianas significantly aid in the closure of the canopy and assist in stabilizing the microclimate below (Schnitzer & Bongers, 2002), allowing shade-tolerant and sun-intolerant species to flourish. By connecting the trees, lianas create a network that offers valuable habitat and a pathway for the movement of arboreal animals (Charles-Dominique & Hladik, 1971; Emmons & Gentry, 1983; Ødegaard, 2000; Schnitzer & Carson, 2010; Koenig et al., 2007). Patterns of liana infection on host trees can also impact the ant community structure (Adams et al., 2017, 2019; Clay et al., 2010; Blüthgen & Fiedler, 2002). The role of lianas in snake predation on parrot nests was demonstrated in a study conducted by Koenig et al. (2007).

The leaves (Aide & Zimmerman, 1990; Chiarello, 1994; Charles & Basset, 2005), flowers (Labeyrie et al., 2001), and fruits (Whitney & Smith, 1998) of lianas provide food for a variety of invertebrate and vertebrate populations (Schnitzer & Bongers, 2002; Moscovice et al., 2007; Martins, 2009; Petroni et al., 2017). For instance, according to Chiarello (1994), lianas account for 41% of the leaves consumed by brown howler monkeys. Similar to this, Martins (2009) noted that lianas constituted 27% of the diet of the brown howling monkey (*Aouatta guariba*) and 33% of the Southern muriqui (*Brachyteles arachnoides*). Chimpanzees on Tanzania's Rubonda Island devoured 35% of their fruit as lianas, according to research by Moscovice et al. (2007). Furthermore, species like *Saba comorensis* can continue to produce fruit throughout the phenological observations, guaranteeing the year-round availability of nutrients for the faunal assemblages (Moscovice et al., 2007).

Lianas have traditionally been used as a food source, medicine, and a means of livelihood among local communities. For instance, according to Tra Bi et al. (2005), indigenous and rural groups used ethnobotanically significant climbers in Cote d'Ivoire, Africa. Significant mentions of ethnobotany and the financial advantages of climbers were found in other research papers (e.g., Gentry, 1992; Mino et al., 1995; Kathriarachchi et al., 2004; Muhwezi et al., 2009). Millions worldwide rely on agro-horticultural vine crops and vines planted from forests for their food. Several common foods are derived from vine products, including grapes (*Vitis* spp.), sweet potatoes (*Ipomoea batatas*), yams (*Dioscorea* spp.), black pepper (*Piper nigrum*), vanilla (*Vanilla planifolia*), melons, squash, and cucumbers. Passion fruit (*Passiflora edulis*) and chayote (*Sechium edule*) are commercially valuable foods in tropical countries (Phillips, 1991). However, the overexploitation of these economically and culturally important species threatens their survival. Considering the extent of faunal dependence and the economic benefits of climbers, developing management tools is crucial to protect climber species from extinction. In recent years, biotechnology has also been emphasized as a means of conserving these species along with the traditional methods (Arnold et al., 2005; Arora et al., 2010; Sharma et al., 2016; McGeoch et al., 2008; Dogar et al., 2018) to meet the growing requirements.

2.8 Functional Traits: Morphology, Anatomy, and Physiology

Plant functional traits are reliable predictors of plants' survival strategies. Climbers have gotten a lot of attention in plant functional trait studies because of their unique characteristics, such as their anatomical properties (Obee, 1934; Ayensu & Stern, 1964; Davis & Evert, 1970; Carlquist, 1981) associated with anomalous cambial activities (Dobbins, 1969, 1971), which have been actively studied for a long time. Most of the research has focused on comparing the wood anatomy of lianas to that of trees (Gartner et al., 1990; Gasson & Dobbins, 1991; Teramura et al., 1991; Chiu & Ewers, 1992; Fisher & Ewers, 1995; Ewers et al., 1997). Climbers' thin, elongated stems that grow inexorably long have always fascinated scientists studying their hydraulic efficiencies in relation to their xylem structure (Scholander et al., 1955, 1957; Ewers & Fisher, 1989; Ewers, 1985; Bell et al., 1988; Braun & Schmid, 1999; Johnson et al., 2013). Climbers' xylem properties, particularly vessel dimensions, are distinct and easily distinguished from those of trees (Ewers & Fisher, 1989; Ewers et al., 1990; Fisher & Ewers, 1992; Caballé, 1993; Jacobsen & Pratt, 2012; Yang & Chen, 2015; Rajput et al., 2021). These differences in xylem dimensions piqued the interest of researchers, and several studies have shown that lianas outperform trees in terms of hydraulic efficiency (Van der Sande et al., 2019; Ganthaler et al., 2019; Bentrup, 2017; De Guzman et al., 2017). However, lianas are more susceptible to freeze–thaw embolism than trees in temperate ecosystems (Cochard et al., 1994; Jiménez-Castillo & Lusk, 2013), limiting climber distribution along the latitudinal gradient. Many studies have been conducted on the evolutionary perspectives of the vascular system in climbers at both the taxonomic and community levels (Pace et al., 2009; Hearn, 2009; Angyalossy et al., 2012, 2015; Cunha Neto et al., 2018).

Climbers are distinguished from other growth habits by their biomechanical properties. Due to the lack of autonomous vertical growth, climbers must make significant adaptations in terms of stem flexibility and architecture, which often vary among climbers of different climbing strategies. Climbers' biomechanical properties have received significantly less attention than their anatomical properties, mostly limited to lianas. Putz and Holbrook (1991) investigated the biomechanical properties of climbers and the biomechanical constraints for climbing among various climbing strategies. Rowe and Speck (1996) demonstrated the shift in young's modulus during ontogeny of a tropical liana, *Condylocarpon guienense*. Few other studies have shown the role of biomechanical properties in determining growth habits among climbers (Isnard et al., 2003, 2005; Ménard et al., 2009), including free-standing lianas (Kennard, 1998) and climbing palms (Isnard et al., 2005; Rowe & Isnard, 2009).

While belowground competition between trees and lianas is influenced by anatomical traits, xylem structure, and associated hydraulic efficiencies, aboveground competition is frequently viewed in terms of leaf trait configuration, which acts as an interface between the plant and its immediate environment. The ecology of morphological traits like leaf size, shape (e.g., Givnish & Vermeij, 1976), margin (e.g., Burnham et al., 2001), thickness, and stomatal density (e.g., Tay & Furukawa,

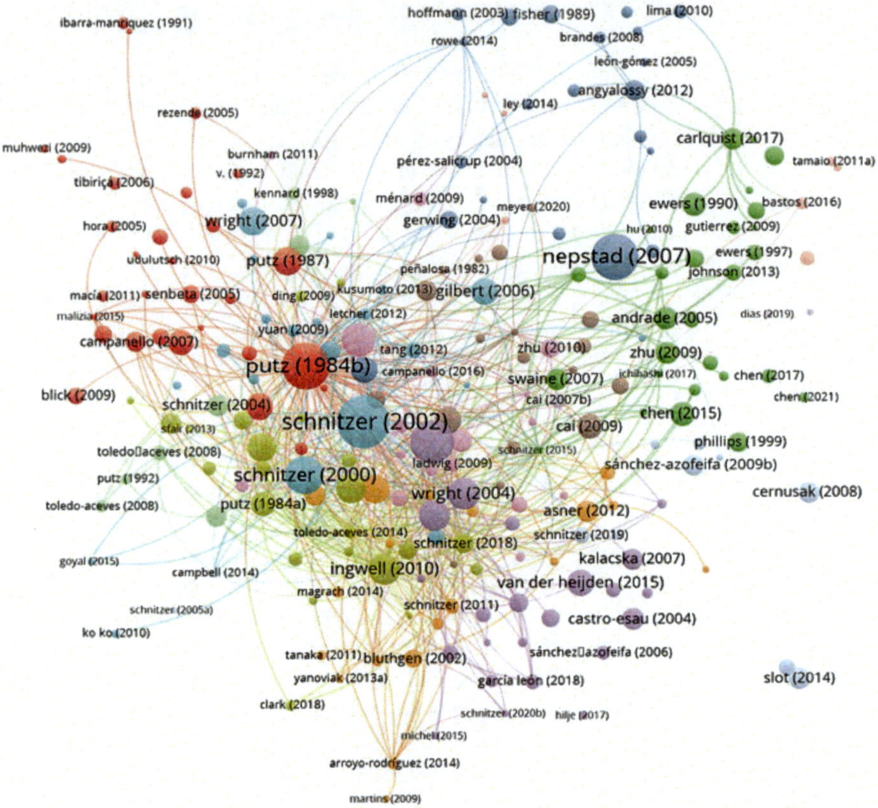

Fig. 2.3 Citation-based cluster analysis of climber-related research articles from 1980 to 2022. The number of citations is indicated by the size of the circles and letters. The analysis is based on the DOIs of 273 research materials from Crossref retrieved and compiled using the default settings of VOSviewer version 1.6.18 (https://www.vosviewer.com) (Van Eck & Waltman, 2010)

2008), physiological/biochemical traits like photosynthetic efficiency (e.g., Avalos & Mulkey, 1999; Zhu & Cao, 2009), leaf nitrogen (e.g., Ackerly, 1992), carbon, and chlorophyll content, anatomical features like mesophyll thickness, chemical traits (e.g., Asner & Martin, 2012), phenological characteristics (e.g., Hegarty, 1990) and their overall relationship with environmental parameters like temperature (e.g., Fetcher, 1981; Guzmán et al., 2018), light (Yuan et al., 2016), and seasonality (e.g., Cai et al., 2009) have all been extensively studied. Many studies have also been conducted to compare the leaf functional strategies of lianas and self-standing plants (e.g., Hegarty, 1990; Avalos et al., 1999; Salzer et al., 2006; Sanches & Válio, 2006; Han et al., 2010; Collins et al., 2016; Werden et al., 2018; Vivek & Parthasarathy, 2018). According to Cai et al. (2009), seasonal variations in leaf-level physiology give lianas a competitive advantage over trees. Vivek and Parthasarathy (2018) reported similar findings from a seasonally dry tropical forest site in India. In general, lianas have drawn attention in functional trait research because of their climbing habits, morphological characteristics, hydraulic efficiency, biomechanical properties, and greater ecological amplitude. However, the global functional trait database continues to underrepresent climbers (Gallagher & Leishman, 2012).

This chapter reviewed past and present climber research across different research themes. From a virtual blank, climbers have undoubtedly become indisputable components of contemporary ecology research in recent times. Figure 2.3 lists some of the most impactful studies that improved our knowledge of the biology of climbers (based on citations). The current trends in climber research and the future perspectives are discussed in Chap. 4.

References

Ackerly, D. D. (1992). Light, leaf age, and leaf nitrogen concentration in a tropical vine. *Oecologia, 89*(4), 596–600.

Adams, B. J., Schnitzer, S. A., & Yanoviak, S. P. (2017). Trees as islands: Canopy ant species richness increases with the size of liana-free trees in a Neotropical forest. *Ecography, 40*(9), 1067–1075.

Adams, B. J., Schnitzer, S. A., & Yanoviak, S. P. (2019). Connectivity explains local ant community structure in a Neotropical forest canopy: A large-scale experimental approach. *Ecology, 100*(6), e02673.

Addo-Fordjour, P., & Owusu-Boadi, K. (2016). Linear edge effects on liana and tree communities in two tropical forest ecosystems in Ghana. *Ecological Research, 31*(5), 709–718.

Addo-Fordjour, P., Anning, A. K., Atakora, E. A., & Agyei, P. S. (2008). Diversity and distribution of climbing plants in a semi-deciduous rain forest, KNUST Botanic Garden, Ghana. *International Journal of Botany.*

Addo-Fordjour, P., El Duah, P., & Agbesi, D. K. K. (2013). Factors influencing liana species richness and structure following anthropogenic disturbance in a tropical forest, Ghana. *International Scholarly Research Notices, 2013.*

Addo-Fordjour, P., Obeng, S., Anning, A. K., & Addo, M. G. (2009). Floristic composition, structure and natural regeneration in a moist semi-deciduous forest following anthropogenic disturbances and plant invasion. *International Journal of Biodiversity and Conservation, 1*(2), 21–37.

Addo-Fordjour, P., Rahmad, Z. B., & Shahrul, A. M. S. (2012). Effects of human disturbance on liana community diversity and structure in a tropical rainforest, Malaysia: Implication for conservation. *Journal of Plant Ecology, 5*(4), 391–399.

Addo-Fordjour, P., Rahmad, Z. B., & Shahrul, A. M. S. (2014). Impacts of forest management on community assemblage and carbon stock of lianas in a tropical lowland forest, Malaysia. *Tropical Conservation Science, 7*(2), 244–259.

Addo-Fordjour, P., Rahmad, Z. B., & Shahrul, A. M. S. (2016). Liana species composition, dominance and host interactions in primary and secondary forests in Malaysia. *Tropical Ecology, 57*(3).

Aide, T. M., & Zimmerman, J. K. (1990). Patterns of insect herbivory, growth, and survivorship in juveniles of a neotropical liana. *Ecology, 71*(4), 1412–1421.

Allen, B. P., Pauley, E. F., & Sharitz, R. R. (1997). Hurricane impacts on liana populations in an old-growth southeastern bottomland forest. *Journal of the Torrey Botanical Society*, 34–42.

Allen, B. P., Sharitz, R. R., & Goebel, P. C. (2005). Twelve years post-hurricane liana dynamics in an old-growth southeastern floodplain forest. *Forest Ecology and Management, 218*(1-3), 259–269.

Alves, L. F., Assis, M. A., van Melis, J., Barros, A. L., Vieira, S. A., Martins, F. R., Joly, C. A., et al. (2012). Variation in liana abundance and biomass along an elevational gradient in the tropical Atlantic Forest (Brazil). *Ecological Research, 27*(2), 323–332.

Alvira, D., Putz, F. E., & Fredericksen, T. S. (2004). Liana loads and post-logging liana densities after liana cutting in a lowland forest in Bolivia. *Forest Ecology and Management, 190*(1), 73–86.

Anbarashan, M., & Parthasarathy, N. (2013). Diversity and ecology of lianas in tropical dry evergreen forests on the Coromandel Coast of India under various disturbance regimes. *Flora-Morphology, Distribution, Functional Ecology of Plants, 208*(1), 22–32.

Angyalossy, V., Angeles, G., Pace, M. R., Lima, A. C., Dias-Leme, C. L., Lohmann, L. G., & Madero-Vega, C. (2012). An overview of the anatomy, development and evolution of the vascular system of lianas. *Plant Ecology & Diversity, 5*(2), 167–182.

Angyalossy, V., Pace, M. R., & Lima, A. C. (2015). Liana anatomy: A broad perspective on structural evolution of the vascular system. *Ecology of Lianas*, 253–287.

Appanah, S., & Putz, F. E. (1984). Climber abundance in virgin dipterocarp forest and the effect of pre-felling climber cutting on logging damage [Peninsular Malaysia]. *Malaysian Forester* (Malaysia).

Arnold, C., Schnitzler, A., Douard, A., Peter, R., & Gillet, F. (2005). Is there a future for wild grapevine (Vitis vinifera subsp. silvestris) in the Rhine Valley? *Biodiversity & Conservation, 14*(6), 1507–1523.

Arora, J., Goyal, S., & Ramawat, K. G. (2010). Enhanced stilbene production in cell cultures of Cayratia trifolia through co-treatment with abiotic and biotic elicitors and sucrose. *In Vitro Cellular & Developmental Biology-Plant, 46*(5), 430–436.

Asner, G. P., & Martin, R. E. (2012). Contrasting leaf chemical traits in tropical lianas and trees: implications for future forest composition. *Ecology Letters, 15*(9), 1001–1007.

Avalos, G., & Mulkey, S. S. (1999). Seasonal changes in liana cover in the upper canopy of a neotropical dry forest. *Biotropica*, 186–192.

Avalos, G., Mulkey, S. S., & Kitajima, K. (1999). Leaf optical properties of trees and lianas in the outer canopy of a tropical dry forest. *Biotropica*, 517–520.

Ayensu, E. S., & Stern, W. L. (1964). Systematic anatomy and ontogeny of the stem in Passifloraceae. *Systematic Plant Studies*.

Baars, R., Kelly, D., & Sparrow, A. D. (1998). Liane distribution within native forest remnants in two regions of the South Island, New Zealand. *New Zealand Journal of Ecology*, 71–85.

Balfour, D. A., & Bond, W. J. (1993). Factors limiting climber distribution and abundance in a southern African forest. *Journal of Ecology*, 93–100.

Bell, D. J., Forseth, I. N., & Teramura, A. H. (1988). Field water relations of three temperate vines. *Oecologia, 74*(4), 537–545.

Bentrup, F. W. (2017). Water ascent in trees and lianas: the cohesion-tension theory revisited in the wake of Otto Renner. *Protoplasma, 254*(2), 627–633.

Birhane, E., Fekensa, Z., Tewolde-Berhan, S., Rannestad, M. M., & Solomon, N. (2020). The diversity and distribution of lianas under various disturbance regimes in Chilimo dry Afromontane Forest, Ethiopia. *Global Ecology and Conservation, 23*, e01045.

Black, H. L., & Harper, K. T. (1979). The adaptive value of buttresses to tropical trees: Additional hypotheses. *Biotropica, 11*(3), 240–240.

Blick, R. A. J., & Burns, K. C. (2011). Liana co-occurrence patterns in a temperate rainforest. *Journal of Vegetation Science, 22*(5), 868–877.

Blüthgen, N., & Fiedler, K. (2002). Interactions between weaver ants Oecophylla smaragdina, homopterans, trees and lianas in an Australian rain forest canopy. *Journal of Animal Ecology, 71*(5), 793–801.

Boom, B. M., & Mori, S. A. (1982). Falsification of two hypotheses on liana exclusion from tropical trees possessing buttresses and smooth bark. *Bulletin of the Torrey Botanical Club*, 447–450.

Braun, P., & Schmid, J. (1999). Sap flow measurements in grapevines (Vitis vinifera L.) 1. Stem morphology and use of the heat balance method. *Plant and Soil, 215*(1), 39–45.

Bremekamp, C. E. B. (1912). Die rotierende Nutation und der Geotropismus der Windepflanzen. *Recueil des travaux botaniques néerlandais, 9*(3), 281–381.

Bruy, D., Ibanez, T., Munzinger, J., & Isnard, S. (2017). Abundance, richness and composition of lianas in forest communities along an elevation gradient in New Caledonia. *Plant Ecology & Diversity, 10*(5-6), 469–481.

Burnham, R. J. (2004). Alpha and beta diversity of lianas in Yasunı, Ecuador. *Forest Ecology and Management, 190*(1), 43–55.

Burnham, R. J., Pitman, N. C., Johnson, K. R., & Wilf, P. (2001). Habitat-related error in estimating temperatures from leaf margins in a humid tropical forest. *American Journal of Botany, 88*(6), 1096–1102.

Caballé, G. U. Y. (1993). Liana structure, function and selection: a comparative study of xylem cylinders of tropical rainforest species in Africa and America. *Botanical Journal of the Linnean Society, 113*(1), 41–60.

Caballé, G., & Martin, A. (2001). Thirteen years of change in trees and lianas in a Gabonese rainforest. *Plant Ecology, 152*(2), 167–173.

Cai, Z. Q., Schnitzer, S. A., & Bongers, F. (2009). Seasonal differences in leaf-level physiology give lianas a competitive advantage over trees in a tropical seasonal forest. *Oecologia, 161*(1), 25–33.

Campanello, P. I., Garibaldi, J. F., Gatti, M. G., & Goldstein, G. (2007). Lianas in a subtropical Atlantic Forest: host preference and tree growth. *Forest Ecology and Management, 242*(2-3), 250–259.

Campanello, P. I., Villagra, M., Garibaldi, J. F., Ritter, L. J., Araujo, J. J., & Goldstein, G. (2012). Liana abundance, tree crown infestation, and tree regeneration ten years after liana cutting in a subtropical forest. *Forest Ecology and Management, 284*, 213–221.

Campbell, E. J. F., & Newbery, D. M. (1993). Ecological relationships between lianas and trees in lowland rain forest in Sabah, East Malaysia. *Journal of Tropical Ecology, 9*(4), 469–490.

Campbell, M. J., Edwards, W., Magrach, A., Alamgir, M., Porolak, G., Mohandass, D., & Laurance, W. F. (2018). Edge disturbance drives liana abundance increase and alteration of liana–host tree interactions in tropical forest fragments. *Ecology and Evolution, 8*(8), 4237–4251.

Cappuccino, N. (2004). Allee effect in an invasive alien plant, pale swallow-wort Vincetoxicum rossicum (Asclepiadaceae). *Oikos, 106*(1), 3–8.

Carlquist, S. (1981). Wood anatomy of Nepenthaceae. *Bulletin of the Torrey Botanical Club*, 324–330.

Carrasco-Urra, F., & Gianoli, E. (2009). Abundance of climbing plants in a southern temperate rain forest: host tree characteristics or light availability? *Journal of Vegetation Science, 20*(6), 1155–1162.

Carse, L. E., Frederickson, T. S., & Licona, J. C. (2000). Liana-tree species associations in a Bolivian dry forest. *Tropical Ecology, 41*(1), 1–10.

Carvalho, E. C. D., Martins, F. R., Oliveira, R. S., Soares, A. A., & Araújo, F. S. (2016). Why is liana abundance low in semiarid climates? *Austral Ecology, 41*(5), 559–571.

Ceballos, S. J., & Malizia, A. (2017). Liana density declined and basal area increased over 12 y in a subtropical montane forest in Argentina. *Journal of Tropical Ecology, 33*(4), 241–248.

Cesar, R. G., Holl, K. D., Girao, V. J., Mello, F. N., Vidal, E., Alves, M. C., & Brancalion, P. H. (2016). Evaluating climber cutting as a strategy to restore degraded tropical forests. *Biological Conservation, 201*, 309–313.

César, R. G., Leite, H. P. P., Martins, J. T., Amarante, K. M., Torres, B. F., Mello, F. N. A., Brancalion, P. H. S., et al. (2018). Shift in abundance from seedling to Juvenile Gives Lianas advantage over trees: A case study in the Atlantic Forest Hotspot. *Tropical Conservation Science, 11*, 1940082918808062.

Chalmers, A. C. (1994). Climbing plants in relation to their supports in a stand of dry rainforest in the Hunter Valley, New South Wales. In *Proc Linn Soc NSW* (Vol. 114, pp. 73–90).

Chaplin, G. E. (1985). An integrated silvicultural solution to weedy climber problems in the Solomon Islands. *The Commonwealth Forestry Review*, 133–139.

Charles, E., & Basset, Y. (2005). Vertical stratification of leaf-beetle assemblages (Coleoptera: Chrysomelidae) in two forest types in Panama. *Journal of Tropical Ecology, 21*(3), 329–336.

Charles-Dominique, P., & Hladik, C. M. (1971). Le Lepilemur du sud de Madagascar: écologie, alimentation et vie sociale. *Revue d'Ecologie, Terre et Vie, 1*, 3–66.

Charles-Dominique, P., Atramentowicz, M., Charles-Dominique, M., Gerard, H., Hladik, A., Hladik, C. M., & Prévost, M. F. (1981). Les mammiferes frugivores arboricoles nocturnes d'une foret guyanaise: Inter-relations plantes-animaux. *Revue d'Ecologie, Terre et Vie, 35*(3), 341–436.

Chen, Y. J., Bongers, F., Cao, K. F., & Cai, Z. Q. (2008). Above-and below-ground competition in high and low irradiance: tree seedling responses to a competing liana Byttneria grandifolia. *Journal of Tropical Ecology, 24*(5), 517–524.

Chettri, A., Barik, S. K., Pandey, H. N., & Lyngdoh, M. K. (2010). Liana diversity and abundance as related to microenvironment in three forest types located in different elevational ranges of the Eastern Himalayas. *Plant Ecology & Diversity, 3*(2), 175–185.

Chhetri, R., & Bhattarai, P. (2013). Floristic composition and diversity in Upper Manaslu Conservation Area, Central Nepal. *Asian Journal of Conservation Biology, 2*(2), 111–121.

Chiarello, A. G. (1994). Diet of the brown howler monkey Alouatta fusca in a semi-deciduous forest fragment of southeastern Brazil. *Primates, 35*(1), 25–34.

Chittibabu, C. V., & Parthasarathy, N. (2001). Liana diversity and host relationships in a tropical evergreen forest in the Indian Eastern Ghats. *Ecological Research, 16*(3), 519–529.

Chiu, S. T., & Ewers, F. W. (1992). Xylem structure and water transport in a twiner, a scrambler, and a shrub of Lonicera (Caprifoliaceae). *Trees, 6*(4), 216–224.

Clay, N. A., Bauer, M., Solis, M., & Yanoviak, S. P. (2010). Arboreal substrates influence foraging in tropical ants. *Ecological Entomology, 35*(4), 417–423.

Cleary, D. F. (2017). Impact of logging on tree, liana and herb assemblages in a Bornean forest. *Journal of Sustainable Forestry, 36*(8), 806–817.

Cleary, D. F. R., Boyle, T. J., Setyawati, T., & Menken, S. B. (2005). The impact of logging on the abundance, species richness and community composition of butterfly guilds in Borneo. *Journal of Applied Entomology, 129*(1), 52–59.

Clinebell, R. R., Phillips, O. L., Gentry, A. H., Stark, N., & Zuuring, H. (1995). Prediction of neotropical tree and liana species richness from soil and climatic data. *Biodiversity & Conservation, 4*(1), 56–90.

Cochard, H., Ewers, F. W., & Tyree, M. T. (1994). Water relations of a tropical vine-like bamboo (Rhipidocladum racemiflorum): root pressures, vulnerability to cavitation and seasonal changes in embolism. *Journal of Experimental Botany, 45*(8), 1085–1089.

Collins, B. S., & Wein, G. R. (1993). Understory vines: Distribution and relation to environment on a southern mixed hardwood site. *Bulletin of the Torrey Botanical Club*, 38–44.

Collins, C. G., Wright, S. J., & Wurzburger, N. (2016). Root and leaf traits reflect distinct resource acquisition strategies in tropical lianas and trees. *Oecologia, 180*(4), 1037–1047.

Cunha Neto, I. L. D., Martins, F. M., Somner, G. V., & Tamaio, N. (2018). Successive cambia in liana stems of Paullinieae and their evolutionary significance in Sapindaceae. *Botanical Journal of the Linnean Society, 186*(1), 66–88.

Darwin, C. (1875). *The movements and habits of climbing plants*. John Murray.

Davis, J. D., & Evert, R. F. (1970). Seasonal cycle of phloem development in woody vines. *Botanical Gazette, 131*(2), 128–138.

De Deurwaerder, H., Okello, J. A., Koedam, N., Schmitz, N., & Steppe, K. (2016). How are anatomical and hydraulic features of the mangroves Avicennia marina and Rhizophora mucronata influenced by siltation? *Trees, 30*(1), 35–45.

De Guzman, M. E., Santiago, L. S., Schnitzer, S. A., & Álvarez-Cansino, L. (2017). Trade-offs between water transport capacity and drought resistance in neotropical canopy liana and tree species. *Tree Physiology, 37*(10), 1404–1414.

Den Dubbelden, K. C., & Oosterbeek, B. (1995). The availability of external support affects allocation patterns and morphology of herbaceous climbing plants. *Functional Ecology*, 628–634.

Dengler, J., Chytrý, M., & Ewald, J. (2008). Phytosociology. In S. E. Jørgensen & B. D. Fath (Eds.), *Encyclopedia of ecology. General ecology* (Vol. 4, pp. 2767–2779). Elsevier.

DeWalt, S. J., Ickes, K., Nilus, R., Harms, K. E., & Burslem, D. F. (2006). Liana habitat associations and community structure in a Bornean lowland tropical forest. *Plant Ecology, 186*(2), 203–216.

Dewalt, S. J., Schnitzer, S. A., & Denslow, J. S. (2000). Density and diversity of lianas along a chronosequence in a central Panamanian lowland forest. *Journal of Tropical Ecology, 16*(1), 1–19.

DeWalt, S. J., Schnitzer, S. A., Chave, J., Bongers, F., Burnham, R. J., Cai, Z., Chuyong, G., Clark, D. B., Ewango, C. E. N., Gerwing, J. J., Gortaire, E., Hart, T., Ibarra-Manríquez, G., Ickes, K., Kenfack, D., Macía, M. J., Jean-Remy, M., Martínez-Ramos, M., Mascaro, J., et al. (2010). Annual rainfall and seasonality predict pan-tropical patterns of liana density and basal area. *Biotropica, 42*(3), 309–317.

Ding, Y., & Zang, R. (2009). Effects of logging on the diversity of lianas in a lowland tropical rain forest in Hainan Island, South China. *Biotropica, 41*(5), 618–624.

Dobbins, D. R. (1969). Studies on the anomalous cambial activity in Doxantha unguis-cati (Bignoniaceae). I. Development of the vascular pattern. *Canadian Journal of Botany, 47*(12), 2101–2106.

Dobbins, D. R. (1971). Studies on the anomalous cambial activity in Doxantha unguis-cati (Bignoniaceae). II. A case of differential production of secondary tissues. *American Journal of Botany, 58*(8), 697–705.

Dogar, R., Yadav, S. K., Nagori, B. P., Mathur, K., & Goyal, M. (2018). A survey on therapeutic utility of Luffa echinata roxb. (Ethnomedicinal uses and Pharmacological activity). *Indian Journal of Drugs, 6*(1), 62–67.

Duncan, W. H. (1967). Woody vines of the southeastern states. *SIDA, Contributions to Botany, 3*(1), 1–76.

Durigon, J., Durán, S. M., & Gianoli, E. (2013). Global distribution of root climbers is positively associated with precipitation and negatively associated with seasonality. *Journal of Tropical Ecology, 29*(4), 357–360.

Eilu, G. (2000). Liana abundance in three tropical rain forests of western Uganda. *Selbyana*, 30–37.

Emmons, L. H., & Gentry, A. H. (1983). Tropical forest structure and the distribution of gliding and prehensile-tailed vertebrates. *The American Naturalist, 121*(4), 513–524.

Ewango, C. E. (2010). *The liana assemblage of a Congolian rainforest: Diversity, structure and dynamics*. Wageningen University and Research.

Ewango, C. E., Bongers, F., Makana, J. R., Poorter, L., & Sosef, M. S. (2015). Structure and composition of the liana assemblage of a mixed rain forest in the Congo Basin. *Plant Ecology and Evolution, 148*(1), 29–42.

Ewers, F. W. (1985). Xylem structure and water conduction in conifer trees, dicot trees, and llanas. *IAWA Journal, 6*(4), 309–317.

Ewers, F. W., & Fisher, J. B. (1989). Variation in vessel length and diameter in stems of six tropical and subtropical lianas. *American Journal of Botany, 76*(10), 1452–1459.

Ewers, F. W., Cochard, H., & Tyree, M. T. (1997). A survey of root pressures in vines of a tropical lowland forest. *Oecologia, 110*(2), 191–196.

Ewers, F. W., Fisher, J. B., & Chiu, S. T. (1990). A survey of vessel dimensions in stems of tropical lianas and other growth forms. *Oecologia, 84*(4), 544–552.

Featherly, H. I. (1941). Silting and forest succession on Deep Fork in southwestern Creek County, Oklahoma. In *Proceedings of the Oklahoma Academy of Science* (pp. 63–64).

Fetcher, N. (1981). Leaf size and leaf temperature in tropical vines. *The American Naturalist, 117*(6), 1011–1014.

Fisher, J. B., & Ewers, F. W. (1992). Xylem pathways in liana stems with variant secondary growth. *Botanical Journal of the Linnean Society, 108*(2), 181–202.

Fisher, J. B., & Ewers, F. W. (1995). Vessel dimensions in liana and tree species of Gnetum (Gnetales). *American Journal of Botany, 82*(11), 1350–1357.

Fox, J. E. D. (1968). Logging damage and the influence of climber cutting prior to logging in the lowland dipterocarp forest of Sabah. *Malayan Forester, 31*, 326–347.

Fox, J. E. D. (1969). Climbers in the lowland dipterocarp forest. *Commonwealth Forestry Review, 48*, 196–198.

Gallagher, R. V., & Leishman, M. R. (2012). A global analysis of trait variation and evolution in climbing plants. *Journal of Biogeography, 39*(10), 1757–1771.

Ganthaler, A., Marx, K., Beikircher, B., & Mayr, S. (2019). Are hydraulic patterns of lianas different from trees? New insights from Hedera helix. *Journal of Experimental Botany, 70*(10), 2811–2822.

García León, M. M., Martínez Izquierdo, L., Mello, F. N. A., Powers, J. S., & Schnitzer, S. A. (2018). Lianas reduce community-level canopy tree reproduction in a Panamanian forest. *Journal of Ecology, 106*(2), 737–745.

Gardette, E. (1996). *The effect of selective timber logging on the diversity of woody climbers at Pasoh.*

Garrido-Perez, E. I., & Burnham, R. J. (2010). The evolution of host specificity in liana-tree interactions. *Puente Biologico, 3*, 145–157.

Gartner, B. L., Bullock, S. H., Mooney, H. A., Brown, V. B., & Whitbeck, J. L. (1990). Water transport properties of vine and tree stems in a tropical deciduous forest. *American Journal of Botany, 77*(6), 742–749.

Gasson, P., & Dobbins, D. R. (1991). Wood anatomy of the Bignoniaceae, with a comparison of trees and lianas. *IAWA Journal, 12*(4), 389–415.

Gentry, A. H. (1974). Coevolutionary patterns in central American Bignoniaceae. *Annals of the Missouri Botanical Garden, 61*(3), 728–759.

Gentry, A. H. (1982). Patterns of neotropical plant species diversity. In *Evolutionary biology* (pp. 1–84). Springer.

Gentry, A. H. (1988). Changes in plant community diversity and floristic composition on environmental and geographical gradients. *Annals of the Missouri Botanical Garden*, 1–34.

Gentry, A. H. (1991). Climbing plants. *The Biology of Vines*, 3.

Gentry, A. H. (1992). Tropical forest biodiversity: Distributional patterns and their conservational significance. *Oikos*, 19–28.

Gentry, A. H. (1993). Tropical forest biodiversity and the potential for new medicinal plants.

Gentry, A. H. (1995). Patterns of diversity and floristic composition in Neotropical montane forests. *Patterns of Diversity and Floristic Composition in Neotropical Montane Forests*, 103–126.

Gentry, A. H. (1985). An ecotaxonomic survey of Panamanian lianas. In W. D'Arey & M. Correa (Eds.), *Historia natural de Panama. Monograph of systematic Botany* (pp. 29–42). Missouri Botanical Garden.

Gerwing, J. J. (2001). Testing liana cutting and controlled burning as silvicultural treatments for a logged forest in the eastern Amazon. *Journal of Applied Ecology, 38*(6), 1264–1276.

Gerwing, J. J., & Uhl, C. (2002). Pre-logging liana cutting reduces liana regeneration in logging gaps in the eastern Brazilian Amazon. *Ecological Applications, 12*(6), 1642–1651.

Gerwing, J. J., Schnitzer, S. A., Burnham, R. J., Bongers, F., Chave, J., DeWalt, S. J., Ewango, C. E. N., Foster, R., Kenfack, D., Martínez-Ramos, M., Parren, M., Parthasarathy, N., Pérez-Salicrup, D. R., Putz, F. E., & Thomas, D. W. (2006). A standard protocol for liana censuses 1. *Biotropica: The Journal of Biology and Conservation, 38*(2), 256–261.

Gianoli, E., Saldaña, A., Jiménez-Castillo, M., & Valladares, F. (2010). Distribution and abundance of vines along the light gradient in a southern temperate rain forest. *Journal of Vegetation Science, 21*(1), 66–73.

Givnish, T. J., & Vermeij, G. J. (1976). Sizes and shapes of liane leaves. *The American Naturalist, 110*(975), 743–778.

Gray, A. (1858). *Manual of the botany of the Northern United States: Including Virginia, Kentucky, and All East of the Mississippi*. Ivison & Phinney.

Guo, Y., Li, G., Kang, B., Wang, D., & Yang, G. (2012). The differential responses of lianas and vines to rainfall gradients in distribution and abundance in Qinling Mountains, China. *Plant Ecology, 213*(11), 1749–1755.

Guo, Y., Zhao, P., Bu, J., & Yue, M. (2019). The differential responses of woody and herbaceous climbers to selective logging and supporter structure in a temperate forest of Xiaolong Mountain, China. *Plant Ecology, 220*(3), 293–304.

Guzmán, Q., Sánchez-Azofeifa, G. A., & Rivard, B. (2018). Differences in leaf temperature between lianas and trees in the neotropical canopy. *Forests, 9*(6), 307.

Han, L., Xie, L. J., Dai, K. J., Yang, Q., & Cai, Z. Q. (2010). Contrasting leaf characteristics of trees and lianas in secondary and mature forests in southwestern China. *Photosynthetica, 48*(4), 559–566.

Hearn, D. J. (2009). Developmental patterns in anatomy are shared among separate evolutionary origins of stem succulent and storage root-bearing growth habits in Adenia (Passifloraceae). *American Journal of Botany, 96*(11), 1941–1956.

Hegarty, E. E. (1990). Leaf life-span and leafing phenology of lianes and associated trees during a rainforest succession. *The Journal of Ecology*, 300–312.

Hegarty, E. E., & Caballé, G. (1991). Distribution and abundance of vines in forest communities. *The Biology of Vines, 313*, 335.

Hegnauer, R. (1975). *Secondary metabolites and crop plants. Crop genetic resources for today and tomorrow* (pp. 249–255). Cambridge University Press.

Hickman, J. E., Wu, S., Mickley, L. J., & Lerdau, M. T. (2010). Kudzu (Pueraria montana) invasion doubles emissions of nitric oxide and increases ozone pollution. *Proceedings of the National Academy of Sciences, 107*(22), 10115–10119.

Hogan, J. A., Mayorquin, S., Rice, K., Thompson, J., Zimmerman, J. K., & Brokaw, N. (2017). Liana dynamics reflect land-use history and hurricane response in a Puerto Rican forest. *Journal of Tropical Ecology, 33*(2), 155–164.

Homeier, J., Englert, F., Leuschner, C., Weigelt, P., & Unger, M. (2010). Factors controlling the abundance of lianas along an altitudinal transect of tropical forests in Ecuador. *Forest Ecology and Management, 259*(8), 1399–1405.

Hu, L. (2011). Distribution and diversity of climbing plants in temperate East Asia. *Biodiversity Science, 19*(5), 567–573.

Ibarra-Manríquez, G., & Martínez-Ramos, M. (2002). Landscape variation of liana communities in a Neotropical rain forest. *Plant Ecology, 160*(1), 91–112.

Ingwell, L. L., Joseph Wright, S., Becklund, K. K., Hubbell, S. P., & Schnitzer, S. A. (2010). The impact of lianas on 10 years of tree growth and mortality on Barro Colorado Island, Panama. *Journal of Ecology, 98*(4), 879–887.

Isnard, S., Speck, T., & Rowe, N. P. (2003). Mechanical architecture and development in Clematis: Implications for canalised evolution of growth forms. *New Phytologist, 158*(3), 543–559.

Isnard, S., Speck, T., & Rowe, N. P. (2005). Biomechanics and development of the climbing habit in two species of the South American palm genus Desmoncus (Arecaceae). *American Journal of Botany, 92*(9), 1444–1456.

Vazquez, J. A., & Givnish, T. J. (1998). Altitudinal gradients in tropical forest composition, structure, and diversity in the Sierra de Manantlán. *Journal of Ecology*, 999–1020.

Jacobs, M. (1976). The study of lianas. *Flora Malesiana Bulletin, 29*(1), 2610–2618.

Jacobsen, A. L., & Pratt, R. B. (2012). No evidence for an open vessel effect in centrifuge-based vulnerability curves of a long-vesselled liana (Vitis vinifera). *New Phytologist, 194*(4), 982–990.

Jamhuri, J., Samantha, L. D., Tee, S. L., Kamarudin, N., Ashton-Butt, A., Zubaid, A., & Azhar, B. (2018). Selective logging causes the decline of large-sized mammals including those in unlogged patches surrounded by logged and agricultural areas. *Biological Conservation, 227*, 40–47.

Janzen, D. H. (1973). Host plants as islands. II. Competition in evolutionary and contemporary time. *The American Naturalist, 107*(958), 786–790.

Janzen, D. H. (1969). Allelopathy by myrmecophytes: The ant Azteca as an allelopathic agent of Cecropia. *Ecology, 50*, 147–153.

Janzen, D. H. (1974). Tropical blackwater rivers, animals, and mast fruiting by the Dipterocarpaceae. *Biotropica, 6*(81), 2618.

Jiménez-Castillo, M., & Lusk, C. H. (2013). Vascular performance of woody plants in a temperate rain forest: Lianas suffer higher levels of freeze–thaw embolism than associated trees. *Functional Ecology, 27*(2), 403–412.

Jiménez-Castillo, M., Wiser, S. K., & Lusk, C. H. (2007). Elevational parallels of latitudinal variation in the proportion of lianas in woody floras. *Journal of Biogeography, 34*(1), 163–168.

Johnson, D. M., Domec, J. C., Woodruff, D. R., McCulloh, K. A., & Meinzer, F. C. (2013). Contrasting hydraulic strategies in two tropical lianas and their host trees. *American Journal of Botany, 100*(2), 374–383.

Jones, I. L., Peres, C. A., Benchimol, M., Bunnefeld, L., & Dent, D. H. (2017). Woody lianas increase in dominance and maintain compositional integrity across an Amazonian dam-induced fragmented landscape. *PLoS One, 12*(10), e0185527.

Kadavul, K., & Parthasarathy, N. (1999). Forest sites on the Kalrayan hills, Eastern Ghats, south India. *Tropical Biodiversity, 6*(3), 197–208.

Kammesheidt, L., Berhaman, A., Tay, J., Abdullah, G., & Azwal, M. (2009). Liana abundance, diversity and tree infestation in the Imbak Canyon conservation area, Sabah, Malaysia. *Journal of Tropical Forest Science*, 265–271.

Kathriarachchi, H. S., Tennakoon, K. U., Gunatilleke, C. V. S., Gunatilleke, I. A. U. N., & Ashton, P. M. S. (2004). Ecology of two selected liana species of utility value in a lowland rain forest of Sri Lanka: Implications for management. *Conservation and Society*, 273–288.

Kennard, D. K. (1998). Biomechanical properties of tree saplings and free-standing lianas as indicators of susceptibility to logging damage. *Forest Ecology and Management, 102*(2–3), 179–191.

Khadanga, S. S., Muthumperumal, C., & Parthasarathy, N. (2015). Changes in Liana diversity over a decade in Indian tropical dry evergreen forests. *Biodiversity in Tropical Ecosystems. New Delhi: Today and Tomorrow's Printers and Publishers*, 61–79.

Killeen, T. J., Jardim, A., Mamani, F., & Rojas, N. (1998). Diversity, composition and structure of a tropical semideciduous forest in the Chiquitania region of Santa Cruz. *Bolivia. Journal of Tropical Ecology, 14*(6), 803–827.

Koenig, S. E., Joseph, M., Wunderle, J. R., & Enkerlin-Hoeflich, E. C. (2007). Vines and canopy contact: A route for snake predation on parrot nests. *Bird Conservation International, 17*(1), 79–91.

Koffi, B. A., Kouame, D., & Yao, Y. C. A. (2016). Structure and composition of the liana assemblage of Azagny National Park in the Southern Cte dIvoire. *International Journal of Biodiversity and Conservation, 8*(9), 206–215.

Koning, C. (1933). C. and Taub. T. Impact buckling of thin bars in the elastic range hinged at both ends. *Luftfahrtforschung, 10*(2), 55–64.

Koningsberger, V. J. (1922). Tropismus und Wachstum. *Recueil des travaux botaniques néerlandais, 19*(1), 1–137.

Labeyrie, E., Pascal, L., Delabie, J., Orivel, J., & Dejean, A. (2001). Protection of Passiflora glandulosa (Passifloraceae) against herbivory: Impact of ants exploiting extra floral. *Sociobiology, 38*(3A).

Ladwig, L. M., & Meiners, S. J. (2009). Impacts of temperate lianas on tree growth in young deciduous forests. *Forest Ecology and Management, 259*(2), 195–200.

Laurance, W. F., Andrade, A. S., Magrach, A., Camargo, J. L., Valsko, J. J., Campbell, M., Fearnside, P. M., Edwards, W., Lovejoy, T. E., & Laurance, S. G. (2014). Long-term changes in liana abundance and forest dynamics in undisturbed Amazonian forests. *Ecology, 95*(6), 1604–1611.

Lebrun, J. (1937). Observations sur la morphologie et lécologie des lianes de la fôret équatoraile du Congo. *Bull Inst Roy Col Belge, 8*(1), 78–87.

Leicht, S. A., & Silander, J. A., Jr. (2006). Differential responses of invasive Celastrus orbiculatus (Celastraceae) and native C. scandens to changes in light quality. *American Journal of Botany, 93*(7), 972–977.

Lieberman, D., Lieberman, M., Peralta, R., & Hartshorn, G. S. (1996). Tropical forest structure and composition on a large-scale altitudinal gradient in Costa Rica. *Journal of Ecology*, 137–152.

Lobos-Catalán, P., & Jiménez-Castillo, M. (2019). Temperature shapes liana diversity pattern along a latitudinal gradient in southern temperate rainforest. *Plant Ecology, 220*(12), 1109–1117.

Londré, R. A., & Schnitzer, S. A. (2006). The distribution of lianas and their change in abundance in temperate forests over the past 45 years. *Ecology, 87*(12), 2973–2978.

Longino, J. T. (1986). A negative correlation between growth and rainfall in a tropical liana. *Biotropica*, 195–200.

Lü, X. T., Tang, J. W., Feng, Z. L., & Li, M. H. (2009). Diversity and aboveground biomass of lianas in the tropical seasonal rain forests of Xishuangbanna, SW China. *Revista de biología tropical, 57*(1–2), 211–222.

Lutz, H. (1943). Injuries to trees caused by Celastrus and Vitis. *Bulletin of the Torrey Botanical Club, 70.*

Macía, M. J. (2011). Spatial distribution and floristic composition of trees and lianas in different forest types of an Amazonian rainforest. *Plant Ecology, 212*(7), 1159–1177.

Magrach, A., Senior, R. A., Rogers, A., Nurdin, D., Benedick, S., Laurance, W. F., Santamaria, L., & Edwards, D. P. (2016). Selective logging in tropical forests decreases the robustness of liana–tree interaction networks to the loss of host tree species. *Proceedings of the Royal Society B: Biological Sciences, 283*(1826), 20153008.

Malizia, A., & Grau, H. R. (2006). Liana–host tree associations in a subtropical montane forest of north-western Argentina. *Journal of Tropical Ecology, 22*(3), 331–339.

Malizia, A., Grau, H. R., & Lichstein, J. W. (2010). Soil phosphorus and disturbance influence liana communities in a subtropical montane forest. *Journal of Vegetation Science, 21*(3), 551–560.

Martínez-Izquierdo, L., García, M. M., Powers, J. S., & Schnitzer, S. A. (2016). Lianas suppress seedling growth and survival of 14 tree species in a Panamanian tropical forest. *Ecology, 97*(1), 215–224.

Martins, M. M. (2009). Lianas as a food resource for brown howlers (Alouatta guariba) and southern muriquis (Brachyteles arachnoides) in a forest fragment. *Animal Biodiversity and Conservation, 32*(1), 51–58.

Mascaro, J., Schnitzer, S. A., & Carson, W. P. (2004). Liana diversity, abundance, and mortality in a tropical wet forest in Costa Rica. *Forest Ecology and Management, 190*(1), 3–14.

McGeoch, L., Gordon, I., & Schmitt, J. (2008). Impacts of land use, anthropogenic disturbance, and harvesting on an African medicinal liana. *Biological Conservation, 141*(9), 2218–2229.

McNab, W. H., & Meeker, M. (1987). Oriental bittersweet: a growing threat to hardwood silviculture in the Appalachians. *Northern Journal of Applied Forestry, 4*(4), 174–177.

Ménard, L., McKey, D., & Rowe, N. (2009). Developmental plasticity and biomechanics of treelets and lianas in Manihot aff. quinquepartita (Euphorbiaceae): A branch-angle climber of French Guiana. *Annals of Botany, 103*(8), 1249–1259.

Mesh, S., Cayetano, D. T., Requena, E., Alvarez, E., Kay, E., Finkral, A., & Putz, F. E. (2017). Logging impacts on liana regeneration and diversity in Belize. *Journal of Tropical Forest Science*, 343–348.

Mino, P. Y., Balslev, H., & Valencia, R. (1995). Useful lianas of the Siona-Secoya indians from Amazonian Ecuador. *Economic botany, 49*(3), 269–275.

Mohandass, D., Campbell, M. J., Beng, K. C., & Davidar, P. (2016). Influence of grazing intensity on swamp plant communities in the tropical montane wetland ecosystems, Nilgiris, Southern India. *Applied Ecology and Environmental Research, 14*(4), 233–268.

Mohandass, D., Campbell, M. J., Hughes, A. C., Mammides, C., & Davidar, P. (2017). The effect of altitude, patch size and disturbance on species richness and density of lianas in montane forest patches. *Acta Oecologica, 83*, 1–14.

Mohandass, D., Davidar, P., Somasundaram, S., Vijayan, L., & Beng, K. C. (2015). Influence of disturbance regime on liana species composition, density and basal area in the tropical montane evergreen forests (sholas) of the Western Ghats, India. *Tropical Ecology, 56*(2), 169–182.

Mohandass, D., Hughes, A. C., Campbell, M., & Davidar, P. (2014). Effects of patch size on liana diversity and distributions in the tropical montane evergreen forests of the Nilgiri Mountains, southern India. *Journal of Tropical Ecology, 30*(6), 579–590.

Molina-Freaner, F., Castillo Gámez, R., Tinoco-Ojanguren, C., & Castellanos, A. E. (2004). Vine species diversity across environmental gradients in northwestern Mexico. *Biodiversity & Conservation, 13*(10), 1853–1874.

Montgomery, G. G., & Sunquist, M. E. (1975). Impact of sloths on Neotropical forest energy flow and nutrient cycling. In *Tropical ecological systems* (pp. 69–98). Springer.

Mori, H., Kamijo, T., & Masaki, T. (2016). Liana distribution and community structure in an old-growth temperate forest: The relative importance of past disturbances, host trees, and microsite characteristics. *Plant Ecology, 217*(10), 1171–1182.

Moscovice, L. R., Issa, M. H., Petrzelkova, K. J., Keuler, N. S., Snowdon, C. T., & Huffman, M. A. (2007). Fruit availability, chimpanzee diet, and grouping patterns on Rubondo Island, Tanzania. *American Journal of Primatology: Official Journal of the American Society of Primatologists, 69*(5), 487–502.

Muhwezi, O., Cunningham, A. B., & Bukenya-Ziraba, R. (2009). Lianas and livelihoods: The role of fibrous forest plants in food security and society around Bwindi Impenetrable National Park, Uganda. *Economic botany, 63*(4), 340–352.

Mumbanza, F. M., Bauters, M., Kearsley, E., Boeckx, P., Lubini, C. A., & Verbeeck, H. (2020). Liana communities exhibit different species composition, diversity and community structure across forest types in the Congo Basin. *Biotropica, 52*(4), 651–663.

Muoghalu, J. I., & Okeesan, O. O. (2005). Climber species composition, abundance and relationship with trees in a Nigerian secondary forest. *African Journal of Ecology, 43*(3), 258–266.

Muthumperumal, C., & Parthasarathy, N. (2009). Angiosperms, climbing plants in tropical forests of southern Eastern Ghats, Tamil Nadu, India. *Check List, 5*(1), 092–111.

Muthumperumal, C., & Parthasarathy, N. (2010). A large-scale inventory of liana diversity in tropical forests of South Eastern Ghats, India. *Systematics and Biodiversity, 8*(2), 289–300.

Muthuramkumar, S., & Parthasarathy, N. (2000). Alpha diversity of lianas in a tropical evergreen forest in the Anamalais, Western Ghats, India. *Diversity and Distributions, 6*(1), 1–14.

Muthuramkumar, S., & Parthasarathy, N. (2001). Tree-liana relationships in a tropical evergreen forest at Varagalaiar, Anamalais, Western Ghats, India. *Journal of tropical ecology, 17*(3), 395–409.

Muthuramkumar, S., Ayyappan, N., Parthasarathy, N., Mudappa, D., Raman, T. S., Selwyn, M. A., & Pragasan, L. A. (2006). Plant community structure in tropical rain forest fragments of the Western Ghats, India 1. *Biotropica: The Journal of Biology and Conservation, 38*(2), 143–160.

Nabe-Nielsen, J. (2001). Diversity and distribution of lianas in a neotropical rain forest, Yasuní National Park, Ecuador. *Journal of Tropical Ecology, 17*(1), 1–19.

Nabe-Nielsen, J., Kollmann, J., & Peña-Claros, M. (2009). Effects of liana load, tree diameter and distances between conspecifics on seed production in tropical timber trees. *Forest Ecology and Management, 257*(3), 987–993.

Nepal, A., & Dorji, U. (2020). Composition and diversity pattern of climbers in tropical forest of Langchenphu, Jomotshangkha Wildlife Sanctuary. *Bhutan Journal of Natural Resources and Development, 7*(1), 11–11.

Nesheim, I., Halvorsen, R., & Nordal, I. (2010). Plant composition in the Maya Biosphere Reserve: Natural and anthropogenic influences. *Plant Ecology, 208*(1), 93–122.

Nurfazliza, K., Nizam, M. S., & Nur Supardi, M. N. (2012). Association of liana communities with their soil properties in a lowland forest of Negeri Sembilan, peninsular Malaysia. *Sains Malaysiana, 41*, 679–690.

Obee, D. J. (1934). Noteworthy anatomical features from the wood of some indigenous Kansas shrubs and lianas. *Transactions of the Kansas Academy of Science (1903-), 37*, 83–89.

Ødegaard, F. (2000). The relative importance of trees versus lianas as hosts for phytophagous beetles (Coleoptera) in tropical forests. *Journal of Biogeography, 27*(2), 283–296.

Ofosu-Bamfo, B., Addo-Fordjour, P., & Belford, E. J. (2019). Does road-edge affect liana community structure and liana-host interactions in evergreen rainforests in Ghana? *Acta Oecologica, 101*, 103476.

Ogle, C. C., Cock, G. L., Arnold, G., & Mickleson, N. (2000). Impact of an exotic vine Clematis vitalba (F. Ranunculaceae) and of control measures on plant biodiversity in indigenous forest, Taihape, New Zealand. *Austral Ecology, 25*(5), 539–551.

Pace, M. R., Lohmann, L. G., & Angyalossy, V. (2009). The rise and evolution of the cambial variant in Bignonieae (Bignoniaceae). *Evolution & Development, 11*(5), 465–479.

Padaki, A., & Parthasarathy, N. (2000). Abundance and distribution of lianas in tropical lowland evergreen forest of Agumbe, central Western Ghats, India. *Tropical Ecology, 41*(2), 143–154.

Pandian, E., & Ravichandran, P. (2019). Diversity and threatened climber plants in tropical forests of Courtallam Hills, Southern Western Ghats, India. *Journal of Tropical Forestry and Environment, 9*(2).

Pandian, E., Parthasarathy, N., & Anbarashan, M. (2017). Liana growth, mortality and recruitment in four inland tropical dry evergreen forest sites of Peninsular India. *Journal of Global Ecology and Environment, 6*(2), 60–72.

Parren, M., & Bongers, F. (2001). Does climber cutting reduce felling damage in southern Cameroon? *Forest Ecology and Management, 141*(3), 175–188.

Parthasarathy, N. (Ed.). (2015). *Biodiversity of lianas* (Vol. 5). Springer.

Pasquini, S. C., Wright, S. J., & Santiago, L. S. (2015). Lianas always outperform tree seedlings regardless of soil nutrients: Results from a long-term fertilization experiment. *Ecology, 96*(7), 1866–1876.

Penfound, W. T., & Howard, J. A. (1940). A phytosociological study of an evergreen oak forest in the vicinity of New Orleans, Louisiana. *American Midland Naturalist*, 165–174.

Pérez-Salicrup, D. R., & De Meijere, W. (2005). Number of lianas per tree and number of trees climbed by lianas at Los Tuxtlas, Mexico 1. *Biotropica: The Journal of Biology and Conservation, 37*(1), 153–156.

Pérez-Salicrup, D. R., Sork, V. L., & Putz, F. E. (2001). Lianas and trees in a liana forest of Amazonian Bolivia 1. *Biotropica, 33*(1), 34–47.

Petroni, L. M., Huffman, M. A., & Rodrigues, E. (2017). Medicinal plants in the diet of woolly spider monkeys (Brachyteles arachnoides, E. Geoffroy, 1806)–a bio-rational for the search of new medicines for human use? *Revista Brasileira de Farmacognosia, 27*, 135–142.

Pfeiffer, N. E. (1926). Microchemical and morphological studies of effect of light on plants. *Botanical Gazette, 81*(2), 173–195.

Phillips, O. (1991). *The ethnobotany and economic botany of tropical vines. The biology of vines* (pp. 427–475). Cambridge University Press.

Phillips, O. L., Martínez, V. R., Arroyo, L., Baker, T. R., Killeen, T., Lewis, S. L., Malhi, Y., Mendoza, A. M., Neill, D., Vargas, P. N., Alexiades, M., Cerón, C., Fiore, A. D., Erwin, T., Jardim, A., Palacios, W., Saldias, M., & Vinceti, B. (2002). Increasing dominance of large lianas in Amazonian forests. *Nature, 418*(6899), 770–774.

Pinard, M. A., & Putz, F. E. (1994). Vine infestation of large remnant trees in logged forest in Sabah, Malaysia: Biomechanical facilitation in vine succession. *Journal of Tropical Forest Science*, 302–309.

Plumier, R. P. C. (1693). Description des Plantes de l'Amérique, avec leurs figures. L'Imprimerie Royale

Pooler, M. R., Dix, R. L., & Feely, J. (2002). Interspecific hybridizations between the native bittersweet, *Celastrus scandens*, and the introduced invasive species, *C. orbiculatus*. *Southeastern Naturalist, 1*(1), 69–76.

Proctor, J., Anderson, J. M., Chai, P., & Vallack, H. W. (1983). Ecological studies in four contrasting lowland rain forests in Gunung Mulu National Park, Sarawak: I. Forest environment, structure and floristics. *The Journal of Ecology*, 237–260.

Putz, F. (1980). Lianas vs. trees. *Biotropica, 12*.

Putz, F. E. (1978). A survey of virgin jungle reserves in Peninsular Malaysia. *A survey of virgin jungle reserves in Peninsular Malaysia*, (73).

Putz, F. E. (1984a). How trees avoid and shed lianas. *Biotropica*, 19–23.

Putz, F. E. (1984b). The natural history of lianas on Barro Colorado Island, Panama. *Ecology, 65*(6), 1713–1724.

Putz, F. E. (1985). Woody vines and forest management in Malaysia. *The Commonwealth Forestry Review*, 359–365.

Putz, F. E., & Chai, P. (1987). Ecological studies of lianas in Lambir national park, Sarawak, Malaysia. *The Journal of Ecology*, 523–531.

Putz, F. E., & Holbrook, N. M. (1991). Biomechanical studies of vines. *The Biology of Vines*, 73–97.

Putz, F. E., & Mooney, H. (1991). *The biology of vines*. Cambridge University Press.

Quigley, M. F., & Plantt, W. J. (1996). Structure and pattern in temperate seasonal forests. *Vegetatio, 123*(2), 117–138.

Rahmad, Z. B., & Akomolafe, G. F. (2018). Distribution, diversity and abundance of ferns in a tropical university campus. *Pertanika Journal of Tropical Agricultural Science, 41*(4).

Rajput, K. S., Gondaliya, A. D., & Moya, R. (2021). Structure of the secondary xylem and development of a cambial variant in Serjania mexicana (Sapindaceae). *IAWA Journal, 43*(1-2), 103–115.

Rashid, M. H., Asaeda, T., & Uddin, M. N. (2010). The allelopathic potential of kudzu (Pueraria montana). *Weed Science, 58*(1), 47–55.

Reddy, M. S., & Parthasarathy, N. (2003). Liana diversity and distribution in four tropical dry evergreen forests on the Coromandel coast of south India. *Biodiversity & Conservation, 12*(8), 1609–1627.

Reddy, M. S., & Parthasarathy, N. (2006). Liana diversity and distribution on host trees in four inland tropical dry evergreen forests of peninsular India. *Tropical Ecology, 47*(1), 109–124.

Rice, K., Brokaw, N., & Thompson, J. (2004). Liana abundance in a Puerto Rican forest. *Forest Ecology and Management, 190*(1), 33–41.

Rich, P. M., Lum, S., Munoz, L. E. D. A., & Quesada, M. A. U. R. I. C. I. O. (1987). Shedding of vines by the palms Welfia georgii and Iriartea gigantea. *Principes, 31*(1), 31–34.

Robertson, P. A., Weaver, G. T., & Cavanaugh, J. A. (1978). Vegetation and tree species patterns near the northern terminus of the southern floodplain forest. *Ecological Monographs, 48*(3), 249–267.

Rodríguez-Quintero, W. D., & Gianoli, E. (2016). Abundance and diversity of lianas in a Neotropical dry forest: The influence of soil moisture. *Plant Ecology and Evolution, 149*(3), 329–334.

Rossell, I. M., & Eggleston, H. (2017). Elevational distribution of temperate lianas along trails in Pisgah National Forest. *Southeastern Naturalist, 16*(3), 443–450.

Rowe, N. P., & Speck, T. (1996). Biomechanical characteristics of the ontogeny and growth habit of the tropical liana Condylocarpon guianense (Apocynaceae). *International Journal of Plant Sciences, 157*(4), 406–417.

Rowe, N., & Isnard, S. (2009). Biomechanics of climbing palms and how they climb. *Plant Signaling & Behavior, 4*(9), 875–877.

Salzer, J., Matezki, S., & Kazda, M. (2006). Nutritional differences and leaf acclimation of climbing plants and the associated vegetation in different types of an Andean montane rainforest. *Oecologia, 147*(3), 417–425.

Sanches, M. C., & Válio, I. F. M. (2006). Leaf optical properties of two liana species Canavalia parviflora Benth. and Gouania virgata Reissk in different light conditions. *Brazilian Journal of Botany, 29*, 319–330.

Schenck, H. (1892). *Contributions to the biology and anatomy of the lianas, in particular the species native to Brazil* (No. 4-5). G Fisher.

Schnitzer, S. A., & Bongers, F. (2002). The ecology of lianas and their role in forests. *Trends in Ecology & Evolution, 17*(5), 223–230.

Schnitzer, S. A., & Bongers, F. (2011). Increasing liana abundance and biomass in tropical forests: Emerging patterns and putative mechanisms. *Ecology Letters, 14*(4), 397–406.

Schnitzer, S. A., & Bongers, F. J. J. M. (2005). Lianas and gap-phase regeneration: Implications for forest dynamics and species diversity. *Forest Climbing Plants of West Africa: Diversity, Ecology and Management*, 59–72.

Schnitzer, S. A., & Carson, W. P. (2010). Lianas suppress tree regeneration and diversity in treefall gaps. *Ecology Letters, 13*(7), 849–857.

Schnitzer, S. A., Kuzee, M. E., & Bongers, F. (2005). Disentangling above-and below-ground competition between lianas and trees in a tropical forest. *Journal of Ecology, 93*(6), 1115–1125.

Schnitzer, S. A., Mangan, S. A., Dalling, J. W., Baldeck, C. A., Hubbell, S. P., Ledo, A., & Yorke, S. R. (2012). Liana abundance, diversity, and distribution on Barro Colorado Island, Panama. *PloS One, 7*(12), e52114.

Schnitzer, S. A., Putz, F. E., Bongers, F., & Kroening, K. (2015). The past, present, and potential future of liana ecology. *Ecology of Lianas*, 1–10.

Schnitzer, S. A., Rutishauser, S., & Aguilar, S. (2008). Supplemental protocol for liana censuses. *Forest Ecology and Management, 255*(3-4), 1044–1049.

Schnitzler, A. (1995). Community ecology of arboreal lianas in gallery forests of the Rhine valley, France. *Ecologica, 16*.

Schnitzler, A., Amigo, J., Hale, B., & Schnitzler, C. (2016). Patterns of climber distribution in temperate forests of the Americas. *Journal of Plant Ecology, 9*(6), 724–733.

Scholander, P., Love, W., & Kanwisher, J. (1955). The rise of sap in tall grapevines. *Plant Physiology, 30*, 93–104.

Scholander, P., Ruud, B., & Leivestad, H. (1957). The rise of sap in a tropical liana. *Plant Physiology, 32*, 1–6.

Senbeta, F., Schmitt, C., Denich, M., Demissew, S., Velk, P. L., Preisinger, H., & Teketay, D. (2005). The diversity and distribution of lianas in the Afromontane rain forests of Ethiopia. *Diversity and Distributions, 11*(5), 443–452.

Sfair, J. C., Rochelle, A. L. C., Rezende, A. A., & Martins, F. R. (2016). Liana avoidance strategies in trees: Combined attributes increase efficiency. *Tropical Ecology, 57*(3), 559–566.

Sfair, J. C., Rochelle, A. L. C., Rezende, A. A., Van Melis, J., de Lara Weiser, V., & Martins, F. R. (2010). Nested liana-tree network in three distinct neotropical vegetation formations. *Perspectives in Plant Ecology, Evolution and Systematics, 12*(4), 277–281.

Sfair, J. C., Rochelle, A. L., Van Melis, J., Rezende, A. A., Weiser, V. D. L., & Martins, F. R. (2015). Theoretical approaches to liana management: A search for a less harmful method. *International Journal of Biodiversity Science, Ecosystem Services & Management, 11*(2), 89–95.

Sharma, S., Shahzad, A., Akhtar, R., & Upadhyay, A. (2016). Micropropagation: A boon for conservation of valuable Vines and Lianas. In *Biotechnological strategies for the conservation of medicinal and ornamental climbers* (pp. 163–193). Springer.

Shigo, A. L. (1986). A new tree biology: Facts, photos, and philosophies on trees and their problems and proper care.

Smith, H. C., & McCay, R. E. (1979). Estimating time required to cut grapevines in young, even-aged hardwood stands. *Southern Journal of Applied Forestry, 3*(3), 125–127.

Steenis, C. (1939). Ecological observations on the genus Pleiocraterium in the Gajo-lands, N. Sumatra. *Rec. Trav. Bot. Neerl, 36*, 446–448.

Sundarapandian, S. M., Muthumperumal, C., & Subashree, K. (2016). Biological invasion of vines, their impacts and management. In *Biodiversity of lianas* (pp. 211–253). Springer.
Swaine, M. D., & Grace, J. (2007). Lianas may be favoured by low rainfall: Evidence from Ghana. *Plant Ecology, 192*(2), 271–276.
Tay, A. C., & Furukawa, A. (2008). Variations in leaf stomatal density and distribution of 53 vine species in Japan. *Plant Species Biology, 23*(1), 2–8.
Teramura, A. H., Gold, W. G., & Forseth, I. N. (1991). *Physiological ecology of mesic, temperate woody vines. The biology of vines* (pp. 245–285). Cambridge University Press.
Tobin, M. F., Wright, A. J., Mangan, S. A., & Schnitzer, S. A. (2012). Lianas have a greater competitive effect than trees of similar biomass on tropical canopy trees. *Ecosphere, 3*(2), 1–11.
Toledo-Aceves, T., & Swaine, M. D. (2008). Above-and below-ground competition between the liana Acacia kamerunensis and tree seedlings in contrasting light environments. *Plant Ecology, 196*(2), 233–244.
Tra Bi, F. H., Kouamé, F. N., & Traoré, D. (2005). Utilisation of climbers in two forest reserves in West Côte d'Ivoire. *Forest Climbing Plants of West Africa: Diversity, Ecology and Management*, 167–181.
Uwalaka, N. O., & Muoghalu, J. I. (2017). Nine years of climber community dynamics in a Nigerian lowland rain forest 31 years after a ground fire. *Biodiversity and Conservation, 26*(4), 997–1008.
Uwalaka, N. O., Khapugin, A. A., & Muoghalu, J. I. (2020). Effect of some environmental factors on liana abundance in a regenerating secondary lowland rainforest in Nigeria three decades after a ground fire. *Ecological Questions, 31*(2), 27–38.
Valladares, F., Gianoli, E., & Saldana, A. (2011). Climbing plants in a temperate rainforest understorey: Searching for high light or coping with deep shade? *Annals of Botany, 108*(2), 231–239.
Van der Heijden, G. M., & Phillips, O. L. (2009). Environmental effects on Neotropical liana species richness. *Journal of Biogeography, 36*(8), 1561–1572.
Van der Heijden, G. M., Healey, J. R., & Phillips, O. L. (2008). Infestation of trees by lianas in a tropical forest in Amazonian Peru. *Journal of Vegetation Science, 19*(6), 747–756.
Van der Sande, M. T., Poorter, L., Schnitzer, S. A., Engelbrecht, B. M., & Markesteijn, L. (2019). The hydraulic efficiency–safety trade-off differs between lianas and trees. *Ecology, 100*(5), e02666.
Van Eck, N., & Waltman, L. (2010). Software survey: VOSviewer, a computer program for bibliometric mapping. *Scientometrics, 84*(2), 523–538.
Vidal, E., Johns, J., Gerwing, J. J., Barreto, P., & Uhl, C. (1997). Vine management for reduced-impact logging in eastern Amazonia. *Forest Ecology and Management, 98*(2), 105–114.
Vivek, P., & Parthasarathy, N. (2015). Liana community and functional trait analysis in tropical dry evergreen forest of India. *Journal of Plant Ecology, 8*(5), 501–512.
Vivek, P., & Parthasarathy, N. (2016). Host-preference of four dominant liana species with contrasting climbing strategies in tropical dry evergreen forest of India. *Journal of Global Ecology and Environment, 4*(2), 86–91.
Vivek, P., & Parthasarathy, N. (2017). Patterns of tree-liana interactions: Distribution and host preference of lianas in a tropical dry evergreen forest in India. *Tropical Ecology, 58*(3), 591–603.
Vivek, P., & Parthasarathy, N. (2018). Contrasting leaf-trait strategies in dominant liana and tree species of Indian tropical dry evergreen forest. *Flora, 249*, 143–149.
Weiser, M. D., Enquist, B. J., Boyle, B., Killeen, T. J., Jørgensen, P. M., Fonseca, G., & Martínez, R. V. (2007). Latitudinal patterns of range size and species richness of New World woody plants. *Global Ecology and Biogeography, 16*(5), 679–688.
Werden, L. K., Waring, B. G., Smith-Martin, C. M., & Powers, J. S. (2018). Tropical dry forest trees and lianas differ in leaf economic spectrum traits but have overlapping water-use strategies. *Tree Physiology, 38*(4), 517–530.

Whitney, K. D., & Smith, T. B. (1998). Habitat use and resource tracking by African Ceratogymna hornbills: Implications for seed dispersal and forest conservation. In *Animal conservation forum* (Vol. 1, No. 2, pp. 107–117). Cambridge University Press.

Wright, S. J., Sun, I. F., Pickering, M., Fletcher, C. D., & Chen, Y. Y. (2015). Long-term changes in liana loads and tree dynamics in a Malaysian forest. *Ecology, 96*(10), 2748–2757.

Yang, S. Z., & Chen, P. H. (2015). Classifying Taiwan lianas with radiating plates of xylem. *Taiwania, 60*(4), 151–159.

Yang, S. Z., Fan, H., Li, K. W., & Ko, T. Y. (2018). How the diversity, abundance, size and climbing mechanisms of woody lianas are related to biotic and abiotic factors in a subtropical secondary forest, Taiwan. *Folia Geobotanica, 53*(1), 77–88.

Yuan, C. M., Liu, W. Y., Tang, C. Q., & Li, X. S. (2009). Species composition, diversity, and abundance of lianas in different secondary and primary forests in a subtropical mountainous area, SW China. *Ecological Research, 24*(6), 1361–1370.

Yuan, C. M., Wu, T., Geng, Y. F., Chai, Y., & Hao, J. B. (2016). Phenotypic plasticity of lianas in response to altered light environment. *Ecological Research, 31*(3), 375–384.

Zhu, S. D., & Cao, K. F. (2009). Hydraulic properties and photosynthetic rates in co-occurring lianas and trees in a seasonal tropical rainforest in southwestern China. *Plant Ecology, 204*(2), 295–304.

3 Ecology of Lianas: Diversity and Distribution

Abstract

It is a central theme in ecology to understand what determines patterns of species diversity and distribution. Plant diversity and abundance are frequently linked to a wide range of underlying environmental gradients and ecological interactions. The lianas are no different. Liana diversity and abundance vary significantly between habitats, landscapes, and even within the tropics. Many factors have been tested at the local and continental scales to address such variations in liana species composition. However, the generality of liana distribution patterns remains unclear, as studies have produced contradictory results. In this chapter, I review and discuss the putative environmental determinants of global liana diversity and distribution. In addition, using India as a case study, I discuss recent trends in increasing liana abundance and biomass in the tropics, as well as potential underlying mechanisms.

Keywords

Environmental determinants · Liana diversity · Anthropogenic disturbance · Soil properties · Edge effect

3.1 Introduction

Lianas are one of the most distinctive features of tropical forest ecosystems. Their ecological significance, particularly in determining the structural and functional diversity of tropical forests, is now more obvious than ever (Schnitzer et al., 2015). According to quantitative ecological studies, woody climbers can account for up to 35% of total woody species diversity and richness in the tropics (Vivek & Parthasarathy, 2015). Furthermore, recent studies have revealed an increased dominance of lianas in tropical forests, although the underlying mechanisms favoring

V. Pandi, *Taxonomy and Ecology of Climbers: Climbing Plants of India*, https://doi.org/10.1007/978-981-19-8645-1_3

lianas are unclear (Schnitzer & Bongers, 2011). Understanding the causes of increased liana dominance will necessitate a better understanding of the factors that control liana community structure, which is still inconclusive, as studies have yielded contradictory results (See Table 3.1). The distinction between tropical and temperate ecosystems (latitudinal variation) in terms of liana diversity and abundance is arguably well established (Gentry, 1991; Schnitzer, 2005; Hu et al., 2010; DeWalt et al., 2015; Schnitzler et al., 2016). Similarly, there is substantial evidence for a negative association between elevation and the diversity and abundance of liana species (Hu et al., 2010; Bruy et al., 2017; Yang et al., 2018). The other key drivers, however, are essentially unequivocal. For example, mean annual rainfall has been identified as a major determinant of liana abundance in the tropics. While some studies found that liana abundance and biomass are positively related to MAP (e.g., Guo et al., 2012; Durigon et al., 2013), others found the opposite (e.g., DeWalt et al., 2010; Swaine & Grace, 2007). Physical features such as soil properties and related parameters follow a similar pattern. Some studies found a strong association between liana community distribution and soil features (Malizia et al., 2010; DeWalt et al., 2006), but see Clinebell et al. (1995) and Perez-Salicrup et al. (2001). The availability of suitable trellis and the age of the forest stand all play important roles in controlling liana diversity and abundance (Van der Heijden & Phillips, 2009a, 2009b; Balfour & Bond, 1993; Yang et al., 2018). However, this association may be weak (see Malizia et al., 2010). Increased forest disturbance and elevated CO_2 were proposed by Schnitzer and Bongers (2011) as one of the putative mechanisms to explain the global increase in liana abundance, but Marvin et al. (2015) found no evidence that elevated CO_2 gives lianas an advantage over trees.

Similarly, a shift in a variety of abiotic factors occurring during the edge or fragmentation process may benefit lianas more than trees (Magnago et al., 2017). In contrast, Ofosu-Bamfo et al. (2019) found only a weak relationship between edge effect and liana composition in a Ghanaian evergreen forest. A slew of studies has recently reported on the edge effect on liana diversity and abundance (Addo-Fordjour & Owusu-Boadi, 2016). Furthermore, several studies have shown that both natural and anthropogenic forest disturbance can increase liana abundance and change liana composition (e.g., Addo-Fordjour et al., 2009, 2012, 2013; Malizia et al., 2010; Mohandass et al., 2015; Mori et al., 2016, but see Nesheim et al. (2010). Moreover, extreme levels of disturbance can have a negative impact on lianas as well. Overall, it is clear that the structural composition of lianas is likely influenced by a variety of factors that differently influence the patterns of liana community structure and assemblage (Fig. 3.1). The sections that follow discuss some of the most common associations between lianas and environmental attributes, as well as their ecological significance.

Table 3.1 A list of studies that have been conducted at the local and continental scales to determine the environmental determinants of the liana community assemblage

	Association			
Parameter	Diversity	Abundance	Growth	Basal area/ Biomass
Rainfall				
Lobos-Catalán & Jiménez-Castillo (2019)	Nil			
Longino (1986)			–	
Clinebell et al. (1995)				
Swaine and Grace (2007)	–			
DeWalt et al. 2010)		–		
Guo et al. (2012)	+	+		
Durigon et al. (2013)	+			
Hu et al. (2010)	+			
Van der Heijden and Phillips (2009a, 2009b)	+			
Temperature				
Lobos-Catalán and Jiménez-Castillo (2019)	–			
Durigon et al. (2013)	–			
Seasonality				
Lobos-Catalán and Jiménez-Castillo (2019)	Nil			
Durigon et al. (2013)	–			
DeWalt et al. (2010)		+		
Van der Heijden and Phillips (2009a, 2009b)	–			
Light				
Baars et al. (1998)	Nil			
Gianoli et al. (2010)	Nil			
Putz (1984a, b)	Nil			
Edaphic factors				
Clinebell et al. (1995)	Nil	Nil		
Balfour and Bond (1993)		+		
Ibarra-Manríquez and Martínez-Ramos (2002)	Nil			
Lawrence and Schlesinger (2001) (soil phosphorous)				+
DeWalt et al. (2006)	+			
Pérez-Salicrup et al. (2001)		–		
Malizia et al. (2010) (soil phosphorous)		+		
Rodríguez-Quintero and Gianoli (2016) (soil water availability)		+		
Rodríguez-Quintero and Gianoli (2016) (soil moisture)	–			

(continued)

Table 3.1 (continued)

Parameter	Association			
	Diversity	Abundance	Growth	Basal area/ Biomass
Altitude/elevation				
Clinebell et al. (1995)				
Balfour and Bond (1993)		+		
Lieberman et al. (1996)		–		
Vazquez and Givnish (1998)	–			
Padaki & Parthasarathy (2000)	–			
Jiménez-Castillo et al. (2007)	–			
Chettri et al. (2010)	–			
Alves et al. (2012)	–			–
Rossell and Eggleston (2017)	–			
Bruy et al. (2017)	–			
Mohandass et al. (2017)	–	–		
Yang et al. (2018)	–			
Latitude				
Molina-Freaner et al. (2004)	–			
Lobos-Catalán and Jiménez-Castillo (2019)	–			
Hu et al. (2010)	–			
Disturbance/tree-fall gap				
Ibarra-Manríquez and Martínez-Ramos (2002) (GAP)		+		
Malizia et al. (2010) (GAP)		+		
Addo-Fordjour et al. (2009)		–		+
Nesheim et al. (2010)	Nil			
Mohandass et al. (2015)	Nil	+		+
Forest structural properties				
Van der Heijden and Phillips (2009a, 2009b)	+			
Balfour and Bond (1993)	+			
Malizia et al. (2010)	–			
Yang et al. (2018)	+	+		
Allen et al. (1997)	+	+		
Baars et al., (1998)	+	+		
Dewalt et al., (2000) (forest stand age)	–	–		
Mohandass et al. (2017) (patch size)	+	+		
Forest edge/fragmentation				
Wu et al. (2010)	+	+		
Campbell et al. (2018)	+	+		
Ofosu-Bamfo et al. (2019)	Nil	Nil		

+ denotes a positive relationship, while – denotes a negative relationship between environmental parameters and liana diversity/abundance/growth rate/basal area or biomass. *Nil* is used in case of no established relationship

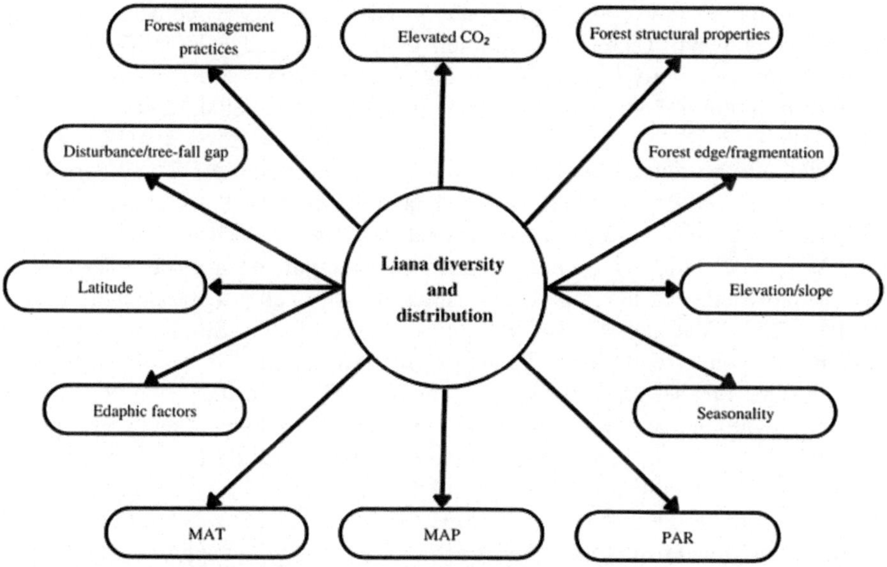

Fig. 3.1 Schematic representation of various environmental and structural attributes that are likely to influence the patterns of liana diversity and distribution. (MAT—mean annual temperature; MAP—mean annual precipitation; PAR—photosynthetically active radiation/light availability)

3.2 Factors Determining Liana Diversity and Abundance

Several studies have reported a sharp decline in liana diversity and abundance along the latitudinal gradients (e.g., Molina-Freaner et al., 2004). Similarly, numerous studies have shown a negative relationship between elevation and liana species diversity (Padaki & Parthasarathy, 2000; Vazquez & Givnish, 1998). The temperature is predicted to play a significant effect in determining liana species diversity in both scenarios (Lobos-Catalán & Jiménez-Castillo, 2019). The inability of lianas to endure colder temperatures could explain their poor establishment at higher elevations (Jiménez-Castillo et al., 2007). Higher altitudes may also have higher rainfall, higher air humidity, and slower tree turnover rates (Marrs et al., 1988), which could explain the drop in liana abundance along the elevational gradients. According to Alves et al. (2012), lowland forests support more lianas than any other forest along the elevational range. Likewise, Balfour and Bond (1993) found higher climber densities in the lower elevational plots.

Lianas in general have wider and longer xylem vessels that are prone to freeze-based embolism (Schnitzer & Bongers, 2002; Rossell & Eggleston, 2017). Despite possessing an effective vascular system, cold temperatures at high elevations might impair liana distribution by encouraging embolism in xylem vessels (Ewers, 1985), reducing the hydraulic efficiency, thus limiting large lianas to lower elevations, and warmer places (Gentry, 1991; Hu et al., 2010). According to Schnitzer (2005), lianas growing at moderately colder temperatures may have vessel elements that are smaller than typical. Therefore, minimum temperature and cold intolerance are among the major environmental factors that seem to have a strong control on the global distribution of lianas (Gentry, 1991; Jiménez-Castillo et al., 2007; Hu et al., 2010).

Rainfall and seasonality are known to influence liana abundance and composition. However, unlike the elevational and latitudinal gradients, it is ambiguous; investigations have produced inconsistent results. The key drivers of liana abundance in tropical climates have been identified as precipitation and seasonality, with liana rising with decreasing rainfall and increasing seasonality (DeWalt et al., 2010; Schnitzer, 2005; Van der Heijden & Phillips, 2009a, 2009b). Similarly, soil water availability is connected to liana abundance (Ibarra-Manríquez & Martínez-Ramos, 2002). Molina-Freaner and Tinoco-Ojanguren (1997) found that liana abundance decreased with decreasing soil water availability among the desert plant communities in Mexico. Considering the large and wider vessels, lianas are sensitive to low water availability (Ewers et al., 1990; Gartner et al., 1990). Following a prolonged drought, the xylem may become dysfunctional due to vessel embolism, impacting individual liana forms. This trend, however, may not apply to all types of lianas. For example, root climbers, a distinct category of climbers, are known to be favorably associated with rainfall (Durigon et al., 2013). Similarly, Guo et al. (2012) revealed that lianas and vines respond differently to rainfall gradients. They showed that liana species diversity was positively related to rainfall, but vine species diversity remained constant, showing that lianas are more sensitive to water availability than their herbaceous counterparts.

Besides temperature, rainfall and seasonality, the soil features are expected to play a major role in controlling liana abundance. Soil pH and phosphorus levels, in particular, have been shown to be positively correlated with liana abundance (e.g., Malizia et al., 2010). Nurfazliza et al. (2012) found that liana diversity and density are influenced by soil pH and soil phosphorous. DeWalt et al. (2006) found that soil phosphorus content is positively correlated with liana species diversity. In contrast, nitrogen concentration has been shown to be inversely related to climber abundance (Balfour & Bond, 1993; Van Daalen, 1984). Similarly, liana diversity may be maintained independently of soil fertility. For instance, Ibarra-Manríquez and Martínez-Ramos (2002) showed that the nutrient-poor forest sites had a higher species diversity.

The role of forest structural characteristics is also an important factor in determining liana abundance and diversity. The availability of suitable trellis, for example, and the properties of host trees have a significant impact on liana distribution patterns (Putz & Chai, 1987; Teramura et al., 1991). Because lianas are structural parasites that rely primarily on trees to reach the forest canopy, maximum liana abundance is frequently associated with high host tree availability (Ibarra-Manríquez & Martínez-Ramos, 2002; Allen et al., 1997). For example, DeWalt et al. (2000) found that lianas were significantly more abundant and diverse in younger forests than in older forests. This could be due to the availability of small-diameter stems that support a wide range of climbing strategies. The proportion of tendril climbers, for example, decreases with increasing stand age (DeWalt et al., 2000). While other guilds have limitations on the maximum support diameter (for example, tendril climbers), root climbers do not (Putz, 1984a, b; Hegarty & Caballé, 1991). In contrast, Alves et al. (2012) found that forest structure is not a good predictor of liana community composition in the Atlantic forest.

The availability of light is also critical for the growth and establishment of climbers (Teramura et al., 1991). The higher liana abundance and biomass in disturbed sites are attributed to a higher proportion of gaps in the site, which provide high light availability. Past canopy disturbance levels and gap formation rates may have an impact on the liana community composition (Ibarra-Manríquez & Martínez-Ramos, 2002). Changes in the forest canopy can affect liana populations by changing the light environment, increasing establishment opportunities, and providing trellis for climbing (Allen et al., 1997). Liana growth is restricted under a closed canopy due to a lack of available resources and a lack of suitable natural trellis (Putz, 1980). In contrast, Gianoli et al. (2010) found no significant changes in liana diversity across the different light environments. The size and frequency of canopy gaps, as well as the type and persistence of canopy disturbance, may influence vine invasion at the community level (Hegarty & Caballé, 1991). Furthermore, the size of the forest patch can affect liana diversity and abundance. Mohandass et al. (2017) found that liana diversity and density correlated positively with patch size, implying that liana adds more complexity to the patch as it expands and is likely to play a role in the forest successional process (Letcher, 2015).

3.3 Synthesis

Overall, it is evident that liana community structure is governed by multiple factors, but this is not consistent across sites. Whether our understanding of lianas as single functional types could be attributed to such differences? It could be possible because each climbing mechanisms have their own constraint for growth and establishment. Structural traits such as trellis availability and stem diameter, for example, may influence the distribution patterns of tendril climbers, which can dominate early successional forests, whereas root climbers are largely free of stem diameter constraints. Furthermore, unlike other climber forms, root climbers have been shown to correlate positively with rainfall and negatively with seasonality. Scramblers, on the other hand, can extend their self-supporting phase and are largely unconstrained by trellis availability. Furthermore, evidence of structural and functional divergence among lianas with different climbing mechanisms, particularly between active and passive climbers, is growing. Scramblers, for example, have been shown to differ from other climber forms in some key functional traits. More importantly, scrambles' functional strategy was more akin to self-standing plants like trees and shrubs than other climbers. For example, Xiao et al. (2021) and Ewers et al. (1990) found that scramblers differ from other climbing forms in stem anatomical, hydraulic, and biomechanical properties, particularly vessel dimensions, which converged with self-standing plants. This could be due to scramblers' extended self-supporting phase, as other climbers barely register a visible self-supporting phase. Congeric comparisons (*Lonicera*) of xylem vessel dimensions revealed that the values of scramblers were comparable to those of shrubs (Chiu & Ewers, 1992). In general, lianas are expected to have a suite of traits that favor higher hydraulic efficiency in order to improve the specific conductivity of their narrow stems. Similarly, their biomechanical properties are typically aligned to support stem flexibility in order to avoid breaking and twisting while climbing (Rowe & Speck, 2004). However, contrary to popular belief, scramblers appear to prefer self-sustaining growth over climbing habits. More evidence-based studies, however, are required to support this assertion. Nonetheless, scramblers' shrub-like strategy may provide them with a selective advantage over other climbers in drier/seasonal forests. Drought-induced embolism may occur in lianas because their wider and longer vessels are sensitive to low water availability. Similarly, the presence of relatively narrow vessels may be advantageous in avoiding vascular dysfunction caused by freeze-induced embolism at higher altitudes with colder temperatures and higher rainfall.

Furthermore, Dias et al. (2021) found that scramblers differ from other climber forms in leaf functional traits, including twiners and tendril climbers. They also revealed that most active climbers use an acquisitive strategy for c-gaining, whereas scramblers use a conservative strategy. This is contrary to popular belief that lianas are highly light-demanding or light-hungry plants (e.g., Putz, 1984a, b). Plants with acquisitive and conservative strategies are typically found at the extremes of the leaf economic spectrum (Kattge et al., 2020), implying functional divergence between scramblers and other climbers. These anatomical, biomechanical, and photosynthetic characteristics are essentially highly sensitive to environmental cues such as light availability, rainfall, and seasonality. Therefore, the structural and functional divergence of scramblers from other climbing forms has important implications for the generality of global liana diversity and distribution. To our knowledge, most studies have tested the environmental determinants of liana diversity and distribution by treating lianas as a single functional group. As a result, generality in liana diversity and distribution patterns has not yet been achieved, as studies at the local and continental scales produced contradictory results, limiting our understanding of their global distribution. Many studies have been conducted to compare the structural and functional differences between lianas and trees, but few have attempted to comprehend the structural and functional differences among lianas. However, recent research findings support the need for decentralized liana research. This came at a critical time, as evidence of liana dominance in tropical forests around the world continues to mount. Understanding the nuances of all parameters' combined efforts (multivariate analysis) while considering all climbing mechanisms as distinct functional types is thus required.

3.4 Increase in Liana Biomass and Abundance: Evidence from India's Seasonally Dry Tropical Forests

Evidence keeps surging on the dominance of lianas across the tropics (e.g., Ingwell et al., 2010; Laurance et al., 2014; Jones et al., 2017). Though few African studies (e.g., Caballé & Martin, 2001; Ewango, 2010) contradict the findings from the Neotropics, long-term monitoring studies from India support the liana dominance hypothesis. The tropical dry evergreen forests (TDEFs) of India are among the liana-dense ecosystems (e.g., Dewalt et al., 2015), where they account for up to 50% of woody species diversity and density (Vivek & Parthasarathy, 2015). Recent studies from six 1-ha TDEF monitoring plots show that liana density has increased by 30% and biomass has increased by 60% since the initial inventory over the last two decades (Table 3.2).

Tree density, on the other hand, had decreased by 30% over the last two decades (Parthasarathy, 2015). Although studies indicate that human-mediated disturbances might play a role in increasing liana abundance and basal area across census plots, the extent or threshold to which the disturbance promotes liana proliferation remains unknown. For example, in site 6, the recent census revealed a significant decline in liana stem density, though there was a significant increase during the first decadal interval, indicating that extreme levels of anthropogenic disturbances can have adverse effects on lianas as well. In general, a minimum to moderate disturbance can induce liana proliferation, through tree-fall gaps and increased light availability. However, the lack of suitable trellis or shortage of host trees caused due to extreme levels of disturbance could adversely affect the liana community assemblage and colonization structure. Increased liana infestation rates may also increase the burden on host trees, causing them to carry more liana loads. The lack of long-term censusing data on lianas across different forest types in India hinders our efforts to investigate the potential role of other environmental factors. Nonetheless, there is compelling evidence for growing liana dominance in India's seasonally dry tropical forest sites. Though the underlying mechanisms are unclear, anthropogenic pressure appears to have played a significant role in India's TDEFs. Whatsoever, the findings from India demand immediate attention in order to decipher the biological mechanisms that favor lianas over trees.

Table 3.2 Consolidated details of changes in liana species richness (SR, ha^{-1}), density (SD, ha^{-1}), and basal area (BA, m^2 ha^{-1}) in various seasonally dry tropical forest sites of India. All study sites are 1-ha contiguous plots

		Census 1			Census 2			Census 3			Net change			
	Census interval	SR	SD	BA	SR	SD	BA	SR	SD	BA	SR	SD	BA	References
Site 1	2003–2013	21	596	1.68	22	760	2.06	–	–	–	1	164	0.38	1, 2
Site 2	2003–2013	26	792	1.09	26	1002	1.76	–	–	–	0	210	0.67	1, 2
Site 3	2003–2013	23	515	0.67	23	726	0.86	–	–	–	0	211	0.19	1, 2
Site 4	2003–2013	27	775	0.44	30	937	1.06	–	–	–	3	162	0.62	3, 4, 5
Site 5	2002–2012–2020	24	812	2.77	20	1259	5.83	18	1262	5.03	- 6	450	3.06	3, 4, 5
Site 6	2003–2012–2020	28	835	1.82	29	940	2.25	27	702	2.75	- 1	- 133	0.93	3, 4, 5

1—Pandian & Parthasarathy, 2016; 2—Pandian et al., 2017; 3—Khadanga et al., 2015; 4—Nath et al., 2022; 5—Nath et al., (unpublished)

References

Addo-Fordjour, P., & Owusu-Boadi, K. (2016). Linear edge effects on liana and tree communities in two tropical forest ecosystems in Ghana. *Ecological Research, 31*(5), 709–718.

Addo-Fordjour, P., Obeng, S., Anning, A. K., & Addo, M. G. (2009). Floristic composition, structure and natural regeneration in a moist semi-deciduous forest following anthropogenic disturbances and plant invasion. *International Journal of Biodiversity and Conservation, 1*(2), 21–37.

Addo-Fordjour, P., Rahmad, Z. B., & Shahrul, A. M. S. (2012). Effects of human disturbance on liana community diversity and structure in a tropical rainforest, Malaysia: Implication for conservation. *Journal of Plant Ecology, 5*(4), 391–399.

Addo-Fordjour, P., El Duah, P., & Agbesi, D. K. K. (2013). *Factors influencing liana species richness and structure following anthropogenic disturbance in a tropical forest, Ghana.* International Scholarly Research Notices.

Allen, B. P., Pauley, E. F., & Sharitz, R. R. (1997). Hurricane impacts on liana populations in an old-growth southeastern bottomland forest. *Journal of the Torrey Botanical Society, 124*, 34–42.

Alves, L. F., Assis, M. A., van Melis, J., Barros, A. L., Vieira, S. A., Martins, F. R., Joly, C. A., et al. (2012). Variation in liana abundance and biomass along an elevational gradient in the tropical Atlantic Forest (Brazil). *Ecological Research, 27*(2), 323–332.

Baars, R., Kelly, D., & Sparrow, A. D. (1998). Liane distribution within native forest remnants in two regions of the South Island, New Zealand. *New Zealand Journal of Ecology*, 71–85.

Balfour, D. A., & Bond, W. J. (1993). Factors limiting climber distribution and abundance in a southern African forest. *Journal of Ecology, 81*, 93–100.

Bruy, D., Ibanez, T., Munzinger, J., & Isnard, S. (2017). Abundance, richness and composition of lianas in forest communities along an elevation gradient in New Caledonia. *Plant Ecology & Diversity, 10*(5–6), 469–481.

Caballé, G., & Martin, A. (2001). Thirteen years of change in trees and lianas in a Gabonese rainforest. *Plant Ecology, 152*(2), 167–173.

Campbell, M. J., Edwards, W., Magrach, A., Alamgir, M., Porolak, G., Mohandass, D., & Laurance, W. F. (2018). Edge disturbance drives liana abundance increase and alteration of liana–host tree interactions in tropical forest fragments. *Ecology and Evolution, 8*(8), 4237–4251.

Chettri, A., Barik, S. K., Pandey, H. N., & Lyngdoh, M. K. (2010). Liana diversity and abundance as related to microenvironment in three forest types located in different elevational ranges of the eastern Himalayas. *Plant Ecology & Diversity, 3*(2), 175–185.

Chiu, S. T., & Ewers, F. W. (1992). Xylem structure and water transport in a twiner, a scrambler, and a shrub of Lonicera (Caprifoliaceae). *Trees, 6*(4), 216–224.

Clinebell, R. R., Phillips, O. L., Gentry, A. H., Stark, N., & Zuuring, H. (1995). Prediction of neotropical tree and liana species richness from soil and climatic data. *Biodiversity and Conservation, 4*(1), 56–90.

Daalen, V. J. C. (1984). Distinguishing features of forest species on nutrient-poor soils in the southern cape. *Bothalia, 15*, 229–239.

Dewalt, S. J., Schnitzer, S. A., & Denslow, J. S. (2000). Density and diversity of lianas along a chronosequence in a central Panamanian lowland forest. *Journal of Tropical Ecology, 16*(1), 1–19.

DeWalt, S. J., Ickes, K., Nilus, R., Harms, K. E., & Burslem, D. F. (2006). Liana habitat associations and community structure in a Bornean lowland tropical forest. *Plant Ecology, 186*(2), 203–216.

DeWalt, S. J., Schnitzer, S. A., Chave, J., Bongers, F., Burnham, R. J., Cai, Z., Chuyong, G., Clark, D. B., Ewango, C. E. N., Gerwing, J. J., Gortaire, E., Hart, T., Ibarra-Manríquez, G., Ickes, K., Kenfack, D., Macía, M. J., Jean-Remy, M., Martínez-Ramos, M., Mascaro, J., Moses, S., Muller-Landau, H. C., Parren, M. P. E., Parthasarathy, N., Pérez-Salicrup, D. R., Putz, F. E.,

Romero-Saltos, H., & Thomas, D. (2010). Annual rainfall and seasonality predict pan-tropical patterns of liana density and basal area. *Biotropica, 42*(3), 309–317.

Dewalt, S. J., Schnitzer, S. A., Alves, L. F., Bongers, F., Burnham, R. J., Cai, Z., Carson, W. P., Chave, J., Chuyong, G. B., Costa, F. R. C., Ewango, C. E. N., Gallagher, R. V., Gerwing, J. J., Amezcua, E. G., Hart, T., Ibarra-Manríquez, G., Ickes, K., Kenfack, D., Letcher, S. G., & Melis, J. V. (2015). Biogeographical patterns of liana abundance and diversity. In S. A. Schnitzer, F. Bongers, R. J. Burnham, & F. E. Putz (Eds.), *Ecology of lianas* (pp. 131–146). Wiley. https://doi.org/10.1002/9781118392409.ch11

Dias, A. S., Oliveira, R., Martins, F., Bongers, F., Anten, N., & Sterck, F. (2021). *Climbing mechanisms as a central trait to understand the ecology of lianas--a global synthesis*. Authorea Preprints.

Durigon, J., Durán, S. M., & Gianoli, E. (2013). Global distribution of root climbers is positively associated with precipitation and negatively associated with seasonality. *Journal of Tropical Ecology, 29*(4), 357–360.

Ewango, C. E. (2010). *The liana assemblage of a Congolian rainforest: Diversity, structure and dynamics*. Wageningen University and Research.

Ewers, F. W. (1985). Xylem structure and water conduction in conifer trees, dicot trees, and llanas. *IAWA Journal, 6*(4), 309–317.

Ewers, F. W., Fisher, J. B., & Chiu, S. T. (1990). A survey of vessel dimensions in stems of tropical lianas and other growth forms. *Oecologia, 84*(4), 544–552.

Gartner, B. L., Bullock, S. H., Mooney, H. A., Brown, V. B., & Whitbeck, J. L. (1990). Water transport properties of vine and tree stems in a tropical deciduous forest. *American Journal of Botany, 77*(6), 742–749.

Gentry, A. H. (1991). Climbing plants. *The Biology of Vines, 3*.

Gianoli, E., Saldaña, A., Jiménez-Castillo, M., & Valladares, F. (2010). Distribution and abundance of vines along the light gradient in a southern temperate rain forest. *Journal of Vegetation Science, 21*(1), 66–73.

Guo, Y., Li, G., Kang, B., Wang, D., & Yang, G. (2012). The differential responses of lianas and vines to rainfall gradients in distribution and abundance in Qinling Mountains, China. *Plant Ecology, 213*(11), 1749–1755.

Hegarty, E. E., & Caballé, G. (1991). Distribution and abundance of vines in forest communities. *The Biology of Vines, 313*, 335.

Hu, L., Li, M., & Li, Z. (2010). Geographical and environmental gradients of lianas and vines in China. *Global Ecology and Biogeography, 19*(4), 554–561.

Ibarra-Manríquez, G., & Martínez-Ramos, M. (2002). Landscape variation of liana communities in a Neotropical rain forest. *Plant Ecology, 160*(1), 91–112.

Ingwell, L. L., Joseph Wright, S., Becklund, K. K., Hubbell, S. P., & Schnitzer, S. A. (2010). The impact of lianas on 10 years of tree growth and mortality on Barro Colorado Island, Panama. *Journal of Ecology, 98*(4), 879–887.

Jiménez-Castillo, M., Wiser, S. K., & Lusk, C. H. (2007). Elevational parallels of latitudinal variation in the proportion of lianas in woody floras. *Journal of Biogeography, 34*(1), 163–168.

Jones, I. L., Peres, C. A., Benchimol, M., Bunnefeld, L., & Dent, D. H. (2017). Woody lianas increase in dominance and maintain compositional integrity across an Amazonian dam-induced fragmented landscape. *PLoS One, 12*(10), e0185527.

Kattge, J., Bönisch, G., Díaz, S., Lavorel, S., Prentice, I. C., Leadley, P., Cuntz, M., et al. (2020). TRY plant trait database–enhanced coverage and open access. *Global Change Biology, 26*(1), 119–188.

Khadanga, S. S., Muthumperumal, C., & Parthasarathy, N. (2015). *Changes in Liana diversity over a decade in Indian tropical dry evergreen forests. Biodiversity in tropical ecosystems* (pp. 61–79). Today and Tomorrow's Printers and Publishers.

Laurance, W. F., Andrade, A. S., Magrach, A., Camargo, J. L., Valsko, J. J., Campbell, M., Fearnside, P. M., Edwards, W., Lovejoy, T. E., & Laurance, S. G. (2014). Long-term changes in liana abundance and forest dynamics in undisturbed Amazonian forests. *Ecology, 95*(6), 1604–1611.

Lawrence, D., & Schlesinger, W. H. (2001). Changes in soil phosphorus during 200 years of shifting cultivation in Indonesia. *Ecology, 82*(10), 2769–2780.

Letcher, S. G. (2015). Patterns of liana succession in tropical forests. *Ecology of Lianas*, 116–130.

Lieberman, D., Lieberman, M., Peralta, R., & Hartshorn, G. S. (1996). Tropical forest structure and composition on a large-scale altitudinal gradient in Costa Rica. *Journal of Ecology*, 137–152.

Lobos-Catalán, P., & Jiménez-Castillo, M. (2019). Temperature shapes liana diversity pattern along a latitudinal gradient in southern temperate rainforest. *Plant Ecology, 220*(12), 1109–1117.

Longino, J. T. (1986). A negative correlation between growth and rainfall in a tropical liana. *Biotropica, 18*, 195–200.

Magnago, L. F. S., Magrach, A., Barlow, J., Schaefer, C. E. G. R., Laurance, W. F., Martins, S. V., & Edwards, D. P. (2017). Do fragment size and edge effects predict carbon stocks in trees and lianas in tropical forests? *Functional Ecology, 31*(2), 542–552.

Malizia, A., Grau, H. R., & Lichstein, J. W. (2010). Soil phosphorus and disturbance influence liana communities in a subtropical montane forest. *Journal of Vegetation Science, 21*(3), 551–560.

Marrs, R. H., Proctor, J., Heaney, A., & Mountford, M. D. (1988). Changes in soil nitrogen-mineralization and nitrification along an altitudinal transect in tropical rain forest in Costa Rica. *The Journal of Ecology, 76*, 466–482.

Marvin, D. C., Winter, K., Burnham, R. J., & Schnitzer, S. A. (2015). No evidence that elevated CO 2 gives tropical lianas an advantage over tropical trees. *Global Change Biology, 21*(5), 2055–2069.

Mohandass, D., Davidar, P., Somasundaram, S., Vijayan, L., & Beng, K. C. (2015). Influence of disturbance regime on liana species composition, density and basal area in the tropical montane evergreen forests (sholas) of the Western Ghats, India. *Tropical Ecology, 56*(2), 169–182.

Mohandass, D., Campbell, M. J., Hughes, A. C., Mammides, C., & Davidar, P. (2017). The effect of altitude, patch size and disturbance on species richness and density of lianas in montane forest patches. *Acta Oecologica, 83*, 1–14.

Molina-Freaner, F., & Tinoco-Ojanguren, C. (1997). Vines of a desert plant community in Central Sonora, Mexico. *Biotropica, 29*(1), 46–56.

Molina-Freaner, F., Castillo Gámez, R., Tinoco-Ojanguren, C., & Castellanos, A. E. (2004). Vine species diversity across environmental gradients in northwestern Mexico. *Biodiversity and Conservation, 13*(10), 1853–1874.

Mori, H., Kamijo, T., & Masaki, T. (2016). Liana distribution and community structure in an old-growth temperate forest: The relative importance of past disturbances, host trees, and microsite characteristics. *Plant Ecology, 217*(10), 1171–1182.

Nath, S., Babu, K. N., Dilshad, K., Dar, A. A., & Parthasarathy, N. (2022). Impact of anthropogenic disturbances on the liana diversity and need for conservation of resource valued species-evidences from Indian tropical dry evergreen forests. *Taiwania, 67*(1), 129–139.

Nesheim, I., Halvorsen, R., & Nordal, I. (2010). Plant composition in the Maya Biosphere Reserve: Natural and anthropogenic influences. *Plant Ecology, 208*(1), 93–122.

Nurfazliza, K., Nizam, M. S., & Nur Supardi, M. N. (2012). Association of liana communities with their soil properties in a lowland forest of Negeri Sembilan, peninsular Malaysia. *Sains Malaysiana, 41*, 679–690.

Ofosu-Bamfo, B., Addo-Fordjour, P., & Belford, E. J. (2019). Does road-edge affect liana community structure and liana-host interactions in evergreen rainforests in Ghana? *Acta Oecologica, 101*, 103476.

Padaki, A., & Parthasarathy, N. (2000). Abundance and distribution of lianas in tropical lowland evergreen forest of Agumbe, Central Western Ghats, India. *Tropical Ecology, 41*(2), 143–154.

Pandian, E., & Parthasarathy, N. (2016). Decadal (2003–2013) changes in liana diversity, abundance and aboveground biomass in four inland tropical dry evergreen forest sites of peninsular India. *Journal of Forestry Research, 27*(1), 133–146.

Pandian, E., Parthasarathy, N., & Anbarashan, M. (2017). Liana growth, mortality and recruitment in four inland tropical dry evergreen forest sites of peninsular India. *Journal of Global Ecology and Environment, 6*(2), 60–72.

Parthasarathy, N. (Ed.). (2015). *Biodiversity of lianas* (Vol. 5). Springer.

Pérez-Salicrup, D. R., Sork, V. L., & Putz, F. E. (2001). Lianas and trees in a liana forest of Amazonian Bolivia 1. *Biotropica, 33*(1), 34–47.

Putz, F. (1980). Lianas vs. trees. *Biotropica, 12, 224.*

Putz, F. E. (1984a). How trees avoid and shed lianas. *Biotropica, 16*, 19–23.

Putz, F. E. (1984b). The natural history of lianas on Barro Colorado Island, Panama. *Ecology, 65*(6), 1713–1724.

Putz, F. E., & Chai, P. (1987). Ecological studies of lianas in Lambir national park, Sarawak, Malaysia. *The Journal of Ecology, 75*, 523–531.

Rodríguez-Quintero, W. D., & Gianoli, E. (2016). Abundance and diversity of lianas in a Neotropical dry forest: The influence of soil moisture. *Plant Ecology and Evolution, 149*(3), 329–334.

Rossell, I. M., & Eggleston, H. (2017). Elevational distribution of temperate lianas along trails in Pisgah National Forest. *Southeastern Naturalist, 16*(3), 443–450.

Rowe, N., & Speck, T. (2004). Hydraulics and mechanics of plants: Novelty, innovation and evolution. In *The evolution of plant physiology* (pp. 297–325). Academic Press.

Schnitzer, S. A. (2005). A mechanistic explanation for global patterns of liana abundance and distribution. *The American Naturalist, 166*(2), 262–276.

Schnitzer, S. A., & Bongers, F. (2002). The ecology of lianas and their role in forests. *Trends in Ecology & Evolution, 17*(5), 223–230.

Schnitzer, S. A., & Bongers, F. (2011). Increasing liana abundance and biomass in tropical forests: Emerging patterns and putative mechanisms. *Ecology Letters, 14*(4), 397–406.

Schnitzer, S. A., Putz, F. E., Bongers, F., & Kroening, K. (2015). The past, present, and potential future of liana ecology. *Ecology of lianas*, 1–10.

Schnitzler, A., Amigo, J., Hale, B., & Schnitzler, C. (2016). Patterns of climber distribution in temperate forests of the Americas. *Journal of Plant Ecology, 9*(6), 724–733.

Swaine, M. D., & Grace, J. (2007). Lianas may be favoured by low rainfall: Evidence from Ghana. *Plant Ecology, 192*(2), 271–276.

Teramura, A. H., Gold, W. G., & Forseth, I. N. (1991). *Physiological ecology of mesic, temperate woody vines. The biology of vines* (pp. 245–285). Cambridge University Press.

Van der Heijden, G. M. F., & Phillips, O. L. (2009a). Liana infestation impacts tree growth in a lowland tropical moist forest. *Biogeosciences, 6*(10), 2217–2226.

Van Der Heijden, G. M., & Phillips, O. L. (2009b). Environmental effects on Neotropical liana species richness. *Journal of Biogeography, 36*(8), 1561–1572.

Vazquez, J. A., & Givnish, T. J. (1998). Altitudinal gradients in tropical forest composition, structure, and diversity in the sierra de Manantlán. *Journal of Ecology*, 999–1020.

Vivek, P., & Parthasarathy, N. (2015). Liana community and functional trait analysis in tropical dry evergreen forest of India. *Journal of Plant Ecology, 8*(5), 501–512.

Wu, Y., Tao, J., Zhao, K., & Hao, J. (2010). Changes of woody vines under the edge effect of Bawangling natural secondary forest in Hainan. *Forestry Science, 46*(5), 1–6.

Xiao, Y., Song, Y., Wu, F. C., Zhang, S. B., & Zhang, J. L. (2021). Divergence of stem biomechanics and hydraulics between Bauhinia lianas and trees. *AoB Plants, 13*(3), plab016.

Yang, S. Z., Fan, H., Li, K. W., & Ko, T. Y. (2018). How the diversity, abundance, size and climbing mechanisms of woody lianas are related to biotic and abiotic factors in a subtropical secondary forest, Taiwan. *Folia Geobotanica, 53*(1), 77–88.

Current Status and Future Prospects 4

Abstract

We can safely assume from Chapter 2 that liana research is expanding rapidly. However, there is a significant disparity between the various research themes investigated. The number of phytosociological studies, for example, is at least 20 times greater than the number of studies on the evolutionary aspects of climbers. Similarly, while many studies on lianas in forest management practices exist, technological applications are underutilized in climber-related research. I analyzed the shift in theme-related research across decades to see if these disparities are dynamic. I filtered the climber research materials published between 1980 and 2020 (see Chapter 2) because the majority of research has been published in the last four decades (>90 percent). The research materials published between 1980 and 2020 were divided into four distinct categories, one for each decade (Fig. 4.1). The research materials under each decade were further classified based on the research theme (See Chapter 2). Although the number of studies increased significantly, there is a clear pattern of a few research themes dominating the study period. Some of the rare and yet important research topics have received little attention. Therefore, in this chapter, I look at potential scientific gaps in global climber research that need to be addressed immediately, as well as future prospects. Furthermore, I highlight lesser-known areas of climber research that merit more attention.

Keywords

Climbing mechanisms · Liana distribution · Liana–tree interactions · Allometric equation · Plant functional traits

V. Pandi, *Taxonomy and Ecology of Climbers: Climbing Plants of India*,
https://doi.org/10.1007/978-981-19-8645-1_4

4.1 Climber Research: Current Status and Future Prospects

I begin with a fundamental discussion of climbing plant terminologies and climbing modes.

Plant growth forms will have ecological and evolutionary implications (Asner & Martin, 2012; Rowe & Speck, 2005). Many, if not the majority, of plants' structural and functional characteristics, are linked to their life forms. For example, a plant's anatomy, physiology, mechanical properties, and overall architecture are determined by whether it is a tree or a climber (Rowe & Speck, 2005). However, we frequently encounter confusing or redundant terminologies used to represent climbers (see Chapters 2 and 5), which may impede cross-study comparisons. Sperotto, Acevedo-Rodríguez, Vasconcelos, and Roque (2020) have recently recommended using standard terminologies for climbing habits because the climbing mechanism is one of the key innovations in the history of plant evolution (Gianoli, 2015). It is therefore critical to implement the use of standardized terminologies for climbers across study sites (Fig. 4.1).

Second, do we need decentralized climber research? Besides lianas and vines, our current understanding is that climbers are a single functional type rather than different groups. Several studies have compared trees and lianas as two distinct functional groups (see Sect. 3.4). However, very few studies have attempted to compare the functional differences among lianas. Lianas with different climbing mechanisms have different ecological requirements and frequently occupy forests at various stages of succession. Only a few studies have provided insights into

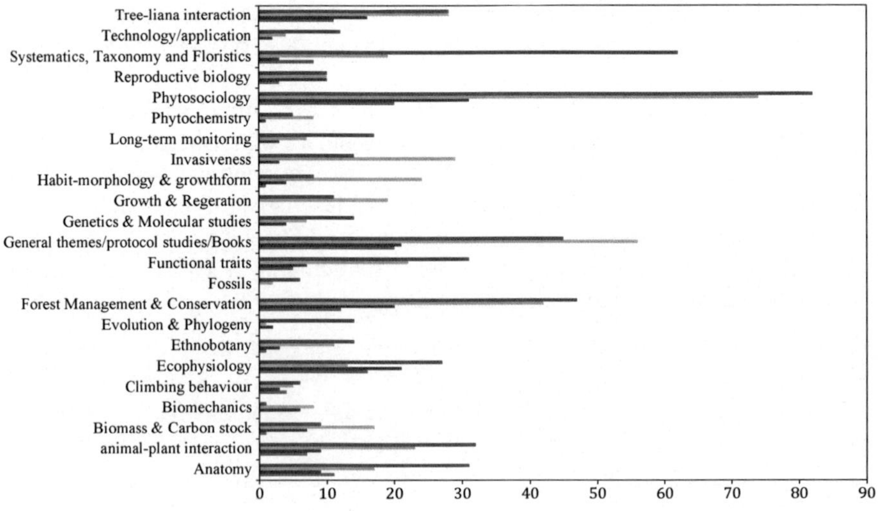

Fig. 4.1 Trends in the number of publications under various research themes over the past four decades (1980–2020). Purple bar represents the time interval from 2010 to 2020; green—2000–2010; brown—1990–2000; blue—1980–1990

functional differences among climbers that use different climbing strategies. For example, Xiao, Song, Wu, Zhang, and Zhang (2021) revealed that scramblers differ from other climbing forms in stem hydraulics and biomechanical properties. Ewers, Fisher, and Chiu (1990) observed that lianas with different climbing habits had similar vessel dimensions, except for scramblers, which had shorter but wider vessels. He also found that the vessel length of scramblers converged with shrubs and trees while diverging from other climbers.

Dias et al. (2021) revealed that lianas with active climbing mechanisms (twiners and tendrils) differ significantly from passively climbing lianas in major structural and functional traits (scramblers). Scramblers' life-history strategies vary considerably from other climbers: scrambling is considered the least specialized climbing mechanism in which the stems lean on the host; circumnutation has not been reported in scramblers, and the retention of the self-supporting phase for an extended period distinguishes them from other climbers. Similarly, the establishment of tendril climbers and root climbers as non-self-supporting plants is impacted by various constraints throughout their life cycle. Nonetheless, the convergence of some of the scramblers' life-history traits with self-supporting plants necessitates special attention. Rowe, Isnard, and Speck (2004) shed more light on the shifting growth forms in climbers. The underlying mechanisms are of biomechanical and anatomical origin; however, the role of the environment in dictating plants' growth strategies cannot be overlooked. Although plant movements' physiological and environmental characteristics are studied, their genetic controls are poorly understood. Molecular and phylogenetic approaches may enhance our understanding of the origin and evolution of climbing behavior in plants (Fig. 4.1).

What controls the diversity and distribution of species form the central theme in ecology. The effect of latitude on climber diversity and distribution is well documented. Other environmental factors, such as temperature and rainfall, have produced contrasting results within the tropics. Similarly, other factors such as disturbance have been shown to promote climber species richness and abundance in some studies while being detrimental in others. This could be attributed to the consideration of climbers as one functional group. For example, root climbers are generally abundant in areas of high rainfall, whereas scramblers may prefer sites with prolonged seasonality. Similarly, the structural requirements of tendril climbers may differ from those of twiners. Intercontinental research involving multiple parameters may provide generality in patterns of climber distribution and richness within tropical ecosystems. Furthermore, establishing an independent relationship between climbers of different climbing mechanisms and environmental factors may depict the underlying factors controlling climber diversity and distribution within the tropics.

Are lianas increasing? Results from the African counterpart raised severe concerns about the generality of the phenomenon across the tropics. While few studies from India support the liana increasing hypothesis, comparisons are hindered by the lack of studies from other parts of the paleotropics. Furthermore, the generality of proposed mechanisms for the global increase in liana abundance is inconclusive. While the underlying mechanisms are still unknown, this will have severe consequences for the structure and dynamics of tropical forests. The increase in the proportion of liana-laden trees in the tropics is another major threat. The antagonistic effects of lianas are well understood, as is their impact on the hosts' carbon stock and sequestration potential. However, the mechanisms of liana–tree interactions remain unclear because most studies focused on the hosts' perspective, employing a spectrum of host traits likely to promote or deter liana colonization. Since the availability and overall architecture of the hosts influence the twiners, tendril climbers, scramblers, and root climbers differently, it will be appropriate to study liana–tree associations that integrate host as well as the climber characteristics. Only a few studies, to my knowledge, have addressed the differential needs in host preferences of lianas using different climbing mechanisms. On the other hand, the criteria for liana infestation on host trees have varied across studies. A few studies looked at the primary contact and trunk infestation rates, while others looked only at the lateral and crown infestation rates, and few studies have included both. Most studies failed to address the methods used in the case of a single liana colonizing multiple hosts, which is common in the tropics. Recent evidence suggests that liana infestation on host trees is becoming more prevalent. A standardized methodology that integrates host, environment, and liana characteristics will aid us in better understanding the patterns of tree–liana interactions and in developing strategies for forest management practices.

Although climbers add significantly to the species diversity and local stem counts, they contribute far less biomass and carbon than trees. Studies have predicted that with their thin stems, lianas may contribute up to 10% of the total woody biomass and carbon stocks in the tropical forests, where they are abundant. In comparison to trees, there are only a few allometric methods for estimating liana biomass (Schnitzer, DeWalt, & Chave, 2006), and those commonly used allometric equations produce inconsistent results (e.g., Addo-Fordjour & Rahmad, 2013; Miao, Koerner, Medjibe, & Poulsen, 2016). Such disparities in outcomes may lead to an under or overestimation of lianas' total aboveground biomass. While researchers suggest the need for species-specific, site-specific, or intracontinental/intercontinental models, the role of forest or vegetation types cannot be overlooked. Despite their importance, the allometric equations for estimating liana biomass are insufficient. Therefore, a comprehensive model to precisely estimate the lianas' biomass is inevitable. Such studies can reveal the actual contribution of lianas to woody plant biomass, as lianas are generally known to have a negative impact on forest carbon stocks and cannot compensate for the tree biomass they displace.

Besides these significant research gaps, the underrepresentation of climbers in functional traits cannot be refuted (Gallagher & Leishman, 2012). Furthermore, despite their critical role in forest structure, function, and productivity, climbers are underrepresented in global vegetation models (Verbeeck & Kearsley, 2016). There is a growing consensus on the negative impacts of lianas on carbon sink potential and their proliferation in tropical forests worldwide, making them a worthwhile inclusion in global vegetation models. A growing trend is applying geospatial technology to ecological research, especially in the control and management of lianas. Long-term monitoring of lianas would be beneficial for understanding how lianas respond to global climate change and anthropogenic pressure, as lianas are often portrayed as better players of drought and disturbance. Lianas play an essential role in tropical forests, but more studies have documented their negative impact; they are often overlooked for their positive effects. When developing forest management plans, it is critical to emphasize the ethnobotany and economic benefits of lianas and their importance as a food resource for faunal assemblages. In terms of fundamental research, the underlying mechanisms of plant movement and the origins and evolution of climbing behavior have long been unclear. Ecological research findings should be supplemented with molecular, phylogenetic, and fossil research to understand better how climbing habits evolved in plants. My recommendation for the future is to conduct more balanced research on climbers, focusing on fundamental and application-based studies.

References

Addo-Fordjour, P., & Rahmad, Z. B. (2013). *Mixed species allometric models for estimating above-ground liana biomass in tropical primary and secondary forests*. International Scholarly Research Notices.

Asner, G. P., & Martin, R. E. (2012). Contrasting leaf chemical traits in tropical lianas and trees: Implications for future forest composition. *Ecology Letters, 15*(9), 1001–1007.

Dias, A. S., Oliveira, R., Martins, F., Bongers, F., Anten, N., & Sterck, F. (2021). Climbing mechanisms as a central trait to understand the ecology of lianas—a global synthesis. Authorea Preprints.

Ewers, F. W., Fisher, J. B., & Chiu, S. T. (1990). A survey of vessel dimensions in stems of tropical lianas and other growth forms. *Oecologia, 84*(4), 544–552.

Gallagher, R. V., & Leishman, M. R. (2012). A global analysis of trait variation and evolution in climbing plants. *Journal of Biogeography, 39*(10), 1757–1771.

Gianoli, E. (2015). Evolutionary implications of the climbing habit in plants. In *Ecology of lianas* (pp. 239–250). Wiley.

Miao, Z., Koerner, S. E., Medjibe, V. P., & Poulsen, J. R. (2016). Wanted. *Biotropica, 48*(5), 561–564.

Rowe, N., & Speck, T. (2005). Plant growth forms: An ecological and evolutionary perspective. *New Phytologist, 166*(1), 61–72.

Rowe, N., Isnard, S., & Speck, T. (2004). Diversity of mechanical architectures in climbing plants: An evolutionary perspective. *Journal of Plant Growth Regulation, 23*(2), 108–128.

Schnitzer, S. A., DeWalt, S. J., & Chave, J. (2006). Censusing and measuring lianas: A quantitative comparison of the common methods. *Biotropica, 38*(5), 581–591.

Sperotto, P., Acevedo-Rodríguez, P., Vasconcelos, T. N., & Roque, N. (2020). Towards a standardization of terminology of the climbing habit in plants. *The Botanical Review, 86*(3), 180–210.

Verbeeck, H., & Kearsley, E. (2016). The importance of including lianas in global vegetation models. *Proceedings of the National Academy of Sciences, 113*(1), E4–E4.

Xiao, Y., Song, Y., Wu, F. C., Zhang, S. B., & Zhang, J. L. (2021). Divergence of stem biomechanics and hydraulics between Bauhinia lianas and trees. *AoB Plants, 13*(3), plab016.

Check for updates

5 Taxonomic Diversity of Climbing Flora in India: A Compendium

Abstract

Understanding the taxonomic diversity of any plant group is a prerequisite for ecological and evolutionary studies. However, climber taxonomic estimates have rarely been studied, and only a few attempts have been made to precisely estimate their extent of diversity. Among the other limitations of estimating climbers are the use of numerous methodologies and different growth form considerations while studying climbers. This chapter presents a novel attempt to estimate the taxonomic diversity of climbing plants found in India. For the first time, a comprehensive list of Indian climbing flora is provided, along with detailed species-level information. Data from over 100 research articles, reviews, Floras, databases, and technical reports revealed the presence of 2600 climber species from 104 spermatophyte plant families. A standard methodology used for this study may provide the most accurate estimate of climber diversity in India, and it can also be replicated elsewhere to estimate climber diversity within a defined geographical area. The species-level information provided in this chapter will serve as a baseline for future research in climber ecology, evolution, and systematics.

Keywords

Angiosperms · Growth form · Phylogenetic diversity · Leaf type · Endemism · IUCN · Climbing mechanisms

5.1 Overview

Our understanding of climbers' taxonomic diversity remains poor despite their growing presence in research. According to recent research, climbers are among the most diverse plant groups, with representation in 171 plant families of

V. Pandi, *Taxonomy and Ecology of Climbers: Climbing Plants of India*,
https://doi.org/10.1007/978-981-19-8645-1_5

Gymnosperms, Pteridophytes, and Angiosperms. Especially families like Convolvulaceae (55 genera/~1850 species), Cucurbitaceae (97 genera/ ~ 990 species), and Menispermaceae (72 genera/~450 species) are exclusively climbers. By numbers, Fabaceae is the most speciose climber family in the paleotropics (Putz and Chai, 1987; Cai et al., 2009; Anbarashan and Parthasarathy, 2013), while Apocynaceae and Fabaceae are the most species-rich climber families in the Neotropics (Acevedo-Rodriguez et al., 2015). The majority of our knowledge about climbing plant diversity comes from floristic studies and ecological inventories. As discussed in previous chapters, many studies excluded climbers from forest inventories due to difficulties in field collections and taxonomic assertions. According to Jacobs (1976), "*collecting climbers is a sport in and of itself.*" Because climbers reproduce vegetatively, forming a dense entanglement, it takes considerable effort to differentiate the ramets and genets; otherwise, the numbers are overestimated.

Furthermore, until recently, the lack of a standard protocol for censusing climbers had kept them off the inventory. The difficulties in assigning the appropriate growth form and climbing mechanisms among climbers add to the complexity. For example, numerous terminologies are used to identify climbers that use various climbing strategies. One classic example is the use of terminologies such as climbing shrubs or scandent shrubs in taxonomic surveys, which is equivalent to scrambler (with or without spines) in ecological inventories (see Table 5.1). Few previous studies included epiphytes, hemi-epiphytes, and rattans among climbers, while others mostly ignored them. Similarly, most taxonomic surveys classify scramblers as shrubs or small trees rather than climbers. The standard protocols on liana inventory methods fairly resolved all issues. However, the ecological inventories primarily focus on lianas, leaving a significant proportion of herbaceous climbers out. This could significantly underestimate global climber diversity. Therefore, climbers are under-documented in ecological inventories or confused by different terminologies used in taxonomic surveys. As a result, our estimates of climber diversity are lower, which may limit our understanding of their ecological and evolutionary significance. The absence of accurate data on climbers is genuinely a global issue; yet, given resource constraints and a lack of transparency in data sharing, we attempted to estimate India's climbing plant diversity.

India, the seventh largest country in the world, roughly occupies 2.2% of the earth's surface. The heterogeneity in the landscape, climate, and diverse range of ecological habitats has contributed significantly to its floristic diversity. India is one of the world's mega biodiversity centers, home to more than 11% of all known flora. According to recent reports, India has over 55,000 plant species, with Angiosperms accounting for ~22,000 of them (Botanical Survey of India, BSI). About 23.20% of Indian flowering plants are endemic to the country, demonstrating India's unique geographical configuration and geological history. Furthermore, India is home to four of the world's 36 biodiversity hotspots. The major Phytogeographical regions of India are the Western Ghats, the Eastern Himalayas, the East Coast, the West Coast, the North-eastern Ranges, the Eastern Plains, the North Deccan, the Northern Plains,

Table 5.1 Various terminologies and growth forms used for selected scrambler species in Indian Floras

Species	Growth form					
	The flora of British India	Flora of the Presidency of Madras	The flora of the Presidency of Bombay	Flora of Karnataka	The excursion flora of Central Tamil Nadu	Flora of Udupi
Ziziphus oenoplia	Straggling shrub often climbing	Straggling or climbing shrub	Straggling shrub often semi-scandent	Scandent and ferruginous shrubs	Shrub or small tree	Scandent shrub
Ziziphus rugosa	Straggling shrub	Straggling shrub	Straggling armed shrub sometimes climbing	Scandent armed shrub	Straggler	Scrambling shrub
Capparis brevispina		Shrub			Shrub or Straggler	
Capparis sepiaria	A spreading wiry branched shrub	Large straggling shrub	Much branched woody climber	Shrub, erect, or climbing	Straggler	
Olax scandens		A rambling or climbing shrub	Climbing and much-branched shrub	Unarmed scandent shrub		
Olax wightiana		A small tree or climbing shrub				
Grewia rhamnifolia	Scandent shrub	A straggling or scandent shrub			Shrub	Trees or erect or straggling shrubs
Toddalia asiatica	A rambling sarmentose shrub	Climbing shrub	Sub-scandent or sarmentose shrub		Armed straggler	
Toddalia asiatica var. *garcilis*				Climbing shrub		
Toddalia asiatica var. *floribunda*				Scandent shrub		
Scutia myrtina	Branches straggling	Straggling shrub	Straggling shrub		Armed straggler	
Salacia chinensis	Straggling tree or large climbing shrub	Straggling shrub	Straggling or climbing shrub		Shrub	Straggling shrub

(continued)

Table 5.1 (continued)

Species	Growth form					
	The flora of British India	Flora of the Presidency of Madras	The flora of the Presidency of Bombay	Flora of Karnataka	The excursion flora of Central Tamil Nadu	Flora of Udupi
Premna corymbosa		Shrub			Scandent shrub	Climbing shrub
Acacia caesia		Climbing shrub	Scandent shrub	Straggler	Straggler	Climbing shrub
Alangium salvifolium	A small tree or a rambling shrub	Small tree or Straggling		Shrub		
Alangium salvifolium var. hexapetalum		Usually climbing		Straggling shrub		Straggling shrub
Grewia oppositifolia		Shrub erect or climbing		Scandent shrub		
Paramygnia monophylla			A climbing evergreen shrub	Straggling shrub		Straggling shrub
Salacia oblonga		Climbing shrub or small tree	Scandent shrub			
Dalbergia volubilis		Large woody climber	Large woody climber			
Caesalpinia mimosoides		Climbing shrub	Climbing shrub	Prickly shrub	Armed stragglers	
Jasminum sambac	Scandent shrub	An erect sometimes climbing shrub	Sub-erect shrub, scarcely climbing		Usually scandent	
Phyllathus reticulatus		Shrub often scandent				
Premna scandens	A tree or a large climber					
Premna villosa	A tree or a large climber					

and the South Deccan. The diversity in biogeography, climate regimes, and rich biodiversity makes India one of the ideal candidates to conduct such studies.

In this chapter, I present a standard methodology for accurately reporting the diversity of climbing Flora, an integrated approach using India as a case study. A comprehensive list of India's climbing Flora and information on their phylogenetic diversity and life-history strategies are provided. The schematic representation of the methodology followed for data sourcing, validation, and curation is given in Fig. 5.1. The detailed methodology is also available in Vivek et al. (2022).

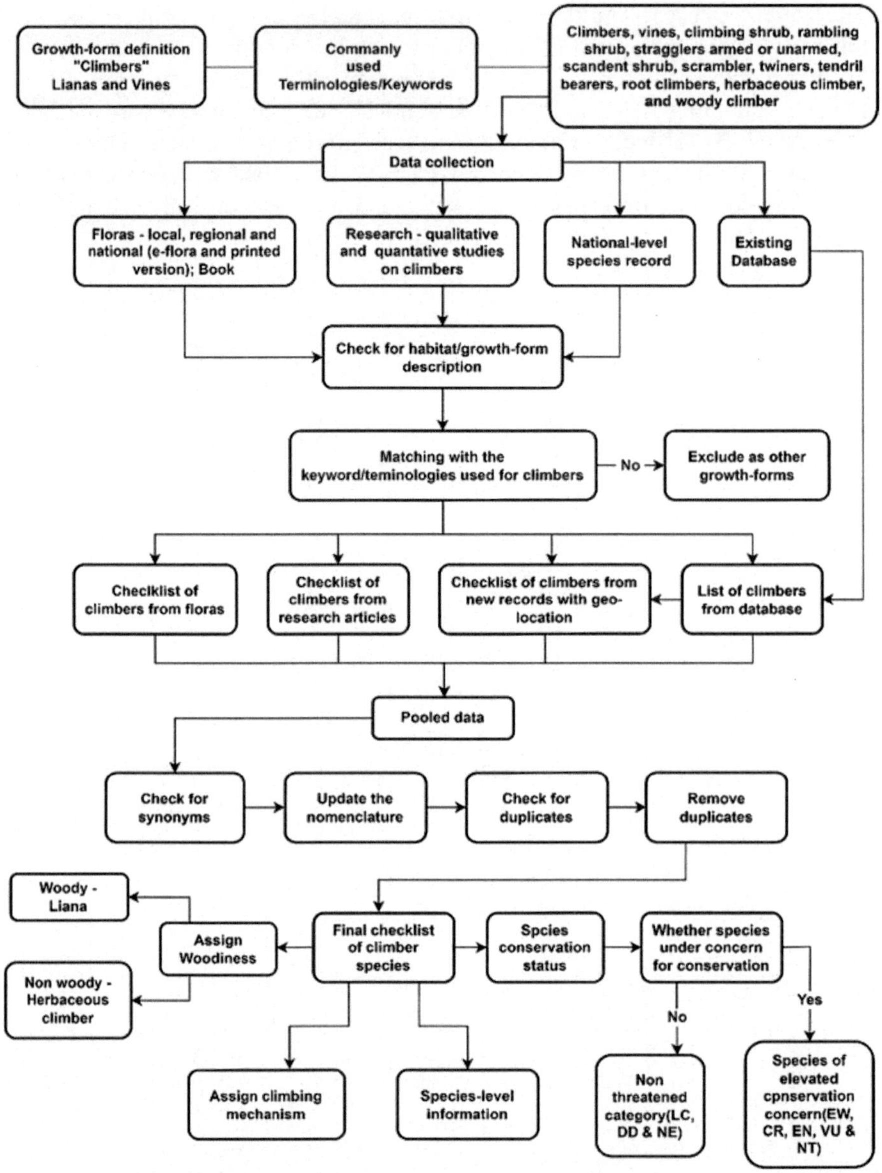

Fig. 5.1 The methodology used to prepare the comprehensive checklist of Indian climbing flora is depicted schematically. Climber terminologies are derived from a variety of sources of literature via rigorous surveys. Information was gathered using 35 Indian Flora (regional and national) and 50+ research articles. New species records published by the Botanical Survey of India and other existing databases were also used as source materials. The final checklist includes species that have been modified to reflect current nomenclature. Climbing mechanisms were assigned for each species, broadly under five categories viz. twiners, tendril climbers, hook climbers, root climbers, and scramblers (armed and unarmed) to facilitate global comparisons. See Vivek et al. (2022) for a detailed methodology. Species-level information was taken from pertinent sources, including research articles, Floras, and online platforms. Species' global distribution data was retrieved from PoWO (www.plantsoftheworldonline.org) (PoWO, 2022).

5.2 An Overview of the Climbing Flora of India

5.2.1 Taxonomic Diversity

The climbing plants of India are phylogenetically diverse (Fig. 5.2). A systematic review of the climbing plants of India revealed the presence of 2624 species representing 585 genera and 104 spermatophyte families (Table 5.2). The climbing flora of India accounts for ~12% and ~ 13% of the known flowering plants and Gymnosperms in India. Among 64 orders in Angiosperms, ~62.5% had climber representations in Indian Flora. Similarly, at the family level, ~25% of the known Angiosperm climber families are represented in India. Among flowering plants, most

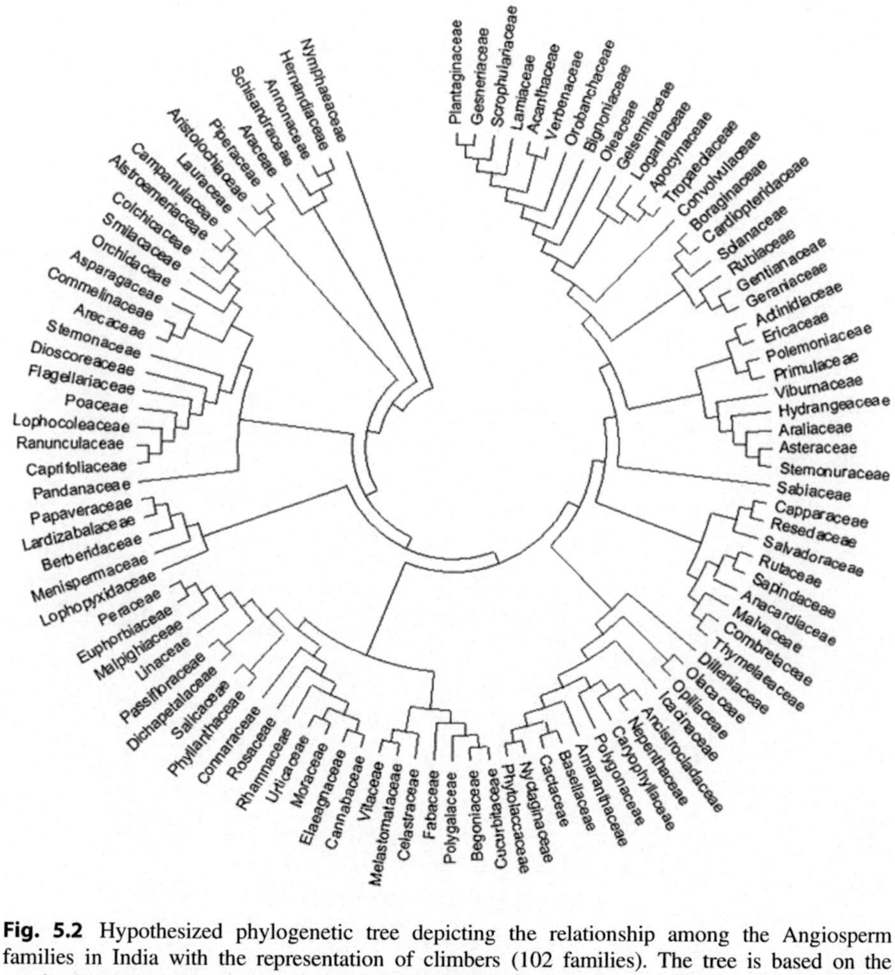

Fig. 5.2 Hypothesized phylogenetic tree depicting the relationship among the Angiosperm families in India with the representation of climbers (102 families). The tree is based on the species-level phylogenetic information of 102 climber species randomly chosen from their representative families (refer to Table 5.1 for the total number of genera and species within each family)

Table 5.2 Indian Spermatophyte plant families with at least one climber species. Climbing mechanism: ST-stem twiner; TC-tendril climber; HC-hook climber; SCR-A-armed scrambler; SCR-UA-unarmed scrambler; RC-root climber

Family	Number of genera	Number of species	Diversity of climbing mechanisms	Types of climbing mechanisms
Fabaceae	81	369	5	HC, TC, SCR-A, SCR-UA, TC
Apocynaceae	78	312	4	ST, SCR-A, SCR-UA, TC
Convolvulaceae	26	217	1	ST
Vitaceae	11	115	1	TC
Cucurbitaceae	39	108	1	TC
Arecaceae	4	71	2	SCR-A, SCR-UA
Annonaceae	14	68	1	SCR-UA
Rosaceae	4	63	1	SCR-A
Rubiaceae	20	61	4	RC, ST, SCR-A, SCR-UA
Celastraceae	13	60	3	ST, SCR-A, SCR-UA
Piperaceae	2	59	1	RC
Menispermaceae	21	56	1	ST
Oleaceae	2	48	2	ST, SCR-UA
Ranunculaceae	4	43	2	SCR-UA, TC
Rhamnaceae	10	40	4	ST, SCR-A, SCR-UA, TC
Capparaceae	6	37	2	SCR-A, SCR-UA
Dioscoreaceae	1	34	1	ST
Lamiaceae	13	34	2	ST, SCR-UA
Passifloraceae	4	34	2	ST, TC
Malvaceae	10	32	2	SCR-A, SCR-UA
Araceae	5	31	3	RC, ST, SCR-UA
Combretaceae	3	30	4	RC, ST, SCR-UA, TC
Rutaceae	8	27	2	SCR-A, SCR-UA
Malpighiaceae	5	26	2	ST, SCR-UA
Asteraceae	12	25	2	ST, SCR-UA
Bignoniaceae	16	25	4	RC, ST, SCR-UA, TC
Moraceae	6	25	3	RC, SCR-A, SCR-UA
Loganiaceae	2	24	3	HC, ST, SCR-UA
Acanthaceae	8	22	2	ST, SCR-UA
Aristolochiaceae	2	21	2	ST, TC
Euphorbiaceae	9	21	2	ST, SCR-UA
Primulaceae	3	21	1	SCR-UA
Smilacaceae	2	20	1	TC

(continued)

Table 5.2 (continued)

Family	Number of genera	Number of species	Diversity of climbing mechanisms	Types of climbing mechanisms
Connaraceae	5	17	2	ST, SCR-UA
Caprifoliaceae	3	16	1	SCR-UA
Solanaceae	4	14	2	SCR-A, SCR-UA
Araliaceae	3	13	3	RC, SCR-A, SCR-UA
Asparagaceae	1	13	1	SCR-A
Icacinaceae	7	13	3	ST, SCR-UA, TC
Gentianaceae	4	12	1	ST
Melastomataceae	7	12	2	ST, SCR-UA
Boraginaceae	3	11	1	SCR-UA
Poaceae	6	10	2	ST, SCR-UA
Phyllanthaceae	4	9	1	SCR-UA
Polygonaceae	5	8	3	ST, SCR-UA, TC
Gnetaceae	1	7	1	ST
Sabiaceae	1	7	1	SCR-UA
Schisandraceae	2	7	1	SCR-UA
Urticaceae	6	7	1	SCR-UA
Ancistrocladaceae	1	6	2	HC, SCR-UA
Campanulaceae	2	6	1	ST
Dilleniaceae	3	6	1	SCR-UA
Sapindaceae	2	6	2	SCR-UA, TC
Lardizabalaceae	2	5	1	SCR-UA
Nyctaginaceae	2	5	1	SCR-A
Olacaceae	2	5	2	SCR-UA, TC
Papaveraceae	1	5	1	SCR-UA
Plantaginaceae	3	5	2	ST, SCR-UA
Schizaeaceae	1	5	1	ST
Dichapetalaceae	1	4	1	SCR-UA
Elaeagnaceae	1	4	2	SCR-A, SCR-UA
Ericaceae	2	4	1	SCR-UA
Hernandiaceae	1	4	1	ST
Linaceae	2	4	1	HC
Verbenaceae	2	4	2	SCR-A, SCR-UA
Amaranthaceae	3	3	2	ST, SCR-UA
Basellaceae	2	3	1	ST
Ephedraceae	1	3	1	SCR-UA
Nepenthaceae	1	3	1	TC
Opiliaceae	1	3	1	SCR-A
Orobanchaceae	2	3	2	ST, SCR-UA
Thymelaeaceae	2	3	1	SCR-UA
Actinidiaceae	1	2	1	SCR-UA
Hydrangeaceae	2	2	2	ST, SCR-UA

(continued)

Table 5.2 (continued)

Family	Number of genera	Number of species	Diversity of climbing mechanisms	Types of climbing mechanisms
Lauraceae	1	2	1	ST
Polygalaceae	2	2	1	SCR-UA
Pteridaceae	2	2	2	ST, SCR-UA
Resedaceae	1	2	1	SCR-UA
Salvadoraceae	1	2	1	SCR-A
Scrophulariaceae	1	2	1	SCR-UA
Tropaeolaceae	1	2	1	ST
Alstroemeriaceae	1	1	1	ST
Anacardiaceae	1	1	1	SCR-UA
Annonaceae	1	1	1	SCR-UA
Begoniaceae	1	1	1	SCR-UA
Berberidaceae	1	1	1	SCR-UA
Cactaceae	1	1	1	SCR-A
Cannabaceae	1	1	1	TC
Cardiopteridaceae	1	1	1	ST
Caryophyllaceae	1	1	1	ST
Colchicaceae	1	1	1	TC
Commelinaceae	1	1	1	ST
Flagellariaceae	1	1	1	SCR-UA
Gelsemiaceae	1	1	1	SCR-UA
Geraniaceae	1	1	1	SCR-UA
Lophocoleaceae	1	1	1	TC
Lophopyxidaceae	1	1	1	SCR-UA
Pandanaceae	1	1	1	RC
Peraceae	1	1	1	ST
Phytolaccaceae	1	1	1	ST
Polemoniaceae	1	1	1	ST
Salicaceae	1	1	1	SCR-UA
Stemonaceae	1	1	1	ST
Stemonuraceae	1	1	1	ST
Viburnaceae	1	1	1	SCR-UA

of the Indian climbing flora is represented within the eudicots (84%). A detailed distribution of Indian climbing flora under different spermatophyte orders and families is given in Table 5.2, Fig. 5.3. Fabaceae (383 species/82 genera), Apocynaceae (333 species/79 genera), and Convolvulaceae (220 species/26 genera) are the most species-rich climber families in India (Table 5.2, Fig. 5.4). Piper and Ipomoea are the most diverse climber genus in India (Fig. 5.5). About 20% of the total climbing flora are endemic to India.

Only 35 climber species from the current survey are listed on the IUCN red list. I categorized climbers under six different climbing mechanisms: stem twiners, armed

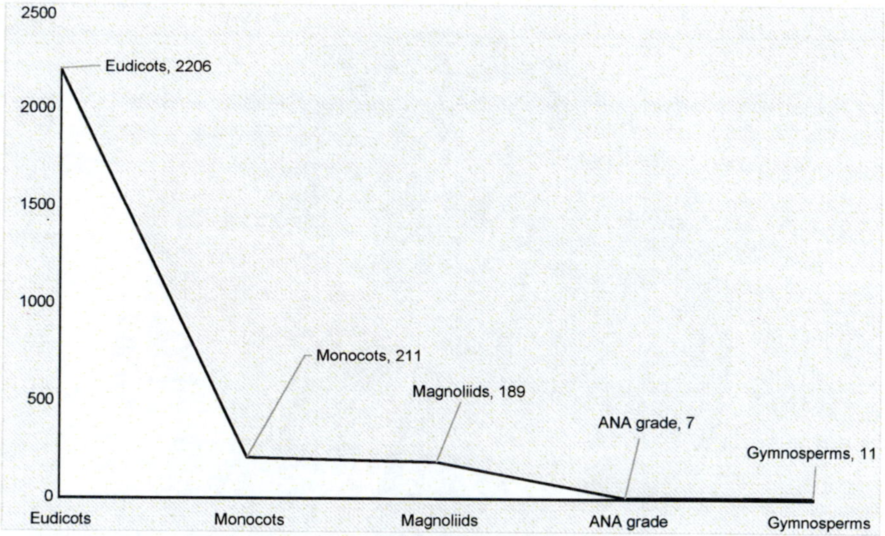

Fig. 5.3 Contribution of Gymnosperms and various subgroups within Angiosperms to the climbing flora of India

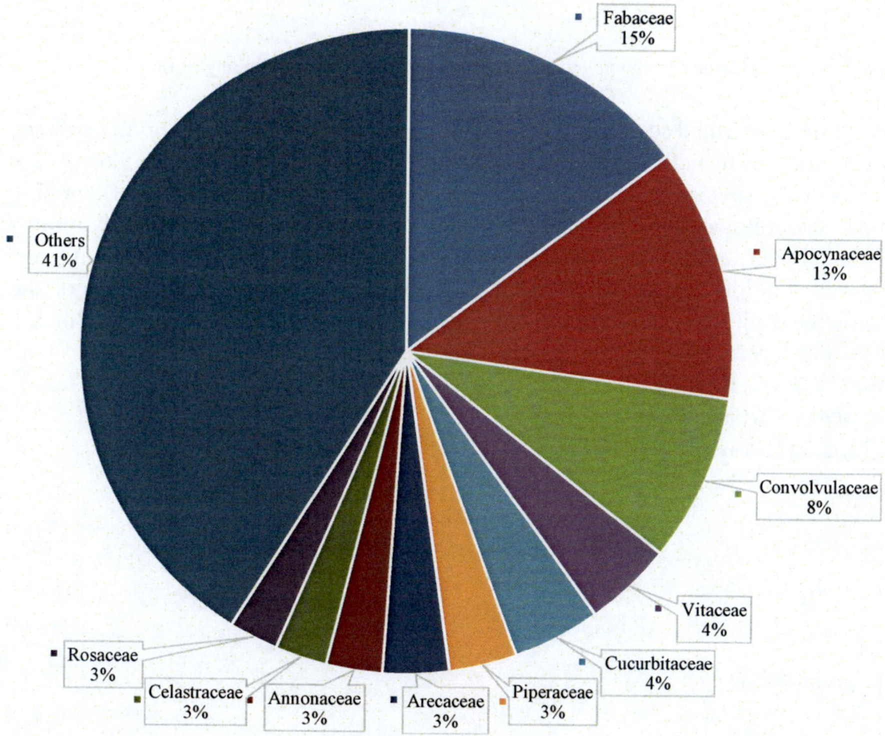

Fig. 5.4 Contribution of top 10 climber families to the species diversity of the Indian climbing flora

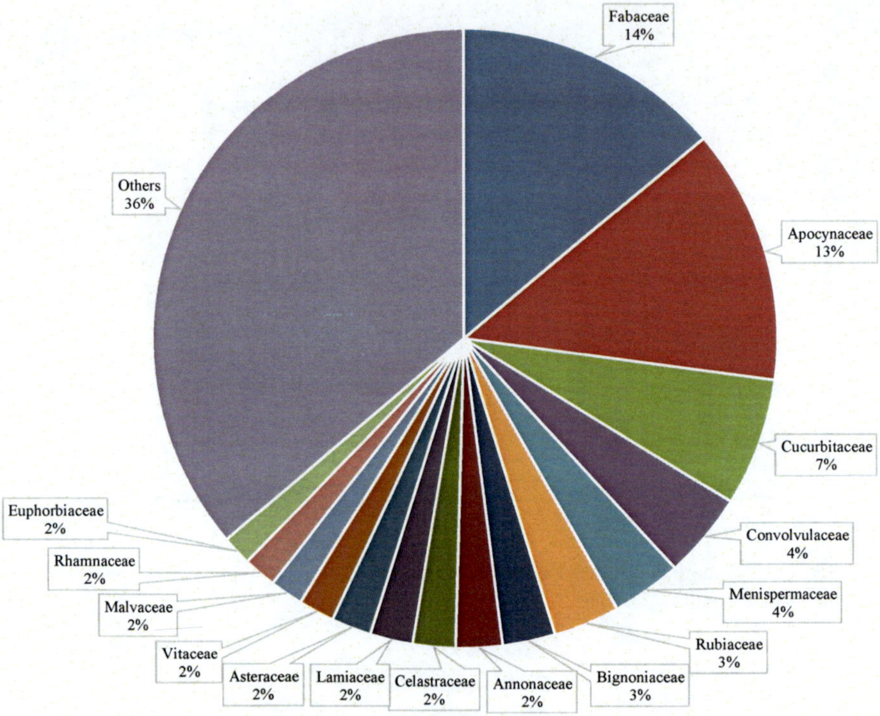

Fig. 5.5 Contribution of top ten genera-rich families to the Indian climbing flora

scramblers, unarmed scramblers, tendril climbers, root climbers, and hook climbers. Most climbers in India are stem twiners and unarmed scramblers (Fig. 5.6). At the family level, species under Fabaceae displayed the maximum number of climbing strategies, followed by Bignoniaceae and Araliaceae (Table 5.2). About two-thirds of the climber families are associated with single climbing strategies, while the rest evolved multiple climbing mechanisms (Table 5.2). Similarly, two-thirds of the enumerated climbers are lianas, and the herbaceous climbers accounted for the rest. The species employing passive climbing strategies (scramblers and hook climbers) are found exclusively woody, while the active climbers were either woody or herbaceous (Plates 5.1, 5.2, 5.3, 5.4, 5.5, 5.6, 5.7, 5.8, 5.9, 5.10, 5.11, 5.12, 5.13, 5.14, 5.15, 5.16, 5.17, 5.18, 5.19, and 5.20).

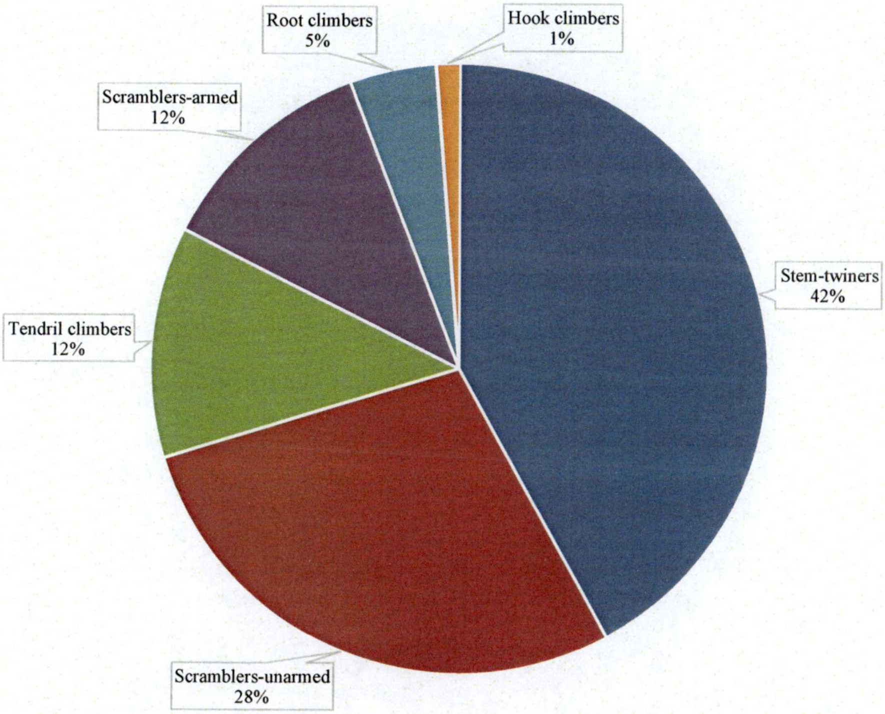

Fig. 5.6 The proportion of Indian climbing flora under various climbing strategies

Plate 5.1 General habit of *Combretum albidum.* Picture credit: Vivek Pandi

Plate 5.2 General habit of *Dregea volubilis.* Picture credit: Vivek Pandi

Plate 5.3 General habit of *Cissus vitiginea*. Picture credit: Vivek Pandi

Plate 5.4 *Jasminum angustifolium (**a**); Ziziphus oenopolia (**b**).* Picture credit: Vivek Pandi

Plate 5.5 *Cansjera rheedi (a); Leptadenia reticulata (**b**); Pothos scandens (**c**).* Picture credit: Vivek Pandi (**a** & **b**); Subbaiah Karuppusamy (**c**)

Plate 5.6 *Cissampelos indica* (***a***); *Celastrus paniculatus* (***b***); *Cissus javana* (***c***). Picture credit: Subbaiah Karuppusamy (**a**); Vivek Pandi (**b** & **c**)

Plate 5.7 *Asparagus racemosus* (***a***); *Toddalia asiatica* (***b***); *Alangium salvifolium* subsp. *hexapetalum* (*c*). Picture credit: Vivek Pandi

Plate 5.8 *Mallotus repandus (**a**); Grewia rhamnifolia (**b**).* Picture credit: Vivek Pandi

Plate 5.9 *Mussaenda frondosa (**a**); Caesalpinia cucullata (**b**).* Picture credit: Vivek Pandi

Plate 5.10 *Artabotrys zeylanicus (**a**); Dioscorea hispida (**b**).* Picture credit: Vivek Pandi (**a**)

Plate 5.11 *Stictocardia tiliifolia* (***a***); *Toddalia asiatica* (***b***); *Piper sp.* (***c***); *Paramygnya monophylla* (***d***). Picture credit: Vivek Pandi

Plate 5.12 Flowers (**a**) and dehisced pods of *Abrus precatorius* exposing the seeds (**b**). Picture credit: Dhatchanamoorthy (**a**) Vivek Pandi (**b**)

Plate 5.13 Flowers of *Ipomea hederifolia* (**a**); *I. pes-tigridis* (**b**); *I. eriocarpha* (**c**); *I. cairica* (**d**); *I. staphylina* (**e**); *I. parasitica* (**f**). Picture credit: Dhatchanamoorthy

Plate 5.14 *Rivea hypocrateriformis* (***a***); *Cissus vitiginea* (***b***). Picture credit: Vivek Pandi

Plate 5.15 Flowers of *Acacia caesia* (**a**); *Acacia pennata* (**b**). Picture credit: Dhatchanamoorthy (**a** & **b**)

Plate 5.16 Flowers of *Gymnema sylvestre* (**a**); *G. inodorum* (**b**). Picture credit: Dhatchanamoorthy (**a**); Subbaiah Karuppusamy (**b**)

Plate 5.17 Flowers of *Jasminum cuspidatum* (**a**); *J. angustifolium* (**b**). Picture credit: Dhatchanamoorthy (**a**); Vivek Pandi (**b**)

Plate 5.18 Flowers of *Passiflora edulis* (**a**); *P. subpeltata* (**b**); *P. foetida* (**c**). Picture credit: Vivek Pandi (**a** & **c**); Dhatchanamoorthy; (**b**)

Plate 5.19 Flowers of *Vallaris solanaceae* (**a**); *Neonotonia wightii* (**b**). Picture credit: Dhatchanamoorthy

Plate 5.20 *Abrus pulchellus* (**a**); *Kedrostis foetidissima* (**b**). Picture credit: Dhatchanamoorthy

5.3 Climbing Flora of India: A Compendium

Acanthus volubilis Wall.
Acanthaceae
Synonyms: *Dilivaria scandens* Nees, *Dilivaria volubilis* (Wall.) Nees
Climbing Mechanism: Scrambler-Unarmed
Distribution (Global): Singapore, Andaman Is., Bangladesh, Malaya, Myanmar, Nicobar Is., Thailand
Distribution (India): Andaman and Nicobar Islands
Leaf Type: Simple and opposite
Inflorescence: Terminal cluster
Fruit Type: Rarely formed, consists of a capsule about 2.5 cm long, ellipsoid, and flattened
Flowering and Fruiting: Throughout the year
IUCN Status: Least Concern
Notes: In Malay culture, powdered seeds are taken with water as a blood-cleansing medicine against ulcers.
Reference: Pl. Asiat. Rar. 2: 56 (1831)

Aphelandra arborea
Acanthaceae
Climbing Mechanism: Scrambler-Unarmed
IUCN Status: Not evaluated

Dipteracanthus suffruticosus
Acanthaceae
Climbing Mechanism: Scrambler-Unarmed
IUCN Status: Not evaluated

Meyenia hawtayneana Nees
Acanthaceae
Synonyms: *Meyenia hawtayneana* (Wall.) Nees, *Thunbergia coerulea* Wight ex Nees, *Thunbergia cordifolia* Wall. *Thunbergia erecta* Nees, *Thunbergia hawtaynii* T. Anderson
Climbing mechanism: Scrambler-Unarmed
Distribution (Global): Assam, Bangladesh, India, Myanmar, Vietnam
Distribution (India): Andhra Pradesh, Maharashtra, Tamil Nadu, Telangana
Leaf type: Simple and opposite 6–9 x 3–4.5 cm, ovate, apex acuminate, base cordate, 3–5 ribbed
Inflorescence: Flowers solitary, axillary
Fruit type: Capsule to 3 x 1.5 cm, globose, beaked; seeds 4, ca 4 mm across, orbicular
Flowering and fruiting: February–April
IUCN status: Not evaluated

Notes: Occurs in the dry deciduous forests of the Western Ghats and the Eastern Ghats
Reference: Tent. Fl. Napal.: 49 (1826)

Nilgirianthus barbatus
Acanthaceae
Climbing Mechanism: Scrambler-Unarmed
IUCN Status: Not evaluated

Strobilanthes lakshminarasimhanii Sameer Patil
Acanthaceae
Climbing Mechanism: Scrambler-Unarmed
Leaf Type: Simple, opposite
Inflorescence: Spikes or capitate heads
Fruit Type: Dry
IUCN Status: Not evaluated
Reference: Sameer Patil. In: Nelumbo 62(2): 113, figs. 1-2 (2020)

Strobilanthes lurida Wight
Acanthaceae
Synonyms: *Didyplosandra lurida* (Wight) Bremek
Common Name: Lurid coneflower
Climbing Mechanism: Scrambler-Unarmed
Distribution (Global): India
Leaf Type: Leaves are opposite
Inflorescence: Spikes
Fruit Type: Capsule, ovoid-ellipsoid
Flowering and Fruiting: December-April
IUCN Status: Not evaluated
Reference: Icon. Pl. Ind. Orient. 4: t. 1515 (1850)

Thelepaepale ixiocephala (Benth.) Bremek
Acanthaceae
Synonyms: *Strobilanthes ixiocephala* Benth. *Strobilanthes eriocephala* T. Anderson., *Strobilanthes glutinosa* J. Graham., *Strobilanthes neesiana* Wight
Common Name: Sky blue karvy
Climbing Mechanism: Scrambler-Unarmed
Distribution (Global): Bangladesh, India
Distribution (India): Goa, Gujarat, Maharashtra, Tamil Nadu
Leaf Type: Simple and opposite
Inflorescence: Axillary or terminal condensed spikes
Flowering and Fruiting: November–January
IUCN Status: Not evaluated
Reference: Flora 32: 557 (1849)

Thunbergia alata Bojer ex Sims
Acanthaceae
Synonyms: *Endomelas alata* (Bojer ex Sims) Raf., *Thunbergia oculata* S. Moore. + 10
Common Name: Black-eyed Susan Vine, Brazilian nightshade
Climbing Mechanism: Scrambler-Unarmed
Distribution (India): Andaman and Nicobar Islands, Andhra Pradesh, Arunachal Pradesh, Gujarat, Madhya Pradesh, Maharashtra, Rajasthan, Tamil Nadu, Telangana
Leaf Type: Opposite decussate
Fruit Type: Depressed-globose capsule
Flowering and Fruiting: December–March
IUCN Status: Not evaluated
Notes: Naturalized. Occurs in lower and higher altitudes of the Western Ghats and the Eastern Ghats
Reference: Bot. Mag. 52: t. 2591 (1825)

Thunbergia bicolor Lindau
Acanthaceae
Synonyms: *Schmidia bicolor* Wight., *Thunbergia wightiana* T. Anderson
Common Name: Bicolor clock vine
Climbing Mechanism: Scrambler-Unarmed
Distribution (India): Tamil Nadu
Leaf Type: Opposite
Inflorescence: Axillary pendate, Raceme
Fruit Type: Capsule, globose
Flowering and Fruiting: November–January
IUCN Status: Not evaluated
Reference: Beibl. Bot. Jahrb. Syst. 41: 42 (1893)

Thunbergia coccinea Wall.
Acanthaceae
Synonyms: *Flemingia coccinea* Buch. -Ham. ex Nees, *Hexacentris acuminata* Nees, *Hexacentris coccinea* (Wall.) Nees, *Hexacentris dentata* Nees, *Thunbergia pendula* Hassk., *Thunbergia quinquenervis* Buch. -Ham. ex Nees
Common Name: Scarlet clock vine
Climbing Mechanism: Scrambler-Unarmed
Distribution (Global): Assam, Bangladesh, China South-Central, East Himalaya, India, Laos, Myanmar, Nepal, Thailand, Tibet, Vietnam, Western Himalaya
Distribution (India): Andaman and Nicobar Islands, Arunachal Pradesh, Assam, Bihar, Odisha, Maharashtra, Meghalaya, Mizoram, Tamil Nadu
Leaf Type: Simple, 3–5 nerved
Inflorescence: Stout racemes
Fruit Type: Dry, capsule
Flowering and Fruiting: November–April

IUCN Status: Not evaluated
Reference: Prodr. Fl. Nepal.: 120 (1825)

Thunbergia erecta (Benth.) T. Anderson
Acanthaceae
Synonyms: *Meyenia erecta* Benth., *Thunbergia ikbaliana* De Wild., *Thunbergia mestdaghi* De Wild
Climbing Mechanism: Scrambler-Unarmed
Distribution (Global): Cameroon, Central African Republic, Congo, Gabon, Ghana, Guinea, Guinea-Bissau, Ivory Coast, Liberia, Nigeria, Rwanda, Senegal, Sierra Leone, Sudan, Tanzania, Uganda, Zaire
Distribution (India): Puducherry
Leaf Type: Elliptic, glabrous
Inflorescence: Solitary
Fruit Type: Capsule, glabrous
IUCN Status: Not evaluated
Notes: Usually found in wet evergreen forests and swamp forests
Reference: J. Proc. Linn. Soc., Bot. 7: 18 (1863)

Thunbergia fragrans Roxb.
Acanthaceae
Synonyms: *Flemingia grandiflora* Roxb. ex Rottler, *Meyenia longiflora* Benth. ex Hohen., *Thunbergia angustifolia* Buch. -Ham. ex Nees., *Thunbergia bodinieri* H. Lév., *Thunbergia cordata* Colla, *Thunbergia fragrans* subsp. *hainanensis* (C.Y.Wu & H.S.Lo) H.P.Tsui, Thunb
Common Name: Sweet clock-vine, White lady
Climbing Mechanism: Stem Twiner
Distribution (Global): Andaman Is., Assam, Bangladesh, Cambodia, China South-Central, China Southeast, East Himalaya, Hainan, India, Laos, Myanmar, Nepal, Philippines, Sri Lanka, Thailand, Vietnam
Distribution (India): Andaman and Nicobar Islands, Andhra Pradesh, Odisha, Madhya Pradesh, Maharashtra, Tamil Nadu
Leaf Type: Simple, 3–5 nerved
Inflorescence: Stout racemes
Fruit Type: Dry, capsule
Flowering and Fruiting: Throughout the year
IUCN Status: Not evaluated
Notes: Frequently found in dry evergreen forests
Reference: Pl. Coromandel 1: 47 (1796)

Thunbergia fragrans var. *hispida*
Acanthaceae
Climbing Mechanism: Stem Twiner
Distribution (Global): India, Sri Lanka
Distribution (India): Peninsular India

Leaf Type: Simple, 3–5 nerved
Inflorescence: Stout racemes
Fruit Type: Dry, capsule
IUCN Status: Not evaluated
Reference: Gamble. In: Fl. Madras 1008. (1924)

Thunbergia fragrans var. *vestita*
Acanthaceae
Climbing Mechanism: Stem Twiner
Distribution (India): Andhra Pradesh, Tamil Nadu
Leaf Type: Simple, 3–5 nerved
Inflorescence: Stout racemes
Fruit Type: Dry, capsule
Flowering and Fruiting: September–February
IUCN Status: Not evaluated
Notes: Found usually in deciduous forests
Reference: Nees. In: Wall., Pl. Asiat. Rar. 3: 78. (1832)

Thunbergia geraniifolia Benth.
Acanthaceae
Synonyms: *Thunbergia chrysops* Hook., *Thunbergia subnymphaeifolia* Lindau
Climbing Mechanism: Scrambler-Unarmed
Distribution (Global): Cameroon, Ghana, Guinea, Ivory Coast, Liberia, Senegal, Sierra Leone, Togo
Leaf Type: Simple, 3–5 nerved
Inflorescence: Stout racemes
Fruit Type: Dry, capsule
IUCN Status: Not evaluated
Reference: Bot. Mag. 70: t. 4119 (1844)

Thunbergia grandiflora (Roxb. ex Rottl.) Roxb.
Acanthaceae
Synonyms: *Pleuremidis grandiflora* (Roxb.) Raf., *Thunbergia adenophora* W.W. Sm., *Thunbergia chinensis* Merr., *Thunbergia cordifolia* Nees, *Thunbergia grandiflora* f. alba Leonard, *Thunbergia grandiflora* f. citrina Leonard, *Thunbergia lacei* Gamble
Common Name: Large-flowered Thunbergia, Bengal trumpet, Blue trumpet vine, Blue sky vine
Climbing Mechanism: Scrambler-Unarmed
Distribution (Global): Assam, Bangladesh, Cambodia, China South-Central, China Southeast, East Himalaya, Hainan, Laos, Malaya, Myanmar, Nepal, Thailand, Vietnam
Distribution (India): Andaman and Nicobar Islands, Arunachal Pradesh, Assam, Bihar, Odisha, Delhi, Gujarat, Kerala, Madhya Pradesh, Maharashtra, Meghalaya, Odisha, Rajasthan, Tamil Nadu, Tripura, Uttar Pradesh, West Bengal

Leaf Type: Simple, 3–5 nerved
Inflorescence: Stout racemes
Fruit Type: Dry, capsule
Flowering and Fruiting: February–July
IUCN Status: Not evaluated
Reference: Bot. Reg. 6: 6 (1820)

Thunbergia hawtayneana Wall
Acanthaceae
Synonyms: *Meyenia hawtayneana* (Wall.) Nees, *Thunbergia coerulea* Wight ex Nees, *Thunbergia cordifolia* Wall., *Thunbergia erecta* Nees, *Thunbergia hawtaynii* T. Anderson
Climbing Mechanism: Scrambler-Unarmed
Distribution (Global): Assam, Bangladesh, India, Myanmar, Vietnam
Distribution (India): Tamil Nadu
Leaf Type: Simple and opposite
Inflorescence: Solitary, axillary
Fruit Type: Capsule, globose
Flowering and Fruiting: February–April
IUCN Status: Not evaluated
Notes: Found across dry deciduous forests of the Western Ghats and the Eastern Ghats
Reference: Tent. Fl. Napal.: 49 (1826)

Thunbergia laevis Nees
Acanthaceae
Synonyms: *Thunbergia fragrans* var. *laevis*
Climbing Mechanism: Stem Twiner
Distribution (Global): Assam, India, Sri Lanka
Distribution (India): Delhi, Goa, Gujarat, Madhya Pradesh, Maharashtra, Rajasthan
Leaf Type: Simple, 3–5 nerved
Inflorescence: Stout racemes
Fruit Type: Dry, capsule
Flowering and Fruiting: July–March
IUCN Status: Not evaluated
Notes: Often found in dry deciduous to moist deciduous forests
Reference: N. Wallich, Pl. Asiat. Rar. 3: 77 (1832)

Thunbergia laurifolia Lindl.
Acanthaceae
Synonyms: *Thunbergia harrisii* Hook.f.
Common Name: Laurel clock vine
Climbing Mechanism: Scrambler-Unarmed
Distribution (Global): Andaman Is., Assam, East Himalaya, Malaya, Myanmar, Nicobar Is., Vietnam

Distribution (India): Andaman and Nicobar Islands, Maharashtra
Leaf Type: In opposite pairs
Inflorescence: Hanging racemes
Fruit Type: Large beaked capsules with a globose basal (fertile) portion
Flowering and Fruiting: October–February
IUCN Status: Not evaluated
Reference: Gard. Chron. 1856: 260 (1856)

Thunbergia lutea T. Anderson
Acanthaceae
Synonyms: *Hexacentris lutea* Lindl., *Thunbergia salweenensis* W.W.Sm.
Common Name: Sikkim clock vine
Climbing Mechanism: Scrambler-Unarmed
Distribution (Global): Bangladesh, China South-Central, East Himalaya, Myanmar, Nepal, Thailand, Tibet
Leaf Type: Leaves ovate or broadly ovate
Inflorescence: Solitary
Fruit Type: Capsule glabrous
Flowering and Fruiting: August–December
IUCN Status: Not evaluated
Reference: J. Linn. Soc., Bot. 9: 448 (1867)

Thunbergia mysorensis (Wight) T. Anderson
Acanthaceae
Synonyms: *Hexacentris mysorensis* Wight
Common Name: Mysore clock vine, Dolls shoes, Brick and butter vine.
Climbing Mechanism: Scrambler-Unarmed
Distribution (Global): India
Distribution (India): Andaman and Nicobar Islands, Kerala, Maharashtra, Tamil Nadu, Western Ghats
Leaf Type: Oppositely arranged, lanceolate
Inflorescence: Racemes axillary
Fruit Type: Capsule
Flowering and Fruiting: December–March
IUCN Status: Not evaluated
Notes: Usually occur in evergreen and semi-evergreen forests
Reference: J. Linn. Soc., Bot. 9: 448 (1867)

Thunbergia tomentosa Wall. ex Nees
Acanthaceae
Climbing Mechanism: Scrambler-Unarmed
Distribution (Global): India
Distribution (India): Tamil Nadu
Leaf Type: Leaves sagittate
Inflorescence: Solitary, axillary

Fruit Type: Capsule depressed-globose
Flowering and Fruiting: January–March
IUCN Status: Not evaluated
Notes: Often found in degraded forests
Reference: N. Wallich, Pl. Asiat. Rar. 3: 78 (1832)

Thunbergia wightiana T. Anderson
Acanthaceae
Climbing Mechanism: Scrambler-Unarmed
Leaf Type: Simple, 3–5 nerved
Inflorescence: Stout racemes
Fruit Type: Dry, capsule
IUCN Status: Not evaluated
Reference: T. Anderson. In: Journ. L. Soc. 9: 448. (1867)

Actinidia callosa Lindl.
Actinidiaceae
Synonyms: *Actinidia alnifolia* Stapf ex Hand. -Mazz., *Actinidia arisanensis* Hayata, *Actinidia callosa* var. *acuminata* C.F. Liang, *Actinidia callosa* var. *formosana* Finet & Gagnep., *Actinidia callosa* var. *pubiramula* C.Y. Wu, *Actinidia formosana* (Finet & Gagnep.)
Common Name: Himalayan kiwi vine
Climbing Mechanism: Scrambler-Unarmed
Distribution (Global): Assam, China North-Central, China South-Central, China Southeast, East Himalaya, India, Malaya, Myanmar, Nepal, Sumatera, Taiwan, Tibet, Vietnam, West Himalaya
Distribution (India): North Eastern India
Leaf Type: Simple and alternate
Inflorescence: Axillary, solitary, cymes, or in fascicles
Fruit Type: Berry
Flowering and Fruiting: June–November
IUCN Status: Not evaluated
Notes: Used in folk medicine
Reference: Intr. Nat. Syst. Bot., ed. 2: 439 (1836)

Actinidia strigosa Hook.f. & Thomson
Actinidiaceae
Climbing Mechanism: Scrambler-Unarmed
Distribution (Global): East Himalaya, Nepal
Distribution (India): West Bengal, North Eastern India
Leaf Type: Simple, alternate, lanceolate-ovate to oblanceolate-elliptic
Inflorescence: Axillary, solitary, cymes, or in fascicles
Fruit Type: berry, globose-ovoid, glabrous, or hairy
IUCN Status: Not evaluated

Notes: Fruits are edible
Reference: J. Proc. Linn. Soc., Bot. 5: 55 (1860)

Bomarea multiflora (L.f.) Mirb.
Alstroemeriaceae
Synonyms: *Alstroemeria bredemeyeriana* Willd. ex Schult. & Schult.f., *Alstroemeria caldasiana* (Herb.) Hemsl., *Alstroemeria caldasii* Kunth, *Alstroemeria floribunda* Kunth, *Alstroemeria multiflora* L.f. +16
Common Name: Vine Alstroemeria
Climbing Mechanism: Stem Twiner
Distribution (Global): Colombia, Ecuador
Distribution (India): Tamil Nadu
IUCN Status: Not evaluated
Reference: Hist. Nat. Pl. 9: 72 (1804)

Brachylepis nervosa Wight & Arn.
Amaranthaceae
Synonyms: *Decalepis nervosa* (Wight & Arn.) Venter
Common Name: Nerved-leaf swallow-root
Climbing Mechanism: Stem Twiner
IUCN Status: Not evaluated
Reference: Contr. Bot. India: 63 (1834)

Cyathula tomentosa (Roth) Moq.
Amaranthaceae
Synonyms: *Achyranthes sequax* Wall., *Achyranthes tomentosa* (Schult.) Roth, *Cyathula sequax* (Wall.) Moq., *Desmochaeta sequax* (Wall.) A. Braun, *Desmochaeta tomentosa* Schult., *Polyscalis sequax* Wall
Common Name: Woolly pasture weed, Cottony chaff flower
Climbing Mechanism: Stem Twiner
Distribution (Global): Assam, Bangladesh, China South-Central, East Himalaya, Myanmar, Nepal, Tibet, West Himalaya
Leaf Type: Alternate, elliptic, pointed
Inflorescence: Spikes
Flowering and Fruiting: January–April
IUCN Status: Not evaluated
Notes: Usually found in evergreen forests
Reference: A.P.de Candolle, Prodr. 13(2): 327 (1849)

Deeringia amaranthoides (Lam.) Merr.
Amaranthaceae
Synonyms: *Achyranthes amaranthoides* Lam., *Celosia amaranthoides* (Lam.) Medik. +10
Common Name: Shrubby Deeringia
Climbing Mechanism: Scrambler-Unarmed

Distribution (Global): Assam, Bangladesh, China South-Central, China Southeast, Christmas I., East Himalaya, Fiji, Hainan, India, Java, Laos, Lesser Sunda Is., Malaya, Marianas, Myanmar, Nepal, New Guinea, New South Wales, Philippines, Queensland, Solomon Is., Sulawesi, Sumatera, Taiwan, Thailand, Tibet, Tonga, Vietnam, West Himalaya, Western Australia
Leaf Type: Alternate
Inflorescence: Racemes
Fruit Type: Globose scarlet berry
Flowering and Fruiting: August–December
IUCN Status: Not evaluated
Notes: Tender leaves are eaten cooked as a vegetable
Reference: Interpr. Herb. Amboin: 211 (1917)

Pegia nitida Colebr.
Anacardiaceae
Synonyms: *Phlebochiton extensum* Wall., *Tapirira extensa* (Wall.) Hook.f. ex Marchand, *Tapirira hirsuta* (Roxb.) Kurz, *Robergia hirsuta* Roxb.
Climbing Mechanism: Scrambler-Unarmed
Distribution (Global): Assam, Bangladesh, China South-Central, China Southeast, East Himalaya, Myanmar, Nepal, Thailand
Distribution (India): Arunachal Pradesh, Assam, Manipur, Meghalaya, Tripura, West Bengal
Leaf Type: Compound
Inflorescence: Terminal panicles
Fruit Type: Drupe, black when ripe
Flowering and Fruiting: February–March
IUCN Status: Not evaluated
Reference: Trans. Linn. Soc. London 15: 364 (1827)

Ancistrocladus attenuatus Dyer
Ancistrocladaceae
Climbing Mechanism: Scrambler-Unarmed
Distribution (Global): Andaman Is., India, Myanmar
Distribution (India): Maharashtra, Andaman and Nicobar Islands
Leaf Type: Simple, alternate
Inflorescence: Inflorescence axillary or terminal, panicles
Fruit Type: Dry, Pods
Flowering and Fruiting: January–May
IUCN Status: Not evaluated
Notes: Common in Moist deciduous
Reference: J. D. Hooker, Fl. Brit. India 1: 300 (1874)

Ancistrocladus griffithii Planch
Ancistrocladaceae
Climbing Mechanism: Hook Climber

Distribution (Global): Andaman Is., Cambodia, Myanmar, Thailand, Vietnam
Leaf Type: Simple, alternate
Inflorescence: Inflorescence axillary or terminal, panicles
Fruit Type: Dry, Pods
IUCN Status: Not evaluated
Reference: Ann. Sci. Nat., Bot., sér. 3, 13: 318 (1849)

Ancistrocladus hamatus (Vahl) Gilg
Ancistrocladaceae
Synonyms: *Ancistrocladus thwaitesii* Tiegh., *Ancistrocladus vahlii* Arn., *Bigamea hamata* (Vahl) Tiegh., *Bigamea thwaitesii* Tiegh., *Wormia hamata* Vahl.
Climbing Mechanism: Scrambler-Unarmed
Distribution (Global): Sri Lanka
Distribution (India): Peninsular India
Leaf Type: Simple, alternate
Inflorescence: Inflorescence axillary or terminal, panicles
Fruit Type: Dry, Pods
IUCN Status: Not evaluated
Reference: H.G.A. Engler & K.A.E. Prantl, Nat. Pflanzenfam. 3(6): 276 (1895)

Ancistrocladus heyneanus Wall. ex J. Graham
Ancistrocladaceae
Climbing Mechanism: Scrambler-Unarmed
Distribution (Global): India
Distribution (India): Kerala, Tamil Nadu, Maharashtra, Karnataka, Puducherry, Western Ghats
Leaf Type: Simple, alternate
Inflorescence: Inflorescence axillary or terminal, panicles
Fruit Type: Dry, Pods
Flowering and Fruiting: January–April
IUCN Status: Not evaluated
Notes: Mostly found in Evergreen forests
Reference: Cat. Pl. Bombay: 28 (1839)

Ancistrocladus tectorius (Lour.) Merr.
Ancistrocladaceae
Synonyms: *Ancistrocladus benomensis* Rischer & G. Bringmann, *Ancistrocladus carallioides* Craib, *Ancistrocladus cochinchinensis* Gagnep., *Ancistrocladus extensus* Wall. ex Planch., *Ancistrocladus hainanensis* Hayata, *Ancistrocladus harmandii* Gagnep.
Climbing Mechanism: Scrambler-Unarmed
Distribution (Global): Andaman Is., Bangladesh, Borneo, Cambodia, Hainan, Laos, Malaya, Myanmar, Sumatera, Thailand, Vietnam
Leaf Type: Simple, alternate
Inflorescence: Inflorescence axillary or terminal, panicles

Fruit Type: Dry, Pods
IUCN Status: Not evaluated
Reference: Lingnan Sci. J. 6: 329 (1928 publ. 1930)

Ancistrocladus wallichii Planch
Ancistrocladaceae
Climbing Mechanism: Hook Climber
Distribution (Global): Andaman Is., Assam, Bangladesh
Distribution (India): Andaman and Nicobar Islands
Leaf Type: Simple, alternate
Inflorescence: Inflorescence axillary or terminal, panicles
Fruit Type: Dry, Pods
IUCN Status: Not evaluated
Reference: Ann. Sci. Nat., Bot., sér. 3, 13: 319 (1849)

Anomianthus dulcis (Dunal) James Sinclair
Annonaceae
Climbing Mechanism: Scrambler-Unarmed
IUCN Status: Not evaluated
Reference: Gard. Bull. Singapore 14: 40 (1953)

Anona odoratissimus R.Br. Champa H.
Annonaceae
Climbing Mechanism: Scrambler-Unarmed
Distribution (India): Bihar, Odisha
IUCN Status: Not evaluated

Artabotrys burmanicus A. DC
Annonaceae
Synonyms: *Artabotrys cubittii* Chatterjee
Climbing Mechanism: Scrambler-Unarmed
Distribution (Global): Assam, Myanmar
Leaf Type: Simple, alternate, exstipulate
Inflorescence: Solitary or in fascicles, recurved hooked leaf opposed
Fruit Type: Wet, berry
IUCN Status: Not evaluated
Reference: Mém. Soc. Phys. Genève 5: 212 (1832)

Artabotrys caudatus Wall. ex Hook.f. & Thomson
Annonaceae
Climbing Mechanism: Scrambler-Unarmed
Distribution (Global): Assam, Bangladesh, Eastern Himalayas
Distribution (India): Arunachal Pradesh, Assam
Leaf Type: Simple, alternate
Inflorescence: Solitary

Fruit Type: Carpels
IUCN Status: Not evaluated
Reference: Fl. Ind. 1: 129 (1855)

Artabotrys crassifolius Hook.f. & Thomson
Annonaceae
Climbing Mechanism: Scrambler-Unarmed
Distribution (Global): Malaya, Myanmar
Leaf Type: Simple, alternate, exstipulate
Inflorescence: Solitary or in fascicles, recurved hooked leaf opposed
Fruit Type: Wet, berry
IUCN Status: Not eevaluated
Reference: J. D. Hooker, Fl. Brit. India 1: 54 (1872)

Artabotrys cubittii Chatterjee
Annonaceae
Synonyms: *Artabotrys burmanicus* A. DC
Climbing Mechanism: Scrambler-Unarmed
Distribution (India): Manipur
Leaf Type: Simple, alternate
Inflorescence: Solitary
Fruit Type: Carpels
IUCN Status: Not eevaluated
Reference: J. Indian Bot. Soc. 19: 1 (1940)

Artabotrys hexapetalus (L.f.) Bhandari
Annonaceae
Synonyms: *Annona hexapetala* L.f., *Annona uncinata* Lam., *Artabotrys hamatus* (Dunal) Blume, *Artabotrys intermedius* Hassk., *Artabotrys odoratissimus* R.Br., *Artabotrys uncatus* (Lour.) Baill., *Artabotrys uncinatus* (Lam.) Merr., *Unona esculenta* Dunal
Common Name: Tail grape
Climbing Mechanism: Scrambler-Unarmed
Distribution (Global): Andaman Is., Assam, Comoros, India, Laos, Myanmar, Sri Lanka, Thailand, Vietnam
Distribution (India): Andhra Pradesh, Eastern Ghats, Gujarat, Kerala, Manipur, Rajasthan, Tamil Nadu, Telangana, West Bengal
Leaf Type: Simple, alternate
Inflorescence: Solitary
Fruit Type: Carpels
Flowering and Fruiting: January–May
IUCN Status: Not evaluated
Notes: Grown as a garden plant. Flowers contain essential oil used in perfumery
Reference: Baileya 12: 149 (1965)

Artabotrys kurzii Hook.f. & Thomson
Annonaceae
Climbing Mechanism: Scrambler-Unarmed
Distribution (Global): Myanmar
Leaf Type: Simple, alternate, exstipulate
Inflorescence: Solitary or in fascicles, recurved hooked leaf opposed
Fruit Type: Wet, berry
IUCN Status: Not evaluated
Reference: J. D. Hooker, Fl. Brit. India 1: 54 (1872)

Artabotrys manoranjanii M. V. Ramana J. Swamy & K. C. Mohan
Annonaceae
Climbing Mechanism: Scrambler-Unarmed
Distribution (Global): Andaman Is.
Distribution (India): Andaman and Nicobar Islands
Leaf Type: Simple, alternate, exstipulate
Inflorescence: Solitary or in fascicles, recurved hooked leaf opposed
Fruit Type: Wet, berry
IUCN Status: Not evaluated
Reference: Nordic J. Bot. 34: 413 (2016)

Artabotrys nicobarianus D. Das
Annonaceae
Climbing Mechanism: Scrambler-Unarmed
Distribution (Global): Andaman Is., Nicobar Is.
Distribution (India): Great Nicobar Island
Leaf Type: Simple, alternate
Inflorescence: Solitary
Fruit Type: Carpels
IUCN Status: Not evaluated
Reference: Bull. Bot. Surv. India 11: 194 (1969 publ. 1971)

Artabotrys pleurocarpus Maingay ex Hook.f. & Thomson
Annonaceae
Climbing Mechanism: Scrambler-Unarmed
Distribution (Global): Malaya
Leaf Type: Simple, alternate, exstipulate
Inflorescence: Solitary or in fascicles, recurved hooked leaf opposed
Fruit Type: Wet, berry
IUCN Status: Not evaluated
Reference: J. D. Hooker, Fl. Brit. India 1: 54 (1872)

Artabotrys sahyadricus Robi, KMP Kumar & Hareesh
Annonaceae
Climbing Mechanism: Scrambler-Unarmed

Distribution (Global): India
Distribution (India): Kerala
Leaf Type: Simple, alternate, exstipulate
Inflorescence: Solitary or in fascicles, recurved hooked leaf opposed
Fruit Type: Wet, berry
IUCN Status: Not evaluated
Reference: Webbia 72: 117 (2017)

Artabotrys speciosus Kurz ex Hook.f. & Thomson
Annonaceae
Climbing Mechanism: Scrambler-Unarmed
Distribution (Global): Andaman Is., Myanmar
Distribution (India): Andaman and Nicobar Islands
Leaf Type: Simple, alternate
Inflorescence: Solitary
Fruit Type: Carpels
IUCN Status: Not evaluated
Reference: J. D. Hooker, Fl. Brit. India 1: 55 (1872)

Artabotrys suaveolens (Blume) Blume
Annonaceae
Synonyms: *Artabotrys corniculatus* (Blanco) Merr., *Artabotrys monogynus* Merr., *Artabotrys parviflorus* Miq., *Artabotrys rolfei* S. Vidal, *Artabotrys trigynus* Merr., *Unona corniculata* Blanco, *Unona suaveolens* Blume
Climbing Mechanism: Scrambler-Unarmed
Distribution (Global): Borneo, Java, Lesser Sunda Is., Malaya, Maluku, Myanmar, New Guinea, Philippines, Sulawesi, Sumatera, Thailand
Leaf Type: Simple, alternate, exstipulate
Inflorescence: Solitary or in fascicles, recurved hooked leaf opposed
Fruit Type: Wet, berry
IUCN Status: Not evaluated
Reference: Fl. Javae, Anon.: 62 (1830)

Artabotrys zeylanicus Hook.f. & Thomson
Annonaceae
Common Name: Ceylon green champa
Climbing Mechanism: Scrambler-Unarmed
Distribution (Global): India, Sri Lanka
Leaf Type: Simple, alternate, exstipulate
Inflorescence: Solitary or in fascicles, recurved hooked leaf opposed
Fruit Type: Wet, berry
Flowering and Fruiting: September–June
IUCN Status: Not evaluated

Notes: Frequently found in evergreen and semi-evergreen forests of the Western Ghats up to 1000 m
Reference: Fl. Ind. 1: 128 (1855)

Artabotrys zeylanicus var. *kottavasaliyana*
Annonaceae
Climbing Mechanism: Scrambler-Unarmed
IUCN Status: Not evaluated

Cymbopetalum brasiliense (Vell.) Benth. ex Baill.
Annonaceae
Synonyms: *Cymbopetalum brasiliense* f. latifolia Huber, *Cymbopetalum odoratissimum* J.J. Rodr., *Trigynaea anastomosans* Rusby, *Uva brasiliensis* (Vell.) Kuntze, *Uvaria brasiliensis* Vell.
Climbing Mechanism: Scrambler-Unarmed
Distribution (Global): Bolivia, Brazil North, Brazil Northeast, Brazil Southeast, Brazil West-Central, Colombia, French Guiana, Guyana, Panamá, Peru, Suriname, Trinidad-Tobago, Venezuela
Distribution (India): Western Ghats
IUCN Status: Least Concern
Reference: Hist. Pl. 1: 240 (1868)

Desmos chinensis Lour.
Annonaceae
Synonyms: *Annona laevigata* Mart., *Artabotrys esquirolii* H. Lév., *Desmos chinensis* var. *brevifolius* (Teijsm. & Binn. ex Boerl.) Bân. +30
Common Name: Sahyadri ylang-ylang
Climbing Mechanism: Scrambler-Unarmed
Distribution (Global): Assam, Bangladesh, Borneo, Cambodia, China South-Central, China Southeast, East Himalaya, Hainan, India, Java, Laos, Lesser Sunda Is., Malaya, Myanmar, Nepal, Philippines, Sri Lanka, Sulawesi, Sumatera, Thailand, Vietnam
Distribution (India): Arunachal Pradesh, Karnataka, Kerala, Maharashtra, Manipur, Mizoram, Tamil Nadu, Tripura, West Bengal
Leaf Type: Compound, imparipinnate
Inflorescence: Extra axillary
Flowering and Fruiting: April–January
IUCN Status: Not evaluated
Notes: Usually found in evergreen and semi-evergreen forests
Reference: Fl. Cochinch.: 352 (1790)

Desmos cochinchinensis Lour.
Annonaceae
Synonyms: *Desmos cochinchinensis* var. *fulvescens* Bân, *Desmos hancei* Merr., *Desmos velutinus* (Hance) Ast, *Unona cabog* Blanco, *Unona cochinchinensis*

(Lour.) DC., *Unona desmos* Raeusch., *Unona discolor* Wall., *Unona fulva* Wall., *Unona odorata* Blanco
Climbing Mechanism: Scrambler-Unarmed
Distribution (Global): Assam, Cambodia, Laos, Malaya, Myanmar, Nicobar Is., Vietnam
Distribution (India): Andaman and Nicobar Islands, Manipur
Leaf Type: Simple, alternate
Inflorescence: Solitary, extra axillary
Fruit Type: dehiscent, peduncled, and arranged in dense clusters
Flowering and Fruiting: April and August
IUCN Status: Not eevaluated
Notes: Medicinally valuable in treating various ailments. Grows mostly in open forests
Reference: Fl. Cochinch.: 352 (1790)

Desmos dumosus (Roxb.) Saff.
Annonaceae
Synonyms: *Dasymaschalon moniliferum* (Merr.) P.T.Li, *Desmos subbiglandulosus* (Miq.) Merr., *Oxymitra monilifera* Merr., *Unona dumosa* Roxb., *Unona subbiglandulosa* Miq.
Climbing Mechanism: Scrambler-Unarmed
Distribution (Global): Assam, Bangladesh, Borneo, China South-Central, China Southeast, East Himalaya, Laos, Malaya, Myanmar, Nicobar Is., Thailand, Vietnam
Distribution (India): North Eastern India, West Bengal
Leaf Type: Simple, alternate
Inflorescence: Solitary, extra axillary
Fruit Type: dark purple fruits have 2–4 smooth seeds with brown seed coats
Flowering and Fruiting: May–January
IUCN Status: Not evaluated
Reference: Bull. Torrey Bot. Club 39: 506 (1912)

Desmos dunalii (Wall. ex Hook.f. & Thomson) Saff.
Annonaceae
Synonyms: *Unona dunalii* Wall. ex Hook.f. & Thomson
Climbing Mechanism: Scrambler-Unarmed
Distribution (Global): Bangladesh, Borneo, Malaya, Myanmar, Thailand
IUCN Status: Not evaluated
Reference: Bull. Torrey Bot. Club 39: 506 (1912)

Desmos viridiflorus Saff.
Annonaceae
Synonyms: *Unona viridiflora* Bedd.
Climbing Mechanism: Scrambler-Unarmed
Distribution (Global): India

Distribution (India): Kerala, Tamil Nadu, Western Ghats
Leaf Type: Solitary
Flowering and Fruiting: April–October
IUCN Status: Not evaluated
Notes: Usually grow in evergreen forests
Reference: Bull. Torrey Bot. Club 39: 506 (1912)

Fissistigma bicolor (Roxb.) Merr.
Annonaceae
Synonyms: *Fissistigma villosum* (Jovet-Ast) Merr., *Melodorum bicolor* (Roxb.) Hook.f. & Thomson, *Uvaria bicolor* Roxb., *Melodorum villosum* Jovet-Ast.
Common Name: Bicolor Uvaria
Climbing Mechanism: Scrambler-Unarmed
Distribution (Global): Assam, Bangladesh, East Himalaya, Myanmar, Vietnam
Distribution (India): Arunachal Pradesh, Manipur, Meghalaya, Tripura, West Bengal
IUCN Status: Not eevaluated
Reference: Philipp. J. Sci. 15: 131 (1919)

Fissistigma fulgens (Hook.f. & Thomson) Merr.
Annonaceae
Synonyms: *Unona cauliflora* Blanco, *Melodorum fulgens* Hook.f. & Thomson, *Myristica finlaysoniana* Wall., *Uvaria fulgens* Wall
Climbing Mechanism: Scrambler-Unarmed
Distribution (Global): Borneo, Malaya, Philippines, Sumatera, Thailand
IUCN Status: Not evaluated
Reference: Philipp. J. Sci. 15: 131 (1919)

Fissistigma lanuginosum (Hook.f. & Thomson) Merr.
Annonaceae
Synonyms: *Melodorum lanuginosum* Hook.f. & Thomson, *Uvaria lanuginosa* Wall., *Uvaria tomentosa* Wall.
Climbing Mechanism: Scrambler-Unarmed
Distribution (Global): Malaya, Thailand
IUCN Status: Not evaluated
Reference: Philipp. J. Sci. 15: 132 (1919)

Fissistigma latifolium (Dunal) Merr.
Annonaceae
Synonyms: *Annona rufa* C. Presl, *Fissistigma borneense* (Miq.) Merr., *Fissistigma parviflorum* (Scheff.) Merr., *Fissistigma rufum* (C. Presl) Merr., *Fissistigma sphaerocarpum* (Miq.) Backer, *Friesodielsia multinervia* (Merr.) Steenis, *Melodorum borneense* Miq.
Climbing Mechanism: Scrambler-Unarmed
Distribution (Global): Borneo, China South-Central, Java, Malaya, Maluku, Philippines, Sulawesi, Sumatera, Thailand, Vietnam

Leaf Type: Simple, alternate
Inflorescence: Solitary, fascicled panicles or cymes
Fruit Type: Carpels
IUCN Status: Not evaluated
Reference: Philipp. J. Sci. 15: 132 (1919)

Fissistigma manubriatum (Hook.f. & Thomson) Merr.
Annonaceae
Synonyms: *Fissistigma korthalsii* (Miq.) Merr., *Melodorum bancanum* Scheff., *Melodorum korthalsii* Miq., *Uvaria manubriata* Wall., *Melodorum manubriatum* Hook.f. & Thomson
Climbing Mechanism: Scrambler-Unarmed
Distribution (Global): Assam, Borneo, Malaya, Myanmar, Sumatera
Distribution (India): Arunachal Pradesh
Leaf Type: Simple, alternate
Inflorescence: Solitary, fascicled panicles or cymes
Fruit Type: Carpels
IUCN Status: Not evaluated
Reference: Philipp. J. Sci. 15: 124 (1919)

Fissistigma polyanthum (Hook.f. & Thomson) Merr.
Annonaceae
Synonyms: *Melodorum polyanthum* Hook.f. & Thomson, *Uvaria polyantha* Wall
Common Name: Many-flowered Uvaria
Climbing Mechanism: Scrambler-Unarmed
Distribution (Global): Assam, Bangladesh, China South-Central, China Southeast, East Himalaya, Hainan, Myanmar, Tibet, Vietnam
Distribution (India): Arunachal Pradesh, Manipur, Meghalaya, Mizoram
Leaf Type: Simple, alternate
Inflorescence: Solitary, fascicled panicles or cymes
Fruit Type: Carpels
Flowering and Fruiting: November–March/April-–August
IUCN Status: Not evaluated
Reference: Philipp. J. Sci. 15: 135 (1919)

Fissistigma rubiginosum (A.DC.) Merr.
Annonaceae
Synonyms: *Melodorum rubiginosum* (A.DC.) Hook.f. & Thomson, *Uvaria fulva* Hook.f. & Thomson, *Uvaria nervosa* Wall., *Uvaria rubiginosa* A. DC
Climbing Mechanism: Scrambler-Unarmed
Distribution (Global): Assam, Bangladesh, Borneo, Cambodia, East Himalaya, Malaya, Myanmar, Vietnam
Distribution (India): Arunachal Pradesh, Meghalaya
Leaf Type: Simple, alternate
Inflorescence: Solitary, fascicled panicles or cymes

Fruit Type: Carpels
Flowering and Fruiting: November–January
IUCN Status: Not evaluated
Reference: Philipp. J. Sci. 15: 135 (1919)

Fissistigma verrucosum (Hook.f. & Thomson) Merr.
Annonaceae
Synonyms: *Melodorum verrucosum* Hook.f. & Thomson
Climbing Mechanism: Scrambler-Unarmed
Distribution (Global): Assam, Bangladesh, Myanmar
Distribution (India): North Eastern India
Leaf Type: Simple, alternate
Inflorescence: Solitary, fascicled panicles or cymes
Fruit Type: Carpels
Flowering and Fruiting: March–January
IUCN Status: Not evaluated
Reference: Philipp. J. Sci. 15: 137 (1919)

Fissistigma wallichii (Hook.f. & Thomson) Merr.
Annonaceae
Synonyms: *Fissistigma oligocarpum* W.T. Wang, *Melodorum wallichii* Hook.f. & Thomson, *Uvaria bicolor* Wall
Climbing Mechanism: Scrambler-Unarmed
Distribution (Global): Assam, Bangladesh, China
Distribution (India): Arunachal Pradesh
Leaf Type: Simple, alternate
Inflorescence: Solitary, fascicled panicles or cymes
Fruit Type: Carpels
Flowering and Fruiting: November–February/July–August
IUCN Status: Not evaluated
Reference: Philipp. J. Sci. 15: 137 (1919)

Friesodielsia biglandulosa (Blume) Steenis
Annonaceae
Synonyms: *Guatteria biglandulosa* Blume, *Monoon biglandulosum* (Blume) Miq., *Oxymitra biglandulosa* (Blume) Scheff., *Richella biglandulosa* (Blume) R.E.Fr., *Polyalthia biglandulosa* (Blume) Hook.f. & Thomson
Climbing Mechanism: Scrambler-Unarmed
Distribution (Global): Borneo, Java, Malaya
IUCN Status: Not Evaluated
Reference: Blumea 12: 358 (1964)

Friesodielsia fornicata (Roxb.) D.Das
Annonaceae

Synonyms: *Oxymitra fornicata* (Roxb.) Hook.f. & Thomson, *Uvaria fornicata* Roxb., *Uvaria roxburghiana* Wall
Common Name: Eastern dwaba-berry
Climbing Mechanism: Scrambler-Unarmed
Distribution (Global): Andaman Is., Assam, Bangladesh, Cambodia, East Himalaya, Myanmar, Nicobar Is., Vietnam
Distribution (India): Arunachal Pradesh
Leaf Type: Simple, alternate
Inflorescence: Solitary, extra axillary
Fruit Type: Carpels
Flowering and Fruiting: May–July/October–December
IUCN Status: Not evaluated
Reference: Bull. Bot. Surv. India 5: 43 (1963)

Friesodielsia glauca (Hook.f. & Thomson) Steenis
Annonaceae
Synonyms: *Friesodielsia argentea* (J. Sinclair) Steenis, *Friesodielsia diadena* (Miq.) Steenis, *Friesodielsia linderifolia* (Ridl.) Steenis, *Oxymitra argentea* J. Sinclair, *Oxymitra diadena* Miq., *Oxymitra glauca* Hook.f. & Thomson, *Oxymitra linderifolia* Ridl.
Climbing Mechanism: Scrambler-Unarmed
Distribution (Global): Borneo, Malaya, Sumatera, Thailand
IUCN Status: Not evaluated
Reference: Blumea 12: 359 (1964)

Friesodielsia khoshooi Vasudeva Rao & Chakrab.
Annonaceae
Synonyms: *Oxymitra fornicata* (Roxb.) Hook. f. & Thomson, *Uvaria fornicata* Roxb.
Climbing Mechanism: Scrambler-Unarmed
Distribution (Global): Nicobar Is.
Distribution (India): Andaman and Nicobar Islands
Leaf Type: Simple, alternate
Inflorescence: Solitary, extra axillary
Fruit Type: Carpels
Flowering and Fruiting: April–May/August
IUCN Status: Not evaluated
Reference: J. Econ. Taxon. Bot. 6: 435 (1985)

Friesodielsia unonifolia (A.DC.) Steenis
Annonaceae
Synonyms: *Guatteria unonifolia* A.DC., *Oxymitra unonifolia* (A.DC.) Hook.f. & Thomson
Climbing Mechanism: Scrambler-Unarmed
Distribution (Global): Myanmar, Thailand

IUCN Status: Not evaluated
Reference: Blumea 12: 361 (1964)

Melodorum bicolor Hook.f. & Thomson
Annonaceae
Synonyms: *Fissistigma bicolor* (Roxb.) Merr., *Fissistigma villosum* (Jovet-Ast) Merr., *Melodorum villosum* Jovet-Ast, *Uvaria bicolor* Roxb.
Common Name: Bicolor Uvaria
Climbing Mechanism: Scrambler-Unarmed
Distribution (Global): Assam, Bangladesh, East Himalaya, Myanmar, Vietnam
Distribution (India): Assam
Leaf Type: Leaves simple, alternate, exstipulate, with strong parallel veins, pubescent to tomentose
Inflorescence: solitary or fascicled panicles or cymes, leaf opposed or axillary
Fruit Type: Berry, edible when ripe
Flowering and Fruiting: March–April/August–October
IUCN Status: Not evaluated
Notes: Predominantly found in evergreen forests
Reference: Philipp. J. Sci. 15: 131 (1919)

Melodorum griffithii Hook.f. & Thomson
Annonaceae
Synonyms: *Fissistigma scandens* Griff.
Climbing Mechanism: Scrambler-Unarmed
Distribution (Global): Malaya, Myanmar, Thailand, Vietnam
Flowering and Fruiting: July–December
IUCN Status: Not evaluated
Reference: Not. Pl. Asiat. 4: 706 (1854)

Melodorum rufinervum Hook.f. & Thomson
Annonaceae
Synonyms: *Fissistigma rufinerve* (Hook.f. & Thomson) Merr.
Climbing Mechanism: Scrambler-Unarmed
Distribution (Global): Assam, Bangladesh, Vietnam
Distribution (India): Assam
IUCN Status: Not evaluated
Reference: Philipp. J. Sci. 15: 136 (1919)

Melodorum verrucosum Hook.f. & Thomson
Annonaceae
Synonyms: *Fissistigma verrucosum* (Hook.f. & Thomson) Merr.
Climbing Mechanism: Scrambler-Unarmed
Distribution (Global): Assam, Bangladesh, Myanmar
Distribution (India): Assam
Leaf Type: Leaves simple, alternate, exstipulate, with strong parallel veins

Inflorescence: Flowers bisexual, solitary or fascicled panicles or cymes, leaf opposed or axillary
Fruit Type: Ripe carpels berried, globose, stalked, seeds many
Flowering and Fruiting: March–January
IUCN Status: Not evaluated
Notes: Mostly found in Subtropical evergreen forests

Mitrella kentii (Blume) Miq.
Annonaceae
Climbing Mechanism: Scrambler-Unarmed
Leaf Type: Simple, alternate
Inflorescence: Flowers in solitary
Fruit Type: Dry, pods
IUCN Status: Not evaluated

Oxymitra filipes Hook.f. & Thomson
Annonaceae
Climbing Mechanism: Scrambler-Unarmed
IUCN Status: Not evaluated
Reference: J. D. Hooker, Fl. Brit. India 1: 71 (1872)

Oxymitra fornicata Hook.f. & Thomson
Annonaceae
Climbing Mechanism: Scrambler-Unarmed
Distribution (India): Assam
IUCN Status: Not evaluated
Reference: Fl. Ind. 1: 146 (1855)

Oxymitra latifolia Hook.f. & Thomson
Annonaceae
Climbing Mechanism: Scrambler-Unarmed
IUCN Status: Not evaluated
Reference: Fl. Ind. 1: 145 (1855)

Oxymitra maclellandii Hook.f. & Thomson
Annonaceae
Climbing Mechanism: Scrambler-Unarmed
IUCN Status: Not evaluated
Reference: J. D. Hooker, Fl. Brit. India 1: 70 (1872)

Oxymitra stenopetala Hook.f. & Thomson
Annonaceae
Climbing Mechanism: Scrambler-Unarmed
IUCN Status: Not evaluated
Reference: J. D. Hooker, Fl. Brit. India 1: 71 (1872)

Phaeanthus ophthalmicus (Roxb. ex G. Don) J.Sinclair
Annonaceae
Synonyms: *Guatteria macropoda* (Miq.) Zipp. ex Burck, *Monoon macropodum* Miq., *Phaeanthus cumingii* Miq., *Phaeanthus ebracteolatus* (C. Presl) Merr., *Phaeanthus macropodus* (Miq.) Diels, *Phaeanthus nigrescens* Elmer, *Phaeanthus nitidus* Merr., *Phaeanthus nutans* Hook
Climbing Mechanism: Scrambler-Unarmed
Distribution (Global): Borneo, Lesser Sunda Is., Malaya, Maluku, New Guinea, Philippines, Sulawesi
IUCN Status: Not evaluated
Reference: Gard. Bull. Singapore 14: 374 (1955)

Pyramidanthe prismatica (Hook.f. & Thomson) Merr.
Annonaceae
Synonyms: *Fissistigma cylindricum* (Maingay ex Hook.f. & Thomson) Merr., *Fissistigma maingayi* (Hook.f. & Thomson) Merr., *Fissistigma prismaticum* (Hook.f. & Thomson) Merr., *Fissistigma rigidum* (Ridl.) Merr., *Melodorum cylindricum* Maingay ex Hook.f. & Thomson
Climbing Mechanism: Scrambler-Unarmed
Distribution (Global): Borneo, Malaya, Sumatera
IUCN Status: Not evaluated
Reference: J. Straits Branch Roy. Asiat. Soc. 84(Spec. No.): 262 (1921)

Unona desmos Raeusch
Annonaceae
Synonyms: *Desmos cochinchinensis* Lour., *Desmos cochinchinensis* var. *fulvescens* Bân., *Desmos hancei* Merr., *Desmos velutinus* (Hance) Ast., *Unona cabog* Blanco., *Unona cochinchinensis* (Lour.) DC., *Unona discolor* Wall., *Unona fulva* Wall., *Unona odorata* Blanco
Climbing Mechanism: Scrambler-Unarmed
Distribution (Global): Assam, Cambodia, Laos, Malaya, Myanmar, Nicobar Is., Vietnam
Distribution (India): Assam
Leaf Type: Simple
Inflorescence: Single axillary flowers
IUCN Status: Not evaluated
Reference: Fl. Cochinch.: 352 (1790)

Unona lawii Hook.f. & Thomson
Annonaceae
Synonyms: *Desmos chinensis* Lour., *Annona laevigata* Mart., *Artabotrys esquirolii* H. Lév., *Desmos chinensis* var. *brevifolius* (Teijsm. & Binn. ex Boerl.) Bân., *Desmos chinensis* var. *latifolia* (Hook.f. & Thomson) Karthik & Moorthy. +26
Climbing Mechanism: Scrambler-Unarmed

Distribution (Global): Assam, Bangladesh, Borneo, Cambodia, China South-Central, China Southeast, East Himalaya, Hainan, India, Java, Laos, Lesser Sunda Is., Malaya, Myanmar, Nepal, Philippines, Sri Lanka, Sulawesi, Sumatera, Thailand, Vietnam
Leaf Type: Ovate-Oblong
Inflorescence: Solitary, axillary
Flowering and Fruiting: October–January
IUCN Status: Not evaluated
Notes: Found mostly in evergreen forests of the Western Ghats
Reference: Fl. Cochinch.: 352 (1790)

Uvaria andamanica King
Annonaceae
Synonyms: *Uvaria lamponga* Scheff.
Climbing Mechanism: Scrambler-Unarmed
Distribution (Global): Andaman Is., Sumatera
Distribution (India): Andaman and Nicobar Islands
Leaf Type: Leaves simple, alternate
Inflorescence: Solitary or cymose fascicled
Fruit Type: Berries, globose
Flowering and Fruiting: December–April
IUCN Status: Not evaluated
Reference: Natuurk. Tijdschr. Ned. -Indië 31: 22 (1869)

Uvaria argentea Blume
Annonaceae
Synonyms: *Anomianthus argenteus* (Blume) Backer., *Cyathostemma argenteum* (Blume) J. Sinclair., *Uva argentea* (Blume) Kuntze
Climbing Mechanism: Scrambler-Unarmed
Distribution (Global): Assam, Bangladesh, Borneo, China South-Central, Java, Malaya, Myanmar, Sumatera, Thailand, Vietnam
Leaf Type: Simple, alternate
Inflorescence: Cymose fascicled
Fruit Type: Wet, globose
IUCN Status: Not evaluated
Reference: Fl. Javae 21-22: 24 (1830)

Uvaria concava Teijsm. & Binn.
Annonaceae
Synonyms: *Unona leytensis* Elmer., *Uva concava* (Teijsm. & Binn.) Kuntze., *Uva membranacea* (Benth.) Kuntze, *Uvaria eucincta* Bedd. ex Dunn., *Uvaria hookeri* King., *Uvaria lauterbachiana* Diels., *Uvaria leytensis* (Elmer) Merr., *Uvaria lurida* Hook.f. & Thomson
Climbing Mechanism: Scrambler-Unarmed

Distribution (Global): Andaman Is., Assam, Bangladesh, Borneo, East Himalaya, India, Java, Lesser Sunda Is., Malaya, New Guinea, Nicobar Is., Philippines, Queensland, Sulawesi, Sumatera, Thailand, Vietnam
Distribution (India): Tamil Nadu
Leaf Type: Simple, alternate
Inflorescence: Cymose fascicled
Fruit Type: Wet, globose
IUCN Status: Not evaluated
Reference: Natuurk. Tijdschr. Ned. -Indië 3: 331 (1852)

Uvaria eucinata Bedd. ex Dunn
Annonaceae
Climbing Mechanism: Scrambler-Unarmed
IUCN Status: Not evaluated

Uvaria eucincta Bedd. ex Dunn
Annonaceae
Climbing Mechanism: Scrambler-Unarmed
Distribution (India): Karnataka, Odisha
Leaf Type: Simple, alternate
Inflorescence: Cymose fascicled
Fruit Type: Wet, globose
IUCN Status: Not evaluated
Reference: Decades kewenses. Plantarum Novarum in Horti Regii Conservatarum. Decades LXXVIII-LXXIX (1914) Decades kewenses. Plantarum Novarum in Horti Regii Conservatarum. Decades LXXVIII-LXXIX. Bulletin of Miscellaneous Information (Royal Gardens, Kew) 1914(5): 181

Uvaria grandiflora Roxb. ex Hornem.
Annonaceae
Synonyms: *Guatteria macrantha* C. Presl., *Unona grandiflora* Lesch. ex DC., *Uva grandiflora* (Lesch. ex DC.) Kuntze., *Uvaria cardinalis* Elmer., *Uvaria flava* Teijsm. & Binn., *Uvaria platypetala* Champ. ex Benth., *Uvaria purpurea* Blume., *Uvaria rhodantha* Hance ex Walp.
Climbing Mechanism: Scrambler-Unarmed
Distribution (Global): Borneo, Cambodia, China Southeast, Hainan, Java, Laos, Malaya, Myanmar, New Guinea, Philippines, Sulawesi, Sumatera, Thailand, Vietnam
Leaf Type: Leaves simple, alternate
Inflorescence: Solitary or cymose fascicled
Fruit Type: Berries, globose
Flowering and Fruiting: March–December
IUCN Status: Not evaluated
Reference: Hort. Bot. Hafn., Suppl.: 141 (1819)

Uvaria griffithii L.L. Zhou, Y.C.F.Su & R.M.K.Saunders
Annonaceae
Synonyms: *Cyathostemma scortechinii* King, *Cyathostemma viridiflorum* Griff., *Cyathostemma viridiflorum* var. *scortechinii* (King) Ridl.
Climbing Mechanism: Scrambler-Unarmed
Distribution (Global): Borneo, Malaya, Sumatera, Thailand
Leaf Type: Simple, alternate
Inflorescence: Cymose fascicled
Fruit Type: Wet, globose
IUCN Status: Not evaluated
Reference: Syst. Biodivers. 7: 255 (2009)

Uvaria hamiltonii Hook.f. & Thomson
Annonaceae
Synonyms: *Uva hamiltonii* Kuntze, *Uvaria grandiflora* Wall
Common Name: Eastern Uvaria
Climbing Mechanism: Scrambler-Unarmed
Distribution (Global): Andaman Is., Assam, Bangladesh, Cambodia, East Himalaya, India, Laos, Myanmar, Nepal, Thailand, Vietnam
Distribution (India): Andaman and Nicobar Islands, Andhra Pradesh, Assam, Bihar, Odisha, Meghalaya, Odisha, West Bengal.
Leaf Type: Leaves simple, alternate
Inflorescence: Solitary or cymose fascicled
Fruit Type: Berries
Flowering and Fruiting: May–September
IUCN Status: Not evaluated
Notes: Occurs mostly along the stream banks of deciduous forests
Reference: Fl. Ind. 1: 96 (1855)

Uvaria hirsuta Jack
Annonaceae
Synonyms: *Guatteria pilosa* G. Don., *Uva hirsuta* (Jack) Kuntze., *Uvaria blumeana* Steud., *Uvaria pilosa* (G.Don) Roxb., *Uvaria subcordata* Miq., *Uvaria trichomalla* Blume., *Uvaria velutina* Blume
Climbing Mechanism: Scrambler-Unarmed
Distribution (Global): Bangladesh, Borneo, Java, Malaya, Myanmar, Thailand, Vietnam
Leaf Type: Simple, alternate
Inflorescence: Cymose fascicled
Fruit Type: Wet, globose
IUCN Status: Not evaluated
Reference: Malayan Misc. 1(5): 46 (1820)

Uvaria hookeri King
Annonaceae

Climbing Mechanism: Scrambler-Unarmed
Distribution (India): Kerala, Maharashtra
Leaf Type: Simple, alternate
Inflorescence: Cymose fascicled
Fruit Type: Wet, globose
IUCN Status: Not evaluated
Reference: Ann. Roy. Bot. Gard. (Calcutta) iv. I. (1893) 28, t. 22

Uvaria kurzii (King) P.T.Li
Annonaceae
Climbing Mechanism: Scrambler-Unarmed
Distribution (Global): Andaman Is., Assam, China South-Central, China Southeast, India, Nicobar Is., Thailand
Leaf Type: Simple, alternate
Inflorescence: Cymose fascicled
Fruit Type: Wet, globose
IUCN Status: Not evaluated
Reference: Acta Phytotax. Sin. 14(1): 106 (1976)

Uvaria littoralis (Blume) Blume
Annonaceae
Synonyms: *Guatteria cordata* Dunal., *Guatteria rufa* Lindl., *Unona camphorata* Blanco., *Unona littoralis* Blume
Uva cordata (Dunal) Kuntze., *Uva synsepala* (Miq.) Kuntze., *Uvaria acrantha* Miq., Uvaria cordata (Dunal) Wall., *Uvaria* gamopetala Zoll., *Uvaria macrophyll*
Climbing Mechanism: Scrambler-Unarmed
Distribution (Global): Andaman Is., Assam, Bangladesh, Borneo, Cambodia, India, Java, Laos, Lesser Sunda Is., Malaya, Maluku, Myanmar, New Guinea, Nicobar Is., Philippines, Solomon Is., Sri Lanka, Sulawesi, Sumatera, Thailand, Vietnam
Leaf Type: Simple, alternate
Inflorescence: Cymose fascicled
Fruit Type: Wet, globose
IUCN Status: Not evaluated
Reference: Fl. Javae 21-22: 26 (1830)

Uvaria lobbiana Hook.f. & Thomson
Annonaceae
Synonyms: *Uva lobbiana* (Hook.f. & Thomson) Kuntze., *Uva ptychocalyx* (Miq.) Kuntze., *Uva subrepanda* (Wall. ex Hook.f. & Thomson) Kuntze., *Uvaria ptychocalyx* Miq.
Climbing Mechanism: Scrambler-Unarmed
Distribution (Global): Andaman Is., Malaya, Myanmar, Sumatera, Thailand
Leaf Type: Simple, alternate
Inflorescence: Cymose fascicled
Fruit Type: Wet, globose

IUCN Status: Not evaluated
Reference: Fl. Ind. 1: 100 (1855)

Uvaria lurida var. *sikkimensis*
Annonaceae
Climbing Mechanism: Scrambler-Unarmed
Leaf Type: Leaves simple, alternate
Inflorescence: Solitary or cymose fascicled
Fruit Type: Berries
Flowering and Fruiting: August–October
IUCN Status: Not evaluated

Uvaria macropoda Hook.f. & Thomson
Annonaceae
Synonyms: *Uva macropoda* Kuntze
Climbing Mechanism: Scrambler-Unarmed
Distribution (Global): Sri Lanka
Distribution (India): Tamil Nadu
Leaf Type: Narrowly elliptic
Inflorescence: Solitary
Flowering and Fruiting: March–June
IUCN Status: Not evaluated
Notes: Found often in semi-evergreen forests of the Western Ghats
Reference: Fl. Ind. 1: 101 (1855)

Uvaria micrantha (A.DC.) Hook.f. & Thomson
Annonaceae
Synonyms: *Anaxagorea sumatrana* Miq., *Cyathostemma micranthum* (A.DC.) J. Sinclair, *Cyathostemma sumatranum* (Miq.) Boerl., *Guatteria micrantha* A. DC., *Polyalthia fruticans* A.DC., *Popowia nitida* King, *Uva micrantha* (A.DC.) Kuntze, *Uva sumatrana* (Miq.) Kuntze
Climbing Mechanism: Scrambler-Unarmed
Distribution (Global): Andaman Is., Borneo, Cambodia, Java, Laos, Lesser Sunda Is., Malaya, Myanmar, New Guinea, Nicobar Is., Philippines, Queensland, Sulawesi, Sumatera, Thailand, Vietnam, Western Australia
Leaf Type: Simple, alternate
Inflorescence: Cymose fascicled
Fruit Type: Wet, globose
IUCN Status: Not evaluated
Reference: Fl. Ind. 1: 103 (1855)

Uvaria narum var. *eunarum*
Annonaceae
Climbing Mechanism: Scrambler-Unarmed
Distribution (Global): India, Sri Lanka

Distribution (India): Peninsular India
Leaf Type: Simple, alternate
Inflorescence: Cymose fascicled
Fruit Type: Wet, globose
IUCN Status: Not evaluated
Reference: Numer. List, 9, 1829

Uvaria narum Wall
Annonaceae
Synonyms: *Hexalobus jussiaeanus* Baill., *Unona narum* Dunal., *Uva narum* (Dunal) Kuntze., *Uvaria flexuosa* B. Heyne ex Wall., *Uvaria malabarica* Oken
Common Name: South-Indian Uvaria
Climbing Mechanism: Scrambler-Unarmed
Distribution (India): Goa, Kerala, Maharashtra, Tamil Nadu, Western Ghats
Leaf Type: Leaves simple, alternate
Inflorescence: Solitary
Flowering and Fruiting: May-October
IUCN Status: Not evaluated
Notes: Found often in moist deciduous to semi-evergreen forests
Reference: Fl. Javae 21-22: 5 (1830)

Uvaria nicobarica Raizada & K.C. Sahni
Annonaceae
Climbing Mechanism: Scrambler-Unarmed
Distribution (Global): Nicobar Is.
Leaf Type: Leaves simple, alternate
Inflorescence: Solitary or cymose fascicled
Flowering and Fruiting: October–November
IUCN Status: Not evaluated
Reference: Indian Forester 87: 101 (1961)

Uvaria rufa Blume
Annonaceae
Synonyms: *Guatteria rufa* Dunal, *Unona setigera* Blanco, *Uva rufa* Kuntze, *Uvaria astrosticta* Miq., *Uvaria bancana* Scheff., *Uvaria branderhorstii* Burck, *Uvaria fauveliana* (Finet & Gagnep.) Pierre ex Ast, *Uvaria ridleyi* King, *Uvaria setigera* (Blanco) Blanco
Climbing Mechanism: Scrambler-Unarmed
Distribution (Global): Andaman Is., Cambodia, China South-Central, Hainan, Java, Laos, Lesser Sunda Is., Malaya, New Guinea, Nicobar Is., Philippines, Queensland, Sumatera, Thailand, Vietnam
Distribution (India): Andaman and Nicobar Islands
Leaf Type: Leaves simple, alternate
Inflorescence: Solitary or cymose fascicled
Fruit Type: Berries

Flowering and Fruiting: May–September
IUCN Status: Not evaluated
Notes: Found often in deciduous to evergreen forests
Reference: Fl. Javae 21-22: 19 (1830)

Uvaria semecarpifolia Hook.f. & Thomson
Annonaceae
Synonyms: *Uva semecarpifolia* Kuntze
Climbing Mechanism: Scrambler-Unarmed
Distribution (Global): Sri Lanka
Leaf Type: Simple, alternate
Inflorescence: Cymose fascicled
Fruit Type: Wet, globose
IUCN Status: Not evaluated
Reference: Fl. Ind. 1: 97 (1855)

Uvaria sphenocarpa Hook.f. & Thomson
Annonaceae
Synonyms: *Uva sphenocarpa* Kuntze
Climbing Mechanism: Scrambler-Unarmed
Distribution (Global): Sri Lanka
Leaf Type: Simple, alternate
Inflorescence: Cymose fascicled
Fruit Type: Wet, globose
IUCN Status: Not evaluated
Reference: Fl. Ind. 1: 98 (1855)

Uvaria zeylanica Dombey ex DC.
Annonaceae
Synonyms: *Guatteria malabarica* Dunal., *Guatteria montana* DC., *Uvaria ceylanica* Heynh., *Uvaria coriacea* Vahl., *Uvaria heyneana* Wight & Arn., *Uvaria paracaroensis* Dennst
Climbing Mechanism: Scrambler-Unarmed
Distribution (Global): India, Sri Lanka
Distribution (India): Kerala, Tamil Nadu, Western Ghats
Leaf Type: Leaves simple, alternate
Inflorescence: Solitary or cymose fascicled
Fruit Type: Berries, globose
Flowering and Fruiting: September–March
IUCN Status: Not evaluated
Reference: Sp. Pl.: 536 (1753)

Adelostemma gracillimum (Wall. ex Wight) Hook.f.
Apocynaceae
Climbing Mechanism: Stem Twiner

Distribution (Global): China South-Central, China Southeast, Myanmar.
Leaf Type: ovate—oblong-ovate, Simple
IUCN Status: Not evaluated
Reference: Hooker's Icon. Pl. 15: 22 (1883)

Aganosma caryophyllata G. Don
Apocynaceae
Climbing Mechanism: Stem Twiner
Distribution (India): Maharashtra, Peninsular India, Bihar, Odisha, Uttar Pradesh
Leaf Type: Simple, elliptic-oblong
Inflorescence: Cymes in dense panicles
Fruit Type: Dry, follicles
Flowering and Fruiting: February-September
IUCN Status: Not evaluated
Reference: Gen. Hist. 4: 77 (1837)

Aganosma cymosa (Roxb.) G. Don
Apocynaceae
Synonyms: *Echites cymosus* Roxb., *Ichnocarpus cymosus* (Roxb.) Mottet.
Common Name: Forest Aganosma
Climbing Mechanism: Stem Twiner
Distribution (Global): Assam, Bangladesh, Cambodia, China South-Central, China Southeast, India, Laos, Sri Lanka, Thailand, Vietnam
Distribution (India): Tamil Nadu, Andhra Pradesh, Western Ghats, Eastern Ghats, Odisha, Puducherry, Andaman and Nicobar Islands, Mizoram, Tripura, Maharashtra
Leaf Type: Simple
Inflorescence: Cyme
Fruit Type: Follicle
Flowering and Fruiting: April–December
IUCN Status: Not evaluated
Reference: Gen. Hist. 4: 77 (1837)

Aganosma cymosa var. *elegans* (G. Don) Hook.f.
Apocynaceae
Synonyms: *Aganosma elegans* G. Don, *Echites elegans* Wall
Climbing Mechanism: Stem Twiner
Distribution (India): Peninsular India
Leaf Type: Simple, elliptic-oblong
Inflorescence: Cymes in dense panicles
Fruit Type: Dry, follicles
IUCN Status: Not evaluated
Notes: Often occur in dry deciduous forests
Reference: Hook. fil. In: Fl. Brit. India 3: 665. (1882)

Aganosma dichotoma K. Schum
Apocynaceae
Synonyms: *Aganosma heynei* (Spreng.) I.M. Turner
Common Name: Clove scented echites
Climbing Mechanism: Stem Twiner
Distribution (Global): India, Peninsular Malaysia, Singapore, and Java
Distribution (India): Telangana, Tamil Nadu, Andhra Pradesh, West Bengal, Madhya Pradesh, Maharashtra, Eastern Ghats
Leaf Type: Simple
Inflorescence: Compound (clustered on terminal panicle)
Flowering and Fruiting: June–September
IUCN Status: Not evaluated
Notes: Flowers used in treating eye diseases
Reference: H.G.A. Engler & K.A.E.Prantl, Nat. Pflanzenfam. 4(2): 173 (1895)

Aganosma gracilis Hook.f.
Apocynaceae
Climbing Mechanism: Stem Twiner
Distribution (Global): India
Distribution (India): West Bengal, Assam, Eastern Himalaya
Leaf Type: Simple, elliptic-oblong
Inflorescence: Cymes in dense panicles
Fruit Type: Dry, follicles
IUCN Status: Not evaluated
Reference: Fl. Brit. India 3: 664 (1882)

Aganosma wallichii G. Don
Apocynaceae
Synonyms: *Aganosma blumei* A.DC., *Aganosma calycina* A.DC., *Echites calycinus* Wall., *Ichnocarpus wallichii* (G. Don) Mottet
Climbing Mechanism: Stem Twiner
Distribution (Global): Java, Malaya, Myanmar, Sumatera, Thailand
Leaf Type: Simple, elliptic-oblong
Inflorescence: Cymes in dense panicles
Fruit Type: Dry, follicles
IUCN Status: Not evaluated
Reference: Gen. Hist. 4: 77 (1837)

Allamanda blanchetii A. DC
Apocynaceae
Synonyms: *Allamanda violacea* Gardner
Common Name: Cherry Allamanda
Climbing Mechanism: Scrambler-Unarmed
Distribution (Global): Brazil, India
Distribution (India): Peninsular India

Flowering and Fruiting: Throughout the year
IUCN Status: Not evaluated
Reference: A.P.de Candolle, Prodr. 8: 319 (1844)

Allamanda schottii Pohl
Apocynaceae
Synonyms: *Allamanda brasiliensis* Schott ex Pohl, *Allamanda cathartica* Schrad., *Allamanda cathartica* var. *schottii* (Pohl) L.H. Bailey & Raffill, *Allamanda magnifica* B.S.Williams, *Allamanda neriifolia* Hook
Common Name: Bush Allamanda, Oleander-leaved Allamand
Climbing Mechanism: Scrambler-Unarmed
Distribution (Global): Argentina Northeast, Brazil South, Brazil Southeast, French Guiana
Distribution (India): Arunachal Pradesh, Telangana, Tamil Nadu, West Bengal, Goa, Gujarat, Madhya Pradesh, Maharashtra, Tripura, Rajasthan, Mizoram, Odisha, Andaman and Nicobar Islands, Kerala
IUCN Status: Not evaluated
Reference: Pl. Bras. Icon. Descr. 1: 73 (1827)

Alyxia fascicularis Benth.
Apocynaceae
Synonyms: *Gynopogon fascicularis* (Wall. ex G. Don) K.Schum., *Hunteria fascicularis* Wall. ex G. Don, *Pulassarium fasciculare* (Wall. ex G. Don) Kuntze
Climbing Mechanism: Stem Twiner
Distribution (Global): Assam, East Himalaya, Thailand, Tibet
Leaf Type: opposite or 3 in a whorl elliptic-oblong or -lanceolate obtusely caudate
Inflorescence: axillary or terminal
Fruit Type: 1 article in each string
IUCN Status: Not evaluated
Reference: Fl. Brit. India 3: 636 (1882)

Alyxia gracilis (Wall. ex A.DC.) Hook.f.
Apocynaceae
Synonyms: *Gynopogon gracilis* (Wall. ex A.DC.) K. Schum., *Hunteria gracilis* Wall. ex A.DC., *Pulassarium gracile* (Wall. ex A.DC.) Kuntze
Climbing Mechanism: Stem Twiner
Distribution (Global): Assam, Bangladesh, Eastern Himalaya
Leaf Type: Whorls of 3
Inflorescence: axillary
IUCN Status: Not evaluated
Reference: Fl. Brit. India 3: 634 (1882)

Alyxia reinwardtii Blume
Apocynaceae

Synonyms: *Alyxia calcicola* Markgr., *Alyxia cinerea* Bakh.f., *Alyxia flavescens* Pierre ex Pit., *Alyxia forbesii* King & Gamble, *Alyxia jasminae* Tsiang & P.T. Li, *Alyxia kerrii* D.J.Middleton, *Alyxia kurzii* Burkill, *Alyxia lucida* Wall., *Alyxia nitens* Kerr.
Climbing Mechanism: Stem Twiner
Distribution (Global): Andaman Is., Bangladesh, Borneo, Cambodia, China South-Central, Java, Laos, Lesser Sunda Is., Malaya, Myanmar, Nicobar Is., Philippines, Sumatera, Thailand, Vietnam
IUCN Status: Not evaluated
Reference: Catalogus: 43 (1823)

Amphineurion marginatum (Roxb.) D.J. Middleton
Apocynaceae
Synonyms: *Aganosma acuminata* G. Don, *Aganosma euloba* Miq., *Aganosma macrocarpa* A.DC., *Aganosma marginata* (Roxb.) G. Don, *Aganosma velutina* A.DC., *Amphineurion acuminatum* (G.Don) Pichon, *Amphineurion velutinum* (A.DC.) Pichon, *Chonemorpha cristata* (Roth) G. Don
Common Name: Long-petal Aganosma, common Aganosma
Climbing Mechanism: Scrambler-Unarmed
Distribution (Global): Andaman Is., Assam, Bangladesh, Borneo, Cambodia, China Southeast, East Himalaya, Hainan, India, Java, Laos, Lesser Sunda Is., Malaya, Maluku, Myanmar, Nepal, Nicobar Is., Philippines, Sulawesi, Sumatera, Thailand, Vietnam
Distribution (India): West Bengal, Andaman and Nicobar Islands
Leaf Type: Simple
Inflorescence: Axillary cyme
Fruit Type: Dry, follicles
IUCN Status: Not evaluated
Reference: Taxon 55: 502 (2006)

Anodendron affine (Hook. & Arn.) Druce
Apocynaceae
Synonyms: *Aganosma laevis* Champ. ex Benth., *Anodendron affine* var. *pingienense* Tsiang & P.T. Li, *Anodendron fangschengense* Tsiang & P.T.Li, *Anodendron laeve* (Champ. ex Benth.) Maxim. ex Franch. & Sav., *Anodendron loheri* Merr.
Climbing Mechanism: Stem Twiner
Distribution (Global): Bangladesh, China South-Central, China Southeast, Hainan, Japan, Laos, Myanmar, Nansei-shoto, Philippines, Taiwan, Thailand, Vietnam
IUCN Status: Not evaluated
Reference: Rep. Bot. Soc. Exch. Club Brit. Isles 1916: 605 (1917)

Anodendron candolleanum Wight
Apocynaceae

Synonyms: *Anodendron rubescens* (Teijsm. & Binn.) Teijsm. & Binn., *Anodendron scandens* (Hassk.) Pichon, *Dendrocharis rubescens* Teijsm. & Binn., *Ecdysanthera rubescens* (Teijsm. & Binn.) Boerl., *Ecdysanthera scandens* Hassk
Climbing Mechanism: Stem Twiner
Distribution (Global): Borneo, Java, Malaya, Philippines, Sumatera, Thailand
IUCN Status: Not evaluated
Reference: Icon. Pl. Ind. Orient. 4: t. 1309 (1848)

Anodendron paniculatum A. DC
Apocynaceae
Synonyms: *Anodendron manubriatum* Merr, *Anodendron rhinosporum* Thwaites
Common Name: Andamanese bowstring Plant, twin-net
Climbing Mechanism: Stem Twiner
Distribution (India): Arunachal Pradesh, Telangana, Tamil Nadu, Andaman and Nicobar Islands, Madhya Pradesh, Bihar, Odisha, Andhra Pradesh, Mizoram, Assam, Eastern Ghats, Puducherry, Western Ghats
Flowering and Fruiting: February–October
IUCN Status: Not evaluated
Notes: Grows on brown soil at higher elevations
Reference: A.P.de Candolle, Prodr. 8: 444, 480 (1844)

Anodendron pauciflorum Hook.f.
Apocynaceae
Climbing Mechanism: Stem Twiner
Distribution (Global): Borneo, Malaya, Sumatera
IUCN Status: Not evaluated
Reference: Fl. Brit. India 3: 669 (1882)

Asclepias campestris Vell
Apocynaceae
Synonyms: *Oxypetalum erectum* subsp. *campestre* (Vell.) Hoehne
Climbing Mechanism: Scrambler-Unarmed
IUCN Status: Not evaluated
Reference: Fl. Flumin. 3: t. 63 (1829)

Atherandra acutifolia Decne.
Apocynaceae
Synonyms: *Atherandra acuminata* Decne., *Atherandra cuspidata* Blume, *Atherandra pubescens* Blume, *Atherostemon javensis* Blume, *Cryptolepis filiformis* Wall
Climbing Mechanism: Stem Twiner
Distribution (Global): Java, Malaya, Myanmar, Sumatera, Thailand, Vietnam
IUCN Status: Not evaluated
Reference: A.P.de Candolle, Prodr. 8: 497 (1844)

Atherolepis wallichii (Wight) Hook.f.
Apocynaceae
Climbing Mechanism: Stem Twiner
IUCN Status: Not evaluated
Reference: Fl. Brit. India 4: 8 (1883)

Baeolepis nervosa (Wight & Arn.) Decne. ex Moq.
Apocynaceae
Synonyms: *Decalepis nervosa* (Wight & Arn.) Venter
Climbing Mechanism: Stem Twiner
Distribution (India): Tamil Nadu, Western Ghats
IUCN Status: Not evaluated
Reference: A.P.de Candolle, Prodr. 13(2): 216 (1849)

Beaumontia grandiflora Wall
Apocynaceae
Synonyms: *Beaumontia longiflora* Hook.f., *Beaumontia longifolia* Lodd. ex G. Don, *Echites grandiflorus* Roxb.
Common Name: Easter-lily-vine, Herald's Trumpet
Climbing Mechanism: Scrambler-Unarmed
Distribution (Global): Assam, Bangladesh, China South-Central, China Southeast, East Himalaya, India, Laos, Myanmar, Nepal, Thailand, Vietnam
Distribution (India): Arunachal Pradesh, Assam, Delhi, Gujarat, Maharashtra, Meghalaya, Mizoram, Tamil Nadu, Tripura, West Bengal
Leaf Type: Simple, opposite, leathery
Inflorescence: Clusters at tips, bell shaped
Fruit Type: Dry, follicles
Flowering and Fruiting: June–December
IUCN Status: Not evaluated
Notes: Grown as ornamental plant in gardens
Reference: Tent. Fl. Napal.: 15, t. 7. (1824)

Beaumontia jerdoniana Wight
Apocynaceae
Common Name: Nepal trumpet flower
Climbing Mechanism: Scrambler-Unarmed
Distribution (Global): Andaman Is., India, Myanmar
Distribution (India): Maharashtra, Tamil Nadu, Western Ghats
Leaf Type: Simple, opposite, leathery
Inflorescence: Clusters at tips, bell shaped
Fruit Type: Dry, follicles
Flowering and Fruiting: November–December
IUCN Status: Not evaluated
Notes: Mostly found in evergreen forests
Reference: Icon. Pl. Ind. Orient. 4: t. 1314 (1848)

Beaumontia khasiana Hook.f.
Apocynaceae
Synonyms: *Beaumontia brevituba* MacGregor & W.W.Sm., *Beaumontia multiflora* Teijsm. & Binn., *Beaumontia shanica* MacGregor & W.W.Sm., *Beaumontia yunnanensis* Tsiang & W.C. Chen
Climbing Mechanism: Scrambler-Unarmed
Distribution (Global): Assam, Bangladesh, China South-Central, Myanmar
Leaf Type: Simple, opposite, leathery
Inflorescence: Clusters at tips, bell shaped
Fruit Type: Dry, follicles
IUCN Status: Not evaluated
Reference: Fl. Brit. India 3: 661 (1882)

Belostemma hirsutum Wall. ex Wight
Apocynaceae
Synonyms: *Vincetoxicum belostemma* (Benth.) Kuntze
Climbing Mechanism: Stem Twiner
Leaf Type: Simple, ovate-oblong, membranous
Inflorescence: Umbel-like cymules
Fruit Type: Dry, follicles
IUCN Status: Not evaluated
Reference: Contr. Bot. India: 52 (1834)

Blyttia spiralis (Forssk.) D.V. Field & J.R.I.Wood
Apocynaceae
Synonyms: *Vincetoxicum spirale* (Forssk.) D.Z.Li
Climbing Mechanism: Stem Twiner
Flowering and Fruiting: August-January.
IUCN Status: Not evaluated
Reference: Kew Bull. 38: 219 (1983)

Brachystelma beddomei Hook.f.
Apocynaceae
Synonyms: *Ceropegia brevitubulata* Bedd.
Climbing Mechanism: Stem Twiner
Leaf Type: Leaf scales opposite, 3-4pairs from the base
Inflorescence: Axillary flowers, umbel
Fruit Type: Dry
Flowering and Fruiting: October–January
IUCN Status: Not evaluated
Reference: Fl. Brit. India 4: 66 (1883)

Brachystelma volubile Hook.f.
Apocynaceae
Synonyms: *Ceropegia volubicaulis* Bruyns

Climbing Mechanism: Stem Twiner
Distribution (India): Andhra Pradesh
Leaf Type: Leaf scales opposite, 3–4pairs from the base
Inflorescence: Axillary flowers, umbel
Fruit Type: Dry
Flowering and Fruiting: August–September
IUCN Status: Not evaluated
Notes: Occurs mostly in deciduous forests
Reference: Fl. Brit. India 4: 65 (1883)

Carissa carandas L.
Apocynaceae
Synonyms: *Arduina carandas* (L.) Baill., *Carissa salicina* Lam., *Echites spinosus* Burm.f., *Jasminonerium carandas* (L.) Kuntze., *Jasminonerium salicinum* (Lam.) Kuntze
Common Name: Christ's thorn, Bengal currants, Jasmine flowered Carissa
Climbing Mechanism: Scrambler-Armed
Distribution (Global): Bangladesh, India.
Distribution (India): Andhra Pradesh, Delhi, Telangana, West Bengal
Leaf Type: Simple, opposite decussate
Inflorescence: Cymes axillary and terminal, corymbose
Fruit Type: Berry
Flowering and Fruiting: March–October
IUCN Status: Not evaluated
Notes: Fruits were used in the ancient Indian herbal system of medicine to treat acidity and indigestion
Reference: Mant. Pl. 1: 52 (1767)

Carissa spinarum L.
Apocynaceae
Synonyms: *Antura edulis* Forssk., *Antura hadiensis* J.F. Gmel., *Arduina brownii* K. Schum., *Arduina campenonii* Drake, *Arduina edulis* (Forssk.) Spreng. +70
Common Name: Num-num, Small Carissa
Climbing Mechanism: Scrambler-Armed
Distribution (Global): Aldabra, Andaman Is., Angola, Bangladesh, Benin, Botswana, Burkina, Burundi, Cambodia, Cameroon, Cape Verde, Caprivi Strip, Central African Republic, Chad, China South-Central, Comoros, Congo, Djibouti, East Himalaya, Egypt, Eritrea, Ethiopia, Ghana, Guinea, Guinea-Bissau, India, Ivory Coast, Kenya, Laos, Madagascar, Malawi, Mali, Mauritius, Mozambique, Myanmar, Namibia, Nepal, New Caledonia, New Guinea, New South Wales, Nicobar Is., Nigeria, Northern Provinces, Northern Territory, Oman, Pakistan, Queensland, Rodrigues, Rwanda, Réunion, Saudi Arabia, Senegal, Seychelles, Socotra, Somalia, Sri Lanka, Sudan, Tanzania, Thailand, Togo, Uganda, Vietnam, West Himalaya, Western Australia, Yemen, Zambia, Zaire, Zimbabwe

Distribution (India): Andhra Pradesh, Odisha, Puducherry, Tamil Nadu, Telangana, Western Ghats
Leaf Type: Simple, opposite decussate
Inflorescence: Terminal or axillary cyme
Fruit Type: Globose, berry
Flowering and Fruiting: February–July.
IUCN Status: Least concern
Notes: Fruits are used in making pickles and also edible when ripe
Reference: Mant. Pl. 2: 559 (1771)

Centrostemma coriaceum Meisn
Apocynaceae
Synonyms: *Hoya coriacea* Blume
Climbing Mechanism: Stem Twiner
Distribution (Global): Borneo, Java, Malaya, Maluku, Philippines, Sumatera, Thailand
IUCN Status: Not evaluated
Reference: Bijdr. Fl. Ned. Ind.: 1061 (1827)

Ceropegia africana R.Br.
Apocynaceae
Climbing Mechanism: Stem Twiner
Distribution (Global): Cape Provinces, KwaZulu-Natal, Lesotho, Northern Provinces
Leaf Type: Simple
Inflorescence: Umbellate cyme
Fruit Type: Dry, follicles
IUCN Status: Not evaluated
Reference: Bot. Reg. 8: t. 626 (1822)

Ceropegia anatii S.R. Yadav, Sardesai & S.P.Gaikwad
Apocynaceae
Climbing Mechanism: Stem Twiner
Leaf Type: Simple
Inflorescence: Umbellate cyme
Fruit Type: Dry, follicles
IUCN Status: Not evaluated

Ceropegia andamanica Sreek., Veenakumari & Prashanth
Apocynaceae
Climbing Mechanism: Stem Twiner
Distribution (Global): Andaman Is.
Leaf Type: Simple
Inflorescence: Umbellate cyme
Fruit Type: Dry, follicles

IUCN Status: Not evaluated
Reference: Blumea 43: 215 (1998)

Ceropegia angustifolia Vahl ex Decne.
Apocynaceae
Climbing Mechanism: Stem Twiner
IUCN Status: Not evaluated

Ceropegia anjanerica Malpure, M.Y. Kamble & S.R.Yadav
Apocynaceae
Common Name: Anjaneri Ceropegia
Climbing Mechanism: Stem Twiner
Distribution (Global): India
Leaf Type: Leaves opposite decussate
Inflorescence: Solitary
Fruit Type: Follicles usually double, ovate, white, and silky
Flowering and Fruiting: August–September
IUCN Status: Endangered
Reference: Curr. Sci. 91: 1141 (2006)

Ceropegia arnottiana Wight
Apocynaceae
Climbing Mechanism: Stem Twiner
Distribution (Global): Myanmar, Thailand
Leaf Type: Simple
Inflorescence: Umbellate cyme
Fruit Type: Dry, follicles
IUCN Status: Not evaluated
Reference: Contr. Bot. India: 32 (1834)

Ceropegia attenuata
Apocynaceae
Climbing Mechanism: Stem Twiner
IUCN Status: Not evaluated

Ceropegia barnesii E.A. Bruce & Chatterjee
Apocynaceae
Climbing Mechanism: Stem Twiner
Distribution (Global): India.
Distribution (India): Tamil Nadu
Leaf Type: Simple
Inflorescence: Umbellate cyme
Fruit Type: Dry, follicles
IUCN Status: Not evaluated

Notes: Occur in high altitudes of the Western Ghats
Reference: Kew Bull. 3: 62 (1948)

Ceropegia beddomei Hook.f.
Apocynaceae
Climbing Mechanism: Stem Twiner
Distribution (Global): India
Leaf Type: Simple
Inflorescence: Umbellate cyme
Fruit Type: Dry, follicles
Flowering and Fruiting: June–September
IUCN Status: Not evaluated
Notes: Mostly found in moist deciduous & semi-evergreen forests. Endemic to Southern Western Ghats
Reference: Fl. Brit. India 4: 75 (1883)

Ceropegia bhatii S.R. Yadav & Shendage
Apocynaceae
Climbing Mechanism: Stem Twiner
Distribution (Global): India
Distribution (India): Karnataka
Leaf Type: Simple
Inflorescence: Umbellate cyme
Fruit Type: Dry, follicles
Flowering and Fruiting: August–November
IUCN Status: Not evaluated
Reference: Kew Bull. 65: 106 (2010)

Ceropegia bulbosa Roxb.
Apocynaceae
Synonyms: *Ceropegia acuminata* Roxb., *Ceropegia brosima* E.A. Bruce & P.R.O. Bally, *Ceropegia bulbosa* var. *esculenta* (Edgew.) Hook.f., *Ceropegia bulbosa* var. *lushii* (Graham) Hook.f., *Ceropegia edulis* Decne., *Ceropegia esculenta* Edgew., *Ceropegia humilis* N.E.Br.
Common Name: Bulbous Ceropegia
Climbing Mechanism: Stem Twiner
Distribution (Global): Bangladesh, Cameroon, Chad, Eritrea, Ethiopia, India, Kenya, Oman, Pakistan, Saudi Arabia, Somalia, Yemen
Distribution (India): Andhra Pradesh, Delhi, Gujarat, Madhya Pradesh, Maharashtra, Tamil Nadu
Leaf Type: Herbaceous, sub-sessile
Inflorescence: Pseudo-umbellate cyme
Fruit Type: Follicles paired
Flowering and Fruiting: July–December
IUCN Status: Not evaluated

Notes: Decoction made from the tubers is used, orally, to ease dropping out urinary bladder stones. Tubers are eaten, either raw or cooked
Reference: Pl. Coromandel 1: 11 (1795)

Ceropegia candelabrum L.
Apocynaceae
Common Name: Goglet flower
Climbing Mechanism: Stem Twiner
Distribution (Global): India, Sri Lanka
Distribution (India): Andhra Pradesh, Gujarat, Kerala, Madhya Pradesh, Maharashtra, Odisha, Tamil Nadu
Leaf Type: Simple, opposite
Inflorescence: Axillary Cymes
Fruit Type: Paired follicles
Flowering and Fruiting: August–January
IUCN Status: Not evaluated
Notes: Occasional in semi-evergreen to moist deciduous forests
Reference: Sp. Pl.: 211 (1753)

Ceropegia ciliata Wight
Apocynaceae
Climbing Mechanism: Stem Twiner
Distribution (Global): India
Distribution (India): Tamil Nadu
Leaf Type: Opposite
Inflorescence: Umbelliform, often branching cymes
Fruit Type: Follicles short
Flowering and Fruiting: August–October
IUCN Status: Not evaluated
Notes: Endemic to Southern Western Ghats
Reference: Icon. Pl. Ind. Orient. 4: t. 1262 (1848)

Ceropegia decaisneana Wight
Apocynaceae
Synonyms: *Ceropegia brevicollis* Hook.f.
Common Name: Decaisne's Ceropegia
Climbing Mechanism: Stem Twiner
Distribution (Global): India, Sri Lanka
Distribution (India): Tamil Nadu
Leaf Type: Ovate-elliptic
Inflorescence: Cymes axillary
Fruit Type: Follicles
Flowering and Fruiting: October–December
IUCN Status: Not evaluated

Notes: Mostly found in evergreen forests
Reference: Icon. Pl. Ind. Orient. 4: t. 1259 (1848)

Ceropegia elegans Wall
Apocynaceae
Synonyms: *Ceropegia ledgeri* N.E.Br., *Ceropegia mysorensis* Wight., *Ceropegia similis* N.E.Br., *Ceropegia sphenanantha* Wight & Arn., *Ceropegia sphenanthera* Decne., *Ceropegia walkerae* Wight
Common Name: Mysore Ceropegi
Climbing Mechanism: Stem Twiner
Distribution (Global): Bangladesh, India, Myanmar, Sri Lanka
Distribution (India): Tamil Nadu
Leaf Type: Ovate, lance shape
Inflorescence: Cymes axillary
Fruit Type: Follicles
Flowering and Fruiting: July–April
IUCN Status: Not evaluated
Notes: Occasional on hills up to 1000 m, mostly on thickets
Reference: Bot. Mag. 57: t. 3015 (1830)

Ceropegia ensifolia Bedd.
Apocynaceae
Synonyms: *Ceropegia albiflora* Hook.f., *Ceropegia ciliata* subsp. *ensifolia* (Bedd.) H. Huber
Common Name: Sword-leaf Ceropegia
Climbing Mechanism: Stem Twiner
Distribution (Global): India.
Distribution (India): Tamil Nadu
Leaf Type: Linear, lance shaped
Inflorescence: Cymes
Fruit Type: Follicles
Flowering and Fruiting: August–October
IUCN Status: Not evaluated
Notes: Endemic to Southern Western Ghats
Reference: Madras J. Lit. Sci., ser. 3, 1: 52 (1864)

Ceropegia evansii McCann
Apocynaceae
Common Name: Evans Ceropegia
Climbing Mechanism: Stem Twiner
Distribution (Global): India
Distribution (India): Kerala, Maharashtra
Leaf Type: Ovate-lanceolate
Fruit Type: Follicles

IUCN Status: Not evaluated
Reference: J. Bombay Nat. Hist. Soc. 45: 209 (1945)

Ceropegia fantastica Sedgw.
Apocynaceae
Common Name: Fantastic Ceropegi
Climbing Mechanism: Stem Twiner
Distribution (Global): India
Distribution (India): Maharashtra
Leaf Type: Ovate-lanceolate
Fruit Type: Follicles
IUCN Status: Not evaluated
Reference: J. Indian Bot. 2: 124 (1921)

Ceropegia fimbriifera Bedd.
Apocynaceae
Climbing Mechanism: Stem Twiner
Distribution (Global): India
Leaf Type: Simple
Inflorescence: Umbellate cyme
Fruit Type: Dry, follicles
Flowering and Fruiting: June–September
IUCN Status: Not evaluated
Reference: Madras J. Lit. Sci., ser. 3, 1: 53 (1864)

Ceropegia huberi Ansari
Apocynaceae
Common Name: Huber's Ceropegia
Climbing Mechanism: Stem Twiner
Distribution (Global): India
Distribution (India): Maharashtra
Leaf Type: Ovate, lance shaped
IUCN Status: Not evaluated
Notes: Endemic to the Western Ghat
Reference: Bull. Bot. Surv. India 10: 219 (1969)

Ceropegia intermedia Wight
Apocynaceae
Common Name: Intermediate Ceropegia
Climbing Mechanism: Stem Twiner
Distribution (Global): India
Distribution (India): Odisha, Tamil Nadu
Leaf Type: Lanceolate
Inflorescence: Axillary racemes
Flowering and Fruiting: June–December

IUCN Status: Not evaluated
Notes: Usually found in semi-evergreen forests and scrub jungles
Reference: Icon. Pl. Ind. Orient. 4: t. 1263 (1848)

Ceropegia jainii Ansari & B.G.P. Kulk
Apocynaceae
Common Name: Jaini Ceropegia
Climbing Mechanism: Stem Twiner
Distribution (Global): India
Leaf Type: Oppositely arranged
Inflorescence: Cymes uniflowered
Flowering and Fruiting: August–October
IUCN Status: Not evaluated
Reference: Bull. Bot. Surv. India 22: 221 (1980 publ. 1982)

Ceropegia juncea Roxb.
Apocynaceae
Common Name: Leafless goglet flower
Climbing Mechanism: Stem Twiner
Distribution (Global): India, Myanmar, Sri Lanka
Distribution (India): Puducherry, Tamil Nadu, Andhra Pradesh
Leaf Type: Opposite pairs
Inflorescence: Umbels/Axillary cymes
Fruit Type: Follicles paired
Flowering and Fruiting: October–January
IUCN Status: Not evaluated
Reference: Pl. Coromandel 1: 12 (1795)

Ceropegia karulensis Punekar, Tamhankar, Lakshmin., Kumaran, A. Raut, S.K. Srivast. & Kavade
Apocynaceae
Synonyms: *Ceropegia sahyadrica* var. *karulensis* (Punekar, Tamhankar, Lakshmin., Kumaran, Raut, S.K. Srivast. & Kavade) Kambale & S.R. Yadav
Common Name: Karul-Ghat Ceropegia
Climbing Mechanism: Stem Twiner
Distribution (Global): India
Distribution (India): Assam, Maharashtra
Leaf Type: Opposite, elliptic-ovate
Inflorescence: Umbel-like cymes
IUCN Status: Not evaluated
Reference: Rheedea 29: 65 (2019)

Ceropegia khasiana Murug., A.A. Mao, Meitei & Kambale
Apocynaceae
Climbing Mechanism: Stem Twiner

Distribution (Global): Assam
Leaf Type: Simple
Inflorescence: Umbellate cyme
Fruit Type: Dry, follicles
IUCN Status: Not evaluated
Reference: Gard. Bull. Singapore 71: 520 (2019)

Ceropegia lawii Hook.f.
Apocynaceae
Synonyms: *Ceropegia lawii* var. *wadhwae* M.R. Almeida., *Ceropegia maharashtrensis* Punekar, Tamhankar, Lakshmin., Kumaran, Raut & S.K. Srivast.
Common Name: Law's Ceropegia, Panchgani Ceropegia
Climbing Mechanism: Stem Twiner
Distribution (Global): India
Leaf Type: Simple
Inflorescence: Umbellate cyme
Fruit Type: Dry, follicles
Flowering and Fruiting: August-September.
IUCN Status: Not evaluated
Reference: Fl. Brit. India 4: 67 (1883)

Ceropegia longifolia Wall.
Apocynaceae
Synonyms: *Ceropegia borii* Raizada., *Ceropegia lanceolata* Wight & Arn.
Common Name: Longleaf Ceropegia
Climbing Mechanism: Stem Twiner
Distribution (Global): Assam, Bangladesh, China South-Central, East Himalaya, Myanmar, Nepal, Tibet, West Himalaya
Distribution (India): Meghalaya, Arunachal Pradesh, Himachal Pradesh, Mizoram, West Bengal
Leaf Type: Lance shaped
Inflorescence: Umbel-like, 4-8-flowered clusters
IUCN Status: Not evaluated
Reference: Pl. Asiat. Rar. 1: 56 (1830)

Ceropegia lucida Wall
Apocynaceae
Common Name: Shining Ceropegia
Climbing Mechanism: Stem Twiner
Distribution (Global): Assam, Bangladesh, East Himalaya, Malaya, Myanmar, Thailand
Leaf Type: Simple, opposite-decussate, exstipulate
Inflorescence: Umbellate Cymes
Fruit Type: Follicles

Flowering and Fruiting: September–November
IUCN Status: Not evaluated
Reference: Pl. Asiat. Rar. 2: 33 (1831)

Ceropegia maccannii Ansari
Apocynaceae
Synonyms: *Ceropegia lawii* var. *maccannii* (Ansari) M.R. Almeida
Common Name: Maccani Ceropegia
Climbing Mechanism: Stem Twiner
Distribution (Global): India
Leaf Type: Lance shaped
Flowering and Fruiting: July–October
IUCN Status: Not evaluated
Notes: Endemic to the Western Ghats
Reference: Bull. Bot. Surv. India 22: 227 (1980 publ. 1982)

Ceropegia macrantha Wight
Apocynaceae
Synonyms: *Ceropegia raizadiana* Babu
Common Name: Large-flowered Ceropegia
Climbing Mechanism: Stem Twiner
Distribution (Global): Assam, Bangladesh, Eastern Himalaya, Nepal, Western Himalaya
Distribution (India): Himachal Pradesh, Madhya Pradesh, Mizoram
Leaf Type: Ovate-lanceolate
Inflorescence: Cymes
IUCN Status: Not evaluated
Reference: Contr. Bot. India: 31 (1834)

Ceropegia maculata Bedd.
Apocynaceae
Common Name: Spotted Ceropegia
Climbing Mechanism: Stem Twiner
Distribution (Global): India
Distribution (India): Tamil Nadu
Leaf Type: Ovate, pointed, or tapering
Inflorescence: Umbels/Sub-umbellate
Flowering and Fruiting: September–November
IUCN Status: Not evaluated
Reference: Madras J. Lit. Sci., ser. 3, 1: 52 (1864)

Ceropegia mahabalei Hemadri & Ansari
Apocynaceae
Common Name: Ralegaon Ceropegia
Climbing Mechanism: Stem Twiner

Distribution (Global): India
Leaf Type: Simple
Inflorescence: Umbellate cyme
Fruit Type: Dry, follicles
Flowering and Fruiting: August–September
IUCN Status: Not evaluated
Notes: Endemic to the Western Ghats
Reference: Indian Forester 97: 105 (1971)

Ceropegia maharashtrensis Punekar, Tamhankar, Lakshmin., Kumaran, A.Raut, & S. K. Srivastava
Apocynaceae
Climbing Mechanism: Stem Twiner
Distribution (India): Maharashtra
Leaf Type: Simple
Inflorescence: Umbellate cyme
Fruit Type: Dry, follicles
IUCN Status: Not evaluated
Reference: Punekar, Tamhankar, Lakshmin., Kumaran, Raut & S. K. Srivast. In: Nelumbo 55: 22. (2013)

Ceropegia mannarana P. Umam. & P. Daniel
Apocynaceae
Climbing Mechanism: Stem Twiner
Distribution (Global): India
Distribution (India): Tamil Nadu, Western Ghats
Leaf Type: Simple
Inflorescence: Umbellate cyme
Fruit Type: Dry, follicles
IUCN Status: Not evaluated
Reference: Fl. Gulf Mannar South. India: 262 (2001)

Ceropegia manoharii Sujanapa, Salim, Anil, & Sasidh.
Apocynaceae
Climbing Mechanism: Stem Twiner
Distribution (Global): India
Distribution (India): Kerala
Leaf Type: Simple, decussate
Inflorescence: Cymes axillary
Fruit Type: Follicles in pairs
Flowering and Fruiting: August–February
IUCN Status: Not evaluated
Notes: Locally threatened due to anthropogenic activities
Reference: J. Bot. Res. Inst. Texas 7: 342 (2013)

Ceropegia media (Huber) Ansari
Apocynaceae
Common Name: Medium Ceropegia
Climbing Mechanism: Stem Twiner
Distribution (Global): India
Distribution (India): Maharashtra
Leaf Type: Oppositely arranged
Flowering and Fruiting: July–October.
IUCN Status: Not evaluated
Reference: Bull. Bot. Surv. India 11: 199 (1969 publ. 1971)

Ceropegia metziana Miq.
Apocynaceae
Synonyms: *Ceropegia gracilis* Bedd.
Climbing Mechanism: Stem Twiner
Distribution (Global): India
Distribution (India): Tamil Nadu
Leaf Type: Ovate-oblong
Fruit Type: Follicle
Flowering and Fruiting: September–November
IUCN Status: Not evaluated
Notes: Endemic to Southern Western Ghats.
Reference: Nieuwe Verh. Eerste Kl. Kon. Ned. Inst. Wetensch. Amsterdam, ser. 3, 5: 11

Ceropegia meyeri Decne.
Apocynaceae
Synonyms: *Ceropegia pubescens* E. Mey
Climbing Mechanism: Stem Twiner
Distribution (Global): Cape Provinces, Caprivi Strip, KwaZulu-Natal, Mozambique, Namibia, Northern Provinces, Zambia, Zimbabwe
Leaf Type: Simple
Inflorescence: Umbellate cyme
Fruit Type: Dry, follicles
IUCN Status: Not evaluated
Reference: A.P.de Candolle, Prodr. 8: 645 (1844)

Ceropegia mizoramensis Ram. Kumar & S. Sharma
Apocynaceae
Synonyms: *Ceropegia longifolia* subsp. *sinensis* H. Huber., *Ceropegia murlensis* Ram.Kumar, & S.Sharma., *Ceropegia profundorum* Hand.-Mazz
Climbing Mechanism: Stem Twiner
Distribution (Global): Assam, China South-Central, China Southeast, East Himalaya
Distribution (India): Mizoram
Leaf Type: Simple

Inflorescence: Umbellate cyme
Fruit Type: Dry, follicles
IUCN Status: Not evaluated
Reference: Notes Roy. Bot. Gard. Edinburgh 8: 17 (1913)

Ceropegia mohanramii S.R. Yadav, M.N.Gavade, & Sardesai
Apocynaceae
Climbing Mechanism: Stem Twiner
Distribution (Global): India
Leaf Type: Simple
Inflorescence: Umbellate cyme
Fruit Type: Dry, follicles
IUCN Status: Not evaluated
Reference: Rheedea 16: 33 (2006)

Ceropegia muliensis W.W.Sm.
Apocynaceae
Climbing Mechanism: Stem Twiner
Distribution (Global): China South-Central
Leaf Type: Simple
Inflorescence: Umbellate cyme
Fruit Type: Dry, follicles
IUCN Status: Not evaluated
Reference: Notes Roy. Bot. Gard. Edinburgh 12: 199 (1920)

Ceropegia nampyana Manudev, Kambale, & Pramod
Apocynaceae
Climbing Mechanism: Stem Twiner
Distribution (Global): India
Distribution (India): Kerala
Leaf Type: Simple
Inflorescence: Umbellate cyme
Fruit Type: Dry, follicles
IUCN Status: Not evaluated
Reference: Int. J. Advanced Res. 4(5): 1408 (2016)

Ceropegia noorjahaniae M.A. Ansari
Apocynaceae
Common Name: Noorjahan Ceropegia
Climbing Mechanism: Stem Twiner
Distribution (Global): India
Leaf Type: Opposite, nearly stalk-less
Flowering and Fruiting: July–August
IUCN Status: Not evaluated
Reference: J. Bombay Nat. Hist. Soc. 69: 250 (1972)

Ceropegia oculata Hook
Apocynaceae
Synonyms: *Ceropegia mahabalei* var. *hemalatae* S.S. Rahangdale & S.R. Rahangdale, *Ceropegia oculata* var. *satpudensis* Punekar, S.D.Jagtap, & Deokule
Common Name: Peacock Ceropegia, Lantern flower
Climbing Mechanism: Stem Twiner
Distribution (Global): India
Distribution (India): Madhya Pradesh, Maharashtra, Tamil Nadu
Leaf Type: Ovate, lance shaped
Flowering and Fruiting: July–October
IUCN Status: Not evaluated
Reference: Bot. Mag. 70: t. 4093 (1844)

Ceropegia oculata var. *satpudensis*
Apocynaceae
Climbing Mechanism: Stem Twiner
Leaf Type: Simple
Inflorescence: Umbellate cyme
Fruit Type: Dry, follicles
IUCN Status: Not evaluated

Ceropegia odorata Nimmo
Apocynaceae
Synonyms: *Ceropegia blatteri* McCann
Common Name: Fragrant Ceropegia
Climbing Mechanism: Stem Twiner
Distribution (Global): India
Distribution (India): Gujarat, Maharashtra
Leaf Type: Simple
Inflorescence: Umbellate cyme
Fruit Type: Dry, follicles
Flowering and Fruiting: August–September
IUCN Status: Critically endangered
Notes: Medicinally valuable and tubers are edible
Reference: Fl. Brit. India 4: 75 (1883)

Ceropegia omissa H. Huber
Apocynaceae
Common Name: Neglected Ceropegia
Climbing Mechanism: Stem Twiner
Distribution (Global): India
Distribution (India): Tamil Nadu
Leaf Type: Lanceolate
Inflorescence: Cymes axillary

Flowering and Fruiting: September–December
IUCN Status: Not evaluated
Notes: Endemic to Southern Western Ghats
Reference: Mem. Soc. Brot. 12: 67 (1957)

Ceropegia procumbens (Gravely & Mayur) Bruyns
Apocynaceae
Synonyms: *Boucerosia procumbens* (Gravely & Mayur.) Plowes., *Caralluma procumbens* Gravely & Mayur., Pendulluma procumbens (Gravely and Mayur.) Plowes
Climbing Mechanism: Stem Twiner
Distribution (Global): India
Distribution (India): Tamil Nadu
Leaf Type: Simple
Inflorescence: Umbellate cyme
Fruit Type: Dry, follicles
IUCN Status: Not evaluated
Reference: Haseltonia 3: 59 (1995)

Ceropegia pullaiahii Kullayiswamy
Apocynaceae
Climbing Mechanism: Stem Twiner
Distribution (Global): India
Distribution (India): Andhra Pradesh
Leaf Type: Simple
Inflorescence: Umbellate cyme
Fruit Type: Dry, follicles
Flowering and Fruiting: June–December
IUCN Status: Not evaluated
Reference: Nordic J. Bot. 31: 166 (2012)

Ceropegia pusilla Wight and Arn.
Apocynaceae
Common Name: Weak Ceropegia
Climbing Mechanism: Stem Twiner
Distribution (Global): India
Leaf Type: Linear, acute
Flowering and Fruiting: June–September
IUCN Status: Not evaluated
Notes: The tubers are edible. Rich in ceropegin and are used in Ayurvedic drug preparation
Reference: Contr. Bot. India: 81 (1834)

Ceropegia ravikumariana Kambale and Gnanasek
Apocynaceae

Climbing Mechanism: Stem Twiner
Distribution (Global): India
Distribution (India): Tamil Nadu
Leaf Type: Simple
Inflorescence: Umbellate cyme
Fruit Type: Dry, follicles
IUCN Status: Not evaluated
Reference: Rheedea 26: 57 (2016)

Ceropegia santapaui Wadhwa and Ansari
Apocynaceae
Common Name: Santapau's Ceropegia
Climbing Mechanism: Stem Twiner
Distribution (Global): India
Distribution (India): Maharashtra
Leaf Type: Lance shaped
Inflorescence: Umbells
Flowering and Fruiting: August–October
IUCN Status: Not evaluated
Reference: Bull. Bot. Surv. India 10: 95 (1968)

Ceropegia schumanniana Swarupan. & Mangaly
Apocynaceae
Climbing Mechanism: Stem Twiner
Distribution (Global): India
Leaf Type: Simple
Inflorescence: Umbellate cyme
Fruit Type: Dry, follicles
IUCN Status: Not evaluated
Reference: Nordic J. Bot. 12: 699 (1992)

Ceropegia spiralis Wight
Apocynaceae
Climbing Mechanism: Stem Twiner
Leaf Type: Simple
Inflorescence: Umbellate cyme
Fruit Type: Dry, follicles
IUCN Status: Not evaluated

Ceropegia thwaitesii Hook.
Apocynaceae
Climbing Mechanism: Stem Twiner
Distribution (Global): India
Distribution (India): Peninsular India
Leaf Type: Ovate-lanceolate

Inflorescence: Cymes
Fruit Type: Follicles
Flowering and Fruiting: November–March
IUCN Status: Not evaluated
Reference: Bot. Mag. 80: t. 4758 (1854)

Ceropegia truncatocoronata (Sedgw.) Kottaim.
Apocynaceae
Synonyms: *Boucerosia nilagiriana* (Kumari & Subba Rao) Plowes., *Boucerosia truncatocoronata* Sedgw., *Caralluma crenulata* Wall., *Caralluma nilagiriana* Kumari & Subba Rao., *Caralluma truncatocoronata* (Sedgw.) Gravely & Mayur., *Ceropegia crenulata* (Wall.) Bruyns
Climbing Mechanism: Stem Twiner
Distribution (Global): India, Myanmar
Distribution (India): Peninsular India
Leaf Type: Simple
Inflorescence: Umbellate cyme
Fruit Type: Dry, follicles
IUCN Status: Not evaluated
Reference: Contr. Bot. India: 34 (1834)

Ceropegia vincifolia Hook.
Apocynaceae
Synonyms: *Ceropegia hirsuta* var. *vincifolia* (Hook.) Hook.f., *Ceropegia polyantha* Blatt. & McCann., *Ceropegia stocksii* Hook.f.
Common Name: Vinca-leaved Ceropegia
Climbing Mechanism: Stem Twiner
Distribution (Global): India
Distribution (India): Maharashtra
Leaf Type: Ovate with a heart-shaped base
Flowering and Fruiting: August–September
IUCN Status: Not evaluated
Reference: Bot. Mag. 66: t. 3740 (1839)

Chilocarpus costatus Miq.
Apocynaceae
Synonyms: *Chilocarpus aurantiacus* Ridl., *Chilocarpus costatus* var. *borneensis* Markgr., *Chilocarpus cuneifolius* Kerr
Chilocarpus diepenhorstii Miq., *Chilocarpus maingayi* Dyer ex Hook.f.
Climbing Mechanism: Stem Twiner
Distribution (Global): Borneo, Malaya, Myanmar, Sumatera, Thailand
IUCN Status: Not evaluated
Reference: Fl. Ned. Ind. 2: 393 (1857)

Chilocarpus decipiens Hook.f.
Apocynaceae
Climbing Mechanism: Stem Twiner
Distribution (Global): Malaya, Sumatera
IUCN Status: Not evaluated
Reference: Fl. Brit. India 3: 627 (1882)

Chilocarpus denudatus Blume
Apocynaceae
Synonyms: *Alstonia micrantha* Ridl., *Chilocarpus alyxifolius* Pierre. +10
Climbing Mechanism: Stem Twiner
Distribution (Global): Borneo, Cambodia, India, Java, Lesser Sunda Is., Malaya, Maluku, Myanmar, New Guinea, Nicobar Is., Sulawesi, Sumatera, Thailand, Vietnam
Distribution (India): Western Ghats
Flowering and Fruiting: March–August
IUCN Status: Not evaluated
Reference: Bijdr. Fl. Ned. Ind.: 1025 (1826)

Chilocarpus vernicosus Blume
Apocynaceae
Synonyms: *Chilocarpus cantleyi* King & Gamble., *Chilocarpus enervis* Hook.f., *Chilocarpus globulifer* Blume., *Chilocarpus nigrescens* King & Gamble
Climbing Mechanism: Stem Twiner
Distribution (Global): Borneo, Malaya, Sumatera
IUCN Status: Not evaluated
Reference: Mus. Bot. 1: 152 (1850)

Chonemorpha fragrans (Moon) Alston
Apocynaceae
Synonyms: *Beluttakaka grandieriana* Pierre, *Beluttakaka griffithii* (Hook.f.) Kuntze, *Beluttakaka macrophylla* (G. Don) Kuntze, *Cercocoma macrantha* Teijsm. and Binn. +20
Common Name: Frangipani vine, Wood vine, Funnel-flower vine
Climbing Mechanism: Scrambler-Unarmed
Distribution (Global): Andaman Is., Assam, Bangladesh, Borneo, Cambodia, China South-Central, China Southeast, East Himalaya, India, Java, Laos, Lesser Sunda Is., Malaya, Myanmar, Nepal, Philippines, Sri Lanka, Sulawesi, Sumatera, Thailand, Tibet, Vietnam, West Himalaya
Distribution (India): Andaman and Nicobar Islands, Arunachal Pradesh, Karnataka, Kerala, Meghalaya, Mizoram, Odisha, Tamil Nadu, Western Ghats
Leaf Type: Elliptic or ovate-orbicular
Inflorescence: Tomenose, paniculate Cymes
Fruit Type: Follicles
Flowering and Fruiting: Throughout the year.

IUCN Status: Not evaluated
Notes: Medicinally valuable in treating various ailments
Reference: Ann. Roy. Bot. Gard. (Peradeniya) 11: 203 (1929)

Chonemorpha verrucosa (Blume) Mabb.
Apocynaceae
Synonyms: *Cercocoma wallichii* (A.DC.) Miq., *Chonemorpha elastica* Merr., *Echites ellipticus* Wall. ex G. Don., *Echites rhynchospermus* Wall., *Echites verticalis* Buch.-Ham. ex Wall., *Rhynchodia capusii* Pierre ex Spire., *Rhynchodia fragrans* Pierre ex Spire
Climbing Mechanism: Scrambler-Unarmed
Distribution (Global): Andaman Is., Assam, Bangladesh, Borneo, Cambodia, China South-Central, China Southeast, East Himalaya, Hainan, Java, Laos, Malaya, Myanmar, Nicobar Is., Sumatera, Thailand, Vietnam
Distribution (India): Mizoram, Andaman and Nicobar Islands
IUCN Status: Not evaluated
Reference: Novon 3: 455 (1993)

Cleghornia acuminata Wight
Apocynaceae
Synonyms: *Baissea acuminata* (Wight) Benth. ex Hook.f.
Cleghornia cymosa Wight.
Climbing Mechanism: Stem Twiner
Distribution (Global): Sri Lanka
Leaf Type: Simple
Inflorescence: Flowers in clusters
Fruit Type: Dry, follicles
IUCN Status: Not evaluated
Reference: Icon. Pl. Ind. Orient. 4: t. 1310 (1848)

Cleghornia malaccensis (Hook.f.) King and Gamble
Apocynaceae
Synonyms: *Baissea malaccensis* Hook.f., *Cleghornia dongnaiensis* Pierre ex Pit., *Giadotrum dongnaiense* (Pierre ex Pit.) Pichon., *Giadotrum malaccense* (Hook. f.) Pichon
Climbing Mechanism: Stem Twiner
Distribution (Global): China South-Central, Laos, Malaya, Thailand, Vietnam
Leaf Type: Simple
Inflorescence: Flowers in clusters
Fruit Type: Dry, follicles
IUCN Status: Not evaluated
Reference: J. Asiat. Soc. Bengal, Pt. 2, Nat. Hist. 74: 491 (1908)

Cosmostigma cordatum (Poir.) M.R. Almeida
Apocynaceae

Synonyms: *Asclepias racemosa* Roxb., *Cosmostigma acuminatum* Wight, *Cosmostigma racemosum* Wight, *Hemidesmus cordatus* (Poir.) Schult., *Periploca cordata* Poir., *Tylophora punctata* Kostel
Common Name: Green milkweed creeper
Climbing Mechanism: Stem Twiner
Distribution (Global): Bangladesh, Cambodia, India, Myanmar, Sri Lanka, Thailand, Vietnam
Distribution (India): Tamil Nadu, Western Ghats
Leaf Type: Opposite
Inflorescence: Incorymb-like or raceme-like
Flowering and Fruiting: April-December
IUCN Status: Not evaluated
Notes: The leaves of this woody climber are used in Indian traditional medicine to cure ulcerous sores
Reference: Fl. Maharashtra 3A: 239 (2001)

Cryptolepis dubia (Burm.f.) M.R. Almeida
Apocynaceae
Synonyms: *Cryptolepis buchananii* R.Br. ex Roem. & Schult., *Chonemorpha reticulata* G. Don., *Cryptolepis reticulata* (Roth) Wall. ex Steud., *Echites cuspidatus* B. Heyne ex Hook.f., *Echites reticulatus* Roth., *Nerium reticulatum* Roxb., *Periploca dubia* Burm.f.
Common Name: Wax leaved climber, Indian sarsaparilla
Climbing Mechanism: Stem Twiner
Distribution (Global): Andaman Is., Assam, Bangladesh, China South-Central, China Southeast, East Himalaya, India, Laos, Myanmar, Nepal, Pakistan, Sri Lanka, Thailand, Vietnam, West Himalaya
Distribution (India): Tamil Nadu
Leaf Type: Oblong or elliptic
Inflorescence: Cymes like panicles
Flowering and Fruiting: March–August
IUCN Status: Not evaluated
Reference: Syst. Veg., ed. 15 bis 4: 409 (1819)

Cryptolepis grandiflora Wight
Apocynaceae
Synonyms: *Cryptolepis wightiana* Wall., *Echites coriaceus* B. Heyne ex Hook.f.
Common Name: Large-flowered Cryptolepis
Climbing Mechanism: Stem Twiner
Distribution (Global): Andaman Is., India
Distribution (India): Tamil Nadu
Leaf Type: Elliptic-oblong, velvety-hairy
Inflorescence: Cymes
Flowering and Fruiting: October–February
IUCN Status: Not evaluated

Notes: Usually found in deciduous to Moist Deciduous Forests. Endemic to Peninsular India
Reference: Icon. Pl. Ind. Orient.: t. 831 (1845)

Cryptolepis sinensis (Lour.) Merr.
Apocynaceae
Synonyms: *Aganosma edithae* Hance, *Apocynum orixense* Rottler ex Hook.f., *Cryptolepis edithae* (Hance) Benth. & Hook.f. ex Maxim., *Cryptolepis elegans* Wall. ex G. Don, *Cryptolepis laxiflora* Blume. + 10
Common Name: Chinese wax-leaf climber, Chinese Cryptolepis
Climbing Mechanism: Stem Twiner
Distribution (Global): Andaman Is., Assam, Bangladesh, Cambodia, China South-Central, China Southeast, East Himalaya, Hainan, India, Java, Laos, Malaya, Myanmar, Philippines, Taiwan, Thailand, Vietnam
Distribution (India): Andaman and Nicobar Islands, Arunachal Pradesh, Meghalaya, Mizoram, Odisha, Telangana, Tripura, West Bengal
Leaf Type: Smaller, oblong to linear
Inflorescence: Cymes
Fruit Type: Follicles
Flowering and Fruiting: June–March
IUCN Status: Not evaluated
Notes: The stems and leaves are used externally for the treatment of snake bites, traumatic injuries, and scabies
Reference: Philipp. J. Sci. 15: 254 (1919)

Cryptostegia grandiflora Roxb. ex R.Br.
Apocynaceae
Synonyms: *Nerium grandiflorum* (Roxb. ex R.Br.) Roxb.
Common Name: Rubber-vine, Purple Allamanda
Climbing Mechanism: Stem Twiner
Distribution (Global): Madagascar
Distribution (India): Andhra Pradesh, Bihar, Odisha, Delhi, Eastern Ghats, Gujarat, Madhya Pradesh, Maharashtra, Odisha, Rajasthan, Tamil Nadu, Telangana, Uttar Pradesh, West Bengal
Leaf Type: Opposite pairs
Flowering and Fruiting: Throughout the year
IUCN Status: Not evaluated
Notes: The Rubber Vine is poisonous it contains glucosides interfering with the heart, and ingesting will provoke stomach and intestinal upset
Reference: Bot. Reg. 5: t. 435 (1819)

Cryptostegia madagascariensis Bojer ex Decne
Apocynaceae

Synonyms: *Cryptostegia glaberrima* Hochr., *Cryptostegia madagascariensis* var. *glaberrima* (Hochr.) Marohasy & P.I. Forst., *Cryptostegia madagascariensis* var. *septentrionalis* Marohasy & P.I.Forst
Climbing Mechanism: Scrambler-Unarmed
Distribution (Global): Madagascar
Distribution (India): Gujarat, Maharashtra, Tamil Nadu
Flowering and Fruiting: June–March
IUCN Status: Not evaluated
Notes: Cultivated in gardens
Reference: A.P.de Candolle, Prodr. 8: 492 (1844)

Cynanchum acutum L.
Apocynaceae
Synonyms: *Solenostemma acutum* (L.) Wehmer, *Vincetoxicum acutum* (L.) Kuntze
Climbing Mechanism: Stem Twiner
Distribution (Global): Afghanistan, Albania, Algeria, Altay, Baleares, Bulgaria, Canary Is., China North-Central, Cyprus, East Aegean Is., Egypt, France, Greece, Inner Mongolia, Iran, Iraq, Italy, Kazakhstan, Kirgizstan, Kriti, Krym, Lebanon–Syria, Libya, Mongolia, Morocco, North Caucasus, Pakistan, Palestine, Portugal, Romania, Saudi Arabia, Sicilia, South European Russi, Spain, Tadzhikistan, Tibet, Transcaucasus, Tunisia, Turkey, Turkey-in-Europe, Turkmenistan, Ukraine, Uzbekistan, West Himalaya, West Siberia, Xinjiang, Yugoslavia
Leaf Type: Simple, opposite
Inflorescence: Axillary umbel-like
Fruit Type: Dry, follicles
IUCN Status: Least concern
Reference: Sp. Pl.: 212 (1753)

Cynanchum alatum Wight & Arn.
Apocynaceae
Synonyms: *Asclepias truncata* Roxb. ex Decne., *Cyathella alata* (Wight & Arn.) C.Y. Wu & D.Z.Li, *Cynanchum mairei* Schltr. ex H. Lév., *Cynoctonum alatum* (Wight & Arn.) Decne., *Vincetoxicum alatum* (Wight & Arn.) Kuntze
Climbing Mechanism: Stem Twiner
Distribution (Global): India, Sri Lanka
Distribution (India): Tamil Nadu
Leaf Type: Elliptic-oblong
Flowering and Fruiting: October–March
IUCN Status: Not evaluated
Notes: Usually found in deciduous forests
Reference: Contr. Bot. India: 57 (1834)

Cynanchum auriculatum Royle ex Wight
Apocynaceae

Synonyms: *Vincetoxicum auriculatum* (Royle ex Wight) Kuntze., *Cynanchum auriculatum* var. *sinense* T. Yamaz., *Cynanchum saccatum* W.T.Wang, *Diploglossum auriculatum* (Royle ex Wight) Meisn., *Endotropis auriculata* (Royle ex Wight) Decne
Common Name: Heart-leaf swallow-wort
Climbing Mechanism: Stem Twiner
Distribution (Global): China South-Central, East Himalaya, Nepal, Pakistan, Tibet, West Himalaya
Distribution (India): Himachal Pradesh, West Bengal
Leaf Type: Simple, opposite
Inflorescence: Axillary umbel-like
Fruit Type: Dry, follicles
Flowering and Fruiting: July–August
IUCN Status: Not evaluated
Reference: Revis. Gen. Pl. 2: 424 (1891)

Cynanchum callialatum Buch. -Ham. ex Wight
Apocynaceae
Synonyms: *Cyathella callialata* (Buch. -Ham. ex Wight) C.Y. Wu & D.Z.Li., *Cynanchum angustifolium* Wight & Arn., *Cynanchum heyneanum* Schult., *Cynanchum robertsoniae* Liede., *Cynoctonum angustifolium* Decne., *Cynoctonum callialata* (Buch.-Ham. ex Wight) Decne
Common Name: Pretty-wing swallow-wort
Climbing Mechanism: Stem Twiner
Distribution (Global): Bangladesh, China South-Central, China Southeast, India, Myanmar, Nepal, Pakistan
Distribution (India): Andhra Pradesh, Odisha, Tamil Nadu, West Bengal
Leaf Type: Elliptic-oblong or linear-oblong, pointed
Inflorescence: Umbel-like Cymes, axillary
Flowering and Fruiting: October–March
IUCN Status: Not evaluated
Notes: Occasionally found on hills
Reference: Contr. Bot. India: 56 (1834)

Cynanchum chinense R.Br.
Apocynaceae
Synonyms: *Cynanchum deltoideum* Hance, *Cynanchum pubescens* Bunge, *Vincetoxicum pubescens* (Bunge) Kuntze
Climbing Mechanism: Stem Twiner
Distribution (Global): China North-Central, China Southeast, Inner Mongolia, Korea, Manchuria, Mongolia, Qinghai
Leaf Type: Simple, opposite
Inflorescence: Axillary umbel-like
Fruit Type: Dry, follicles

IUCN Status: Not evaluated
Reference: Mem. Wern. Nat. Hist. Soc. 1: 44 (1811)

Cynanchum corymbosum Wight
Apocynaceae
Synonyms: *Cyathella corymbosa* (Wight) C.Y. Wu & D.Z.Li, *Cynoctonum corymbosum* (Wight) Decne., *Vincetoxicum corymbosum* (Wight) Benth. & Hook.f. ex B.D. Jacks
Climbing Mechanism: Stem Twiner
Distribution (Global): Assam, Bangladesh, Cambodia, China South-Central, China Southeast, East Himalaya, Laos, Malaya, Myanmar, Nicobar Is., Thailand, Vietnam
Distribution (India): Andaman and Nicobar Islands, Tripura, West Bengal
Leaf Type: Simple, opposite
Inflorescence: Axillary, umbel-like
Fruit Type: Dry, follicles
Flowering and Fruiting: September–October
IUCN Status: Not evaluated
Reference: Contr. Bot. India: 56 (1834)

Cynanchum dalhousiae Wight
Apocynaceae
Synonyms: *Vincetoxicum dalhousiae* (Wight) Kuntze
Common Name: Dalhousie swallow-wort
Climbing Mechanism: Stem Twiner
Distribution (Global): Nepal, Pakistan, West Himalaya
Distribution (India): Himachal Pradesh
Leaf Type: Simple, opposite
Inflorescence: Axillary, umbel-like
Fruit Type: Dry, follicles
Flowering and Fruiting: July–August
IUCN Status: Not evaluated
Reference: Contr. Bot. India: 55 (1834)

Cynanchum heydei Hook.f.
Apocynaceae
Synonyms: *Vincetoxicum heydei* (Hook.f.) Kuntze
Climbing Mechanism: Stem Twiner
Leaf Type: Simple, opposite
Inflorescence: Axillary umbel-like
Fruit Type: Dry, follicles
IUCN Status: Not evaluated
Reference: Fl. Brit. India 4: 25 (1883)

Cynanchum ovalifolium Wight
Apocynaceae
Synonyms: *Cyathella formosana* (Maxim.) C.Y. Wu & D.Z.Li, *Cyathella formosana* var. *ovalifolia* (Y.Tsiang & P.T.Li) C.Y.Wu & D.Z.Li, *Cynanchum crassifolium* Hatus., *Cynanchum formosanum* (Maxim.) Hemsl. + 16
Climbing Mechanism: Stem Twiner
Distribution (Global): Borneo, Java, Malaya, Maluku, New Guinea, Philippines, Queensland, Sulawesi, Sumatera, Taiwan, Thailand
Leaf Type: Simple, opposite
Inflorescence: Axillary umbel-like
Fruit Type: Dry, follicles
IUCN Status: Not evaluated
Reference: Contr. Bot. India: 57 (1834)

Cynanchum tunicatum (Retz.) Alston
Apocynaceae
Synonyms: *Asclepias tunicata* (Retz.) Roxb., *Cynanchum inconspicuum* Griff., *Cynanchum pauciflorum* R.Br., *Cynanchum truncatum* R.Br., *Cynoctonum pauciflorum* Decne., *Periploca tunicata* Retz., *Vincetoxicum tunicatum* (Retz.) Kuntze
Common Name: White milk wine
Climbing Mechanism: Stem Twiner
Distribution (Global): India, Sri Lanka
Distribution (India): Andhra Pradesh, Maharashtra, Tamil Nadu
Leaf Type: Decussate, narrowly cordiform
Inflorescence: Cymes umbel-like, solitary
Fruit Type: Follicles, fusiform
Flowering and Fruiting: October–February
IUCN Status: Not evaluated
Notes: Occasional on hills above 800 m in open deciduous forests
Reference: H. Trimen, Handb. Fl. Ceylon 6(Suppl.): 194 (1931)

Cynanchum viminale (L.) L.
Apocynaceae
Synonyms: *Cynanchum aphyllum* L., *Sarcostemma viminale* (L.) R.Br., *Euphorbia viminalis* L.
Common Name: Moon creeper, Moon plant
Climbing Mechanism: Scrambler-Unarmed
Distribution (Global): Aldabra, Angola, Benin, Botswana, Burkina, Burundi, Cape Provinces, Comoros, Djibouti, Equatorial Guinea, Eritrea, Ethiopia, Free State, Ghana, India, Kenya, KwaZulu-Natal, Lesotho, Madagascar, Malawi, Malaya, Mali, Mozambique, Myanmar, Namibia, New Caledonia, New South Wales, Nigeria, Northern Provinces, Northern Territory, Oman, Pakistan, Philippines, Queensland, Rodrigues, Rwanda, Réunion, Saudi Arabia, Seychelles, Socotra,

Somalia, South Australia, Sri Lanka, Sudan, Swaziland, Tanzania, Thailand, Uganda, Western Australia, Yemen, Zambia, Zaire, Zimbabwe
Leaf Type: Simple, opposite
Inflorescence: Axillary umbel-like
Fruit Type: Dry, follicles
Flowering and Fruiting: August–May
IUCN Status: Not evaluated
Notes: Occasionally found in dry deciduous forest and scrub jungle on rocky surfaces
Reference: Mant. Pl. 2: 392 (1771)

Cynanchum wallichii Wight
Apocynaceae
Synonyms: *Cyathella wallichii* (Wight) C.Y. Wu & D.Z.Li, *Cynoctonum wallichii* (Wight) Decne., *Vincetoxicum wallichii* (Wight) Kuntze
Climbing Mechanism: Stem Twiner
Distribution (India): Arunachal Pradesh
Leaf Type: Simple, opposite
Inflorescence: Axillary umbel-like
Fruit Type: Dry, follicles
IUCN Status: Not evaluated
Reference: Contr. Bot. India: 56 (1834)

Decalepis arayalpathra (J. Joseph & V. Chandras.) Venter
Apocynaceae
Synonyms: *Janakia arayalpathra* J. Joseph & V. Chandras
Climbing Mechanism: Stem Twiner
Distribution (Global): India
Distribution (India): Peninsular India
Leaf Type: Round-attenuate
Inflorescence: Axillary cymes
Fruit Type: Follicles, linear
Flowering and Fruiting: February–September
IUCN Status: Not evaluated
Notes: The tuberous roots of the plant are highly aromatic. Medicinally valuable and used to treat various ailments
Reference: Taxon 46: 712 (1997)

Decalepis hamiltonii Wight & Arn.
Apocynaceae
Synonyms: *Apocynum reticulatum* Wall., *Streptocaulon hamiltonii* Wight
Common Name: Swallow-root
Climbing Mechanism: Stem Twiner
Distribution (Global): India
Distribution (India): Andhra Pradesh, Eastern Ghats, Tamil Nadu, Telangana

Leaf Type: Simple, opposite
Inflorescence: Cymes, trichotomously branched
Fruit Type: Follicles
Flowering and Fruiting: May–December
IUCN Status: Endangered
Notes: The tubers are medicinal and often used for making pickles
Reference: Contr. Bot. India: 64 (1834)

Decalepis nervosa (Wight & Arn.) Venter
Apocynaceae
Synonyms: *Baeolepis nervosa* (Wight & Arn.) Decne. ex Moq., *Brachylepis nervosa* Wight & Arn.
Common Name: Nerved-leaf swallow-root
Climbing Mechanism: Stem Twiner
Distribution (Global): India
Distribution (India): Tamil Nadu
Leaf Type: Simple, opposite
Inflorescence: Condensed cymes
Fruit Type: Follicles narrowly lance shaped
Flowering and Fruiting: December–April
IUCN Status: Not evaluated
Notes: Medicinally valuable climber used for treating various ailments
Reference: Taxon 46: 712 (1997)

Dischidia bengalensis Colebr.
Apocynaceae
Synonyms: *Dischidia cuneifolia* Wall., *Dischidia loeseneriana* Schltr., *Dischidia pseudobenghalensis* Costantin, *Dischidia spatulata* Blume
Climbing Mechanism: Scrambler-Unarmed
Distribution (Global): Andaman Is., Assam, Bangladesh, Borneo, East Himalaya, Java, Malaya, Myanmar, Nepal, Nicobar Is., Sumatera, Thailand, Vietnam
Distribution (India): Assam, West Bengal
IUCN Status: Not evaluated
Reference: Trans. Linn. Soc. London 12: 357 (1818)

Dischidia complex Griff.
Apocynaceae
Synonyms: *Dischidia shelfordii* H. Pearson
Climbing Mechanism: Scrambler-Unarmed
Distribution (Global): Borneo, Malaya
IUCN Status: Not evaluated
Reference: Not. Pl. Asiat. 4: 50 (1854)

Dischidia major (Vahl) Merr.
Apocynaceae

Synonyms: *Collyris major* Vahl, *Dischidia bauerlenii* Schltr., *Dischidia clavata* Wall., *Dischidia merguiensis* Becc., *Dischidia pubiflora* Schltr., *Dischidia rafflesiana* Wall., *Dischidia singaporensis* Ridl., *Dischidia timorensis* Decne.
Climbing Mechanism: Scrambler-Unarmed
Distribution (Global): Andaman Is., Assam, Bangladesh, Borneo, Cambodia, Java, Laos, Lesser Sunda Is., Malaya, Myanmar, New Guinea, Nicobar Is., Philippines, Queensland, Solomon Is., Sulawesi, Sumatera, Thailand, Vietnam
IUCN Status: Not evaluated
Reference: Interpr. Herb. Amboin.: 437 (1917)

Dischidia nummularia R.Br.
Apocynaceae
Synonyms: *Dischidia aemula* Schltr., *Dischidia beiningiana* Schltr., *Dischidia copelandii* Schltr., *Dischidia decipiens* Schltr., *Dischidia dirhiza* Schltr., *Dischidia gaudichaudii* Decne. + 12
Climbing Mechanism: Scrambler-Unarmed
Distribution (Global): Andaman Is., Assam, Bangladesh, Bismarck Archipelago, Borneo, Cambodia, China South-Central, China Southeast, East Himalaya, Hainan, Java, Laos, Lesser Sunda Is., Malaya, Maluku, Myanmar, New Guinea, Philippines, Queensland, Solomon Is., Sulawesi, Sumatera, Thailand, Vietnam
Distribution (India): Andaman and Nicobar Islands
IUCN Status: Not evaluated
Reference: Prodr. Fl. Nov. Holland.: 461 (1810)

Dittoceras andersonii Hook.f.
Apocynaceae
Synonyms: *Heterostemma andersonii* (Hook.f.) Rodda
Climbing Mechanism: Stem Twiner
Distribution (India): West Bengal
IUCN Status: Not evaluated
Reference: Hooker's Icon. Pl. 15: t. 1422 (1883)

Dregea lanceolata (Cooke) Santapau & Wagh
Apocynaceae
Synonyms: *Wattakaka lanceolata* (T. Cooke) Kerr
Climbing Mechanism: Stem Twiner
Distribution (India): Maharashtra
IUCN Status: Not evaluated
Reference: Bull. Bot. Surv. India 5: 106 (1964)

Dregea volubilis (L.f.) Benth. ex Hook.f.
Apocynaceae
Synonyms: *Asclepias volubilis*, *Wattakaka volubilis*
Common Name: Sneeze wort, Cotton milk plant, Green milkweed climber, Green wax flower, Sneezing silk

Climbing Mechanism: Stem Twiner
Distribution (India): Andaman and Nicobar Islands, Arunachal Pradesh, Bihar, Odisha, Delhi, Gujarat, Tamil Nadu, Telangana, West Bengal, Western Ghats
IUCN Status: Not evaluated
Reference: Fl. Brit. India 4: 46 (1883)

Epigynum auritum (C.K. Schneid.) Tsiang & P.T.Li
Apocynaceae
Synonyms: *Epigynum lachnocarpum* Pichon, *Trachelospermum auritum* C.K. Schneid., *Trachelospermum curtisii* King & Gamble
Climbing Mechanism: Scrambler-Unarmed
Distribution (Global): Assam, Cambodia, China South-Central, Laos, Malaya, Myanmar, Thailand
IUCN Status: Not evaluated
Reference: Acta Phytotax. Sin. 11: 397 (1973)

Epigynum griffithianum Wight
Apocynaceae
Synonyms: *Argyronerium odoratum* Pit., *Epigynum curtisii* King & Gamble, *Epigynum maingayi* Hook.f., *Epigynum odoratum* (Pit.) Kerr, *Epigynum perakense* King & Gamble, *Legouixia amabilis* Van Heurck & Müll.Arg.
Climbing Mechanism: Scrambler-Unarmed
Distribution (Global): Cambodia, India, Laos, Malaya, Myanmar, Thailand, Vietnam
IUCN Status: Not evaluated
Reference: Icon. Pl. Ind. Orient. 4: t. 1308 (1848)

Finlaysonia obovata Wall
Apocynaceae
Synonyms: *Gurua obovata* (Wall.) Buch. -Ham. ex Voigt, *Tabernaemontana cirrhosa* Blanco
Climbing Mechanism: Stem Twiner
Distribution (Global): Andaman Is., Bangladesh, Caroline Is., India, Java, Lesser Sunda Is., Malaya, Maluku, Myanmar, Nicobar Is., Northern Territory, Philippines, Queensland, Thailand, Vietnam
Distribution (India): Great, Nicobar, Island, Odisha, West Bengal
IUCN Status: Not evaluated
Reference: Pl. Asiat. Rar. 2: 48 (1831)

Finlaysonia wallichii (Wight) Venter
Apocynaceae
Synonyms: *Atherolepis wallichii* (Wight) Hook.f., *Atherandra wallichii* (Wight) Benth. & Hook.f., *Hemidesmus wallichii* Wight
Climbing Mechanism: Stem Twiner
Distribution (Global): Myanmar

Distribution (India): Telangana
IUCN Status: Not evaluated
Reference: Ann. Missouri Bot. Gard. 88: 565 (2001)

Genianthus crassifolius (Wight) Hook.f.
Apocynaceae
Synonyms: *Genianthus blumei* (Decne.) Boerl., *Secamone crassifolia* (Wight) K. Schum., *Secamone fulva* (Blume) Koord., *Toxocarpus blumei* Decne., *Toxocarpus crassifolius* Wight
Climbing Mechanism: Scrambler-Unarmed
Distribution (Global): Bangladesh, Borneo, East Himalaya, Java, Malaya, Myanmar, Sumatera, Thailand
IUCN Status: Not evaluated
Reference: Fl. Brit. India 4: 16 (1883)

Genianthus horei Vasudeva Rao
Apocynaceae
Climbing Mechanism: Stem Twiner
Distribution (Global): Nicobar Is.
Distribution (India): Great Nicobar Island
IUCN Status: Not evaluated
Reference: J. Econ. Taxon. Bot. 6: 732 (1985)

Genianthus laurifolius (Roxb.) Hook.f.
Apocynaceae
Synonyms: *Asclepias laurifolia* Roxb., *Secamone laurifolia* (Roxb.) K. Schum., *Toxocarpus crassifolius* Dalz. & Gibs.
Toxocarpus laurifolius (Roxb.) Wight & Arn.
Common Name: Laurel-leaf milkweed
Climbing Mechanism: Stem Twiner
Distribution (India): Andaman and Nicobar Islands, Arunachal Pradesh, Maharashtra, West Bengal
Leaf Type: Leaves are elliptic
Inflorescence: Cymes
IUCN Status: Not evaluated
Reference: Fl. Brit. India 4: 16 (1883)

Genianthus maingayi Hook.f.
Apocynaceae
Synonyms: *Secamone maingayi* (Hook.f.) Rodda
Climbing Mechanism: Stem Twiner
Distribution (Global): Borneo, Malaya, Sumatera
IUCN Status: Not evaluated
Reference: Fl. Brit. India 4: 16 (1883)

Gongronema nepalense (Wall.) Decne.
Apocynaceae
Climbing Mechanism: Scrambler-Unarmed
Distribution (India): Arunachal Pradesh, Meghalaya, Mizoram, West Bengal
IUCN Status: Not evaluated
Reference: Prodr. [A. P. de Candolle] 8: 624. 1844 [mid-Mar 1844]

Gongronema thomsonii (Hook.f.) K.M. Matthew
Apocynaceae
Synonyms: *Gymnema thomsonii* Hook.f.
Climbing Mechanism: Scrambler-Unarmed
Distribution (Global): East Himalaya
Distribution (India): West Bengal
IUCN Status: Not evaluated
Reference: Kew Bull. 34: 68 (1979)

Goniostemma acuminata Wight
Apocynaceae
Climbing Mechanism: Scrambler-Unarmed
IUCN Status: Not evaluated
Reference: Wight. In: Contrib. 62. (1834).

Gymnema acuminatum Wall
Apocynaceae
Synonyms: *Asclepias acuminata* (Wall.) Roxb., *Marsdenia acuminata* (Wall.) I.M. Turner
Climbing Mechanism: Stem Twiner
Distribution (Global): Assam, Bangladesh, Cambodia, Malaya, Myanmar, Thailand, Vietnam
Distribution (India): West Bengal
IUCN Status: Not evaluated
Reference: Tent. Fl. Napal.: 50 (1826)

Gymnema cuspidatum (Thunb.) Kuntze
Apocynaceae
Synonyms: *Bidaria celsicola* H. Huber, *Bidaria cuspidata* (Thunb.) H. Huber, *Bidaria cuspidata* var. *stenoloba* (Hook.f.) H.Huber, *Bidaria pergularioides* Thwaites, *Cynanchum cuspidatum* Thunb., *Gymnema pergularioides* (Thwaites) Hook.f., *Gymnema stenolobum* Hook.f.
Climbing Mechanism: Stem Twiner
Distribution (Global): India, Sri Lanka
Distribution (India): Maharashtra
Flowering and Fruiting: December–April
IUCN Status: Not evaluated
Reference: Revis. Gen. Pl. 2: 420 (1891)

Gymnema decaisneanum Wight
Apocynaceae
Synonyms: *Gymnema hirsutum* Wight & Arn.
Gymnema mohanramii Karthik. & Moorthy
Gymnema mohanramii var. *decaisneanum* (Wight) Karthik & Moorthy
Climbing Mechanism: Stem Twiner
Distribution (Global): India, Myanmar
Distribution (India): Tamil Nadu
IUCN Status: Not evaluated
Reference: Icon. Pl. Ind. Orient. 4: t. 1271 (1848)

Gymnema elegans Wight & Arn.
Apocynaceae
Synonyms: *Bidaria elegans* (Wight & Arn.) Decne.
Climbing Mechanism: Stem Twiner
Distribution (Global): India
Distribution (India): Tamil Nadu
Flowering and Fruiting: October–January
IUCN Status: Not evaluated
Reference: Contr. Bot. India: 46 (1834)

Gymnema inodorum (Lour.) Decne.
Apocynaceae
Synonyms: *Asclepias tingens* Buch. -Ham. ex Roxb., *Asclepias tingens* Roxb., *Bidaria inodora* (Lour.) Decne., *Bidaria pubiflora* Miq., *Bidaria syringifolia* (Decne.) Decne., *Bidaria tingens* (Roxb.) Decne., *Cynanchum inodorum* Lour., *Gymnema pubiflora* (Miq.) Hook
Common Name: Scentless cowplant
Climbing Mechanism: Stem Twiner
Distribution (Global): Andaman Is., Assam, Bangladesh, Borneo, China South-Central, China Southeast, East Himalaya, Hainan, India, Java, Malaya, Myanmar, Nepal, Philippines, Thailand, Vietnam, West Himalaya
Distribution (India): Tamil Nadu
IUCN Status: Not evaluated
Reference: A.P.de Candolle, Prodr. 8: 551 (1844)

Gymnema khandalense Santapau
Apocynaceae
Synonyms: *Bidaria khandalensis* (Santapau) A.P. Jagtap & N.P. Singh
Climbing Mechanism: Stem Twiner
Distribution (India): Maharashtra, Western Ghats
Leaf Type: Opposite
Inflorescence: Axillary umbellate cymes
Flowering and Fruiting: October–November

IUCN Status: Endangered
Reference: Kew Bull. 3: 486 (1948 publ. 1949)

Gymnema kollimalayanum A. Ramach. & M.B. Viswan.
Apocynaceae
Synonyms: *Gymnema latifolium* Wall. ex Wight
Climbing Mechanism: Stem Twiner
Distribution (India): Tamil Nadu
IUCN Status: Not evaluated
Reference: Adansonia, sér. 3, 31: 408 (2009)

Gymnema lactiferum (L.) R.Br. ex Schult.
Apocynaceae
Synonyms: *Asclepias lactifera* L., *Gymnema malayanum* Griff., *Gymnema nitens* Blume, *Gymnema zeylanicum* Decne., *Marsdenia lacti*fera (L.) I.M. Turner, *Marsdenia lactifera* var. *thwaitesii* (Hook.f.) I.M. Turner
Climbing Mechanism: Stem Twiner
Distribution (Global): Assam, Borneo, India, Malaya, Myanmar, Sri Lanka, Sumatera
Distribution (India): Tamil Nadu
IUCN Status: Not evaluated
Reference: J.J. Roemer & J.A.Schultes, Syst. Veg., ed. 15 bis 6: 57 (1820)

Gymnema latifolium Wall. ex Wight
Apocynaceae
Synonyms: *Bidaria latifolia* (Wall. ex Wight) P.T. Li, *Gymnema khandalense* Santapau, *Gymnema kollimalayanum* A.Ramach. & M.B. Viswan., *Bidaria khandalensis* (Santapau) A.P.Jagtap & N.P.Singh
Climbing Mechanism: Stem Twiner
Distribution (Global): Andaman Is., Assam, Bangladesh, China South-Central, China Southeast, East Himalaya, India, Myanmar, Nicobar Is., Thailand, Vietnam
Distribution (India): Andaman and Nicobar Islands
IUCN Status: Not evaluated
Reference: Contr. Bot. India: 45 (1834)

Gymnema lushaiense M.A. Rahman & Wilcock
Apocynaceae
Climbing Mechanism: Stem Twiner
Distribution (Global): India
Distribution (India): Mizoram, Assam
IUCN Status: Not evaluated
Reference: J. Econ. Taxon. Bot. 13: 181 (1989)

Gymnema molle Wall. ex Wight
Apocynaceae

Climbing Mechanism: Stem Twiner
Distribution (Global): Bangladesh, Myanmar
IUCN Status: Not evaluated
Reference: Contr. Bot. India: 45 (1834)

Gymnema montanum Hook.f.
Apocynaceae
Synonyms: *Asclepias montana* Roxb., *Bidaria elegans* Dalzell & A. Gibson, *Bidaria montana* (Roxb.) M.A. Rahman & Wilcock, *Periploca montana* (Roxb.) Rottler ex Wight
Climbing Mechanism: Stem Twiner
Distribution (Global): India
Distribution (India): Tamil Nadu
Flowering and Fruiting: April–May
IUCN Status: Not evaluated
Reference: Fl. Brit. India 4: 31 (1883)

Gymnema montanum var. *beddomei* Hook.f.
Apocynaceae
Climbing Mechanism: Stem Twiner
Distribution (India): Tamil Nadu
IUCN Status: Not evaluated
Reference: Hook. fil. In: Fl. Brit. India 4: 32. (1883)

Gymnema sylvestre (Retz.) R.Br. ex Sm.
Apocynaceae
Synonyms: *Apocynum alterniflorum* Lour., *Asclepias geminata* Roxb., *Conocalpis umbellata* Bojer ex Decne., *Cynanchum senegalense* Sieber ex Decne., *Cynanchum subvolubile* Schumach. & Thonn., *Gymnema affine* Decne., *Gymnema alterniflorum* (Lour.) Merr.
Common Name: Gurmar, Cowplant, Australian cowplant, Periploca of the woods, Small Indian ipecacuanha
Climbing Mechanism: Stem Twiner
Distribution (Global): Angola, Assam, Bangladesh, Benin, Botswana, Burkina, Burundi, Cambodia, Cameroon, Cape Provinces, Caprivi Strip, Central African Repu, Chad, China South-Central, China Southeast, Comoros, Congo, Eritrea, Ethiopia, Gabon, Ghana, Guinea, Guinea-Bissau, Hainan, India, Ivory Coast, Kenya, KwaZulu-Natal, Laos, Madagascar, Malawi, Malaya, Mali, Mauritania, Mozambique, Namibia, Nansei-shoto, Niger, Nigeria, Northern Provinces, Northern Territory, Queensland, Rwanda, Saudi Arabia, Senegal, Sri Lanka, Sudan, Taiwan, Tanzania, Togo, Uganda, Vietnam, Western Australia, Yemen, Zambia, Zaire, Zimbabwe
Distribution (India): Andhra Pradesh, Bihar, Odisha, Eastern Ghats, Goa, Gujarat, Jharkhand, Madhya Pradesh, Maharashtra, Odisha, Puducherry, Rajasthan, Tamil Nadu, Telangana, West Bengal

Leaf Type: Simple
Inflorescence: Umbellate cymes
Fruit Type: Paired follicle
Flowering and Fruiting: July–January
IUCN Status: Not evaluated
Reference: A. Rees, Cycl. 17: no. 4 (1811)

Hemidesmus indicus (L.) R.Br. ex Schult.
Apocynaceae
Synonyms: *Hemidesmus pubescens* Wight & Arn., *Periploca cordata* Dennst., *Periploca indica* L., *Periploca malabarica* Burm. ex Decne., *Asclepias pseudosarsa* Roxb., *Cosmostigma cordatum* (Poir.) M.R. Almeida, *Hemidesmus cordatus* (Poir.) Schult., *Hemidesmus wallich*
Common Name: Indian sarsaparilla
Climbing Mechanism: Stem Twiner
Distribution (Global): Assam, Bangladesh, Cambodia, East Himalaya, India, Laos, Malaya, Myanmar, Pakistan, Sri Lanka, Vietnam
Distribution (India): Andhra Pradesh, Bihar, Odisha, Eastern Ghats, Goa, Gujarat, Jharkhand, Kerala, Madhya Pradesh, Maharashtra, Odisha, Puducherry, Rajasthan, Tamil Nadu, Telangana, Tripura, Uttar Pradesh, West Bengal, Western Ghats
Leaf Type: Simple
Inflorescence: Axillary fascicles
Fruit Type: Follicles
IUCN Status: Not evaluated
Reference: W.T. Aiton, Hortus Kew., ed. 2, 2: 75 (1811)

Heterostemma alatum Wight & Arn.
Apocynaceae
Climbing Mechanism: Stem Twiner
Distribution (Global): Assam, East Himalaya, India, Myanmar, Nepal, West Himalaya
Distribution (India): Arunachal Pradesh, Himachal Pradesh, Maharashtra, Meghalaya, Mizoram, Tripura, West Bengal
IUCN Status: Not evaluated
Reference: Contr. Bot. India: 42 (1834)

Heterostemma barikiana P. Agnihotri, D.Husain, P.Katiyar, D.Sahoo, Rodda & T. Husain
Apocynaceae
Climbing Mechanism: Stem Twiner
Distribution (India): Manipur
IUCN Status: Not evaluated
Reference: P. Agnihotri, D. Husain, P. Katiyar, D. Sahoo, Rodda & T. Husain. In: Phytotaxa 419(2): 232, as "barikiana". (2019)

Heterostemma dalzellii Hook.f.
Apocynaceae
Synonyms: *Heterostemma wallichii* Dalzell & A. Gibson
Common Name: Dalzell's Heterostemma
Climbing Mechanism: Stem Twiner
Distribution (Global): India
Distribution (India): Maharashtra, Goa
IUCN Status: Not evaluated
Reference: Fl. Brit. India 4: 48 (1883)

Heterostemma decanense (Talbot) Swarupan. & Mangaly
Apocynaceae
Climbing Mechanism: Stem Twiner
Distribution (India): Maharashtra, Telangana
Leaf Type: Opposite
Flowering and Fruiting: August–January
IUCN Status: Not evaluated
Reference: Swarupan., Mangaly. In: Bot. J. L. Soc., 101(2): 255. (1989)

Heterostemma disciflorum (Hook.f.) Swarupan. & Mangaly
Apocynaceae
Synonyms: *Oianthus disciflorus* Hook.f.
Climbing Mechanism: Stem Twiner
Distribution (Global): India
Distribution (India): Peninsular India
IUCN Status: Not evaluated
Reference: Bot. J. Linn. Soc. 101: 256 (1989)

Heterostemma stellatum Hook.f.
Apocynaceae
Climbing Mechanism: Stem Twiner
Distribution (Global): Assam, East Himalaya
IUCN Status: Not evaluated
Reference: Fl. Brit. India 4: 47 (1883)

Heterostemma tanjorense Wight & Arn.
Apocynaceae
Synonyms: Stapelia involucrata Hook.f.
Climbing Mechanism: Stem Twiner
Distribution (Global): India, Sri Lanka
Distribution (India): Andaman and Nicobar Islands, Andhra Pradesh, Maharashtra, Odisha, Tamil Nadu
IUCN Status: Not evaluated
Reference: Contr. Bot. India: 42 (1834)

Heterostemma urceolatum Dalzell
Apocynaceae
Synonyms: *Oianthus urceolatus* (Dalzell) Benth.
Climbing Mechanism: Stem Twiner
Distribution (Global): India
Distribution (India): Maharashtra
IUCN Status: Not evaluated
Reference: Hooker's J. Bot. Kew Gard. Misc. 4: 295 (1852)

Heterostemma vasudevanii Swarupan. & Mangaly
Apocynaceae
Climbing Mechanism: Stem Twiner
Distribution (Global): India
Distribution (India): Peninsular India
Flowering and Fruiting: August–December
IUCN Status: Not evaluated
Reference: Bot. J. Linn. Soc. 101: 257 (1989)

Heterostemma wallichii
Apocynaceae
Climbing Mechanism: Stem Twiner
IUCN Status: Not evaluated

Holostemma ada-kodien Schult.
Apocynaceae
Synonyms: *Cynanchum annularium* (Roxb.) Liede & Khanum
Common Name: Holostemma creeper
Climbing Mechanism: Stem Twiner
Distribution (India): Andhra Pradesh, Eastern Ghats, Goa, Kerala, Madhya Pradesh, Odisha, Rajasthan, Telangana, West Bengal
Leaf Type: Leaves are ovate to heart shaped
Inflorescence: Axillary paduncled cymes
Fruit Type: Follicles
Flowering and Fruiting: September–November
IUCN Status: Not evaluated
Reference: J.J. Roemer & J.A.Schultes, Syst. Veg., ed. 15 bis 6: 95 (1820)

Holostemma rheedei Wall.
Apocynaceae
Synonyms: *Cynanchum annularium* (Roxb.) Liede & Khanum
Climbing Mechanism: Stem Twiner
Leaf Type: Simple, opposite, heart shaped
Inflorescence: Simple, white-pink flowers
Fruit Type: Dry, folicles

IUCN Status: Not evaluated
Reference: Pl. Asiat. Rar. 2: 51 (1831)

Hoya acuminata (Wight) Benth. ex Hook.f.
Apocynaceae
Synonyms: *Hoya griffithiana* Decne. ex Hook.f., *Pterostelma acuminata* Wight.
Climbing Mechanism: Stem Twiner
Distribution (Global): Assam, Bangladesh, East Himalaya, Myanmar
Leaf Type: Simple, opposite, with sap
Inflorescence: Axillary cymes, umbellate
Fruit Type: Dry, follicles
IUCN St9atus: Not evaluated
Reference: Fl. Brit. India 4: 53 (1883)

Hoya alexicaca (Jacq.) Moon
Apocynaceae
Synonyms: *Asclepias alexicaca* Jacq., *Asclepias pendula* Roxb., *Hoya iconum* Santapau, *Hoya pendula* (Roxb.) Wight & Arn.
Climbing Mechanism: Stem Twiner
Distribution (Global): Bangladesh, India, Myanmar
Distribution (India): Maharashtra, Odisha, Tamil Nadu
Leaf Type: Simple, opposite, with sap
Inflorescence: Axillary cymes, umbellate
Fruit Type: Dry, follicles
IUCN Status: Not evaluated
Reference: Cat. Pl. Ceylon: 21 (1824)

Hoya arnottiana Wight
Apocynaceae
Common Name: Porcelain flower
Wax plant, Arnott's wax flower
Climbing Mechanism: Stem Twiner
Distribution (Global): Assam, Bangladesh, East Himalaya, Laos, Nepal, Thailand, Vietnam
Distribution (India): Arunachal Pradesh, Assam
Leaf Type: Simple, opposite, with sap
Inflorescence: Axillary cymes, umbellate
Fruit Type: Dry, follicles
IUCN Status: Not evaluated
Reference: Contr. Bot. India: 37 (1834)

Hoya carnosa (L.f.) R.Br.
Apocynaceae
Synonyms: *Hoya carnosa* var. *gushanica* W. Xu, *Hoya chinensis* (Lour.) Traill, *Hoya compacta* C.M. Burton, *Hoya crassifolia* (J.Jacq.) Haw., *Hoya intermedia*

A.C.Sm., *Hoya laurifolia* Miq., *Hoya motoskei* Teijsm. & Binn., *Hoya picta* Siebold ex C. Morren
Common Name: Wax plant, Wax flower, Porcelain flower
Climbing Mechanism: Stem Twiner
Distribution (Global): China South-Central, China Southeast, Hainan, Japan, Laos, Nansei-shoto, Taiwan
Distribution (India): Maharashtra
Leaf Type: Simple, opposite, with sap
Inflorescence: Axillary cymes, umbellate
Fruit Type: Dry, follicles
IUCN Status: Not evaluated
Reference: Asclepiadeae: 16 (1810)

Hoya coronaria Blume
Apocynaceae
Synonyms: *Eriostemma coronaria* (Blume) Kloppenb. & Gilding, *Eriostemma obtusifolioides* Gilding & T. Green, *Hoya velutina* Wight
Climbing Mechanism: Stem Twiner
Distribution (Global): Borneo, Java, Malaya, Philippines, Sulawesi, Sumatera, Thailand
Leaf Type: Simple, opposite, with sap
Inflorescence: Axillary cymes, umbellate
Fruit Type: Dry, follicles
IUCN Status: Not evaluated
Reference: Bijdr. Fl. Ned. Ind.: 1063 (1827)

Hoya diversifolia Blume
Apocynaceae
Synonyms: *Eriostemma zollingerianum* (Miq.) Kloppenb., *Hoya alba* Kostel., *Hoya bulusanensis* Elmer, *Hoya coriacea* Zoll. ex Miq., *Hoya crassipes* Turcz., *Hoya diversifolia* subsp. *elnidicus* (Kloppenb.) Kloppenb., *Hoya elnidicus* Kloppenb., *Hoya esculenta* Tsiang
Common Name: Mangrove wax plant, Andaman wax flower
Climbing Mechanism: Stem Twiner
Distribution (Global): Andaman Is., Borneo, Cambodia, Hainan, Java, Laos, Lesser Sunda Is., Malaya, Myanmar, Philippines, Sulawesi, Sumatera, Thailand, Vietnam
Leaf Type: Simple, opposite, with sap
Inflorescence: Axillary cymes, umbellate
Fruit Type: Dry, follicles
IUCN Status: Not evaluated
Reference: Bijdr. Fl. Ned. Ind.: 1064 (1827)

Hoya edeni King ex Hook.f.
Apocynaceae

Climbing Mechanism: Stem Twiner
Distribution (Global): Assam, East Himalaya, Nepal
Distribution (India): Arunachal Pradesh
Leaf Type: Simple, opposite, with sap
Inflorescence: Axillary cymes, umbellate
Fruit Type: Dry, follicles
IUCN Status: Not evaluated
Reference: Fl. Brit. India 4: 53 (1883)

Hoya elliptica Hook.f.
Apocynaceae
Climbing Mechanism: Stem Twiner
Distribution (Global): Borneo, Malaya, Sumatera, Thailand
Leaf Type: Simple, opposite, with sap
Inflorescence: Axillary cymes, umbellate
Fruit Type: Dry, follicles
IUCN Status: Not evaluated
Reference: Fl. Brit. India 4: 58 (1883)

Hoya fusca Wall.
Apocynaceae
Common Name: Dusky wax flower, Dusky porcelain flower
Climbing Mechanism: Stem Twiner
Distribution (Global): Assam, Bangladesh, Cambodia, China South-Central, China Southeast, East Himalaya, Hainan, India, Laos, Myanmar, Nepal, Thailand, Tibet, Vietnam
Distribution (India): Arunachal Pradesh
Leaf Type: Simple, opposite, with sap
Inflorescence: Axillary cymes, umbellate
Fruit Type: Dry, follicles
IUCN Status: Not evaluated
Reference: Pl. Asiat. Rar. 1: 68 (1830)

Hoya globulosa Hook.f.
Apocynaceae
Synonyms: *Hoya villosa* Costantin
Climbing Mechanism: Stem Twiner
Distribution (Global): Assam, Bangladesh, China South-Central, China Southeast, East Himalaya, Hainan, Laos, Nicobar Is., Vietnam, West Himalaya
Distribution (India): Andaman and Nicobar Islands, Arunachal Pradesh, Assam, Tripura
Leaf Type: Simple, opposite, with sap
Inflorescence: Axillary cymes, umbellate
Fruit Type: Dry, follicles

IUCN Status: Not evaluated
Reference: Gard. Chron., n.s., 17: 732 (1882)

Hoya griffithii Hook.f.
Apocynaceae
Synonyms: *Hoya kwangsiensis* Y. Tsiang & P.T.Li, *Hoya lancilimba* Merr., *Hoya tsoii* Merr.
Climbing Mechanism: Stem Twiner
Distribution (Global): Assam, China South-Central, China Southeast, Hainan, Laos, Vietnam
Distribution (India): Maharashtra, Mizoram
Leaf Type: Simple, opposite, with sap
Inflorescence: Axillary cymes, umbellate
Fruit Type: Dry, follicles
IUCN Status: Not evaluated
Reference: Fl. Brit. India 4: 59 (1883)

Hoya imperialis Lindl.
Apocynaceae
Synonyms: *Calotropis sussuela* G. Don
Climbing Mechanism: Stem Twiner
Distribution (Global): Borneo, Java, Malaya, Philippines, Sumatera, Thailand
Leaf Type: Simple, opposite, with sap
Inflorescence: Axillary cymes, umbellate
Fruit Type: Dry, follicles
IUCN Status: Not evaluated
Reference: Edwards's Bot. Reg. 32: t. 68 (1846)

Hoya kanyakumariana A.N. Henry & Swamin.
Apocynaceae
Climbing Mechanism: Stem Twiner
Distribution (Global): India
Leaf Type: Opposite
Inflorescence: Lateral umbellate solitary cymes
Flowering and Fruiting: August–December
IUCN Status: Not evaluated
Reference: J. Bombay Nat. Hist. Soc. 75: 462 (1978 publ. 1979)

Hoya lacunosa Blume
Apocynaceae
Synonyms: *Hoya suaveolens* Miq., *Otostemma lacunosum* (Blume) Blume
Climbing Mechanism: Stem Twiner
Distribution (Global): Borneo, Java, Malaya, Philippines, Sumatera, Thailand
Leaf Type: Simple, opposite, with sap
Inflorescence: Axillary cymes, umbellate

Fruit Type: Dry, follicles
IUCN Status: Not evaluated
Reference: Bijdr. Fl. Ned. Ind.: 1063 (1827)

Hoya lanceolata Lindl.
Apocynaceae
Common Name: Lanceleaf wax flower
Climbing Mechanism: Stem Twiner
Distribution (Global): Assam, Bangladesh, China South-Central, East Himalaya, Myanmar, Nepal, Vietnam, West Himalaya
Distribution (India): Arunachal Pradesh, West Bengal
Leaf Type: Simple, opposite, with sap
Inflorescence: Axillary cymes, umbellate
Fruit Type: Dry, follicles
IUCN Status: Not evaluated
Reference: Prodr. Fl. Nepal.: 130 (1825)

Hoya linearis Wall. ex D. Don
Apocynaceae
Climbing Mechanism: Stem Twiner
Distribution (Global): Assam, China South-Central, East Himalaya, Laos, Myanmar, Nepal, Vietnam
Distribution (India): Arunachal Pradesh
Leaf Type: Simple, opposite, with sap
Inflorescence: Axillary cymes, umbellate
Fruit Type: Dry, follicles
IUCN Status: Not evaluated
Reference: Prodr. Fl. Nepal.: 130 (1825)

Hoya lobbii Hook.f.
Apocynaceae
Synonyms: Hoya lazaroi Kloppenb. & Siar
Climbing Mechanism: Stem Twiner
Distribution (Global): Assam, Cambodia, Thailand
Distribution (India): Arunachal Pradesh
Leaf Type: Simple, opposite, with sap
Inflorescence: Axillary cymes, umbellate
Fruit Type: Dry, follicles
IUCN Status: Not evaluated
Reference: Fl. Brit. India 4: 54 (1883)

Hoya longifolia Wall. ex Wight
Apocynaceae
Synonyms: *Acanthostemma kuhlii* Bl., *Hoya shepherdii* Short ex Hook.
Climbing Mechanism: Stem Twiner

Distribution (Global): Assam, China South-Central, East Himalaya, India, Myanmar, Nepal, Nicobar Is., Thailand, West Himalaya
Distribution (India): Arunachal Pradesh
Leaf Type: Simple, opposite, with sap
Inflorescence: Axillary cymes, umbellate
Fruit Type: Dry, follicles
IUCN Status: Not evaluated
Reference: Contr. Bot. India: 36 (1834)

Hoya multiflora Blume
Apocynaceae
Synonyms: *Asclepias carnosa* Blanco, *Asclepias stellata* Burm. ex Decne., *Centrostemma cyrtoceras* Meisn., *Centrostemma elegans* Blume, *Centrostemma floribunda* Bosse, *Centrostemma laurifolium* Blume, *Centrostemma lindleyanum* Decne., *Centrostemma micranthum* Blume
Climbing Mechanism: Stem Twiner
Distribution (Global): Assam, Bangladesh, Borneo, Cambodia, China South-Central, China Southeast, Java, Laos, Malaya, Myanmar, New Guinea, Philippines, Sulawesi, Sumatera, Thailand, Vietnam
Leaf Type: Simple, opposite, with sap
Inflorescence: Axillary cymes, umbellate
Fruit Type: Dry, follicles
IUCN Status: Not evaluated
Reference: Catalogus: 49 (1823)

Hoya obcordata Hook.f.
Apocynaceae
Synonyms: *Hoya obreniformis* King
Climbing Mechanism: Stem Twiner
Distribution (Global): East Himalaya
Distribution (India): West Bengal
Leaf Type: Simple, opposite, with sap
Inflorescence: Axillary cymes, umbellate
Fruit Type: Dry, follicles
IUCN Status: Not evaluated
Reference: Fl. Brit. India 4: 56 (1883)

Hoya oblanceolata Hook.f.
Apocynaceae
Climbing Mechanism: Stem Twiner
Distribution (Global): Assam, Sumatera
Leaf Type: Simple, opposite, with sap
Inflorescence: Axillary cymes, umbellate
Fruit Type: Dry, follicles

IUCN Status: Not evaluated
Reference: Fl. Brit. India 4: 57 (1883)

Hoya ovalifolia Wight & Arn.
Apocynaceae
Climbing Mechanism: Stem Twiner
Distribution (Global): India, Sri Lanka
Distribution (India): Maharashtra
Leaf Type: Simple, opposite, with sap
Inflorescence: Axillary cymes, umbellate
Fruit Type: Dry, follicles
IUCN Status: Not evaluated
Reference: Contr. Bot. India: 37 (1834)

Hoya parasitica Wall. ex Traill
Apocynaceae
Climbing Mechanism: Stem Twiner
Distribution (India): Andaman and Nicobar Islands, Arunachal Pradesh, Odisha, Tripura
Leaf Type: Simple, opposite, with sap
Inflorescence: Axillary cymes, umbellate
Fruit Type: Dry, follicles
IUCN Status: Not evaluated

Hoya parviflora Wight
Apocynaceae
Synonyms: *Hoya variifolia* Ridl.
Climbing Mechanism: Stem Twiner
Distribution (Global): Bangladesh, Malaya, Myanmar, Sumatera, Thailand, Vietnam
Leaf Type: Simple, opposite, with sap
Inflorescence: Axillary cymes, umbellate
Fruit Type: Dry, follicles
IUCN Status: Not evaluated
Reference: Contr. Bot. India: 37 (1834)

Hoya pauciflora Wight
Apocynaceae
Synonyms: *Hoya wightiana* Thwaites
Common Name: South-Indian wax flower
Climbing Mechanism: Stem Twiner
Distribution (Global): India, Sri Lanka
Distribution (India): Arunachal Pradesh, Tamil Nadu
Leaf Type: Simple, opposite, with sap
Inflorescence: Axillary cymes, umbellate
Fruit Type: Dry, follicles

Flowering and Fruiting: November–December
IUCN Status: Not evaluated
Reference: Icon. Pl. Ind. Orient. 4: t. 1356 (1848)

Hoya pendula Wight & Arn.
Apocynaceae
Common Name: Pendulous wax flower
Climbing Mechanism: Stem Twiner
Distribution (India): Andhra Pradesh
Leaf Type: Simple, opposite, with sap
Inflorescence: Axillary cymes, umbellate
Fruit Type: Dry, follicles
IUCN Status: Not evaluated
Reference: Contr. Bot. India: 36 (1834)

Hoya polyneura Hook.f.
Apocynaceae
Climbing Mechanism: Stem Twiner
Distribution (Global): Assam, China South-Central, East Himalaya, Laos, Myanmar, Nepal, Tibet
Distribution (India): Arunachal Pradesh
Leaf Type: Simple, opposite, with sap
Inflorescence: Axillary cymes, umbellate
Fruit Type: Dry, follicles
IUCN Status: Not evaluated
Reference: Fl. Brit. India 4: 54 (1883)

Hoya retusa Warb.
Apocynaceae
Synonyms: *Hoya tsiangiana* P.T.Li
Climbing Mechanism: Stem Twiner
Distribution (India): Maharashtra
Leaf Type: Simple, opposite, with sap
Inflorescence: Axillary cymes, umbellate
Fruit Type: Dry, follicles
IUCN Status: Not evaluated
Reference: Repert. Spec. Nov. Regni Veg. 3: 344 (1907)

Hoya revoluta Wight ex Decne.
Apocynaceae
Climbing Mechanism: Stem Twiner
Distribution (Global): Borneo, Malaya, Sumatera, Thailand
Leaf Type: Simple, opposite, with sap
Inflorescence: Axillary cymes, umbellate
Fruit Type: Dry, follicles

Flowering and Fruiting: November–December
IUCN Status: Not evaluated
Reference: Fl. Brit. India 4: 55 (1883)

Hoya wallichii (Wight) C.M. Burton
Apocynaceae
Synonyms: *Physostelma wallichii* Wight
Climbing Mechanism: Stem Twiner
Distribution (Global): Borneo, Malaya
Leaf Type: Simple, opposite, with sap
Inflorescence: Axillary cymes, umbellate
Fruit Type: Dry, follicles
IUCN Status: Not evaluated
Reference: Hoyan 18(1: 2): 5 (1996)

Hoya wightii Hook.f.
Apocynaceae
Synonyms: *Hoya pallida* Dalzell & A. Gibson
Climbing Mechanism: Stem Twiner
Distribution (Global): India
Distribution (India): Maharashtra, Tamil Nadu
Leaf Type: Simple, opposite, with sap
Inflorescence: Axillary cymes, umbellate
Fruit Type: Dry, follicles
Flowering and Fruiting: March–April
IUCN Status: Not evaluated
Reference: Fl. Brit. India 4: 59 (1883)

Hoya wightii subsp. *palniensis* Thwaites
Apocynaceae
Climbing Mechanism: Stem Twiner
Leaf Type: Simple, opposite, with sap
Inflorescence: Axillary cymes, umbellate
Fruit Type: Dry, follicles
IUCN Status: Not evaluated

Ichnocarpus frutescens (L.) W.T. Aiton
Apocynaceae
Synonyms: *Aganosma affinis* (Roem. & Schult.) G. Don, *Apocynum crassifolium* Salisb., *Apocynum frutescens* L., *Beluttakaka malabarica* Kuntze, *Carruthersia daronensis* Elmer, *Chonemorpha bantamensis* G.Don, *Chonemorpha malabarica* G.Don + 31
Common Name: Black creeper
Climbing Mechanism: Stem Twiner

Distribution (Global): Andaman Is., Assam, Bangladesh, Bismarck Archipelago, Borneo, Cambodia, China South-Central, China Southeast, East Himalaya, Hainan, India, Java, Laos, Lesser Sunda Is., Malaya, Maluku, Myanmar, Nepal, New Guinea, Nicobar Is., Northern Territory, Pakistan, Philippines, Queensland, Sri Lanka, Sulawesi, Sumatera, Thailand, Vietnam, West Himalaya, Western Australia

Distribution (India): Andaman and Nicobar Islands, Andhra Pradesh, Arunachal Pradesh, Bihar, Odisha, Delhi, Eastern Ghats, Himachal Pradesh, Jharkhand, Kerala, Madhya Pradesh, Meghalaya, Mizoram, Odisha, Puducherry, Rajasthan, Tamil Nadu, Telangana

Leaf Type: Simple

Inflorescence: Cymes

Fruit Type: Follicles

Flowering and Fruiting: August–March

IUCN Status: Not evaluated

Reference: Hortus Kew. 2: 69 (1811)

Kamettia caryophyllata (Roxb.) Nicolson & Suresh

Apocynaceae

Synonyms: *Aganosma roxburghii* G. Don, *Echites caryophyllatus* Roxb., *Ellertonia rheedei* Wight, *Ichnocarpus roxburghii* Mottet, *Kamettia malabarica* Kostel

Climbing Mechanism: Stem Twiner

Distribution (Global): India.

Distribution (India): Kerala, Maharashtra, Tamil Nadu

Leaf Type: Leaves 3-4 at nodes, oblong, apex abruptly acuminate, nerves to 14 pairs

Inflorescence: Flowers in terminal peduncled cymes

Fruit Type: Follicles, divaricate, linear, terete

Flowering and Fruiting: September–January

IUCN Status: Not evaluated

Notes: Usually found in evergreen and semi-evergreen forests

Reference: Taxon 35: 354 (1986)

Leptadenia reticulata (Retz.) Wight & Arn.

Apocynaceae

Synonyms: *Asclepias javanica* Burm. ex Decne., *Asclepias suberosa* Roxb., *Asclepias volubilis* Hook.f., *Asclepias zeylanica* Burm. ex Decne., *Curinila rheedei* Schult., *Curnilia sarmentosa* Raf. +10

Common Name: Cork swallow-wort

Climbing Mechanism: Stem Twiner

Distribution (Global): Bangladesh, Cambodia, India, Laccadive Is., Myanmar, Nepal, Sri Lanka

Distribution (India): Andhra Pradesh, Bihar, Odisha, Delhi, Eastern Ghats, Gujarat, Madhya Pradesh, Maharashtra, Odisha, Puducherry, Rajasthan, Tamil Nadu, Telangana, Western Ghats

Leaf Type: Leaves are leathery, ovate, pointed, hairless above, and finely velvety below
Inflorescence: Flowers are greenish-yellow, in lateral or subaxillary cymes, often with small hairs
Fruit Type: Follicles paired, bluntly acute at both ends
Flowering and Fruiting: September–December
IUCN Status: Not evaluated
Notes: Usually found in moist and dry deciduous forests
Reference: R. Wight, Cat. Indian Pl.: 89 (1834)

Lygisma angustifolia (Wight) Hook.f.
Apocynaceae
Climbing Mechanism: Stem Twiner
IUCN Status: Not evaluated

Mandevilla boliviensis (Hook.f.) Woodson
Apocynaceae
Synonyms: *Dipladenia bella* Pittier, *Dipladenia boliviensis* J.J. Veitch, *Mandevilla bella* (Pittier) Woodson, *Mandevilla cereola* Woodson, *Mandevilla permixta* Woodson, *Mandevilla pittieri* Woodson
Common Name: Dipladenia white, Mandevilla white
Climbing Mechanism: Scrambler-Unarmed
Distribution (Global): Bolivia, Brazil Northeast, Brazil Southeast, Colombia, Costa Rica, Ecuador, Peru, Venezuela
Distribution (India): Peninsular India
Leaf Type: Shining green oblong, slender pointed leaves
Inflorescence: orange-yellow throat in axillary recemes
Fruit Type: Follicles
Flowering and Fruiting: October–March
IUCN Status: Not evaluated
Reference: Ann. Missouri Bot. Gard. 20: 716 (1933)

Mandevilla laxa (Ruiz & Pav.) Woodson
Apocynaceae
Synonyms: *Amblyanthera bridgesii* Müll.Arg., *Amblyanthera suaveolens* (Lindl.) Müll.Arg., *Echites glandulosus* Poir., *Echites laxus* Ruiz & Pav., *Echites suaveolens* (Lindl.) A.DC., *Mandevilla bangii* Rusby, *Mandevilla bridgesii* (Müll.Arg.) Woodson
Common Name: Chilean jasmine
Climbing Mechanism: Scrambler-Unarmed
Distribution (Global): Argentina Northeast, Argentina Northwest, Bolivia, Peru
Distribution (India): Tamil Nadu
IUCN Status: Not evaluated
Reference: Ann. Missouri Bot. Gard. 19: 68 (1932)

Marsdenia floribunda (Brongn.) Schltr.
Apocynaceae
Synonyms: *Stephanotis floribunda* Jacques. *Ceropegia stephanotis* Schult., *Isaura alliacea* Steud., *Isaura allicia* Comm. ex Poir., *Marsdenia floribunda* (Jacques) Schltr., *Marsdenia isaura* Choux, *Stephanotis isaura* Decne
Common Name: Madagascar jasmine, Bridal bouquet, wax flower
Climbing Mechanism: Stem Twiner
Distribution (Global): Madagascar
Leaf Type: leathery oval leaves opposite, and clusters of pure white, waxy
Flowering and Fruiting: Throughout the year.
IUCN Status: Vulnerable
Reference: Ann. Soc. Hort. Paris 15: 28 (1834)

Marsdenia lucida Edgew. ex Madden
Apocynaceae
Common Name: Shining Marsdenia
Climbing Mechanism: Stem Twiner
Distribution (Global): East Himalaya, India, Nepal, West Himalaya
Distribution (India): Himachal Pradesh
Leaf Type: Leaves are elliptic or ovate pointed or bluntly tapering
Inflorescence: Flower-cymes corymb-like, very shortly stalked. Flowers are borne in many-flowered corymbs. Flower-cluster-stalk stout
IUCN Status: Not evaluated
Reference: J. Asiat. Soc. Bengal 17(1): 370 (1848)

Marsdenia maingayi (Hook.f.) P.I. Forst.
Apocynaceae
Synonyms: *Jasminanthes maingayi* (Hook.f.) Rodda., *Stephanotis maingayi* Hook.f.
Climbing Mechanism: Stem Twiner
Distribution (Global): Borneo, Malaya
Leaf Type: Leaf lamina elliptic to ovate, chartaceous
Inflorescence: Inflorescence extra-axillary, racemiform
Fruit Type: A single follicle, cylindrical-oblong
IUCN Status: Not evaluated
Notes: Usually found along the forest margin and open forest gaps of disturbed lowland forests
Reference: Phytotaxa 405: 137 (2019)

Marsdenia raziana Yogan. & Subr.
Apocynaceae
Climbing Mechanism: Stem Twiner
Distribution (Global): India
Distribution (India): Peninsular India
Leaf Type: Leaves opposite, oblong, acuminate, base rounded or acute, membranous, glabrous, and glossy

Inflorescence: Flowers brownish-pink, in 10–15 flowered umbellate racemes
Flowering and Fruiting: January–March
IUCN Status: Not evaluated
Reference: Proc. Indian Acad. Sci., B 83: 147 (1976)

Marsdenia roylei Wight
Apocynaceae
Synonyms: *Pergularia roylei* (Wight) D. Dietr.
Common Name: Royle's Pergularia
Climbing Mechanism: Stem Twiner
Distribution (Global): Assam, Bangladesh, East Himalaya, India, Myanmar, Nepal, Pakistan, West Himalaya
Distribution (India): Arunachal Pradesh, Assam, Himachal Pradesh, West Bengal
Flowering and Fruiting: June–August
IUCN Status: Not evaluated
Reference: Contr. Bot. India: 40 (1834)

Marsdenia tenacissima (Roxb.) Moon
Apocynaceae
Synonyms: *Asclepias echinata* Hook.f.
Asclepias tenacissima Roxb., *Asclepias tomentosa* Hook.f., *Gymnema tenacissimum* (Roxb.) Spreng., *Pergularia tenacissima* (Roxb.) D.Dietr., *Pergularia tomentosa* Span
Common Name: Rajmahal hemp, Devil's tongue, Bush banana
Climbing Mechanism: Stem Twiner
Distribution (Global): Assam, Bangladesh, Cambodia, China South-Central, East Himalaya, India, Laos, Myanmar, Nepal, Sri Lanka, Thailand, Vietnam, West Himalaya
Distribution (India): Andhra Pradesh, Bihar, Odisha, Eastern Ghats, Gujarat, Madhya Pradesh, Maharashtra, Odisha, Rajasthan, Tamil Nadu, Telangana, West Bengal
Leaf Type: eaves ovate, base deeply heart shaped with rounded sinus, tip long-pointed, basal veins 5–7, lateral veins 2 or 3 pairs
Inflorescence: Flower-clusters cymes are much branched, broader than long, many flowered. Clusters are carried on stalks
Fruit Type: Merocarps velvety, pubescent
Flowering and Fruiting: August–December
IUCN Status: Not evaluated
Notes: Usually found in dry deciduous forests of the Western Ghats and the Eastern Ghats
Reference: Cat. Pl. Ceylon: 21 (1824)

Marsdenia thyrsiflora Hook.f.
Apocynaceae
Synonyms: *Marsdenia tinctoria* Hook.f. & Thomson

Climbing Mechanism: Stem Twiner
Distribution (Global): Assam, Bangladesh, Cambodia, East Himalaya, Malaya, Thailand
Distribution (India): West Bengal
IUCN Status: Not evaluated
Reference: Fl. Brit. India 4: 37 (1883)

Marsdenia tinctoria R.Br.
Apocynaceae
Synonyms: *Asclepias tinctoria* (R.Br.) Roxb., *Cynanchum tingens* Buch. -Ham. ex Hook.f.
Marsdenia akkar Blanco, *Marsdenia globifera* Tsiang, *Marsdenia monostachya* Wall. ex Hook.f.
Marsdenia tagudinia Blanco, *Marsdenia tinctoria* var. *tomentosa* Masam. ex Tsiang
Climbing Mechanism: Stem Twiner
Distribution (Global): Assam, Bangladesh, Borneo, Cambodia, China South-Central, China Southeast, East Himalaya, Hainan, Japan, Java, Laos, Malaya, Myanmar, Nansei-shoto, Nepal, Philippines, Sri Lanka, Taiwan, Thailand, Tibet, Vietnam
Distribution (India): Arunachal Pradesh, Maharashtra, West Bengal
Leaf Type: Leaf blade oblong to broadly ovate, thin papery, bluish when dry, base rounded to cordate
Inflorescence: Inflorescences dense, globose to long racemelike clusters of cymules
Fruit Type: Follicles oblong-lanceolate
Flowering and Fruiting: March-December
IUCN Status: Not evaluated
Reference: Asclepiadeae: 17 (1810)

Marsdenia tirunelvelica A.N. Henry & Subr.
Apocynaceae
Climbing Mechanism: Stem Twiner
Distribution (Global): India
Distribution (India): Tamil Nadu
Leaf Type: Leaves, elliptic, acuminate, round to subcordate at base, glaucous beneath
Inflorescence: Flowers creamy, in short peduncled umbels
Flowering and Fruiting: March–April
IUCN Status: Not evaluated
Notes: Often occur in evergreen forests
Reference: J. Bombay Nat. Hist. Soc. 73: 186 (1976)

Metastelma parviflorum
Apocynaceae
Climbing Mechanism: Stem Twiner
IUCN Status: Not evaluated

Micrechites andamanicus M. Gangop. & Chakrab.
Apocynaceae
Synonyms: *Ichnocarpus andamanica* (M. Gangop. & Chakrab.) Rasingam, Diwakar & R.P. Pandey
Climbing Mechanism: Stem Twiner
Distribution (Global): Andaman Is.
IUCN Status: Not evaluated
Reference: J. Econ. Taxon. Bot. 16: 42 (1992)

Micrechites parkinsonii M. Gangop. & Chakrab.
Apocynaceae
Climbing Mechanism: Stem Twiner
IUCN Status: Not evaluated

Micrechites polyanthus (Blume) Miq.
Apocynaceae
Climbing Mechanism: Scrambler-Unarmed
IUCN Status: Not evaluated

Myriopteron extensum (Wight & Arn.) K. Schum.
Apocynaceae
Climbing Mechanism: Scrambler-Unarmed
Distribution (India): Assam, Mizoram, West Bengal
IUCN Status: Not evaluated

Odontadenia macrantha (Roem. & Schult.) Markgr.
Apocynaceae
Synonyms: *Odontadenia speciosa* Benth
Common Name: Odontadenia
Climbing Mechanism: Stem Twiner
Distribution (Global): Tropical America
Distribution (India): Peninsular India
Flowering and Fruiting: October–January
IUCN Status: Not evaluated
Reference: Meded. Afd. Handelsmus. Kolon. Inst. 30: 461 (1937)

Odontadenia semidigyna (P.J. Bergius) J.F. Morales
Apocynaceae
Synonyms: *Angadenia sylvestris* (A.DC.) Miers, *Cylicadenia harrisii* (Hook.) Lem., *Dipladenia brearleyana* T. Moore. + 21
Climbing Mechanism: Stem Twiner
Distribution (Global): Belize, Brazil North, Brazil West-Central, Colombia, Costa Rica, Ecuador, French Guiana, Guatemala, Guyana, Honduras, Mexico Southeast, Mexico Southwest, Nicaragua, Panamá, Peru, Suriname, Trinidad-Tobago, Venezuela

Distribution (India): Tamil Nadu
IUCN Status: Not evaluated
Reference: Novon 26: 162 (2018)

Oianthus beddomei Hook.f.
Apocynaceae
Synonyms: *Oianthus beddomei* Hook.f.
Climbing Mechanism: Stem Twiner
Distribution (Global): India
Distribution (India): Peninsular India
Flowering and Fruiting: August–September
IUCN Status: Not evaluated
Reference: Bot. J. Linn. Soc. 101: 254 (1989)

Oianthus disciflorus Hook.f.
Apocynaceae
Synonyms: *Oianthus disciflorus* Hook.f.
Climbing Mechanism: Stem Twiner
Distribution (Global): India
Distribution (India): Andhra Pradesh, Maharashtra
Flowering and Fruiting: July–October
IUCN Status: Not evaluated
Reference: Bot. J. Linn. Soc. 101: 256 (1989)

Oianthus urceolatus Benth.
Apocynaceae
Synonyms: *Heterostemma urceolatum* Dalzell
Climbing Mechanism: Stem Twiner
Distribution (Global): India
Distribution (India): Maharashtra
Flowering and Fruiting: July–August
IUCN Status: Not evaluated
Reference: Hooker's Icon. Pl. 12: t. 1191 (1876)

Oxystelma esculentum (L.f.) Sm.
Apocynaceae
Synonyms: *Asclepias rosea* Roxb., *Marsdenia brockmaniana* W. Fitzg. + 8
Climbing Mechanism: Stem Twiner
Distribution (Global): Egypt to NE. Tanzania, SW. Yemen, Sinai to S. China and W. Malesia, N. Australia
Distribution (India): Andhra Pradesh, Bihar, Odisha, Madhya Pradesh, Maharashtra, Mizoram, Odisha, Rajasthan, Telangana, West Bengal, Western Ghats
Flowering and Fruiting: August–April
IUCN Status: Least concern

Notes: Found mostly in scrub jungles
Reference: A. Rees, Cycl. 25: no. 1 (1813)

Oxystelma secamone H. Karst.
Apocynaceae
Climbing Mechanism: Stem Twiner
Distribution (India): Delhi, Gujarat, Tamil Nadu, Uttar Pradesh
IUCN Status: Not evaluated
Reference: Deut. Fl.: 1031 (1883)

Parameria laevigata (Juss.) Moldenke
Apocynaceae
Common Name: Smooth Parameria
Climbing Mechanism: Scrambler-Unarmed
Distribution (India): Andaman and Nicobar Islands
IUCN Status: Least concern
Reference: First published in Revista Sudamer. Bot. 6: 176 (1940)

Parameria polyneura Hook.f.
Apocynaceae
Synonyms: *Micrechites ovalifolia* Ridl., *Parameria griffithii* Pierre.
Climbing Mechanism: Scrambler-Unarmed
IUCN Status: Not evaluated
Reference: Fl. Brit. India 3: 660 (1882)

Parsonsia alboflavescens (Dennst.) Mabb.
Apocynaceae
Synonyms: *Aganosma concanensis* Hook., *Aganosma laevigata* (Moon) J. Graham, *Apocynum reticulatum* L. + 39
Common Name: Spiral vined silkpod
Climbing Mechanism: Stem Twiner
Distribution (Global): Andaman Is., Assam, Bangladesh, Bismarck Archipelago, Borneo, Cambodia, China Southeast, Hainan, India, Java, Laos, Malaya, Maluku, Myanmar, Nansei-shoto, New Guinea, Nicobar Is., Northern Territory, Philippines, Solomon Is., Sri Lanka, Sulawesi, Sumatera, Taiwan, Thailand, Vietnam
Distribution (India): Andaman and Nicobar Islands, Arunachal Pradesh, Maharashtra, Mizoram, Tamil Nadu, West Bengal, Western Ghats
Flowering and Fruiting: March–June
IUCN Status: Not evaluated
Notes: Distributed mostly in semi-evergreen and evergreen forests
Reference: Taxon 26: 532 (1977)

Parsonsia inodora (Lour.) M.R. Almeida & S.M. Almeida
Apocynaceae

Synonyms: *Gymnema inodorum* (Lour.) Decne., *Aganosma inodora* Lour., *Asclepias daemia* Blanco, *Bidaria elegans* Dalz. & Gibs., *Bidaria inodora* (Lour.) Decne., *Cynanchum inodorum* Lour., *Heligme rheedii* Náves, *Parsonsia rheedii* F. Villar.
Climbing Mechanism: Stem Twiner
Distribution (India): Andaman and Nicobar Islands, Goa
Flowering and Fruiting: March–April
IUCN Status: Not evaluated
Reference: J. Bombay Nat. Hist. Soc. 90: 428 (1993 publ. 1994)

Pentalinon luteum (L.) B.F. Hansen & Wunderlin
Apocynaceae
Synonyms: *Angadenia jamaicensis* (Griseb.) Lippold, *Apocynum speciosissimum* Mill., *Asketanthera obtusifolia* Alain, *Chariomma domingensis* (Jacq.) Miers + 40
Common Name: Hammock viper's tail, Lice bush, Wild Allamanda, Yellow mandevilla
Climbing Mechanism: Scrambler-Unarmed
Distribution (India): Tamil Nadu
Flowering and Fruiting: Throughout the year
IUCN Status: Not evaluated
Reference: Taxon 35: 167 (1986)

Pentanura khasiana Kurz
Apocynaceae
Climbing Mechanism: Stem Twiner
Distribution (India): Arunachal Pradesh
IUCN Status: Not evaluated
Reference: Forest Fl. Burma 2: 196 (1877)

Pentasachme fasciculatum (Buch. -Ham. ex Wight) M.R. Almeida
Apocynaceae
Climbing Mechanism: Stem Twiner
Distribution (India): Telangana
IUCN Status: Not evaluated

Pentatropis capensis (L.f.) Bullock
Apocynaceae
Synonyms: *Asclepias microphylla* Roth, *Colostephanus capensis* (L.f.) Harv., *Cynanchum acuminatum* Thunb., *Cynanchum capense* L.f., *Cynoctonum capense* (L.f.) E. Mey., *Pentatropis microphylla* (Roth) Wight & Arn., *Vincetoxicum capense* (L.f.) Kuntze
Common Name: Salt killer vine
Climbing Mechanism: Stem Twiner
Distribution (Global): Bangladesh, India, Pakistan, Sri Lanka

Distribution (India): Andhra Pradesh, Gujarat, Maharashtra, Odisha, Tamil Nadu, Telangana, West Bengal
Flowering and Fruiting: March–November
IUCN Status: Not evaluated
Notes: Mostly found in deciduous forests
Reference: Kew Bull. 10: 284 (1955)

Pentatropis cynanchoides R.Br.
Apocynaceae
Climbing Mechanism: Stem Twiner
Distribution (Global): Afghanistan, Burkina, Chad, Comoros, Djibouti, Egypt, Eritrea, Ethiopia, Guinea, Gulf States, India, Iran, Kenya, Madagascar, Mali, Mauritania, Nigeria, Oman, Pakistan, Palestine, Saudi Arabia, Senegal, Sinai, Somalia, Sudan, Tanzania, Uganda, Yemen
IUCN Status: Not evaluated
Reference: D. Oliver & auct. suc. (eds.), Fl. Trop. Afr. 4: 380 (1902)
First published in D. Oliver & auct. suc. (eds.), Fl. Trop. Afr. 4: 380 (1902)

Pentatropis nivalis (J.F. Gmel.) D.V. Field & J.R.I. Wood
Apocynaceae
Synonyms: *Asclepias forsskalii* Schult., *Asclepias nivalis* J.F. Gmel., *Asclepias nivea* Forssk., *Oxystelma caudatum* Buch.-Ham. ex Wall. + 10
Common Name: White milkweed
Climbing Mechanism: Stem Twiner
Distribution (India): Maharashtra
Leaf Type: Variable leaves
Inflorescence: Cyme
Flowering and Fruiting: January–October
IUCN Status: Not evaluated
Notes: Frequently found on the hedges of cultivated fields
Reference: Kew Bull. 38: 215 (1983)

Pergularia brunoniana (Wight & Arn.) D.Dietr.
Apocynaceae
Climbing Mechanism: Stem Twiner
IUCN Status: Not evaluated
Reference: Syn. Plant. 2: 894 (1840)

Pergularia calesiana (Wight) Buch. -Ham. ex D. Dietr.
Apocynaceae
Climbing Mechanism: Stem Twiner
IUCN Status: Not evaluated
Reference: Fl. Brit. India 4: 37 (1883)

Pergularia daemia (Forssk.) Chiov.
Apocynaceae
Synonyms: *Asclepias daemia* Forssk., *Doemia forsskalii* Schult
Common Name: Trellis Vine
Climbing Mechanism: Stem Twiner
Distribution (Global): Afghanistan, Andaman Is., Angola, Assam, Bangladesh, Benin, Botswana, Burkina, Burundi, Cameroon, Cape Provinces, Caprivi Strip, Central African Repu, Chad, Djibouti, Equatorial Guinea, Eritrea, Ethiopia, Gambia, Ghana, Guinea, Guinea-Bissau, Gulf of Guinea Is., India, Iran, Ivory Coast, Kenya, Liberia, Malawi, Mali, Mauritania, Mozambique, Myanmar, Namibia, Nepal, Niger, Nigeria, Oman, Pakistan, Rwanda, Saudi Arabia, Senegal, Sinai, Somalia, Sri Lanka, Sudan, Tanzania, Uganda, West Himalaya, Yemen, Zambia, Zaire, Zimbabwe
Distribution (India): Andhra Pradesh, Assam, Delhi, Gujarat, Himachal Pradesh, Kerala, Madhya Pradesh, Maharashtra, Odisha, Rajasthan, Tamil Nadu, Telangana, Tripura, Uttar Pradesh, West Bengal
Leaf Type: Simple, opposite decussate
Inflorescence: Cyme
Fruit Type: Follicle
Flowering and Fruiting: August–March
IUCN Status: Least concern
Notes: Frequently found in scrub and open deciduous forests
Reference: Res. Sci. Somalia Ital. 1: 115 (1916)

Pergularia exilis (Colebr.) Spreng.
Apocynaceae
Climbing Mechanism: Stem Twiner
IUCN Status: Not evaluated
Reference: Syst. Veg., 1: 844 (1824)

Pergularia hamiltonii (Wight) D. Dietr.
Apocynaceae
Climbing Mechanism: Stem Twiner
Flowering and Fruiting: April–July
IUCN Status: Not evaluated
Reference: Syn. Plant. 2: 894 (1840)

Pergularia odoratissima Sm.
Apocynaceae
Climbing Mechanism: Stem Twiner
IUCN Status: Not evaluated
Reference: Icon. Pict. Pl. Rar. 3: t. 16 (1793)

Periploca calophylla (Wight) Falc.
Apocynaceae

Synonyms: *Streptocaulon calophyllum* Wight
Climbing Mechanism: Stem Twiner
Distribution (Global): Assam, China South-Central, China Southeast, East Himalaya, India, Myanmar, Nepal, Tibet, Vietnam, West Himalaya
Distribution (India): Arunachal Pradesh, Himachal Pradesh, Meghalaya, Sikkim, West Bengal
IUCN Status: Not evaluated
Reference: Proc. Linn. Soc. London 1: 115 (1841)

Pottsia laxiflora (Blume) Kuntze
Apocynaceae
Synonyms: *Euthodon paniculata* Griff., *Parapottsia hookeriana* (Wight) Miq., *Parapottsia laxiflora* (Blume) Miq., *Parsonsia ovata* Wall. ex G. Don, *Pottsia cantonensis* Hook. & Arn., *Pottsia hookeriana* Wight, *Pottsia inodora* Pit. + 6
Climbing Mechanism: Stem Twiner
Distribution (Global): Assam, Bangladesh, Cambodia, China South-Central, China Southeast, Hainan, Java, Laos, Lesser Sunda Is., Malaya, Myanmar, Sumatera, Thailand, Vietnam
Distribution (India): Arunachal Pradesh, Meghalaya
IUCN Status: Not evaluated
Reference: Revis. Gen. Pl. 2: 416 (1891)

Pycnorhachis maingayi Hook.f.
Apocynaceae
Synonyms: *Pycnorhachis benthamiana* Baill.
Climbing Mechanism: Stem Twiner
Distribution (Global): Malaya
IUCN Status: Not evaluated
Reference: Fl. Brit. India 4: 63 (1883)

Raphistemma pulchellum (Roxb.) Wall.
Apocynaceae
Synonyms: *Asclepias pulchella* Roxb., *Oxystelma pulchellum* (Roxb.) D. Dietr., *Pergularia campanulata* Buch. -Ham. ex Wall.
Climbing Mechanism: Stem Twiner
Distribution (India): Arunachal Pradesh, Bihar, Odisha, West Bengal
Leaf Type: Simple, ovate, glabrous, deeply cordate at base
Inflorescence: Peduncles
Fruit Type: Dry, follicles
Flowering and Fruiting: September–October
IUCN Status: Not evaluated
Reference: Pl. Asiat. Rar. 2: 50 (1831)

Sarcolobus carinatus Griff.
Apocynaceae

Synonyms: *Sarcolobus virulentus* Griff.
Climbing Mechanism: Stem Twiner
Distribution (Global): Andaman Is., Bangladesh, India, Myanmar, Nicobar Is., Thailand
Distribution (India): Andaman and Nicobar Islands, Andhra Pradesh, Bihar, Odisha, West Bengal
Leaf Type: Simple, opposite
Inflorescence: Corymbose cymes
Fruit Type: Dry, follicles
Flowering and Fruiting: March–September
IUCN Status: Not evaluated
Reference: Asiat. Res. 12: 570 (1816)

Sarcolobus globosus Wall.
Apocynaceae
Synonyms: *Sarcolobus globosus* subsp. *globosus.*, *Sarcolobus globosus* subsp. *peregrinus* (Blanco) Rintz.
Climbing Mechanism: Stem Twiner
Distribution (Global): Andaman Is., Bangladesh, Borneo, India, Java, Laos, Lesser Sunda Is., Malaya, Maluku, Myanmar, New Guinea, Nicobar Is., Philippines, Sulawesi, Sumatera, Thailand, Vietnam
Distribution (India): Great Nicobar Island, Odisha, West Bengal
Leaf Type: Simple, opposite
Inflorescence: Corymbose cymes
Fruit Type: Dry, follicles
IUCN Status: Not evaluated
Reference: Asiat. Res. 12: 568 (1816)

Sarcostemma acidum (Roxb.) Voigt
Apocynaceae
Synonyms: *Asclepias acida* Roxb., *Asclepias aphylla* Roxb. ex Wight., *Sarcostemma brevistigma* Wight & Arn., *Cynanchum acidum* (Roxb.) Oken.
Common Name: Leafless East-Indian vine
Climbing Mechanism: Scrambler-Unarmed
Distribution (Global): Bangladesh, China Southeast, Hainan, India, Myanmar, Nepal, Pakistan, Thailand, Vietnam
Distribution (India): Great Nicobar Island, Odisha, West Bengal
Leaf Type: Leafless
Inflorescence: cymes, borne at branch ends or in axils, white-yellowish flowers
Fruit Type: Lance-shaped pods
Flowering and Fruiting: March–November
IUCN Status: Not evaluated
Reference: Allg. Naturgesch. 3(2): 1032 (1841)

Sarcostemma intermedium Decne.
Apocynaceae
Synonyms: *Cynanchum sarcomedium* Meve & Liede
Common Name: Leafless climbing milkweed
Climbing Mechanism: Scrambler-Unarmed
Distribution (Global): India
Distribution (India): Gujarat, Puducherry, Rajasthan, Telangana
Leaf Type: Leafless
Inflorescence: umbels at branch-ends, 5–15 flowered
Fruit Type: Smooth and long, winged seeds
Flowering and Fruiting: August–December
IUCN Status: Not evaluated
Reference: Kew Bull. 67: 753 (2012)

Sarcostemma secamone (L.) Bennett
Apocynaceae
Synonyms: *Oxystelma secamone* (L.) H. Karst., *Periploca secamone* Thunb., *Periploca secamone* L.
Secamone thunbergii E. Mey., *Secamone alpini* Schult.
Climbing Mechanism: Scrambler-Unarmed
Distribution (Global): Cape Provinces, Free State, Kenya, KwaZulu-Natal, Malawi, Mozambique, Northern Provinces, Swaziland, Tanzania, Uganda, Zambia, Zimbabwe
Distribution (India): Arunachal Pradesh, Telangana
Leaf Type: Simple, opposite leaves
Inflorescence: Paired, axillary, solitary, or in lax racemes
Fruit Type: Follicles, inflated
Flowering and Fruiting: September–April
IUCN Status: Not evaluated
Notes: Found usually along banks of rivers and streams
Reference: J.J. Roemer & J.A.Schultes, Syst. Veg., ed. 15 bis 6: 125 (1820)

Sarcostemma stocksii Hook.f.
Apocynaceae
Synonyms: *Sarcostemma stocksii* Hook.f., *Sarcostemma viminale* subsp. *stocksii* (Hook.f.) Ali.
Climbing Mechanism: Scrambler-Unarmed
Distribution (Global): India, Pakistan
Distribution (India): Peninsular India
Leaf Type: Leaves reduced
Inflorescence: Umbels
Fruit Type: Dry, follicles
IUCN Status: Not evaluated
Reference: Kew Bull. 67: 754 (2012)

Sarcostemma viminale subsp. *brunonianum* (Wight & Arn.) P.I. Forst
Apocynaceae
Synonyms: *Sarcostemma brunonianum* Wight & Arn., *Cynanchum viminale* subsp. *brunonianum* (Wight & Arn.) Meve & Liede
Climbing Mechanism: Scrambler-Unarmed
Distribution (Global): India, Malaya, Myanmar, New Caledonia, New South Wales, Northern Territory, Philippines, Queensland, Sri Lanka, Thailand, Western Australia
Distribution (India): Maharashtra, Rajasthan, Telangana, Western Ghats
Leaf Type: Leafless
Inflorescence: Umbels lateral, 10–15 flowered, green
Fruit Type: Follicles
Flowering and Fruiting: March–June
IUCN Status: Not evaluated
Notes: Found in dry deciduous forests of the Western Ghats and the Eastern Ghats
Reference: Kew Bull. 67: 754 (2012)

Secamone andamanica Goel & Vasudeva Rao
Apocynaceae
Climbing Mechanism: Stem Twiner
IUCN Status: Not evaluated
Reference: J. Bombay Nat. Hist. Soc. 85: 161 (1988)

Secamone emetica (Retz.) R.Br. ex Schult.
Apocynaceae
Synonyms: *Periploca emetica* Retz.
Climbing Mechanism: Stem Twiner
Distribution (Global): India, Sri Lanka
Distribution (India): Andhra Pradesh, Eastern Ghats, Gujarat, Kerala, Puducherry, Tamil Nadu, Telangana
Leaf Type: Simple, opposite
Inflorescence: Axillary cyme, trichotomously branched
Fruit Type: Paired follicles
Flowering and Fruiting: August–March
IUCN Status: Not evaluated
Notes: Mostly found in deciduous forests and scrub jungles
Reference: A. Rees, Cycl. 32: no. 2 (1815)

Seshagiria sahyadrica Ansari & Hemadri
Apocynaceae
Synonyms: *Cynanchum sahyadricum* (Ansari & Hemadri) Liede & Khanum
Climbing Mechanism: Stem Twiner
Distribution (Global): India, Laccadive Is.
Distribution (India): Maharashtra
Flowering and Fruiting: June–September

IUCN Status: Not evaluated
Reference: Taxon 65: 480 (2016)

Stelmacrypton khasianum Baill.
Apocynaceae
Synonyms: *Decalepis khasiana* (Kurz) Ionta ex Kambale., *Finlaysonia khasiana* (Kurz) Venter., *Hanghomia marseillei* Gagnep. & Thénint., *Pentanura khasiana* Kurz, *Periploca acuminata* M.A. Rahman & Wilcock., *Periploca khasiana* (Kurz) Benth. ex Hook.f.
Common Name: Khasi swallow root
Climbing Mechanism: Stem Twiner
Distribution (Global): Assam, Bangladesh, China South-Central, China Southeast, East Himalaya, Laos, Myanmar
Distribution (India): Mizoram, Tripura
Leaf Type: Simple, ovate
Inflorescence: Simple, single
Fruit Type: Dry, follicles
Flowering and Fruiting: May–September
IUCN Status: Not evaluated
Reference: Taxon 65: 876 (2016)

Streptocaulon juventas (Lour.) Merr.
Apocynaceae
Synonyms: *Apocynum juventas* Lour., *Streptocaulon griffithii* Hook.f., *Streptocaulon tomentosum* Wight., *Tylophora juventas* (Lour.) Woodson
Climbing Mechanism: Stem Twiner
Distribution (Global): Cambodia, China South-Central, China Southeast, Laos, Malaya, Myanmar, Thailand, Vietnam
Leaf Type: Simple
Inflorescence: Thyrsoid
Fruit Type: Dry, follicles
IUCN Status: Not evaluated
Reference: Trans. Amer. Philos. Soc., n.s., 24(2): 315 (1935)

Streptocaulon kleinii Wight & Arn.
Apocynaceae
Climbing Mechanism: Stem Twiner
Distribution (Global): Cambodia, India, Laos, Thailand, Vietnam
Distribution (India): Tamil Nadu
Leaf Type: Simple
Inflorescence: Thyrsoid
Fruit Type: Dry, follicles
IUCN Status: Not evaluated
Reference: R. Wight, Contr. Bot. India: 65 (1834)

Streptocaulon sylvestre Wight
Apocynaceae
Climbing Mechanism: Stem Twiner
Distribution (Global): Assam, Bangladesh, East Himalaya, India
Distribution (India): Bihar, Odisha
Leaf Type: Simple
Inflorescence: Thyrsoid
Fruit Type: Dry, follicles
IUCN Status: Not evaluated
Reference: Contr. Bot. India: 65 (1834)

Streptocaulon wallichii Wight
Apocynaceae
Synonyms: *Periploca calumpitensis* Llanos., *Streptocaulon obtusum* Turcz., *Streptocaulon baumii* Decne.
Climbing Mechanism: Stem Twiner
Distribution (Global): Cambodia, Malaya, Philippines, Thailand, Vietnam
Leaf Type: Simple
Inflorescence: Thyrsoid
Fruit Type: Dry, follicles
IUCN Status: Not evaluated
Reference: Contr. Bot. India: 65 (1834)

Strophanthus caudatus (L.) Kurz
Apocynaceae
Synonyms: *Apocynum floristratum* Noronha., *Echites caudatus* L., *Strophanthus griffithii* Wight., *Strophanthus horsfieldianus* Miq. + 12
Climbing Mechanism: Scrambler-Unarmed
Distribution (Global): Andaman Is., Borneo, Cambodia, China Southeast, Java, Laos, Lesser Sunda Is., Malaya, Maluku, Myanmar, New Guinea, Nicobar Is., Philippines, Sulawesi, Sumatera, Thailand, Vietnam
Distribution (India): Andaman and Nicobar Islands, Maharashtra
Leaf Type: Single, opposite
Inflorescence: Clusters on long stalks
Fruit Type: Dry, follicles
IUCN Status: Not evaluated
Reference: J. Asiat. Soc. Bengal, Pt. 2, Nat. Hist. 46: 257 (1877)

Strophanthus gratus (Wall. & Hook.) Baill.
Apocynaceae
Synonyms: *Strophanthus pierreanus* De Wild
Climbing Mechanism: Scrambler-Unarmed
Distribution (Global): Cameroon, Gabon, Nigeria
Distribution (India): Tamil Nadu
Leaf Type: Single, opposite

Inflorescence: Clusters on long stalks
Fruit Type: Dry, follicles
IUCN Status: Not evaluated
Reference: J. Bot. (Morot) 7: 299 (1893)

Strophanthus singaporianus (Wall. ex G. Don) Gilg
Apocynaceae
Synonyms: *Cercocoma singaporiana* Wall. ex G. Don., *Strophanthus brevicaudatus* Wight.
Climbing Mechanism: Scrambler-Unarmed
Distribution (Global): Borneo, Malaya
Leaf Type: Single, opposite
Inflorescence: Clusters on long stalks
Fruit Type: Dry, follicles
IUCN Status: Not evaluated
Reference: H.G.A. Engler, Monogr. Afrik. Pflanzen-Fam. 7: 11 (1903)

Strophanthus wallichii A. DC.
Apocynaceae
Synonyms: *Nerium caudatum* Roxb., *Strophanthus caudatus* Kurz., *Strophanthus robustus* (Pierre ex Gilg) Pierre ex Pit.
Climbing Mechanism: Scrambler-Unarmed
Distribution (Global): Andaman Is., Assam, Bangladesh, China South-Central, East Himalaya, India, Laos, Malaya, Myanmar, Thailand, Vietnam
Distribution (India): Andaman and Nicobar Islands, Arunachal Pradesh, Bihar, Odisha, Meghalaya, Odisha, Tripura, West Bengal
Leaf Type: Single, opposite
Inflorescence: Clusters on long stalks
Fruit Type: Dry, follicles
Flowering and Fruiting: April–January
IUCN Status: Not evaluated
Reference: A.P.de Candolle, Prodr. 8: 418 (1844)

Strophanthus wightianus Wall. ex Wight
Apocynaceae
Common Name: Wight's twisted flower
Climbing Mechanism: Scrambler-Unarmed
Distribution (Global): India
Distribution (India): Tamil Nadu
Leaf Type: Elliptic-Ovate
Inflorescence: Branched Cymes
Fruit Type: Follicles, stuot
Flowering and Fruiting: March–January
IUCN Status: Not evaluated

Notes: Found mostly in scrub jungles and sacred groves along the sea coasts
Reference: Icon. Pl. Ind. Orient. 4: t. 1301 (1848)

Telosma cordata (Burm.f.) Merr.
Apocynaceae
Synonyms: *Apocynum odoratissimum* Lour. ex Pritz., *Pergularia minor* Andrews., *Oxystelma ovatum* P.T. Li & S.Z. Huang. + 10
Common Name: Chinese violet, Fragrant Telosma, Pakalana vine, Cowslip creeper, Tonkin creeper
Climbing Mechanism: Stem Twiner
Distribution (Global): Bangladesh, Cambodia, China Southeast, India, Laos, Myanmar, Pakistan, Thailand, Vietnam, West Himalaya
Distribution (India): Maharashtra, Rajasthan, Tripura
Leaf Type: Heart shaped, opposite
Inflorescence: Cymes
Flowering and Fruiting: May–December
IUCN Status: Not evaluated
Reference: Philipp. J. Sci. 19: 372 (1921)

Telosma pallida (Roxb.) W.G. Craib
Apocynaceae
Synonyms: *Asclepias pallida* Roxb., *Pergularia montana* Decne., *Pergularia coromandeliana* Decne., *Pergularia pallida* (Roxb.) Wight & Arn.
Common Name: Telosma vine
Climbing Mechanism: Stem Twiner
Distribution (Global): Assam, Bangladesh, East Himalaya, India, Myanmar, Nepal, Pakistan, Taiwan, Thailand, Vietnam, West Himalaya
Distribution (India): Andhra Pradesh, Bihar, Odisha, Delhi, Gujarat, Madhya Pradesh, Maharashtra, Odisha, Tamil Nadu, Telangana, Uttar Pradesh, West Bengal
Flowering and Fruiting: May–December
IUCN Status: Not evaluated
Notes: Often found in scrub jungles
Reference: Bull. Misc. Inform. Kew 1911: 418 (1911)

Telosma puberula (Miq.) Kerr
Apocynaceae
Synonyms: *Pergularia puberula* Miq.
Climbing Mechanism: Stem Twiner
Distribution (Global): Java, Malaya, Thailand
IUCN Status: Not evaluated
Reference: W.G. Craib, Fl. Siam. 3(1): 32 (1951)

Toxocarpus beddomei Gamble
Apocynaceae

Climbing Mechanism: Stem Twiner
Distribution (Global): India
Distribution (India): Tamil Nadu, Western Ghats
Leaf Type: Leaves opposite
Fruit Type: Follicular mericarps
Flowering and Fruiting: March–May
IUCN Status: Not evaluated
Notes: Endemic to Southern Western Ghats
Reference: Bull. Misc. Inform. Kew 1922: 119 (1922)

Toxocarpus concanensis Hook.f.
Apocynaceae
Climbing Mechanism: Stem Twiner
Distribution (Global): India
Distribution (India): Andaman and Nicobar Islands
IUCN Status: Not evaluated
Reference: Fl. Brit. India 4: 14 (1883)

Toxocarpus griffithii Decne.
Apocynaceae
Synonyms: *Secamone griffithii* (Decne.) Klack., *Toxocarpus roxburghii* Griff.
Climbing Mechanism: Stem Twiner
Distribution (Global): Java, Malaya, Myanmar
IUCN Status: Not evaluated
Reference: Blumea 55: 239 (2010)

Toxocarpus himalensis Falc. ex Hook.f.
Apocynaceae
Climbing Mechanism: Stem Twiner
Distribution (Global): Assam, Bangladesh, China South-Central, China Southeast, East Himalaya, Myanmar, West Himalaya
Distribution (India): Andaman and Nicobar Islands, Arunachal Pradesh, Meghalaya, Mizoram, Sikkim, West Bengal
IUCN Status: Not evaluated
Reference: Fl. Brit. India 4: 13 (1883)

Toxocarpus kleinii Wight & Arn.
Apocynaceae
Synonyms: *Secamone kleinii* (Wight & Arn.) K. Schum.
Common Name: Klein's bowfruit vine.
Climbing Mechanism: Stem Twiner
Distribution (Global): Andaman Is., India, Sri Lanka
Distribution (India): Andhra Pradesh, Maharashtra, Tamil Nadu
Leaf Type: Leaves decussate
Inflorescence: Cymes dichasial, lax

Fruit Type: Follicles paired, elongate, fusiform
Flowering and Fruiting: July
IUCN Status: Not evaluated
Reference: R. Wight, Contr. Bot. India: 61 (1834)

Toxocarpus kurzii Hook.f.
Apocynaceae
Climbing Mechanism: Stem Twiner
Distribution (Global): Andaman Is.
Distribution (India): Andaman and Nicobar Islands
IUCN Status: Not evaluated
Reference: Fl. Brit. India 4: 15 (1883)

Toxocarpus longistigma Wight & Arn. ex Steud.
Apocynaceae
Synonyms: *Toxocarpus roxburghii* Wight & Arn., *Asclepias longistigma* Roxb.
Climbing Mechanism: Stem Twiner
Distribution (Global): India
Distribution (India): Arunachal Pradesh
Flowering and Fruiting: March–July
IUCN Status: Not evaluated
Reference: Nomencl. Bot., ed. 2, 2: 694 (1841)

Toxocarpus palghatensis Gamble
Apocynaceae
Climbing Mechanism: Stem Twiner
Distribution (Global): India
Distribution (India): Tamil Nadu
Leaf Type: Leaves opposite
Inflorescence: Axillary, villous racemose cymes
Fruit Type: Follicles
Flowering and Fruiting: Throughout the year
IUCN Status: Not evaluated
Reference: Bull. Misc. Inform. Kew 1922: 119 (1922)

Toxocarpus roxburghii Wight & Arn.
Apocynaceae
Climbing Mechanism: Stem Twiner
Distribution (India): Andhra Pradesh
IUCN Status: Not evaluated
Reference: Wight & Arn. In: Wight Contr. Bot. India 61. (1834)

Toxocarpus villosus (Blume) Decne.
Apocynaceae
Synonyms: *Secamone villosa* Blume

Climbing Mechanism: Stem Twiner
Distribution (Global): Cambodia, China South-Central, China Southeast, Java, Laos, Lesser Sunda Is., Myanmar, Thailand, Vietnam
IUCN Status: Not evaluated
Reference: A.P.de Candolle, Prodr. 8: 506 (1844)

Trachelospermum asiaticum (Siebold & Zucc.) Nakai
Apocynaceae
Synonyms: *Malouetia asiatica* Siebold & Zucc., *Microchonea lucida* Pierre, *Parechites borneanus* Miq., *Trachelospermum asiaticum* var. *liukiuense* (Hatus.) Hatus., *Trachelospermum bessonii* Pierre ex Pit., *Trachelospermum borneanum* (Miq.) Boerl. + 10
Climbing Mechanism: Scrambler-Unarmed
Distribution (Global): Assam, Bangladesh, Borneo, China South-Central, China Southeast, Hainan, India
IUCN Status: Not evaluated
Reference: T. Mori, Enum. Pl. Corea: 293 (1922)

Trachelospermum assamense Woodson
Apocynaceae
Climbing Mechanism: Scrambler-Unarmed
Distribution (Global): Assam, East Himalaya
Distribution (India): West Bengal
IUCN Status: Not evaluated
Reference: Sunyatsenia 3: 80 (1936)

Trachelospermum axillare Hook.f.
Apocynaceae
Synonyms: *Melodinus chaffanjonii* H. Lév., *Periploca astacus* H.Lév.
Climbing Mechanism: Scrambler-Unarmed
Distribution (Global): China South-Central, China Southeast, East Himalaya, Nepal, Thailand, Tibet, Vietnam, West Himalaya
Distribution (India): Arunachal Pradesh, Meghalaya, West Bengal
IUCN Status: Not evaluated
Reference: Fl. Brit. India 3: 668 (1882)

Trachelospermum jasminoides (Lindl.) Lem.
Apocynaceae
Synonyms: KB
Nerium divaricatum Thunb., *Parechites adnascens* Hance, *Parechites thunbergii* A. Gray, *Rhyncospermum jasminoides* Lindl., *Rhyncospermum pulchrum* Jacob-Makoy, *Trachelospermum divaricatum* Kanitz
Common Name: Confederate jasmine, Star jasmine
Climbing Mechanism: Scrambler-Unarmed

Distribution (Global): China North-Central, China South-Central, China Southeast, Hainan, Japan, Korea, Taiwan, Tibet, Vietnam
Distribution (India): Delhi
Leaf Type: Leaves are opposite, oval to lance shaped
IUCN Status: Not evaluated
Reference: Jard. Fleur. 1: t. 61 (1851)

Trachelospermum lucidum (D. Don) K.Schum.
Apocynaceae
Synonyms: *Alstonia lucida* D. Don, *Blaberopus lucidus* (D.Don) A.DC., *Echites triangularis* Buch.-Ham. ex A.DC., *Ichnocarpus fragrans* Wall. ex G. Don. + 12
Common Name: Shining Star Jasmine
Climbing Mechanism: Scrambler-Unarmed
Distribution (Global): Assam, China South-Central, China Southeast, East Himalaya, Laos, Nepal, Pakistan, Taiwan, Thailand, Tibet, Vietnam, West Himalaya
Distribution (India): Arunachal Pradesh, Himachal Pradesh, Meghalaya, Mizoram, West Bengal
Leaf Type: Elliptic, lance shaped
Flowering and Fruiting: April–July
IUCN Status: Not evaluated
Reference: H.G.A. Engler & K.A.E.Prantl, Nat. Pflanzenfam. 4(2): 173 (1895)

Treutlera insignis Hook.f.
Apocynaceae
Climbing Mechanism: Stem Twiner
Distribution (Global): Assam, East Himalaya, Nepal
Distribution (India): West Bengal
IUCN Status: Not evaluated
Reference: Hooker's Icon. Pl. 15: t. 1425 (1883)

Tylophora asthmatica (L.f.) Wight & Arn.
Apocynaceae
Synonyms: *Vincetoxicum indicum* var. *indicum*, *Asclepias prolifera* Rottler ex Ainslie, *Tylophora flavescens* Ridl. + 22
Climbing Mechanism: Stem Twiner
Distribution (Global): Andaman Is., Assam, Bangladesh, Borneo, Cambodia, India, Laccadive Is., Laos, Malaya, Myanmar, Nicobar Is., Sri Lanka, Sumatera, Thailand, Vietnam
Distribution (India): Bihar, Odisha, Gujarat
Leaf Type: Simple, cordiform
Inflorescence: Axillary umbel
Fruit Type: Dry, follicles
IUCN Status: Not evaluated
Reference: Wight, Arn. In: Wright, Contrib. 51. (1834)

Tylophora cordifolia Thwaites
Apocynaceae
Synonyms: *Vincetoxicum cordifolium* (Thwaites) Kuntze., *Tylophora thwaitesii* K. Schum.
Climbing Mechanism: Stem Twiner
Distribution (Global): India, Sri Lanka
Leaf Type: Leaves opposite, small, ovate-cordate
Inflorescence: Solitary
Fruit Type: Follicles are usually solitary
Flowering and Fruiting: April–August
IUCN Status: Not evaluated
Notes: Usually found in deciduous forests
Reference: Revis. Gen. Pl. 2: 424 (1891)

Tylophora dalzellii Hook.f.
Apocynaceae
Synonyms: *Vincetoxicum dalzellii* (Hook.f.) Kuntze., *Tylophora carnosa* Dalzell & A. Gibson
Common Name: Dalzell ipecac
Climbing Mechanism: Stem Twiner
Distribution (Global): India
Distribution (India): Goa, Gujarat, Maharashtra
Leaf Type: Oppositely arranged leaves
Inflorescence: Umbellate cyme
Fruit Type: Follicles
Flowering and Fruiting: June–August
IUCN Status: Not evaluated
Notes: Usually found in deciduous forests
Reference: Revis. Gen. Pl. 2: 424 (1891)

Tylophora fasciculata Thwaites
Apocynaceae
Synonyms: *Vincetoxicum iphisia* Meve & Liede., *Iphisia multiflora* Wight & Arn., *Tylophora iphisia* Decne., *Tylophora multiflora* (Wight & Arn.) Alston., *Vincetoxicum multiflorum* (Wight & Arn.) Kuntze
Common Name: Brown-flowered ipecac
Climbing Mechanism: Stem Twiner
Distribution (Global): India, Sri Lanka
Distribution (India): Andhra Pradesh, Bihar, Odisha, Gujarat, Odisha
Leaf Type: Ovate
Inflorescence: Stalked cymes
Fruit Type: Follice, ellipsoid
Flowering and Fruiting: April–December
IUCN Status: Not evaluated

Notes: Usually found in dry deciduous forests and also in grasslands
Reference: Phytotaxa 369: 152 (2018)

Tylophora flexuosa R.Br.
Apocynaceae
Synonyms: *Hoya flexuosa* (R.Br.) Spreng., *Tylophora flexuosa* R.Br.
Climbing Mechanism: Stem Twiner
Distribution (Global): Andaman Is., Assam, Bangladesh, Borneo, Cambodia, Caroline Is., China North-Central, China South-Central, China Southeast, Hainan, India, Java, Laos, Lesser Sunda Is., Malaya, Maluku, Myanmar, New Guinea, Nicobar Is., Northern Territory, Philippines, Queensland, Sri Lanka, Sulawesi, Sumatera, Taiwan, Thailand, Vietnam, Western Australia
Distribution (India): West Bengal
Leaf Type: Simple, cordiform
Inflorescence: Axillary umbel
Fruit Type: Dry, follicles
IUCN Status: Not evaluated
Notes: Found mostly in scrub jungles
Reference: Revis. Gen. Pl. 2: 424 (1891)

Tylophora glabra Costantin
Apocynaceae
Synonyms: *Vincetoxicum renchangii* (Tsiang) Meve & Liede., *Tylophora longipedicellata* Y. Tsiang & P.T.Li
Tylophora renchangii Tsiang
Climbing Mechanism: Stem Twiner
Distribution (Global): China Southeast, Hainan, Vietnam
Distribution (India): Andaman and Nicobar Islands
Leaf Type: Simple, cordiform
Inflorescence: Axillary umbel
Fruit Type: Dry, follicles
IUCN Status: Not evaluated
Reference: Phytotaxa 369: 159 (2018)

Tylophora govanii (Wight & Arn.) Decne.
Apocynaceae
Synonyms: *Vincetoxicum govanii* (Wight & Arn.) Meve & Liede., *Iphisia govanii* Wight & Arn.
Common Name: Govan's swallow-wort
Climbing Mechanism: Stem Twiner
Distribution (Global): India, West Himalaya
Leaf Type: Simple, cordiform
Inflorescence: Axillary umbel
Fruit Type: Dry, follicles

IUCN Status: Not evaluated
Reference: Phytotaxa 369: 146 (2018)

Tylophora himalaica Hook.f.
Apocynaceae
Synonyms: *Vincetoxicum himalaicum* (Hook.f.) Kuntze
Climbing Mechanism: Stem Twiner
Distribution (Global): Assam, East Himalaya, India
Distribution (India): Arunachal Pradesh, West Bengal
Leaf Type: Simple, cordiform
Inflorescence: Axillary umbel
Fruit Type: Dry, follicles
IUCN Status: Not evaluated
Reference: Revis. Gen. Pl. 2: 424 (1891)

Tylophora hirsuta Wight
Apocynaceae
Synonyms: *Vincetoxicum hirsutum* (Wall.) Kuntze
Diplolepis apiculata Lindl., Diplolepis ovata Lindl., *Tylophora ovata* var. *lanyuensis* (Y.C. Liu & F.Y.Lu) S.S.Ying. + 12
Common Name: Hairy ipecac
Climbing Mechanism: Stem Twiner
Distribution (Global): Assam, Bangladesh, Cambodia, China South-Central, China Southeast, East Himalaya, Hainan, India, Java, Malaya, Myanmar, Nepal, Pakistan, Taiwan, Thailand, Vietnam, West Himalaya
Distribution (India): Arunachal Pradesh, Himachal Pradesh, Rajasthan, West Bengal
Leaf Type: Simple, cordiform
Inflorescence: Axillary umbel
Fruit Type: Dry, follicles
Flowering and Fruiting: April–August
IUCN Status: Not evaluated
Reference: Revis. Gen. Pl. 2: 424 (1891)

Tylophora indica (Burm.f.) Merr.
Apocynaceae
Synonyms: *Vincetoxicum indicum* (Burm.f.) Mabb., *Cynanchum indicum* Burm.f.
Common Name: Panacea twiner, Emetic swallow-wort, Emetic swallow-wort, Indian ipecacuanha
Climbing Mechanism: Stem Twiner
Distribution (Global): Andaman Is., Assam, Bangladesh, Borneo, Cambodia, India, Laccadive Is., Laos, Malaya, Myanmar, Nicobar Is., Sri Lanka, Sumatera, Thailand, Vietnam
Distribution (India): Andaman and Nicobar Islands, Andhra Pradesh, Delhi, Goa, Gujarat, Kerala, Maharashtra, Odisha, Puducherry, Tamil Nadu, Telangana, Uttar Pradesh, West Bengal

Leaf Type: Simple, opposite decussate
Inflorescence: fascicles in cymes
Fruit Type: Follicles
Flowering and Fruiting: May–December
IUCN Status: Not evaluated
Notes: Distributed mainly in scrub and deciduous forests
Reference: Mabberley's Pl.-Book, ed. 4: 1102 (2017)

Tylophora indica var. *glabra* (Decne.) H. Huber
Apocynaceae
Synonyms: *Vincetoxicum indicum* var. *glabrum* (Decne.) A. Kidyoo
Cynanchum flavum Thunb., *Tylophora asthmatica* var. *glabra* Decne., *Tylophora flava* Trimen
Common Name: Emetic swallow wort
Climbing Mechanism: Stem Twiner
Distribution (Global): India, Sri Lanka
Distribution (India): Peninsular India
Leaf Type: Opposite
Inflorescence: Umbellate cymes
Flowering and Fruiting: May–December
IUCN Status: Not evaluated
Reference: Taiwania 63: 30 (2018)

Tylophora iphisia Decne
Apocynaceae
Climbing Mechanism: Stem Twiner
Leaf Type: Simple, cordiform
Inflorescence: Axillary umbel
Fruit Type: Dry, follicles
IUCN Status: Not evaluated
Reference: Decne. In: DC. Prod. 8: 610. (1844)

Tylophora longifolia Wight
Apocynaceae
Synonyms: *Vincetoxicum longifolium* (Wight) Kuntze
Climbing Mechanism: Stem Twiner
Distribution (Global): Assam, Bangladesh, China South-Central, India, Malaya, Myanmar
Distribution (India): Arunachal Pradesh, Mizoram, West Bengal
Leaf Type: Simple, cordiform
Inflorescence: Axillary umbel
Fruit Type: Dry, follicles
IUCN Status: Not evaluated
Reference: Revis. Gen. Pl. 2: 424 (1891)

Tylophora multiflora (Wight & Arn.) Alston
Apocynaceae
Climbing Mechanism: Stem Twiner
Distribution (India): Tamil Nadu
Leaf Type: Simple, cordiform
Inflorescence: Axillary umbel
Fruit Type: Dry, follicles
IUCN Status: Not evaluated
Reference: Alston. In: Trimen, HandBook Fl. Sri Lanka, 6: Suppl., 195. (1931)

Tylophora nicobarica Murugan & M.Y. Kamble
Apocynaceae
Synonyms: *Vincetoxicum nicobaricum* (Murugan & M.Y. Kamble) Meve & Liede
Climbing Mechanism: Stem Twiner
Distribution (Global): Nicobar Is.
Distribution (India): Andaman and Nicobar Islands
Leaf Type: Simple, cordiform
Inflorescence: Axillary umbel
Fruit Type: Dry, follicles
IUCN Status: Not evaluated
Reference: Phytotaxa 369: 157 (2018)

Tylophora ovata (Lindl.) Hook. ex Steud
Apocynaceae
Climbing Mechanism: Stem Twiner
Distribution (India): Tamil Nadu, Telangana, Western Ghats
Leaf Type: Simple, cordiform
Inflorescence: Axillary umbel
Fruit Type: Dry, follicles
IUCN Status: Not evaluated
Reference: Hook. In: Nomencl. Bot. ed. 2, 2: 726. (1841)

Tylophora pauciflora Hohen. ex Hook.f.
Apocynaceae
Climbing Mechanism: Stem Twiner
Distribution (India): Andhra Pradesh, Tamil Nadu
Leaf Type: Simple, cordiform
Inflorescence: Axillary umbel
Fruit Type: Dry, follicles
IUCN Status: Not evaluated
Reference: Hohen. ex Hook. fil. In: Fl. Brit Ind. 4: 44. (1883)

Tylophora rotundifolia Buch. -Ham. ex Wight
Apocynaceae

Synonyms: *Vincetoxicum rotundifolium* (Buch. -Ham. ex Wight) Kuntze., *Tylophora trichophylla* Tsiang
Common Name: Roundleaf ipecac
Climbing Mechanism: Stem Twiner
Distribution (Global): Assam, China Southeast, East Himalaya, Hainan, India, Nepal, Thailand
Distribution (India): Andhra Pradesh, Gujarat, Madhya Pradesh, Maharashtra, Odisha, Telangana
Leaf Type: Ovate or orbicular, cordate
Inflorescence: Umbels erect, axillary
Fruit Type: Follicle, ellipsoid
Flowering and Fruiting: May–September
IUCN Status: Not evaluated
Notes: Often found in shola and evergreen forests
Reference: Revis. Gen. Pl. 2: 425 (1891)

Tylophora subramanii A.N. Henry
Apocynaceae
Synonyms: *Vincetoxicum subramanii* (A.N. Henry) Meve & Liede
Common Name: Subramani's ipecac
Climbing Mechanism: Stem Twiner
Distribution (Global): India
Distribution (India): Tamil Nadu
Leaf Type: Simple, cordiform
Inflorescence: Axillary umbel
Fruit Type: Dry, follicles
Flowering and Fruiting: March–April
IUCN Status: Not evaluated
Notes: Endemic to Southern Western Ghats
Reference: Phytotaxa 369: 164 (2018)

Tylophora zeylanica Decne
Apocynaceae
Synonyms: *Vincetoxicum zeylanicum* (Decne.) Meve & Liede., *Cynanchum micranthum* Thunb., *Tylophora micrantha* (Thunb.) Thwaites., *Vincetoxicum micranthum* (Thunb.) Kuntze
Climbing Mechanism: Stem Twiner
Distribution (Global): India
Distribution (India): Andaman and Nicobar Islands
Leaf Type: Simple, cordiform
Inflorescence: Axillary umbel
Fruit Type: Dry, follicles
Flowering and Fruiting: February–June
IUCN Status: Not evaluated

Notes: Often occur in evergreen forests
Reference: Phytotaxa 369: 167 (2018)

Urceola brachysepala Hook.f.
Apocynaceae
Synonyms: *Carruthersia imberbis* Elmer., *Chavannesia brachysepala* (Hook.f.) Pichon., *Chavannesia imberbis* (Elmer) Pichon., *Chavannesia philippinensis* (Merr.) Pichon., *Urceola acuteacuminata* Boerl., *Urceola imberbis* (Elmer) Merr., *Urceola maingayi* Hook.f.
Climbing Mechanism: Scrambler-Unarmed
Distribution (Global): Borneo, Java, Malaya, Philippines, Sumatera
Leaf Type: Simple
Inflorescence: Cymes paniculate
Fruit Type: Dry, follicles
IUCN Status: Not evaluated
Reference: Fl. Brit. India 3: 659 (1882)

Urceola elastica Roxb.
Apocynaceae
Synonyms: *Tabernaemontana elastica* (Roxb.) Spreng., *Tabernaemontana polyantha* Miq.
Climbing Mechanism: Scrambler-Unarmed
Distribution (Global): Borneo, Java, Malaya, Sumatera
Leaf Type: Simple
Inflorescence: Cymes paniculate
Fruit Type: Dry, follicles
IUCN Status: Not evaluated
Reference: Asiat. Res. 5: 169 (1798)

Urceola lakhimpurensis (S.K. Srivast. & Mehrotra) Karthik. & Moorthy
Apocynaceae
Synonyms: *Ecdysanthera lakhimpurensis* S.K. Srivast. & Mehrotra
Climbing Mechanism: Stem Twiner
Distribution (Global): Assam
Leaf Type: Simple
Inflorescence: Cymes paniculate
Fruit Type: Dry, follicles
IUCN Status: Not evaluated
Reference: Fl. Pl. India: 140 (2009)

Urceola lucida (A.DC.) Benth. ex Kurz
Apocynaceae
Synonyms: *Chavannesia esculenta* A.DC., *Chavannesia lucida* A.DC., *Echites esculentus* Wall., *Echites lucidus* Wall. ex G. Don, *Urceola esculenta* (A.DC.)

Benth. ex Kurz., *Urceòla reticulata* King & Gamble., *Xylinabaria esculenta* (A. DC.) Pierre
Climbing Mechanism: Scrambler-Unarmed
Distribution (Global): Laos, Malaya, Myanmar, Sumatera, Thailand
Leaf Type: Simple
Inflorescence: Cymes paniculate
Fruit Type: Dry, follicles
IUCN Status: Not evaluated
Reference: J. Asiat. Soc. Bengal, Pt. 2, Nat. Hist. 46: 255 (1877)

Urceola micrantha (Wall. ex G. Don) Mabb.
Apocynaceae
Synonyms: *Chavannesia montana* (M.R. Hend.) Pichon., *Cudicia gyrandra* Buch.-Ham. ex Dillwyn., *Ecdysanthera annamensis* Vernet., *Ecdysanthera brachiata* A.DC., *Ecdysanthera cambodiensis* Pierre., *Ecdysanthera langbiani* Vernet., *Ecdysanthera linearicarpa* Pierre
Climbing Mechanism: Scrambler-Unarmed
Distribution (Global): Assam, Bangladesh, Cambodia, China South-Central, China Southeast, East Himalaya, Hainan, Laos, Malaya, Myanmar, Nansei-shoto, Nepal, Taiwan, Thailand, Tibet, Vietnam
Leaf Type: Simple
Inflorescence: Cymes paniculate
Fruit Type: Dry, follicles
IUCN Status: Not evaluated
Reference: Novon 4: 151 (1994)

Urceola torulosa Hook.f.
Apocynaceae
Synonyms: *Chavannesia torulosa* (Hook.f.) Pichon., *Echites monilifer* Wall., *Urceola longisepala* Merr.
Urceola malaccensis Hook.f.
Climbing Mechanism: Scrambler-Unarmed
Distribution (Global): Borneo, Malaya, Sumatera
Leaf Type: Simple
Inflorescence: Cymes paniculate
Fruit Type: Dry, follicles
IUCN Status: Not evaluated
Reference: Fl. Brit. India 3: 659 (1882)

Urceola tournieri (Pierre) Mabb.
Apocynaceae
Synonyms: *Ecdysanthera tournieri* Pierre., *Parabarium burmanicum* Lý., *Parabarium hookeri* Pierre ex Spire., *Parabarium tournieri* (Pierre) Pierre
Climbing Mechanism: Scrambler-Unarmed

Distribution (Global): China South-Central, East Himalaya, Laos, Myanmar, Nepal, Thailand, Vietnam
Distribution (India): Mizoram, West Bengal
Leaf Type: Simple
Inflorescence: Cymes paniculate
Fruit Type: Dry, follicles
IUCN Status: Not evaluated
Reference: Novon 4: 151 (1994)

Utleria salicifolia Bedd. ex Hook.f.
Apocynaceae
Synonyms: *Decalepis salicifolia* (Bedd. ex Hook.f.) Bruyns.
Common Name: Willow-leaved swallow-root
Climbing Mechanism: Scrambler-Unarmed
Distribution (Global): India, Laccadive Is.
Leaf Type: Linear-lanceolate
Inflorescence: Dichotomous cymes, axillary
Flowering and Fruiting: June
IUCN Status: Critically endangered
Reference: Taxon 65: 499 (2016)

Vallaris glabra (L.) Kuntze
Apocynaceae
Synonyms: *Echites dichotomus* Roxb., *Echites hircosus* Roxb., *Emericia pergularia* (Burm.f.) Roem. & Schult., *Pergularia glabra* L., *Vallaris dichotoma* Wall., *Vallaris indica* J.F. Gmel., *Vallaris ovalis* Miq., *Vallaris pergulana* Burm.f.
Climbing Mechanism: Scrambler-Unarmed
Distribution (Global): Java, Lesser Sunda Is., Sumatera
Leaf Type: Simple
Inflorescence: Umbellate racemes in clusters
Fruit Type: Dry, follicles
IUCN Status: Not evaluated
Reference: Revis. Gen. Pl. 2: 417 (1891)

Vallaris solanacea (Roth) Kuntze
Apocynaceae
Synonyms: *Peltanthera solanacea* Roth, *Vallaris assamensis* Griff., *Vallaris heynei* Spreng
Common Name: Ramsar, Bread flower
Climbing Mechanism: Scrambler-Unarmed
Distribution (Global): Assam, Bangladesh, Cambodia, East Himalaya, Hainan, India, Laos, Myanmar, Nepal, Pakistan, Sri Lanka, Thailand, Vietnam, West Himalaya
Distribution (India): Andaman and Nicobar Islands, Andhra Pradesh, Arunachal Pradesh, Delhi, Eastern Ghats, Gujarat, Himachal Pradesh, Madhya Pradesh,

Maharashtra, Mizoram, Odisha, Rajasthan, Tamil Nadu, Telangana, Tripura, Uttar Pradesh, West Bengal
Leaf Type: Elliptic, oblong
Flowering and Fruiting: March–November
IUCN Status: Not evaluated
Reference: Revis. Gen. Pl. 2: 417 (1891)

Vincetoxicum capparidifolium (Wight & Arn.) Kuntze
Apocynaceae
Synonyms: *Asclepias tenuis* Hook.f., *Tylophora capparidifolia* Wight & Arn.
Climbing Mechanism: Stem Twiner
Distribution (India): Tamil Nadu
Leaf Type: Simple
Inflorescence: Cymes in clusters
Fruit Type: Dry, pods
IUCN Status: Not evaluated
Reference: Revis. Gen. Pl. 2: 424 (1891)

Vincetoxicum dalzellii Kuntze
Apocynaceae
Synonyms: *Tylophora carnosa* Dalzell & A. Gibson, *Tylophora dalzellii* Hook.f.
Common Name: Dalzell ipecac
Climbing Mechanism: Stem Twiner
Distribution (Global): India
Distribution (India): Tamil Nadu
Leaf Type: Oppositely arranged leaves
Flowering and Fruiting: June–August
IUCN Status: Not evaluated
Reference: Revis. Gen. Pl. 2: 424 (1891)

Vincetoxicum hirsutum Kuntze
Apocynaceae
Climbing Mechanism: Stem Twiner
Distribution (India): Tamil Nadu
Leaf Type: Simple
Inflorescence: Cymes in clusters
Fruit Type: Dry, pods
IUCN Status: Not evaluated

Willughbeia cirrhifera Abeyw.
Apocynaceae
Synonyms: *Ancylocladus ceylanicus* (Wight) Kuntze., *Chilocarpus ceylanicus* Wight., *Willughbeia ceylanica* (Wight) Thwaites.
Climbing Mechanism: Tendril climber
Distribution (Global): Sri Lanka

IUCN Status: Vulnerable
Reference: Ceylon J. Sci., Biol. Sci. 2: 84 (1959)

Willughbeia edulis Roxb.
Apocynaceae
Synonyms: *Ambelania edulis* (Roxb.) J. Presl, *Ancylocladus cochinchinensis* Pierre, *Ancylocladus curtisianus* Pierre, *Ancylocladus edulis* (Roxb.) Kuntze, *Pacouria roxburghii* Kostel., *Willughbeia cochinchinensis* (Pierre) K. Schum., *Willughbeia curtisiana* (Pierre) K
Climbing Mechanism: Tendril climber
Distribution (Global): Assam, Bangladesh, Cambodia, Laos, Malaya, Myanmar, Nicobar Is., Thailand, Vietnam
Distribution (India): Arunachal Pradesh, Great Nicobar Island, Mizoram, Tripura
Leaf Type: Oblong, acuminate
Inflorescence: Aillary cyme
Flowering and Fruiting: May
IUCN Status: Not evaluated
Reference: Pl. Coromandel 3: 77 (1820)

Willughbeia grandiflora Dyer ex Hook.f.
Apocynaceae
Synonyms: *Ancylocladus glaucinus* Pierre., *Ancylocladus grandiflorus* (Dyer ex Hook.f.) Kuntze., *Willughbeia glaucina* (Pierre) K. Schum
Climbing Mechanism: Tendril climber
Distribution (Global): Borneo, Malaya, Thailand
IUCN Status: Not evaluated
Reference: Fl. Brit. India 3: 625 (1882)

Epipremnum aureum (Linden & André) G.S. Bunting
Araceae
Synonyms: *Epipremnum mooreense* Nadeaud, *Pothos aureus* Linden & André, *Rhaphidophora aurea* (Linden & André) Birdsey, *Scindapsus aureus* (Linden & André) Engl.
Common Name: Money plant, Golden pothos, Ceylon creeper, Hunter's robe, Ivy arum, House plant, Silver vine, Solomon Islands ivy, Marble queen, Taro vine
Climbing Mechanism: Root climber
Distribution (Global): Society Is.
Distribution (India): Tamil Nadu
IUCN Status: Not evaluated
Reference: Ann. Missouri Bot. Gard. 50: 28 (1964)

Epipremnum giganteum (Roxb.) Schott
Araceae
Synonyms: *Monstera gigantea* (Roxb.) Schott, *Pothos giganteus* Roxb., *Rhaphidophora gigantea* (Roxb.) Ridl., *Scindapsus giganteus* (Roxb.) Schott

Climbing Mechanism: Root climber
Distribution (Global): Cambodia, Malaya, Myanmar, Thailand, Vietnam
IUCN Status: Not evaluated
Reference: Bonplandia (Hannover) 5: 45 (1857)

Epipremnum pinnatum (L.) Engl.
Araceae
Synonyms: *Epipremnum angustilobum* K. Krause, *Epipremnum crassifolium* Engl., *Epipremnum elegans* Engl., *Epipremnum formosanum* Hayata, *Epipremnum glaucicephalum* Elmer, *Epipremnum merrillii* Engl. & K.Krause, *Epipremnum mirabile* Schott, *Epipremnum robinsonii* K.Kr
Common Name: Centipede tongavine, Dragon-tail plant
Climbing Mechanism: Root climber
Distribution (Global): Andaman Is., Assam, Bangladesh, Bismarck Archipelago, Borneo, Cambodia, Caroline Is., China South-Central, China Southeast, Cook Is., Fiji, Hainan, India, Java, Laos, Lesser Sunda Is., Malaya, Maluku, Marshall Is., Myanmar, Nansei-shoto, New Caledonia, New Guinea, Nicobar Is., Northern Territory, Philippines, Queensland, Samoa, Santa Cruz Is., Solomon Is., Sulawesi, Sumatera, Taiwan, Thailand, Tonga, Vanuatu, Vietnam, Wallis-Futuna Is.
Distribution (India): Assam, Odisha, Tamil Nadu, Western Ghats
IUCN Status: Not evaluated
Reference: Pflanzenr., IV, 23B: 60 (1908)

Monstera deliciosa Liebm.
Araceae
Climbing Mechanism: Scrambler-Unarmed
Distribution (India): Gujarat, Odisha, Rajasthan, Tamil Nadu
IUCN Status: Not evaluated

Pothos armatus C.E.C. Fisch.
Araceae
Climbing Mechanism: Stem Twiner
Distribution (Global): India
Distribution (India): Tamil Nadu
Leaf Type: Simple, leathery, rounded at base
Inflorescence: Solitary, peduncle erect or spread
Fruit Type: Wet, berry
Flowering and Fruiting: February–March
IUCN Status: Not evaluated
Notes: Mostly found in evergreen forests
Reference: Bull. Misc. Inform. Kew 1929: 126 (1929)

Pothos boyceanus C. Rajkumar, Shaju, Nazarudeen & Prakashk
Araceae

Climbing Mechanism: Root climber
Distribution (Global): India
Leaf Type: Simple, leathery, rounded at base
Inflorescence: Solitary, peduncle erect or spread
Fruit Type: Wet, berry
IUCN Status: Not evaluated
Reference: Taiwania 65: 114 (2020)

Pothos chinensis (Raf.) Merr.
Araceae
Synonyms: *Pothos balansae* Engl., *Pothos cathcartii* Schott, *Pothos chinensis* var. *lotienensis* C.Y. Wu & H.Li, *Pothos seemannii* Schott, *Pothos warburgii* Engl., *Pothos yunnanensis* Engl., *Tapanava chinensis* Raf.
Climbing Mechanism: Stem Twiner
Distribution (Global): Assam, Bangladesh, Cambodia, China South-Central, China Southeast, East Himalaya, Hainan, India, Laos, Myanmar, Nansei-shoto, Nepal, Taiwan, Thailand, Tibet, Vietnam, West Himalaya
Leaf Type: Simple, leathery, rounded at base
Inflorescence: Solitary, peduncle erect or spread
Fruit Type: Wet, berry
IUCN Status: Not evaluated
Reference: J. Arnold Arbor. 29: 210 (1948)

Pothos crassipedunculatus Sivad. & N. Mohanan
Araceae
Climbing Mechanism: Stem Twiner
Distribution (Global): India
Distribution (India): Peninsular India
Leaf Type: Simple, leathery, rounded at base
Inflorescence: Solitary, peduncle erect or spread
Fruit Type: Wet, berry
Flowering and Fruiting: December–May
IUCN Status: Not evaluated
Notes: Mostly found in evergreen forests
Reference: Pl. Syst. Evol. 168: 221 (1989)

Pothos curtisii Hook.f.
Araceae
Synonyms: *Pothos kunstleri* Hook.f., *Pothos latifolius* Hook.f., *Pothos maingayi* Hook.f., *Pothos peninsularis* Alderw.
Climbing Mechanism: Stem Twiner
Distribution (Global): India, Malaya, Sumatera, Thailand
Leaf Type: Simple, leathery, rounded at base
Inflorescence: Solitary, peduncle erect or spread
Fruit Type: Wet, berry

IUCN Status: Not evaluated
Reference: Fl. Brit. India 6: 554 (1893)

Pothos junghuhnii de Vriese
Araceae
Synonyms: *Pothos junghuhnianus* Schott, *Pothos longipedunculatus* Engl., *Pothos macrophyllus* de Vriese, *Pothos roxburghii* Schott, *Pothos vrieseanus* Schott.
Climbing Mechanism: Stem Twiner
Distribution (Global): Assam, Bangladesh, Borneo, East Himalaya, Java, Myanmar, Sumatera
Leaf Type: Simple, leathery, rounded at base
Inflorescence: Solitary, peduncle erect or spread
Fruit Type: Wet, berry
IUCN Status: Not evaluated
Reference: F.A.W. Miquel, Pl. Jungh.: 104 (1851)

Pothos keralensis A.G. Pandurangan & V.J.Nair
Araceae
Climbing Mechanism: Stem Twiner
Distribution (Global): India.
Distribution (India): Peninsular India.
Leaf Type: Simple, leathery, rounded at base
Inflorescence: Solitary, peduncle erect or spread
Fruit Type: Wet, berry
Flowering and Fruiting: February–May
IUCN Status: Not evaluated
Notes: Mostly found in evergreen forests
Reference: Indian J. Forest. 17: 64 (1994)

Pothos lancifolius Hook.f.
Araceae
Synonyms: *Pothos penicilliger* Gagnep.
Climbing Mechanism: Stem Twiner
Distribution (Global): Malaya, Vietnam
Leaf Type: Simple, leathery, rounded at base
Inflorescence: Solitary, peduncle erect or spread
Fruit Type: Wet, berry
IUCN Status: Not evaluated
Reference: Fl. Brit. India 6: 554 (1893)

Pothos macrocephalus Scort. ex Hook.f.
Araceae
Climbing Mechanism: Stem Twiner
Distribution (Global): Malaya, Nicobar Is., Sumatera, Thailand
Distribution (India): Andaman and Nicobar Islands, Arunachal Pradesh

Leaf Type: Simple, leathery, rounded at base
Inflorescence: Solitary, peduncle erect or spread
Fruit Type: Wet, berry
IUCN Status: Not evaluated
Reference: Fl. Brit. India 6: 553 (1893)

Pothos roxburghii de Vriese
Araceae
Synonyms: *Pothos neoroxburghii* P.C. Boyce
Climbing Mechanism: Stem Twiner
Distribution (Global): Assam, Bangladesh, East Himalaya, India, Malaya, Myanmar, Thailand, West Himalaya
Distribution (India): Sikkim
Leaf Type: Simple, leathery, rounded at base
Inflorescence: Solitary, peduncle erect or spread
Fruit Type: Wet, berry
IUCN Status: Not evaluated
Reference: F.A.W. Miquel, Pl. Jungh.: 103 (1851)

Pothos scandens L.
Araceae
Synonyms: *Batis hermaphrodita* Blanco, *Podospadix angustifolia* Raf., *Pothos angustifolius* (Raf.) C.Presl, *Pothos chapelieri* Schott, *Pothos cognatus* Schott + 18
Common Name: Climbing aroid
Climbing Mechanism: Stem Twiner
Distribution (Global): Andaman Is., Assam, Bangladesh, Borneo, Cambodia, China South-Central, Comoros, East Himalaya, India, Java, Laos, Lesser Sunda Is., Madagascar, Malaya, Maluku, Myanmar, New Guinea, Nicobar Is., Philippines, Seychelles, Sri Lanka, Sulawesi, Sumatera, Thailand, Tibet, Vietnam
Distribution (India): Andaman and Nicobar Islands, Arunachal Pradesh, Assam, Gujarat, Jharkhand, Karnataka, Kerala, Odisha, Tamil Nadu, Tripura, West Bengal, Western Ghats
Leaf Type: Simple, leathery, rounded at base
Inflorescence: Solitary, peduncle erect or spread
Fruit Type: Wet, berry
Flowering and Fruiting: May–November
IUCN Status: Not evaluated
Notes: Mostly found in evergreen forests
Reference: Sp. Pl.: 968 (1753)

Pothos thomsonianus Schott
Araceae
Climbing Mechanism: Stem Twiner
Distribution (Global): India

Distribution (India): Peninsular India
Leaf Type: Simple, leathery, rounded at base
Inflorescence: Solitary, peduncle erect or spread
Fruit Type: Wet, berry
Flowering and Fruiting: January–March
IUCN Status: Not evaluated
Notes: Mostly found in evergreen forests
Reference: Aroideae: 24 (1856)

Pothos tirunelveliensis Sasikala & Reema Kumari
Araceae
Climbing Mechanism: Stem Twiner
Distribution (Global): India
Distribution (India): Kerala
Leaf Type: Simple, leathery, rounded at base
Inflorescence: Solitary, peduncle erect or spread
Fruit Type: Wet, berry
IUCN Status: Not evaluated
Reference: Sci. Res. Reporter 3: 152 (2013)

Rhaphidophora beccarii (Engl.) Engl.
Araceae
Synonyms: *Epipremnum beccarii* Engl., *Rhaphidophora borneensis* Engl., *Rhaphidophora fluminea* Ridl.
Climbing Mechanism: Root Climber
Distribution (Global): Andaman Is., Borneo, Malaya, Sumatera, Thailand
Distribution (India): Arunachal Pradesh
Leaf Type: Simple, broad leaves
Inflorescence: Solitary at tips
Fruit Type: Wet, berry
IUCN Status: Not evaluated
Reference: Bot. Jahrb. Syst. 1: 181 (1880)

Rhaphidophora calophylla
Araceae
Climbing Mechanism: Root Climber
IUCN Status: Not evaluated

Rhaphidophora decursiva (Roxb.) Schott
Araceae
Synonyms: *Monstera decursiva* (Roxb.) Schott, *Monstera multijuga* K. Koch ex Ender, *Pothos decursivus* Roxb., *Rhaphidophora affinis* Schott, *Rhaphidophora eximia* Schott, *Rhaphidophora grandis* Schott, *Scindapsus decursivus* (Roxb.) Schott.
Climbing Mechanism: Root Climber

Distribution (Global): Andaman Is., Assam, Bangladesh, Cambodia, China South-Central, China Southeast, East Himalaya, Hainan, India, Laos, Myanmar, Nepal, Nicobar Is., Sri Lanka, Thailand, Tibet, Vietnam, West Himalaya
Distribution (India): Andhra Pradesh, Arunachal Pradesh, Bihar, Odisha, Gujarat
Leaf Type: Simple, broad leaves
Inflorescence: Solitary at tips
Fruit Type: Wet, berry
Flowering and Fruiting: June–November
IUCN Status: Not evaluated
Reference: Bonplandia (Hannover) 5: 45 (1857)

Rhaphidophora glauca (Wall.) Schott
Araceae
Synonyms: *Monstera glauca* K. Koch ex Ender, *Pothos glaucus* Wall., *Pothos wallichii* Steud., *Scindapsus glaucus* (Wall.) Schott.
Climbing Mechanism: Root Climber
Distribution (Global): Assam, Bangladesh, East Himalaya, Laos, Myanmar, Nepal, Thailand, Tibet, Vietnam, West Himalaya
Distribution (India): Arunachal Pradesh, Odisha
Leaf Type: Simple, broad leaves
Inflorescence: Solitary at tips
Fruit Type: Wet, berry
Flowering and Fruiting: October–November
IUCN Status: Not evaluated
Reference: Bonplandia (Hannover) 5: 45 (1857)

Rhaphidophora hongkongensis Schott
Araceae
Synonyms: *Pothos obliquus* Wall. ex Hook.f.
Climbing Mechanism: Root Climber
Distribution (Global): Assam, Bangladesh, China South-Central, China Southeast, Hainan, Laos, Myanmar, Taiwan, Thailand, Vietnam
Distribution (India): Arunachal Pradesh
Leaf Type: Simple, broad leaves
Inflorescence: Solitary at tips
Fruit Type: Wet, berry
IUCN Status: Not evaluated
Reference: Bonplandia (Hannover) 5: 44 (1857)

Rhaphidophora hookeri Schott
Araceae
Climbing Mechanism: Root Climber
Distribution (Global): Assam, Bangladesh, China South-Central, China Southeast, East Himalaya, Laos, Myanmar, Nepal, Thailand, Tibet, Vietnam, West Himalaya

Distribution (India): Arunachal Pradesh, Odisha
Leaf Type: Simple, broad leaves
Inflorescence: Solitary at tips
Fruit Type: Wet, berry
Flowering and Fruiting: January–June
IUCN Status: Not evaluated
Reference: Prodr. Syst. Aroid.: 381 (1860)

Rhaphidophora korthalsii Schott
Araceae
Synonyms: *Epipremnum multicephalum* Elmer., *Pothos bifarius* Wall. ex Hook.f., *Pothos celatocaulis* N.E.Br., *Rhaphidophora celatocaulis* (N.E.Br.) Alderw., *Rhaphidophora copelandii* Engl., *Rhaphidophora grandifolia* K. Krause, *Rhaphidophora grandis* Ridl. + 7
Climbing Mechanism: Root Climber
Distribution (Global): Bismarck Archipelago, Borneo, Caroline Is., East Himalaya, Java, Malaya, Maluku, Nansei-shoto, New Guinea, Philippines, Solomon Is., Sulawesi, Sumatera, Thailand, Vanuatu
Distribution (India): Arunachal Pradesh
Leaf Type: Simple, broad leaves
Inflorescence: Solitary at tips
Fruit Type: Wet, berry
IUCN Status: Not evaluated
Reference: Ann. Mus. Bot. Lugduno-Batavi 1: 129 (1863)

Rhaphidophora lancifolia Schott
Araceae
Climbing Mechanism: Root Climber
Distribution (Global): Assam, Bangladesh, China South-Central, China Southeast
Distribution (India): Tripura
Leaf Type: Simple, broad leaves
Inflorescence: Solitary at tips
Fruit Type: Wet, berry
IUCN Status: Not evaluated
Reference: Prodr. Syst. Aroid.: 380 (1860)

Rhaphidophora peepla (Roxb.) Schott
Araceae
Synonyms: *Monstera peepla* (Roxb.) Schott, *Pothos peepla* Roxb., *Pothos pipla* Schult. & Schult.f., *Rhaphidophora manipurensis* Engl. & K. Krause, *Scindapsus peepla* (Roxb.) Schott.
Climbing Mechanism: Root Climber
Distribution (Global): Assam, Bangladesh, Cambodia, China South-Central, East Himalaya, Laos, Myanmar, Nepal, Thailand, Vietnam
Distribution (India): Tripura, Arunachal Pradesh

Leaf Type: Simple, broad leaves
Inflorescence: Solitary at tips
Fruit Type: Wet, berry
IUCN Status: Not evaluated
Reference: Bonplandia (Hannover) 5: 45 (1857)

Rhaphidophora pertusa (Roxb.) Schott
Araceae
Synonyms: *Monstera pertusa* (Roxb.) Schott, *Pothos pertusus* Roxb., *Rhaphidophora lacera* Hassk., *Scindapsus peepla* Thwaites, *Scindapsus pertusus* (Roxb.) Schott.
Climbing Mechanism: Root Climber
Distribution (Global): Andaman Is., Bangladesh, India, Maldives, Myanmar, Sri Lanka, Thailand
Distribution (India): Andaman and Nicobar Islands, Andhra Pradesh, Odisha
Leaf Type: Simple, broad leaves
Inflorescence: Solitary at tips
Fruit Type: Wet, berry
Flowering and Fruiting: January–March
IUCN Status: Not evaluated
Reference: Bonplandia (Hannover) 5: 45 (1857)

Rhaphidophora puberula Engl.
Araceae
Synonyms: *Rhaphidophora batoeensis* Engl. & K. Krause, *Rhaphidophora gracilipes* Hook.f., *Rhaphidophora hallieri* Alderw., *Rhaphidophora kunstleri* Hook.f., *Rhaphidophora pilosula* Alderw., *Rhaphidophora scaberula* Alderw., *Rhaphidophora scortechinii* Hook.f.
Climbing Mechanism: Root Climber
Distribution (Global): Assam, Borneo, Lesser Sunda Is., Malaya, Sumatera
Leaf Type: Simple, broad leaves
Inflorescence: Solitary at tips
Fruit Type: Wet, berry
IUCN Status: Not evaluated
Reference: Bot. Jahrb. Syst. 1: 180 (1880)

Scindapsus hederaceus Miq.
Araceae
Synonyms: *Pothos hederaceus* Zoll. & Moritzi., *Scindapsus inquinatus* Schott., *Scindapsus poilanei* Gagnep., *Scindapsus pothoides* Miq.
Climbing Mechanism: Stem Twiner
Distribution (Global): Borneo, Cambodia, Java, Laos, Lesser Sunda Is., Malaya, Philippines, Sumatera, Thailand, Vietnam
IUCN Status: Least Concern
Reference: Fl. Ned. Ind. 3: 185 (1856)

Scindapsus officinalis (Roxb.) Schott
Araceae
Synonyms: Monstera officinalis (Roxb.) Schott., Pothos officinalis Roxb., *Scindapsus annamicus* Gagnep.
Climbing Mechanism: Root Climber
Distribution (Global): Andaman Is., Assam, Bangladesh, Cambodia, East Himalaya, India, Laos, Myanmar, Nepal, Thailand, Vietnam, West Himalaya
Distribution (India): Andaman and Nicobar Islands, Andhra Pradesh, Arunachal Pradesh, Bihar, Odisha, Odisha, Sikkim, Tripura, Uttar Pradesh, West Bengal
Flowering and Fruiting: June–January
IUCN Status: Not evaluated
Notes: Often found in damp valleys
Reference: H.W. Schott & S.L.Endlicher, Melet. Bot.: 21 (1832)

Scindapsus scortechinii Hook.f.
Araceae
Climbing Mechanism: Root Climber
Distribution (Global): Bangladesh, Malaya, Thailand
IUCN Status: Not evaluated
Reference: Fl. Brit. India 6: 541 (1893)

Aralia devendrae Pusalkar
Araliaceae
Climbing Mechanism: Stem Twiner
Distribution (Global): West Himalaya
Distribution (India): Jammu and Kashmir
IUCN Status: Not evaluated
Reference: Taiwania 54: 226 (2009)

Eleutherococcus trifoliatus (L.) S.Y. Hu
Araliaceae
Synonyms: *Acanthopanax aculeatus* (Aiton) Witte, *Acanthopanax sepium* Seem., *Acanthopanax spinifolius* Merr., *Acanthopanax trifoliatus* (L.) Voss, *Aralia trifoliata* (L.) Meyen, *Eleutherococcus spinifolius* (Merr.) S.Y. Hu, *Eleutherococcus trifoliatus* var. *intege*
Climbing Mechanism: Scrambler-Armed
Distribution (Global): Assam, China South-Central, China Southeast, Myanmar, Nepal, Philippines, Taiwan, Thailand, Vietnam
Distribution (India): Manipur
Leaf Type: Leaves are compound, 3–5 foliolate
Inflorescence: Axillary or terminal umbels
Flowering and Fruiting: October–February
IUCN Status: Not evaluated
Reference: J. Arnold Arbor. 61: 110 (1980)

Hedera canariensis Willd.
Araliaceae
Synonyms: *Hedera canariensis* var. *arborescens* Hibberd, *Hedera grandifolia* Hibberd, *Hedera helix* subsp. *canariensis* (Willd.) Cout., *Hedera sevillana* Sprenger, *Hedera viridis* (Hibberd) G. Nicholson
Common Name: Canary Island Ivy
Climbing Mechanism: Root Climber
Distribution (Global): Canary Is.
IUCN Status: Not evaluated
Reference: Mag. Neuesten Entdeck. Gesammten Naturk. Ges. Naturf. Freunde Berlin 2: 171 (1808)

Hedera helix L.
Araliaceae
Synonyms: *Hedera communis, Hedera cordata.*
Common Name: Common Ivy, English ivy, European ivy
Climbing Mechanism: Root Climber
Distribution (Global): Albania, Austria, Baleares, Baltic States, Belarus, Belgium, Bulgaria, Central European Rus, Corse, Czechoslovakia, Denmark, East Aegean Is., France, Germany, Great Britain, Greece, Hungary, Ireland, Italy, Kriti, Netherlands, North Caucasus, Northwest European R, Norway, Poland, Portugal, Romania, Sardegna, Sicilia, South European Russi, Spain, Sweden, Switzerland, Turkey, Turkey-in-Europe, Ukraine, Yugoslavia
Distribution (India): Maharashtra, Tamil Nadu, North Eastern India
IUCN Status: Not evaluated
Reference: Sp. Pl.: 202 (1753)

Hedera nepalensis K. Koch
Araliaceae
Synonyms: *Cissus wallichiana* Turcz., *Hedera cinerea* (Hibberd) Bean, *Hedera helix* var. *chrysocarpa* DC., nom. illeg., *Hedera helix* var. *cinerea* Hibberd, *Hedera helix* var. *himalaica* Hibberd, *Hedera himalaica* (Hibberd) Carrière
Common Name: Himalayan Ivy
Climbing Mechanism: Root Climber
Distribution (India): Arunachal Pradesh, Himachal Pradesh, Maharashtra, Manipur, Meghalaya, Mizoram, West Bengal
Leaf Type: 3–5 lobed
Inflorescence: Umbels
IUCN Status: Not evaluated
Reference: K. Koch. In: Hort. Dendrol.: 284. (1853)

Schefflera clarkeana Craib
Araliaceae
Synonyms: *Hedera macrophylla* Wall., *Heptapleurum clarkeanum* (Craib) Lowry & G.M. Plunkett., *Heptapleurum ellipticum* var. *macrophyllum* (C.B.Clarke) Lowry

& G.M.Plunkett., *Heptapleurum venulosum* var. *macrophyllum* C.B.Clarke., *Schefflera elliptica* var. *macrophylla* (C.B.)
Climbing Mechanism: Scrambler-Unarmed
Distribution (Global): Bangladesh, East Himalaya, Myanmar, Thailand
Distribution (India): Andhra Pradesh, Arunachal Pradesh, Mizoram
Leaf Type: Digitately 5–6 foliolate
Inflorescence: Panicles
Fruit Type: Wet, berry, globose
Flowering and Fruiting: July–September
IUCN Status: Not evaluated
Reference: Bull. Misc. Inform. Kew 1919: 229 (1919)

Schefflera elliptica (Blume) Harms
Araliaceae
Synonyms: *Paratropia elliptica* (Blume) Miq., *Sciodaphyllum ellipticum* Blume., *Heptapleurum ellipticum* (Blume) Seem
Common Name: Elliptic-leaved Schefflera
Climbing Mechanism: Scrambler-Unarmed
Distribution (Global): Andaman Is., Assam, Bangladesh, Bismarck Archipelago, Borneo, Cambodia, Caroline Is., China South-Central, China Southeast, Christmas I., East Himalaya, India, Java, Laos, Lesser Sunda Is., Malaya, Myanmar, Nepal, New Guinea, Nicobar Is., Philippines, Queensland, Sulawesi, Sumatera, Taiwan, Thailand, Vietnam, West Himalaya
Distribution (India): Andhra Pradesh, Arunachal Pradesh, Mizoram, Telangana
Leaf Type: Digitately compound leaves
Inflorescence: Panicle of umbels at the end of branches
Fruit Type: Ovoid to ellipsoid or round
Flowering and Fruiting: March–July
IUCN Status: Least Concern
Reference: J. Bot. 3: 78 (1865)

Schefflera emarginata (Moon) Harms
Araliaceae
Synonyms: *Aralia praemorsa* Jacob-Makoy., *Hedera emarginata* Moon., *Paratropia emarginata* (Moon) Regel., *Heptapleurum emarginatum* (Moon) Seem
Climbing Mechanism: Scrambler-Unarmed
Distribution (Global): Sri Lanka
Leaf Type: Digitately 5–6 foliolate
Inflorescence: Panicles
Fruit Type: Wet, berry, globose
IUCN Status: Not evaluated
Reference: J. Bot. 3: 80 (1865)

Schefflera pubigera (Brongn. ex Planch.) Frodin
Araliaceae
Synonyms: *Heptapleurum ellipticum* var. *ellipticum*, *Aralia moorei* F. Muell., *Schefflera fukienensis* Merr., *Schefflera minimiflora* Ridl. + 24
Climbing Mechanism: Scrambler-Unarmed
Distribution (Global): Andaman Is., Assam, Bangladesh, Bismarck Archipelago, Borneo, Cambodia, Caroline Is., China South-Central, China Southeast, Christmas I., East Himalaya, Java, Laos, Lesser Sunda Is., Malaya, Myanmar, Nepal, New Guinea, Nicobar Is., Philippines, Queensland, Sulawesi, Sumatera, Taiwan, Thailand, Vietnam, West Himalaya
Distribution (India): West Bengal
Leaf Type: Digitately 5–6 foliolate
Inflorescence: Panicles
Fruit Type: Wet, berry, globose
Flowering and Fruiting: November–April
IUCN Status: Not evaluated
Reference: D.G. Frodin & R.H.A.Govaerts, World Checkl. & Bibliogr. Araliaceae: 368 (2003 publ. 2004)

Schefflera pueckleri (K. Koch) Frodin
Araliaceae
Synonyms: *Heptapleurum calyptratum* (Hook.f. & Thomson) Y.F. Deng., *Aralia elliptica* K.Koch., *Aralia pulchra* Van Houtte ex H.Jaeger., *Heptapleurum pulchrum* (Van Houtte ex H.Jaeger) Voss., *Paratropia pulchra* Decne. & Planch., *Paratropia wallichiana* Planch.
Climbing Mechanism: Scrambler-Unarmed
Distribution (Global): Assam, Bangladesh, Cambodia, China South-Central, Laos, Myanmar, Thailand, Tibet, Vietnam
Leaf Type: Digitately 5–6 foliolate
Inflorescence: Panicles
Fruit Type: Wet, berry, globose
IUCN Status: Data Deficient
Reference: Med. Fl. China 7: 469 (2018)

Schefflera roxburghii Gamble
Araliaceae
Synonyms: *Heptapleurum digitatum* (G. Don ex Loudon) Lowry & G.M. Plunkett., *Actinophyllum digitatum* Wall. ex Loudon., *Aralia digitata* Roxb., *Paratropia digitata* Voigt., *Schefflera venulosa* var. *roxburghii* (Gamble) K.K. Khanna., *Sciodaphyllum digitatum* G.Don ex Loudon
Climbing Mechanism: Scrambler-Unarmed
Distribution (Global): Assam, East Himalaya, India, Myanmar, Nepal
Distribution (India): Andhra Pradesh, Madhya Pradesh, Maharashtra, Tamil Nadu
Leaf Type: Leaves alternate, digitate
Inflorescence: Flowers in panicles of umbels or compound racemes, usually terminal

Fruit Type: Drupes 5–6 angled
Flowering and Fruiting: May–September
IUCN Status: Least Concern
Notes: Mostly found in evergreen forests
Reference: Novon 28: 151 (2020)

Schefflera stellata (Gaertn.) Baill.
Araliaceae
Synonyms: *Heptapleurum stellatum* Gaertn., *Hedera terebinthinacea* Vahl., *Hedera terebinthinacea* Vahl., *Heptapleurum acutangulum* Gaertn., *Paratropia terebinthinacea* (Vahl) Arn.
Common Name: Starry Schefflera
Climbing Mechanism: Scrambler-Unarmed
Distribution (Global): India, Sri Lanka
Distribution (India): Andhra Pradesh, Kerala, Odisha, Tamil Nadu, Western Ghats
Leaf Type: Leaves digitate, alternate distichous
Inflorescence: Panicles terminal and axillary, greenish
Fruit Type: Drup, globose, red when ripe
Flowering and Fruiting: April–June
IUCN Status: Not evaluated
Notes: Mostly distributed in moist deciduous forests
Reference: Fruct. Sem. Pl. 2: t. 178, f. 3 (1791)

Schefflera venulosa (Wight & Arn.) Harms
Araliaceae
Synonyms: *Heptapleurum venulosum* (Wight & Arn.) Seem., *Hedera venulosa* (Wight & Arn.) Bedd., *Paratropia venulosa* Wight & Arn.
Common Name: Schefflera vine
Climbing Mechanism: Scrambler-Unarmed
Distribution (Global): Bangladesh, India
Distribution (India): Andhra Pradesh, Arunachal Pradesh, Assam, Goa, Karnataka, Madhya Pradesh, Meghalaya, Mizoram, Odisha, Tamil Nadu, Tripura, Western Ghats
Leaf Type: Leaves are digitately compound
Inflorescence: Umbels
Fruit Type: Drupe, globose, 5-angled
Flowering and Fruiting: April–September
IUCN Status: Not evaluated
Notes: Mostly found in evergreen forests
Reference: J. Bot. 3: 80 (1865)

Schefflera wallichiana (Wight & Arn.) Harms
Araliaceae
Synonyms: *Heptapleurum wallichianum* (Wight & Arn.) Seem., *Aralia wallichiana* (Wight & Arn.) D.Dietr., *Paratropia wallichiana* Wight & Arn.

Climbing Mechanism: Scrambler-Unarmed
Distribution (Global): India
Distribution (India): Arunachal Pradesh, Kerala, Meghalaya, Tamil Nadu, Western Ghats
Leaf Type: Digitately, compound, alternate, stipulate
Inflorescence: Umbels on elongate racemes arranged in panicles on short axis, Pale green flowers
Fruit Type: Berry, globose, 5-angled
Flowering and Fruiting: April–May
IUCN Status: Not evaluated
Notes: Mostly found in evergreen forests
Reference: J. Bot. 3: 80 (1865)

Calamus acanthospathus Griff.
Arecaceae
Synonyms: *Calamus feanus* Becc., *Calamus feanus* var. *medogensis* S.J. Pei & S.Y. Chen, *Calamus montanus* T.Anderson, *Calamus yunnanensis* S.J.Pei & S.Y. Chen, *Calamus yunnanensis* var. *densiflorus* S.J.Pei & S.Y.Chen, *Calamus yunnanensis* var. *intermedius* S.J.Pei & S.Y.
Climbing Mechanism: Scrambler-Armed
Distribution (Global): Assam, China South-Central, East Himalaya, Laos, Myanmar, Nepal, Thailand, Tibet, Vietnam
Distribution (India): North Eastern India
Leaf Type: Leaves pinnatisect
Inflorescence: Spadicles elongate, branched
Fruit Type: Wet, subglobose
IUCN Status: Not evaluated
Reference: Calcutta J. Nat. Hist. 5: 39 (1845)

Calamus andamanicus Kurz
Arecaceae
Synonyms: *Palmijuncus andamanicus* (Kurz) Kuntze, *Calamus semierectus* Renuka & Vijayak
Climbing Mechanism: Scrambler-Armed
Distribution (Global): Andaman Is., Nicobar Is.
Distribution (India): Andaman and Nicobar Islands
Leaf Type: Leaves pinnatisect
Inflorescence: Spadicles elongate, branched
Fruit Type: Wet, subglobose
IUCN Status: Not evaluated
Reference: J. Asiat. Soc. Bengal, Pt. 2, Nat. Hist. 43: 211 (1874)

Calamus baratangensis Renuka & Vijayak
Arecaceae
Synonyms: *Calamus basui* Renuka & Vijayak

Climbing Mechanism: Scrambler-Armed
Distribution (Global): Andaman Is.
Leaf Type: Leaves pinnatisect
Inflorescence: Spadicles elongate, branched
Fruit Type: Wet, subglobose
IUCN Status: Not evaluated
Reference: Rheedea 4: 141 (1994)

Calamus basui Renuka & Vijayak
Arecaceae
Climbing Mechanism: Scrambler-Armed
Leaf Type: Leaves pinnatisect
Inflorescence: Spadicles elongate, branched
Fruit Type: Wet, subglobose
IUCN Status: Not evaluated
Reference: Rheedea 4: 120 (1994)

Calamus brandisii Becc.
Arecaceae
Climbing Mechanism: Scrambler-Armed
Distribution (Global): India
Distribution (India): Tamil Nadu
Leaf Type: Leaves pinnatisect
Inflorescence: Spadicles elongate, branched
Fruit Type: Wet, subglobose
Flowering and Fruiting: October–May
IUCN Status: Not evaluated
Notes: Frequently distributed in evergreen forests
Reference: J. D. Hooker, Fl. Brit. India 6: 448 (1892)

Calamus caesius Blume
Arecaceae
Synonyms: *Calamus glaucescens* Blume, *Calamus tapa* Becc., *Palmijuncus glaucescens* Kuntze, *Palmijuncus caesius* (Blume) Kuntze, *Rotang caesius* (Blume) Baill
Climbing Mechanism: Scrambler-Armed
Distribution (Global): Borneo, Malaya, Philippines, Sumatera, Thailand
Leaf Type: Leaves pinnatisect
Inflorescence: Spadicles elongate, branched
Fruit Type: Wet, subglobose
IUCN Status: Not evaluated
Reference: Rumphia 3: 57 (1847)

Calamus castaneus Griff.
Arecaceae

Synonyms: *Calamus griffithianus* Mart. ex Walp., *Palmijuncus griffithianus* (Mart. ex Walp.) Kuntze, *Calamus inopinatus* Furtado
Climbing Mechanism: Scrambler-Armed
Distribution (Global): Malaya, Sumatera, Thailand
Leaf Type: Leaves pinnatisect
Inflorescence: Spadicles elongate, branched
Fruit Type: Wet, subglobose
IUCN Status: Not evaluated
Reference: Calcutta J. Nat. Hist. 5: 28 (1845)

Calamus delessertianus Becc.
Arecaceae
Climbing Mechanism: Scrambler-Armed
Leaf Type: Leaves pinnatisect
Inflorescence: Spadicles elongate, branched
Fruit Type: Wet, subglobose
Flowering and Fruiting: October–April
IUCN Status: Not evaluated
Notes: Often distributed in evergreen and semi-evergreen forests
Reference: Ann. Roy. Bot. Gard. (Calcutta) 11(1): 276 (1908)

Calamus densiflorus Becc.
Arecaceae
Synonyms: *Calamus neglectus* Becc.
Climbing Mechanism: Scrambler-Armed
Distribution (Global): Malaya, Sumatera, Thailand
Leaf Type: Leaves pinnatisect
Inflorescence: Spadicles elongate, branched
Fruit Type: Wet, subglobose
IUCN Status: Not evaluated
Reference: J. D. Hooker, Fl. Brit. India 6: 445 (1893)

Calamus diepenhorstii Miq.
Arecaceae
Synonyms: *Calamus diepenhorstii* var. *major* J. Dransf., *Calamus singaporensis* Becc., *Calamus pacificus* Ridl., *Palmijuncus diepenhorstii* (Miq.) Kuntze
Climbing Mechanism: Scrambler-Armed
Distribution (Global): Borneo, Malaya, Philippines, Sumatera, Thailand
Leaf Type: Leaves pinnatisect
Inflorescence: Spadicles elongate, branched
Fruit Type: Wet, subglobose
IUCN Status: Not evaluated
Reference: Fl. Ned. Ind., Eerste Bijv.: 594 (1861)

Calamus digitatus Becc.
Arecaceae
Climbing Mechanism: Scrambler-Armed
Distribution (Global): Sri Lanka
Leaf Type: Leaves pinnatisect
Inflorescence: Spadicles elongate, branched
Fruit Type: Wet, subglobose
IUCN Status: Not evaluated
Reference: J. D. Hooker, Fl. Brit. India 6: 442 (1892)

Calamus dilaceratus Becc.
Arecaceae
Climbing Mechanism: Scrambler-Armed
Distribution (Global): Nicobar Is.
Distribution (India): Great Nicobar Island
Leaf Type: Leaves pinnatisect
Inflorescence: Spadicles elongate, branched
Fruit Type: Wet, subglobose
IUCN Status: Not evaluated
Reference: Rec. Bot. Surv. India 2: 198 (1902)

Calamus dransfieldii Renuka
Arecaceae
Climbing Mechanism: Scrambler-Armed
Distribution (India): Peninsular India
Leaf Type: Leaves pinnatisect
Inflorescence: Spadicles elongate, branched
Fruit Type: Wet, subglobose
Flowering and Fruiting: November–January
IUCN Status: Not evaluated
Reference: Kew Bull. 42: 433 (1987)

Calamus erectus Roxb.
Arecaceae
Synonyms: *Calamus collinus* Griff., *Calamus erectus* var. *schizospathus* (Griff.) Becc., *Calamus macrocarpus* Griff. ex Walp., *Calamus schizospathus* Griff., *Palmijuncus collinus* (Griff.) Kuntze, *Palmijuncus erectus* (Roxb.) Kuntze, *Palmijuncus macrocarpus* (Griff.)
Common Name: Cane fruit.
Climbing Mechanism: Scrambler-Armed
Distribution (Global): Assam, Bangladesh, China South-Central, East Himalaya, Laos, Myanmar, Nepal, Thailand
Leaf Type: Leaves pinnatisect
Inflorescence: Spadicles elongate, branched
Fruit Type: Wet, subglobose

Flowering and Fruiting: December–May
IUCN Status: Not evaluated
Reference: Fl. Ind. ed. 1832, 3: 774 (1832)

Calamus flagellum Griff. ex Mart.
Arecaceae
Synonyms: *Calamus flagellum* var. *furvifurfuraceus* S.J. Pei & S.Y.Chen, *Calamus jenkinsianus* Griff., *Calamus polygamus* Roxb., *Calamus karinensis* (Becc.) S.J. Pei & S.Y.Chen, *Palmijuncus flagellum* (Griff. ex Walp.) Kuntze, *Palmijuncus polygamus* (Roxb.) Kuntze
Climbing Mechanism: Scrambler-Armed
Distribution (Global): Assam, Bangladesh, Cambodia, China South-Central, China Southeast, East Himalaya, Laos, Myanmar, Thailand, Tibet, Vietnam
Distribution (India): North Eastern India
Leaf Type: Leaves pinnatisect
Inflorescence: Spadicles elongate, branched
Fruit Type: Wet, subglobose
IUCN Status: Not evaluated
Reference: Ann. Bot. Syst. 3: 484 (1852)

Calamus floribundus Griff.
Arecaceae
Synonyms: *Calamus mishmeensis* Griff., *Palmijuncus mishmeensis* (Griff.) Kuntze, *Palmijuncus floribundus* (Griff.) Kuntze
Climbing Mechanism: Scrambler-Armed
Distribution (Global): Assam, Bangladesh, East Himalaya, Myanmar

Distribution (India): Arunachal Pradesh, Tripura
Leaf Type: Leaves pinnatisect
Inflorescence: Spadicles elongate, branched
Fruit Type: Wet, subglobose
IUCN Status: Not evaluated
Reference: Calcutta J. Nat. Hist. 5: 56 (1845)

Calamus gamblei Becc.
Arecaceae
Synonyms: *Calamus dransfieldii* Renuka, *Calamus laccifer* Lakshmana & Renuka, *Calamus neelagiricus* Renuka, *Calamus prasinus* Lakshmana & Renuka, *Calamus renukae* J. Jacob, N.Mohanan & Kariyappa, *Calamus shendurunii* Anto, Renuka & Sreek
Climbing Mechanism: Scrambler-Armed
Distribution (Global): India
Distribution (India): Karnataka, Tamil Nadu
Leaf Type: Leaves pinnatisect

Inflorescence: Spadicles elongate, branched
Fruit Type: Wet, subglobose
Flowering and Fruiting: March–April
IUCN Status: Not evaluated
Reference: J. D. Hooker, Fl. Brit. India 6: 453 (1893)

Calamus guruba Buch. -Ham. ex Mart.
Arecaceae
Synonyms: *Calamus mastersianus* Griff., *Daemonorops guruba* (Buch. -Ham. ex Mart.) Mart., *Daemonorops guruba* var. *hamiltonianus* Mart., *Daemonorops guruba* var. *mastersianus* (Griff.) Mart., *Palmijuncus guruba* (Buch. -Ham. ex Mart.) Kuntze
Climbing Mechanism: Scrambler-Armed
Distribution (Global): Assam, Bangladesh, East Himalaya, India
Distribution (India): Odisha, Tripura, North Eastern India
Leaf Type: Leaves pinnatisect
Inflorescence: Spadicles elongate, branched
Fruit Type: Wet, subglobose
Flowering and Fruiting: March–December
IUCN Status: Not evaluated
Notes: Frequently found along the streams in the hills
Reference: Hist. Nat. Palm. 3: 211 (1838)

Calamus hookerianus Becc.
Arecaceae
Climbing Mechanism: Scrambler-Armed
Distribution (Global): India
Leaf Type: Leaves pinnatisect
Inflorescence: Spadicles elongate, branched
Fruit Type: Wet, subglobose
Flowering and Fruiting: October–April
IUCN Status: Not evaluated
Reference: Ann. Roy. Bot. Gard. (Calcutta) 11(1): 226 (1908)

Calamus hypoleucus Kurz
Arecaceae
Climbing Mechanism: Scrambler-Armed
Leaf Type: Leaves pinnatisect
Inflorescence: Spadicles elongate, branched
Fruit Type: Wet, subglobose
IUCN Status: Not evaluated
Reference: J. Asiat. Soc. Bengal, Pt. 2, Nat. Hist. 43: 208 (1874)

Calamus insignis Griff.
Arecaceae

Synonyms: *Calamus spathulatus* Becc., *Palmijuncus insignis* (Griff.) Kuntze, *Calamus subspathulatus* Ridl.
Climbing Mechanism: Scrambler-Armed
Distribution (Global): Malaya, Sumatera, Thailand
Leaf Type: Leaves pinnatisect
Inflorescence: Spadicles elongate, branched
Fruit Type: Wet, subglobose
IUCN Status: Not evaluated
Reference: Calcutta J. Nat. Hist. 5: 59 (1845)

Calamus karnatakensis Renuka & Lakshmana
Arecaceae
Climbing Mechanism: Scrambler-Armed
Distribution (India): Peninsular India
Leaf Type: Leaves pinnatisect
Inflorescence: Spadicles elongate, branched
Fruit Type: Wet, subglobose
Flowering and Fruiting: October–May
IUCN Status: Not evaluated
Reference: R. I. C. Bull. 9: 10 (1990)

Calamus lacciferus Lakshmana & Renuka
Arecaceae
Climbing Mechanism: Scrambler-Armed
Distribution (India): Peninsular India
Leaf Type: Leaves pinnatisect
Inflorescence: Spadicles elongate, branched
Fruit Type: Wet, subglobose
IUCN Status: Not evaluated
Reference: Lakshmana & Renuka. In: J. Econ. Taxon. Bot. 14: 707. (1990)

Calamus lakshmanae Renuka
Arecaceae
Climbing Mechanism: Scrambler-Armed
Distribution (Global): India
Distribution (India): Peninsular India
Leaf Type: Leaves pinnatisect
Inflorescence: Spadicles elongate, branched
Fruit Type: Wet, subglobose
IUCN Status: Not evaluated
Reference: J. Econ. Taxon. Bot. 14: 703 (1990)

Calamus latifolius Roxb.
Arecaceae

Synonyms: *Calamus dumetorum* Ridl., *Calamus gregisectus* Burret, *Calamus humilis* Roxb., *Calamus kerrianus* Becc., *Calamus loeiensis* Hodel, *Calamus macracanthus* T. Anderson, *Calamus palustris* Griff., *Calamus palustris* var. *cochinchinensis* Becc., *Palmijuncus humi*
Climbing Mechanism: Scrambler-Armed
Distribution (Global): Andaman Is., Assam, Bangladesh, Cambodia, East Himalaya, Laos, Malaya, Myanmar, Nepal, Nicobar Is., Thailand, Vietnam
Distribution (India): Andhra Pradesh, Arunachal Pradesh, Bihar, Odisha, Sikkim
Leaf Type: Leaves pinnatisect
Inflorescence: Spadicles elongate, branched
Fruit Type: Wet, subglobose
Flowering and Fruiting: December–March
IUCN Status: Not evaluated
Reference: Hort. Bengal.: 73 (1814)

Calamus leptospadix Griff.
Arecaceae
Synonyms: *Palmijuncus leptospadix* (Griff.) Kuntze
Climbing Mechanism: Scrambler-Armed
Distribution (Global): Assam, Bangladesh, East Himalaya, Myanmar, Nepal
Distribution (India): West Bengal, North Eastern India
Leaf Type: Leaves pinnatisect
Inflorescence: Spadicles elongate, branched
Fruit Type: Wet, subglobose
IUCN Status: Not evaluated
Reference: Calcutta J. Nat. Hist. 5: 49 (1845)

Calamus longisetus Griff.
Arecaceae
Synonyms: *Calamus tigrinus* Kurz, *Palmijuncus longisetus* (Griff.) Kuntze, *Palmijuncus tigrinus* (Kurz) Kuntze
Climbing Mechanism: Scrambler-Armed
Distribution (Global): Andaman Is., Assam, Bangladesh, Malaya, Myanmar, Nicobar Is., Thailand
Distribution (India): Andaman and Nicobar Islands
Leaf Type: Leaves pinnatisect
Inflorescence: Spadicles elongate, branched
Fruit Type: Wet, subglobose
IUCN Status: Not evaluated
Reference: Calcutta J. Nat. Hist. 5: 36 (1845)

Calamus mahanandensis S. Mondal, S. K. Basu & M. Chowdhury
Arecaceae
Climbing Mechanism: Scrambler-Armed
Distribution (Global): East Himalaya

Leaf Type: Leaves pinnatisect
Inflorescence: Spadicles elongate, branched
Fruit Type: Wet, subglobose
IUCN Status: Not evaluated
Reference: Turczaninowia 23(3): 93 (2020)

Calamus manan Miq.
Arecaceae
Synonyms: *Calamus giganteus* Becc., *Rotang manan* (Miq.) Baill., *Palmijuncus manan* (Miq.) Kuntze
Climbing Mechanism: Scrambler-Armed
Distribution (Global): Borneo, Malaya, Sumatera, Thailand
Leaf Type: Leaves pinnatisect
Inflorescence: Spadicles elongate, branched
Fruit Type: Wet, subglobose
IUCN Status: Not evaluated
Reference: Fl. Ned. Ind., Eerste Bijv.: 595 (1861)

Calamus meghalayensis A.J. Hend.
Arecaceae
Synonyms: *Calamus floribundus* var. *depauperatus* Becc.
Climbing Mechanism: Scrambler-Armed
Distribution (Global): Assam
Leaf Type: Leaves pinnatisect
Inflorescence: Spadicles elongate, branched
Fruit Type: Wet, subglobose
IUCN Status: Not evaluated
Reference: Taiwania 52: 155 (2007)

Calamus melanacanthus Mart.
Arecaceae
Climbing Mechanism: Scrambler-Armed
Leaf Type: Leaves pinnatisect
Inflorescence: Spadicles elongate, branched
Fruit Type: Wet, subglobose
IUCN Status: Not evaluated
Reference: Hist. Nat. Palm., 3: 211, 1838

Calamus metzianus Schltdl.
Arecaceae
Synonyms: *Palmijuncus rivalis* (Thwaites ex Trimen) Kuntze, *Calamus rivalis* Thwaites ex Trimen.
Climbing Mechanism: Scrambler-Armed
Distribution (Global): India, Sri Lanka
Distribution (India): Peninsular India

Leaf Type: Leaves pinnatisect
Inflorescence: Spadicles elongate, branched
Fruit Type: Wet, subglobose
Flowering and Fruiting: October–May
IUCN Status: Not evaluated
Reference: Linnaea 26: 727 (1855)

Calamus nagbettai R.R. Fernald & Dey
Arecaceae
Climbing Mechanism: Scrambler-Armed
Distribution (Global): India
Distribution (India): Peninsular India
Leaf Type: Leaves pinnatisect
Inflorescence: Spadicles elongate, branched
Fruit Type: Wet, subglobose
Flowering and Fruiting: December–January
IUCN Status: Not evaluated
Reference: Indian Forester 96: 223 (1970)

Calamus nambariensis Becc.
Arecaceae
Climbing Mechanism: Scrambler-Armed
Leaf Type: Leaves pinnatisect
Inflorescence: Spadicles elongate, branched
Fruit Type: Wet, subglobose
IUCN Status: Not evaluated
Reference: Ann. Roy. Bot. Gard. (Calcutta) 11(1): 433 (1908)

Calamus neelagiricus Renuka
Arecaceae
Climbing Mechanism: Scrambler-Armed
Distribution (India): Peninsular India
Leaf Type: Leaves pinnatisect
Inflorescence: Spadicles elongate, branched
Fruit Type: Wet, subglobose
Flowering and Fruiting: April–May
IUCN Status: Not evaluated
Reference: Rheedea 7: 69 (1997)

Calamus nicobaricus Becc.
Arecaceae
Synonyms: *Calamus pseudorivalis* Becc.
Climbing Mechanism: Scrambler-Armed
Distribution (Global): Nicobar Is.
Leaf Type: Leaves pinnatisect

Inflorescence: Spadicles elongate, branched
Fruit Type: Wet, subglobose
IUCN Status: Not evaluated
Reference: J. D. Hooker, Fl. Brit. India 6: 446 (1892)

Calamus ovoideus Thwaites ex Trimen
Arecaceae
Synonyms: *Palmijuncus ovoideus* (Thwaites ex Trimen) Kuntze
Climbing Mechanism: Scrambler-Armed
Distribution (Global): Sri Lanka
Leaf Type: Leaves pinnatisect
Inflorescence: Spadicles elongate, branched
Fruit Type: Wet, subglobose
IUCN Status: Not evaluated
Reference: J. Bot. 23: 269 (1885)

Calamus oxleyanus Teijsm. & Binn. ex Miq.
Arecaceae
Synonyms: *Calamus diffusus* Becc., *Calamus fernandezii* H. Wendl., *Calamus leiospathus* Bartlett, *Daemonorops fasciculata* Mart. + 3
Climbing Mechanism: Scrambler-Armed
Distribution (Global): Malaya, Sumatera
Leaf Type: Leaves pinnatisect
Inflorescence: Spadicles elongate, branched
Fruit Type: Wet, subglobose
IUCN Status: Least Concern
Reference: Palm. Archip. Ind.: 17 (1868)

Calamus pachystemonus Thwaites
Arecaceae
Synonyms: *Palmijuncus pachystemonus* (Thwaites) Kuntze
Calamus gracilis Thwaites
Climbing Mechanism: Scrambler-Armed
Distribution (Global): Sri Lanka
Leaf Type: Leaves pinnatisect
Inflorescence: Spadicles elongate, branched
Fruit Type: Wet, subglobose
IUCN Status: Not evaluated
Reference: Enum. Pl. Zeyl.: 431 (1864)

Calamus palustris Griff.
Arecaceae
Synonyms: *Calamus dumetorum* Ridl., *Calamus kerrianus* Becc., *Calamus loeiensis* Hodel, *Calamus palustris* var. *amplissimus* Becc., *Calamus palustris* var.

cochinchinensis Becc., *Calamus palustris* var. *malaccensis* Becc., *Palmijuncus palustris* (Griff.) Kuntze
Climbing Mechanism: Scrambler-Armed
Distribution (India): Andaman and Nicobar Islands
Leaf Type: Leaves pinnatisect
Inflorescence: Spadicles elongate, branched
Fruit Type: Wet, subglobose
IUCN Status: Not evaluated
Reference: Calcutta J. Nat. Hist. 5: 60 (1845)

Calamus penicillatus Roxb.
Arecaceae
Synonyms: *Calamus martianus* Becc., *Calamus penicillatus* Mart., *Calamus penangensis* Ridl., *Palmijuncus penicillatus* (Roxb.) Kuntze
Climbing Mechanism: Scrambler-Armed
Distribution (Global): Malaya
Leaf Type: Leaves pinnatisect
Inflorescence: Spadicles elongate, branched
Fruit Type: Wet, subglobose
IUCN Status: Not evaluated
Reference: Fl. Ind. ed. 1832, 3: 781 (1832)

Calamus prasinus Lakshmana & Renuka
Arecaceae
Climbing Mechanism: Scrambler-Armed
Distribution (India): Peninsular India
Leaf Type: Leaves pinnatisect
Inflorescence: Spadicles elongate, branched
Fruit Type: Wet, subglobose
Flowering and Fruiting: November–December
IUCN Status: Not evaluated
Notes: Cane is used in furniture industry
Reference: J. Econ. Taxon. Bot. 14: 705 (1991)

Calamus pseudoerectus S. Mondal, S.K. Basu & M. Chowdhury
Arecaceae
Climbing Mechanism: Scrambler-Armed
Distribution (India): West Bengal
Leaf Type: Leaves pinnatisect
Inflorescence: Spadicles elongate, branched
Fruit Type: Wet, subglobose
IUCN Status: Not evaluated

Calamus pseudorivalis Becc.
Arecaceae

Climbing Mechanism: Scrambler-Armed
Distribution (India): Andaman and Nicobar Islands
Leaf Type: Leaves pinnatisect
Inflorescence: Spadicles elongate, branched
Fruit Type: Wet, subglobose
IUCN Status: Not evaluated
Reference: Ann. Roy. Bot. Gard. (Calcutta) 11(1): 222 (1908)

Calamus pseudotenuis Becc.
Arecaceae
Common Name: Slender rattan cane
Climbing Mechanism: Scrambler-Armed
Distribution (Global): India, Sri Lanka
Distribution (India): Tamil Nadu, Goa
Leaf Type: Leaves pinnatisect
Inflorescence: Spadicles elongate, branched
Fruit Type: Wet, subglobose
Flowering and Fruiting: October–February
IUCN Status: Not evaluated
Notes: Occasional in evergreen forests across ghats. Cane is used in manufacturing furniture and baskets
Reference: J. D. Hooker, Fl. Brit. India 6: 445 (1892)

Calamus quinquenervius Roxb.
Arecaceae
Synonyms: Palmijuncus quinquenervius (Roxb.) Kuntze
Climbing Mechanism: Scrambler-Armed
Distribution (Global): Bangladesh
Leaf Type: Leaves pinnatisect
Inflorescence: Spadicles elongate, branched
Fruit Type: Wet, subglobose
IUCN Status: Not evaluated
Reference: Fl. Ind. ed. 1832, 3: 777 (1832)

Calamus radiatus Thwaites
Arecaceae
Synonyms: *Palmijuncus radiatus* (Thwaites) Kuntze
Climbing Mechanism: Scrambler-Armed
Distribution (Global): Sri Lanka
Leaf Type: Leaves pinnatisect
Inflorescence: Spadicles elongate, branched
Fruit Type: Wet, subglobose
IUCN Status: Not evaluated
Reference: Enum. Pl. Zeyl.: 431 (1864)

Calamus renukae J.Jacob, N.Mohanan & Kariyappa
Arecaceae
Climbing Mechanism: Scrambler-Armed
Distribution (India): Peninsular India
Leaf Type: Leaves pinnatisect
Inflorescence: Spadicles elongate, branched
Fruit Type: Wet, subglobose
Flowering and Fruiting: Throughout the year
IUCN Status: Not evaluated
Reference: Rheedea 18: 29 (2008)

Calamus rheedei Griff.
Arecaceae
Synonyms: *Calamus travancoricus* Bedd. ex Becc., *Palmijuncus rheedei* (Griff.) Kuntze, *Daemonorops rheedei* (Griff.) Mart.
Climbing Mechanism: Scrambler-Armed
Distribution (Global): India
Distribution (India): Tamil Nadu
Leaf Type: Leaves pinnatisect
Inflorescence: Spadicles elongate, branched
Fruit Type: Wet, subglobose
IUCN Status: Not evaluated
Reference: Calcutta J. Nat. Hist. 5: 73 (1845)

Calamus semierectus Renuka & Vijayak
Arecaceae
Climbing Mechanism: Scrambler-Armed
Leaf Type: Leaves pinnatisect
Inflorescence: Spadicles elongate, branched
Fruit Type: Wet, subglobose
IUCN Status: Not evaluated
Reference: Rheedea 4: 122 (1994)

Calamus shendurunii Anto, Renuka & Sreek
Arecaceae
Climbing Mechanism: Scrambler-Armed
Distribution (India): Peninsular India
Leaf Type: Leaves pinnatisect
Inflorescence: Spadicles elongate, branched
Fruit Type: Wet, subglobose
Flowering and Fruiting: Throughout the year
IUCN Status: Not evaluated
Reference: Rheedea 11: 37 (2001)

Calamus stoloniferus Renuka
Arecaceae
Climbing Mechanism: Scrambler-Armed
Distribution (India): Peninsular India
Leaf Type: Leaves pinnatisect
Inflorescence: Spadicles elongate, branched
Fruit Type: Wet, subglobose
IUCN Status: Not evaluated
Reference: Renuka. In: J. Econ. Taxon. Bot. 14: 701. (1990)

Calamus tenuis Roxb.
Arecaceae
Synonyms: *Calamus amarus* Lour., *Calamus delessertianus* Becc., *Calamus heliotropium* Buch. -Ham. ex Mart., *Calamus royleanus* Griff., *Calamus stolonifer* Teijsm. & Binn., *Palmijuncus amarus* (Lour.) Kuntze, *Palmijuncus heliotropium* (Buch. -Ham. ex Mart.) Kuntze
Climbing Mechanism: Scrambler-Armed
Distribution (Global): Assam, Bangladesh, East Himalaya, India, Myanmar, West Himalaya
Distribution (India): Uttar Pradesh, West Bengal
Leaf Type: Leaves pinnatisect
Inflorescence: Spadicles elongate, branched
Fruit Type: Wet, subglobose
IUCN Status: Least Concern
Reference: Fl. Ind. ed. 1832, 3: 780 (1832)

Calamus thwaitesii Becc.
Arecaceae
Synonyms: *Calamus thwaitesii* var. *canaranus* Becc.
Common Name: Rattan cane
Climbing Mechanism: Scrambler-Armed
Distribution (Global): India, Sri Lanka
Distribution (India): Goa, Tamil Nadu
Leaf Type: Leaves pinnatisect
Inflorescence: Spadicles elongate, branched
Fruit Type: Wet, subglobose
Flowering and Fruiting: December–January
IUCN Status: Not evaluated
Notes: Cane is used in manufacturing walking sticks and suspension bridges
Reference: J. D. Hooker, Fl. Brit. India 6: 441 (1892)

Calamus travancoricus Bedd. ex Becc.
Arecaceae
Climbing Mechanism: Scrambler-Armed
Distribution (India): Peninsular India

Leaf Type: Leaves pinnatisect
Inflorescence: Spadicles elongate, branched
Fruit Type: Wet, subglobose
Flowering and Fruiting: October–June
IUCN Status: Not evaluated
Reference: J. D. Hooker, Fl. Brit. India 6: 452 (1893)

Calamus unifarius H. Wendl.
Arecaceae
Synonyms: Palmijuncus unifarius (H. Wendl.) Kuntze
Climbing Mechanism: Scrambler-Armed
Distribution (Global): Java, Lesser Sunda Is., Nicobar Is., Sumatera
Distribution (India): Great Nicobar Island
Leaf Type: Leaves pinnatisect
Inflorescence: Spadicles elongate, branched
Fruit Type: Wet, subglobose
IUCN Status: Not evaluated
Reference: Bot. Zeitung (Berlin) 17: 158 (1859)

Calamus unifarius var. *pentong*
Arecaceae
Climbing Mechanism: Scrambler-Armed
Leaf Type: Leaves pinnatisect
Inflorescence: Spadicles elongate, branched
Fruit Type: Wet, subglobose
IUCN Status: Not evaluated
Reference: H. Wendl. In: Bot. Zeitung (Berlin) 17: 158. (1859)

Calamus vattayila Renuka
Arecaceae
Synonyms: *Calamus pseudofeanus* S.K. Basu
Climbing Mechanism: Scrambler-Armed
Distribution (Global): India.
Distribution (India): Tamil Nadu
Leaf Type: Leaves pinnatisect
Inflorescence: Spadicles elongate, branched
Fruit Type: Wet, subglobose
Flowering and Fruiting: May–August
IUCN Status: Not evaluated
Reference: Curr. Sci. 56: 1012 (1987)

Calamus viminalis Willd.
Arecaceae
Synonyms: *Calamus fasciculatus* Roxb., *Calamus fasciculatus* subvar. *andamanicus* Becc., *Calamus fasciculatus* subvar. *bengalensis* Becc., *Calamus*

fasciculatus subvar. *cochinchinensis* Becc., *Calamus fasciculatus* subvar. *pinangianus* Becc., *Calamus litoralis* Blume
Climbing Mechanism: Scrambler-Armed
Distribution (Global): Andaman Is., Assam, Bangladesh, Cambodia, China South-Central, East Himalaya, India, Java, Laos, Lesser Sunda Is., Malaya, Myanmar, Nicobar Is., Thailand, Vietnam
Distribution (India): Andaman and Nicobar Islands, Andhra Pradesh
Leaf Type: Leaves pinnatisect
Inflorescence: Spadicles elongate, branched
Fruit Type: Wet, subglobose
Flowering and Fruiting: September–April
IUCN Status: Not evaluated
Reference: Sp. Pl., ed. 4, 2: 203 (1799)

Calamus wightii Griff.
Arecaceae
Synonyms: *Calamus huegelianus* Mart. ex Walp., *Calamus melanolepis* (Mart.) H. Wendl., *Daemonorops melanolepis* Mart., *Palmijuncus huegelianus* (Mart. ex Walp.) Kuntze, *Palmijuncus melanolepis* (Mart.) Kuntze, *Palmijuncus wightii* (Griff.) Kuntze
Climbing Mechanism: Scrambler-Armed
Distribution (Global): India
Leaf Type: Leaves pinnatisect
Inflorescence: Spadicles elongate, branched
Fruit Type: Wet, subglobose
Flowering and Fruiting: May–July
IUCN Status: Not evaluated
Reference: Palms Brit. E. Ind.: t. 216C (1850)

Daemonorops angustifolia (Griff.) Mart.
Arecaceae
Synonyms: *Calamus melanochaetes* (Blume) Miq., *Acanthophoenix grandis* André, *Calamus acanthopis* Griff., *Calamus angustifolius* Griff. + 80
Climbing Mechanism: Scrambler-Armed
Distribution (Global): Andaman Is., Assam, Bangladesh, Cambodia, China South-east, East Himalaya, Hainan, Java, Laos, Lesser Sunda Is., Malaya, Myanmar, New Guinea, Philippines, Sumatera, Taiwan, Thailand, Vietnam
Leaf Type: Pinnately compound
Inflorescence: Erect
Fruit Type: Globose
Flowering and Fruiting: April–May
IUCN Status: Not evaluated
Notes: Usually found in evergreen forests of Shoal Bay, South Andamans
Reference: Verh. Kon. Akad. Wetensch., Afd. Natuurk. 11(5): 28 (1868)

Daemonorops aurea Renuka & Vijayak
Arecaceae
Climbing Mechanism: Scrambler-Armed
IUCN Status: Not evaluated
Reference: Renuka & Vijayak. In: Rheedea 4: 122. (1994)

Daemonorops calicarpa (Griff.) Mart.
Arecaceae
Synonyms: *Calamus calicarpus* Griff. *Calamus petiolaris* Griff., *Daemonorops calicarpa* (Griff.) Mart., *Daemonorops microthamnus* Becc., *Daemonorops petiolaris* (Griff.) Mart., *Palmijuncus calicarpus* (Griff.) Kuntze, *Palmijuncus petiolaris* (Griff.) Kuntze
Climbing Mechanism: Scrambler-Armed
Distribution (Global): Malaya, Sumatera
Leaf Type: Pinnate, cirrate, very large
Inflorescence: Head inflorescence
Fruit Type: rounded, tipped by a short beak
IUCN Status: Not evaluated
Reference: Calcutta J. Nat. Hist. 5: 92 (1845)

Daemonorops didymophylla Becc.
Arecaceae
Synonyms: *Calamus gracilipes* Miq. *Calamus cochleatus* Miq. + 10
Climbing Mechanism: Scrambler-Armed
Distribution (Global): Borneo, Malaya, Sumatera, Thailand
Leaf Type: Pinnately compound, Opposite or alternate pairs
Fruit Type: Ripe fruit ovoid with endosperm
IUCN Status: Not evaluated
Notes: Produces a medium-quality cane, commercially used
Reference: Verh. Kon. Akad. Wetensch., Afd. Natuurk. 11(5): 28 (1868)

Daemonorops draco (Willd.) Blume
Arecaceae
Synonyms: *Calamus draco* Willd., *Calamus draconis* Oken, *Palmijuncus draco* (Willd.) Kuntze
Climbing Mechanism: Scrambler-Armed
Distribution (Global): Sumatera
IUCN Status: Not evaluated
Notes: Produces high-quality cane, commercially used
Reference: Sp. Pl., ed. 4, 2: 203 (1799)

Daemonorops geniculata (Griff.) Mart.
Arecaceae
Synonyms: *Calamus geniculatus* Griff., *Calamus collarifer* (Becc.) W.J. Baker, *Calamus plagiocyclus* (Burret) W.J. Baker, *Daemonorops acanthobola* Becc.,

Daemonorops collarifera Becc., *Daemonorops geniculata* (Griff.) Mart., *Daemonorops plagiocycla* Burret
Climbing Mechanism: Scrambler-Armed
Distribution (Global): Borneo, Malaya, Sumatera, Thailand
Leaf Type: Pinnately compound
Fruit Type: Rounded, endosperm ruminate
IUCN Status: Not evaluated
Reference: Calcutta J. Nat. Hist. 5: 67 (1845)

Daemonorops grandis (Griff.) Mart.
Arecaceae
Synonyms: *Calamus melanochaetes* (Blume) Miq., *Acanthophoenix grandis* André, *Calamus acanthopis* Griff., *Calamus angustifolius* Griff. +80
Climbing Mechanism: Scrambler-Armed
IUCN Status: Not evaluated
Reference: Verh. Kon. Akad. Wetensch., Afd. Natuurk. 11(5): 28 (1868)

Daemonorops hirsuta Blume
Arecaceae
Synonyms: *Calamus hirsutus* (Blume) Miq. *Palmijuncus hirsutus* (Blume) Kuntze
Climbing Mechanism: Scrambler-Armed
Distribution (Global): Borneo, Malaya, Sumatera
IUCN Status: Not evaluated
Reference: Verh. Kon. Akad. Wetensch., Afd. Natuurk. 11(5): 28 (1868)

Daemonorops jenkinsiana (Griff.) Mart.
Arecaceae
Synonyms:
Climbing Mechanism: Scrambler-Armed
Distribution (India): Bihar, Odisha, Sikkim
IUCN Status: Not evaluated
Reference: Mart. In: Hist. Nat. Palm. 3: 327 (1853)

Daemonorops kurziana Hook.f. ex Becc.
Arecaceae
Synonyms:
Climbing Mechanism: Scrambler-Armed
Distribution (India): Andaman and Nicobar Islands
IUCN Status: Not evaluated
Reference: J. D. Hooker. In: Fl. Brit. India 6: 463. (1893)

Daemonorops leptopus (Griff.) Mart.
Arecaceae
Synonyms: *Calamus leptopus* Griff., *Daemonorops congesta* Ridl., *Daemonorops leptopus* (Griff.) Mart., *Palmijuncus leptopus* (Griff.) Kuntze

Climbing Mechanism: Scrambler-Armed
Distribution (Global): Malaya, Thailand
IUCN Status: Not evaluated
Reference: Calcutta J. Nat. Hist. 5: 73 (1845)

Daemonorops lewisiana (Griff.) Mart.
Arecaceae
Climbing Mechanism: Scrambler-Armed
IUCN Status: Not evaluated
Reference: Mart. In: Hist. Nat. Palm. 3: 327. (1853)

Daemonorops manii Becc.
Arecaceae
Climbing Mechanism: Scrambler-Armed
Distribution (India): Andaman and Nicobar Islands
IUCN Status: Not evaluated
Reference: J. D. Hooker. In: Fl. Brit. India 6: 463. (1893)

Daemonorops rarispinosa Renuka & Vijayak
Arecaceae
Climbing Mechanism: Scrambler-Armed
IUCN Status: Not evaluated
Reference: Renuka & Vijayak. In: Rheedea 4: 125. (1994)

Daemonorops verticillaris (Griff.) Mart.
Arecaceae
Synonyms: *Calamus verticillaris* Griff. *Daemonorops setigera* Ridl., *Daemonorops stipitata* Furtado., *Daemonorops verticillaris* var. *straminea* Furtado, *Palmijuncus verticillaris* (Griff.) Kuntze
Climbing Mechanism: Scrambler-Armed
Distribution (Global): Malaya, Sumatera, Thailand
Leaf Type: Pinnate
IUCN Status: Not evaluated
Reference: Calcutta J. Nat. Hist. 5: 63 (1845)

Daemonorops wrightmyoensis Renuka & Vijayak
Arecaceae
Climbing Mechanism: Scrambler-Armed
IUCN Status: Not evaluated
Reference: Renuka & Vijayak. In: Rheedea 4: 125. (1994)

Korthalsia laciniosa (Griff.) Mart.
Arecaceae
Synonyms: *Calamosagus harinifolius* Griff., *Calamosagus laciniosus* Griff., *Calamosagus wallichiifolius* Griff., *Korthalsia andamanensis* Becc., *Korthalsia*

grandis Ridl., *Korthalsia scaphigera* Kurz, *Korthalsia teysmannii* Miq., *Korthalsia wallichiifolia* (Griff.)
Climbing Mechanism: Scrambler-Armed
Distribution (Global): Andaman Is., Cambodia, Java, Laos, Malaya, Myanmar, Nicobar Is., Philippines, Sumatera, Thailand, Vietnam
Distribution (India): Andaman and Nicobar Islands
Leaf Type: leaflets, equidistant, cuneate-rhomboid, irregularly toothed, dark green above, pale below
Inflorescence: Inflorescence large, twice or thrice branched, flower arranged in longitudinal rows
Fruit Type: Fruit widely depressed ovate, orange-red, endosperm ruminant
Flowering and Fruiting: April–November
IUCN Status: Not evaluated
Notes: Widely spread all over South Andaman and Nicobar Islands
Reference: Hist. Nat. Palm. 3: 211 (1845)

Korthalsia rogersii Becc.
Arecaceae
Climbing Mechanism: Scrambler-Armed
Distribution (Global): Andaman Is.
Leaf Type: Leaflets, rhomboid, sheath light green, sparingly armed with spines, praemorse, versatile, long yellowish green
Inflorescence: Partial inflorescence with upto 5 rachillae lax
Fruit Type: Obovoid with persistent calyx, ruminant endosperm
IUCN Status: Not evaluated
Reference: nn. Roy. Bot. Gard. (Calcutta) 12(2): 131 (1918)

Plectocomia assamica Griff.
Arecaceae
Climbing Mechanism: Scrambler-Armed
Distribution (Global): Assam, China South-Central, East Himalaya, Myanmar
Distribution (India): North Eastern India
Leaf Type: Pinnate, leaf sheaths without knees, covered spines
Inflorescence: Axillary spikes
Fruit Type: Wet, globose
IUCN Status: Not evaluated
Reference: Calcutta J. Nat. Hist. 5: 97 (1845)

Plectocomia bractealis Becc.
Arecaceae
Climbing Mechanism: Scrambler-Armed
Distribution (Global): Assam
Leaf Type: Pinnate, leaf sheaths without knees, covered spines
Inflorescence: Axillary spikes
Fruit Type: Wet, globose

IUCN Status: Not evaluated
Reference: Ann. Roy. Bot. Gard. (Calcutta) 12(2): 40 (1918)

Plectocomia himalayana Griff.
Arecaceae
Synonyms: *Plectocomia andersonii* W. Bull, *Plectocomia montana* Hook.f. & Thomson, *Plectocomia montana* Griff. ex T. Anderson
Climbing Mechanism: Scrambler-Armed
Distribution (Global): Assam, Bangladesh, China South-Central, East Himalaya, Laos, Nepal, Thailand
Distribution (India): North Eastern India
Leaf Type: Pinnate, leaf sheaths without knees, covered spines
Inflorescence: Axillary spikes
Fruit Type: Wet, globose
IUCN Status: Least Concern
Reference: Calcutta J. Nat. Hist. 5: 100 (1845)

Plectocomia khasyana Griff.
Arecaceae
Climbing Mechanism: Scrambler-Armed
Distribution (Global): Assam
Leaf Type: Pinnate, leaf sheaths without knees, covered spines
Inflorescence: Axillary spikes
Fruit Type: Wet, globose
IUCN Status: Not evaluated
Reference: Calcutta J. Nat. Hist. 5: 98 (1845)

Plectocomia macrostachya Kurz
Arecaceae
Climbing Mechanism: Scrambler-Armed
Distribution (Global): Myanmar
Leaf Type: Pinnate, leaf sheaths without knees, covered spines
Inflorescence: Axillary spikes
Fruit Type: Wet, globose
IUCN Status: Not evaluated
Reference: J. Asiat. Soc. Bengal, Pt. 2, Nat. Hist. 43: 207 (1874)

Calamus rotang L.
Arecaceae+B397E403
Synonyms: *Calamus roxburghii* Griff., *Calamus scipionum* Lam., *Calamus monoecus* Roxb., *Palmijuncus monoecus* (Roxb.) Kuntze, *Rotang linnaei* Baill., *Rotanga calamus* Crantz, *Draco rotang* Crantz
Common Name: Common rattan
Climbing Mechanism: Scrambler-Armed
Distribution (Global): India, Sri Lanka

Distribution (India): Andhra Pradesh, Gujarat, Odisha, Tamil Nadu, Telangana
Leaf Type: Leaves pinnatisect
Inflorescence: Spadicles elongate, branched
Fruit Type: Wet, subglobose
Flowering and Fruiting: Throughout the year
IUCN Status: Not evaluated
Notes: Cane is used for making furniture and baskets
Reference: Sp. Pl.: 325 (1753)

Aristolochia assamica
Aristolochiaceae
Climbing Mechanism: Stem Twiner
IUCN Status: Not evaluated

Aristolochia atropurpurea Parish ex Hook.f.
Aristolochiaceae
Climbing Mechanism: Stem Twiner
Distribution (Global): Myanmar
Distribution (India): Rajasthan
Leaf Type: Simple, lobed
Inflorescence: Simple, racemes in leaf-axils
Fruit Type: Dry, capsule
IUCN Status: Not evaluated
Reference: Fl. Brit. India 5: 76 (1886)

Aristolochia bracteolata Lam.
Aristolochiaceae
Synonyms: *Aristolochia bracteata* Retz., *Aristolochia crenata* Ehrenb. ex Duch., *Aristolochia abyssinica* Klotzsch, *Aristolochia mauritiana* Pers., *Aristolochia maurorum* Klotzsch, *Aristolochia sempervirens* Forssk
Common Name: Bracteated birthwort, Worm killer
Climbing Mechanism: Stem Twiner
Distribution (Global): Assam, Bangladesh, Burkina, Cameroon, Central African Repu, Chad, Djibouti, Eritrea, Ethiopia, Gulf States, India, Kenya, Mali, Myanmar, Niger, Nigeria, Oman, Pakistan, Saudi Arabia, Somalia, Sri Lanka, Sudan, Tanzania, Uganda, Yemen
Distribution (India): Telangana, Andhra Pradesh, Maharashtra, Tamil Nadu, Madhya Pradesh
Leaf Type: Simple, lobed
Inflorescence: Simple, racemes in leaf-axils
Fruit Type: Dry, capsule
Flowering and Fruiting: Throughout the year
IUCN Status: Not evaluated
Notes: Leaf paste is used in treating itches and insect bites
Reference: Encycl. 1: 258 (1783)

Aristolochia cymbifera Mart.
Aristolochiaceae
Synonyms: *Ambuya labiosa* Raf., *Aristolochia abbreviata* Mart. ex Mast., *Aristolochia cymbifera* var. *abbreviata* Duch., *Aristolochia cymbifera* var. *genuina* Duch., *Aristolochia cymbifera* var. *labiosa* (Ker Gawl.) Duch., *Aristolochia galeata* Moritzi
Climbing Mechanism: Stem Twiner
Distribution (Global): Brazil Southeast

.
Distribution (India): Peninsular India
Leaf Type: Simple, lobed
Inflorescence: Simple, racemes in leaf-axils
Fruit Type: Dry, capsule
IUCN Status: Not evaluated
Reference: Nov. Gen. Sp. Pl. 1: 76 (1824)

Aristolochia fimbriata Cham.
Aristolochiaceae
Synonyms: *Aristolochia bonplandii* Ten., *Aristolochia ciliata* Hook., *Aristolochia ciliosa* Benth., *Aristolochia insignis* Verschaff., *Howardia fimbriata* Klotzsch.
Climbing Mechanism: Stem Twiner
Distribution (Global): Argentina Northeast, Bolivia, Brazil South, Paraguay, Uruguay
Distribution (India): Peninsular India
Leaf Type: Simple, lobed
Inflorescence: Simple, racemes in leaf-axils
Fruit Type: Dry, capsule
Flowering and Fruiting: September–October
IUCN Status: Not evaluated
Reference: Linnaea 7: 210 (1832)

Aristolochia gourigangaica N.C. Nair
Aristolochiaceae
Synonyms: *Aristolochia dilatata* N.E.Br.
Climbing Mechanism: Stem Twiner
Leaf Type: Simple, lobed
Inflorescence: Simple, racemes in leaf-axils
Fruit Type: Dry, capsule
IUCN Status: Not evaluated
Reference: Bull. Bot. Surv. India 10: 332 (1969)

Aristolochia griffithii Hook.f. & Thomson ex Duch.
Aristolochiaceae

Synonyms: *Isotrema griffithii* (Hook.f. & Thomson ex Duch.) C.E.C. Fisch., *Siphisia griffithii* Klotzsch ex Duch.
Climbing Mechanism: Stem Twiner
Distribution (Global): Assam, East Himalaya, Myanmar, Nepal, Tibet
Distribution (India): North Eastern India
Leaf Type: Simple, lobed
Inflorescence: Simple, racemes in leaf-axils
Fruit Type: Dry, capsule
IUCN Status: Not evaluated
Reference: A.P.de Candolle, Prodr. 15(1): 437 (1864)

Aristolochia gurinderii Ravi Kumar Umesh Kumar Tiwari & N. Balachandran
Aristolochiaceae
Climbing Mechanism: Stem Twiner
Distribution (India): Andaman and Nicobar Islands
Leaf Type: Simple, lobed
Inflorescence: Simple, racemes in leaf-axils
Fruit Type: Dry, capsule
IUCN Status: Not evaluated
Reference: K. Ravik., U. L. Tiwari & N. Balach. In: Phytotaxa 172(2): 118. (2014)

Aristolochia indica L.
Aristolochiaceae
Synonyms: *Aristolochia indica* var. *magna* Benth.
Aristolochia lanceolata Wight
Aristolochia maysorensis Fisch. ex Duch.
Aristolochia pandurata Wall. ex Duch.
Common Name: Indian birthwort
Climbing Mechanism: Stem Twiner
Distribution (Global): Andaman Is., Assam, Bangladesh, East Himalaya, India, Myanmar, Nepal, Sri Lanka, Vietnam
Distribution (India): Peninsular India, Odisha, West Bengal, Gujarat, Rajasthan, Madhya Pradesh
Leaf Type: Simple, lobed
Inflorescence: Simple, racemes in leaf-axils
Fruit Type: Dry, capsule
Flowering and Fruiting: July–February
IUCN Status: Not evaluated
Reference: Sp. Pl.: 960 (1753)

Aristolochia inflata Kunth
Aristolochiaceae
Synonyms: *Aristolochia gibbosa* Duch., *Aristolochia odoratissima* Benth., *Aristolochia podocarpa* Bertol., *Howardia benthamii* Klotzsch *Aristolochia torta* Willd. ex Duch., *Howardia inflata* (Kunth) Klotzsch.

Climbing Mechanism: Stem Twiner
Distribution (Global): Colombia, Costa Rica, El Salvador, Guatemala, Honduras, Mexico Gulf, Mexico Southeast, Mexico Southwest, Nicaragua, Panamá, Venezuela
Leaf Type: Simple, lobed
Inflorescence: Simple, racemes in leaf-axils
Fruit Type: Dry, capsule
IUCN Status: Not evaluated
Reference: F.W.H.von Humboldt, A.J.A. Bonpland & C.S. Kunth, Nov. Gen. Sp. 2: 145 (1817)

Aristolochia jackii Steud.
Aristolochiaceae
Synonyms: *Aristolochia hastata* Jack, *Aristolochia tripartita* Backer, *Aristolochia ungulifolia* Mast.
Climbing Mechanism: Stem Twiner
Distribution (Global): Borneo, Java, Malaya, Nicobar Is., Philippines, Sumatera
Leaf Type: Simple, lobed
Inflorescence: Simple, racemes in leaf-axils
Fruit Type: Dry, capsule
IUCN Status: Not evaluated
Reference: Nomencl. Bot., ed. 2, 1: 141 (1840)

Aristolochia krisagathra Sivar. & Pradeep
Aristolochiaceae
Climbing Mechanism: Stem Twiner
Distribution (Global): India
Distribution (India): Peninsular India
Leaf Type: Simple, lobed
Inflorescence: Simple, racemes in leaf-axils
Fruit Type: Dry, capsule
Flowering and Fruiting: September–January
IUCN Status: Not evaluated
Notes: Often found in semi-evergreen forests
Reference: Pl. Syst. Evol. 163: 31 (1989)

Aristolochia labiata Willd.
Aristolochiaceae
Synonyms: *Aristolochia brasiliensis* Mart., *Aristolochia brasiliensis* var. *galeata* (Mart.) Hoehne, *Aristolochia brasiliensis* var. *macrophylla* Duch., *Aristolochia brasiliensis* var. *parviflora* Duch., *Aristolochia galeata* Mart., *Aristolochia macrorrhyncha* Hoehne
Climbing Mechanism: Stem Twiner
Distribution (Global): Brazil North, Brazil Northeast, Brazil South, Brazil Southeast, Brazil West-Central

Leaf Type: Simple
Fruit Type: Dry dehiscent capsule
Flowering and Fruiting: Throughout the year
IUCN Status: Not evaluated
Reference: Mém. Soc. Imp. Naturalistes Moscou 2: 101 (1809)

Aristolochia littoralis Parodi
Aristolochiaceae
Synonyms: *Aristolochia elegans* Mast., *Aristolochia elegans* var. *hassleriana* (Chodat) Hassl., *Aristolochia hassleriana* Chodat
Climbing Mechanism: Stem Twiner
Distribution (Global): Argentina Northeast, Argentina Northwest, Bolivia, Brazil North, Brazil Northeast, Brazil South, Brazil Southeast, Brazil West-Central, Colombia, Ecuador, Paraguay, Peru
Distribution (India): Madhya Pradesh, Maharashtra
Leaf Type: Simple, lobed
Inflorescence: Simple, racemes in leaf-axils
Fruit Type: Dry, capsule
Flowering and Fruiting: April–July
IUCN Status: Not evaluated
Notes: Cultivated and used as an ornamental plant
Reference: Anales Soc. Ci. Argent. 5: 155 (1878)

Aristolochia platanifolia (Klotzsch) Duch.
Aristolochiaceae
Synonyms: *Isotrema platanifolium* (Klotzsch) X.X. Zhu, S.Liao & J.S.Ma, *Siphisia platanifolia* Klotzsch.
Climbing Mechanism: Stem Twiner
Distribution (Global): Assam, East Himalaya
Distribution (India): Arunachal Pradesh
Leaf Type: Simple, lobed
Inflorescence: Simple, racemes in leaf-axils
Fruit Type: Dry, capsule
Flowering and Fruiting: October–April
IUCN Status: Not evaluated
Reference: A.P.de Candolle, Prodr. 15(1): 437 (1864)

Aristolochia punjabensis Lace
Aristolochiaceae
Synonyms: *Isotrema punjabense* (Lace) X.X. Zhu, S.Liao & J.S.Ma
Climbing Mechanism: Stem Twiner
Distribution (Global): Pakistan, West Himalaya
Distribution (India): Himachal Pradesh
Leaf Type: Simple, lobed
Inflorescence: Simple, racemes in leaf-axils

Fruit Type: Dry, capsule
IUCN Status: Not evaluated
Reference: Bull. Misc. Inform. Kew 1911: 273 (1911)

Aristolochia ringens Vahl
Aristolochiaceae
Synonyms: *Aristolochia turbacensis* Kunth
Howardia cymbifera Klotzsch, *Howardia galeata* Klotzsch, *Howardia ringens* (Vahl) Klotzsch.
Common Name: Gaping Dutchman's Pipe, Pipe Vine
Climbing Mechanism: Stem Twiner
Distribution (Global): Argentina Northwest, Bolivia, Colombia, Costa Rica, Ecuador, El Salvador, Panamá, Peru, Venezuela
Distribution (India): Tamil Nadu, Maharashtra
Leaf Type: Simple, lobed
Inflorescence: Simple, racemes in leaf-axils
Fruit Type: Dry, capsule
Flowering and Fruiting: January–June
IUCN Status: Not evaluated
Reference: Symb. Bot. 3: 99 (1794)

Aristolochia saccata Wall.
Aristolochiaceae
Synonyms: *Isotrema saccatum* (Wall.) X.X. Zhu, S. Liao & J.S. Ma, *Siphisia angustifolia* Klotzsch, *Siphisia saccata* (Wall.) Klotzsch.
Climbing Mechanism: Stem Twiner
Distribution (Global): Assam, Bangladesh, East Himalaya, Myanmar, Nepal, Thailand, Vietnam
Distribution (India): Arunachal Pradesh, Mizoram
Leaf Type: Simple, lobed
Inflorescence: Simple, racemes in leaf-axils
Fruit Type: Dry, capsule
IUCN Status: Not evaluated
Reference: Pl. Asiat. Rar. 2: 2 (1830)

Aristolochia tagala Cham.
Aristolochiaceae
Synonyms: *Aristolochia acuminata* Roxb., *Aristolochia anguicida* Sieber ex Duch., *Aristolochia angulosa* Wall. ex Duch., *Aristolochia eschscholtzii* Ledeb. ex Duch., *Aristolochia japonica* Miq. + 8
Climbing Mechanism: Stem Twiner
Distribution (India): Andaman and Nicobar Islands, Andhra Pradesh, Arunachal Pradesh, Assam, Bihar, Odisha, Karnataka, Maharashtra, Tamil Nadu, Western Ghats
Leaf Type: Simple, lobed

Inflorescence: Simple, racemes in leaf-axils
Fruit Type: Dry, capsule
Flowering and Fruiting: Throughout the year
IUCN Status: Not evaluated
Notes: Found on thickets and in semi-evergreen forests
Reference: Linnaea 7: 207 (1832)

Aristolochia tomentosa
Aristolochiaceae
Climbing Mechanism: Stem Twiner
IUCN Status: Not evaluated

Asparagus aethiopicus L.
Asparagaceae
Synonyms: *Asparagopsis aethiopica* (L.) Kunth, *Asparagopsis lancea* (Thunb.) Kunth, *Asparagus aculeatus* Voss, *Asparagus laetus* Salisb., *Asparagus lanceus* Thunb., *Asparagus maximus* Voss, *Protasparagus aethiopicus* (L.) Oberm.
Climbing Mechanism: Scrambler-Armed
Distribution (Global): Cape Provinces, Northern Provinces
Leaf Type: Scale leaves in clusters, linear-falcate
Inflorescence: Axillary racemes
Fruit Type: Wet, berry
IUCN Status: Least Concern
Reference: Mant. Pl.: 1 (1770)

Asparagus dumosus Baker
Asparagaceae
Synonyms: Protasparagus dumosus (Baker) Kamble
Climbing Mechanism: Scrambler-Armed
Distribution (Global): India, Pakistan
Distribution (India): Gujarat
Leaf Type: Scale leaves in clusters, linear-falcate
Inflorescence: Axillary racemes
Fruit Type: Wet, berry
IUCN Status: Not evaluated
Reference: J. Linn. Soc., Bot. 14: 608 (1875)

Asparagus falcatus L.
Asparagaceae
Synonyms: *Protasparagus falcatus* (L.) Oberm.
Climbing Mechanism: Scrambler-Armed
Distribution (Global): Cape Provinces, Comoros, Ethiopia, Gulf States, India, Kenya, KwaZulu-Natal, Mozambique, Northern Provinces, Rwanda, Saudi Arabia, Somalia, Sri Lanka, Swaziland, Tanzania, Uganda, Yemen, Zimbabwe
Leaf Type: Scale leaves in clusters, linear-falcate

Inflorescence: Axillary racemes
Fruit Type: Wet, berry
IUCN Status: Not evaluated
Reference: Sp. Pl.: 313 (1753)

Asparagus filicinus Buch. -Ham. ex D. Don
Asparagaceae
Synonyms: *Asparagus qinghaiensis* Y. Wan, *Protasparagus filicinus* (Buch. -Ham. ex D. Don) Kamble, *Protasparagus filicinus* var. *brevifolius* (Baker) Kamble, *Protasparagus filicinus* var. *brevipes* (Baker) Kamble, *Protasparagus filicinus* var. *microclada* (Hook.f.) Kamble
Common Name: Fern Asparagus
Climbing Mechanism: Scrambler-Armed
Distribution (Global): Assam, Bangladesh, China North-Central, China South-Central, China Southeast, East Himalaya, Myanmar, Nepal, Thailand, Vietnam, West Himalaya
Distribution (India): Arunachal Pradesh
Leaf Type: Scale leaves in clusters, linear-falcate
Inflorescence: Axillary racemes
Fruit Type: Wet, berry
IUCN Status: Data Deficient
Reference: Prodr. Fl. Nepal.: 49 (1825)

Asparagus fysonii J.F. Macbr.
Asparagaceae
Synonyms: *Asparagus asiaticus* Wight, *Asparagus subulatus* Steud. ex Baker, *Protasparagus fysonii* (J.F. Macbr.) Kamble
Climbing Mechanism: Scrambler-Armed
Distribution (Global): India
Distribution (India): Western Ghats
Leaf Type: Scale leaves in clusters, linear-falcate
Inflorescence: Axillary racemes
Fruit Type: Wet, berry
Flowering and Fruiting: March–July
IUCN Status: Not evaluated
Reference: Contr. Gray Herb. 56: 17 (1918)

Asparagus gonoclados Baker
Asparagaceae
Synonyms: *Protasparagus gonoclados* (Baker) Kamble
Climbing Mechanism: Scrambler-Armed
Distribution (Global): India, Sri Lanka
Distribution (India): Andhra Pradesh, Gujarat, Tamil Nadu, Telangana
Leaf Type: Scale leaves in clusters, linear-falcate
Inflorescence: Axillary racemes

Fruit Type: Wet, berry
Flowering and Fruiting: August–January
IUCN Status: Not evaluated
Notes: Common in thickets of deciduous forests
Reference: J. Linn. Soc., Bot. 14: 627 (1875)

Asparagus karthikeyanii (Kamble) M.R. Almeida
Asparagaceae
Synonyms: *Protasparagus karthikeyanii* Kamble
Climbing Mechanism: Scrambler-Armed
Distribution (Global): India
Leaf Type: Scale leaves in clusters, linear-falcate
Inflorescence: Axillary racemes
Fruit Type: Wet, berry
IUCN Status: Not evaluated
Reference: Fl. Maharashtra 5A: 166 (2009)

Asparagus laevissimus Steud. ex Baker
Asparagaceae
Synonyms: *Protasparagus laevissimus* (Steud. ex Baker) Kamble
Climbing Mechanism: Scrambler-Armed
Distribution (Global): India
Distribution (India): Andhra Pradesh, Tamil Nadu
Leaf Type: Scale leaves in clusters, linear-falcate
Inflorescence: Axillary racemes
Fruit Type: Wet, berry
Flowering and Fruiting: May–August
IUCN Status: Not evaluated
Reference: J. Linn. Soc., Bot. 14: 623 (1875)

Asparagus neglectus
Asparagaceae
Climbing Mechanism: Scrambler-Armed
IUCN Status: Not evaluated

Asparagus officinalis L.
Asparagaceae
Synonyms: *Asparagus altilis* (L.) Asch., *Asparagus altilis* subsp. *oxycarpus* (Steven) K. Richt., *Asparagus caspius* Schult. & Schult.f., *Asparagus caspius* Hohen., *Asparagus collinus* Schur, *Asparagus esculentus* Salisb., *Asparagus fiori* Sennen, *Asparagus hedecarp*
Common Name: Garden asparagus
Climbing Mechanism: Scrambler-Armed
Distribution (Global): Afghanistan, Albania, Altay, Austria, Belarus, Bulgaria, Central European Rus, Corse, Czechoslovakia, France, Germany, Greece, Hungary,

Iran, Kazakhstan, Krasnoyarsk, Krym, Lebanon-Syria, Mongolia, North Caucasus, Pakistan, Poland, Sicilia, South European Russi, Switzerland, Transcaucasus, Turkey, Turkey-in-Europe, Ukraine, West Siberia, Xinjiang
Distribution (India): Himachal Pradesh, Tamil Nadu
Leaf Type: Scale leaves in clusters, linear-falcate
Inflorescence: Axillary racemes
Fruit Type: Wet, berry
IUCN Status: Least Concern
Reference: Sp. Pl.: 313 (1753)

Asparagus racemosus var. *javanicus*
Asparagaceae
Climbing Mechanism: Scrambler-Armed
Distribution (India): Andhra Pradesh, Gujarat
Leaf Type: Scale leaves in clusters, linear-falcate
Inflorescence: Axillary racemes
Fruit Type: Wet, berry
Flowering and Fruiting: June–January
IUCN Status: Not evaluated

Asparagus racemosus Willd.
Asparagaceae
Synonyms: *Asparagopsis abyssinica* Kunth, *Asparagopsis acerosa* Kunth, *Asparagopsis brownei* Kunth, *Asparagopsis decaisnei* Kunth, *Asparagopsis floribunda* Kunth, *Asparagopsis hohenackeri* Kunth, *Asparagopsis javanica* Kunth, *Asparagopsis retrofracta* Schweinf. ex
Common Name: Indian Asparagus, Spiny Asparagus, Wild Asparagus
Climbing Mechanism: Scrambler-Armed
Distribution (Global): Andaman Is., Angola, Assam, Bangladesh, Burkina, Cambodia, Central African Repu, East Himalaya, Equatorial Guinea, Eritrea, Ethiopia, Gabon, Ghana, Guinea, Guinea-Bissau, India, Ivory Coast, Java, Kenya, Lesser Sunda Is., Liberia, Madagascar, Malaya, Maldives, Mali, Maluku, Mozambique, Myanmar, Nepal, Nigeria, Northern Territory, Oman, Pakistan, Queensland, Rwanda, Sierra Leone, Somalia, Sri Lanka, Sudan, Tanzania, Thailand, Tibet, Togo, Uganda, Vietnam, West Himalaya, Western Australia, Yemen, Zimbabwe
Distribution (India): Andaman and Nicobar Islands, Andhra Pradesh, Arunachal Pradesh, Bihar, Odisha, Delhi, Goa, Gujarat, Himachal Pradesh, Jharkhand, Kerala, Madhya Pradesh, Maharashtra, Odisha, Puducherry, Rajasthan, Sikkim, Tamil Nadu, Telangana, Uttar Pradesh, West Bengal, Western Ghats
Leaf Type: Simple
Inflorescence: Axillary, racemes
Fruit Type: Berry
Flowering and Fruiting: June–September

IUCN Status: Not evaluated
Reference: Sp. Pl., ed. 4, 2: 152 (1799)

Asparagus recemosus var. *subacerosa*
Asparagaceae
Climbing Mechanism: Scrambler-Armed
Distribution (India): Sikkim
Leaf Type: Scale leaves in clusters, linear-falcate
Inflorescence: Axillary racemes
Fruit Type: Wet, berry
IUCN Status: Not evaluated

Asparagus rottleri Baker
Asparagaceae
Synonyms: *Asparagus floribundus* Royle, *Protasparagus rottleri* (Baker) Kamble
Climbing Mechanism: Scrambler-Armed
Distribution (Global): India
Leaf Type: Scale leaves in clusters, linear-falcate
Inflorescence: Axillary racemes
Fruit Type: Wet, berry
IUCN Status: Not evaluated
Reference: J. Linn. Soc., Bot. 14: 611 (1875)

Asparagus scandens Thunb.
Asparagaceae
Synonyms: *Asparagopsis scandens* (Thunb.) Kunth, *Asparagus pectinatus* Redouté, *Myrsiphyllum scandens* (Thunb.) Oberm.
Climbing Mechanism: Scrambler-Armed
Distribution (Global): Cape Provinces
Distribution (India): Gujarat
Leaf Type: Scale leaves in clusters, linear-falcate
Inflorescence: Axillary racemes
Fruit Type: Wet, berry
IUCN Status: Not evaluated
Reference: Prodr. Pl. Cap.: 66 (1794)

Asparagus setaceus (Kunth) Jessop
Asparagaceae
Synonyms: *Asparagopsis setacea* Kunth, *Asparagus lujae* De Wild., *Asparagus plumosus* Baker, *Asparagus tenuissimus* Van Geert, *Asparagus zanzibaricus* Baker, *Protasparagus plumosus* (Baker) Oberm., *Protasparagus setaceus* (Kunth) Oberm.
Common Name: Lacy Asparagus fern
Climbing Mechanism: Scrambler-Armed

Distribution (Global): Botswana, Cape Provinces, Comoros, Ethiopia, Free State, Kenya, KwaZulu-Natal, Lesotho, Malawi, Mozambique, Northern Provinces, Swaziland, Tanzania, Zambia, Zimbabwe
Distribution (India): Tamil Nadu
Leaf Type: Scale leaves in clusters, linear-falcate
Inflorescence: Axillary racemes
Fruit Type: Wet, berry
Flowering and Fruiting: June–January
IUCN Status: Not evaluated
Reference: Bothalia 9: 51 (1966)

Asparagus virgatus Baker
Asparagaceae
Synonyms: *Asparagus sylvaticus* Burch. ex Baker, *Protasparagus virgatus* (Baker) Oberm.
Climbing Mechanism: Scrambler-Armed
Distribution (Global): Angola, Cape Provinces, Free State, KwaZulu-Natal, Malawi, Mozambique, Namibia, Northern Provinces, Swaziland, Tanzania, Yemen, Zambia, Zimbabwe
Leaf Type: Scale leaves in clusters, linear-falcate
Inflorescence: Axillary racemes
Fruit Type: Wet, berry
IUCN Status: Not evaluated
Reference: Refug. Bot. 3: t. 214 (1870)

Blumea repanda (Roxb.) Hand. -Mazz.
Asteraceae
Synonyms: *Blumea eberhardtii* Gagnep., *Blumea procera* DC., *Blumea semivestita* DC., *Conyza procera* Wall., *Conyza repanda* Roxb. + 4
Climbing Mechanism: Scrambler-Unarmed
Distribution (Global): Andaman Is., Assam, Bangladesh, China South-Central, East Himalaya, Java, Laos, Myanmar, Nepal, New Caledonia, Vietnam
IUCN Status: Not evaluated
Reference: Symb. Sin. 7: 1378 (1936)

Blumea riparia (Blume) DC.
Asteraceae
Synonyms: *Baccharis nitida* Wall., *Blumea nitida* Wall. ex C.B. Clarke, *Conyza longissima* Zoll. & Moritzi, *Conyza riparia* Blume +3
Climbing Mechanism: Scrambler-Unarmed
Distribution (Global): Andaman Is., Assam, Borneo, Cambodia, China South-Central, China Southeast, East Himalaya, Hainan, India, Java, Laos, Malaya, Maluku, Myanmar, Nepal, New Caledonia, New Guinea, Nicobar Is., Philippines, Queensland, Solomon Is., Sulawesi, Sumatera, Taiwan, Thailand, Vietnam

Distribution (India): Arunachal Pradesh, Manipur, West Bengal
Flowering and Fruiting: November–February
IUCN Status: Not evaluated
Reference: Prodr. 5: 444 (1836)

Cissampelopsis ansteadii (Tadul. & Jacob) C. Jeffrey & Y.L. Chen
Asteraceae
Synonyms: *Senecio ansteadii* Tadul. & Jacob
Climbing Mechanism: Stem Twiner
Distribution (Global): India
Leaf Type: Alternate
Inflorescence: Axillary and terminal corymbs
Fruit Type: Achenes glabrous.
Flowering and Fruiting: March–April
IUCN Status: Not evaluated
Reference: Kew Bull. 39: 341 (1984)

Cissampelopsis calcadensis C. Jeffrey & Y.L. Chen
Asteraceae
Synonyms: *Senecio calcadensis* Ramaswami
Climbing Mechanism: Stem Twiner
Distribution (Global): India
Distribution (India): Peninsular India
Leaf Type: Ovate-triangular
Inflorescence: Capitula, axillary panicles
Fruit Type: Cypselas
Flowering and Fruiting: February
IUCN Status: Not evaluated
Reference: Kew Bull. 39: 341 (1984)

Cissampelopsis corifolia C. Jeffrey & Y.L. Chen
Asteraceae
Climbing Mechanism: Stem Twiner
Distribution (Global): Assam, China South-Central, East Himalaya, Myanmar, Nepal, Thailand, Tibet, Vietnam
Distribution (India): Western Ghats
Leaf Type: Ovate to broadly ovate
Inflorescence: Capitulate, Panicle broad, florets
Fruit Type: Cypselas
Flowering and Fruiting: September–March
IUCN Status: Not evaluated
Reference: Kew Bull. 39: 342 (1984)

Cissampelopsis corymbosa (Wall. ex DC.) C.Jeffrey & Y.L. Chen
Asteraceae

Synonyms: *Senecio corymbosus* Wall. ex DC., *Staehelina corymbosa* Wight ex DC.
Climbing Mechanism: Stem Twiner
Distribution (Global): India, Sri Lanka, Vietnam
Distribution (India): Tamil Nadu
Leaf Type: Ovate
Inflorescence: Panicles road, Capitula large, numerous in usually dense panicles
Fruit Type: Cypselas
Flowering and Fruiting: December–February
IUCN Status: Not evaluated
Reference: Kew Bull. 39: 341 (1984)

Cissampelopsis vivekananthanii Gopal & Chitra
Asteraceae
Climbing Mechanism: Stem Twiner
Distribution (Global): India
Distribution (India): Tamil Nadu
Leaf Type: Simple, orbicular/cordate
Inflorescence: Corymbose cyme
Fruit Type: Wet, drupe
IUCN Status: Not evaluated
Reference: J. Econ. Taxon. Bot. 32: 599 (2008)

Cissampelopsis volubilis (Blume) Miq.
Asteraceae
Synonyms: *Cacalia volubilis* Blume, *Ligularia araneosa* DC., *Senecio arachnoideus* Wall., *Senecio araneosus* DC., *Senecio blumei* DC., *Senecio chionopappus* Sch.Bip., *Senecio hoi* Dunn, *Senecio subaraneosus* Sch.Bip., *Vernonia esquirolii* Vaniot
Climbing Mechanism: Stem Twiner
Distribution (Global): Assam, Borneo, China South-Central, China Southeast, East Himalaya, Hainan, Java, Laos, Malaya, Myanmar, Sulawesi, Sumatera, Thailand, Vietnam
Distribution (India): Manipur, Tamil Nadu.
Leaf Type: Ovate or ovate-triangular
Inflorescence: Panicles narrow, capitula numerous per panicle
Fruit Type: Cypselas
Flowering and Fruiting: October–February
IUCN Status: Not evaluated
Reference: Fl. Ned. Ind. 2: 103 (1856)

Cissampelopsis walkeri (Arn.) C.Jeffrey & Y.L. Chen
Asteraceae
Synonyms: *Senecio corymbosus* var. *walkeri* (Arn.) Grierson, *Senecio walkeri* Arn.
Climbing Mechanism: Stem Twiner
Distribution (Global): India, Sri Lanka

Distribution (India): Arunachal Pradesh, Tamil Nadu, Western Ghats
Leaf Type: Alternate, ovate
Inflorescence: Corymbose terminal, panicle broad, capitula numerous per panicle
Fruit Type: Cypselas/Achenes
Flowering and Fruiting: January–April
IUCN Status: Not evaluated
Reference: Kew Bull. 39: 341 (1984)

Decaneuropsis blanda (DC.) H. Rob. & Skvarla
Asteraceae
Synonyms: *Cacalia blanda* Kuntze
Conyza blanda Wall.
Gymnanthemum blandum Steetz
Vernonia blanda DC.
Vernonia blandula C.B. Clarke
Vernonia tavoyana C.E.C. Fisch.
Climbing Mechanism: Scrambler-Unarmed
Distribution (Global): Assam, China South-Central, China Southeast, Laos, Myanmar, Thailand, Tibet
IUCN Status: Not evaluated
Reference: Proc. Biol. Soc. Washington 120: 364 (2007)

Decaneuropsis vagans (DC.) H. Rob. & Skvarla
Asteraceae
Synonyms: *Cacalia vagans* Kuntze, *Conyza vagans* Wall., *Gymnanthemum vagans* Steetz, *Vernonia vagans* DC.
Climbing Mechanism: Scrambler-Unarmed
Distribution (Global): Assam, East Himalaya
IUCN Status: Not evaluated
Reference: Proc. Biol. Soc. Washington 120: 365 (2007)

Delairea odorata Lem.
Asteraceae
Synonyms: *Senecio mikanioides* Otto ex Walp., *Senecio scandens* DC.
Climbing Mechanism: Stem Twiner
Distribution (Global): Cape Provinces, KwaZulu-Natal, Lesotho
Distribution (India): Tamil Nadu.
IUCN Status: Not evaluated
Reference: Ann. Sci. Nat., Bot., sér. 3, 1: 380 (1844)

Microglossa pyrifolia (Lam.) Kuntze
Asteraceae
Climbing Mechanism: Scrambler-Unarmed
Distribution (India): Mizoram
IUCN Status: Not evaluated

Mikania cordata (Burm.f.) B.L. Rob.
Asteraceae
Climbing Mechanism: Stem Twiner
Distribution (India): Arunachal Pradesh, Assam, Tamil Nadu, Tripura
Leaf Type: Simple, opposite, ovate or triangular
Inflorescence: 4 small heads, greenish-white
Fruit Type: Dry, achenes
IUCN Status: Not evaluated

Mikania micrantha Kunth
Asteraceae
Climbing Mechanism: Stem Twiner
Distribution (India): Arunachal Pradesh, Assam, Kerala, Manipur, Meghalaya, Sikkim, Tamil Nadu, Uttar Pradesh, West Bengal
Leaf Type: Simple, opposite, ovate or triangular
Inflorescence: 5 small heads, greenish-white
Fruit Type: Dry, achenes
IUCN Status: Not evaluated

Mikania scandens (L.) Willd.
Asteraceae
Climbing Mechanism: Stem Twiner
Leaf Type: Simple, opposite, ovate or triangular
Inflorescence: Six small heads, greenish-white
Fruit Type: Dry, achenes
IUCN Status: Not evaluated

Piptocarpha riedelii (Sch.Bip.) Baker
Asteraceae
Synonyms: *Carphobolus riedelii* Sch.Bip., *Vernonia conferta* Sch.Bip. ex Baker, *Vernonia scandens* Less. ex Baker
Climbing Mechanism: Scrambler-Unarmed
Distribution (Global): Brazil Northeast
IUCN Status: Not evaluated
Reference: C.F.P.von Martius & auct. suc. (eds.), Fl. Bras. 6(2): 124 (1873)

Prenanthes scandens Hook.f. & Thomson
Asteraceae
Synonyms: *Notoseris scandens* (Hook.f. ex Benth. & Hook.f.) N. Kilian
Climbing Mechanism: Scrambler-Unarmed
Distribution (India): Arunachal Pradesh, West Bengal
IUCN Status: Not evaluated
Reference: Gen. Pl. 2: 527 (1873)

Pseudogynoxys chenopodioides (Kunth) Cabrera
Asteraceae
Synonyms: *Cacalia scandens* Zeyh. ex DC., *Gynoxys berlandieri* DC., *Gynoxys berlandieri* var. *cordifolia* DC., *Gynoxys berlandieri* var. *cuneata* DC., *Gynoxys cordifolia* var. *neaei* DC., *Pseudogynoxys berlandieri* (DC.) Cabrera +5
Climbing Mechanism: Stem Twiner
Distribution (Global): Belize, Brazil South, Brazil Southeast, Colombia, El Salvador, Guatemala, Honduras, Mexico Central, Mexico Gulf, Mexico Northeast, Mexico Northwest, Mexico Southeast, Mexico Southwest, Nicaragua, Peru, Venezuela, Venezuelan Antilles.
IUCN Status: Not evaluated
Reference: Brittonia 7: 56 (1950)

Senecio confusus Burtt
Asteraceae
Synonyms: *Pseudogynoxys chenopodioides* (Kunth) Cabrera., *Cacalia scandens* Zeyh. ex DC., *Gynoxys berlandieri* DC., *Gynoxys berlandieri* var. *cordifolia* DC., *Gynoxys berlandieri* var. *cuneata* DC., *Gynoxys cordifolia* var. *neaei* DC., *Pseudogynoxys berlandieri* (DC.) Cabrer
Common Name: Mexican flame vine, Orange glow vine
Climbing Mechanism: Scrambler-Unarmed
Distribution (Global): Belize, Brazil South, Brazil Southeast, Colombia, El Salvador, Guatemala, Honduras, Mexico Central, Mexico Gulf, Mexico Northeast, Mexico Northwest, Mexico Southeast, Mexico Southwest, Nicaragua, Peru, Venezuela, Venezuelan Antilles
Distribution (India): Peninsular India
Leaf Type: Thick and Alternate leaves
IUCN Status: Not evaluated
Reference: Brittonia 7: 56 (1950)

Senecio wightianus DC. ex Wight
Asteraceae
Synonyms: *Senecio scandens* var. *scandens*., *Cacalia wightiana* Wall. ex DC. + 12
Climbing Mechanism: Scrambler-Unarmed
Distribution (Global): Assam, Cambodia, China North-Central, China South-Central, China Southeast, East Himalaya, Hainan, India, Japan, Laos, Myanmar, Nepal, Philippines, Qinghai, Sri Lanka, Sulawesi, Taiwan, Thailand, Tibet, Vietnam, West Himalaya
Distribution (India): Arunachal Pradesh, Mizoram
IUCN Status: Not evaluated
Reference: Wight, Contr. Bot. India 22. 1834

Tarlmounia elliptica (DC.) H. Rob., S.C. Keeley, Sandhyarani & Karuppusamy, Skvarla & R. Chan
Asteraceae

Synonyms: *Cacalia elaeagnifolia* Kuntze., *Strobocalyx elliptica* Sch.Bip., *Conyza elaeagnifolia* Wall., *Vernonia elaeagnifolia* DC., *Eupatorium elaeagnifolium* DC., *Vernonia elliptica* DC., *Strobocalyx elaeagnifolia* Sch.Bip.
Common Name: Curtain creeper, Vernonia creeper
Climbing Mechanism: Stem Twiner
Distribution (Global): Cambodia, China South-Central, Laos, Malaya, Myanmar, Thailand, Vietnam
Distribution (India): Tamil Nadu, Telangana
Flowering and Fruiting: January–April
IUCN Status: Not evaluated
Reference: Proc. Biol. Soc. Washington 121: 32 (2008)

Vernonia anceps C.B. Clarke ex Hook.f.
Asteraceae
Synonyms: *Uniyala anceps* (C.B. Clarke ex Hook.f.) H. Rob. & Skvarla., *Gymnanthemum anceps* (C.B. Clarke ex Hook.f.) H. Rob.
Climbing Mechanism: Scrambler-Unarmed
Distribution (Global): India, Sri Lanka
Distribution (India): Rajasthan
Leaf Type: Simple, alternate
Inflorescence: Corymbose panicle
Fruit Type: Dry, achenes
IUCN Status: Not evaluated
Reference: Proc. Biol. Soc. Washington 122: 153 (2009)

Vernonia gratiosa Hance
Asteraceae
Synonyms: *Decaneuropsis gratiosa* (Hance) H. Rob. & Skvarla., *Cacalia gratiosa* Kuntze
Climbing Mechanism: Scrambler-Unarmed
Distribution (Global): China Southeast, Taiwan
Leaf Type: Simple, alternate
Inflorescence: Corymbose panicle
Fruit Type: Dry, achenes
IUCN Status: Not evaluated
Reference: Proc. Biol. Soc. Washington 120: 365 (2007)

Vernonia thomsonii Hook.f.
Asteraceae
Synonyms: *Cacalia thomsonii* Kuntze
Climbing Mechanism: Scrambler-Unarmed
Distribution (Global): Bangladesh, Myanmar
Leaf Type: Simple, alternate
Inflorescence: Corymbose panicle
Fruit Type: Dry, achenes

IUCN Status: Not evaluated
Reference: Fl. Brit. India 3: 232 (1881)

Anredera baselloides (Kunth) Baill.
Basellaceae
Synonyms: *Anredera weberbaueri* (Ulbr.) Soukup, *Boussingaultia baselloides* Kunth, *Boussingaultia weberbaueri* Ulbr., *Tournonia leptostachys* Moq.
Climbing Mechanism: Stem Twiner
Distribution (Global): Ecuador, Peru
IUCN Status: Not evaluated
Reference: Hist. Pl. 9: 147 (1888)

Anredera cordifolia (Ten.) Steenis
Basellaceae
Synonyms: *Anredera americana* J.St.-Hil., *Anredera cordifolia* subsp. *gracilis* (Miers) Xifreda & Argimón, *Boussingaultia cordata* Spreng., *Boussingaultia cordifolia* Ten., *Boussingaultia gracilis* Miers, *Boussingaultia gracilis* f.
Common Name: Madeira vine, Mignonette vine
Climbing Mechanism: Stem Twiner
Distribution (Global): Argentina Northeast, Argentina Northwest, Bolivia, Brazil South, Brazil Southeast, Colombia, Ecuador, Paraguay, Peru, Uruguay, Venezuela
Distribution (India): Tamil Nadu
Flowering and Fruiting: June–October
IUCN Status: Not evaluated
Reference: Fl. Males. 5: 303 (1957)

Basella alba L.
Basellaceae
Synonyms: *Basella alba* var. *cordifolia* (Lam.) M.R. Almeida, *Basella cananifolia* Buch. -Ham. ex Wall., *Basella cordifolia* Lam., *Basella crassifolia* Salisb., *Basella japonica* Burm.f., *Basella lucida* L., *Basella nigra* Lour., *Basella oleracea* Alef., *Basella ramos*
Common Name: Indian Spinach, Malabar Spinach
Climbing Mechanism: Stem Twiner
Distribution (Global): Bangladesh, Borneo, Cambodia, India, Java, Laos, Lesser Sunda Is., Malaya, Maluku, Myanmar, New Guinea, Philippines, Sri Lanka, Sulawesi, Sumatera, Thailand, Vietnam
Distribution (India): Gujarat, Himachal Pradesh, Madhya Pradesh, Tamil Nadu, Telangana, Uttar Pradesh, West Bengal
Leaf Type: Simple
Inflorescence: Axillary or terminal spike
Flowering and Fruiting: Throughout the year
IUCN Status: Not evaluated
Reference: Sp. Pl.: 272 (1753)

Begonia mannii
Begoniaceae
Climbing Mechanism: Scrambler-Unarmed
IUCN Status: Not evaluated

Berberis rotundifolia Poepp. & Endl.
Berberidaceae
Synonyms: *Berberis heterophylla* var. *pluriflora* Reiche, *Berberis philippii* Ahrendt, *Berberis polymorpha* Phil., *Berberis umbellata* Phil.
Climbing Mechanism: Scrambler-Unarmed
Distribution (Global): Chile Central, Chile South
IUCN Status: Not evaluated
Reference: Nov. Gen. Sp. Pl. 2: 63 (1839)

Adenocalymma comosum (Cham.) DC.
Bignoniaceae
Synonyms: *Bignonia comosa*, *Adenocalymma nitidum*, *Adenocalymma acutissimum*
Common Name: Yellow trumpet vine
Climbing Mechanism: Stem Twiner
Distribution (Global): Brazil
Distribution (India): Delhi, Maharashtra
Leaf Type: Simple, Deeply lobed
Inflorescence: Simple, white
Fruit Type: Wet, Capsule
IUCN Status: Not evaluated
Reference: Prodr. 9: 201 (1845)

Amphilophium buccinatorium (DC.) L.G. Lohmann
Bignoniaceae
Synonyms: *Bignonia buccinatoria* (DC.) Mairet ex Hemsl., *Bignonia cherere* Lindl., *Bignonia giesbreghtii* C. Heller ex Peyr, *Distictis buccinatoria* (DC.) A.H. Gentry, *Phaedranthus buccinatorius* (DC.) Miers, *Phaedranthus exsertus* (DC.) Miers, *Phaedranthus lindleyan*
Climbing Mechanism: Scrambler-Unarmed
Distribution (Global): Mexico Central, Mexico Northeast, Mexico Southeast, Mexico Southwest
Leaf Type: 2 or 3 foliolate
IUCN Status: Not evaluated
Reference: Ann. Missouri Bot. Gard. 99: 401 (2014)

Amphilophium paniculatum (L.) Kunth
Bignoniaceae
Synonyms: *Amphilophium glanduliferum* Mart. ex DC., *Amphilophium macrophyllum* Kunth, *Amphilophium molle* Schltdl. & Cham., *Amphilophium*

mollicomum Pittier, *Amphilophium mutisii* Kunth, *Amphilophium paniculatum* var. *imatacense* A.H. Gentry, *Amphilophium paniculatum*
Common Name: Duckbill Creeper, Cuello Creeper
Climbing Mechanism: Scrambler-Unarmed
Distribution (Global): Argentina Northeast, Argentina Northwest, Belize, Bolivia, Brazil North, Brazil Northeast, Brazil South, Brazil Southeast, Brazil West-Central, Colombia, Costa Rica, Ecuador, El Salvador, French Guiana, Guatemala, Guyana, Honduras, Leeward Is., Mexico Gulf, Mexico Northeast, Mexico Northwest, Mexico Southeast, Mexico Southwest, Nicaragua, Panamá, Paraguay, Peru, Puerto Rico, Suriname, Trinidad-Tobago, Venezuela, Windward Is.
Leaf Type: 3 or 3 foliolate
Flowering and Fruiting: September–November
IUCN Status: Not evaluated
Reference: F.W.H.von Humboldt, A.J.A. Bonpland & C.S. Kunth, Nov. Gen. Sp. Pl. 3: 148 (1818 publ. 1819)

Anemopaegma chamberlaynii (Sims) Bureau & K. Schum.
Bignoniaceae
Synonyms: *Anemopaegma chamberlaynii* var. *brachybotrys* (DC.) Bureau & K. Schum., *Anemopaegma chamberlaynii* var. *tenerior* Bureau & K. Schum., *Anemopaegma citrifolium* (DC.) Baill., *Anemopaegma longipes* K. Schum., *Anemopaegma racemosum* Mart. ex DC., *Anemopaegma racem*
Common Name: Trumpet vine, Yellow trumpet vine
Climbing Mechanism: Scrambler-Unarmed
Distribution (Global): Bolivia, Brazil Northeast, Brazil South, Brazil Southeast, Brazil West-Central, Paraguay

.
Distribution (India): Maharashtra
IUCN Status: Not evaluated
Reference: C.F.P.von Martius & auct. suc. (eds.), Fl. Bras. 8(2): 128 (1896)

Bignonia aequinoctialis L.
Bignoniaceae
Synonyms: *Anemopaegma tonduzianum* Kraenzl., *Arrabidaea guatemalensis* K. Schum. & Loes., *Arrabidaea isthmica* Standl., *Arrabidaea pseudochica* Kraenzl., *Bignonia aequinoctialis* var. *hirtella* (Benth.) J.F. Morales +25
Climbing Mechanism: Scrambler-Unarmed
Distribution (Global): Belize, Bolivia, Brazil North, Brazil Northeast, Brazil West-Central, Colombia, Costa Rica, Cuba, Dominican Republic, Ecuador, El Salvador, French Guiana, Guatemala, Guyana, Haiti, Honduras, Leeward Is., Mexico Gulf, Mexico Northwest, Mexico Southeast, Mexico Southwest, Nicaragua, Panamá, Peru, Puerto Rico, Suriname, Trinidad-Tobago, Venezuela, Windward Is.

Distribution (India): Bihar, Odisha, Tamil Nadu
Leaf Type: Odd-pinnate, opposite, terminal one large and symmetric
Inflorescence: Flowers 5-merous in paicles
Fruit Type: Dry, capsule
IUCN Status: Not evaluated
Reference: Sp. Pl.: 623 (1753)

Bignonia binata Thunb.
Bignoniaceae
Synonyms: *Bignonia eximia* Morong, *Bignonia umbellata* DC., *Bignonia umbellulata* DC., *Clytostoma binatum* (Thunb.) Sandwith, *Clytostoma floridum* Miers ex Bureau & K. Schum.
Climbing Mechanism: Scrambler-Unarmed
Distribution (Global): Argentina Northeast, Bolivia, Brazil South, Brazil Southeast, Brazil West-Central, Paraguay
Leaf Type: Odd-pinnate, opposite, terminal one large and symmetric
Inflorescence: Flowers 5-merous in paicles
Fruit Type: Dry, capsule
Flowering and Fruiting: December–April
IUCN Status: Not evaluated
Reference: Pl. Bras. 3: 35 (1821)

Bignonia callistegioides Cham.
Bignoniaceae
Synonyms: *Bignonia lindleyi* DC., *Bignonia picta* Lindl., *Bignonia speciosa* Graham, *Clytostoma callistegioides* (Cham.) Bureau ex Griseb. + 3
Climbing Mechanism: Tendril Climber
Distribution (Global): Argentina Northeast, Argentina Northwest, Bolivia, Brazil South, Paraguay, Uruguay
Distribution (India): Bihar, Odisha
Leaf Type: Odd-pinnate, opposite, terminal one large and symmetric
Inflorescence: Flowers 5-merous in paicles
Fruit Type: Dry, capsule
IUCN Status: Not evaluated
Reference: Linnaea 7: 712 (1833)

Bignonia capreolata L.
Bignoniaceae
Synonyms: *Anisostichus capreolata* (L.) Bureau, *Batocydia capreolata* Mart. ex DC., *Bignonia capreolata* var. *atrosanguinea* Hook.f., *Bignonia capreolata* f. lutea Heineke, Doxantha capreolata (L.) Miers.
Common Name: Crossvine
Climbing Mechanism: Stem Twiner

Distribution (Global): Alabama, Arkansas, Florida, Georgia, Illinois, Indiana, Kentucky, Louisiana, Maryland, Mississippi, Missouri, North Carolina, Ohio, Oklahoma, South Carolina, Tennessee, Texas, Virginia, West Virginia
Distribution (India): Gujarat, Maharashtra
Leaf Type: Odd-pinnate, opposite, terminal one large and symmetric
Inflorescence: Flowers 5-merous in paicles
Fruit Type: Dry, capsule
IUCN Status: Not evaluated
Reference: Sp. Pl.: 624 (1753)

Bignonia magnifica W. Bull
Bignoniaceae
Synonyms: *Arrabidaea magnifica* (W. Bull) Sprague ex Steenis
Saritaea magnifica (W. Bull) Dugand
Common Name: Glow Vine, Purple Bignonia, Purple funnel vine
Climbing Mechanism: Stem Twiner
Distribution (Global): Colombia, Ecuador, Panamá, Venezuela
Distribution (India): Tamil Nadu
Leaf Type: Odd-pinnate, opposite, terminal one large and symmetric
Inflorescence: Flowers 5-merous in paicles
Fruit Type: Dry, capsule
Flowering and Fruiting: November–June
IUCN Status: Not evaluated
Reference: Gard. Chron., n.s., 12: 73 (1879)

Campsis × *tagliabuana* (Vis.) Rehder
Bignoniaceae
Common Name: Galen trumpet creeper
Climbing Mechanism: Scrambler-Unarmed
Leaf Type: Compound, odd-pinnate
Inflorescence: Clusters at tips, trumpet shaped
Fruit Type: Dry, capsule
IUCN Status: Not evaluated
Reference: J. Arnold Arbor. 13: 340 (1932)

Campsis grandiflora (Thunb.) K. Schum.
Bignoniaceae
Synonyms: *Bignonia chinensis* Lam., *Bignonia grandiflora* Thunb., *Campsis adrepens* Lour., *Campsis chinensis* (Lam.) Voss, *Gelseminum grandiflorum* (Thunb.) Kuntze, *Incarvillea chinensis* Spreng. ex DC. +5
Common Name: Chinese trumpet creeper, Chinese trumpet vine
Climbing Mechanism: Scrambler-Unarmed
Distribution (Global): China North-Central, China Southeast, Japan
Distribution (India): Delhi, Maharashtra, Rajasthan, Tamil Nadu, Uttar Pradesh
Leaf Type: Compound, odd-pinnate

Inflorescence: Clusters at tips, trumpet shaped
Fruit Type: Dry, capsule
Flowering and Fruiting: June–January
IUCN Status: Not evaluated
Reference: H.G.A. Engler & K.A.E. Prantl, Nat. Pflanzenfam. 4(3b): 230 (1894)

Campsis radicans (L.) Seem.
Bignoniaceae
Synonyms: *Bignonia coccinea* Steud., *Bignonia florida* Salisb., *Bignonia radicans* L., *Bignonia radicans* var. *coccinea* Pursh, *Bignonia radicans* var. *flammea* Pursh, *Bignonia radicans* var. *minor* Castigl., *Campsis curtisii* Seem., *Campsis radicans* var. *praecox* (H.J.)
Common Name: Trumpet creeper, Trumpet vine, Cow-itch vine
Climbing Mechanism: Scrambler-Unarmed
Distribution (Global): Alabama, Arkansas, Connecticut, Delaware, District of Columbia, Florida, Georgia, Illinois, Indiana, Iowa, Kansas, Kentucky, Louisiana, Maine, Mississippi, Missouri, Nebraska, New Hampshire, New Jersey, New York, North Carolina, North Dakota, Ohio, Oklahoma, Pennsylvania, South Carolina, South Dakota, Tennessee, Texas, Utah, Virginia, West Virginia, Wisconsin
Distribution (India): Maharashtra, Tamil Nadu
Leaf Type: Compound, odd-pinnate
Inflorescence: Clusters at tips, trumpet shaped
Fruit Type: Dry, capsule
Flowering and Fruiting: September–February
IUCN Status: Not evaluated
Reference: Monogr. Bignon. 2(Atlas): 16 (1864)

Clytostoma callistegioides (Cham.) Baill.
Bignoniaceae
Synonyms: *Bignonia callistegioides* Cham., *Bignonia lindleyi* DC., *Bignonia picta* Lindl., *Bignonia speciosa* Graham, *Clytostoma uniflorum* K. Schum. ex Bureau & K. Schum., *Cuspidaria callistegioides* (Cham.) DC., *Pithecoctenium callistegioides* (Cham.) Niederl.
Climbing Mechanism: Stem Twiner
Distribution (Global): Argentina Northeast, Argentina Northwest, Bolivia, Brazil South, Paraguay, Uruguay
Distribution (India): Maharashtra
IUCN Status: Not evaluated
Reference: Linnaea 7: 712 (1833)

Distictis × *riversii*
Bignoniaceae
Climbing Mechanism: Stem Twiner
IUCN Status: Not evaluated

Dolichandra unguis-cati (L.) L.G. Lohmann
Bignoniaceae
Synonyms: *Batocydia exoleta* Mart. ex DC., *Batocydia unguis* Mart. ex DC., *Batocydia unguis-cati* (L.) Mart. ex Britton & P. Wilson, *Bignonia acutistipula* Schltdl., *Bignonia californica* Brandegee, *Bignonia catharinensis* Schenck, *Bignonia dasyonyx* S.F. Blake
Common Name: Cat's claw, Cat's claw creeper
Climbing Mechanism: Root Climber
Distribution (Global): Argentina Northeast, Argentina Northwest, Bahamas, Belize, Bolivia, Brazil North, Brazil Northeast, Brazil South, Brazil Southeast, Brazil West-Central, Colombia, Costa Rica, Cuba, Dominican Republic, Ecuador, El Salvador, French Guiana, Guatemala, Guyana, Haiti, Honduras, Jamaica, Leeward Is., Mexico Central, Mexico Gulf, Mexico Northeast, Mexico Northwest, Mexico Southeast, Mexico Southwest, Nicaragua, Panamá, Paraguay, Peru, Puerto Rico, Southwest Caribbean, Suriname, Trinidad-Tobago, Uruguay, Venezuela, Venezuelan Antilles, Windward Is.
Distribution (India): Tamil Nadu
Leaf Type: Compound leaves
IUCN Status: Not evaluated
Reference: O. Hokche, P.E. Berry & O. Huber (eds.), Nuevo Cat. Fl. Vasc. Venezuela: 273 (2008)

Mansoa alliacea (Lam.) A.H. Gentry
Bignoniaceae
Synonyms: *Adenocalymma alliaceum* (Lam.) Miers, *Adenocalymma obovatum* Urb., *Adenocalymma pachypus* (K. Schum.) Bureau & K. Schum. + 10
Common Name: Garlic vine
Climbing Mechanism: Stem Twiner
Distribution (Global): Bolivia, Brazil North, Brazil Northeast, Colombia, Ecuador, French Guiana, Guyana, Leeward Is., Peru, Suriname, Trinidad-Tobago, Windward Is.
Distribution (India): Tamil Nadu
Leaf Type: Leaves are bright green, slightly coriaceous, opposite, apex mucronate, margin entire, surface glabrous
Inflorescence: Flowers are violet in color and grow in terminal or axillary racemosa inflorescence
Flowering and Fruiting: Throughout the year
IUCN Status: Not evaluated
Reference: Ann. Missouri Bot. Gard. 66: 782 (1979 publ. 1980)

Mansoa hymenaea (DC.) A.H. Gentry
Bignoniaceae
Synonyms: *Adenocalymma alboviolaceum* Loes., *Adenocalymma ciliolatum* S.F. Blake, *Adenocalymma hosmeca* Pittier, *Adenocalymma laevigatum* Mart.

ex DC., *Adenocalymma laevigatum* Bureau & K. Schum., *Adenocalymma macrocarpum* Donn.Sm., *Adenocalymma pohlianum* Bureau & K
Climbing Mechanism: Stem Twiner
Distribution (Global): Belize, Brazil North, Brazil Northeast, Brazil Southeast, Colombia, Costa Rica, Ecuador, El Salvador, French Guiana, Guatemala, Honduras, Mexico Central, Mexico Gulf, Mexico Northeast, Mexico Northwest, Mexico Southeast, Mexico Southwest, Nicaragua, Panamá, Peru, Suriname, Venezuela, Windward Is.
IUCN Status: Not evaluated
Reference: Ann. Missouri Bot. Gard. 66: 782 (1979 publ. 1980)

Nycticalos thomsonae Hook.f.
Bignoniaceae
Climbing Mechanism: Stem Twiner
Distribution (India): Tripura
IUCN Status: Not evaluated

Nyctocalos cuspidatum (Blume) Miq.
Bignoniaceae
Synonyms: *Beaumontia trifoliata* Steenis, *Gelseminum cuspidatum* (Blume) Kuntze, *Nyctocalos thomsonii* Hook.f., *Tecoma cuspidata* Blume.
Climbing Mechanism: Stem Twiner
Distribution (Global): Assam, Bangladesh, Maluku, Sulawesi, Thailand, Vietnam
Distribution (India): Mizoram, Tamil Nadu
IUCN Status: Not evaluated
Notes: Often found in open and exposed forests
Reference: Ann. Mus. Bot. Lugduno-Batavi 3: 249 (1867)

Pandorea jasminoides (Lindl.) K. Schum.
Bignoniaceae
Synonyms: *Bignonia jasminoides* A. Cunn. ex DC., *Pandorea jasminoides* var. *alba* Rehder, *Gelseminum jasminoides* (Lindl.) Kuntze, *Tecoma jasminoides* Lindl.
Common Name: Bower plant, Pandora vine
Climbing Mechanism: Stem Twiner
Distribution (Global): Queensland
Distribution (India): Maharashtra, Tamil Nadu
Leaf Type: Pinnately compound
Fruit Type: Elliptical fruits containing many seeds
Flowering and Fruiting: December–May
IUCN Status: Not evaluated
Reference: H.G.A. Engler & K.A.E. Prantl, Nat. Pflanzenfam. 4(3b): 230 (1894)

Pandorea pandorana (Andrews) Steenis
Bignoniaceae

Synonyms: *Bignonia pandorana* Andrews, *Tecoma pandorana* (Andrews) Skeels, *Campsis pan*dorana (Andrews) Steenis
Climbing Mechanism: Stem Twiner
Distribution (Global): Lesser Sunda Is., Maluku, New Caledonia, New Guinea, New South Wales, Norfolk Is., Northern Territory, Queensland, Solomon Is., Tasmania, Vanuatu, Victoria
Distribution (India): Maharashtra
Leaf Type: Imparipinnate
Fruit Type: Rugose
IUCN Status: Not evaluated
Reference: Bull. Jard. Bot. Buitenzorg, sér. 3, 10: 198 (1928)

Pleonotoma jasminifolia (Kunth) Miers
Bignoniaceae
Synonyms: *Bignonia jasminifolia* Kunth, *Bignonia tetragonocaulis* DC., *Pleonotoma tetragonocaulis* (DC.) Miers.
Climbing Mechanism: Scrambler-Unarmed
Distribution (Global): Brazil North, Colombia, Venezuela
Distribution (India): Tamil Nadu
Leaf Type: Pinnate, leaf sheaths without knees, covered spines
Inflorescence: Axillary spikes
Fruit Type: Wet, globose
IUCN Status: Not evaluated
Reference: Proc. Roy. Hort. Soc. London 3: 184 (1863)

Podranea ricasoliana (Tanfani) Sprague
Bignoniaceae
Synonyms: *Pandorea ricasoliana* (Tanfani) K. Schum., *Tecomaria ricasoliana* (Tanfani) Kraenzl., *Tecoma ricasoliana* Tanfani
Common Name: Port St. John's creeper, Pink trumpet tree
Climbing Mechanism: Stem Twiner
Distribution (Global): Cape Provinces, KwaZulu-Natal, Malawi, Mozambique, Zambia
Distribution (India): Tamil Nadu, Telangana
Leaf Type: Imparipinnate
Inflorescence: Panicle
Flowering and Fruiting: November–June
IUCN Status: Not evaluated
Reference: W.H. Harvey & auct. suc. (eds.), Fl. Cap. 4(2): 450 (1904)

Pyrostegia venusta (Ker Gawl.) Miers
Bignoniaceae
Synonyms: *Bignonia ignea* Vell., *Bignonia tecomiflora* Rusby, *Bignonia tubulosa* Klotzsch, *Bignonia venusta* Ker Gawl., *Jacaranda echinata* Spreng., *Pyrostegia amabilis* Miers +13

Common Name: Flame vine, Golden shower, Orange trumpet vine
Climbing Mechanism: Scrambler-Unarmed
Distribution (Global): Argentina Northeast, Bolivia, Brazil North, Brazil Northeast, Brazil South, Brazil Southeast, Brazil West-Central, Colombia, Costa Rica, Ecuador, El Salvador, Guatemala, Guyana, Honduras, Mexico Central, Mexico Gulf, Mexico Northwest, Mexico Southeast, Mexico Southwest, Panamá, Paraguay, Peru, Suriname, Venezuela
Distribution (India): Delhi, Gujarat, Madhya Pradesh, Maharashtra, Odisha, Rajasthan, Tamil Nadu, Telangana, Tripura
Leaf Type: Compound
Fruit Type: Capsule
Flowering and Fruiting: December–June
IUCN Status: Not evaluated
Reference: Proc. Roy. Hort. Soc. London 3: 188 (1863)

Tanaecium jaroba
Bignoniaceae
Climbing Mechanism: Tendril Climber
IUCN Status: Not evaluated

Cordia diffusa K.C. Jacob
Boraginaceae
Synonyms: *Gerascanthus diffusus* (K.C. Jacob) Borhidi
Climbing Mechanism: Scrambler-Unarmed
Distribution (Global): India
Distribution (India): Tamil Nadu
Leaf Type: Simple, opposite
Inflorescence: Cymes, flowers clustered
Fruit Type: Wet, pulp fleshy
Flowering and Fruiting: February–June
IUCN Status: Not evaluated
Reference: J. Bombay Nat. Hist. Soc. 45: 78 (1944)

Tournefortia heyneana Wall. ex G. Don
Boraginaceae
Synonyms: *Tournefortia reticosa* Wight.
Climbing Mechanism: Scrambler-Unarmed
Distribution (Global): India
Distribution (India): Tamil Nadu
Leaf Type: Leaves alternate
Inflorescence: Dichotomous scorpioid cymes
Fruit Type: Drupes
Flowering and Fruiting: August–February
IUCN Status: Not evaluated

Notes: Endemic to Southern Western Ghats
Reference: Gen. Hist. 4: 369 (1837)

Tournefortia hookeri C.B. Clarke
Boraginaceae
Synonyms: *Tournefortia viridiflora* Gamble
Climbing Mechanism: Scrambler-Unarmed
Distribution (Global): East Himalaya, Myanmar
Distribution (India): West Bengal
Leaf Type: Simple
IUCN Status: Not evaluated
Reference: J. D. Hooker, Fl. Brit. India 4: 147 (1883)

Tournefortia montana Lour.
Boraginaceae
Synonyms: *Messerschmidia montana* (Lour.) Roem. & Schult.
Common Name: Mountain soldierbush
Climbing Mechanism: Scrambler-Unarmed
Distribution (Global): Assam, Bangladesh, Cambodia, China South-Central, East Himalaya, Laos, Vietnam
Distribution (India): Arunachal Pradesh
Leaf Type: Simple
IUCN Status: Not evaluated
Reference: Fl. Cochinch.: 122 (1790)

Tournefortia ovata Wall.
Boraginaceae
Climbing Mechanism: Scrambler-Unarmed
Distribution (Global): Andaman Is., Myanmar, Nicobar Is.
Distribution (India): Andaman and Nicobar Islands, Arunachal Pradesh
Leaf Type: Simple
IUCN Status: Not evaluated
Reference: Gen. Hist. 4: 369 (1837)

Tournefortia reticosa Wight
Boraginaceae
Climbing Mechanism: Scrambler-Unarmed
Distribution (India): Tamil Nadu
Leaf Type: Simple
IUCN Status: Not evaluated
Reference: Wight. In: Ic. t. 1386. (1848)

Tournefortia roxburghii C.B. Clarke
Boraginaceae

Synonyms: *Tournefortia montana* var. *griffithii* (C.B. Clarke) I.M. Johnst., *Heliotropium paniculatum* Roxb., *Heliotropium roxburghiana* Voigt, *Heliotropium roxburghii* Spreng.
Climbing Mechanism: Scrambler-Unarmed
Distribution (Global): Bangladesh
Distribution (India): Tripura
Leaf Type: Simple
IUCN Status: Not evaluated
Reference: J. Arnold Arbor. 32: 117 (1953)

Tournefortia tetrandra Blume
Boraginaceae
Synonyms: *Heliotropium biblianum* Craven., *Heliotropium scandens* Noronha, *Tetrandra glabra* (Hassk.) Miq., *Tetrandra wallichii* Miq., *Tetrandra zollingeri* Miq., *Tournefortia tetrandra* Wall., *Tournefortia wall*ichii A.DC., *Tournefortia glabra* (Hassk.) Zoll. & Moritzi.
Climbing Mechanism: Scrambler-Unarmed
Distribution (Global): Borneo, Java, Malaya, New Guinea, Nicobar Is., Philippines, Sumatera
Distribution (India): Great Nicobar Island
Leaf Type: Simple
IUCN Status: Not evaluated
Reference: Blumea 50: 379 (2005)

Tournefortia viridiflora Wall.
Boraginaceae
Synonyms: *Heliotropium viridiflorum* Lehm., *Heliotropium viridiflorum* Lehm.
Climbing Mechanism: Scrambler-Unarmed
Distribution (Global): Assam, Bangladesh, Myanmar
Distribution (India): Meghalaya
Leaf Type: Simple
IUCN Status: Not evaluated
Reference: Pl. Asperif. Nucif.: 30 (1818)

Tournefortia wightii C.B. Clarke
Boraginaceae
Climbing Mechanism: Scrambler-Unarmed
Distribution (Global): India
Distribution (India): Tamil Nadu
Leaf Type: Simple
IUCN Status: Not evaluated
Notes: Occurs mostly in evergreen forests
Reference: J. D. Hooker, Fl. Brit. India 4: 146 (1883)

Trichodesma khasianum C.B. Clarke
Boraginaceae
Synonyms: *Boraginella khasiana* (C.B. Clarke) Kuntze
Climbing Mechanism: Scrambler-Unarmed
Distribution (Global): Assam, Bangladesh, Myanmar, Vietnam
Distribution (India): Mizoram
IUCN Status: Not evaluated
Reference: J. D. Hooker, Fl. Brit. India 4: 154 (1883)

Pereskia aculeata Mill.
Cactaceae
Synonyms: *Cactus lucidus* Salisb., *Cactus pereskia* L., *Pereskia acardia* J. Parm. ex Pfeiff., *Pereskia aculeata* var. *brasiliensis* (Pfeiff.) Borg, *Pereskia aculeata* var. *longispina* (Haw.) DC., *Pereskia aculeata* var. *rubescens* Pfeiff. + 8
Common Name: Barbados gooseberry, Yellow vine
Climbing Mechanism: Scrambler-Armed
Distribution (Global): Argentina Northeast, Bolivia, Brazil Northeast, Brazil South, Brazil Southeast, Brazil West-Central, Colombia, French Guiana, Guyana, Panamá, Paraguay, Suriname, Venezuela
Distribution (India): Peninsular India
Flowering and Fruiting: June–March
IUCN Status: Least Concern
Reference: Gard. Dict. ed. 8: s.p. (1768)

Campanula pallida Wall.
Campanulaceae
Synonyms: *Campanula colorata* Wall., *Campanula himalayensis* Klotzsch, *Campanula hoffmeisteri* Klotzsch, *Campanula microcarpa* C.Y. Wu, *Campanula moorcroftiana* Wall., *Campanula nepetifolia* H. Lév. & Vaniot, *Campanula nervosa* Royle, *Campanula pallida* var. *tibetica*
Common Name: Pale bellflower, Purple bellflower
Climbing Mechanism: Stem Twiner
Distribution (Global): Afghanistan, Bangladesh, China South-Central, East Himalaya, India, Laos, Myanmar, Nepal, Pakistan, Thailand, Tibet, Vietnam, West Himalaya
Flowering and Fruiting: October–December
IUCN Status: Not evaluated
Reference: Asiat. Res. 13: 375 (1820)

Codonopsis affinis Hook.f. & Thomson
Campanulaceae
Climbing Mechanism: Stem Twiner
Distribution (Global): Assam, East Himalaya, Myanmar, Nepal, Tibet, West Himalaya
Distribution (India): West Bengal

IUCN Status: Not evaluated
Reference: J. Proc. Linn. Soc., Bot. 2: 12 (1857)

Codonopsis gracilis Hook.f. & Thomson
Campanulaceae
Synonyms: *Leptocodon gracilis* (Hook.f. & Thomson) Lem.
Common Name: Slender bonnet bellflower
Climbing Mechanism: Stem Twiner
Distribution (Global): China South-Central, East Himalaya, Myanmar, Nepal, Tibet
Leaf Type: Alternate, rarely those on branchlets opposite
Inflorescence: Capsule, semiglobose
Flowering and Fruiting: August–October
IUCN Status: Not evaluated
Notes: Found in Eastern Himalayas (Nepal to Burma) at altitudes of 2000–2500 m
Reference: Ill. Himal. Pl.: t. 16 A (1855)

Codonopsis javanica (Blume) Hook.f. & Thomson
Campanulaceae
Synonyms: *Campanula javanica* (Blume) D. Dietr., *Campanumoea javanica* Blume
Climbing Mechanism: Stem Twiner
Distribution (Global): Assam, Bangladesh, China North-Central, China South-Central, China Southeast, East Himalaya, Hainan, Japan, Java, Laos, Myanmar, Nansei-shoto, Sumatera, Taiwan, Thailand, Vietnam
Distribution (India): Arunachal Pradesh, Meghalaya, Mizoram
IUCN Status: Not evaluated
Reference: Ill. Himal. Pl.: t. 16 B (1855)

Codonopsis rotundifolia
Campanulaceae
Climbing Mechanism: Stem Twiner
IUCN Status: Not evaluated

Codonopsis viridis Wall.
Campanulaceae
Synonyms: *Campanula viridis* (Wall.) Spreng., *Codonopsis griffithii* C.B. Clarke, *Codonopsis viridis* var. *hirsuta* Chipp, *Glosocomia viridis* (Wall.) Rupr., *Wahlenbergia viridis* (Wall.) A. DC.
Common Name: Green bellflower vine
Climbing Mechanism: Stem Twiner
Distribution (Global): Assam, East Himalaya, Myanmar, Nepal, Pakistan, Tibet, West Himalaya
Distribution (India): Himachal Pradesh, Meghalaya, West Bengal
Leaf Type: Simple, fascicled
Inflorescence: Solitary, terminal
Fruit Type: Dry, capsule

IUCN Status: Not evaluated
Reference: W. Roxburgh, Fl. Ind. 2: 103 (1824)

Celtis tournefortii Lam.
Cannabaceae
Climbing Mechanism: Tendril Climber
Distribution (Global): Bulgaria, Cyprus, East Aegean Is., Greece, Iran, Iraq, Kriti, Krym, Lebanon-Syria, Sicilia, Transcaucasus, Turkey, Yugoslavia
IUCN Status: Least Concern
Reference: Encycl. 4: 138 (1797)

Cadaba farinosa Forssk.
Capparaceae
Synonyms: *Stroemia farinosa* (Forssk.) Vahl
Climbing Mechanism: Scrambler-Unarmed
Distribution (Global): Angola, Benin, Burkina, Burundi, Cameroon, Central African Repu, Chad, Congo, Djibouti, Egypt, Eritrea, Ethiopia, Gambia, Ghana, India, Kenya, Mali, Mauritania, Morocco, Mozambique, Niger, Nigeria, Oman, Pakistan, Rwanda, Saudi Arabia, Senegal, Socotra, Somalia, Sudan, Tanzania, Uganda, Western Sahara, Yemen, Zaire
Leaf Type: Compound, 3-foliolate, obovate
Inflorescence: Corymbose racemes
Fruit Type: Wet, berry
IUCN Status: Least Concern
Reference: Fl. Aegypt. -Arab.: 68 (1775)

Cadaba fruticosa (L.) Druce
Capparaceae
Synonyms: *Cadaba indica* Lam., *Mozambe levigata* Raf., *Stroemia tetrandra* Vahl, *Cleome fruticosa* L.
Common Name: Indian Cadaba
Climbing Mechanism: Scrambler-Unarmed
Distribution (Global): Bangladesh, India, Sri Lanka
Distribution (India): Andhra Pradesh, Gujarat, Maharashtra, Rajasthan, Tamil Nadu, Telangana
Leaf Type: Compound, 3-foliolate, obovate
Inflorescence: Corymbose racemes
Fruit Type: Wet, berry
IUCN Status: Not evaluated
Reference: Rep. Bot. Exch. Club Soc. Brit. Isles 3: 415 (1913 publ. 1914)

Cadaba trifoliata Wight & Arn.
Capparaceae
Synonyms: *Cadaba triphylla* Wight, *Desmocarpus missionis* Wall., *Stroemia trifoliata* Roxb.

Common Name: Three-leaved Cadaba
Climbing Mechanism: Scrambler-Unarmed
Distribution (Global): India, Sri Lanka
Distribution (India): Kerala
Leaf Type: Trifoliate, alternate
Inflorescence: Racemes corymbose
Flowering and Fruiting: February–November
IUCN Status: Not evaluated
Notes: Occurs in both deciduous and evergreen forests
Reference: R. Wight, Cat. Ind. Pl.: 7 (1833)

Capparidastrum sola (J.F. Macbr.) Cornejo & Iltis
Capparaceae
Synonyms: *Capparidastrum sola* (J.F. Macbr.) Cornejo & Iltis, *Capparis acutifolia* J.F. Macbr., *Capparis sola* J.F. Macbr., *Capparis sola* var. *longiracemosa* Dugand
Climbing Mechanism: Scrambler-Armed
Distribution (Global): Bolivia, Colombia, Ecuador, Peru, Suriname, Venezuela
IUCN Status: Not evaluated
Reference: Global Fl. 4: 142 (2018)

Capparis acutifolia subsp. *sabiifolia* (Hook.f. & Thomson) M. Jacobs
Capparaceae
Climbing Mechanism: Scrambler-Armed
Distribution (India): Manipur
Leaf Type: Simple
Inflorescence: Corymbose racemes
Fruit Type: Wet, berry
IUCN Status: Not evaluated
Reference: Blumea 12: 433, non Buch. -Ham. ex DC. 1824. (1965)

Capparis acutifolia subsp. *viminea* M. Jacobs
Capparaceae
Climbing Mechanism: Scrambler-Armed
IUCN Status: Not evaluated

Capparis acutifolia Sweet
Capparaceae
Synonyms: *Capparis chinensis* G. Don, *Capparis kikuchii* Hayata, *Capparis leptophylla* Hayata, *Capparis membranacea* Gardner & Champ., *Capparis tenuifolia* Hayata
Common Name: Chinese caper
Climbing Mechanism: Scrambler-Armed
Distribution (Global): China Southeast, Laos, Taiwan, Thailand, Vietnam
Distribution (India): Arunachal Pradesh, Manipur, Mizoram

Leaf Type: Simple, alternate
Inflorescence: Supra-axillary rows
Fruit Type: Globose or subglobose, blackish when ripe
Flowering and Fruiting: April–October
IUCN Status: Not evaluated
Notes: Often found in scrub jungles
Reference: Hort. Brit., ed. 2: 585 (1830)

Capparis assamica Hook.f. & Thomson
Capparaceae
Synonyms: *Capparis gallatlyi* King
Common Name: Assam caper
Climbing Mechanism: Scrambler-Armed
Distribution (Global): Assam, Bangladesh, China South-Central, China Southeast, East Himalaya, Hainan, Laos, Myanmar, Thailand, Tibet, Vietnam
Distribution (India): Arunachal Pradesh, Assam, Meghalaya
Leaf Type: Simple, alternate
Inflorescence: Nearly racemes
Fruit Type: Globose, rec when ripe
Flowering and Fruiting: March–April
IUCN Status: Not evaluated
Reference: J. D. Hooker, Fl. Brit. India 1: 177 (1872)

Capparis baducca L.
Capparaceae
Common Name: Rheed's caper
Climbing Mechanism: Scrambler-Armed
Leaf Type: Simple
Inflorescence: Corymbose racemes
Fruit Type: Wet, berry
IUCN Status: Not evaluated
Reference: L. In: Sp. Pl. 504 p. p. (1753)

Capparis brevispina DC.
Capparaceae
Synonyms: *Capparis wallichiana* Wight & Arn.
Common Name: Indian caper
Climbing Mechanism: Scrambler-Armed
Distribution (Global): Assam, India, Sri Lanka
Distribution (India): Maharashtra, Puducherry, Tamil Nadu
Leaf Type: Simple
Inflorescence: Corymbose racemes
Fruit Type: Wet, berry
Flowering and Fruiting: January–September
IUCN Status: Not evaluated

Notes: Occasionally found in clumps on plains and scrub jungles
Reference: Prodr. 1: 246 (1824)

Capparis cantoniensis Lour.
Capparaceae
Synonyms: *Capparis ambigua* Kurz, *Capparis celebica* Miq., *Capparis gomeziana* Steud., *Capparis hasseltiana* Miq., *Capparis pumila* Champ. ex Benth., *Capparis punctata* Wall., *Capparis salaccensis* Blume, *Capparis sciaphila* Hance, *Olofuton racemosum* Raf.
Climbing Mechanism: Scrambler-Armed
Distribution (Global): Andaman Is., Assam, Bangladesh, China South-Central, China Southeast, East Himalaya, Hainan, Java, Lesser Sunda Is., Maluku, Myanmar, Philippines, Sulawesi, Sumatera, Thailand, Vietnam
Distribution (India): Mizoram, Tripura
Leaf Type: Simple, alternate
Inflorescence: Corymbs in axillary
IUCN Status: Not evaluated
Notes: Distributed mostly in hill forests
Reference: Fl. Cochinch.: 330 (1790)

Capparis cleghornii Dunn
Capparaceae
Common Name: Cleghorn caper
Climbing Mechanism: Scrambler-Armed
Distribution (Global): India
Distribution (India): Kerala, Maharashtra
Leaf Type: Alternate
Inflorescence: Corymbs, rarely solitary
Flowering and Fruiting: February–October
IUCN Status: Not evaluated
Notes: Found commonly along the margins of semi-evergreen forests
Reference: J.S. Gamble, Fl. Madras: 146 (1915)

Capparis danielii Murugan, R. Manik., S.P. Nithya, B. Karthik & Arisdason
Capparaceae
Climbing Mechanism: Scrambler-Armed
Leaf Type: Simple
Inflorescence: Corymbose racemes
Fruit Type: Wet, berry
IUCN Status: Not evaluated
Reference: Murugan, R. Manik., S. P. Nithya, B. Karthik & Arisdason. In: Phytotaxa 472(3): 284. (2020)

Capparis decidua (Forssk.) Edgew
Capparaceae

Common Name: Bare Caper, Caper berry, Leafless caper-bush
Climbing Mechanism: Scrambler-Armed
Distribution (India): Gujarat, Maharashtra
Leaf Type: Simple
Inflorescence: Corymbose racemes
Fruit Type: Wet, berry
IUCN Status: Least Concern
Notes: Occurs mostly in dry deciduous forests of the Western Ghats
Reference: J. Linn. Soc. Lond. Bot. 6: 184. (1862)

Capparis divaricata Lam.
Capparaceae
Synonyms: *Capparis horrida* Banks ex Wight & Arn., *Capparis stylosa* DC.
Common Name: Spreading Caper
Climbing Mechanism: Scrambler-Armed
Distribution (Global): India, Sri Lanka
Distribution (India): Andhra Pradesh
Leaf Type: Simple
Inflorescence: Corymbose racemes
Fruit Type: Wet, berry
Flowering and Fruiting: March–November
IUCN Status: Not evaluated
Notes: Occasional in scrub or deciduous forests up to 300 m
Reference: Encycl. 1: 606 (1785)

Capparis floribunda Lepr. ex Walp.
Capparaceae
Synonyms: *Capparis tomentosa* Lam., *Capparis alexandrae* Chiov., *Capparis biloba* Hutch. & Dalziel. + 10
Common Name: Woolly caper bush
Climbing Mechanism: Scrambler-Armed
Distribution (Global): Angola, Benin, Botswana, Burkina, Burundi, Cameroon, Cape Provinces, Caprivi Strip, Central African Repu, Chad, Eritrea, Ethiopia, Gabon, Gambia, Ghana, Guinea, Gulf of Guinea Is., Ivory Coast, Kenya, KwaZulu-Natal, Malawi, Mali, Mauritania, Mauritius, Mozambique, Namibia, Niger, Nigeria, Northern Provinces, Rwanda, Saudi Arabia, Senegal, Somalia, Sudan, Swaziland, Tanzania, Uganda, Yemen, Zambia, Zaire, Zimbabwe
Distribution (India): Andaman and Nicobar Islands, Andhra Pradesh, Eastern Ghats, Kerala, Odisha, Tamil Nadu
Leaf Type: Simple
Inflorescence: Corymbose racemes
Fruit Type: Wet, berry
Flowering and Fruiting: January–October
IUCN Status: Not evaluated

Notes: Occurs mostly in scrub jungles
Reference: Encycl. 1: 606 (1785)

Capparis fusifera Dunn
Capparaceae
Synonyms: *Capparis tomentella* Dunn
Common Name: Small-flowered Caper
Climbing Mechanism: Scrambler-Armed
Distribution (Global): India
Distribution (India): Tamil Nadu
Leaf Type: Simple
Inflorescence: Corymbose racemes
Fruit Type: Wet, berry
Flowering and Fruiting: April–June
IUCN Status: Not evaluated
Reference: Bull. Misc. Inform. Kew 1914: 377 (1914)

Capparis grandiflora Wall. ex Hook.f. & Thomson
Capparaceae
Synonyms: *Capparis pyrifolia* Wight & Arn.
Climbing Mechanism: Scrambler-Armed
Distribution (Global): India
Distribution (India): Kerala
Leaf Type: Simple
Inflorescence: Corymbose racemes
Fruit Type: Wet, berry
Flowering and Fruiting: March–April
IUCN Status: Not evaluated
Notes: Distributed mostly in hill forests
Reference: J. D. Hooker, Fl. Brit. India 1: 174 (1872)

Capparis grandis L.f.
Capparaceae
Synonyms: *Capparis auricans* (Kurz) Craib, *Capparis bisperma* Roxb., *Capparis disperma* Walp., *Capparis maxima* B. Heyne ex Roth, *Capparis obovata* Buch.-Ham. ex DC., *Capparis racemifera* DC.
Climbing Mechanism: Scrambler-Armed
Distribution (Global): India, Myanmar, Sri Lanka, Thailand, Vietnam
Leaf Type: Simple
Inflorescence: Corymbose racemes
Fruit Type: Wet, berry
Flowering and Fruiting: Throughout the year
IUCN Status: Not evaluated

Notes: Fresh leaves are cooked and eaten as vegetable soup to treat skin eruptions. Fresh leaves are crushed and the pulp is applied to treat insect bites
Reference: Suppl. Pl.: 263 (1782)

Capparis kollimalayana M.B. Viswan
Capparaceae
Climbing Mechanism: Scrambler-Armed
Distribution (Global): India
Distribution (India): Tamil Nadu
Leaf Type: Simple
Inflorescence: Corymbose racemes
Fruit Type: Wet, berry
IUCN Status: Not evaluated
Reference: Kew Bull. 55: 245 (2000)

Capparis micracantha DC.
Capparaceae
Synonyms: *Capparis micracantha* subsp. *korthalsiana* (Miq.) M.Jacobs, *Capparis micracantha* subsp. *Micracantha*, *Capparis micracantha* var. *microcarpa* Fici.
Climbing Mechanism: Scrambler-Armed
Distribution (Global): Andaman Is., Bangladesh, Borneo, Cambodia, China South-Central, China Southeast, Hainan, Java, Laos, Lesser Sunda Is., Malaya, Maluku, Myanmar, Philippines, Sulawesi, Sumatera, Thailand, Vietnam
Distribution (India): Andaman and Nicobar Islands
Leaf Type: Simple
Inflorescence: Corymbose racemes
Fruit Type: Wet, berry
IUCN Status: Not evaluated
Reference: Prodr. 1: 247 (1824)

Capparis moonii Wight
Capparaceae
Common Name: Large caper
Climbing Mechanism: Scrambler-Armed
Distribution (Global): India, Sri Lanka
Distribution (India): Goa, Karnataka, Kerala, Tamil Nadu, Western Ghats
Leaf Type: Simple
Inflorescence: Corymbose racemes
Fruit Type: Wet, berry
Flowering and Fruiting: December–October
IUCN Status: Not evaluated
Notes: Often found in evergreen forests
Reference: Ill. Ind. Bot. 1: 35 (1840)

Capparis multiflora Hook.f. & Thomson
Capparaceae
Climbing Mechanism: Scrambler-Armed
Distribution (Global): Assam, China South-Central, East Himalaya, Myanmar, Nepal, Tibet
Distribution (India): North Eastern India
Leaf Type: Simple, alternate
Fruit Type: Berry
Flowering and Fruiting: February–June
IUCN Status: Not evaluated
Reference: J. D. Hooker, Fl. Brit. India 1: 178 (1872)

Capparis pyrifolia Lam.
Capparaceae
Synonyms: *Capparis acuminata* Willd., *Capparis dasypetala* Turcz., *Capparis foetida* Blume, *Capparis kerrii* Craib, *Capparis oxyphylla* Miq., *Capparis zeylanica* Wight & Arn., *Oligloron zeylanica* Raf.
Climbing Mechanism: Scrambler-Armed
Distribution (Global): Cambodia, Java, Laos, Lesser Sunda Is., Sumatera, Thailand, Vietnam
Leaf Type: Simple
Inflorescence: Corymbose racemes
Fruit Type: Wet, berry
IUCN Status: Not evaluated
Reference: Encycl. 1: 606 (1785)

Capparis rotundifolia Rottler
Capparaceae
Synonyms: *Capparis longispina* Hook.f. & Thomson, *Capparis orbiculata* Wall. ex Hook.f. & Thomson, *Capparis rotundifolia* var. *longispina* (Hook.f. & Thomson) M.R. Almeida
Common Name: Round leaf caper
Climbing Mechanism: Scrambler-Armed
Distribution (India): Karnataka, Puducherry, Tamil Nadu
Leaf Type: Alternate, circular
Flowering and Fruiting: April–August
IUCN Status: Not evaluated
Notes: Occasional in Eastern Ghats, coastal regions, and moist deciduous forests of the Western Ghats
Reference: Neue Schriften Ges. Naturf. Freunde Berlin 4: 185 (1803)

Capparis roxburghii DC.
Capparaceae
Synonyms: *Capparis aguba* Banks ex DC., *Capparis bisperma* Wight ex Hook.f. & Thomson, *Capparis corymbosa* Roxb.

Climbing Mechanism: Scrambler-Armed
Distribution (Global): India, Sri Lanka
Distribution (India): Andhra Pradesh, Arunachal Pradesh, Eastern Ghats, Odisha, Tamil Nadu
Leaf Type: Simple
Inflorescence: Corymbose racemes
Fruit Type: Wet, berry
Flowering and Fruiting: March–October
IUCN Status: Not evaluated
Notes: Rare in deciduous forests up to 600 m
Reference: Prodr. 1: 247 (1824)

Capparis sepiaria L.
Capparaceae
Common Name: Wild caper bush
Climbing Mechanism: Scrambler-Armed
Distribution (Global): Andaman Is., Angola, Bangladesh, Benin, Borneo, Burkina, Burundi, Cambodia, Cameroon, Cape Provinces, Chad, China Southeast, Eritrea, Ethiopia, Hainan, India, Java, Kenya, KwaZulu-Natal, Laos, Lesser Sunda Is., Malawi, Malaya, Mali, Maluku, Mauritania, Mozambique, Myanmar, New Guinea, Nicobar Is., Northern Provinces, Northern Territory, Philippines, Queensland, Rwanda, Senegal, Somalia, Sri Lanka, Sudan, Sulawesi, Swaziland, Tanzania, Thailand, Uganda, Vietnam, Western Australia, Zambia, Zaire, Zimbabwe
Distribution (India): Andaman and Nicobar Islands, Andhra Pradesh, Delhi, Eastern Ghats, Gujarat, Karnataka, Kerala, Madhya Pradesh, Odisha, Puducherry, Rajasthan, Tamil Nadu, Telangana
Leaf Type: Simple
Inflorescence: Corymbose racemes
Fruit Type: Wet, berry
Flowering and Fruiting: March–October
IUCN Status: Least Concern
Notes: Often found in scrub jungle and dry deciduous forests
Reference: Syst. Nat. ed. 10, 2: 1071 (1759)

Capparis sepiaria var. *fischeri* (Pax) DeWolf
Capparaceae
Synonyms: *Capparis cerasiflora* Gilg, *Capparis djurica* Gilg & Gilg-Ben., *Capparis fischeri* Pax.
Climbing Mechanism: Scrambler-Armed
Distribution (Global): Benin, Burkina, Cameroon, Chad, Eritrea, Ethiopia, Kenya, Mali, Mauritania, Senegal, Somalia, Sudan, Tanzania
Leaf Type: Simple
Inflorescence: Corymbose racemes
Fruit Type: Wet, berry

IUCN Status: Not evaluated
Reference: Kew Bull. 16: 81 (1962)

Capparis shevaroyensis Sundararagh
Capparaceae
Synonyms: *Capparis parviflora* Hook.f. & Thomson
Climbing Mechanism: Scrambler-Armed
Distribution (Global): India
Distribution (India): Kerala, Tamil Nadu
Leaf Type: Simple
Inflorescence: Corymbose racemes
Fruit Type: Wet, berry
Flowering and Fruiting: August–October
IUCN Status: Not evaluated
Notes: Found mostly in the degraded moist deciduous forests
Reference: Kew Bull. 37: 72 (1982)

Capparis sikkimensis Kurz
Capparaceae
Climbing Mechanism: Scrambler-Armed
Distribution (Global): Assam, East Himalaya, Myanmar, Tibet
Distribution (India): West Bengal
Leaf Type: Simple
Inflorescence: Corymbose racemes
Fruit Type: Wet, berry
IUCN Status: Not evaluated
Reference: J. Asiat. Soc. Bengal, Pt. 2, Nat. Hist. 43: 181 (1874)

Capparis spinosa L.
Capparaceae
Common Name: Caperbush, Caper, Caperberry
Climbing Mechanism: Scrambler-Armed
Distribution (Global): Afghanistan, Albania, Algeria, Baleares, Caroline Is., Chad, Cook Is., Corse, Cyprus, East Aegean Is., Egypt, Fiji, France, Greece, Gulf States, India, Iran, Iraq, Italy, Kazakhstan, Kirgizstan, Kriti, Krym, Lebanon-Syria, Lesser Sunda Is., Libya, Marianas, Marshall Is., Mongolia, Morocco, Nauru, Nepal, New Caledonia, New Guinea, Niue, North Caucasus, Oman, Pakistan, Palestine, Philippines, Pitcairn Is., Portugal, Samoa, Sardegna, Saudi Arabia, Sicilia, Sinai, Society Is., Solomon Is., Spain, Tadzhikistan, Tonga, Transcaucasus, Tuamotu, Tubuai Is., Tunisia, Turkey, Turkmenistan, Uzbekistan, Vanuatu, Wallis-Futuna Is., West Himalaya, Xinjiang, Yemen, Yugoslavia
Distribution (India): Goa, Maharashtra
Leaf Type: Simple
Inflorescence: Corymbose racemes

Fruit Type: Wet, berry
IUCN Status: Least Concern
Notes: Usually found in hill forests
Reference: Sp. Pl.: 503 (1753)

Capparis tenera Dalzell
Capparaceae
Synonyms: *Capparis disticha* Kurz, *Capparis oxyphylla* Wall., *Capparis tetrasperma* Thwaites
Climbing Mechanism: Scrambler-Armed
Distribution (Global): Andaman Is., Assam, China South-Central, India, Laos, Myanmar, Sri Lanka, Thailand, Tibet
Distribution (India): Andaman and Nicobar Islands, Assam, Kerala, Maharashtra, Manipur, Mizoram, West Bengal
Leaf Type: Alternate, distichous
Inflorescence: Supra axillary
Flowering and Fruiting: February–June
IUCN Status: Not evaluated
Notes: Mostly found in evergreen forests
Reference: Hooker's J. Bot. Kew Gard. Misc. 2: 41 (1850)

Capparis trinervia Hook.f. & Thomson
Capparaceae
Synonyms: *Capparis kunstleri* King
Climbing Mechanism: Scrambler-Armed
Distribution (Global): Laos, Malaya, Myanmar, Sumatera, Thailand, Vietnam
Leaf Type: Simple
Inflorescence: Corymbose racemes
Fruit Type: Wet, berry
IUCN Status: Not evaluated
Reference: J. D. Hooker, Fl. Brit. India 1: 175 (1872)

Capparis versicolor Griff.
Capparaceae
Synonyms: *Capparis koi* Merr. & Chun, *Capparis larutensis* King, *Capparis nhatrangensis* Gagnep.
Climbing Mechanism: Scrambler-Armed
Distribution (Global): China Southeast, Hainan, Malaya, Myanmar, Thailand, Vietnam
Leaf Type: Simple
Inflorescence: Corymbose racemes
Fruit Type: Wet, berry
IUCN Status: Not evaluated
Reference: Not. Pl. Asiat. 4: 577 (1854)

Maerua apetala (Spreng.) M.Jacobs
Capparaceae
Synonyms: *Capparis apetala* B. Heyne ex Roth, *Capparis linifolia* Roxb., *Crateva apetala* (B. Heyne ex Roth) Spreng., *Niebuhria apetala* (B. Heyne ex Roth) Dunn, *Niebuhria linifolia* Voigt.
Common Name: Petal-less caper
Climbing Mechanism: Scrambler-Unarmed
Distribution (Global): India
Distribution (India): Andhra Pradesh
Leaf Type: Leaves are found only on young shoots, trifoliate. Leaflets are linear-oblong
Inflorescence: Flowers are fragrant, creamish tinged with purple, borne in corymbs at branch ends
Fruit Type: Berry is oblong-ovoid, 1-2-ridged
Flowering and Fruiting: April–July
IUCN Status: Not evaluated
Notes: Found all over Eastern Ghats
Reference: Blumea 12: 207 (1964)

Maerua arenaria Hook.f. & Thomson
Capparaceae
Common Name: Desert caper, Desert maerua
Climbing Mechanism: Scrambler-Unarmed
Distribution (Global): India, Pakistan, Sri Lanka
Distribution (India): Delhi, Madhya Pradesh, Rajasthan
Leaf Type: Leaves are oblong-ovate, broad, entire
Inflorescence: Flowers are usually in dense, corymb-like racemes, greenish-white, flower-stalklate
Fruit Type: Fruit cylindrical, globose, tuberculate
IUCN Status: Not evaluated
Reference: J. D. Hooker, Fl. Brit. India 1: 171 (1872)

Maerua oblongifolia (Forssk.) A. Rich
Capparaceae
Synonyms: *Capparis heteroclita* Roxb., *Capparis heterogenea* Roxb. ex Steud., *Capparis mithridatica* Forssk., *Capparis oblongifolia* Forssk., *Crateva oblongifolia* (Forssk.) Spreng., *Maerua albomarginata* Gilg & Gilg-Ben. +20
Climbing Mechanism: Scrambler-Unarmed
Distribution (Global): Benin, Burkina, Cameroon, Central African Repu, Chad, Djibouti, Ghana, India, Kenya, Mali, Mauritania, Nigeria, Oman, Saudi Arabia, Senegal, Somalia, Sudan, Uganda, Yemen, Zaire
Distribution (India): Andhra Pradesh, Eastern Ghats, Gujarat, Karnataka, Maharashtra, Odisha, Puducherry, Tamil Nadu, Telangana, West Bengal
Leaf Type: Leaves Simple, alternate distichous, entire

Inflorescence: Flowers in short dense corymbose racemes; pedicels stout, mostly 5–9 mm long in flower, up to 15–18 mm long in fruit
Fruit Type: A moniliform berry, twisted; seeds many, globose, tuberculate
Flowering and Fruiting: January–June
IUCN Status: Not evaluated
Notes: Usually found in deciduous forests of the Western Ghats
Reference: Tent. Fl. Abyss. 1: 32 (1847)

Roydsia obtusifolia Hook.f. & Thomson
Capparaceae
Synonyms: *Stixis obtusifolia* (Hook.f. & Thomson) Baill.
Climbing Mechanism: Scrambler-Unarmed
Leaf Type: Simple, glabrous
Inflorescence: Cymose
Fruit Type: Wet, drupe
IUCN Status: Not evaluated
Reference: J. D. Hooker, Fl. Brit. India 1: 180 (1872)

Steriphoma paradoxum (Jacq.) Endl.
Capparaceae
Synonyms: *Morisonia paradoxa* (Jacq.) Christenh. & Byng., *Capparis paradoxa* Jacq., *Hermupoa elliptica* (DC.) Kuntze., *Steriphoma cleomoides* Spreng. + 10
Climbing Mechanism: Scrambler-Unarmed
Distribution (Global): Colombia, Costa Rica, Guatemala, Mexico Southeast, Panamá, Trinidad-Tobago, Venezuela, Venezuelan Antilles
Leaf Type: Simple, alternate
Inflorescence: Single axillary flowers
Fruit Type: Wet, drupe
IUCN Status: Least Concern
Reference: Global Fl. 4: 141 (2018)

Diervilla florida Siebold & Zucc.
Caprifoliaceae
Synonyms: *Calysphyrum floridum* Bunge, *Calysphyrum pauciflorum* Bunge, *Calysphyrum roseum* (Lindl.) C.A. Mey. + 20
Climbing Mechanism: Scrambler-Unarmed
Distribution (Global): China North-Central, China Southeast, Inner Mongolia, Japan, Korea, Manchuria, Primorye
Distribution (India): Peninsular India
Leaf Type: Opposite
Fruit Type: Smooth capsule
Flowering and Fruiting: April–August
IUCN Status: Not evaluated
Notes: Cultivated as an ornamental plant
Reference: Ann. Sci. Nat., Bot., sér. 2, 11: 241 (1839)

Leycesteria dibangalliensis S.K. Das & G.S. Giri
Caprifoliaceae
Climbing Mechanism: Scrambler-Unarmed
Distribution (India): Arunachal Pradesh
IUCN Status: Not evaluated

Lonicera acuminata Wall.
Caprifoliaceae
Synonyms: *Caprifolium acuminatum* (Wall.) Kuntze, *Caprifolium fuschsioides* Kuntze, *Caprifolium henryi* (Hemsl.) Kuntze. + 20
Common Name: Vine honeysuckle
Climbing Mechanism: Scrambler-Unarmed
Distribution (Global): Assam, China North-Central, China South-Central, China Southeast, East Himalaya, Java, Lesser Sunda Is., Myanmar, Nepal, Philippines, Sumatera, Tibet, Vietnam
Distribution (India): North Eastern India, West Bengal
Leaf Type: Ciliate leaves are opposite, arranged alternately along the stems
Inflorescence: Inflorescence axillary, terminal many flowered heads corymb like, also often axillary 2-flowered peduncles also clustered, on the upper part of the branchlet
Fruit Type: Berries are blue-black, ovoid
Flowering and Fruiting: May–July
IUCN Status: Not evaluated
Reference: W. Roxburgh, Fl. Ind. 2: 176 (1824)

Lonicera caprifolioides K. Koch
Caprifoliaceae
Synonyms: *Lonicera fragrantissima* subsp. *fragrantissima*., *Abelia splendens* K. Koch, *Caprifolium fragrantissimum* Kuntze, *Caprifolium standishii* (Carrière) Kuntze, *Chamaecerasus niagerilli* Dippel, *Lonicera caprifolioïdes* K. Koch, *Lonicera fortunei* Dippel. + 5
Common Name: Winter honeysuckle
Climbing Mechanism: Scrambler-Unarmed
Distribution (Global): China North-Central, China South-Central, China Southeast
Distribution (India): Peninsular India
IUCN Status: Not evaluated

Lonicera caprifolium L.
Caprifoliaceae
Synonyms: *Caprifolium atropurpureum* K. Koch, *Caprifolium germanicum* Delarbre, *Caprifolium hortense* Lam., *Caprifolium italicum* Medik., *Caprifolium magnevilleae* Bosse. +10
Common Name: Italian woodbine, Italian honeysuckle, Perfoliate honeysuckle
Climbing Mechanism: Scrambler-Unarmed

Distribution (Global): Albania, Austria, Bulgaria, Czechoslovakia, Germany, Greece, Hungary, Italy, North Caucasus, Romania, Transcaucasus, Turkey, Turkey-in-Europe, Yugoslavia
Distribution (India): Tamil Nadu
Leaf Type: Simple, opposite leaves, which are oval to obovate
Inflorescence: Sessile inflorescences
Fruit Type: A fleshy, red, few- to many-seeded berry
Flowering and Fruiting: May–June
IUCN Status: Not evaluated
Reference: Sp. Pl.: 173 (1753)

Lonicera confusa DC.
Caprifoliaceae
Synonyms: *Caprifolium confusum* (DC.) G. Gordon, *Lonicera confusa* var. *glabrocalyx* R.H. Miao & X.J. Liu, *Lonicera dasystyla* Rehder, *Lonicera multiflora* Champ. ex Benth., *Lonicera periclymenum* Lour., *Lonicera telfairii* Hook. & Arn., *Nintooa confusa* Sweet
Common Name: Soft-leafed honeysuckle
Climbing Mechanism: Scrambler-Unarmed
Distribution (Global): China South-Central, China Southeast, Hainan, Laos, Nepal, Thailand, Vietnam
Distribution (India): Maharashtra, Tamil Nadu
Leaf Type: Leaves are opposite, oval-lanceolate
Inflorescence: Inflorescence consists of 2-flowered axillary cymes with fragrant white flowers
Fruit Type: Fruit is globose berry turning black when ripe.
Flowering and Fruiting: March–October
IUCN Status: Not evaluated
Notes: Usually found in mountain slopes, and along roadsides and riversides
Reference: Prodr. 4: 333 (1830)

Lonicera etrusca Santi
Caprifoliaceae
Common Name: Etruscan honeysuckle
Climbing Mechanism: Scrambler-Unarmed
Distribution (Global): Albania, Algeria, Bulgaria, Corse, Cyprus, East Aegean Is., France, Greece, Italy, Kriti, Lebanon-Syria, Libya, Morocco, North Caucasus, Palestine, Portugal, Sardegna, Sicilia, Spain, Switzerland, Transcaucasus, Tunisia, Turkey, Turkey-in-Europe, Yugoslavia
Distribution (India): Tamil Nadu
IUCN Status: Not evaluated
Reference: Viagg. Montamiata: 113 (1795)

Lonicera glabrata Wall.
Caprifoliaceae

Common Name: Himalayan honeysuckle
Climbing Mechanism: Scrambler-Unarmed
Distribution (Global): Assam, East Himalaya, Myanmar, Nepal, Tibet
Distribution (India): Arunachal Pradesh, Manipur, Sikkim, Tripura, West Bengal
Leaf Type: Leaves cordate-oblong, obovate, acute
Inflorescence: Flowers spikes axillary, short, often panicled
Fruit Type: Very large clusters of shiny black berries
Flowering and Fruiting: May–October
IUCN Status: Not evaluated
Reference: W. Roxburgh, Fl. Ind. 2: 175 (1824)

Lonicera hildebrandiana Collett & Hemsl.
Caprifoliaceae
Common Name: Giant Burmese honeysuckle
Climbing Mechanism: Scrambler-Unarmed
Distribution (Global): SE Asia, China
IUCN Status: Not evaluated

Lonicera japonica Thunb.
Caprifoliaceae
Synonyms: *Caprifolium brachypodum* G. Gordon, *Caprifolium japonicum* (Thunb.) Dum.Cours., *Caprifolium roseum* Lam., *Lonicera aureoreticulata* T. Moore, *Lonicera brachypoda* DC., *Lonicera chinensis* P. Watson, *Lonicera cochinchinensis* G. Don, *Lonicera confusa* Miq.
Common Name: Japanese honeysuckle
Climbing Mechanism: Scrambler-Unarmed
Distribution (Global): China North-Central, China South-Central, China Southeast, Japan, Korea, Manchuria, Taiwan
Distribution (India): Maharashtra, Tamil Nadu
Leaf Type: Opposite, simple oval leaves
Inflorescence: Flowers in peduncled pairs, white to whitish-yellow zygomorphic tubular flowers, peduncle solitary in upper leaf axils
Fruit Type: The fruit is a globose dark blueberry
Flowering and Fruiting: April–November
IUCN Status: Not evaluated
Reference: J.A. Murray (ed.), Syst. Veg. ed. 14: 216 (1784)

Lonicera japonica var. *chinensis*
Caprifoliaceae
Common Name: Japanese honeysuckle
Climbing Mechanism: Scrambler-Unarmed
IUCN Status: Not evaluated

Lonicera leschenaultii Wall.
Caprifoliaceae

Synonyms: *Lonicera macrantha* (D. Don) Spreng., *Caprifolium leschenaultii* (Wall.) Kuntze, *Caprifolium macranthum* D. Don, *Lonicera calvescens* (Chun & F.C. How) P.S. Hsu & H.J. Wang, *Lonicera esquirolii* H. Lév., *Lonicera fulvotomentosa* P.S. Hsu & S.C. Cheng.
Common Name: Long-flower honeysuckle
Climbing Mechanism: Scrambler-Unarmed
Distribution (Global): Assam, China South-Central, China Southeast, East Himalaya, Myanmar, Nepal, Taiwan, Tibet, Vietnam
Distribution (India): Tamil Nadu
Leaf Type: Elliptic leaves, oblong or oblong-lanceolate, pointed, heart-shaped base
Inflorescence: Flowers pale white or pale yellow, in axillary peduncled in pairs of dense terminal panicles
Fruit Type: fruit globose, black when ripe
Flowering and Fruiting: April–September
IUCN Status: Not evaluated
Reference: Syst. Veg. ed. 16, 4(2): 82 (1827)

Lonicera ligustrina Wall.
Caprifoliaceae
Synonyms: Xylosteon ligustrinum (Wall.) D.Don
Climbing Mechanism: Scrambler-Unarmed
Distribution (Global): Assam, China North-Central, China South-Central, China Southeast, East Himalaya, Nepal
Distribution (India): Tamil Nadu, Western Ghats
Leaf Type: Leaves opposite, lanceolate-ellipticals, sometimes leaves in first pair are variable
Inflorescence: Flowers in short pubescent, peduncled pairs
Fruit Type: Fruits berry usually confluent into a single small spherical fruit, purplish-red, black when ripe, juicy
Flowering and Fruiting: May–October
IUCN Status: Not evaluated
Notes: Occurs in high altitude evergreen forests
Reference: W. Roxburgh, Fl. Ind. 2: 179 (1824)

Lonicera macrantha (D. Don) Spreng.
Caprifoliaceae
Synonyms: Synonym
Climbing Mechanism: Scrambler-Unarmed
Distribution (India): Arunachal Pradesh, Manipur, Mizoram, Meghalaya, Tripura, West Bengal
IUCN Status: Not evaluated

Lonicera sempervirens L.
Caprifoliaceae

Synonyms: *Caprifolium sempervirens* (L.) Moench, *Kantemon angustifolium* Raf., *Lonicera angustifolia* Raf., *Lonicera caroliniana* Marshall, *Lonicera coccinea* Pers., *Lonicera connata* Meerb., *Lonicera flammea* Salisb., *Lonicera fuchsioides* Dippel.
Common Name: Trumpet honeysuckle
Climbing Mechanism: Scrambler-Unarmed
Distribution (Global): Alabama, Arkansas, Connecticut, Delaware, District of Columbia, Florida, Georgia, Indiana, Iowa, Kansas, Kentucky, Louisiana, Maine, Maryland, Massachusetts, Michigan, Mississippi, Missouri, New Hampshire, New Jersey, New York, North Carolina, Ohio, Oklahoma, Pennsylvania, Rhode I., South Carolina, Tennessee, Texas, Vermont, Virginia, West Virginia
Distribution (India): Maharashtra, Tamil Nadu
Leaf Type: Leaves opposite, obovate to elliptic, or lanceolate, leathery, margin entire
Inflorescence: Stem terminates an inflorescence anywhere between the base and shoot of the climber. Flowers axillary usually in whorled 3 pairs
Fruit Type: Fruits scarlet red, fleshy, ellipsoid
Flowering and Fruiting: June–October
IUCN Status: Not evaluated
Reference: Sp. Pl.: 173 (1753)

Lonicera × *brownii* (Regel) Carriere
Caprifoliaceae
Synonyms: *Caprifolium* × *puniceum* G. Kirchn., *Lonicera* × *brownii* var. *fuchsioides* (Rehder) Rehder, *Lonicera* × *brownii* var. *punicea* (G. Kirchn.) Rehder, *Lonicera* × *brownii* var. *youngii* (K. Koch) Rehder, *Lonicera ciliosa* f. *youngii* (K. Koch) Dippel.
Common Name: Dropmore scarler honeysuckle
Climbing Mechanism: Scrambler-Unarmed
Distribution (India): Tamil Nadu
Leaf Type: Leaves are Simple, opposite, obovate, obtuse, entire margin
Inflorescence: Orange
Red/Burgundy flowers with spike inflorescence
Fruit Type: Bright red berries
Flowering and Fruiting: April–June
IUCN Status: Not evaluated
Reference: J. Gén. Hort. 11: 123 (1856)

Cardiopteris quinqueloba (Hassk.) Hassk.
Cardiopteridaceae
Synonyms: *Cardiopteris hamulosa* Griff., *Cardiopteris javanica* Blume., *Cardiopteris lobata* Wall. ex Benn. & R.Br., *Cardiopteris platycarpa* Gagnep., *Cardiopteris rumphii* Baill., *Cardiopteris subhamata* Wall. ex Baill., *Peripterygium platycarpum* (Gagnep.) Sleumer.
Climbing Mechanism: Stem Twiner

Distribution (Global): Assam, Bangladesh, Borneo, Cambodia, China South-Central, China Southeast, East Himalaya
Distribution (India): Arunachal Pradesh, Manipur, Mizoram, West Bengal
Leaf Type: Simple, alternate, spiral
Inflorescence: Axillary panicled cymes
Flowering and Fruiting: May–March
IUCN Status: Not evaluated
Notes: Usually found in forest clearings, secondary forests, and open forests
Reference: Retzia 1: 64 (1855)

Brachystemma calycinum D. Don
Caryophyllaceae
Synonyms: *Arenaria nepalensis* Spreng.
Climbing Mechanism: Stem Twiner
Distribution (Global): Assam, Cambodia, China South-Central, East Himalaya, Laos, Myanmar, Nepal, Thailand, Tibet, Vietnam
Distribution (India): Arunachal Pradesh, Manipur, Mizoram, West Bengal
Leaf Type: Simple, opposite
Inflorescence: Axillary or terminal
Fruit Type: Capsule
IUCN Status: Not evaluated
Reference: Prodr. Fl. Nepal.: 216 (1825)

Arnicratea grahamii (Wight) N. Hallé
Celastraceae
Synonyms: *Reissantia grahamii* (Wight) Ding Hou, *Hippocratea ellipticarpa* Merrill, *Hippocratea grahamii* Wight, *Hippocratea lanceolata* Buch. -Ham. ex Wall., *Hippocratea megalocarpa* Merrill, *Hippocratea obtusifolia* R. Grah. ex Wight.
Common Name: Running straggler
Climbing Mechanism: Scrambler-Unarmed
Distribution (India): Maharashtra
Flowering and Fruiting: November–May
IUCN Status: Not evaluated
Notes: Found alonf the Ghats in semi-evergreen forests
Reference: Bull. Mus. Natl. Hist. Nat., B, Adansonia 6: 14 (1984)

Celastrus hindsii Benth.
Celastraceae
Synonyms: *Celastrus approximatus* Craib., *Celastrus axillaris* Ridl., *Celastrus cantonensis* Hance., *Celastrus merrillii* Tardieu., *Celastrus oblongifolius* Hayata., *Celastrus racemulosus* Hassk., *Celastrus venulosus* Wall. Ex. M.-A. Lawson., *Celastrus xizangensis* Y.R.Li
Climbing Mechanism: Scrambler-Unarmed

Distribution (Global): Assam, Bangladesh, Borneo, China South-Central, China Southeast, East Himalaya, Hainan, Java, Malaya, Maluku, Marianas, Myanmar, Ogasawara-shoto, Sulawesi, Sumatera, Taiwan, Thailand, Tibet, Vietnam
Distribution (India): Manipur, Mizoram
Leaf Type: Simple, alternate, spiral
Inflorescence: Axillary or terminal paniculate cymes
Fruit Type: Capsules, subglobose
Flowering and Fruiting: July–October
IUCN Status: Not evaluated
Notes: Found in primary forests, moist and shaded places, along stream-banks and hill-slopes
Reference: Hooker's J. Bot. Kew Gard. Misc. 3: 334 (1851)

Celastrus hookeri Prain
Celastraceae
Climbing Mechanism: Scrambler-Unarmed
Distribution (Global): Assam, China South-Central, East Himalaya, Myanmar, Nepal, Pakistan, Tibet, Vietnam
Distribution (India): Manipur
Leaf Type: Simple
Inflorescence: Panicles
Fruit Type: Dry, capsule
IUCN Status: Not evaluated
Reference: J. Asiat. Soc. Bengal, Pt. 2, Nat. Hist. 73: 197 (1904)

Celastrus monospermus Roxb.
Celastraceae
Synonyms: *Catha benthamii* Gardner & Champ., *Catha monosperma* Benth., *Celastrus annamensis* Tardieu., *Celastrus benthamii* (Gardner & Champ.) Rehder & E.H. Wilson., *Celastrus championii* Benth., *Celastrus laoticus* Pit., *Monocelastrus monosperma* (Roxb.) F.T. Wang & T. Tan.
Climbing Mechanism: Scrambler-Unarmed
Distribution (Global): Assam, Bangladesh, Borneo, China South-Central, China Southeast, East Himalaya, Hainan, Laos, Myanmar, Pakistan, Thailand, Vietnam
Distribution (India): Assam, Tripura, West Bengal
Leaf Type: Simple
Inflorescence: Panicles
Fruit Type: Dry, capsule
IUCN Status: Not evaluated
Reference: Fl. Ind. 2: 394 (1824)

Celastrus paniculatus subsp. *aggregatus*
Celastraceae
Climbing Mechanism: Scrambler-Unarmed
Distribution (Global): India

Distribution (India): Tamil Nadu
Leaf Type: Simple
Inflorescence: Panicles
Fruit Type: Dry, capsule
Flowering and Fruiting: April–October
IUCN Status: Not evaluated
Notes: Occur in the lower elevations in moist deciduous forests and tropical wet evergreen forests
Reference: Kew Bull. 46: 540 (1991)

Celastrus paniculatus subsp. *angladeanus*
Celastraceae
Climbing Mechanism: Scrambler-Unarmed
Distribution (Global): India
Distribution (India): Tamil Nadu
Leaf Type: Simple
Inflorescence: Panicles
Fruit Type: Dry, capsule
IUCN Status: Not evaluated
Reference: Taiwania 62: 311 (2017)

Celastrus paniculatus Willd.
Celastraceae
Synonyms: *Catha paniculata* (Willd.) Scheidw., *Ceanothus paniculatus* (Willd.) B. Heyne ex Roth, *Scutia paniculata* (Willd.) G. Don
Common Name: Black oil plant, Climbing staff plant, Intellect plant
Climbing Mechanism: Scrambler-Unarmed
Distribution (Global): Andaman Is., Assam, Bangladesh, Cambodia, China South-Central, China Southeast, Christmas I., East Himalaya, Hainan, India, Java, Laos, Lesser Sunda Is., Malaya, Myanmar, Nepal, Philippines, Sri Lanka, Sulawesi, Sumatera, Taiwan, Thailand, Vietnam, West Himalaya
Distribution (India): Andaman and Nicobar Islands, Andhra Pradesh, Arunachal Pradesh, Assam, Bihar, Odisha, Eastern Ghats, Goa, Gujarat, Himachal Pradesh, Jharkhand, Kerala, Madhya Pradesh, Maharashtra, Manipur, Meghalaya, Mizoram, Odisha, Rajasthan, Tamil Nadu, Telangana, Uttar Pradesh
Leaf Type: simple, alternate
Inflorescence: Terminal drooping panicles
Fruit Type: Capsule, globose
Flowering and Fruiting: March–December
IUCN Status: Not evaluated
Notes: Leaves are emmenagogue whereas seeds are acrid, bitter, hot, appetizer, laxative, emetic, aphrodisiac, powerful brain tonic, cause burning sensation.
Reference: Sp. Pl., ed. 4, 1: 1125 (1798)

Celastrus stylosus Wall
Celastraceae
Climbing Mechanism: Stem Twiner
Distribution (Global): Assam, Bangladesh, China South-Central, China Southeast, East Himalaya, India, Java, Lesser Sunda Is., Myanmar, Nepal, Thailand, Vietnam
Distribution (India): Arunachal Pradesh, Assam, Rajasthan
Leaf Type: Simple
Inflorescence: Panicles
Fruit Type: Dry, capsule
IUCN Status: Not evaluated
Reference: W. Roxburgh, Fl. Ind. 2: 401 (1824)

Euonymus echinatus Wall
Celastraceae
Synonyms: *Euonymus arboricola* Hayata, *Euonymus cinereus* M.A. Lawson, *Euonymus echinatus* var. *trichocarpus* (Hayata) S.S. Ying, *Euonymus fungosus* Ohwi, *Euonymus mupinensis* Loes. & Rehder, *Euonymus scandens* Graham, *Euonymus subsessilis* Sprague, *Euonymus trichoca*
Common Name: Climbing spindle bush
Climbing Mechanism: Scrambler-Unarmed
Distribution (Global): Assam, Bangladesh, China North-Central, China South-Central, China Southeast, East Himalaya, Hainan, India, Myanmar, Nansei-shoto, Nepal, Taiwan, Thailand, Tibet, West Himalaya
Distribution (India): Arunachal Pradesh, Assam, Himachal Pradesh, West Bengal
IUCN Status: Least Concern
Reference: W. Roxburgh, Fl. Ind. 2: 410 (1824)

Euonymus frigidus Wall
Celastraceae
Synonyms: *Euonymus amygdalifolius* Franch., *Euonymus assamicus* Blakelock, *Euonymus austrotibetanus* Y.R.Li, *Euonymus burmanicus* Merr., *Euonymus crinitus* Pamp., *Euonymus dasydictyon* Loes. & Rehder, *Euonymus fugongensis* Y.R.Li, *Euonymus gongshanensis* Y.R.Li
Climbing Mechanism: Scrambler-Unarmed
Distribution (Global): Assam, China North-Central, China South-Central, China Southeast, East Himalaya, Inner Mongolia, Myanmar, Nepal, Qinghai, Tibet, West Himalaya
Distribution (India): West Bengal
IUCN Status: Not evaluated
Reference: W. Roxburgh, Fl. Ind. 2: 409 (1824)

Euonymus theifolius Wall. ex M.A. Lawsen
Celastraceae
Synonyms: *Euonymus paravagans* Z.M. Gu & C.Y. Cheng

Pragmotessara theifolia (Wall. ex M.A. Lawson) Pierre
Climbing Mechanism: Scrambler-Unarmed
Distribution (Global): Assam, Bangladesh, China South-Central, East Himalaya, Myanmar, Nepal, Thailand, Tibet
Distribution (India): Assam, West Bengal, Mizoram
IUCN Status: Not evaluated
Reference: J. D. Hooker, Fl. Brit. India 1: 612 (1875)

Euonymus vagans Wall
Celastraceae
Synonyms: *Euonymus echinatus* var. *vagans* (Wall.) R.C. Srivast.
Euonymus jinfoshanensis Z.M. Gu, *Euonymus jinggangshanensis* M.X. Nie, *Euonymus pseudosootepensis* Y.R. Li & S.G.Wu
Climbing Mechanism: Scrambler-Unarmed
Distribution (Global): Assam, Bangladesh, China South-Central, China Southeast, East Himalaya, Laos, Myanmar, Nepal, Tibet
Distribution (India): Assam, West Bengal, Manipur
IUCN Status: Least Concern
Reference: W. Roxburgh, Fl. Ind. 2: 412 (1824)

Gymnosporia emarginata (Willd.) Thwaites
Celastraceae
Synonyms: *Catha emarginata* (Willd.) G. Don, *Celastrus emarginatus* Willd., *Celastrus semiarillatus* Turcz., *Elaeodendron horizontale* Turcz., *Maytenus emarginata* (Willd.) Ding Hou, *Schinzia inconspicua* Dennst., *Staphylea vepretum* Roxb. ex Wight & Arn.
Common Name: Red spike thorn
Climbing Mechanism: Scrambler-Armed
Distribution (Global): India, Malaya, Queensland, Sri Lanka
Leaf Type: Simple
Inflorescence: Axillary or Terminal cymes
Fruit Type: Obovoid globose capsule
IUCN Status: Not evaluated
Reference: Enum. Pl. Zeyl.: 409 (1864)

Gymnosporia royleana Wall. ex M.A. Lawson
Celastraceae
Synonyms: *Celastrus royleanus* Wall. ex M.A. Lawson, *Celastrus spinosus* Royle ex Boiss., *Euonymus spinosus* Griff., *Maytenus royleana* (Wall. ex M.A. Lawson) Cufod., *Maytenus yimenensis* H. Shao
Common Name: Royle's spike thorn
Climbing Mechanism: Scrambler-Armed
Distribution (Global): Afghanistan, India, Pakistan, Tibet, West Himalaya
IUCN Status: Not evaluated
Reference: J. D. Hooker, Fl. Brit. India 1: 620 (1875)

Hippocratea andamanica King
Celastraceae
Synonyms: *Loeseneriella andamanica* (King) H.B. Naithani & S. Biswas
Climbing Mechanism: Scrambler-Unarmed
Distribution (India): Andaman and Nicobar Islands
IUCN Status: Not evaluated
Reference: J. Asiat. Soc. Bengal, Pt. 2, Nat. Hist. 65: 359 (1896)

Hippocratea arborea Roxb.
Celastraceae
Synonyms: *Reissantia arborea* (Roxb.) H. Hara
Climbing Mechanism: Scrambler-Unarmed
Distribution (India): North Eastern India, Bihar, Odisha
IUCN Status: Not evaluated
Reference: Pl. Coromandel 3: 3 (1811)

Hippocratea bourdillonii Gamble
Celastraceae
Climbing Mechanism: Scrambler-Unarmed
IUCN Status: Not evaluated

Hippocratea cumingii M.A. Lawson
Celastraceae
Synonyms: *Loeseneriella cumingii* (M.A. Lawson) Ding Hou
Climbing Mechanism: Scrambler-Unarmed
IUCN Status: Not evaluated
Reference: J. D. Hooker, Fl. Brit. India 1: 624 (1875)

Hippocratea lobbii M.A. Lawson
Celastraceae
Synonyms: *Reissantia indica* var. *indica*
Climbing Mechanism: Scrambler-Unarmed
IUCN Status: Not evaluated
Reference: J. D. Hooker, Fl. Brit. India 1: 624 (1875)

Hippocratea majumdarii Chakrab. & M. Gangop
Celastraceae
Synonyms: *Salacia majumdarii* (Chakrab. & M. Gangop.) B.D. Naithani
Climbing Mechanism: Scrambler-Unarmed
IUCN Status: Not evaluated
Reference: J. Econ. Taxon. Bot. 14: 122 (1990)

Hippocratea nicobarica Kurz
Celastraceae
Synonyms: *Loeseneriella nicobarica* (Kurz) H.B. Naithani & S. Biswas

Climbing Mechanism: Scrambler-Unarmed
IUCN Status: Not evaluated
Reference: J. Asiat. Soc. Bengal, Pt. 2, Nat. Hist. 44: 203 (1875)

Hippocratea obtusifolia var. *barbata* (F. Muell.) Benth.
Celastraceae
Climbing Mechanism: Scrambler-Unarmed
IUCN Status: Not evaluated

Loeseneriella africana (Willd.) R. Wilczek
Celastraceae
Synonyms: *Calypso africana* (Willd.) G. Don, *Hippocratea africana* (Willd.) Loes. ex Engl., *Salacia africana* (Willd.) DC., *Tonsella africana* Willd.
Climbing Mechanism: Scrambler-Unarmed
Distribution (Global): Angola, Benin, Botswana, Burkina, Cameroon, Caprivi Strip, Central African Repu, Chad, Congo, Ethiopia, Gabon, Gambia, Ghana, Guinea, Guinea-Bissau, India, Ivory Coast, Kenya, Laos, Liberia, Madagascar, Malawi, Mali, Mauritania, Mozambique, Myanmar, Niger, Nigeria, Northern Provinces, Rwanda, Senegal, Sierra Leone, Sri Lanka, Sudan, Tanzania, Thailand, Togo, Uganda, Zambia, Zaire, Zimbabwe
Distribution (India): Karnataka, Tamil Nadu
Leaf Type: Simple, opposite, broad leaves
Inflorescence: Flowers 3–? in simple axillary dichasial sometimes corymbose cymes with pubescent or glabrescent branches
Fruit Type: Loculicidal capsule fruit
IUCN Status: Not evaluated
Reference: Fl. Congo Belge 9: 154 (1960)

Loeseneriella arnottiana (Wight) A.C.Sm.
Celastraceae
Synonyms: *Pristimera arnottiana* (Wight) R.H. Archer., *Hippocratea arnottiana* Wight, *Loeseneriella arnottiana* (Wight) A.C.Sm., *Prionostemma arnottiana* (Wight) N. Hallé, *Salacia terminalis* Thwaites
Climbing Mechanism: Scrambler-Unarmed
Distribution (Global): India, Sri Lanka
Distribution (India): Kerala, Western Ghats
IUCN Status: Not evaluated
Reference: Molec. Phylogen. Evol. 59: 328 (2011)

Loeseneriella bourdillonii (Gamble) Ramamoorthy
Celastraceae
Synonyms: *Hippocratea bourdillonii* Gamble
Climbing Mechanism: Scrambler-Unarmed
Distribution (Global): India
Distribution (India): Kerala, Western Ghats

Leaf Type: Leaves, oblong, apex abruptly acuminate, base rounded, margins recurved, sub-entire
Inflorescence: Flowers in pedunculate cymes, quadrangular
Fruit Type: Loculicidal capsule
Flowering and Fruiting: March–May
IUCN Status: Not evaluated
Notes: Often found in evergreen and semi-evergreen forests
Reference: J. Biol. Sci. (Bombay) 8(2): 56 (1965 publ. 1966)

Loeseneriella macrantha (Korth.) A.C.Sm.
Celastraceae
Synonyms: *Hippocratea macrantha* Korth.
Climbing Mechanism: Scrambler-Unarmed
Distribution (Global): Andaman Is., Assam, Bangladesh, Bismarck Archipelago, Borneo, Caroline Is., Java, Malaya, Maluku, Myanmar, New Guinea, Nicobar Is., Solomon Is., Sri Lanka, Sulawesi, Sumatera, Thailand, Vanuatu
Distribution (India): Peninsular India
IUCN Status: Critically endangered
Reference: Amer. J. Bot. 28: 439 (1941)

Loeseneriella obtusifolia (Roxb.) A.C.Sm.
Celastraceae
Synonyms: *Loeseneriella africana* var. *obtusifolia* (Roxb.) N. Hallé., *Hippocratea grandiflora* Wall., *Hippocratea obtusifolia* Roxb., *Hippocratea rigida* Hampe ex M.A. Lawson, *Hippocratea tortuosa* Wall., *Loeseneriella indochinensis* Tardieu
Climbing Mechanism: Scrambler-Unarmed
Distribution (Global): India, Laos, Myanmar, Sri Lanka, Thailand
Distribution (India): Andhra Pradesh, Eastern Ghats, Kerala, Mizoram, Tamil Nadu, Telangana, Western Ghats
Leaf Type: Leaves, ovate-lanceolate or elliptic-oblong, apex obtuse or rounded, coarsely crenate, coriaceous
Inflorescence: Panicles axillary or terminal
Flowering and Fruiting: July–September
IUCN Status: Not evaluated
Notes: Occurs mostly in dry deciduous forests
Reference: Fl. Gabon 29: 255 (1986)

Loeseneriella ovata (Lam.) M.R. Almeida
Celastraceae
Synonyms: *Hippocratea volubilis* L. *Bejuco pendulus* Loefl., *Hippocratea acutiflora* Moc. & Sessé ex DC., *Hippocratea affinis* Cambess., *Hippocratea anafensis* Urb., *Hippocratea aubletiana* Miers. + 40
Climbing Mechanism: Scrambler-Unarmed
Distribution (Global): Argentina Northeast, Argentina Northwest, Belize, Bolivia, Brazil North, Brazil Northeast, Brazil South, Brazil Southeast, Brazil West-

Central, Colombia, Costa Rica, Ecuador, El Salvador, Florida, French Guiana, Guatemala, Guyana, Honduras, Leeward Is., Mexico Central, Mexico Gulf, Mexico Northeast, Mexico Northwest, Mexico Southeast, Mexico Southwest, Nicaragua, Panamá, Paraguay, Peru, Southwest Caribbean, Suriname, Trinidad-Tobago, Venezuela, Windward Is.
Distribution (India): Maharashtra
Leaf Type: Leaf blade elliptic to ovate or obovate
Inflorescence: Brownish tomentulose or puberulent
Fruit Type: Capsules
Flowering and Fruiting: June–December
IUCN Status: Not evaluated
Reference: Sp. Pl.: 1191 (1753)

Maytenus acuminata (L.f.) Loes.
Celastraceae
Synonyms: *Ilex livida* E. Mey., *Ilex lucida* E. Mey. ex C. Presl, *Catha acuminata* (L.f.) C. Presl, *Catha rupestris* (Eckl. & Zeyh.) C.Presl. *Celastrus acuminatus* L.f. + 16
Climbing Mechanism: Scrambler-Armed
Distribution (Global): Bangladesh, Burundi, Cameroon, Cape Provinces, Free State, Gabon, Kenya, KwaZulu-Natal, Lesotho, Malawi, Mozambique, Northern Provinces, Rwanda, Swaziland, Tanzania, Uganda, Zambia, Zaire, Zimbabwe
Leaf Type: Leaves not fasciculate, petiolate; lamina dark green and ± glossy above, paler beneath, ovate to narrowly lanceolate or rarely elliptic or elliptic-oblong
Inflorescence: Cymes 0.3–3 cm long, monochasial after 1st to 3rd branching or pseudumbellate to fasciculate, axillary, sessile (fasciculate)
Fruit Type: Capsule green or reddish-green to yellow or orange or red, 4–11 mm long, 3-lobed (or globose to 2-lobed due to abortion), with lobes rounded, smooth, not fleshy
Flowering and Fruiting: October–February
IUCN Status: Least Concern
Notes: Often occur in forest understorey and margin of upland or riverine forest
Reference: H.G.A. Engler, Nat. Pflanzenfam. ed. 2, 20b: 138 (1942)

Maytenus heyneana (Roth) D.C.S. Raju & Babu
Celastraceae
Synonyms: *Gymnosporia heyneana* (Roth) M.A. Lawson. *Catha heyneana* (Roth) Walp., *Celastrus heyneanus* Roth, *Celastrus serrulatus* Roth.
Climbing Mechanism: Scrambler-Armed
Distribution (Global): India
Distribution (India): Tamil Nadu
Leaf Type: Simple, alternate-spiral
Inflorescence: In axillary or terminal cymes; white
Fruit Type: An obcordate capsule; red; seeds 2 per cell, aril fleshy
Flowering and Fruiting: Throughout the year

IUCN Status: Not evaluated
Notes: Commonly found along the slopes, shola border, and deciduous forests from plains to 800 m
Reference: J. D. Hooker, Fl. Brit. India 1: 620 (1875)

Maytenus rothiana Bennet & K.C. Sahni
Celastraceae
Synonyms: *Gymnosporia rothiana* (Walp.) M.A. Lawson. *Catha rothiana* Walp., *Celastrus rothianus* Wight & Arn., *Gymnosporia wightiana* R.S. Rao, *Maytenus wightiana* Babu
Common Name: Roth's spike thorn
Climbing Mechanism: Scrambler-Armed
Distribution (Global): India
Distribution (India): Maharashtra
Leaf Type: Leaves alternate, simple, broadly ovate, margin serrate, acute apex
Inflorescence: Flowers 2–3 mm across, greenish-white, in fascicles of short cymes at leafless nodes; Fruiting pedicels
Fruit Type: Capsule 1.5 x 1.5 cm, obovate, trigonous, 3-valved
Flowering and Fruiting: June–August
IUCN Status: Not evaluated
Notes: Predominantly occur in the evergreen forests of the southern Western Ghats
Reference: J. D. Hooker, Fl. Brit. India 1: 620 (1875)

Microtropis densiflora Wight
Celastraceae
Climbing Mechanism: Scrambler-Unarmed
Distribution (India): Arunachal Pradesh, Tamil Nadu
IUCN Status: Endangered

Plagiopteron suaveolens Griff.
Celastraceae
Synonyms: *Plagiopteron chinensis* X.X. Chen, *Plagiopteron fragrans* Griff.
Climbing Mechanism: Scrambler-Unarmed
Distribution (Global): Assam, China Southeast, Myanmar, Thailand, Vietnam
Distribution (India): Assam
IUCN Status: Not evaluated
Reference: Calcutta J. Nat. Hist. 4: 244 (1843)

Pristimera arnottiana (Wight) R.H. Archer
Celastraceae
Synonyms: *Hippocratea arnottiana* Wight, *Loeseneriella arnottiana* (Wight) A.C. Sm., *Prionostemma arnottiana* (Wight) N. Hallé, *Salacia terminalis* Thwaites
Climbing Mechanism: Scrambler-Armed
Distribution (Global): India, Sri Lanka
Distribution (India): Tamil Nadu

IUCN Status: Not evaluated
Reference: Molec. Phylogen. Evol. 59: 328 (2011)

Pristimera grahamii (Wight) A.C.Sm.
Celastraceae
Synonyms: *Reissantia grahamii* (Wight) Ding Hou
Climbing Mechanism: Scrambler-Armed
Flowering and Fruiting: November–May
IUCN Status: Not evaluated
Notes: Found along the Ghats in semi-evergreen forests
Reference: J. Arnold Arbor. 26: 178 (1945)

Reissantia arborea (Roxb.) H. Hara
Celastraceae
Synonyms: *Hippocratea arborea* Roxb., *Pristimera arborea* (Roxb.) A.C.Sm., *Hippocratea disperma* Wall.
Climbing Mechanism: Scrambler-Unarmed
Distribution (Global): Assam, China South-Central, China Southeast, East Himalaya, Myanmar, Nepal
Distribution (India): Arunachal Pradesh, Tripura
Leaf Type: Simple, opposite
Fruit Type: Capsule
Flowering and Fruiting: November–September
IUCN Status: Not evaluated
Reference: J. Jap. Bot. 40: 327 (1965)

Reissantia grahamii (Wight) Ding Hou
Celastraceae
Synonyms: *Arnicratea grahamii* (Wight) N. Hallé, *Hippocratea ellipticarpa* Merr., *Hippocratea grahamii* Wight, *Hippocratea lanceolata* Buch. -Ham. ex Wall., *Hippocratea megalocarpa* Merr., *Hippocratea obtusifolia* Graham ex Wight, *Hippocratea salacioides* Korth.
Climbing Mechanism: Scrambler-Unarmed
Distribution (Global): Andaman Is., Assam, Bangladesh, Borneo, India, Java, Maluku, Myanmar, New Guinea, Philippines, Sulawesi, Sumatera, Thailand
Distribution (India): Eastern Ghats, Kerala, Tamil Nadu
Leaf Type: Simple, decussate, ovate-obovate
Inflorescence: Cymes rustly, sometimes with branches
Fruit Type: Dry, samaroid
Flowering and Fruiting: January–May
IUCN Status: Not evaluated
Notes: Found usually along the ghats in semi-evergreen forests
Reference: Blumea 12: 33 (1963)

Reissantia indica (Willd.) N. Hallé
Celastraceae
Synonyms: *Hippocratea indica* Willd., *Pristimera indica* (Willd.) A.C.Sm., *Hippocratea lobbii* M. Laws., *Hippocratea disperma* Vahl, *Hippocratea euonymoides* Vahl, *Hippocratea fuscescens* Kurz, *Hippocratea obtusifolia* Wall., *Hippocratea pallida* Craib.
Climbing Mechanism: Scrambler-Unarmed
Distribution (Global): Angola, Assam, Bangladesh, Benin, Borneo, Botswana, Cambodia, Cameroon, Caprivi Strip, Central African Repu, China Southeast, Congo, Ethiopia, Gabon, Ghana, Guinea, Guinea-Bissau, Gulf of Guinea Is., Hainan, India, Ivory Coast, Java, Laos, Lesser Sunda Is., Liberia, Malawi, Malaya, Mozambique, Myanmar, Namibia, Nigeria, Northern Provinces, Philippines, Senegal, Sierra Leone, Sri Lanka, Sudan, Sulawesi, Sumatera, Tanzania, Thailand, Uganda, Vietnam, Zambia, Zaire, Zimbabwe
Distribution (India): Andhra Pradesh, Arunachal Pradesh, Eastern Ghats, Kerala, Maharashtra, Odisha, Puducherry, Tamil Nadu, Tripura
Leaf Type: Simple, decussate, ovate-obovate
Inflorescence: Cymes rustly, sometimes with branches
Fruit Type: Dry, samaroid
Flowering and Fruiting: April–December
IUCN Status: Not evaluated
Notes: Frequently found in semi-evergreen and moist deciduous forests
Reference: Mém. Inst. Franç. Afrique Noire 64: 85 (1962)

Salacia agasthiamalana Udayan, Regy Yohannan & Pradeep
Celastraceae
Climbing Mechanism: Scrambler-Unarmed
Distribution (Global): India
Distribution (India): Kerala, Tamil Nadu
Leaf Type: Simple, subopposite
Inflorescence: Axillary umbels
Fruit Type: Wet, berry, globose
Flowering and Fruiting: September–July
IUCN Status: Not evaluated
Notes: Found mostly in evergreen forests
Reference: Edinburgh J. Bot. 69: 255 (2012)

Salacia beddomei Gamble
Celastraceae
Climbing Mechanism: Scrambler-Unarmed
Distribution (Global): India
Distribution (India): Karnataka, Kerala, Tamil Nadu, Western Ghats
Leaf Type: Simple, subopposite
Inflorescence: Axillary umbels
Fruit Type: Wet, berry, globose

IUCN Status: Not evaluated
Reference: Bull. Misc. Inform. Kew 1916: 133 (1916)

Salacia brunoniana Wight & Arn.
Celastraceae
Climbing Mechanism: Scrambler-Unarmed
Distribution (Global): India
Leaf Type: Simple, subopposite
Inflorescence: Axillary umbels
Fruit Type: Wet, berry, globose
Flowering and Fruiting: January–February
IUCN Status: Not evaluated
Reference: Prodr. Fl. Ind. Orient. 1: 105 (1834)

Salacia chinensis L. Celastraceae
Synonyms: *Johnia coromandeliana* Roxb., *Johnia serrata* Buch. -Ham., Salacia *chinensis* var. *latifolia* (Wall. ex M.A. Lawson) Chakrab. & M. Gangop., *Salacia coromandeliana* Teijsm. & Binn. +12
Common Name: Chinese Salacia, Holly-leaved berried spindle tree
Climbing Mechanism: Scrambler-Unarmed
Distribution (Global): Andaman Is., Assam, Bangladesh, Bismarck Archipelago, Borneo, Cambodia, Caroline Is., China Southeast, East Himalaya, India, Java, Laos, Lesser Sunda Is., Malaya, Maluku, Myanmar, New Guinea, Nicobar Is., Northern Territory, Philippines, Queensland, Solomon Is., Sri Lanka, Sulawesi, Sumatera, Thailand, Vietnam, Western Australia
Distribution (India): Andaman and Nicobar Islands, Andhra Pradesh, Eastern Ghats, Goa, Karnataka, Kerala, Odisha, Puducherry, Tamil Nadu, Western Ghats
Leaf Type: Simple, alternate
Inflorescence: Axillary umbels
Fruit Type: Berry, round or ovate red colored when ripe
Flowering and Fruiting: February–October
IUCN Status: Not evaluated
Notes: Found in stream banks of moist deciduous forests
Reference: Mant. Pl. 2: 293 (1771)

Salacia exsculpta Korth.
Celastraceae
Synonyms: *Salacia rubra* M.A. Lawson., *Salacia wrayi* King
Climbing Mechanism: Scrambler-Unarmed
Distribution (Global): Andaman Is., Malaya, Sumatera
Distribution (India): Andaman and Nicobar Islands
Leaf Type: Simple, subopposite
Inflorescence: Axillary umbels
Fruit Type: Wet, berry, globose

IUCN Status: Not evaluated
Reference: Flora 31: 579 (1848)

Salacia fruticosa Wall.
Celastraceae
Common Name: Woody Salacia
Climbing Mechanism: Scrambler-Unarmed
Distribution (Global): Bangladesh, India
Distribution (India): Karnataka, Kerala, Tamil Nadu, Western Ghats
Leaf Type: Pinnately compound (Paripinnate)
Inflorescence: Cymes
Fruit Type: Berry, spherical red with 1–3 seeds
Flowering and Fruiting: February–May
IUCN Status: Not evaluated
Notes: Usually found in evergreen and semi-evergreen forests
Reference: J. D. Hooker, Fl. Brit. India 1: 628 (1875)

Salacia gambleana Whiting & Kaul
Celastraceae
Synonyms: *Salacia talbotii* Gamble
Climbing Mechanism: Scrambler-Unarmed
Distribution (Global): India
Leaf Type: Simple, subopposite
Inflorescence: Axillary umbels
Fruit Type: Wet, berry, globose
IUCN Status: Not evaluated
Reference: Bull. Misc. Inform. Kew 1940: 300 (1941)

Salacia jenkinsii Kurz
Celastraceae
Climbing Mechanism: Scrambler-Unarmed
Distribution (Global): India
Leaf Type: Simple, opposite, decussate
Inflorescence: Dichotomous cymes
Fruit Type: Capsule
IUCN Status: Not evaluated
Reference: J. Asiat. Soc. Bengal, Pt. 2, Nat. Hist. 44: 203 (1875)

Salacia khasiana Purkayastha
Celastraceae
Climbing Mechanism: Scrambler-Unarmed
Distribution (Global): India
Distribution (India): Arunachal Pradesh, Mizoram
Leaf Type: Simple, subopposite
Inflorescence: Axillary umbels

Fruit Type: Wet, berry, globose
IUCN Status: Not evaluated
Reference: Indian Forester 64: 277 (1938)

Salacia latifolia Wall.
Celastraceae
Climbing Mechanism: Scrambler-Unarmed
Distribution (India): Andaman and Nicobar Islands
Leaf Type: Simple, subopposite
Inflorescence: Axillary umbels
Fruit Type: Wet, berry, globose
IUCN Status: Not evaluated

Salacia leptoclada Tul.
Celastraceae
Synonyms: *Salacia baumannii* Loes., *Salacia wardii* I. Verd.
Climbing Mechanism: Scrambler-Unarmed
Distribution (Global): Burundi, Comoros, Ghana, Ivory Coast, Kenya, KwaZulu-Natal, Madagascar, Mozambique, Tanzania, Togo, Zimbabwe
Leaf Type: Simple, subopposite
Inflorescence: Axillary umbels
Fruit Type: Wet, berry, globose
IUCN Status: Not evaluated
Reference: Ann. Sci. Nat., Bot., sér. 4, 8: 96 (1857)

Salacia macrosperma Wight
Celastraceae
Climbing Mechanism: Scrambler-Unarmed
Distribution (Global): India, Nicobar Is.
Distribution (India): Great, Nicobar, Island, Karnataka, Kerala, Tamil Nadu
Leaf Type: Simple, subopposite
Inflorescence: Axillary umbels
Fruit Type: Wet, berry, globose
Flowering and Fruiting: January–September
IUCN Status: Not evaluated
Notes: Usually found in evergreen and semi-evergreen forests
Reference: Spicil. Neilgherr. 1: 28 (1846)

Salacia maingayi M.A. Lawson
Celastraceae
Synonyms: *Salacia lobbii* M.A. Lawson., Salacia *megasperma* Ridl.
Climbing Mechanism: Scrambler-Unarmed
Distribution (Global): Borneo, Malaya
Leaf Type: Simple, subopposite
Inflorescence: Axillary umbels

Fruit Type: Wet, berry, globose
IUCN Status: Not evaluated
Reference: J. D. Hooker, Fl. Brit. India 1: 626 (1875)

Salacia malabarica Gamble
Celastraceae
Common Name: Malabar Salacia
Climbing Mechanism: Scrambler-Unarmed
Distribution (Global): Bangladesh, India
Distribution (India): Karnataka, Kerala, Tamil Nadu, Western Ghats
Leaf Type: Opposite
Inflorescence: Axillary fascicles, Greenish yellow petals
Fruit Type: Berry, hard and spherical
Flowering and Fruiting: February–April
IUCN Status: Not evaluated
Notes: Usually found in evergreen forests
Reference: Bull. Misc. Inform. Kew 1916: 133 (1916)

Salacia megacarpa N.V. Page & Nandikar
Celastraceae
Climbing Mechanism: Scrambler-Unarmed
Distribution (Global): India
Leaf Type: Simple, subopposite
Inflorescence: Axillary umbels
Fruit Type: Wet, berry, globose
IUCN Status: Not evaluated
Reference: Nordic J. Bot. 38(4)-e02647: 2 (2020)

Salacia membranacea M.A. Lawson
Celastraceae
Climbing Mechanism: Scrambler-Unarmed
Distribution (Global): India
Leaf Type: Simple, subopposite
Inflorescence: Axillary umbels
Fruit Type: Wet, berry, globose
IUCN Status: Not evaluated
Reference: J. D. Hooker, Fl. Brit. India 1: 627 (1875)

Salacia miqueliana Loes.
Celastraceae
Synonyms: *Salacia macrophylla* Miq., *Thermophila macrophylla* Miers.
Climbing Mechanism: Scrambler-Unarmed
Distribution (Global): Brazil North, French Guiana, Suriname
Leaf Type: Simple, subopposite
Inflorescence: Axillary umbels

Fruit Type: Wet, berry, globose
IUCN Status: Not evaluated
Reference: H.G.A. Engler & K.A.E. Prantl, Nat. Pflanzenfam. 3(5): 230 (1893)

Salacia multiflora (Lam.) DC.
Celastraceae
Synonyms: *Hippocratea multiflora* Lam., *Tonsella multiflora* (Lam.) Vahl., *Hylenaea multiflora* (Lam.) Miers., *Tontelea multiflora* (Lam.) Vahl.
Climbing Mechanism: Scrambler-Unarmed
Distribution (Global): Bolivia, Brazil North, Brazil Northeast, Colombia, Costa Rica, Ecuador, French Guiana, Guyana, Nicaragua, Panamá, Peru, Suriname, Venezuela
Leaf Type: Simple, subopposite
Inflorescence: Axillary umbels
Fruit Type: Wet, berry, globose
IUCN Status: Least Concern
Reference: Prodr. 1: 570 (1824)

Salacia oblonga Wall.
Celastraceae
Common Name: Oblong leaf Salacia
Climbing Mechanism: Scrambler-Unarmed
Distribution (Global): India, Sri Lanka
Distribution (India): Andhra Pradesh, Eastern Ghats, Goa, Tamil Nadu
Leaf Type: Simple, subopposite
Inflorescence: Axillary umbels
Fruit Type: Wet, berry, globose
Flowering and Fruiting: March–May
IUCN Status: Vulnerable
Notes: Usually found in evergreen and semi-evergreen forests
Reference: Prodr. Fl. Ind. Orient. 1: 106 (1834)

Salacia ovalis Korth.
Celastraceae
Synonyms: *Salacia integrifolia* Merr.
Climbing Mechanism: Scrambler-Unarmed
Distribution (Global): Java, Maluku, Philippines, Sulawesi, Sumatera
Leaf Type: Simple, subopposite
Inflorescence: Axillary umbels
Fruit Type: Wet, berry, globose
IUCN Status: Not evaluated
Reference: Flora 31: 579 (1848)

Salacia reticulata Wight
Celastraceae

Climbing Mechanism: Scrambler-Unarmed
Distribution (Global): Andaman Is., Sri Lanka
Distribution (India): Eastern Ghats, Karnataka, Kerala
Leaf Type: Opposite or alternate
Inflorescence: Flowers on axillary tubercles (Cyme)
Fruit Type: Tuberculate fruits
Flowering and Fruiting: February–August
IUCN Status: Not evaluated
Reference: Ill. Ind. Bot. 1: 134 (1839)

Salacia roxburghii Wall.
Celastraceae
Climbing Mechanism: Scrambler-Unarmed
Distribution (India): Arunachal Pradesh, Assam
Leaf Type: Simple, subopposite
Inflorescence: Axillary umbels
Fruit Type: Wet, berry, globose
IUCN Status: Not evaluated

Salacia rubra
Celastraceae
Climbing Mechanism: Scrambler-Unarmed
IUCN Status: Not evaluated

Salacia salacioides (Roxb.) R.S. Rao & Hemadri
Celastraceae
Synonyms: *Johnia salacioides* Roxb.
Climbing Mechanism: Scrambler-Unarmed
Distribution (Global): Andaman Is., Assam, Bangladesh, East Himalaya, Myanmar
Distribution (India): Mizoram
Leaf Type: Simple, subopposite
Inflorescence: Axillary umbels
Fruit Type: Wet, berry, globose
Flowering and Fruiting: March
IUCN Status: Not evaluated
Notes: Usually found in evergreen forests
Reference: J. Bombay Nat. Hist. Soc. 67: 358 (1970)

Salacia talbotii Baker f.
Celastraceae
Climbing Mechanism: Scrambler-Unarmed
Distribution (India): Karnataka
Leaf Type: Simple, subopposite
Inflorescence: Axillary umbels
Fruit Type: Wet, berry, globose

IUCN Status: Not evaluated
Reference: A.B. Rendle & al., Cat. Pl. Oban: 19 (1913)

Salacia tortuosa Griff.
Celastraceae
Climbing Mechanism: Scrambler-Unarmed
Distribution (Global): Andaman Is., Myanmar
Distribution (India): Andaman and Nicobar Islands
Leaf Type: Simple, subopposite
Inflorescence: Axillary umbels
Fruit Type: Wet, berry, globose
IUCN Status: Not evaluated
Reference: Not. Pl. Asiat. 4: 471 (1854)

Salacia viminea Wall.
Celastraceae
Climbing Mechanism: Scrambler-Unarmed
Distribution (Global): Cambodia, Laos, Malaya, Myanmar, Sumatera, Thailand, Vietnam
Leaf Type: Simple, subopposite
Inflorescence: Axillary umbels
Fruit Type: Wet, berry, globose
IUCN Status: Not evaluated
Reference: J. D. Hooker, Fl. Brit. India 1: 627 (1875)

Gloriosa superba L.
Colchicaceae
Synonyms: *Eugone superba* (L.) Salisb., *Gloriosa angulata* Schumach., *Gloriosa cirrhifolia* Stokes, *Gloriosa doniana* Schult. & Schult.f., *Gloriosa lutea* Anon., *Gloriosa nepalensis* G. Don, *Gloriosa rockefelleriana* Stehlé & M. Stehlé, *Gloriosa rothschildiana* O'Bri
Common Name: Climbing lily, Flame lily, Superb lily
Climbing Mechanism: Tendril Climber
Distribution (Global): Andaman Is., Angola, Assam, Bangladesh, Benin, Borneo, Botswana, Burkina, Burundi, Cabinda, Cambodia, Cameroon, Cape Provinces, Caprivi Strip, Central African Repu, Chad, China South-Central, Congo, East Himalaya, Equatorial Guinea, Ethiopia, Gabon, Ghana, Guinea, Guinea-Bissau, Gulf of Guinea Is., India, Ivory Coast, Java, Kenya, Laccadive Is., Laos, Lesser Sunda Is., Liberia, Madagascar, Malawi, Malaya, Maldives, Mozambique, Myanmar, Nepal, Nigeria, Northern Provinces, Pakistan, Rwanda, Senegal, Seychelles, Sierra Leone, Sri Lanka, Sudan, Sulawesi, Sumatera, Tanzania, Thailand, Togo, Uganda, Vietnam, West Himalaya, Zambia, Zaire, Zimbabwe
Distribution (India): Andaman and Nicobar Islands, Andhra Pradesh, Assam, Gujarat, Himachal Pradesh, Jharkhand, Kerala, Madhya Pradesh, Maharashtra,

Odisha, Puducherry, Sikkim, Tamil Nadu, Telangana, Tripura, Uttar Pradesh, West Bengal, Western Ghats
Flowering and Fruiting: July–December
IUCN Status: Least Concern
Reference: Sp. Pl.: 305 (1753)

Combretum acuminatum Roxb.
Combretaceae
Synonyms: *Combretum costatum* Roxb., *Combretum neurophyllum* Miq., *Combretum sarcopterum* Thwaites, *Combretum stenopetalum* Van Heurck & Müll.Arg., *Combretum vanheurckii* Müll.Arg., *Embryogonia acuminata* (Roxb.) Blume, *Terminalia borneensis* Ridl.
Climbing Mechanism: Scrambler-Unarmed
Distribution (Global): Assam, Bangladesh, Borneo, Christmas I., India, Java, Laos, Malaya, Maluku, Myanmar, New Guinea, Philippines, Sri Lanka, Sulawesi, Vietnam
Distribution (India): North Eastern India, Tamil Nadu
Leaf Type: Simple
Inflorescence: Racemes, panicles spikes
Fruit Type: Dry, 4–5 winged
IUCN Status: Not evaluated
Notes: Usually found in primary forests, sal forests often along streams
Reference: Fl. Ind. ed. 1832 2: 228 (1832)

Combretum alatum Craib
Combretaceae
Climbing Mechanism: Stem Twiner
Distribution (Global): Myanmar, Thailand
Leaf Type: Simple
Inflorescence: Racemes, panicles spikes
Fruit Type: Dry, 4–5 winged
IUCN Status: Not evaluated
Reference: Bull. Misc. Inform. Kew 1930: 162 (1930)

Combretum albidum G. Don
Combretaceae
Synonyms: *Combretum albidum* var. cooperi (Haines) H.O. Saxena & Brahmam., *Combretum heyneanum* Wall., *Combretum laxum* Roxb., *Combretum ovalifolium* Roxb. ex G. Don., *Combretum roxburghii* G. Don., *Combretum thwaitesianum* Van Heurck & Müll.Arg., *Combretum wightianum* Thwa
Common Name: Buffalo calf plant, Oval-leaved wheel creeper
Climbing Mechanism: Stem Twiner
Distribution (Global): India, Sri Lanka
Distribution (India): Andhra Pradesh, Eastern Ghats, Kerala, Madhya Pradesh, Maharashtra, Odisha, Puducherry, Tamil Nadu, Telangana, Western Ghats

Leaf Type: Simple, opposite decussate
Inflorescence: Spikes compound, axillary
Fruit Type: Winged, red, globose
Flowering and Fruiting: February–May
IUCN Status: Not evaluated
Notes: Usually found in semi-evergreen forests, along the river banks
Reference: Trans. Linn. Soc. London 15: 429 (1827)

Combretum albidum var. *cooperi*
Combretaceae
Climbing Mechanism: Stem Twiner
Distribution (India): Odisha, Peninsular India
Leaf Type: Simple
Inflorescence: Racemes, panicles spikes
Fruit Type: Dry, 4–5 winged
IUCN Status: Not evaluated

Combretum album Pers.
Combretaceae
Synonyms: *Combretum decandrum* Roxb., *Combretum klossii* Ridl., *Pentaptera roxburghii* (Spreng.) Tul.
Common Name: White Combretum.
Climbing Mechanism: Stem Twiner
Distribution (Global): Andaman Is., Assam, Bangladesh, China South-Central, China Southeast, East Himalaya, India, Laos, Myanmar, Nepal, Pakistan, Thailand, Vietnam, West Himalaya
Distribution (India): Tripura, West Bengal
Leaf Type: Opposite
Inflorescence: Laxy compound spikes
Fruit Type: Glossy, cylindrical
Flowering and Fruiting: Throughout the year
IUCN Status: Not evaluated
Notes: Often found abundant in open forests
Reference: Syst. Veg. ed. 16, 2: 331 (1825)

Combretum apetalum Wall. ex Kurz
Combretaceae
Synonyms: *Combretum virgatum* Wall.
Climbing Mechanism: Stem Twiner
Distribution (Global): Bangladesh, Myanmar
Leaf Type: Simple
Inflorescence: Racemes, panicles spikes
Fruit Type: Dry, 4–5 winged
IUCN Status: Not evaluated
Reference: J. Asiat. Soc. Bengal, Pt. 2, Nat. Hist. 46: 55 (1877)

Combretum chinense Roxb.
Combretaceae
Synonyms: *Combretum annulatum* Craib
Climbing Mechanism: Stem Twiner
Distribution (Global): Andaman Is., Assam, Myanmar, Thailand
Leaf Type: Simple
Inflorescence: Racemes, panicles spikes
Fruit Type: Dry, 4–5 winged
IUCN Status: Not evaluated
Reference: Trans. Linn. Soc. London 15: 417 (1827)

Combretum coccineum (Sonn.) Lam.
Combretaceae
Synonyms: *Combretum pachycladum* Baker, *Combretum purpureum* Vahl, *Cristaria coccinea* Sonn.
Grislea coccinea (Sonn.) Britton, *Poivrea coccinea* DC., *Poivrea rufipes* Tul.
Common Name: Scarlet comb, Flame vine
Climbing Mechanism: Stem Twiner
Distribution (Global): Madagascar, Myanmar
Distribution (India): Gujarat, Maharashtra, Tamil Nadu
Leaf Type: Ovate
Inflorescence: Racemes, flowers are red
IUCN Status: Not evaluated
Notes: A native of Madagascar and Mauritius, cultivated as ornamental plant
Reference: Encycl. 1: 734 (1785)

Combretum constrictum (Benth.) M.A. Lawson
Combretaceae
Synonyms: *Combretum infundibuliforme* Engl., *Poivrea constricta* Benth.
Common Name: Powderpuff Combretum
Climbing Mechanism: Stem Twiner
Distribution (Global): Kenya, Mozambique, Nigeria, Somalia, Tanzania
Distribution (India): Tamil Nadu
Leaf Type: Almost opposite
Inflorescence: Head-like spikes
IUCN Status: Not evaluated
Reference: D. Oliver & auct. suc. (eds.), Fl. Trop. Afr. 2: 423 (1871)

Combretum indicum (L.) DeFilipps
Combretaceae
Synonyms: *Kleinia quadricolor* Crantz, *Mekistus sinensis* Lour. ex Gomes Mach., *Quisqualis ebracteata* P. Beauv., *Quisqualis glabra* Burm.f. + 10
Common Name: Rangoon creeper, Burma creeper, Chinese honeysuckle
Climbing Mechanism: Stem Twiner
Distribution (Global): Pan tropical

Distribution (India): Assam, Kerala, Telangana
Leaf Type: Simple
Inflorescence: Racemes, panicles spikes
Fruit Type: Dry, 4–5 winged
Flowering and Fruiting: Throughout the year
IUCN Status: Not evaluated
Notes: Grown as ornamental plant
Reference: Useful Pl. Dominica: 277 (1998)

Combretum latifolium Blume
Combretaceae
Synonyms: *Combretum cyclophyllum* Steud., *Combretum extensum* Roxb. ex G. Don. + 10
Common Name: Large-leaved climbing bushwillow
Climbing Mechanism: Stem Twiner
Distribution (Global): Andaman Is., Assam, Bangladesh, Borneo, Cambodia, China South-Central, India, Java, Laos, Lesser Sunda Is., Malaya, Myanmar, New Guinea, Nicobar Is., Philippines, Sri Lanka, Sulawesi, Sumatera, Thailand, Vietnam
Distribution (India): Andaman and Nicobar Islands, Andhra Pradesh, Eastern Ghats, Goa, Karnataka, Kerala, Maharashtra, Meghalaya, Mizoram, Odisha, Tamil Nadu, Telangana, Western Ghats
Leaf Type: Simple
Inflorescence: Racemes, panicles spikes
Fruit Type: Dry, 4–5 winged
Flowering and Fruiting: April–December
IUCN Status: Not evaluated
Notes: Usually found in semi-evergreen and evergreen forests
Reference: Bijdr. Fl. Ned. Ind.: 641 (1826)

Combretum malabaricum (Bedd.) Sujana, Ratheesh & Anil
Combretaceae
Synonyms: *Quisqualis indica* Blanco., *Quisqualis malabarica* Bedd., *Quisqualis spinosa* Náves ex Fern. -Vill.
Climbing Mechanism: Stem Twiner
Distribution (Global): India
Distribution (India): Kerala
Leaf Type: Simple
Inflorescence: Racemes, panicles spikes
Fruit Type: Dry, 4–5 winged
IUCN Status: Not evaluated
Reference: J. Bot. Res. Inst. Texas 6: 453 (2012)

Combretum ovalifolium Roxb.
Combretaceae

Climbing Mechanism: Stem Twiner
Distribution (India): Andhra Pradesh, Bihar, Odisha, Gujarat, Rajasthan, Tamil Nadu
Leaf Type: Simple
Inflorescence: Racemes, panicles spikes
Fruit Type: Dry, 4–5 winged
IUCN Status: Not evaluated
Reference: Roxb. In: Hort. Beng. (28) (1814) nom. nud., Fl. Ind. 2: 226. (1832)

Combretum pilosum Roxb. ex G. Don
Combretaceae
Synonyms: *Combretum insigne* Van Heurck & Müll.Arg., *Combretum laetum* Wall., *Combretum spinescens* Wall., *Poivrea pilosa* Wight & Arn.
Climbing Mechanism: Stem Twiner
Distribution (Global): Assam, Bangladesh, Cambodia, China South-Central, Hainan, Laos, Myanmar, Thailand, Vietnam
Distribution (India): Arunachal Pradesh, Assam, Manipur, Meghalaya, Mizoram
Leaf Type: Opposite or subopposite
Inflorescence: Terminal and axillary, densely compound spikes
Fruit Type: Pink or vivid pink, glossy
IUCN Status: Not evaluated
Reference: Trans. Linn. Soc. London 15: 434 (1827)

Combretum punctatum Blume
Combretaceae
Synonyms: *Combretum punctatum* var. *punctatum*, *Combretum punctatum* var. *squamosum* (Roxb. ex G. Don) M. Gangop. & Chakrab.
Climbing Mechanism: Stem Twiner
Distribution (Global): Andaman Is., Assam, Bangladesh, Borneo, Cambodia, China South-Central, China Southeast, East Himalaya, Hainan, Java, Laos, Malaya, Myanmar, Nepal, Nicobar Is., Philippines, Sumatera, Thailand, Vietnam
Distribution (India): Arunachal Pradesh, Meghalaya, Mizoram, Tripura, West Bengal
Leaf Type: Simple
Inflorescence: Racemes, panicles spikes
Fruit Type: Dry, 4–5 winged
IUCN Status: Not evaluated
Reference: Bijdr. Fl. Ned. Ind.: 640 (1826)

Combretum punctatum var. *squamosum* (Roxb. ex G. Don) M. Gangop. & Chakrab.
Combretaceae
Climbing Mechanism: Stem Twiner
Distribution (India): Arunachal Pradesh
Leaf Type: Simple
Inflorescence: Racemes, panicles spikes

Fruit Type: Dry, 4–5 winged
IUCN Status: Not evaluated
Reference: M. Gangop., Chakrab. In: J. Econ. Taxon. Bot., 17(3): 680. (1993)

Combretum pyrifolium Kurz
Combretaceae
Synonyms: *Myrobalanus pyrifolia* (Kurz) Kuntze
Climbing Mechanism: Stem Twiner
Distribution (Global): Myanmar
Leaf Type: Simple
Inflorescence: Racemes, panicles spikes
Fruit Type: Dry, 4–5 winged
IUCN Status: Not evaluated
Reference: J. Asiat. Soc. Bengal, Pt. 2, Nat. Hist. 43: 188 (1874)

Combretum razianum K.G. Bhat
Combretaceae
Climbing Mechanism: Stem Twiner
Distribution (Global): India
Distribution (India): Kerala
Leaf Type: Simple
Inflorescence: Racemes, panicles spikes
Fruit Type: Dry, 4–5 winged
Flowering and Fruiting: January–March
IUCN Status: Not evaluated
Reference: Indian J. Forest. 28: 210 (2005)

Combretum recurvatum Sujana, Ratheesh, & Anil
Combretaceae
Climbing Mechanism: Stem Twiner
Distribution (Global): India
Distribution (India): Kerala
Leaf Type: Simple
Inflorescence: Racemes, panicles spikes
Fruit Type: Dry, 4–5 winged
Flowering and Fruiting: January–May
IUCN Status: Not evaluated
Notes: Usually found in semi-evergreen forests
Reference: J. Bot. Res. Inst. Texas 6: 453 (2012)

Combretum sanjappae Chakrab. & Lakra
Combretaceae
Climbing Mechanism: Stem Twiner
Distribution (Global): Andaman Is.
Leaf Type: Simple

Inflorescence: Racemes, panicles spikes
Fruit Type: Dry, 4–5 winged
IUCN Status: Not evaluated
Reference: J. Econ. Taxon. Bot. 26: 502 (2002)

Combretum shivannae Gholave, Kambale, Lekhak & S.R. Yadav
Combretaceae
Climbing Mechanism: Stem Twiner
Distribution (Global): India.
Distribution (India): Peninsular India
Leaf Type: Sub-opposite to opposite
Inflorescence: Terminal axillary, spicate
Flowering and Fruiting: December–February
IUCN Status: Not evaluated
Notes: Usually found in evergreen forests
Reference: Kew Bull. 70(3)-33: 1 (2015)

Combretum squamosum Roxb. ex G. Don
Combretaceae
Climbing Mechanism: Stem Twiner
Distribution (India): Andaman and Nicobar Islands
Leaf Type: Simple
Inflorescence: Racemes, panicles spikes
Fruit Type: Dry, 4–5 winged
IUCN Status: Not evaluated

Combretum tetragonocarpum Kurz
Combretaceae
Synonyms: *Combretum mokimii* M.G. Gangop. & Chakrab., *Combretum tetragonocarpum* var. *mokimii* (M.G. Gangop. & Chakrab.) M.G. Gangop. & Chakrab.
Climbing Mechanism: Stem Twiner
Distribution (Global): Myanmar
Leaf Type: Simple
Inflorescence: Racemes, panicles spikes
Fruit Type: Dry, 4–5 winged
IUCN Status: Not evaluated
Reference: J. Asiat. Soc. Bengal, Pt. 2, Nat. Hist. 41: 306 (1872)

Combretum tetralophum C.B. Clarke
Combretaceae
Synonyms: *Combretum tetragonocarpum* var. *tetralophum* (C.B. Clarke) M.G. Gangop. & Chakrab., *Combretum wrayi* King
Climbing Mechanism: Stem Twiner

Distribution (Global): Borneo, Caroline Is., Java, Laos, Malaya, New Guinea, Nicobar Is., Sulawesi, Sumatera, Thailand, Vietnam
Leaf Type: Simple
Inflorescence: Racemes, panicles spikes
Fruit Type: Dry, 4–5 winged
IUCN Status: Not evaluated
Reference: J. D. Hooker, Fl. Brit. India 2: 454 (1878)

Combretum trifoliatum Vent.
Combretaceae
Synonyms: *Cacoucia lucida* (Blume) Hassk., *Cacoucia trifoliata* (Vent.) DC., *Combretum bellum* Steud., *Combretum lucidum* Blume., *Combretum subalternans* Wall., *Combretum undulatum* Wall., *Embryogonia lucida* (Blume) Blume., *Terminalia lancifolia* Griff.
Climbing Mechanism: Stem Twiner
Distribution (Global): Borneo, Cambodia, Java, Laos, Lesser Sunda Is., Malaya, Myanmar, New Guinea, Queensland, Sulawesi, Sumatera, Thailand, Vietnam
Leaf Type: Simple
Inflorescence: Racemes, panicles spikes
Fruit Type: Dry, 4–5 winged
IUCN Status: Not evaluated
Reference: Choix Pl.: t. 58 (1808)

Combretum wallichii DC.
Combretaceae
Synonyms: *Combretum wallichii* var. *deciduum* M. Gangop. & Chakrab., *Combretum wallichii* var. *flagrocarpum* (C.B. Clarke) M.G. Gangop. & Chakrab., *Combretum wallichii* var. *ternatum* (C.B. Clarke) Govaerts, *Combretum wallichii* var. *wallichii*
Climbing Mechanism: Stem Twiner
Distribution (Global): Andaman Is., Assam, Bangladesh, Borneo, China South-Central, China Southeast, East Himalaya, Malaya, Myanmar, Nepal, Sumatera, Vietnam
Distribution (India): Arunachal Pradesh, Assam, Tripura
Leaf Type: Opposite or subopposite
Inflorescence: Terminal and axillary
Fruit Type: Purple or red, gloss
Flowering and Fruiting: March–November
IUCN Status: Not evaluated
Reference: Prodr. 3: 21 (1828)

Combretum wallichii var. *flagrocarpum* (C.B. Clarke) M.G. Gangop. & Chakrab.
Combretaceae
Climbing Mechanism: Stem Twiner
Leaf Type: Simple

Inflorescence: Racemes, panicles spikes
Fruit Type: Dry, 4–5 winged
IUCN Status: Not evaluated
Reference: M. Gangop., Chakrab. In: J. Econ. Taxon. Bot., 17(3): 681. (1993)

Combretum wallichii var. *griffithii* (Van Heurck & Müll.Arg.) M.G. Gangop. & Chakrab.
Combretaceae
Synonyms: *Combretum griffithii* Van Heurck & Müll.Arg.
Climbing Mechanism: Stem Twiner
Distribution (Global): Assam, Bangladesh, Borneo, China South-Central, East Himalaya, Laos, Myanmar, Thailand, Vietnam
Leaf Type: Simple
Inflorescence: Racemes, panicles spikes
Fruit Type: Dry, 4–5 winged
IUCN Status: Not evaluated
Reference: Observ. Bot. Descript. Pl. Nov. Herb. Van Heurckiani 2: 231 (1871)

Combretum wallichii var. *porterianum* (C.B. Clarke) M.G. Gangop. & Chakrab.
Combretaceae
Synonyms: *Combretum porterianum* (C.B. Clarke) Wall. ex Craib
Climbing Mechanism: Stem Twiner
Distribution (Global): Assam, Bangladesh, Malaya, Thailand
Leaf Type: Simple
Inflorescence: Racemes, panicles spikes
Fruit Type: Dry, 4–5 winged
IUCN Status: Not evaluated
Reference: Fl. Siam. 1: 618 (1931)

Getonia floribunda Roxb.
Combretaceae
Synonyms: *Calycopteris floribunda* (Roxb.) Lam. ex Poir.
Calycopteris nutans (Roxb.) Kurz
Getonia nitida B. Heyne ex Roth
Getonia nutans Roxb. ex DC.
Common Name: Paper flower climber
Climbing Mechanism: Scrambler-Unarmed
Distribution (Global): Andaman Is., Assam, Bangladesh, Cambodia, China South-Central, India, Laos, Malaya, Myanmar, Thailand, Vietnam

.

Distribution (India): Andhra Pradesh, Tamil Nadu, Telangana, West Bengal, Western Ghats
IUCN Status: Not evaluated
Reference: Pl. Coromandel 1: 61 (1795)

Quisqualis densiflora Wall. ex Miq.
Combretaceae
Climbing Mechanism: Scrambler-Unarmed
Leaf Type: Simple, opposite, elliptic to oblong
Inflorescence: Axillary or terminal spikes
Fruit Type: Dry, 5 winged
IUCN Status: Not evaluated
Reference: Fl. Serres Jard. Eur. 6: 348 (1851)

Quisqualis malabarica Bedd.
Combretaceae
Climbing Mechanism: Scrambler-Unarmed
Distribution (India): Tamil Nadu, Western Ghats
Leaf Type: Alternate or opposite
Inflorescence: Terminal cymes
Flowering and Fruiting: January–April
IUCN Status: Not evaluated
Reference: Icon. Pl. Ind. Or. 1: 33 (1874)

Streptolirion volubile Edgew.
Commelinaceae
Common Name: Climbing twisted lily
Climbing Mechanism: Stem Twiner
Distribution (Global): Assam, China North-Central, China South-Central, China Southeast, East Himalaya, Korea, Laos, Manchuria, Myanmar, Nepal, Primorye, Thailand, Tibet, Vietnam, West Himalaya
Distribution (India): Arunachal Pradesh
Leaf Type: Simple, cordate
Inflorescence: racemes, panicles
Fruit Type: Dry, capsule
Flowering and Fruiting: July–August
IUCN Status: Not evaluated
Reference: Proc. Linn. Soc. London 1: 254 (1845)

Cnestis palala (Lour.) Merr.
Connaraceae
Synonyms: *Thysanus palala* Lour.
Common Name: Stinging-Hair Cnestis.
Climbing Mechanism: Scrambler-Unarmed
Distribution (Global): Andaman Is., Borneo, Cambodia, Hainan, Laos, Malaya, Myanmar, Nicobar Is., Philippines, Sulawesi, Sumatera, Thailand, Vietnam
Leaf Type: Pinnately compound
Inflorescence: Axillary raceme or panicle
Fruit Type: Follicle

IUCN Status: Not evaluated
Reference: J. Straits Branch Roy. Asiat. Soc. 85: 201 (1922)

Connarus andamanicus M.S. Mondal
Connaraceae
Climbing Mechanism: Scrambler-Unarmed
Distribution (Global): Andaman Is., Nicobar Is.
Leaf Type: Compound, imparipinnate
Inflorescence: Terminal and axillary panicles
Fruit Type: Dry, follicles
IUCN Status: Not evaluated
Reference: J. Econ. Taxon. Bot. 15: 459 (1991)

Connarus championii Thwaites
Connaraceae
Climbing Mechanism: Scrambler-Unarmed
Distribution (Global): Sri Lanka
Leaf Type: Compound, imparipinnate
Inflorescence: Terminal and axillary panicles
Fruit Type: Dry, follicles
IUCN Status: Not evaluated
Reference: Enum. Pl. Zeyl.: 80 (1858)

Connarus kingii Schellenb.
Connaraceae
Climbing Mechanism: Scrambler-Unarmed
Distribution (Global): Andaman Is., Assam, Myanmar, Nicobar Is.
Leaf Type: Compound, imparipinnate
Inflorescence: Terminal and axillary panicles
Fruit Type: Dry, follicles
IUCN Status: Not evaluated
Reference: Candollea 2: 96 (1925)

Connarus monocarpus Wight & Arn.
Connaraceae
Climbing Mechanism: Scrambler-Unarmed
Distribution (India): Kerala, Maharashtra, Tamil Nadu
Leaf Type: Compound, imparipinnate leaflets 3 or 5
Inflorescence: Terminal, pyramidal paniculate cymes
Fruit Type: Follicle, obovoid
Flowering and Fruiting: April–June
IUCN Status: Not evaluated
Notes: Usually found in semi-evergreen and moist deciduous forests
Reference: Prodr. Fl. Ind. Orient., 143, 1834

Connarus nicobaricus King
Connaraceae
Climbing Mechanism: Scrambler-Unarmed
Distribution (Global): Andaman Is., Nicobar Is.
Distribution (India): Great Nicobar Island
Leaf Type: Compound, imparipinnate
Inflorescence: Terminal and axillary panicles
Fruit Type: Dry, follicles
IUCN Status: Not evaluated
Reference: J. Asiat. Soc. Bengal, Pt. 2, Nat. Hist. 66: 6 (1897)

Connarus paniculatus Roxb.
Connaraceae
Synonyms: *Connarus hainanensis* Merr., *Connarus harmandianus* Pierre., *Connarus paniculatus* subsp. *tonkinensis* (Lecomte) Y.M. Shui., *Connarus pentandrus* Roxb., *Connarus rufulus* Pierre., *Connarus tonkinensis* Lecomte.
Climbing Mechanism: Scrambler-Unarmed
Distribution (Global): Andaman Is., Assam, Bangladesh, Cambodia, East Himalaya, Hainan, India, Laos, Malaya, Myanmar, Nicobar Is., Thailand, Vietnam
Distribution (India): Andhra Pradesh, Eastern Ghats, Great, Nicobar, Island, Meghalaya, Mizoram, Tamil Nadu, Tripura
Leaf Type: Compound, imparipinnate
Inflorescence: Terminal and axillary panicles
Fruit Type: Dry, follicles
Flowering and Fruiting: October–May
IUCN Status: Not evaluated
Notes: Mostly found in evergreen forests
Reference: Fl. Ind. ed. 1832, 3: 139 (1832)

Connarus parameswaranii Ramam. & Rajan
Connaraceae
Climbing Mechanism: Scrambler-Unarmed
Distribution (Global): India
Distribution (India): Peninsular India
Leaf Type: Compound, imparipinnate
Inflorescence: Terminal and axillary panicles
Fruit Type: Dry, follicles
Flowering and Fruiting: January–May
IUCN Status: Not evaluated
Reference: J. Bombay Nat. Hist. Soc. 85: 390 (1988)

Connarus planchonianus Schellenb.
Connaraceae
Climbing Mechanism: Scrambler-Unarmed
Distribution (Global): Malaya, Myanmar, Nicobar Is., Sumatera, Thailand

Distribution (India): Great Nicobar Island
Leaf Type: Compound, imparipinnate
Inflorescence: Terminal and axillary panicles
Fruit Type: Dry, follicles
IUCN Status: Not evaluated
Reference: Bull. Misc. Inform. Kew 1927: 375 (1927)

Connarus sclerocarpus (Wight & Arn.) Schellenb.
Connaraceae
Synonyms: *Rourea sclerocarpa* Wight & Arn. ex Hook.f.
Climbing Mechanism: Scrambler-Unarmed
Distribution (Global): India
Distribution (India): Karnataka, Kerala, Tamil Nadu, Western Ghats
Leaf Type: Opposite
Inflorescence: Panicles
Fruit Type: Not seen
Flowering and Fruiting: September-April.
IUCN Status: Not evaluated
Notes: Mostly found in evergreen forests
Reference: Candollea 2: 98 (1925)

Connarus semidecandrus Jack
Connaraceae
Synonyms: *Connarus amplifolius* Pierre, *Connarus asiaticus* Willd., *Connarus balsahanensis* Elmer, *Connarus borneensis* Merr., *Connarus ellipticus* G. Schellenb. +20
Climbing Mechanism: Scrambler-Unarmed
Distribution (Global): Andaman Is., Bangladesh, Bismarck Archipelago, Borneo, Cambodia, Caroline Is., Java, Laos, Lesser Sunda Is., Malaya, Myanmar, New Guinea, Nicobar Is., Philippines, Solomon Is., Sulawesi, Sumatera, Thailand, Vietnam
Distribution (India): Andaman and Nicobar Islands
Leaf Type: Compound, imparipinnate
Inflorescence: Terminal and axillary panicles
Fruit Type: Dry, follicles
IUCN Status: Not evaluated
Reference: Malayan Misc. 2(7): 39 (1822)

Connarus wightii Hook.f.
Connaraceae
Synonyms: *Connarus ritchiei* Hook.f.
Climbing Mechanism: Scrambler-Unarmed
Distribution (Global): India
Distribution (India): Goa, Tamil Nadu, Western Ghats
Leaf Type: Alternate

Inflorescence: Panicle
Flowering and Fruiting: March–May
IUCN Status: Not evaluated
Notes: Endemic to Southern Western Ghats
Reference: Fl. Brit. India 2: 51 (1876)

Ellipanthus calophyllus Kurz
Connaraceae
Climbing Mechanism: Scrambler-Unarmed
Distribution (Global): Andaman Is., Myanmar
IUCN Status: Not evaluated
Reference: J. Asiat. Soc. Bengal, Pt. 2, Nat. Hist. 41: 305 (1872)

Rourea commutata Planch.
Connaraceae
Synonyms: *Rourea minor* (Gaertn.) Alston
Climbing Mechanism: Scrambler-Unarmed
Distribution (India): Andaman and Nicobar Islands
Leaf Type: Simple, opposite/alternate
Inflorescence: Fascicles
Fruit Type: Dry
IUCN Status: Not evaluated
Reference: Linnaea 23: 420 (1850)

Rourea minor (Gaertn.) Alston
Connaraceae
Synonyms: *Cnestis acuminata* Wall., *Cnestis erecta* Blanco, *Cnestis florida* Jack, *Cnestis glabra* Blanco, *Cnestis monadelpha* DC., *Connarus javanicus* Blume, *Connarus obliquus* (C. Presl) C. Presl, *Connarus roxburghii* Hook. & Arn., *Connarus santaloides* Vahl.
Climbing Mechanism: Scrambler-Unarmed
Distribution (Global): Andaman Is., Angola, Assam, Bangladesh, Benin, Borneo, Burkina, Burundi, Cambodia, Cameroon, Central African Repu, China South-Central, China Southeast, Congo, East Himalaya, Fiji, Gabon, Ghana, Guinea, Guinea-Bissau, India, Ivory Coast, Java, Kenya, Laos, Lesser Sunda Is., Liberia, Madagascar, Malaya, Mali, Maluku, Myanmar, New Caledonia, New Guinea, Nicobar Is., Nigeria, Niue, Philippines, Samoa, Senegal, Sierra Leone, Solomon Is., Sri Lanka, Sudan, Sulawesi, Sumatera, Taiwan, Tanzania, Thailand, Togo, Tonga, Uganda, Vanuatu, Vietnam, Wallis-Futuna Is., Zambia, Zaire
Distribution (India): Eastern Ghats, Goa, Kerala, Manipur, Meghalaya, Mizoram, Tamil Nadu, Western Ghats
Leaf Type: Simple, opposite/alternate
Inflorescence: Fascicles
Fruit Type: Dry
Flowering and Fruiting: December–August

IUCN Status: Not evaluated
Reference: H. Trimen, Handb. Fl. Ceylon 6(Suppl.): 67 (1931)

Rourea prainiana Talbot
Connaraceae
Synonyms: *Roureopsis scortechinii* King, *Santaloides prainiana* (Talbot) G. Schellenb., *Tali prainiana* (Talbot) M.R. Almeida
Climbing Mechanism: Scrambler-Unarmed
Distribution (Global): India, Malaya, Thailand
Distribution (India): Peninsular India
Leaf Type: Simple, opposite/alternate
Inflorescence: Fascicles
Fruit Type: Dry
Flowering and Fruiting: December–August
IUCN Status: Not evaluated
Reference: For. Fl. Bombay 1: 368 (1909)

Taeniochlaena griffithii Hook.f.
Connaraceae
Synonyms: *Rourea acutipetala* Miq., *Roureopsis acutipetala* (Miq.) Leenh., *Santalodes acutipetalum* Kuntze., *Santalodes acutipetalum* Kuntze
Climbing Mechanism: Scrambler-Unarmed
Distribution (Global): Borneo, Malaya, Sumatera, Thailand, Vietnam
IUCN Status: Not evaluated
Reference: Fl. Ned. Ind., Eerste Bijv.: 528 (1861)

Tali minor (Gaertn.) M.R. Almeida
Connaraceae
Synonyms: *Rourea minor* (Gaertn.) Alston., *Cnestis acuminata* Wall., *Santalodes roxburghii* Kuntze. + 80
Climbing Mechanism: Stem Twiner
Distribution (Global): Andaman Is., Angola, Assam, Bangladesh, Benin, Borneo, Burkina, Burundi, Cambodia, Cameroon, Central African Repu, China South-Central, China Southeast, Congo, East Himalaya, Fiji, Gabon, Ghana, Guinea, Guinea-Bissau, India, Ivory Coast, Java, Kenya, Laos, Lesser Sunda Is., Liberia, Madagascar, Malaya, Mali, Maluku, Myanmar, New Caledonia, New Guinea, Nicobar Is., Nigeria, Niue, Philippines, Samoa, Senegal, Sierra Leone, Solomon Is., Sri Lanka, Sudan, Sulawesi, Sumatera, Taiwan, Tanzania, Thailand, Togo, Tonga, Uganda, Vanuatu, Vietnam, Wallis-Futuna Is., Zambia, Zaire
Distribution (India): Maharashtra
Leaf Type: Glabrescent
Inflorescence: Axillary, fascicled
Fruit Type: Obliquely ovoid follicle, red when mature
IUCN Status: Not evaluated

Notes: Occur mostly in moist deciduous forests
Reference: Fl. Maharashtra 1: 283 (1996)

Tali prainiana (Talbot) M.R. Almeida
Connaraceae
Synonyms: *Rourea prainiana* Talbot., *Roureopsis scortechinii* Kin., *Santaloides prainiana* (Talbot) G. Schellenb.
Climbing Mechanism: Stem Twiner
Distribution (Global): India, Malaya, Thailand
Distribution (India): Maharashtra
IUCN Status: Not evaluated
Reference: For. Fl. Bombay 1: 368 (1909)

Aniseia martinicensis (Jacq.) Choisy
Convolvulaceae
Synonyms: *Aniseia carnea* Griseb., *Aniseia cernua* Moric., *Aniseia cernua* var. *glabra* Choisy, *Aniseia cernua* f. *parviflora* Chodat & Hassl., *Aniseia emarginata* (Vahl) Hassk., *Aniseia ensifolia* Choisy, *Aniseia ensifolia* var. *minor* Choisy
Climbing Mechanism: Stem Twiner
Distribution (Global): Argentina Northeast, Belize, Bolivia, Brazil North, Brazil Northeast, Brazil South, Brazil Southeast, Brazil West-Central, Colombia, Costa Rica, Cuba, Dominican Republic, Ecuador, El Salvador, Florida, French Guiana, Guatemala, Guyana, Haiti, Honduras, Jamaica, Leeward Is., Mexico Gulf, Mexico Southeast, Mexico Southwest, Nicaragua, Panamá, Paraguay, Peru, Puerto Rico, Suriname, Trinidad-Tobago, Venezuela, Windward Is.
Distribution (India): Kerala, Maharashtra
Flowering and Fruiting: August–March
IUCN Status: Least Concern
Notes: Occurs in marshy areas
Reference: Mém. Soc. Phys. Genève 8: 66 (1837)

Argyreia arakuensis N.P. Balakr.
Convolvulaceae
Climbing Mechanism: Stem Twiner
Distribution (Global): India
Distribution (India): Andhra Pradesh, Eastern Ghats
Leaf Type: Simple
Inflorescence: Axillary cyme
Fruit Type: Wet, berry
Flowering and Fruiting: August–January
IUCN Status: Not evaluated
Reference: Bull. Bot. Surv. India 3: 163 (1962)

Argyreia argentata var. *venusta*
Convolvulaceae

Climbing Mechanism: Stem Twiner
Distribution (India): Peninsular India
Leaf Type: Simple
Inflorescence: Axillary cyme
Fruit Type: Wet, berry
IUCN Status: Not evaluated
Reference: Fl. Ned. Ind. 2: 588 (1857)

Argyreia argentea (Roxb.) Sweet
Convolvulaceae
Climbing Mechanism: Stem Twiner
Distribution (India): Arunachal Pradesh, Tripura, Tamil Nadu, Assam, Mizoram
Leaf Type: Simple
Inflorescence: Axillary cyme
Fruit Type: Wet, berry
IUCN Status: Not evaluated
Reference: Sweet. In: Hort. Brit., ed. 2; 373. (1830)

Argyreia bella Raizada
Convolvulaceae
Synonyms: *Ipomoea euantha* J.R.I. Wood & Scotland, *Lettsomia bella* C.B. Clarke, *Mouroucoa bella* (C.B. Clarke) Kuntze
Climbing Mechanism: Stem Twiner
Distribution (Global): India, Myanmar, Nepal
Distribution (India): Peninsular India, Madhya Pradesh, Orissa
Leaf Type: Simple
Inflorescence: Axillary cyme
Fruit Type: Wet, berry
Flowering and Fruiting: August–March
IUCN Status: Not evaluated
Reference: Indian Forester 93: 754 (1967)

Argyreia boseana Santapau & V. Patel
Convolvulaceae
Climbing Mechanism: Stem Twiner
Distribution (Global): India
Distribution (India): Maharashtra
Leaf Type: Simple
Inflorescence: Axillary cyme
Fruit Type: Wet, berry
IUCN Status: Not evaluated
Reference: Trans. Bose Res. Inst. Calcutta 22: 35 (1958)

Argyreia bracteata Choisy
Convolvulaceae

Synonyms: *Ipomoea austroindica* J.R.I. Wood & Scotland, Lettsomia bracteata (Choisy) Roberty
Common Name: Woodrose
Climbing Mechanism: Stem Twiner
Distribution (Global): India
Distribution (India): Peninsular India
Leaf Type: Simple
Inflorescence: Axillary cyme
Fruit Type: Wet, berry
Flowering and Fruiting: July–April
IUCN Status: Not evaluated
Notes: Common as undergrowth of deciduous forest
Reference: Mém. Soc. Phys. Genève 6: 412 (1833 publ. 1834)

Argyreia capitiformis (Poir.) Ooststr.
Convolvulaceae
Synonyms: *Argyreia capitata* Choisy, *Argyreia rufohirsuta* H. Lév., *Argyreia strigosa* (Roth) Roberty, *Argyreia verrucosohispida* Y.Y. Qian, *Convolvulus capitatus* Vahl, *Convolvulus capitiformis* Poir., *Convolvulus confertus* Steud., *Convolvulus setosus* Wall.
Climbing Mechanism: Stem Twiner
Distribution (Global): Andaman Is., Assam, Bangladesh, Cambodia, China South-Central, China Southeast, East Himalaya, Hainan, India, Java, Laos, Malaya, Myanmar, Sri Lanka, Sumatera, Thailand, Vietnam
Distribution (India): Arunachal Pradesh, Meghalaya, Tripura, Maharashtra, Andaman and Nicobar Islands, North Eastern India
Leaf Type: Simple
Inflorescence: Axillary cyme
Fruit Type: Wet, berry
IUCN Status: Not evaluated
Reference: Fl. Males. 6: 941 (1972)

Argyreia choisyana Regel & K
Convolvulaceae
Synonyms: *Ipomoea descolei* O'Donell
Climbing Mechanism: Stem Twiner
Distribution (India): Peninsular India
Leaf Type: Simple
Inflorescence: Axillary cyme
Fruit Type: Wet, berry
IUCN Status: Not evaluated
Reference: Index Seminum (LE, Petropolitanus) 1858: 40 (1859)

Argyreia coacta Alston
Convolvulaceae

Climbing Mechanism: Stem Twiner
Distribution (Global): Sri Lanka
Distribution (India): Tamil Nadu
Leaf Type: Simple
Inflorescence: Axillary cyme
Fruit Type: Wet, berry
IUCN Status: Not evaluated
Reference: Ann. Roy. Bot. Gard. (Peradeniya) 11: 209 (1929)

Argyreia coonoorensis W.W.Sm. & Ramos
Convolvulaceae
Synonyms: *Ipomoea coonoorensis* (W.W.Sm. & Ramaswami) J.R.I. Wood & Scotland
Climbing Mechanism: Stem Twiner
Distribution (Global): India
Distribution (India): Tamil Nadu
Leaf Type: Simple
Inflorescence: Axillary cyme
Fruit Type: Wet, berry
IUCN Status: Not evaluated
Reference: Rec. Bot. Surv. India 6: 30 (1914)

Argyreia cuneata Ker Gawl.
Convolvulaceae
Synonyms: *Convolvulus cuneatus* Willd., *Convolvulus cuneiformis* Buch. -Ham. ex Wall., *Ipomoea atrosanguinea* Hook., *Ipomoea cuneata* (Willd.) J.R.I. Wood & Scotland, *Lettsomia cuneata* (Willd.) Roxb., *Lettsomia cuneata* Roxb. ex J.V. Thomps., *Rivea cuneata* (Willd.) W
Common Name: Purple convolvulus
Climbing Mechanism: Stem Twiner
Distribution (Global): India
Distribution (India): Gujarat, Maharashtra, Peninsular India
Leaf Type: Simple
Inflorescence: Axillary cyme
Fruit Type: Berry, yellowish-brown
Flowering and Fruiting: August–February
IUCN Status: Not evaluated
Notes: Occasional on hills above 750 m and on degraded slopes
Reference: Bot. Reg. 8: t. 661 (1822)

Argyreia cymosa Sweet
Convolvulaceae
Synonyms: *Ipomoea cymulosa* J.R.I. Wood & Scotland, *Lettsomia cymosa* Roxb.
Climbing Mechanism: Stem Twiner
Distribution (Global): India, Nepal

Distribution (India): Tamil Nadu, Andhra Pradesh, Maharashtra, Odisha, Eastern Ghats
Leaf Type: Simple
Inflorescence: Axillary cyme
Fruit Type: Wet, berry
Flowering and Fruiting: September–February
IUCN Status: Not evaluated
Notes: Mostly found in deciduous forests and scrub jungles
Reference: Hort. Brit.: 289 (1826)

Argyreia daltonii C.B. Clarke
Convolvulaceae
Synonyms: *Ipomoea daltonii* (C.B. Clarke) J.R.I. Wood & Scotland
Climbing Mechanism: Stem Twiner
Distribution (Global): India
Distribution (India): Tamil Nadu, Andhra Pradesh, Madhya Pradesh, Bihar, Odisha, Puducherry, Eastern Ghats
Leaf Type: Simple
Inflorescence: Axillary cyme
Fruit Type: Wet, berry
Flowering and Fruiting: October–February
IUCN Status: Not evaluated
Notes: Frequently found growing on thickets
Reference: J. D. Hooker, Fl. Brit. India 4: 190 (1883)

Argyreia elliptica Arn. ex Choisy
Convolvulaceae
Synonyms: *Argyreia laurifolia* (Roxb.) Voigt
Argyreia strigosa subsp. *obovata* (C.B. Clarke) Panigrahi & Murti
Convolvulus ellipticus (Roth) Spreng.
Convolvulus laurifolius Roxb.
Ipomoea elliptica Roth
Ipomoea laurifolia (Roxb.) Sweet
Climbing Mechanism: Stem Twiner
Distribution (Global): India, Nepal, Sri Lanka
Distribution (India): Tamil Nadu, Goa, Maharashtra, Eastern Ghats
Leaf Type: Simple
Inflorescence: Axillary cyme
Fruit Type: Wet, berry
Flowering and Fruiting: September–March
IUCN Status: Not evaluated
Notes: Mostly occur on hills above 900 m growing on thickets
Reference: Mém. Soc. Phys. Genève 6: 417 (1833 publ. 1834)

Argyreia fulgens Choisy
Convolvulaceae
Synonyms: *Convolvulus fulgens* Wall., *Ipomoea fulgens* (Choisy) J.R.I. Wood & Scotland
Climbing Mechanism: Stem Twiner
Distribution (Global): India
Leaf Type: Simple
Inflorescence: Axillary cyme
Fruit Type: Wet, berry
Flowering and Fruiting: December–March
IUCN Status: Not evaluated
Notes: Occur in moist deciduous forests
Reference: Mém. Soc. Phys. Genève 6: 415 (1833 publ. 1834)

Argyreia hirsuta Arn.
Convolvulaceae
Synonyms: *Argyreia hirsuta* var. *coacta* C.B. Clarke, *Ipomoea coacta* (C.B. Clarke) J.R.I. Wood & Scotland, *Ipomoea villifolia* J.R.I. Wood & Scotland, *Rivea hirsuta* (Wight & Arn.) Wight, *Rivea zeylanica* var. *hirsuta* (Wight & Arn.) Thwaites
Climbing Mechanism: Stem Twiner
Distribution (Global): India, Sri Lanka
Distribution (India): Tamil Nadu, Andhra Pradesh, Madhya Pradesh, Maharashtra, Eastern Ghats
Leaf Type: Simple
Inflorescence: Axillary cyme
Fruit Type: Wet, berry
Flowering and Fruiting: August–February
IUCN Status: Not evaluated
Notes: Mostly occur in deciduous forests
Reference: Madras J. Lit. Sci. 5: 18 (1837)

Argyreia hookeri C.B. Clarke
Convolvulaceae
Synonyms: *Argyreia populifolia* Choisy, *Ipomoea himalayana* J.R.I. Wood & Scotland, *Rivea hookeri* (C.B. Clarke) Hallier f.
Climbing Mechanism: Stem Twiner
Distribution (Global): Andaman Is., Assam, Bangladesh, East Himalaya, Myanmar, Nepal, Thailand
Distribution (India): Maharashtra, Peninsular India, North Eastern India, Andaman and Nicobar Islands
Leaf Type: Simple
Inflorescence: Axillary cyme
Fruit Type: Wet, berry

IUCN Status: Not evaluated
Reference: J. D. Hooker, Fl. Brit. India 4: 185 (1883)

Argyreia imbricata Santapau & V. Patel
Convolvulaceae
Synonyms: *Argyreia aggregata* (Roxb.) Arn. ex Choisy, *Argyreia osyrensis* var. *aggregata* (Roxb.) K.K.N. Nair, *Convolvulus aggregatus* (Roxb.) Steud., *Convolvulus imbricatus* Spreng., *Ipomoea imbricata* Roth, *Lettsomia aggregata* Roxb., *Lettsomia mysorensis* C.
Climbing Mechanism: Stem Twiner
Distribution (India): Tamil Nadu, Maharashtra
Leaf Type: Simple
Inflorescence: Axillary cyme
Fruit Type: Wet, berry
Flowering and Fruiting: August–January
IUCN Status: Not evaluated
Reference: Trans. Bose Res. Inst. Calcutta 22: 40 (1958)

Argyreia involucrata C.B. Clarke
Convolvulaceae
Synonyms: *Ipomoea dalzellii* J.R.I. Wood & Scotland
Climbing Mechanism: Stem Twiner
Distribution (Global): India
Distribution (India): Andhra Pradesh, Goa, Madhya Pradesh, Maharashtra, Tamil Nadu, Eastern Ghats
Leaf Type: Simple
Inflorescence: Axillary cyme
Fruit Type: Wet, berry
Flowering and Fruiting: August–February
IUCN Status: Not evaluated
Reference: J. D. Hooker, Fl. Brit. India 4: 187 (1883)

Argyreia kleiniana Raizada
Convolvulaceae
Synonyms: *Argyreia populifolia* Choisy, *Convolvulus zeylanicus* Hook. ex Steud., *Ipomoea kleiniana* Roem. & Schult., *Rivea zeylanica* var. *populifolia* Thwaites
Climbing Mechanism: Stem Twiner
Distribution (Global): India, Sri Lanka
Distribution (India): Tamil Nadu, Andhra Pradesh, Madhya Pradesh, Andaman and Nicobar Islands
Leaf Type: Simple
Inflorescence: Axillary cyme
Fruit Type: Wet, berry
Flowering and Fruiting: August–December
IUCN Status: Not evaluated

Notes: Occasional on hills above 500 m on thickets
Reference: Indian Forester 92: 302 (1966)

Argyreia kondaparthiensis P. Daniel & Vajr.
Convolvulaceae
Synonyms: *Argyreia choisyana* (Wight) Wight ex C.B. Clarke, *Argyreia kondaparthiensis* var. *wightii* (C.B. Clarke) Parmar, *Batatas choisyana* Wight, *Ipomoea choisyana* (Wight) Hallier f., *Ipomoea kondaparthiensis* (P. Daniel & Vajr.) J.R.I. Wood & Scotland
Climbing Mechanism: Stem Twiner
Distribution (Global): India
Distribution (India): Peninsular India
Leaf Type: Simple
Inflorescence: Axillary cyme
Fruit Type: Wet, berry
Flowering and Fruiting: April–December
IUCN Status: Not evaluated
Notes: Found in high-altitude grasslands and shola forest margins
Reference: J. Econ. Taxon. Bot. 3: 675 (1982)

Argyreia lakshminarasimhanii S. Shalini, Sujana, Arisdason & D. Maity
Convolvulaceae
Climbing Mechanism: Stem Twiner
Leaf Type: Simple
Inflorescence: Axillary cyme
Fruit Type: Wet, berry
IUCN Status: Not evaluated
Reference: S. Shalini, Sujana, Arisdason & D. Maity. In: Rheedea 30(4): 444, figs. 1-4. (2020)

Argyreia lanceolata Choisy
Convolvulaceae
Synonyms: *Convolvulus erythrocarpus* Wall., *Ipomoea lancifolia* J.R.I. Wood & Scotland, *Lettsomia lanceolata* (Choisy) Kerr.
Climbing Mechanism: Stem Twiner
Distribution (Global): Cambodia, India, Laos, Myanmar, Thailand, Vietnam

.

Leaf Type: Simple
Inflorescence: Axillary cyme
Fruit Type: Wet, berry
IUCN Status: Not evaluated
Reference: Mém. Soc. Phys. Genève 6: 421 (1833 publ. 1834)

Argyreia lawii C.B. Clarke
Convolvulaceae
Synonyms: *Ipomoea lawii* (C.B. Clarke) J.R.I. Wood & Scotland
Climbing Mechanism: Stem Twiner
Distribution (Global): India
Distribution (India): Tamil Nadu, Maharashtra
Leaf Type: Simple
Inflorescence: Axillary cyme
Fruit Type: Wet, berry
IUCN Status: Not evaluated
Reference: J. D. Hooker, Fl. Brit. India 4: 190 (1883)

Argyreia leschenaultii Thwaites
Convolvulaceae
Synonyms: *Argyreia zeylanica* (Gaertn.) Voigt
Climbing Mechanism: Stem Twiner
Distribution (India): Peninsular India
Leaf Type: Simple
Inflorescence: Axillary cyme
Fruit Type: Wet, berry
Flowering and Fruiting: September–March
IUCN Status: Not evaluated
Notes: Mostly found in deciduous forests and scrub jungles
Reference: Enum. Pl. Zeyl.: 209 (1860)

Argyreia mastersii (Prain) Raizada
Convolvulaceae
Synonyms: *Ipomoea mastersii* (Prain) J.R.I. Wood & Scotland, Lettsomia mastersii Prain
Climbing Mechanism: Stem Twiner
Distribution (Global): Assam, China South-Central, Myanmar, Thailand
Distribution (India): Mizoram
Leaf Type: Simple
Inflorescence: Axillary cyme
Fruit Type: Wet, berry
IUCN Status: Not evaluated
Reference: Indian Forester 93: 754 (1967)

Argyreia mollis (Burm.f.) Choisy
Convolvulaceae
Synonyms: *Lettsomia hispida* Hook. ex Choisy B.D. Jacks., *Mouroucoa championii* (Benth.) Kuntze, *Argyreia argentata* Miq., *Argyreia championii* Benth., *Argyreia obtecta* (Choisy) C.B. Clarke, *Convolvulus mollis* Burm.f., *Convolvulus sericeus* L., *Ipomoea bracteata*
Climbing Mechanism: Stem Twiner

Distribution (Global): Andaman Is., Bangladesh, Cambodia, Hainan, Java, Laos, Lesser Sunda Is., Malaya, Myanmar, Sumatera, Thailand, Vietnam
Distribution (India): Andaman and Nicobar Islands
Leaf Type: Simple
Inflorescence: Axillary cyme
Fruit Type: Wet, berry
IUCN Status: Not evaluated
Reference: Mém. Soc. Phys. Genève 6: 421 (1833 publ. 1834)

Argyreia nasirii D.F. Austin
Convolvulaceae
Synonyms: *Lettsomia thomsonii* C. B. Cl., *Argyreia capitata* Brand., *Mouroucoa thomsonii* (C. B. Cl.) Kuntze
Climbing Mechanism: Stem Twiner
Distribution (India): Himachal Pradesh
Leaf Type: Simple
Inflorescence: Axillary cyme
Fruit Type: Wet, berry
IUCN Status: Not evaluated
Reference: Fl. W. Pakistan 126: 4 (1979)

Argyreia nellygherya Choisy
Convolvulaceae
Synonyms: *Ipomoea nellygherya* (Choisy) J.R.I. Wood & Scotland
Climbing Mechanism: Stem Twiner
Distribution (Global): India
Distribution (India): Western Ghats, Tamil Nadu
Leaf Type: Simple
Inflorescence: Axillary cyme
Fruit Type: Wet, berry
Flowering and Fruiting: August–January
IUCN Status: Not evaluated
Notes: Occurs in shola forests
Reference: Mém. Soc. Phys. Genève 6: 414 (1833 publ. 1834)

Argyreia nervosa (Burm.f.) Bojer
Convolvulaceae
Synonyms: *Convolvulus nervosus* Burm.f., *Convolvulus speciosus* L.f., *Ipomoea nervosa* (Burm.f.) J.R.I. Wood & Scotland, *Ipomoea speciosa* (L.f.) Pers., *Lettsomia nervosa* (Burm.f.) Roxb., *Rivea nervosa* (Burm.f.) Hallier f., *Samudra speciosa* (L.f.) Raf. *Argyreia speciosa* (L.f.) Sweet
Common Name: Elephant creeper, Woolly morning glory
Climbing Mechanism: Stem Twiner
Distribution (Global): Assam, Bangladesh, India, Myanmar, Nepal

Distribution (India): Arunachal Pradesh, Tamil Nadu, Andhra Pradesh, Goa, Gujarat, Madhya Pradesh, Delhi, Maharashtra, North Eastern India, Western Ghats
Leaf Type: Simple
Inflorescence: Axillary cyme
Fruit Type: Wet, berry
Flowering and Fruiting: December–June
IUCN Status: Not evaluated
Notes: Grown as a medicinal plant
Reference: Hortus Maurit.: 224 (1837)

Argyreia osyrensis (Roth) Choisy
Convolvulaceae
Synonyms: *Argyreia aggregata* (Roxb.) Arn. ex Choisy
Argyreia brachypoda (Kerr) Ooststr., *Argyreia imbricata* (Roth) Santapau & Patel, *Convolvulus aggregatus* (Roxb.) Steud., *Convolvulus imbricatus* (Roth) Spreng., *Convolvulus orixensis* J.G. Klein & Rottler
Climbing Mechanism: Stem Twiner
Distribution (Global): Cambodia, Hainan, India, Laos, Myanmar, Sri Lanka, Sumatera, Thailand, Vietnam
Distribution (India): Tamil Nadu, Odisha, Andaman and Nicobar Islands
Leaf Type: Simple
Inflorescence: Axillary cyme
Fruit Type: Wet, berry
Flowering and Fruiting: September–April
IUCN Status: Not evaluated
Notes: Found on plains across scrub jungles
Reference: A.P.de Candolle, Prodr. 9: 334 (1845)

Argyreia pierreana Bois
Convolvulaceae
Synonyms: *Argyreia liliiflora* C.Y. Wu, *Argyreia seguinii* Vaniot ex H. Lév., *Ipomoea pierreana* (Bois) J.R.I. Wood & Scotland, *Lettsomia seguinii* H. Lév.
Climbing Mechanism: Stem Twiner
Distribution (Global): China South-Central, China Southeast, Laos, Myanmar, Thailand, Vietnam
Leaf Type: Simple
Inflorescence: Axillary cyme
Fruit Type: Wet, berry
IUCN Status: Not evaluated
Reference: Rev. Hort. (Paris) 78: 562 (1906)

Argyreia pilosa Wight & Arn.
Convolvulaceae
Synonyms: *Ipomoea pilosula* J.R.I. Wood & Scotland

Climbing Mechanism: Stem Twiner
Distribution (Global): India
Distribution (India): Tamil Nadu, Maharashtra, Andhra Pradesh, Gujarat, Madhya Pradesh, Odisha
Leaf Type: Simple
IUCN Status: Not evaluated
Reference: Madras J. Lit. Sci. 5: 18 (1837)

Argyreia pomacea Choisy
Convolvulaceae
Synonyms: *Argyreia nellygherya* Choisy, *Argyreia leschenaultii* Choisy, *Argyreia zeylanica* Voigt, *Convolvulus choisyanus* Wall., *Convolvulus pomaceus* Wall., *Convolvulus strigosus* Russell ex Wall., *Lettsomia pomacea* (Wall. ex Choisy) Roxb.
Climbing Mechanism: Stem Twiner
Distribution (India): Western Ghats, Tamil Nadu
Leaf Type: Simple
Inflorescence: Axillary cyme
Fruit Type: Wet, berry
Flowering and Fruiting: August-April.
IUCN Status: Not evaluated
Notes: Usually found in dry deciduous forests
Reference: Convolv. Orient. 32. 1834 [Aug 1834]

Argyreia populifolia Choisy
Convolvulaceae
Synonyms: *Convolvulus fastigiatus* Herb. Roxb. ex Wall., *Convolvulus zeylanicus* (Gaertn.) Hook. ex Steud., *Ipomoea fastigiata* Wall. ex Choisy, *Ipomoea populifolia* (Choisy) Hall. fil., *Ipomoea zeylanica* Gaertn., *Rivea zeylanica* (Gaertn.) Thw.
Climbing Mechanism: Stem Twiner
Distribution (India): Tamil Nadu, Pondicherry, Eastern Ghats
Leaf Type: Simple
Inflorescence: Axillary cyme
Fruit Type: Wet, berry
Flowering and Fruiting: August–January
IUCN Status: Not evaluated
Reference: A.P.de Candolle, Prodr. 9: 329 (1845)

Argyreia roxburghii (Wall.) Arn. ex Choisy
Convolvulaceae
Climbing Mechanism: Stem Twiner
Distribution (India): Andhra Pradesh, Bihar, Odisha, North Eastern India
Leaf Type: Simple
Inflorescence: Axillary cyme

Fruit Type: Wet, berry
IUCN Status: Not evaluated
Reference: Choisy. In: Conv. Or. 37. (1834)

Argyreia roxburghii var. *ampla* (Wall.) C.B. Clarke
Convolvulaceae
Synonyms: *Argyreia ampla* Choisy, *Argyreia burneyi* Gage, *Argyreia multiflora* Voigt, *Convolvulus roxburghii* Wall., *Ipomoea multiflora* Roxb., *Ipomoea roxburghii* Sweet, *Rivea multiflora* Hallier f.
Climbing Mechanism: Stem Twiner
Distribution (Global): Assam, Bangladesh, East Himalaya, Myanmar, Nepal, Thailand
Distribution (India): Peninsular India
Leaf Type: Simple
Inflorescence: Axillary cyme
Fruit Type: Wet, berry
Flowering and Fruiting: August–January
IUCN Status: Not evaluated
Notes: Rarely found on lateritic soil
Reference: Mém. Soc. Phys. Genève 6: 419 (1833 publ. 1834)

Argyreia rubens Raizada
Convolvulaceae
Synonyms: *Ipomoea rubens* Choisy
Climbing Mechanism: Stem Twiner
Distribution (India): Peninsular India
Leaf Type: Simple
Inflorescence: Axillary cyme
Fruit Type: Wet, berry
Flowering and Fruiting: August–January
IUCN Status: Not evaluated
Reference: Indian Forester 93: 754 (1967)

Argyreia sericea Dalzell
Convolvulaceae
Synonyms: *Ipomoea concanica* J.R.I. Wood & Scotland
Climbing Mechanism: Stem Twiner
Distribution (Global): India
Distribution (India): Telangana, Tamil Nadu, Andhra Pradesh, Madhya Pradesh, Maharashtra, Uttar Pradesh, Gujarat, Rajasthan
Leaf Type: Simple
Inflorescence: Axillary cyme
Fruit Type: Wet, berry
Flowering and Fruiting: July–February
IUCN Status: Not evaluated

Notes: Usually found in deciduous forests
Reference: Bombay Fl.: 169 (1861)

Argyreia setosa Arn. ex Choisy
Convolvulaceae
Synonyms: *Ipomoea baccata* J.R.I. Wood & Scotland, *Lettsomia setosa* Roxb., *Mouroucoa setosa* (Roxb.) Kuntze
Climbing Mechanism: Stem Twiner
Distribution (Global): Assam, India, Myanmar, Nepal, Pakistan, West Himalaya
Distribution (India): Tamil Nadu, Andhra Pradesh, Odisha
Leaf Type: Simple
Inflorescence: Axillary cyme
Fruit Type: Wet, berry
Flowering and Fruiting: July–February
IUCN Status: Not evaluated
Notes: Usually found in moist deciduous forests
Reference: Hort. Brit., ed. 2: 373 (1830)

Argyreia sikkimensis Ooststr.
Convolvulaceae
Synonyms: *Ipomoea sikkimensis* (C.B. Clarke) J.R.I. Wood & Scotland, *Mouroucoa sikkimensis* (C.B. Clarke) Kuntze, *Lettsomia sikkimensis* C.B. Clarke.
Climbing Mechanism: Stem Twiner
Distribution (Global): Assam, Bangladesh, East Himalaya, Myanmar, Thailand
Distribution (India): Arunachal Pradesh, Mizoram
Leaf Type: Simple
Inflorescence: Axillary cyme
Fruit Type: Wet, berry
IUCN Status: Not evaluated
Reference: Blumea 7: 178 (1952)

Argyreia splendens (Hornem.) Sweet
Convolvulaceae
Synonyms: *Amphione splendens* (Hornem.) Raf., *Convolvulus splendens* Hornem., *Ipomoea splendens* (Roxb.) Sims, *Lettsomia splendens* Roxb.
Climbing Mechanism: Stem Twiner
Distribution (Global): Assam, Bangladesh, China South-Central, Myanmar
Distribution (India): Arunachal Pradesh, Meghalaya, Tripura, Mizoram
Leaf Type: Simple
Inflorescence: Axillary cyme
Fruit Type: Wet, berry
IUCN Status: Not evaluated
Reference: Hort. Brit.: 289 (1826)

Argyreia srinivasanii Subba Rao & Kumari
Convolvulaceae
Climbing Mechanism: Stem Twiner
Distribution (Global): India
Distribution (India): Peninsular India
Leaf Type: Simple
Inflorescence: Axillary cyme
Fruit Type: Wet, berry
Flowering and Fruiting: October
IUCN Status: Not evaluated
Notes: Usually found in deciduous forests and scrub jungles
Reference: Fl. Visakhapatnam Distr. 1: 556 (2003)

Argyreia strigosa Roberty
Convolvulaceae
Synonyms: *Argyreia capitiformis* (Poir.) Ooststr.
Climbing Mechanism: Stem Twiner
Distribution (India): Goa, Gujarat, Madhya Pradesh, Maharashtra, Rajasthan, Peninsular India
Leaf Type: Simple
Inflorescence: Axillary cyme
Fruit Type: Wet, berry
Flowering and Fruiting: January–February
IUCN Status: Not evaluated
Reference: Candollea 14: 44 (1952)

Argyreia uniflora Sweet
Convolvulaceae
Synonyms: Rivea hypocrateriformis (Desr.) Choisy
Climbing Mechanism: Stem Twiner
Distribution (India): North and Central India
Leaf Type: Simple
Inflorescence: Axillary cyme
Fruit Type: Wet, berry
IUCN Status: Not evaluated
Reference: Hort. Brit.: 289 (1826)

Argyreia venusta Choisy
Convolvulaceae
Synonyms: *Ipomoea formosa* J.R.I. Wood & Scotland
Climbing Mechanism: Stem Twiner
Distribution (Global): Assam, East Himalaya, Myanmar, Nepal
Distribution (India): Tamil Nadu
Leaf Type: Simple
Inflorescence: Axillary cyme

Fruit Type: Wet, berry
IUCN Status: Not evaluated
Reference: Mém. Soc. Phys. Genève 6: 419 (1833 publ. 1834)

Argyreia wallichii Choisy
Convolvulaceae
Synonyms: *Argyreia wallichii* var. *coriacea* C.B. Clarke
Ipomoea staplesii J.R.I. Wood & Scotland
Ipomoea staplesii var. *coriacea* (C.B. Clarke) J.R.I. Wood & Scotland
Lettsomia nervosa C.B. Clarke
Climbing Mechanism: Stem Twiner
Distribution (Global): Assam, China South-Central, East Himalaya, Laos, Myanmar, Thailand
Distribution (India): North Eastern India, Andaman and Nicobar Islands
Leaf Type: Simple
Inflorescence: Axillary cyme
Fruit Type: Wet, berry
IUCN Status: Not evaluated
Reference: Mém. Soc. Phys. Genève 6: 421 (1833 publ. 1834)

Argyreia zeylanica Kurz
Convolvulaceae
Synonyms: *Argyreia kurzii* (C.B. Clarke) Boerl.
Climbing Mechanism: Stem Twiner
Distribution (India): Tamil Nadu
Leaf Type: Simple
Inflorescence: Axillary cyme
Fruit Type: Wet, berry
IUCN Status: Not evaluated
Reference: Forest Fl. Burma 2: 215 (1877)

Bonamia semidigyna (Roxb.) Hallier f.
Convolvulaceae
Synonyms: *Breweria roxburghii* Choisy, *Breweria semidigyna* (Roxb.) Kuntze, *Convolvulus semidigynus* Roxb., *Ipomoea semidigyna* (Roxb.) Sweet
Climbing Mechanism: Stem Twiner
Distribution (Global): Bangladesh, Borneo, Cambodia, Comoros, India, Java, Laos, Madagascar, Malaya, Maluku, Myanmar, New Guinea, Philippines, Sri Lanka, Sulawesi, Sumatera, Thailand, Vietnam
Distribution (India): Andhra Pradesh, Goa, Maharashtra, Tamil Nadu
Flowering and Fruiting: November–March
IUCN Status: Not evaluated
Reference: Bot. Jahrb. Syst. 16: 528 (1893)

Breweria cordata Blume
Convolvulaceae
Synonyms: *Bonamia semidigyna* var. *semidigyna*
Climbing Mechanism: Stem Twiner
IUCN Status: Not evaluated
Reference: Bijdr.: 722 (1826)

Breweria latifolia Terrac.
Convolvulaceae
Climbing Mechanism: Stem Twiner
IUCN Status: Not evaluated

Breweriopsis elegans (Choisy) Roberty
Convolvulaceae
Synonyms: *Bonamia elegans* (Choisy) Hallier f.
Climbing Mechanism: Stem Twiner
IUCN Status: Not evaluated
Reference: Candollea 14: 31 (1952)

Calystegia hederacea Wall.
Convolvulaceae
Synonyms: *Volvulus hederaceus* (Wall.) Kuntze, *Convolvulus wallichianus* Spreng.
Climbing Mechanism: Stem Twiner
Distribution (Global): Afghanistan, Altay, Bangladesh, China North-Central, China South-Central, China Southeast, East Himalaya, Ethiopia, Hainan, India, Inner Mongolia, Japan, Kazakhstan, Khabarovsk, Korea, Manchuria, Mongolia, Myanmar, Nepal, Pakistan, Primorye, Qinghai, Tadzhikistan, Tibet, West Himalaya, Xinjiang
IUCN Status: Not evaluated
Reference: W. Roxburgh, Fl. Ind. 2: 94 (1824)

Convolvulus arvensis L.
Convolvulaceae
Synonyms: *Convolvulus ambigens* House. + 40
Common Name: Field bindweed
Climbing Mechanism: Stem Twiner
Distribution (Global): Pan tropical
Distribution (India): Andhra Pradesh, Bihar, Odisha, Delhi, Himachal Pradesh, Jharkhand, Maharashtra, Rajasthan, Tamil Nadu, Telangana, Uttar Pradesh
Leaf Type: Alternate, spirally arranged
Fruit Type: Capsule ovoid-globose
Flowering and Fruiting: August–February
IUCN Status: Not evaluated
Reference: Sp. Pl.: 153 (1753)

Convolvulus auricomus Bhandari
Convolvulaceae
Synonyms: *Convolvulus glomeratus* var. *glomeratus*., *Convolvulus auricomus* var. *ferruginosus* Bhandari., Convolvulus *auricomus* var. *volubilis* (C.B. Clarke) Bhandari., *Convolvulus congestus* R.Br., *Convolvulus faurotii* Franch., *Convolvulus glomeratus* var. *volubilis* C.B.C
Common Name: Clustered bindweed
Climbing Mechanism: Stem Twiner
Distribution (Global): Afghanistan, Algeria, Chad, Djibouti, Egypt, Eritrea, Ethiopia, Gulf States, India, Iran, Kenya, Lebanon-Syria, Niger, Nigeria, Oman, Pakistan, Palestine, Saudi Arabia, Sinai, Socotra, Somalia, Sudan, Yemen
Distribution (India): Gujarat
Leaf Type: Ovate or lance-shaped-oblong
Inflorescence: Borne in heads of 4–10
Fruit Type: Capsules, ovoid
Flowering and Fruiting: December–August
IUCN Status: Not evaluated
Notes: The plant is used as fodder for goats and camels, also used as a purgative

Convolvulus auriculifolius Eastw.
Convolvulaceae
Synonyms: *Calystegia occidentalis* subsp. *fulcrata* (A. Gray) Brummitt., *Calystegia fulcrata* (A. Gray) Brummitt., *Calystegia fulcrata* subsp. *gracilenta* (Greene) Brummitt., *Convolvulus fulcratus* (A. Gray) Greene., *Convolvulus gracilentus* Greene
Climbing Mechanism: Stem Twiner
Distribution (Global): California
Leaf Type: Simple, petiolate
Inflorescence: Axillary cyme
Fruit Type: Dry, capsule
IUCN Status: Not evaluated
Reference: Kew Bull. 29: 502 (1974)

Convolvulus microcalyx C.B. Clarke
Convolvulaceae
Climbing Mechanism: Stem Twiner
Distribution (Global): Assam
Leaf Type: Simple, petiolate
Inflorescence: Axillary cyme
Fruit Type: Dry, capsule
IUCN Status: Not evaluated
Reference: J. D. Hooker, Fl. Brit. India 4: 220 (1883)

Convolvulus prostratus Forssk.
Convolvulaceae

Synonyms: *Convolvulus austroaegyptiacus* Abdallah & Sa'ad, *Convolvulus austroaegyptiacus* var. *cancerianus* (Abdallah & Sa'ad) Alfarhan, *Convolvulus cancerianus* Abdallah & Sa'ad, *Convolvulus deserti* Hochst. & Steud. ex Baker & Rendle, *Convolvulus evolvuloides* Bois.
Common Name: Prostrate bindweed
Climbing Mechanism: Stem Twiner
Distribution (Global): Afghanistan, Algeria, Burkina, Cape Verde, Chad, Djibouti, Egypt, Gulf States, India, Iran, Iraq, Libya, Mauritania, Morocco, Nepal, Niger, Oman, Pakistan, Saudi Arabia, Senegal, Sinai, Somalia, Sudan, West Himalaya, Yemen
Distribution (India): Tamil Nadu
Leaf Type: Subsessile
Inflorescence: Cymes
Fruit Type: Capsule, dark brown, Subglobose
Flowering and Fruiting: November–March
IUCN Status: Not evaluated
Notes: Medicinally valuable and used for treating various ailments
Reference: Fl. Aegypt. -Arab.: 203 (1775)

Convolvulus rufescens Choisy
Convolvulaceae
Climbing Mechanism: Stem Twiner
Distribution (Global): India
Distribution (India): Tamil Nadu
Leaf Type: Simple, petiolate
Inflorescence: Axillary cyme
Fruit Type: Dry, capsule
IUCN Status: Not evaluated
Reference: Mém. Soc. Phys. Genève 6: 479 (1833 publ. 1834)

Cuscuta approximata Bab.
Convolvulaceae
Synonyms: *Cuscuta epithymum* subsp. *approximata* (Bab.) Rouy, *Cuscuta planiflora* subsp. *approximata* (Bab.) Nyman, *Cuscuta planiflora* var. *approximata* (Bab.) Engelm., *Cuscuta planiflora* subsp. *approximata* (Bab.) H. Lindb.
Climbing Mechanism: Stem Twiner
Distribution (Global): Tropical and temperate regions
Distribution (India): Andhra Pradesh
Leaf Type: Leafless, parasitic vine
Inflorescence: Lateral fascicles/short raceme
Fruit Type: Dry, capsule
Flowering and Fruiting: October–March
IUCN Status: Not evaluated
Reference: Ann. Mag. Nat. Hist. 13: 253 (1844)

Cuscuta campestris Yunck.
Convolvulaceae
Synonyms: *Cuscuta arvensis* Beyr. ex Engelm., *Cuscuta arvensis* var. *calycina* (Engelm.) Engelm., *Cuscuta gymnocarpa* subsp. *deflexa* Buia, *Cuscuta kamelorum* Pavlov, *Cuscuta pentagona* var. *calycina* Engelm., *Cuscuta pentagona* var. *subulata* Yunck.
Common Name: Golden dodder, Common dodder, Field dodder, Prairie dodder
Climbing Mechanism: Stem Twiner
Distribution (Global): Pan-tropical and temperate regions
Distribution (India): Madhya Pradesh, Tamil Nadu
Leaf Type: No leaves
Inflorescence: Compact spherical clusters
Fruit Type: Greenish-yellow fruits
Flowering and Fruiting: November–April
IUCN Status: Not evaluated
Reference: Mem. Torrey Bot. Club 18: 138 (1932)

Cuscuta chinensis Lam.
Convolvulaceae
Synonyms: *Grammica chinensis* (Lam.) Hadac & Chrtek, Pentake chinense (Lam.) Raf.
Common Name: Dodder, Chinese dodder
Climbing Mechanism: Stem Twiner
Distribution (Global): Afghanistan, Arizona, Assam, Bangladesh, China North-Central, China South-Central, China Southeast, Eritrea, Ethiopia, Ghana, India, Inner Mongolia, Iran, Iraq, Japan, Kazakhstan, Kirgizstan, Korea, Madagascar, Manchuria, Mexico Central, Mexico Northeast, Mexico Northwest, Mexico Southwest, Mongolia, Nansei-shoto, New Mexico, New South Wales, Northern Territory, Pakistan, Primorye, Qinghai, Queensland, Saudi Arabia, Socotra, Sri Lanka, Sudan, Tadzhikistan, Taiwan, Texas, Thailand, Tibet, Turkmenistan, Utah, Uzbekistan, Vietnam, West Himalaya, Western Australia, Xinjiang
Distribution (India): Tamil Nadu, Telangana, Uttar Pradesh
Leaf Type: No leaves
Inflorescence: Lateral fasicles, sessile flowers
Fruit Type: Capsule, globose
Flowering and Fruiting: October–April
IUCN Status: Not evaluated
Notes: Usually found in moist and dry deciduous forests
Reference: Encycl. 2: 229 (1786)

Cuscuta chinensis var. *ciliaris* (Hohen. ex Boiss.) Engelm.
Convolvulaceae
Climbing Mechanism: Stem Twiner
Distribution (India): Gujarat
Leaf Type: No leaves

Flowering and Fruiting: May–October
IUCN Status: Not evaluated
Reference: Engelm. In: (1859). (1859)

Cuscuta hyalina Roth
Convolvulaceae
Synonyms: *Cuscuta acutissima* Buchinger ex Engelm., *Cuscuta arabica* Wight, *Cuscuta boissieri* Stocks, *Cuscuta epitribulum* Schinz, *Cuscuta oxypetala* Boiss., *Grammica hyalina* (Roth) Holub.
Climbing Mechanism: Stem Twiner
Distribution (Global): Afghanistan, Angola, Bangladesh, Botswana, Burundi, Cape Provinces, Cape Verde, Chad, Eritrea, Ethiopia, India, Kenya, Namibia, Northern Provinces, Pakistan, Rwanda, Sudan, Tanzania, Turkey, Uganda, Zimbabwe
Distribution (India): Andhra Pradesh, Gujarat, Madhya Pradesh, Tamil Nadu
Leaf Type: Leafless, parasitic vine
Inflorescence: Lateral fascicles/short raceme
Fruit Type: Dry, capsule
Flowering and Fruiting: September–March
IUCN Status: Not evaluated
Reference: J.J. Roemer & J.A. Schultes, Syst. Veg., ed. 15 bis 6: 203 (1820)

Cuscuta krishnae Udayan & Robi
Convolvulaceae
Climbing Mechanism: Stem Twiner
Distribution (Global): India
Distribution (India): Tamil Nadu
Leaf Type: Leafless, parasitic vine
Inflorescence: Lateral fascicles/short raceme
Fruit Type: Dry, capsule
IUCN Status: Not evaluated
Reference: Webbia 71: 45 (2016)

Cuscuta reflexa Roxb.
Convolvulaceae
Synonyms: *Kadurias reflexa* (Roxb.) Raf., *Monogynella reflexa* (Roxb.) Holub.
Common Name: Common dodder
Climbing Mechanism: Stem Twiner
Distribution (Global): Afghanistan, Assam, Bangladesh, China South-Central, China Southeast, East Himalaya, India, Java, Laos, Myanmar, Nepal, Pakistan, Sri Lanka, Thailand, Tibet, Vietnam, West Himalaya
Distribution (India): Andaman and Nicobar Islands, Andhra Pradesh, Assam, Delhi, Gujarat, Himachal Pradesh, Jharkhand, Madhya Pradesh, Maharashtra, Meghalaya, Mizoram, Tamil Nadu, Telangana, Tripura, Uttar Pradesh, West Bengal
Leaf Type: No leaves

Inflorescence: Lateral racemes, Sessile flowers
Fruit Type: Capsule, succulent
Flowering and Fruiting: October–April
IUCN Status: Least Concern
Notes: Medicinally valuable in treating various ailments
Reference: Pl. Coromandel 2: 3 (1799)

Dinetus grandiflorus (Wall.) Staples
Convolvulaceae
Synonyms: *Convolvulus hamiltonii* Spreng., *Dinetopsis grandiflora* (Wall.) Roberty, *Ipomoea cuspidata* D. Don, *Ipomoea nepalensis* F. Dietr., *Porana grandiflora* Wall., *Porana stenoloba* Kurz.
Climbing Mechanism: Stem Twiner
Distribution (Global): East Himalaya, Nepal, Tibet
IUCN Status: Not evaluated
Reference: Novon 3: 199 (1993)

Dinetus malabaricus Staples
Convolvulaceae
Synonyms: *Porana malabarica* C.B. Clarke, *Porana racemosa* Dalzell & A. Gibson
Common Name: Malabar snow creeper
Climbing Mechanism: Stem Twiner
Distribution (Global): India
Distribution (India): Tamil Nadu
IUCN Status: Not evaluated
Reference: Blumea 51: 441 (2006)

Dinetus racemosus (Roxb.) Buch. -Ham. ex Sweet
Convolvulaceae
Climbing Mechanism: Stem Twiner
Distribution (India): Tamil Nadu, Mizoram
IUCN Status: Not evaluated
Reference: Sweet. In: Brit. Fl. Gard. 2: pl. 127. (1825).

Erycibe albiflora Hallier f.
Convolvulaceae
Climbing Mechanism: Stem Twiner
Distribution (India): Meghalaya
Leaf Type: Simple, elliptic, base rounded
Inflorescence: Panicle cymes
Fruit Type: Wet, berry
IUCN Status: Not evaluated
Reference: Bull. Herb. Boissier 5: 1052 (1897)

Erycibe expansa Wall. ex G. Don
Convolvulaceae
Synonyms: *Erycibe ferruginea* C.Y. Wu, *Erycibe ferruginosa* Griff., *Erycibe paniculata* var. *expansa* (Wall. ex G. Don) DC.
Climbing Mechanism: Stem Twiner
Distribution (Global): China South-Central, Malaya, Myanmar, Thailand
Distribution (India): Andaman and Nicobar Islands.
Leaf Type: Simple, elliptic, base rounded
Inflorescence: Panicle cymes
Fruit Type: Wet, berry
IUCN Status: Not evaluated
Reference: Gen. Hist. 4: 392 (1837)

Erycibe glomerata Blume
Convolvulaceae
Climbing Mechanism: Stem Twiner
Distribution (Global): Borneo, Java, Sumatera
Leaf Type: Simple, elliptic, base rounded
Inflorescence: Panicle cymes
Fruit Type: Wet, berry
IUCN Status: Not evaluated
Reference: Bijdr.: 1047 (1826)

Erycibe griffithii C.B. Clarke
Convolvulaceae
Synonyms: *Erycibe cuprea* Gagnep.
Climbing Mechanism: Stem Twiner
Distribution (Global): Andaman Is., Malaya, Myanmar, Nicobar Is., Thailand, Vietnam
Distribution (India): Andaman and Nicobar Islands
Leaf Type: Simple, elliptic, base rounded
Inflorescence: Panicle cymes
Fruit Type: Wet, berry
IUCN Status: Not evaluated
Reference: J. D. Hooker, Fl. Brit. India 4: 182 (1883)

Erycibe maingayi C.B. Clarke
Convolvulaceae
Climbing Mechanism: Stem Twiner
Distribution (Global): Borneo, Malaya, Sumatera
Leaf Type: Simple, elliptic, base rounded
Inflorescence: Panicle cymes
Fruit Type: Wet, berry
IUCN Status: Not evaluated
Reference: J. D. Hooker, Fl. Brit. India 4: 182 (1883)

Erycibe malaccensis C.B. Clarke
Convolvulaceae
Synonyms: *Erycibe dubia* Elmer
Climbing Mechanism: Stem Twiner
Distribution (Global): Malaya, Myanmar, Philippines, Sulawesi
Leaf Type: Simple, elliptic, base rounded
Inflorescence: Panicle cymes
Fruit Type: Wet, berry
IUCN Status: Not evaluated
Reference: J. D. Hooker, Fl. Brit. India 4: 182 (1883)

Erycibe paniculata Roxb.
Convolvulaceae
Synonyms: *Catonia elliptica* Vahl, *Catonia glauca* Vahl, *Erimatalia rheedii* Schult., *Erycibe albiflora* Hallier f., *Erycibe rheedii* Anon., *Erycibe wightiana* J. Graham
Climbing Mechanism: Stem Twiner
Distribution (Global): Assam, Bangladesh, India, Sri Lanka
Distribution (India): Andaman and Nicobar Islands, Andhra Pradesh, Arunachal Pradesh, Bihar, Odisha, Eastern Ghats, Goa, Maharashtra, Meghalaya, Odisha, Tamil Nadu, Uttar Pradesh, Western Ghats
Leaf Type: Simple, elliptic, base rounded
Inflorescence: Panicle cymes
Fruit Type: Wet, berry
Flowering and Fruiting: November–March
IUCN Status: Not evaluated
Reference: Pl. Coromandel 2: 31 (1802)

Erycibe peguensis Prain
Convolvulaceae
Synonyms: *Lettsomia peguensis* C. B. Cl., *Mouroucoa peguensis* (C. B. Cl.) Kuntze
Climbing Mechanism: Stem Twiner
Distribution (Global): Andaman Is., Assam, Bangladesh, India, Myanmar, Nicobar Is., Thailand
Distribution (India): Tripura, Mizoram
Leaf Type: Simple, elliptic, base rounded
Inflorescence: Panicle cymes
Fruit Type: Wet, berry
IUCN Status: Not evaluated
Reference: J. Asiat. Soc. Bengal, Pt. 2, Nat. Hist. 63: 83 (1894)

Erycibe princei Wall.
Convolvulaceae
Climbing Mechanism: Stem Twiner
Leaf Type: Simple, elliptic, base rounded
Inflorescence: Panicle cymes

Fruit Type: Wet, berry
IUCN Status: Not evaluated
Reference: Wall. In: Cat. n. 1335 (1829), et ex Choisy, Ann. Sc. Nat. Ser. II. 1: 224, descr. (1834)

Erycibe schmidtii Craib
Convolvulaceae
Climbing Mechanism: Stem Twiner
Distribution (India): Meghalaya
Leaf Type: Simple, elliptic, base rounded
Inflorescence: Panicle cymes
Fruit Type: Wet, berry
IUCN Status: Not evaluated
Reference: Craib. In: Bot. Tidsskr. 32: 352. (1916)

Erycibe subspicata Wall. ex G. Don
Convolvulaceae
Synonyms: *Erycibe paniculata* var. *subspicat*a (Wall. ex G. Don) Choisy ex DC.
Climbing Mechanism: Stem Twiner
Distribution (Global): Assam, Bangladesh, Cambodia, China South-Central, China Southeast, India, Laos, Myanmar, Thailand, Vietnam
Distribution (India): Arunachal Pradesh
Leaf Type: Simple, elliptic, base rounded
Inflorescence: Panicle cymes
Fruit Type: Wet, berry
IUCN Status: Not evaluated
Reference: Gen. Hist. 4: 392 (1837)

Erycibe wightiana J. Graham
Convolvulaceae
Synonyms: *Erycibe paniculata* Roxb., *Erycibe albiflora*, *Erycibe paniculata* var. *wightiana*.
Common Name: Panicled Erycibe
Climbing Mechanism: Stem Twiner
Distribution (India): Tamil Nadu
Leaf Type: Simple, elliptic, base rounded
Inflorescence: Panicle cymes
Fruit Type: Wet, berry
IUCN Status: Not evaluated
Reference: Cat. Pl. Bombay: 137 (1839)

Hewittia malabarica (L.) Suresh
Convolvulaceae
Synonyms: *Aniseia afzelii* G. Don, *Aniseia bracteata* Hassk., *Argyreia malabarica* (L.) Choisy, *Bonamia volkensii* Dammer, *Calystegia keriana* Sweet, *Convolvulus*

bicolor Vahl, *Convolvulus bracteatus* Vahl, *Convolvulus hederaceus* Blanco, *Convolvulus involucratus* W
Common Name: Malabar bindweed
Climbing Mechanism: Stem Twiner
Distribution (Global): Angola, Assam, Bangladesh, Benin, Borneo, Burundi, Cambodia, Cameroon, Cape Provinces, Central African Repu, Chad, China South-Central, China Southeast, Ethiopia, Gabon, Gambia, Ghana, Guinea, Hainan, India, Ivory Coast, Java, Kenya, KwaZulu-Natal, Laos, Lesser Sunda Is., Liberia, Madagascar, Malawi, Malaya, Mali, Maluku, Mozambique, Myanmar, New Guinea, Nigeria, Northern Provinces, Philippines, Rwanda, Senegal, Sierra Leone, Somalia, South China Sea, Sri Lanka, Sudan, Sumatera, Swaziland, Taiwan, Tanzania, Thailand, Togo, Uganda, Vietnam, Zambia, Zaire, Zimbabwe
Distribution (India): Andaman and Nicobar Islands, Maharashtra, Tamil Nadu
Flowering and Fruiting: October–January
IUCN Status: Not evaluated
Reference: D.H. Nicolson, C.R. Suresh & K.S. Manilal, Interpr. Van Rheede, Hort. Malab.: 88 (1988)

Hewittia scandens (J. König ex Milne) Mabb.
Convolvulaceae
Synonyms: *Hewittia malabarica* (L.) Suresh, *Convolvulus malabaricus*, *Hewittia bicolor*, *Convolvulus scandens*
Common Name: Malabar bindweed
Climbing Mechanism: Stem Twiner
Distribution (India): Andaman and Nicobar Islands, Andhra Pradesh, Eastern Ghats, West Bengal
IUCN Status: Not evaluated
Reference: K.S. Manilal, Bot. Hist. Hort. Malabaricus: 84 (1980)

Ipomoea aculeata Blume
Convolvulaceae
Synonyms: *Convolvulus aculeatus* (Blume) D. Dietr.
Climbing Mechanism: Stem Twiner
Distribution (Global): Bangladesh, Cambodia, Hainan, India, Java, Laos, Lesser Sunda Is., Myanmar, Philippines, Queensland, Sumatera, Thailand, Vietnam
Leaf Type: Simple, palmately 3–5 fid, 5 nerved at base
Inflorescence: Axillary, umbellate cymes/panicles
Fruit Type: Dry, capsule
IUCN Status: Not evaluated
Reference: Bijdr.: 715 (1826)

Ipomoea alba L. Convolvulaceae
Synonyms: *Chonemorpha convolvuloides* G. Don, *Bonanox indica* Raf., *Bonanox riparia* Raf., *Calonyction aculeatum* (L.) House, *Calonyction aculeatum*

f. apopetalum Allard, *Calonyction album* (L.) House, *Calonyction bona-nox* (Spreng.) Bojer, *Calonyction macrantholeu*
Common Name: Moon flower, Moon vine
Climbing Mechanism: Stem Twiner
Distribution (Global): Argentina Northeast, Argentina Northwest, Bahamas, Belize, Bolivia, Brazil North, Brazil Northeast, Brazil South, Brazil Southeast, Brazil West-Central, Central American Pac, Colombia, Costa Rica, Cuba, Dominican Republic, Ecuador, El Salvador, Florida, French Guiana, Galápagos, Guatemala, Guyana, Haiti, Honduras, Jamaica, Leeward Is., Mexico Central, Mexico Gulf, Mexico Northeast, Mexico Northwest, Mexico Southeast, Mexico Southwest, Nicaragua, Panamá, Paraguay, Peru, Puerto Rico, Suriname, Trinidad-Tobago, Uruguay, Venezuela, Windward Is.
Distribution (India): Andaman and Nicobar Islands, Andhra Pradesh, Arunachal Pradesh, Gujarat, Madhya Pradesh, Maharashtra, Mizoram, Odisha, Rajasthan, Tamil Nadu
Leaf Type: Simple, palmately 3–5 fid, 5 nerved at base
Inflorescence: Axillary, umbellate cymes/panicles
Fruit Type: Dry, capsule
Flowering and Fruiting: October–December
IUCN Status: Least Concern
Reference: Sp. Pl.: 161 (1753)

Ipomoea aquatica Forssk.
Convolvulaceae
Synonyms: *Convolvulus repens* Vahl, *Ipomoea reptans* Poir., *Batatas incurva* Benth., *Convolvulus adansonii* Desr., *Convolvulus clappertonii* Spreng., *Convolvulus incurvus* Schum. & Thonn., *Convolvulus rostratus* Zipp. ex Span., *Ipomoea clappertonii* R. Br.
Common Name: Swamp cabbage, Water morning glory, Water spinach
Climbing Mechanism: Stem Twiner
Distribution (Global): Angola, Assam, Bangladesh, Borneo, Botswana, Burundi, Cambodia, Cameroon, Caprivi Strip, Caroline Is., Central African Repu, Chad, China South-Central, China Southeast, Comoros, Congo, Eritrea, Ethiopia, Fiji, Gabon, Gambia, Ghana, Gilbert Is., Guinea, Gulf States, India, Ivory Coast, Java, Kenya, KwaZulu-Natal, Laos, Lesser Sunda Is., Liberia, Madagascar, Malawi, Malaya, Mali, Maluku, Marianas, Mauritania, Mozambique, Myanmar, Namibia, Nepal, New Guinea, Nigeria, Northern Territory, Oman, Pakistan, Philippines, Queensland, Saudi Arabia, Senegal, Sierra Leone, Somalia, Sri Lanka, Sudan, Sulawesi, Sumatera, Tanzania, Thailand, Togo, Uganda, Vanuatu, Vietnam, West Himalaya, Western Australia, Yemen, Zambia, Zaire, Zimbabwe
Distribution (India): Andaman and Nicobar Islands, Andhra Pradesh, Kerala, Tamil Nadu, Telangana, Uttar Pradesh
Leaf Type: Alternate
Inflorescence: Solitary, cymes
Fruit Type: Capsule

Flowering and Fruiting: November–March
IUCN Status: Least Concern
Reference: Fl. Aegypt. -Arab.: 44 (1775)

Ipomoea arachnosperma Welw.
Convolvulaceae
Synonyms: *Ipomoea pilosa*
Common Name: Hairy little bell
Climbing Mechanism: Stem Twiner
Distribution (India): Himachal Pradesh, Maharashtra
Leaf Type: Trilobed
Fruit Type: Capsule
IUCN Status: Not evaluated
Reference: Apont.: 588 (1859)

Ipomoea asarifolia (Desr.) Roem. & Schult.
Convolvulaceae
Synonyms: *Amphione asarifolia* (Desr.) Raf., *Convolvulus asarifolius* Desr., *Convolvulus beladambu* (Schult.) Spreng., *Convolvulus flagelliformis* Roxb., *Convolvulus rugosus* Rottler, *Convolvulus urbicus* Salzm. ex Choisy, *Ipomoea beladamboe* Roem. & Schult.
Common Name: Ginger-leaf morning glory
Climbing Mechanism: Stem Twiner
Distribution (Global): Angola, Argentina Northeast, Brazil North, Brazil Northeast, Brazil Southeast, Brazil West-Central, Cambodia, Cameroon, Cape Verde, Chad, Colombia, Costa Rica, Cuba, Ecuador, French Guiana, Gabon, Gambia, Ghana, Guatemala, Guinea, Honduras, India, Ivory Coast, Jamaica, Java, Laos, Lesser Sunda Is., Mali, Mexico Southeast, Mozambique, Nicaragua, Niger, Nigeria, Paraguay, Peru, Senegal, Sri Lanka, Sudan, Thailand, Venezuela, Vietnam, Windward Is., Zambia
Distribution (India): Andhra Pradesh, Maharashtra, Odisha, Tamil Nadu
Leaf Type: Alternate
Inflorescence: Solitary or simple or compound cymes
Fruit Type: Capsule
Flowering and Fruiting: October–December
IUCN Status: Not evaluated
Reference: Syst. Veg., ed. 15 bis 4: 251 (1819)

Ipomoea barlerioides (Choisy) Benth. ex C.B. Clarke
Convolvulaceae
Synonyms: *Aniseia barlerioides* Choisy
Convolvulus barlerioides Buch. -Ham. ex Wall.
Common Name: Pink morning glory
Climbing Mechanism: Stem Twiner
Distribution (Global): India, Myanmar, Nepal, West Himalaya

Distribution (India): Andhra Pradesh, Madhya Pradesh, Maharashtra, Tamil Nadu, Telangana
Leaf Type: Simple, palmately 3–5 fid, 5 nerved at base
Inflorescence: Axillary, umbellate cymes/panicles
Fruit Type: Dry, capsule
Flowering and Fruiting: June–December
IUCN Status: Not evaluated
Reference: J. D. Hooker, Fl. Brit. India 4: 201 (1883)

Ipomoea batatas (L.) Lam.
Convolvulaceae
Synonyms: *Batatas betacea* Lindl., *Batatas edulis* (Thunb.) Choisy, *Batatas wallii* C. Morren, *Batatas xanthorhiza* Bojer, *Convolvulus apiculata* M. Martens & Galeotti, *Convolvulus attenuatus* M. Martens & Galeotti, *Convolvulus batata* Vell., *Convolvulus batatas* L.
Common Name: Sweet potato
Climbing Mechanism: Stem Twiner
Distribution (Global): Belize, Colombia, Costa Rica, Ecuador, El Salvador, Guatemala, Honduras, Mexico Gulf, Mexico Northeast, Mexico Southeast, Mexico Southwest, Nicaragua, Panamá, Venezuela
Distribution (India): Arunachal Pradesh, Gujarat, Jharkhand, Madhya Pradesh, Tamil Nadu, Telangana
Leaf Type: Simple, palmately 3–5 fid, 5 nerved at base
Inflorescence: Axillary, umbellate cymes/panicles
Fruit Type: Dry, capsule
IUCN Status: Data Deficient
Reference: Tabl. Encycl. 1: 465 (1793)

Ipomoea beladamboe Roem. & Schult.
Convolvulaceae
Synonyms: *Ipomoea asarifolia* (Desr.) Roem. & Schult.
Climbing Mechanism: Stem Twiner
Leaf Type: Simple, palmately 3–5 fid, 5 nerved at base
Inflorescence: Axillary, umbellate cymes/panicles
Fruit Type: Dry, capsule
IUCN Status: Not evaluated
Reference: Syst. Veg., ed. 15 bis 4: 233 (1819)

Ipomoea biflora (L.) Pers.
Convolvulaceae
Synonyms: *Aniseia biflora* (L.) Choisy, *Aniseia calycina* Choisy, *Convolvulus bifidus* Christm., *Convolvulus biflorus* L., *Convolvulus calycinus* Roxb., *Convolvulus forskalii* Spreng., *Convolvulus hardwickii* Spreng., *Convolvulus plebeius* (R.Br.) Spreng.
Common Name: Chinese morning glory

Climbing Mechanism: Stem Twiner
Distribution (Global): Angola, Botswana, Burundi, Cape Provinces, Chad, China South-Central, China Southeast, East Himalaya, Egypt, Eritrea, Ethiopia, India, Java, Kenya, KwaZulu-Natal, Laos, Lesser Sunda Is., Malawi, Mauritania, Mozambique, Myanmar, Namibia, Nansei-shoto, New South Wales, Niger, Northern Provinces, Northern Territory, Oman, Philippines, Queensland, Rwanda, Saudi Arabia, Socotra, Somalia, South Australia, Sudan, Sulawesi, Swaziland, Taiwan, Tanzania, Thailand, Uganda, Vietnam, West Himalaya, Western Australia, Yemen, Zambia, Zimbabwe
Distribution (India): Tamil Nadu
Leaf Type: Heart shaped
IUCN Status: Not evaluated
Reference: Syn. Pl. 1: 183 (1805)

Ipomoea bracteata Cav.
Convolvulaceae
Synonyms: *Convolvulus bractiflorus* Sessé & Moc., *Convolvulus obvallatus* Spreng., *Exogonium bracteatum* (Cav.) Choisy ex G. Don, *Exogonium bracteatum* var. *pubescens* (B.L. Rob. & Greenm.) House, *Exogonium olivae* Bárcena, *Exogonium spicatum* (Kunth) Choisy
Climbing Mechanism: Stem Twiner
Distribution (Global): Mexico Central, Mexico Gulf, Mexico Northeast, Mexico Northwest, Mexico Southwest
Leaf Type: Simple, palmately 3–5 fid, 5 nerved at base
Inflorescence: Axillary, umbellate cymes/panicles
Fruit Type: Dry, capsule
IUCN Status: Not evaluated
Reference: Icon. 5: 51 (1799)

Ipomoea bracteata var. *lobata*
Convolvulaceae
Climbing Mechanism: Stem Twiner
Leaf Type: Simple, palmately 3–5 fid, 5 nerved at base
Inflorescence: Axillary, umbellate cymes/panicles
Fruit Type: Dry, capsule
IUCN Status: Not evaluated

Ipomoea caespitosa Hallier f. Convolvulaceae
Synonyms: *Hewittia caespitosa* (Roxb.) Steud., *Ipomoea caespitosa* (Roxb.) Hall. fil., *Ipomoea linifolia* Bl., *Ipomoea philippinensis* Choisy, *Ipomoea setulosa* Zoll. & Mor., *Lepistemon decurrens* Hand. -Mazz., *Merremia caespitosa* (Roxb.) Hall.
Climbing Mechanism: Stem Twiner
Distribution (India): Bihar, Odisha
Leaf Type: Simple, alternate
Inflorescence: Solitary, axillary

Fruit Type: Capsular
Flowering and Fruiting: August–December
IUCN Status: Not evaluated

Ipomoea cairica (L.) Sweet
Convolvulaceae
Synonyms: *Convolvulus lupulifolia* Griff., *Convolvulus lymphaticus* Vell., *Convolvulus lymphaticus* Vell., *Convolvulus mucronatus* G. Forst., *Convolvulus paniculatus* Náves ex Fern. -Vill., *Convolvulus pendulus* (R.Br.) Spreng., *Convolvulus quinquelobus* Vahl.
Common Name: Railway Creeper, Cairo morning glory, Messina creeper, Ivy-leaved morning glory
Climbing Mechanism: Stem Twiner
Distribution (Global): Angola, Bangladesh, Botswana, Burundi, Cameroon, Cape Provinces, China South-Central, China Southeast, Congo, Equatorial Guinea, Eritrea, Ethiopia, Gabon, Gambia, Ghana, Guinea, Guinea-Bissau, Gulf of Guinea Is., Hainan, India, Ivory Coast, Kenya, KwaZulu-Natal, Laos, Liberia, Madagascar, Malawi, Mauritius, Mozambique, Myanmar, Nansei-shoto, Nepal, Nigeria, Northern Provinces, Ogasawara-shoto, Oman, Palestine, Rodrigues, Rwanda, Réunion, Senegal, Sierra Leone, Somalia, South China Sea, Sri Lanka, Sudan, Swaziland, Taiwan, Tanzania, Thailand, Togo, Uganda, Vietnam, West Himalaya, Yemen, Zambia, Zaire, Zimbabwe
Distribution (India): Andhra Pradesh, Delhi, Gujarat, Himachal Pradesh, Kerala, Madhya Pradesh, Maharashtra, Odisha, Rajasthan, Tamil Nadu, Uttar Pradesh, Western Ghats
Leaf Type: Simple, palmately 3–5 fid, 5 nerved at base
Inflorescence: Axillary, umbellate cymes/panicles
Fruit Type: Dry, capsule
IUCN Status: Least Concern
Reference: Hort. Brit.: 287 (1826)

Ipomoea calophylla Fenzl
Convolvulaceae
Synonyms: *Ipomoea aitonii* Lindl.
Climbing Mechanism: Stem Twiner
Leaf Type: Simple, palmately 3–5 fid, 5 nerved at base
Inflorescence: Axillary, umbellate cymes/panicles
Fruit Type: Dry, capsule
IUCN Status: Not evaluated
Reference: Flora 27: 312 (1844)

Ipomoea carajasensis D.F. Austin
Convolvulaceae
Synonyms: *Ipomoea maurandioides* Meisn.
Climbing Mechanism: Stem Twiner

Leaf Type: Simple, palmately 3–5 fid, 5 nerved at base
Inflorescence: Axillary, umbellate cymes/panicles
Fruit Type: Dry, capsule
IUCN Status: Not evaluated
Reference: Acta Amazonica 11: 291 (1981)

Ipomoea carnea Jacq.
Convolvulaceae
Synonyms: *Convolvulus carneus* (Jacq.) Spreng.
Common Name: Bush morning glory
Climbing Mechanism: Stem Twiner
Distribution (Global): Argentina Northeast, Argentina Northwest, Belize, Bolivia, Brazil North, Brazil Northeast, Brazil South, Brazil Southeast, Brazil West-Central, Colombia, Costa Rica, Ecuador, El Salvador, Guatemala, Guyana, Honduras, Mexico Central, Mexico Gulf, Mexico Northeast, Mexico Northwest, Mexico Southeast, Mexico Southwest, Nicaragua, Panamá, Paraguay, Peru, Venezuela
Distribution (India): Delhi, Maharashtra, Mizoram, Odisha, Rajasthan
Leaf Type: Simple, palmately 3–5 fid, 5 nerved at base
Inflorescence: Axillary, umbellate cymes/panicles
Fruit Type: Dry, capsule
IUCN Status: Not evaluated
Reference: Enum. Syst. Pl.: 13 (1760)

Ipomoea cheirophylla O'Donell
Convolvulaceae
Synonyms: *Ipomoea digitata*
Common Name: Palm-leaf morning glory
Climbing Mechanism: Stem Twiner
Distribution (Global): Argentina Northeast, Argentina Northwest, Bolivia, Brazil West-Central, Paraguay
Distribution (India): Tamil Nadu
Leaf Type: Simple, palmately 3–5 fid, 5 nerved at base
Inflorescence: Axillary, umbellate cymes/panicles
Fruit Type: Dry, capsule
IUCN Status: Not evaluated
Reference: Lilloa 29: 141 (1959)

Ipomoea cholulensis Kunth
Convolvulaceae
Synonyms: *Convolvulus cholulensis* (Kunth) Spreng., *Ipomoea parviflora* Sessé & Moc., *Quamoclit cholulensis* (Kunth) G. Don,
Quamoclit indivisa var. *pubescens* (Schltdl. & Cham.) Hallier f.
Climbing Mechanism: Stem Twiner

Distribution (Global): Colombia, Costa Rica, Ecuador, El Salvador, Guatemala, Honduras, Mexico Central, Mexico Gulf, Mexico Northeast, Mexico Northwest, Mexico Southeast, Mexico Southwest, Nicaragua, Panamá, Venezuela
Leaf Type: Simple, palmately 3–5 fid, 5 nerved at base
Inflorescence: Axillary, umbellate cymes/panicles
Fruit Type: Dry, capsule
IUCN Status: Not evaluated
Reference: F.W.H.von Humboldt, A.J.A. Bonpland & C.S. Kunth, Nov. Gen. Sp. 3: 112 (1819)

Ipomoea clarkei Hook.f.
Convolvulaceae
Synonyms: *Ipomoea stocksii* C. B. Cl.
Climbing Mechanism: Stem Twiner
Distribution (India): Maharashtra, Tamil Nadu
Leaf Type: Simple, palmately 3–5 fid, 5 nerved at base
Inflorescence: Axillary, umbellate cymes/panicles
Fruit Type: Dry, capsule
Flowering and Fruiting: September–December
IUCN Status: Not evaluated
Reference: Fl. Brit. India 4: 734 (1885)

Ipomoea coptica (L.) Roth ex Roem. & Schult.
Convolvulaceae
Synonyms: *Operculina coptica* (L.) House, *Convolvulus copticus* L., *Convolvulus stipulatus* Desr., *Convolvulus thonningii* Schumach., *Ipomoea coptica* var. *malvifolia* Hall. fil., *Ipomoea dissecta* Willd., *Ipomoea multisecta* Welw., *Sanilum copticum* (L.) Rafin.
Common Name: Egyptian morning glory, Alamo vine
Climbing Mechanism: Stem Twiner
Distribution (Global): Angola, Botswana, Burundi, Cameroon, Cape Provinces, Cape Verde, Caprivi Strip, Central African Repu, Chad, Congo, Ghana, Guinea, India, Kenya, KwaZulu-Natal, Madagascar, Malawi, Mali, Mauritania, Mozambique, Namibia, New Guinea, Niger, Nigeria, Northern Provinces, Northern Territory, Queensland, Senegal, Somalia, Sri Lanka, Sudan, Tanzania, Thailand, Togo, Uganda, Western Australia, Zambia, Zaire, Zimbabwe
Distribution (India): Andhra Pradesh, Delhi, Maharashtra, Rajasthan, Tamil Nadu
Leaf Type: Simple, palmately 3–5 fid, 5 nerved at base
Inflorescence: Axillary, umbellate cymes/panicles
Fruit Type: Dry, capsule
IUCN Status: Least Concern
Reference: J.J. Roemer & J.A. Schultes, Syst. Veg., ed. 15 bis 4: 208 (1819)

Ipomoea cynanchifolia Meisn.
Convolvulaceae

Synonyms: *Ipomoea peckoltii* var. *major* Meisn.
Climbing Mechanism: Stem Twiner
Distribution (Global): Bolivia, Brazil Northeast, Brazil South, Brazil Southeast, Brazil West-Central
Leaf Type: Simple, palmately 3–5 fid, 5 nerved at base
Inflorescence: Axillary, umbellate cymes/panicles
Fruit Type: Dry, capsule
IUCN Status: Least Concern
Reference: C.F.P.von Martius & auct. suc. (eds.), Fl. Bras. 7: 274 (1869)

Ipomoea deccana D.F. Austin
Convolvulaceae
Synonyms: *Ipomoea bracteata* Wight, *Rivea bracteata* Hall. fil.
Common Name: Deccan morning glory
Climbing Mechanism: Stem Twiner
Distribution (India): Kerala, Maharashtra, Tamil Nadu
Leaf Type: Simple, palmately 3–5 fid, 5 nerved at base
Inflorescence: Axillary, umbellate cymes/panicles
Fruit Type: Dry, capsule
Flowering and Fruiting: November–January
IUCN Status: Not evaluated
Reference: M.D. Dassanayake & al. (eds.), Revis. Handb. Fl. Ceylon 1: 324 (1980)

Ipomoea denticulata Choisy
Convolvulaceae
Synonyms: *Convolvulus angustifolius* (Jacq.) Desr., *Convolvulus blumii* D. Dietr., *Convolvulus capensis* Willd., *Convolvulus filicaulis* Vahl, *Convolvulus hastatus* Desr., *Convolvulus sonneratii* W. Wood, *Ipomoea angustifolia* Jacq., *Ipomoea blumii* Steud
Climbing Mechanism: Stem Twiner
Distribution (India): Andaman and Nicobar Islands
Leaf Type: Simple, palmately 3–5 fid, 5 nerved at base
Inflorescence: Axillary, umbellate cymes/panicles
Fruit Type: Dry, capsule
IUCN Status: Not evaluated
Reference: Mém. Soc. Phys. Genève 6: 447 (1833 publ. 1834)

Ipomoea emarginata Kuntze
Convolvulaceae
Synonyms: *Merremia emarginata* (Burm.f.) Hallier f.
Common Name: Kidney leaf morning glory
Climbing Mechanism: Stem Twiner
Distribution (India): Peninsular India
Leaf Type: Simple, palmately 3–5 fid, 5 nerved at base
Inflorescence: Axillary, umbellate cymes/panicles

Fruit Type: Dry, capsule
IUCN Status: Not evaluated
Reference: Revis. Gen. Pl. 2: 443 (1891)

Ipomoea eriocarpa R.Br.
Convolvulaceae
Synonyms: *Convolvulus curtana* Buch. -Ham. ex Wall., *Convolvulus eriocarpus* (R. Br.) Spreng., *Convolvulus hispidus* Vahl, *Convolvulus rampania* Buch. -Ham. ex Wall., *Convolvulus sessiliflorus* (Roth) Spreng., *Convolvulus sphaerocephalus* Roxb., *Convolvulus tamnifoli*
Common Name: Tiny morning glory, Woolly fruited morning glory
Climbing Mechanism: Stem Twiner
Distribution (Global): Afghanistan, Angola, Assam, Benin, Botswana, Burundi, Cambodia, Cameroon, Cape Verde, Central African Repu, Chad, China South-Central, Comoros, Congo, East Himalaya, Egypt, Ethiopia, Gambia, Ghana, Guinea, Guinea-Bissau, Gulf of Guinea Is., India, Ivory Coast, Java, Kenya, Lesser Sunda Is., Liberia, Madagascar, Malawi, Malaya, Mali, Mauritania, Mozambique, Myanmar, Nepal, New Guinea, Nigeria, Northern Provinces, Northern Territory, Pakistan, Philippines, Queensland, Rwanda, Saudi Arabia, Senegal, Sierra Leone, Somalia, Sri Lanka, Sudan, Sulawesi, Sumatera, Tadzhikistan, Tanzania, Togo, Uganda, Vietnam, West Himalaya, Western Australia, Zambia, Zaire, Zimbabwe
Distribution (India): Andaman and Nicobar Islands, Andhra Pradesh, Delhi, Gujarat, Himachal Pradesh, Madhya Pradesh, Maharashtra, Odisha, Tamil Nadu, Telangana, Uttar Pradesh
Leaf Type: Simple, alternate
Inflorescence: Cymes
Fruit Type: Capsule
Flowering and Fruiting: October–January
IUCN Status: Not evaluated
Reference: Prodr. Fl. Nov. Holland.: 484 (1810)

Ipomoea fistulosa Mart. ex Choisy
Convolvulaceae
Synonyms: *Ipomoea carnea* subsp. *fistulosa* (Mart. ex Choisy) D.F. Austin
Common Name: Bush morning glory, Morning glory tree bush
Climbing Mechanism: Stem Twiner
Distribution (India): Assam, Gujarat, Tripura
Leaf Type: Cordate to deltoid large leaves to 5 inches' long
Inflorescence: Terminal clusters
IUCN Status: Not evaluated
Reference: A.P.de Candolle, Prodr. 9: 349 (1845)

Ipomoea glenieii Thwaites ex C.B. Clarke
Convolvulaceae

Synonyms: *Ipomoea aquatica* var. *aquatica*
Climbing Mechanism: Stem Twiner
Leaf Type: Simple, palmately 3–5 fid, 5 nerved at base
Inflorescence: Axillary, umbellate cymes/panicles
Fruit Type: Dry, capsule
IUCN Status: Not evaluated
Reference: J. D. Hooker, Fl. Brit. India 4: 208 (1883)

Ipomoea gomezii C.B. Clarke
Convolvulaceae
Synonyms: *Decalobanthus mammosus* (Lour.) A.R. Simões & Staples
Climbing Mechanism: Stem Twiner
Leaf Type: Simple, palmately 3–5 fid, 5 nerved at base
Inflorescence: Axillary, umbellate cymes/panicles
Fruit Type: Dry, capsule
IUCN Status: Not evaluated
Reference: J. D. Hooker, Fl. Brit. India 4: 211 (1883)

Ipomoea hederifolia L.
Convolvulaceae
Synonyms: *Convolvulus acutangulus* (Ruiz & Pav.) Spreng., *Convolvulus angulatus* (Lam.) Spreng., *Convolvulus coccineus* var. *hederifolius* (L.) Kuntze, *Convolvulus hederifolius* (L.) Spreng., *Convolvulus luteolus* (Jacq.) Spreng., *Convolvulus phoeniceus* (Roxb. ex Rottler) Spreng.
Common Name: Ivy-leaf morning-glory, Red morning glory, Red star glory, Scarlet creeper, Star Ipomoea
Climbing Mechanism: Stem Twiner
Distribution (Global): Alabama, Argentina Northwest, Arizona, Belize, Bermuda, Bolivia, Brazil North, Brazil Northeast, Brazil South, Brazil Southeast, Brazil West-Central, Cayman Is., Colombia, Costa Rica, Cuba, Dominican Republic, Ecuador, El Salvador, Florida, French Guiana, Georgia, Guatemala, Guyana, Haiti, Honduras, Jamaica, Leeward Is., Louisiana, Mexico Central, Mexico Gulf, Mexico Northeast, Mexico Northwest, Mexico Southeast, Mexico Southwest, Mississippi, New Mexico, Nicaragua, Panamá, Paraguay, Peru, Puerto Rico, Suriname, Texas, Trinidad-Tobago, Venezuela, Windward Is.
Distribution (India): Andhra Pradesh, Assam, Gujarat, Kerala, Madhya Pradesh, Maharashtra, Mizoram, Odisha, Rajasthan, Tamil Nadu, Telangana, Tripura, Uttar Pradesh
Leaf Type: Simple
Inflorescence: Cymes
Fruit Type: Capsule
Flowering and Fruiting: October–December
IUCN Status: Not evaluated
Reference: Syst. Nat. ed. 10, 2: 925 (1759)

Ipomoea hirtifolia R.C. Fang & S.H. Huang
Convolvulaceae
Climbing Mechanism: Stem Twiner
Distribution (Global): Tibet
Distribution (India): Andaman and Nicobar Islands
Leaf Type: Simple, palmately 3–5 fid, 5 nerved at base
Inflorescence: Axillary, umbellate cymes/panicles
Fruit Type: Dry, capsule
IUCN Status: Not evaluated
Reference: Fl. Xizang. 4: 28 (1985)

Ipomoea hispida Zuccagni
Convolvulaceae
Synonyms: *Convolvulus hispidus*, *Ipomoea eriocarpa*
Climbing Mechanism: Stem Twiner
Distribution (India): Bihar, Odisha
Leaf Type: Simple, palmately 3–5 fid, 5 nerved at base
Inflorescence: Axillary, umbellate cymes/panicles
Fruit Type: Dry, capsule
IUCN Status: Not evaluated
Reference: Cent. Observ. Pl. 1: 15 (1806)

Ipomoea horsfalliae Hook.
Convolvulaceae
Synonyms: *Convolvulus horsfalliae* (Hook.) D.Dietr.
Common Name: Cardinal creeper, Horsfall's morning glory, Kuhio vine, Lady doorly's morning glory, Prince's vine
Climbing Mechanism: Stem Twiner
Distribution (Global): Guyana, Suriname
Distribution (India): Maharashtra, Rajasthan, Tamil Nadu
Leaf Type: Palmate
Fruit Type: Berries
IUCN Status: Not evaluated
Reference: Bot. Mag. 61: t. 3315 (1834)

Ipomoea illustris Prain
Convolvulaceae
Climbing Mechanism: Stem Twiner
Distribution (India): Andaman and Nicobar Islands, Andhra Pradesh, Goa, Gujarat
Leaf Type: Simple, palmately 3–5 fid, 5 nerved at base
Inflorescence: Axillary, umbellate cymes/panicles
Fruit Type: Dry, capsule
IUCN Status: Not evaluated
Reference: Bengal Pl. 2: 735 (1903)

Ipomoea imperati (Vahl) Griseb.
Convolvulaceae
Synonyms: *Batatas acetosifolia* (Vahl) Choisy, *Batatas incurva* (Schumach. & Thonn.) Benth., *Batatas littoralis* (L.) Choisy, *Batatas sinuatus* (Petagna) Guss., *Convolvulus acetosifolius* Vahl, *Convolvulus arenarius* Vahl, *Convolvulus auritus* Roem. & Schult.
Common Name: Beach morning glory
Climbing Mechanism: Stem Twiner
Distribution (Global): Alabama, Algeria, Andaman Is., Angola, Azores, Bahamas, Baleares, Belize, Benin, Brazil North, Brazil Northeast, Brazil South, Brazil Southeast, Cameroon, Canary Is., Cape Verde, Cayman Is., China Southeast, Colombia, Congo, Costa Rica, Cuba, Cyprus, Dominican Republic, East Aegean Is., Ecuador, Egypt, Equatorial Guinea, Florida, French Guiana, Gabon, Galápagos, Gambia, Georgia, Ghana, Guatemala, Guinea, Gulf of Guinea Is., Guyana, Hainan, Haiti, Hawaii, Honduras, Italy, Ivory Coast, Jamaica, Java, Kazan-retto, Kriti, Lebanon-Syria, Leeward Is., Liberia, Libya, Louisiana, Madeira, Malaya, Mexico Gulf, Mexico Northeast, Mexico Northwest, Mexico Southeast, Mexico Southwest, Mississippi, Morocco, Mozambique, Nansei-shoto, Nicaragua, Nigeria, Northern Territory, Ogasawara-shoto, Palestine, Panamá, Philippines, Puerto Rico, Queensland, Senegal, Sierra Leone, Sinai, South Carolina, Sri Lanka, Suriname, Taiwan, Tanzania, Texas, Thailand, Togo, Trinidad-Tobago, Turkey, Venezuela, Vietnam, Windward Is., Zaire
Leaf Type: Simple, palmately 3–5 fid, 5 nerved at base
Inflorescence: Axillary, umbellate cymes/panicles
Fruit Type: Dry, capsule
IUCN Status: Not evaluated
Reference: Cat. Pl. Cub.: 203 (1866)

Ipomoea indica (Burm.) Merr.
Convolvulaceae
Synonyms: *Convolvulus acuminatus* Vahl, *Convolvulus africanus* Nicolson bis, *Convolvulus amoenus* (Blume) D. Dietr., *Convolvulus bogotensis* Humb. & Bonpl. ex Willd., *Convolvulus bogotensis* f. albiflorus Kuntze, *Convolvulus bogotensis* f. lilacinus Kuntze
Common Name: Blue dawn flower, Blue morning glory, Island Morning Glory, Oceanblue morning Glory
Climbing Mechanism: Stem Twiner
Distribution (Global): Argentina Northeast, Argentina Northwest, Bahamas, Belize, Brazil North, Brazil Northeast, Brazil South, Brazil Southeast, Brazil West-Central, Cayman Is., Central American Pac, Colombia, Costa Rica, Cuba, Dominican Republic, Florida, French Guiana, Guatemala, Guyana, Haiti, Honduras, Jamaica, Leeward Is., Mexican Pacific Is., Mexico Central, Mexico Gulf, Mexico Northeast, Mexico Southeast, Mexico Southwest, Nicaragua, Panamá, Paraguay, Puerto Rico, Suriname, Texas, Trinidad-Tobago, Uruguay, Venezuela, Venezuelan Antilles, Windward Is.

Distribution (India): Gujarat, Himachal Pradesh, Maharashtra, Puducherry, Rajasthan, Tamil Nadu
Leaf Type: Simple, palmately 3–5 fid, 5 nerved at base
Inflorescence: Axillary, umbellate cymes/panicles
Fruit Type: Dry, capsule
IUCN Status: Not evaluated
Reference: Interpr. Herb. Amboin.: 445 (1917)

Ipomoea kingii Diels
Convolvulaceae
Climbing Mechanism: Stem Twiner
Distribution (India): Arunachal Pradesh, Mizoram
Leaf Type: Simple, palmately 3–5 fid, 5 nerved at base
Inflorescence: Axillary, umbellate cymes/panicles
Fruit Type: Dry, capsule
IUCN Status: Not evaluated

Ipomoea laciniata Balf.f.
Convolvulaceae
Synonyms: *Ipomoea kotschyana* Hochst. ex Choisy
Common Name: Cut-leaved morning glory
.
Climbing Mechanism: Stem Twiner
Leaf Type: Simple, palmately 3–5 fid, 5 nerved at base
Inflorescence: Axillary, umbellate cymes/panicles
Fruit Type: Dry, capsule
IUCN Status: Not evaluated
Reference: Proc. Roy. Soc. Edinburgh 12: 82 (1883)

Ipomoea laxiflora H.J. Chowdhery & Debta
Convolvulaceae
Synonyms: *Ipomoea triloba* L.
Climbing Mechanism: Stem Twiner
Leaf Type: Simple, palmately 3–5 fid, 5 nerved at base
Inflorescence: Axillary, umbellate cymes/panicles
Fruit Type: Dry, capsule
IUCN Status: Not evaluated
Reference: Indian J. Forest. 32: 120 (2009)

Ipomoea littoralis Blume
Convolvulaceae
Synonyms: *Convolvulus denticulatus* Desr., *Convolvulus laevigatus* Steud., *Convolvulus maritimus* D. Dietr., *Convolvulus thalassicus* Steud., *Ipomoea carnea* G. Forst., *Ipomoea choisyana* W. Wight, *Ipomoea cymosa* Baker, *Ipomoea denticulata* (Desr.) Choisy

Climbing Mechanism: Stem Twiner
Distribution (Global): Bangladesh, Borneo, Caroline Is., Cook Is., Fiji, Gilbert Is., Hawaii, India, Java, Kazan-retto, Lesser Sunda Is., Madagascar, Malaya, Maluku, Marianas, Marquesas, Marshall Is., Myanmar, Nansei-shoto, Nauru, New Caledonia, New Guinea, Nicobar Is., Niue, Ogasawara-shoto, Philippines, Queensland, Samoa, Seychelles, Society Is., Solomon Is., South China Sea, Sri Lanka, Sulawesi, Sumatera, Taiwan, Thailand, Tonga, Tuamotu, Tubuai Is., Vanuatu, Vietnam, Wallis-Futuna Is.
Leaf Type: Simple, palmately 3–5 fid, 5 nerved at base
Inflorescence: Axillary, umbellate cymes/panicles
Fruit Type: Dry, capsule
IUCN Status: Least Concern
Reference: Bijdr. Fl. Ned. Ind.: 713 (1826)

Ipomoea lobata (Cerv.) Thell.
Convolvulaceae
Synonyms: *Ipomoea mina* Voss, *Ipomoea versicolor* Meisn., *Mina cordata* Micheli, *Mina lobata* Cerv., *Quamoclit lobata* (Cerv.) House, *Quamoclit mina* G. Don. Quamoclit *pallescens* Brongn. ex Neumann, *Convolvulus mina* Kuntze
Climbing Mechanism: Stem Twiner
Distribution (Global): Mexico Southwest
Leaf Type: Simple, palmately 3–5 fid, 5 nerved at base
Inflorescence: Axillary, umbellate cymes/panicles
Fruit Type: Dry, capsule
IUCN Status: Not evaluated
Reference: Vierteljahrsschr. Naturf. Ges. Zürich 64: 775 (1919)

Ipomoea marginata (Desr.) Verdc.
Convolvulaceae
Synonyms: *Batatas abyssinica* A. Rich., *Convolvulus diversifolius* Schumach. & Thonn., *Convolvulus incrassatus* Wall., *Convolvulus javanicus* Burm.f., *Convolvulus javanicus* Spreng., *Convolvulus marginatus* Desr., *Convolvulus sepiarius* Wall., *Convolvulus stipulaceu*
Common Name: Purple heart glory, Arrow-leaf morning glory, Hedge bind-weed, spotted-leaved Ipomoea
Climbing Mechanism: Stem Twiner
Distribution (Global): Angola, Bangladesh, Benin, Burundi, Cameroon, Cape Verde, Central African Repu, Chad, Eritrea, Ethiopia, Gambia, Ghana, Guinea, Hainan, India, Ivory Coast, Java, Kenya, Lesser Sunda Is., Malawi, Malaya, Mali, Mauritania, Mozambique, Myanmar, New Guinea, Nigeria, Queensland, Senegal, Somalia, Sri Lanka, Sudan, Sulawesi, Tanzania, Thailand, Togo, Uganda, Vietnam, Zaire, Zimbabwe
Distribution (India): Kerala, Maharashtra, Tamil Nadu
Leaf Type: Alternately arranged ovate-heart-shaped leave
Inflorescence: Racemose inflorescence, flowers darker red-purple

Fruit Type: Capsule, long, globose
Flowering and Fruiting: August–November
IUCN Status: Not evaluated
Notes: Distributed mostly near swamps and moist places near the coast
Reference: Fl. Indica: 50 (1768)

Ipomoea mauritiana Jacq.
Convolvulaceae
Synonyms: *Apopleumon bignonioides* (Sims) Raf., Batatas bignonioides (Sims) G. Don. + 30
Common Name: Giant potato, Mauritanian convolvulus
Climbing Mechanism: Stem Twiner
Distribution (Global): Pantropical
Distribution (India): Andhra Pradesh, Arunachal Pradesh, Goa, Great Nicobar Island, Gujarat, Maharashtra, Odisha, Rajasthan, Tamil Nadu, Telangana
Leaf Type: Usually palmately 5–7-divided to or beyond middle, rarely entire or shallowly lobed
Inflorescence: Inflorescence axillary, few to many-flowered cymes; peduncles
Fruit Type: Capsule, ovoid
Flowering and Fruiting: August–September
IUCN Status: Not evaluated
Reference: Collectanea 4: 216 (1791)

Ipomoea maxima Don ex Sweet
Convolvulaceae
Synonyms: *Ipomoea alba*. *Chonemorpha convolvuloides* G. Don, *Bonanox indica* Raf., *Bonanox riparia* Raf., *Calonyction aculeatum* (L.) House, *Calonyction aculeatum* f. apopetalum Allard, *Calonyction album* (L.) House, *Calonyction bona-nox* (Spreng.) Bojer. + 40
Common Name: Moon Vine, Moonflower
Climbing Mechanism: Stem Twiner
Distribution (Global): Pantropical
Distribution (India): Uttar Pradesh
Leaf Type: leaves are entire or three lobed
Inflorescence: Flowers solitary or few in cymes
Fruit Type: Capsule
Flowering and Fruiting: October–December
IUCN Status: Not evaluated
Notes: Usually found in the moist deciduous forests of the Eastern Ghats & the Western Ghats
Reference: Sp. Pl.: 161 (1753)

Ipomoea molleri Gand.
Convolvulaceae
Climbing Mechanism: Stem Twiner

Distribution (Global): Gulf of Guinea Is.
Distribution (India): Western Ghats
Leaf Type: Simple, palmately 3–5 fid, 5 nerved at base
Inflorescence: Axillary, umbellate cymes/panicles
Fruit Type: Dry, capsule
IUCN Status: Not evaluated
Reference: Bull. Soc. Bot. France 65: 61 (1918)

Ipomoea mombassana Vatke
Convolvulaceae
Climbing Mechanism: Stem Twiner
Distribution (Global): Kenya, Sudan, Tanzania
Distribution (India): Peninsular India
Leaf Type: Leaves, simple, triangular-ovate, deeply cordate to sagitate
Inflorescence: Axillary, 1–3 flowered
Fruit Type: Capsule, globose
Flowering and Fruiting: October–December
IUCN Status: Not evaluated
Notes: Often found in degraded dry deciduous forests
Reference: Linnaea 43: 515 (1882)

Ipomoea muricata (L.) Jacq.
Convolvulaceae
Synonyms: *Ipomoea capillacea* (Kunth) G. Don. *Convolvulus capillaceus* Kunth, *Diatrema muricata* Raf., *Ipomoea aristulata* M. Martens & Galeotti, *Ipomoea armata* Roem. & Schult., *Ipomoea muricata* f. alba Woodson & Seibert, *Ipomoea muricatisepala* Matuda
Common Name: Lovely morning glory
Climbing Mechanism: Stem Twiner
Distribution (Global): Arizona, Colombia, Colorado, Costa Rica, Ecuador, El Salvador, Guatemala, Honduras, Mexico Central, Mexico Gulf, Mexico Northeast, Mexico Northwest, Mexico Southeast, Mexico Southwest, New Mexico, Nicaragua, Panamá, Peru, Texas, Venezuela
Distribution (India): Delhi, Gujarat, Tamil Nadu, Uttar Pradesh
Leaf Type: Sessile, incised and appearing compound
Inflorescence: Axillary, flowers solitary on peduncles
Fruit Type: Globose capsule
Flowering and Fruiting: July–September
IUCN Status: Not evaluated
Notes: Found in pine forests, oak woodlands, and plains
Reference: Gen. Hist. 4: 267 (1837)

Ipomoea nil (L.) Roth
Convolvulaceae

Synonyms: *Convolvuloides triloba* Moench, *Convolvulus caeruleus* (Roxb. ex Ker-Gawl.) Spreng., *Convolvulus coelestis* G. Forst., *Convolvulus dillenii* Desr., *Convolvulus hederaceus* L., *Convolvulus hederifolius* Salisb. + 40
Common Name: White-edge morning glory
Climbing Mechanism: Stem Twiner
Distribution (Global): Pantropical
Distribution (India): Andaman and Nicobar Islands, Andhra Pradesh, Arunachal Pradesh, Delhi, Gujarat, Himachal Pradesh, Madhya Pradesh, Maharashtra, Mizoram, Odisha, Rajasthan, Tamil Nadu, Telangana, Uttar Pradesh
Leaf Type: Leaves are simple, Alternate distichous, 3-lobed, rarely entire, broadly ovate or nearly circular
Inflorescence: Inflorescences axillary, 1- to few flowered, Flowers solitary or in cymes
Fruit Type: Capsules, ovoid
Flowering and Fruiting: November–January
IUCN Status: Not evaluated
Reference: Catal. Bot. 1: 36 (1797)

Ipomoea obscura (L.) Ker Gawl.
Convolvulaceae
Synonyms: *Convolvulus curassavicus* (Balb. ex Roem. & Schult.) Colla, *Convolvulus gonatodes* Steud. ex A. Rich., *Convolvulus insuavis* (Blume) D. Dietr., *Convolvulus obscurus* L., *Convolvulus ocularis* (Bartl.) Endl., *Convolvulus pilifer* Steud. + 25
Common Name: Wild petunia, Yellow Ipomoea, Obscure morning glory
Climbing Mechanism: Stem Twiner
Distribution (Global): Angola, Assam, Bangladesh, Borneo, Botswana, Burundi, Cambodia, Cameroon, Cape Provinces, Caprivi Strip, Central African Repu, Chad, China South-Central, China Southeast, Christmas I., Comoros, Djibouti, Eritrea, Ethiopia, Fiji, Free State, Gambia, Ghana, Hainan, India, Ivory Coast, Java, Kenya, KwaZulu-Natal, Laos, Lesser Sunda Is., Madagascar, Malawi, Malaya, Maluku, Mozambique, Myanmar, Namibia, Nepal, New Guinea, Nigeria, Northern Provinces, Oman, Philippines, Rwanda, Saudi Arabia, Sierra Leone, Society Is., Socotra, Solomon Is., Somalia, South China Sea, Sri Lanka, Sudan, Sulawesi, Sumatera, Swaziland, Taiwan, Tanzania, Thailand, Togo, Uganda, Vietnam, Yemen, Zambia, Zimbabwe
Distribution (India): Andaman and Nicobar Islands, Andhra Pradesh, Assam, Delhi, Kerala, Madhya Pradesh, Maharashtra, Odisha, Puducherry, Rajasthan, Tamil Nadu, Telangana, Uttar Pradesh
Leaf Type: Simple, alternate distichous., cordiform
Inflorescence: In solitary or sub-umbellate cymes; white or yellow, tinged pink
Fruit Type: Capsule globose, glabrous, tipped with persistent style base
Flowering and Fruiting: August–March
IUCN Status: Not evaluated
Reference: Bot. Reg. 3: t. 239 (1817)

Ipomoea ochracea (Lindl.) G. Don
Convolvulaceae
Synonyms: *Convolvulus ochraceus* Lindl., *Convolvulus ochraceus* Vahl ex Choisy, *Convolvulus trichocalyx* Schumach. & Thonn., *Ipomoea afra* Choisy, *Ipomoea clarkei* Hook.f., *Ipomoea conica* Didr., *Ipomoea curtisii* House, *Ipomoea kentrocarpa* Hochst. ex A. Rich
Climbing Mechanism: Stem Twiner
Distribution (Global): Angola, Botswana, Burundi, Cameroon, Cape Verde, Central African Repu, Chad, Congo, Eritrea, Ethiopia, Gabon, Ghana, Guinea, Guinea-Bissau, Ivory Coast, Kenya, Malawi, Mauritania, Mauritius, Mozambique, Nigeria, Rwanda, Senegal, Sierra Leone, Sudan, Tanzania, Togo, Uganda, Zambia, Zaire, Zimbabwe
Distribution (India): Peninsular India
Leaf Type: Leaves broadly
ovate, cordate with rounded
lobes at base, entire at margins
Inflorescence: Inflorescences a cyme, 1–3 flowered
Fruit Type: Capsules ovoid, acute with persistent style, glabrous, brown, 4-valved, 2-celled
Flowering and Fruiting: August–September
IUCN Status: Least Concern
Reference: Hort. Brit., ed. 2: 371 (1830)

Ipomoea pallida Santapau & V. Patel
Convolvulaceae
Climbing Mechanism: Stem Twiner
Distribution (India): Maharashtra, Rajasthan
Leaf Type: Simple, palmately 3–5 fid, 5 nerved at base
Inflorescence: Axillary, umbellate cymes/panicles
Fruit Type: Dry, capsule
IUCN Status: Not evaluated
Reference: Trans. Bose Res. Inst. Calcutta 22: 38 (1958)

Ipomoea parasitica (Kunth) G. Don
Convolvulaceae
Synonyms: *Convolvulus circinnatus* Roem. & Schult., *Convolvulus parasiticus* Kunth, *Ipomoea perlonga* B.L. Rob., *Pharbitis parasitica* (Kunth) V.M.Badillo
Common Name: Yellow-throated morning glory, Yellow-throated Ipomoea
Climbing Mechanism: Stem Twiner
Distribution (Global): Bolivia, Brazil Northeast, Brazil Southeast, Brazil West-Central, Colombia, Costa Rica, El Salvador, Guatemala, Honduras, Mexico Central, Mexico Gulf, Mexico Northeast, Mexico Northwest, Mexico Southeast, Mexico Southwest, Nicaragua, Peru, Venezuela
Distribution (India): Telangana

Leaf Type: Leaves are alternate, simple, ovate, 5–15 × 5–15 cm, blunt or long-pointed at the tip, heart shaped
Inflorescence: Inflorescence, Beautiful funnel-shaped flowers are borne in several-flowered cymes
Fruit Type: Capsule ovoid to globose, glabrous
Flowering and Fruiting: September–January
IUCN Status: Not evaluated
Reference: Gen. Hist. 4: 275 (1837)

Ipomoea pes-tigridis L.
Convolvulaceae
Synonyms: *Convolvuloides palmata* Moench, *Convolvulus bryoniifolius* Salisb., *Convolvulus pes-tigridis* (L.) Spreng.
Common Name: Tiger-foot Ipomoea, Tiger foot morning glory
Climbing Mechanism: Stem Twiner
Distribution (Global): Angola, Bangladesh, Benin, Borneo, Botswana, Cambodia, Caprivi Strip, Chad, China South-Central, China Southeast, East Himalaya, Guinea-Bissau, Hainan, India, Java, Kenya, Lesser Sunda Is., Malawi, Malaya, Mali, Maluku, Mauritius, Mozambique, Myanmar, Namibia, Nepal, New Guinea, Niger, Nigeria, Northern Provinces, Pakistan, Philippines, Somalia, South China Sea, Sri Lanka, Sudan, Sulawesi, Sumatera, Taiwan, Tanzania, Thailand, Vietnam, West Himalaya, Zambia, Zaire, Zimbabwe
Distribution (India): Andaman, Nicobar, Islands, Andhra Pradesh, Delhi, Goa, Gujarat, Himachal Pradesh, Madhya Pradesh, Maharashtra, Odisha, Puducherry, Rajasthan, Tamil Nadu, Telangana, Uttar Pradesh
Leaf Type: Leaf-blade palmately 7–9-lobed
Inflorescence: Inflorescence, subsessile in axillary, capitate clusters
Fruit Type: Capsule brown, glabrous
Flowering and Fruiting: September–December
IUCN Status: Not evaluated
Reference: Sp. Pl.: 162 (1753)

Ipomoea pileata Roxb.
Convolvulaceae
Synonyms: *Convolvulus guinensis* D. Dietr., *Convolvulus involucratus* Spreng., *Convolvulus pileatus* (Roxb.) Spreng., *Convolvulus trichocalyx* Zoll. & Moritzi, *Elythrostamna convolvulacea* Bojer ex Desjardins, *Ipomoea involucrata* Anon.
Climbing Mechanism: Stem Twiner
Distribution (Global): Angola, Assam, Bangladesh, Borneo, Botswana, Cambodia, Central African Repu, Chad, China South-Central, China Southeast, Comoros, Gambia, Hainan, India, Java, Kenya, Laos, Madagascar, Malawi, Malaya, Mozambique, Myanmar, Namibia, Nigeria, Northern Provinces, Philippines, Senegal, Sri Lanka, Sudan, Tanzania, Thailand, Togo, Vietnam, Zambia, Zaire, Zimbabwe
Distribution (India): Kerala, Maharashtra, Tamil Nadu

Leaf Type: ovate-acuminate, mucronate, base cordate, glabrescent above and pubescent below
Inflorescence: Peduncle often longer than the petiole
Fruit Type: Capsule ovoid
Flowering and Fruiting: November–January
IUCN Status: Not evaluated
Notes: Usually found in moist deciduous forests of the Western Ghats
Reference: Fl. Ind. 2: 94 (1824)

Ipomoea platanifolia
Convolvulaceae
Climbing Mechanism: Stem Twiner
IUCN Status: Not evaluated

Ipomoea poranoides C.B. Clarke in Hook.f.
Convolvulaceae
Synonyms: *Merremia poranoides* (C.B. Clarke) Hallier f., *Ipomoea courchetii* Gagnep., *Ipomoea longipedunculata* C.Y. Wu, *Merremia longipedunculata* R.C. Fang.
Climbing Mechanism: Stem Twiner
Distribution (Global): Assam, China South-Central, East Himalaya, India, Thailand, Vietnam
Leaf Type: Simple, palmately 3–5 fid, 5 nerved at base
Inflorescence: Axillary, umbellate cymes/panicles
Fruit Type: Dry, capsule
IUCN Status: Not evaluated
Reference: Bull. Herb. Boissier 5: 375 (1897)

Ipomoea pulchella Roth
Convolvulaceae
Synonyms: *Ipomoea cairica* (L.) Sweet., *Batatas cavanillesii* (Schult.) G. Don, *Batatas pulchella* (Roth) Bojer, *Batatas senegalensis* (Lam.) G. Don, *Convolvulus bellus* Spreng., *Convolvulus cairicus* L., *Convolvulus cavanillesii* (Schult.) Spreng. + 40
Common Name: Cairo morning glory, Messina creeper, Coastal morning glory, Railway creeper, Mile a minute vine
Climbing Mechanism: Stem Twiner
Distribution (Global): Pantropical
Distribution (India): Gujarat, Rajasthan
Leaf Type: Palmately divided into 5–7, Ovate
Inflorescence: Flower axillary in 1-few flowered cymes
Fruit Type: Capsule, subglobose
Flowering and Fruiting: October–April
IUCN Status: Not evaluated

Notes: Usually found along roadsides, grassy fields, wastelands, dry slopes, and railway tracks
Reference: Hort. Brit.: 287 (1826)

Ipomoea purpurea (L.) Roth
Convolvulaceae
Synonyms: *Cleiemera cuspidata* Raf., *Convolvuloides leucosperma* Moench, *Convolvuloides purpurea* (L.) Moench, *Convolvulus eriocaulos* Roem. & Schult., *Convolvulus glandulifer* Spreng. +30
Common Name: Common morning glory, Tall morning glory, Purple morning glory, Garden morning glory
Climbing Mechanism: Stem Twiner
Distribution (Global): Pantropical
Distribution (India): Arunachal Pradesh, Delhi, Himachal Pradesh, Kerala, Madhya Pradesh, Maharashtra, Rajasthan, Tamil Nadu, Uttar Pradesh
Leaf Type: Leaf, ovate, entire or tri-lobed
Inflorescence: Flowers are borne on solitary or in 1–5 flowered cymes
Fruit Type: Capsules
Flowering and Fruiting: July–September
IUCN Status: Not evaluated
Notes: Mostly found in wetlands and coastal habitats
Reference: Bot. Abh. Beobacht.: 27 (1787)

Ipomoea quamoclit L.
Convolvulaceae
Climbing Mechanism: Stem Twiner
IUCN Status: Not evaluated

Ipomoea rubens Choisy
Convolvulaceae
Climbing Mechanism: Stem Twiner
IUCN Status: Least Concern

Ipomoea rubriflora O'Donell
Convolvulaceae
Synonyms: *Ipomoea hederifolia* Anon.
Common Name: Redstar, Red morning glory, Scarlet creeper
Climbing Mechanism: Stem Twiner
Distribution (Global): Argentina Northeast, Argentina Northwest, Bolivia, Paraguay
Leaf Type: Alternate, 3-lobed leaves are heart shaped at the base
Inflorescence: Cymes few-flowered, axillary
Fruit Type: Capsule strongly rostrate terminating in a persistent, the fruiting pedicel erec
Flowering and Fruiting: October–December

IUCN Status: Not evaluated
Reference: Lilloa 29: 79 (1959)

Ipomoea rumicifolia Choisy
Convolvulaceae
Synonyms: *Convolvulus lasiospermus* Vis., *Ipomoea gnaphalosperma* Hochst. ex Choisy
Common Name: Dock-leaf morning glory
Climbing Mechanism: Stem Twiner
Distribution (Global): Angola, Eritrea, Ethiopia, India, Senegal, Sudan
Distribution (India): Andhra Pradesh, Tamil Nadu, Western Ghats
Leaf Type: Leaves are broadly arrow shaped to ovate-heart shaped
Inflorescence: Flowers are borne in leaf axils, in 1–2 flowered clusters
IUCN Status: Not evaluated
Reference: Mém. Soc. Phys. Genève 6: 447 (1833 publ. 1834)

Ipomoea salicifolia Herb.Madr. ex Choisy
Convolvulaceae
Synonyms: *Stictocardia tiliifolia* (Desr.) Hallier f., *Amphione tiliifolia* (Desr.) Raf., *Argyreia campanulata* Alston. +20
Climbing Mechanism: Stem Twiner
Distribution (Global): Bangladesh, Borneo, Caroline Is., Christmas I., Cook Is., Fiji, Hainan, Howland-Baker Is., India, Java, Kazan-retto, Lesser Sunda Is., Malaya, Maluku, Marianas, Myanmar, Nansei-shoto, New Caledonia, New Guinea, Northern Territory, Ogasawara-shoto, Philippines, Queensland, Samoa, Society Is., Sri Lanka, Sulawesi, Sumatera, Taiwan, Thailand, Tonga, Tubuai Is., Vanuatu, Vietnam
Leaf Type: Leaves simple, cordate to cordate-ovate
Inflorescence: Flowers axillary, solitary or in 2–6 flowered cymes
Fruit Type: Non-capsular fruits enclosed in enlarged, leathery sepals
Flowering and Fruiting: December–April
IUCN Status: Not evaluated
Reference: Bot. Jahrb. Syst. 18: 159 (1893)

Ipomoea salsettensis Santapau & V. Patel
Convolvulaceae
Climbing Mechanism: Stem Twiner
Distribution (Global): India
Distribution (India): Maharashtra
Leaf Type: Leaves broadly ovate or orbicular, entire or 3–5 lobed, base cordate, apex acuminate or mucronate
Inflorescence: Flowers 3–6 in umbellate cymes, rarely uniflorate
Fruit Type: Capsules depressed-globose
Flowering and Fruiting: September–December
IUCN Status: Not evaluated

Notes: Mostly grows on roadside bushes.
Reference: Trans. Bose Res. Inst. Calcutta 22: 34 (1958)

Ipomoea soluta Kerr
Convolvulaceae
Synonyms: *Adamboe bicolor* Raf., *Argyreia campanulata* (L.) Alston, *Convolvulus campanulatus* (L.) Spreng., *Rivea campanulata* (L.) House, *Stictocardia campanulata* (L.) Merr.
Climbing Mechanism: Stem Twiner
Distribution (Global): Andaman Is., Bangladesh, Borneo, Cambodia, China South-Central, India, Java, Laos, Lesser Sunda Is., Malaya, Maluku, Myanmar, New Guinea, Philippines, Solomon Is., Sri Lanka, Sulawesi, Sumatera, Thailand, Vietnam
Distribution (India): Peninsular India
Leaf Type: Alternately arranged ovate leaves
Inflorescence: Flowers occur in branched, many-flowered cymes, flowers are white to pale violet, with a deep purple throat
Fruit Type: Capsule, globose, glabrous
Flowering and Fruiting: October–January
IUCN Status: Not evaluated
Notes: Commonly found in moist deciduous forests and mangrove forests
Reference: Sp. Pl.: 160 (1753)

Ipomoea staphylina Roem. & Schult.
Convolvulaceae
Synonyms: *Batatas multiflora* Bojer ex Choisy, *Convolvulus kleinii* Spreng., *Convolvulus racemosus* Roem. & Schult., *Ipomoea racemosa* Roth.
Common Name: Clustered morning glory
Climbing Mechanism: Stem Twiner
Distribution (Global): Assam, Bangladesh, India
Distribution (India): Andhra Pradesh, Maharashtra, Tamil Nadu, Telangana
Leaf Type: Leaves are simple, broadly ovate, base heart shaped, tip pointed
Inflorescence: Pink funnel-shaped flowers 2 cm long, borne on axillary or subterminal panicles of cymes arise in leaf-axils
Fruit Type: A subglobose capsule
Flowering and Fruiting: January–April
IUCN Status: Not evaluated
Notes: Usually found in deciduous forests
Reference: Syst. Veg., ed. 15 bis 4: 249 (1819)

Ipomoea stocksii C.B. Clarke
Convolvulaceae
Synonyms: *Ipomoea diversifolia* R.Br., *Convolvulus diversifolius* (R.Br.) Spreng., *Ipomoea laciniata* (Dalzell) C.B. Clarke, *Pharbitis laciniata* Dalzell
Climbing Mechanism: Stem Twiner

Distribution (Global): India, Northern Territory, Philippines, Queensland, Western Australia
Distribution (India): Maharashtra, Rajasthan
Leaf Type: Leaves orbicular in outline, digitate, with 5 coarsely and irregularly dentate to pinnatifid segments
Inflorescence: Inflorescences axillary, mostly 1-, sometimes to 3-flowered
Fruit Type: Capsule subglobose, glabrous
IUCN Status: Not evaluated
Notes: Usually found in low altitude grasslands
Reference: Prodr. Fl. Nov. Holland.: 487 (1810)

Ipomoea tiliacea (Willd.) Choisy
Convolvulaceae
Synonyms: *Convolvulus acuminatus* Salzm. ex Meisn., *Convolvulus biflorus* Sessé & Moc., *Convolvulus essequebensis* Spreng., *Convolvulus fastigiatus* Willd.
Climbing Mechanism: Stem Twiner
Distribution (Global): Bahamas, Belize, Bolivia, Brazil North, Brazil Northeast, Brazil South, Brazil Southeast, Cayman Is., Colombia, Costa Rica, Cuba, Dominican Republic, Ecuador, El Salvador, French Guiana, Guatemala, Guyana, Haiti, Honduras, Jamaica, Leeward Is., Mexico Central, Mexico Gulf, Mexico Northeast, Mexico Southeast, Mexico Southwest, Nicaragua, Panamá, Peru, Puerto Rico, Suriname, Trinidad-Tobago, Venezuela, Windward Is.
Leaf Type: Simple, palmately 3–5 fid, 5 nerved at base
Inflorescence: Axillary, umbellate cymes/panicles
Fruit Type: Dry, capsule
IUCN Status: Least Concern
Reference: A.P.de Candolle, Prodr. 9: 375 (1845)

Ipomoea trichosperma Blume
Convolvulaceae
Synonyms: *Ipomoea aspera* (Choisy) Vatke., *Calonyction asperum* Choisy, *Convolvulus asper* Wall.
Climbing Mechanism: Stem Twiner
Distribution (Global): Andaman Is., Bangladesh, Cambodia, Laos, Myanmar, Thailand, Vietnam
Distribution (India): Maharashtra
Leaf Type: Leaves
ovate-cordate, cordate
at base, caudate-acuminate at apex, rarely entire
Inflorescence: Flowers nocturnal, solitary-axillary
Fruit Type: Fruits not seen
Flowering and Fruiting: October–November
IUCN Status: Not evaluated
Reference: Linnaea 43: 508 (1882)

Ipomoea tricolor Cav.
Convolvulaceae
Synonyms: *Convolvulus rubrocaeruleus* (Hook.) D.Dietr., *Convolvulus venustus* Spreng., *Ipomoea hookeri* G. Don, *Ipomoea rubrocaerulea* Hook., *Ipomoea schiedeana* Ham., *Ipomoea violacea* Lunan, *Pharbitis rubrocaerulea* (Hook.) Planch., *Pharbitis tricolor* (Cav.) Chitt
Climbing Mechanism: Stem Twiner
Distribution (Global): Mexico Central, Mexico Gulf, Mexico Northeast, Mexico Southeast, Mexico Southwest
Distribution (India): Madhya Pradesh, Rajasthan
Leaf Type: Leaves petiolate, alternate, ovate, cordate with rather angular, nearly rounded auricles
Inflorescence: Inflorescence of pedunculate, few-flowered axillary cymes
Fruit Type: Capsule, ovoid, glabrous, rostrate
Flowering and Fruiting: October–February
IUCN Status: Not evaluated
Notes: Widely cultivated in temperate and subtropical regions
Reference: Icon. 3: 5 (1795)

Ipomoea triloba L.
Convolvulaceae
Synonyms: *Amphione lobata* Raf., *Convolvulus heterophyllus* Sessé & Moc., *Convolvulus ipomoea* Vell., *Convolvulus mariannensis* (Choisy) Gaudich. ex Saff., *Convolvulus sloanei* Spreng., *Convolvulus subquinquelobus* Wood, *Convolvulus trilobus* (L.) Desr.
Common Name: Little bell, Little bell morning glory, Pink convolvulus, Three-lobed morning glory
Climbing Mechanism: Stem Twiner
Distribution (Global): Aruba, Bahamas, Belize, Brazil North, Brazil Northeast, Brazil South, Brazil Southeast, Brazil West-Central, Cayman Is., Colombia, Costa Rica, Cuba, Dominican Republic, El Salvador, Galápagos, Guatemala, Haiti, Honduras, Jamaica, Leeward Is., Mexico Central, Mexico Gulf, Mexico Northeast, Mexico Northwest, Mexico Southeast, Mexico Southwest, Netherlands Antilles, Panamá, Puerto Rico, Trinidad-Tobago, Turks-Caicos Is., Venezuela, Windward Is.
Distribution (India): Gujarat, Maharashtra, Rajasthan, Tamil Nadu, Telangana
Leaf Type: Leaves are heart shaped, fine-tipped, entire or 3-lobed
Inflorescence: Flowers aggregated in umbellate cymes
Fruit Type: Capsule subglobose, densely hispid
Flowering and Fruiting: September–March
IUCN Status: Least Concern
Notes: Often found in degraded forest areas
Reference: Sp. Pl.: 161 (1753)

Ipomoea tropica Santapau & Patel
Convolvulaceae
Synonyms: *Ipomoea biflora* (L.) Pers. *Aniseia biflora* (L.) Choisy, *Aniseia calycina* Choisy, *Convolvulus bifidus* Christm., *Convolvulus biflorus* L., *Convolvulus calycinus* Roxb., *Convolvulus forskalii* Spreng. +24
Common Name: Chinese morning glory
Climbing Mechanism: Stem Twiner
Distribution (Global): Angola, Botswana, Burundi, Cape Provinces, Chad, China South-Central, China Southeast, East Himalaya, Egypt, Eritrea, Ethiopia, India, Java, Kenya, KwaZulu-Natal, Laos, Lesser Sunda Is., Malawi, Mauritania, Mozambique, Myanmar, Namibia, Nansei-shoto, New South Wales, Niger, Northern Provinces, Northern Territory, Oman, Philippines, Queensland, Rwanda, Saudi Arabia, Socotra, Somalia, South Australia, Sudan, Sulawesi, Swaziland, Taiwan, Tanzania, Thailand, Uganda, Vietnam, West Himalaya, Western Australia, Yemen, Zambia, Zimbabwe
Distribution (India): Peninsular India
Leaf Type: simple, petiolate, lamina entire, triangular to ovate
Inflorescence: Cymose, 1–3 flowered
Fruit Type: Capsules
Flowering and Fruiting: September–December
IUCN Status: Not evaluated
Reference: Syn. Pl. 1: 183 (1805)

Ipomoea tuberculata Ker Gawl.
Convolvulaceae
Synonyms: *Ipomoea cairica* (L.) Sweet., *Batatas cavanillesii* (Schult.) G. Don, *Batatas pulchella* (Roth) Bojer, *Batatas senegalensis* (Lam.) G. Don, *Convolvulus bellus* Spreng., *Convolvulus cairicus* L., *Convolvulus cavanillesii* (Schult.) Spreng.
Climbing Mechanism: Stem Twiner
Distribution (Global): Pantropical
Distribution (India): Madhya Pradesh, Tamil Nadu
Leaf Type: The alternately arranged leaves are divided into five or seven narrow lobes
Inflorescence: Flowers purple, pink or rarely pinkish white, solitary or in groups of 2–3
Fruit Type: 4-valved globular capsule
Flowering and Fruiting: Throughout the year
IUCN Status: Not evaluated
Notes: Mostly found on roadsides, grassy fields, wastelands, dry slopes, and railway tracks
Reference: Hort. Brit.: 287 (1826)

Ipomoea turbinata Lag. ex Choisy
Convolvulaceae

Synonyms: *Ipomoea muricata* (L.) Jacq. *Bonanox muricata* (L.) Raf., *Calonyction longiflorum* Hassk., *Calonyction muricatum* (L.) G. Don, *Calonyction pseudomuricatum* (Bernh. ex Link) G. Don, *Convolvulus colubrinus* Blanco, *Convolvulus muricatus* L.
Common Name: Purple moonflower, Clove bean, Lavender moonvine, Lavender moonflower
Climbing Mechanism: Stem Twiner
Distribution (Global): Pantropical
Distribution (India): Andhra Pradesh, Himachal Pradesh, Madhya Pradesh, Maharashtra, Odisha, Rajasthan, Tamil Nadu, Telangana
Leaf Type: Leaves are simple, smooth, soft and entire, the base is heart shaped, lobed
Inflorescence: Inflorescence of 1–2 (–5)-flowered, pedunculate cymes
Fruit Type: Capsule ovoid, glabrous, rostrate, the persistent style
Flowering and Fruiting: April–May
IUCN Status: Not evaluated
Reference: Pl. Hort. Schoenbr. 3: 40 (1798)

Ipomoea violacea L.
Convolvulaceae
Synonyms: Synonym
Climbing Mechanism: Stem Twiner
Distribution (India): Delhi, Maharashtra, Mizoram, Odisha, Tamil Nadu
Leaf Type: Simple, palmately 3–5 fid, 5 nerved at base
Inflorescence: Axillary, umbellate cymes/panicles
Fruit Type: Dry, capsule
IUCN Status: Not evaluated

Ipomoea wightii (Wall.) Choisy
Convolvulaceae
Synonyms: *Convolvulus wightii* Wall.
Climbing Mechanism: Stem Twiner
Distribution (Global): Bangladesh, Burundi, Cape Provinces, Central African Repu, Ethiopia, India, Kenya, KwaZulu-Natal, Madagascar, Malawi, Mozambique, Northern Provinces, Rwanda, Somalia, Sri Lanka, Sudan, Tanzania, Uganda, Zambia, Zaire, Zimbabwe
Distribution (India): Andhra Pradesh, Odisha, Rajasthan, Tamil Nadu, Telangana
Leaf Type: Leaves, orbicular in outline, cordate at base, shallowly 3-lobed
Inflorescence: Very rarely solitary
Fruit Type: Capsule globose, slightly to densely bristly above or with faint white tomentum
Flowering and Fruiting: December–January
IUCN Status: Least Concern
Notes: Often occur in grasslands and moist deciduous forests near streams
Reference: Mém. Soc. Phys. Genève 6: 470 (1833 publ. 1834)

Jacquemontia ovalifolia (Choisy) Hallier f.
Convolvulaceae
Synonyms: *Convolvulus coeruleus* Schumach. & Thonn., *Convolvulus ovalifolius* Vahl, *Convolvulus ovalifolius* Vahl ex H. West, Convolvulus quinquepartitus Vahl, *Ipomoea oleracea* Welw., *Ipomoea ovalifolia* Choisy, *Ipomoea quinquepartita* (Vahl) Roem. & Schult.
Climbing Mechanism: Stem Twiner
Distribution (Global): Angola, Ethiopia, Ghana, Guinea, Kenya, Madagascar, Mozambique, Somalia, Tanzania, Togo, Uganda, Zaire
Distribution (India): Tamil Nadu
Leaf Type: Simple, cordiform
Inflorescence: Axillary cymes, corymbose
Fruit Type: Dry, capsule, subglobose
Flowering and Fruiting: December–July
IUCN Status: Not evaluated
Reference: Bot. Jahrb. Syst. 16: 543 (1893)

Jacquemontia paniculata (Burm.f.) Hallier f.
Convolvulaceae
Synonyms: *Convolvulus paniculatus* (Burm.f.) Kuntze, *Ipomoea paniculata* Burm.f.
Common Name: Mauve clustervine
Climbing Mechanism: Stem Twiner
Distribution (Global): Assam, Bangladesh, Borneo, Cambodia, China South-Central, China Southeast, Christmas I., Comoros, Hainan, India, Java, Kenya, Laos, Lesser Sunda Is., Madagascar, Malaya, Maldives, Maluku, Mozambique, Myanmar, New Caledonia, New Guinea, Nicobar Is., Northern Territory, Philippines, Queensland, Solomon Is., South China Sea, Sri Lanka, Sulawesi, Sumatera, Taiwan, Tanzania, Thailand, Vanuatu, Vietnam, Western Australia
Distribution (India): Andaman and Nicobar Islands, Andhra Pradesh, Gujarat, Madhya Pradesh, Maharashtra, Odisha, Puducherry, Rajasthan, Tamil Nadu, Telangana
Leaf Type: eaves ovate, villous or glabrous adaxially, base shallowly cordate or rounded
Inflorescence: Inflorescences umbelliform-cymose
Fruit Type: Capsule globose, straw-colored
Flowering and Fruiting: October–December
IUCN Status: Not evaluated
Reference: Bot. Jahrb. Syst. 18: 95 (1893)

Jacquemontia pentantha G. Don
Convolvulaceae
Climbing Mechanism: Stem Twiner
Distribution (India): Delhi, Gujarat, Maharashtra, Rajasthan, Tamil Nadu
Leaf Type: Simple, cordiform
Inflorescence: Axillary cymes, corymbose

Fruit Type: Dry, capsule, subglobose
IUCN Status: Not evaluated

Lepistemon binectariferum (Wall.) Kuntze
Convolvulaceae
Synonyms: *Polemonium obscurum* Blanco.
Climbing Mechanism: Stem Twiner
Distribution (Global): Philippines (Luzon, Mindanao), and Hainan
Leaf Type: leaf blade cordate-ovate, 5-18 X 5-15 cm, base deeply cordate, margin entire, angulate, or shallowly 3–7 lobed
Inflorescence: Cymes axillary, few to many flowered; peduncle short or absent. Pedicel ca. 7 mm, glabrous or pilose
Fruit Type: Capsule globose to ovoid, 6–8 mm, apiculate
Flowering and Fruiting: December–January
IUCN Status: Not evaluated

Lepistemon leiocalyx Stapf
Convolvulaceae
Synonyms: *Ipomoea leiocalyx* (Stapf) J.R.I. Wood & Scotland
Climbing Mechanism: Stem Twiner
Distribution (Global): Bangladesh, India
Distribution (India): Tamil Nadu
Leaf Type: Leaves alternate, broadly ovate, often 3 lobed, 8–10 by 4–7 cm, base deeply cordate, apex gradually acuminate
Inflorescence: Flowers in dense axillary cymes, bisexual, 5-merous. Calyx-lobes 5, subequal
Fruit Type: Capsule 4-valved, seeds glabrous
Flowering and Fruiting: January–March
IUCN Status: Not evaluated
Reference: Bull. Misc. Inform. Kew 1895: 113 (1895)

Lepistemon verdcourtii P. Mathew & S.D. Biju
Convolvulaceae
Climbing Mechanism: Stem Twiner
IUCN Status: Not evaluated

Lettsomia adpressa Miq.
Convolvulaceae
Climbing Mechanism: Stem Twiner
IUCN Status: Not evaluated

Lettsomia barbata C.B. Clarke
Convolvulaceae
Synonyms: *Argyreia barbata* (Wall.) Raizada. *Convolvulus barbatus* Wall., *Ipomoea barbata* (Wall.) Choisy, *Mouroucoa barbata* (Wall.) Kuntze

Climbing Mechanism: Stem Twiner
Distribution (Global): Bangladesh, Myanmar
IUCN Status: Not evaluated
Reference: Indian Forester 93: 754 (1967)

Lettsomia bella C.B. Clarke
Convolvulaceae
Synonyms: *Argyreia bella* (C.B. Clarke) Raizada., *Ipomoea euantha* J.R.I. Wood & Scotland, *Mouroucoa bella* (C.B. Clarke) Kuntze
Climbing Mechanism: Stem Twiner
Distribution (Global): India, Myanmar, Nepal
Distribution (India): Bihar, Odisha
IUCN Status: Not evaluated
Reference: Indian Forester 93: 754 (1967)

Lettsomia bracteosa C.B. Clarke
Convolvulaceae
Synonyms: *Argyreia bracteosa* (C.B. Clarke) Raizada., *Mouroucoa bracteosa* (C.B. Clarke) Kuntze
Climbing Mechanism: Stem Twiner
Distribution (Global): Myanmar
IUCN Status: Not evaluated
Reference: Indian Forester 93: 754 (1967)

Lettsomia elliptica Wight
Convolvulaceae
Synonyms: *Argyreia elliptica* (Roth) Choisy. *Argyreia laurifolia* (Roxb.) Voigt, *Argyreia strigosa* subsp. *obovata* (C.B. Clarke) Panigrahi & Murti, *Convolvulus ellipticus* (Roth) Spreng., *Convolvulus laurifolius* Roxb., *Ipomoea elliptica* Roth.
Common Name: Oval leaved silverweed
Climbing Mechanism: Stem Twiner
Distribution (Global): India, Nepal, Sri Lanka
Leaf Type: Leaves are elliptic-ovate to obovate, 5–9 cm long, 3–5 cm broad, smooth above, velvety below
Inflorescence: Flowers are pink-rose coloured and are borne in lax flat-topped clusters
Fruit Type: Fruit is an orange colored berry
Flowering and Fruiting: October–February
IUCN Status: Not evaluated
Notes: Mostly found along the margins of shola forests
Reference: Mém. Soc. Phys. Genève 6: 417 (1833 publ. 1834)

Lettsomia hancorniifolia C.B. Clarke
Convolvulaceae

Synonyms: *Argyreia hancorniifolia* Gardner ex Thwaites., *Ipomoea hancorniifolia* (Gardner ex Thwaites) J.R.I. Wood & Scotland, *Mouroucoa hancorniifolia* (Gardner ex Thwaites) Kuntze
Climbing Mechanism: Stem Twiner
Distribution (Global): Sri Lanka
IUCN Status: Not evaluated
Reference: Enum. Pl. Zeyl.: 210 (1860)

Lettsomia kurzii C.B. Clarke
Convolvulaceae
Synonyms: *Argyreia kurzii* (C.B. Clarke) Boerl., *Argyreia zeylanica* Kurz, *Ipomoea kurzii* (C.B. Clarke) J.R.I. Wood & Scotland, *Mouroucoa kurzii* (C.B. Clarke) Kuntze
Climbing Mechanism: Stem Twiner
Distribution (Global): Myanmar
IUCN Status: Not evaluated
Reference: Handl. Fl. Ned. Ind. 2: 513 (1899)

Lettsomia maingayi C.B. Clarke
Convolvulaceae
Synonyms: *Argyreia maingayi* (C.B. Clarke) Hoogland., *Ipomoea maingayi* (C.B. Clarke) J.R.I. Wood & Scotland, *Mouroucoa maingayi* (C.B. Clarke) Kuntze
Climbing Mechanism: Stem Twiner
Distribution (Global): Malaya, Thailand
Leaf Type: Leaves ovate, elliptic-oblong, or obovate, base rounded or shortly attenuate, lateral veins 5 or 6 pairs
Inflorescence: Inflorescence densely capitate
Fruit Type: Berry ovoid, pink
IUCN Status: Not Evaluated
Reference: Blumea 7: 185 (1952)

Lettsomia mysorensis C.B. Clarke
Convolvulaceae
Synonyms: *Argyreia osyrensis* (Roth) Choisy., *Argyreia aggregata* (Roxb.) Arn. ex Choisy, *Argyreia brachypoda* (Kerr) Ooststr., *Argyreia imbricata* (Roth) Santapau & Patel, *Convolvulus aggregatus* (Roxb.) Steud. + 10
Common Name: Shingled silverweed
Climbing Mechanism: Stem Twiner
Distribution (Global): Cambodia, Hainan, India, Laos, Myanmar, Sri Lanka, Sumatera, Thailand, Vietnam
Leaf Type: Leaves are ovate or broadly ovate to nearly circular
Inflorescence: Flowers are pink, tubular-bell shaped are borne in head-like cluster, cymes densely white tomentoses
Fruit Type: Berry is enclosed by sepals, red, round

Flowering and Fruiting: December–May.
IUCN Status: Not evaluated
Notes: Most found along the margins of grasslands
Reference: A.P.de Candolle, Prodr. 9: 334 (1845)

Lettsomia penangiana Miq.
Convolvulaceae
Climbing Mechanism: Stem Twiner
IUCN Status: Not evaluated

Lettsomia rubens C.B. Clarke
Convolvulaceae
Synonyms: *Ipomoea rubens* Choisy., *Convolvulus lilacinus* D. Dietr. +20
Climbing Mechanism: Stem Twiner
Distribution (Global): Angola, Assam, Bangladesh, Benin, Borneo, Botswana, Burundi, Cameroon, Central African Repu, Chad, Comoros, Congo, Ethiopia, Ghana, Guinea, Gulf of Guinea Is., India, Ivory Coast, Java, Kenya, Lesser Sunda Is., Madagascar, Malawi, Mozambique, Namibia, Niger, Nigeria, Philippines, Rwanda, Senegal, Somalia, Sudan, Sumatera, Tanzania, Thailand, Uganda, Vietnam, Zambia, Zaire, Zimbabwe
Leaf Type: Leaves simple, alternate, broadly ovate, deeply cordate with rounded auriclesat base
Inflorescence: Inflorescence axillary formed of 1–13-fowered cymes, cymes subumbellate, Flowers pink with dark at center
Flowering and Fruiting: December–March
IUCN Status: Not evaluated
Notes: Occasionally found in fresh water swamps near the coast
Reference: Mém. Soc. Phys. Genève 6: 463 (1833 publ. 1834)

Lettsomia rubicunda C.B. Clarke
Convolvulaceae
Synonyms: *Argyreia rubicunda* Choisy., *Lettsomia laxiflora* Prain, *Lettsomia rubicunda* (Choisy) C.B. Clarke, *Mouroucoa rubicunda* (Choisy) Kuntze
Climbing Mechanism: Stem Twiner
Distribution (Global): Malaya, Myanmar
Leaf Type: Leaves elliptic or sometimes ovate-elliptic
Inflorescence: Inflorescences axillary, cymosely branched at apex, several- to many-flowered, cyme 6–12 cm diam
Fruit Type: Fruit ovoid, pinkish white or purplish red, base enclosed by cupular to shortly funnelform calyx
IUCN Status: Not evaluated
Reference: Mém. Soc. Phys. Genève 6: 426 (1833 publ. 1834)

Lettsomia setosa Roxb.
Convolvulaceae

Synonyms: *Argyreia setosa* (Roxb.) Sweet., *Ipomoea bac*cata J.R.I. Wood & Scotland, *Lettsomia setosa* Roxb., *Mouroucoa setosa* (Roxb.) Kuntze
Climbing Mechanism: Stem Twiner
Distribution (Global): Assam, India, Myanmar, Nepal, Pakistan, West Himalaya
Distribution (India): Bihar, Odisha
Leaf Type: Leaves alternate, broadly ovate to orbicular, acute or acuminate, base cordate, sparsely hirsute
Inflorescence: Flowers in axillary corybiform panicles; bracts densely strigose
Fruit Type: Berry red, fruiting calyx brown
Flowering and Fruiting: July–December
IUCN Status: Not evaluated
Notes: Distributed mostly along the evergreen forests of the Western Ghats
Reference: Hort. Brit., ed. 2: 373 (1830)

Lettsomia sikkimensis C.B. Clarke
Convolvulaceae
Synonyms: *Argyreia sikkimensis* (C.B. Clarke) Ooststr., *Ipomoea sikkimensis* (C.B. Clarke) J.R.I.Wood & Scotland, *Mouroucoa sikkimensis* (C.B.Clarke) Kuntze
Climbing Mechanism: Stem Twiner
Distribution (Global): Assam, Bangladesh, East Himalaya, Myanmar, Thailand
Leaf Type: Leaves large elliptic shortly acuminate at both ends or subobtuse sparsely hairy beneath
Inflorescence: Peduncles long, corymbs small few-fld, bracts linear-oblong, sepals 1/2 in.
IUCN Status: Not evaluated
Reference: Blumea 7: 178 (1952)

Lettsomia thomsonii C.B. Clarke
Convolvulaceae
Synonyms: *Argyreia thomsonii* (C.B. Clarke) Babu. *Argyreia capitata* Brandis, *Argyreia nasirii* D.F. Austin, *Ipomoea thomsonii* (C.B. Clarke) J.R.I. Wood & Scotland, *Mouroucoa thomsonii* (C.B. Clarke) Kuntze
Climbing Mechanism: Stem Twiner
Distribution (Global): East Himalaya, Nepal, Pakistan, West Himalaya
Distribution (India): Bihar, Odisha
IUCN Status: Not evaluated
Reference: Herb. Fl. Dehra Dun: 320, 661 (1977)

Merremia aegyptia (L.) Urb.
Convolvulaceae
Synonyms: *Batatas pentaphylla* L., *Convolvulus aegyptius* (L.) L., *Convolvulus aphyllus* Viv., *Convolvulus coriaceus* Choisy, *Convolvulus cujanensis* Bowdich, *Convolvulus hirsutus* Roxb. + 20
Common Name: Hairy woodrose, Hairy Merremia

Climbing Mechanism: Stem Twiner

Distribution (Global): Angola, Argentina Northeast, Argentina Northwest, Aruba, Bahamas, Belize, Benin, Bolivia, Brazil North, Brazil Northeast, Brazil South, Brazil Southeast, Brazil West-Central, Cameroon, Chad, Colombia, Costa Rica, Cuba, Dominican Republic, Ecuador, Equatorial Guinea, Eritrea, Ethiopia, French Guiana, Galápagos, Gambia, Ghana, Guatemala, Guinea, Guinea-Bissau, Guyana, Haiti, Jamaica, Leeward Is., Mali, Mexico Gulf, Mexico Northwest, Mexico Southeast, Mexico Southwest, Mozambique, Netherlands Antilles, Nicaragua, Niger, Nigeria, Panamá, Paraguay, Peru, Puerto Rico, Senegal, Sierra Leone, Somalia, Sudan, Suriname, Togo, Trinidad-Tobago, Venezuela, Venezuelan Antilles, Windward Is., Zambia, Zaire, Zimbabwe

Distribution (India): Andhra Pradesh, Gujarat, Madhya Pradesh, Maharashtra, Odisha, Puducherry, Rajasthan, Tamil Nadu, Telangana

Leaf Type: Leaves are alternate, digitately 5-foliate

Inflorescence: Flowers are borne in few flowered, lax, hairy, raceme-like cymes

Fruit Type: Capsules are 8 mm long, ovoid, papery, covered by silky, silky sepals, 4 celled, 4 valved

Flowering and Fruiting: December–March

IUCN Status: Not evaluated

Reference: Bot. J. Linn. Soc. 183: 573 (2017)

Merremia cissoides (Lam.) Hallier f.

Convolvulaceae

Climbing Mechanism: Stem Twiner

IUCN Status: Not evaluated

Merremia crispatula Prain

Convolvulaceae

Synonyms: *Operculina petaloidea* (Choisy) Ooststr., *Convolvulus crispatulus* Wall., *Ipomoea petaloidea* Choisy, *Ipomoea xanthantha* Kurz, *Merremia petaloidea* (Choisy) Burkill, *Operculina petaloidea* var. *pauciflor*a (C.B. Clarke) Parmar

Climbing Mechanism: Stem Twiner

Distribution (Global): Bangladesh, Cambodia, India, Myanmar, Thailand, Vietnam, West Himalaya

Leaf Type: Leaves ovate, broadly elliptic, to oblong-linear, 5.8–11.3 by 2.7–5.8 cm, pubescent when young, later glabrate, base rounded to truncate, apex obtuse or acute, mucronulate; lateral veins 7–9 per side; petiole 1.5–2.6 cm, ridged to narrowly winged

Inflorescence: Inflorescence axillary or terminal on branches, cymose, 1–9-flowered; peduncles terete, 0.9–2 cm; bracts linear-lanceolate, 7–20 mm, deciduous; pedicels clavate, striate-angulate, 10–18 mm

Fruit Type: Capsule c. 15 mm diam., tan, cupped by accrescent calyx

Flowering and Fruiting: February–April

IUCN Status: Not evaluated

Notes: Rarely found in exposed areas in evergreen

forests, along stream banks, and roadsides
Reference: Blumea 3: 369 (1939)

Merremia dissecta (Jacq.) Hallier f.
Convolvulaceae
Synonyms: *Distimake dissectus* (Jacq.) A.R. Simões & Staples., *Convolvulus dissectus* Jacq., *Convolvulus palmatus* Mill., *Convolvulus variabilis* Weinm. ex Steud., *Distimake dissectus* var. *chacoensis* (O'Donell) Petrongari & Sim. -Bianch.
Common Name: Alamo vine, Noyau vine
Climbing Mechanism: Stem Twiner
Distribution (Global): Pantropical.
Distribution (India): Andhra Pradesh, Gujarat, Maharashtra, Rajasthan, Tamil Nadu, Telangana, Uttar Pradesh
Leaf Type: The leaves are alternate, spiral, simple, with very hairy stalks 2–7 cm long
Fruit Type: The fruit is a dehiscent 5-valved capsule, globular, papery, smooth and non-fles
IUCN Status: Not evaluated
Notes: Usually grows in open forest but also in monsoon forest or disturbed areas or on the margins of rain forest
Reference: Bot. J. Linn. Soc. 183: 574 (2017)

Merremia gangetica Cufod.
Convolvulaceae
Synonyms: *Merremia gangetica*, *Convolvulus reniformis*, *Ipomoea reniformis*
Climbing Mechanism: Stem Twiner
Distribution (Global): Pan tropical species found in Southeast Asia, India, Nepal, Sri Lanka, Australia, and tropical Africa
Distribution (India): Telangana
Leaf Type: The leaves are small, kidney shaped to somewhat heart shaped
Inflorescence: Solitary or 2–3 flowered axillary cymes
Fruit Type: The capsule is rounded and about 5 mm in diameter
Flowering and Fruiting: October–March
IUCN Status: Least Concern
Notes: Frequently found on the banks of canals and pools. Also found in wastelands, waterlogged areas, roadsides, and grasslands
Reference: Bull. Jard. Bot. Natl. Belg. 39(Suppl.): XXX (1969)

Merremia gemella (Burm.f.) Hallier f.
Convolvulaceae
Synonyms: *Convolvulus gemellus* Burm.f., *Ipomoea gemella* (Burm.f.) Roth.
Climbing Mechanism: Stem Twiner

Distribution (Global): Bangladesh, East Himalaya, India, Java, Laos, Lesser Sunda Is., Malaya, Maluku, Myanmar, New Guinea, Northern Territory, Philippines, Queensland, Sri Lanka, Taiwan, Thailand, Vietnam, Western Australia
Distribution (India): Maharashtra, Tamil Nadu
Leaf Type: Glabrous or short pilose, base broadly cordate, proximal margin entire, undulate or coarsely crenate
Inflorescence: Inflorescences umbelliform or forked with short raceme-like branches, few flowered; peduncle 2.5–10 cm; bracts early deciduous, minute. Pedicel 3–6 mm
Fruit Type: Capsule depressed-globose, ca. 7 mm, coarsely wrinkled. Fruits globular, about 5–8 mm diam., sepals persistent, reflexed, not enclosing the fruit
Flowering and Fruiting: Throughtout the year
IUCN Status: Not evaluated
Notes: Usually found associated with thickets
Reference: Bot. Jahrb. Syst. 16: 552 (1893)

Merremia hederacea (Burm.f.) Hallier f.
Convolvulaceae
Synonyms: *Convolvulus flavus* Willd., *Evolvulus hederaceus* Burm.f.
Climbing Mechanism: Stem Twiner
Distribution (Global): Assam, Bangladesh, Benin, Borneo, Cambodia, Cameroon, Caroline Is., Central African Repu, Chad, China South-Central, China Southeast, Christmas I., Congo, East Himalaya, Ethiopia, Gabon, Ghana, Guinea, Guinea-Bissau, Hainan, India, Ivory Coast, Java, Kenya, Laos, Lesser Sunda Is., Liberia, Madagascar, Malaya, Mali, Marianas, Mauritius, Mozambique, Myanmar, Nansei-shoto, Nepal, New Guinea, Niger, Nigeria, Northern Territory, Ogasawara-shoto, Pakistan, Philippines, Queensland, Rwanda, Réunion, Senegal, Sierra Leone, Somalia, Sri Lanka, Sudan, Sumatera, Taiwan, Tanzania, Thailand, Togo, Vietnam, West Himalaya, Western Australia, Zambia, Zaire, Zimbabwe
Distribution (India): Andhra Pradesh, Assam, Gujarat, Himachal Pradesh, Kerala, Madhya Pradesh, Maharashtra, Odisha, Rajasthan, Tamil Nadu, Telangana, Tripura, Uttar Pradesh
Leaf Type: Leaves Simple trilobed, alternate distichous 2–4 x 1.5–3 cm, ovate, rarely 3-lobed, base cordate, apex acuminate-apiculate, glabrous or sparsely pubescent
Inflorescence: Solitary cymes many-flowered, axillary; peduncle 2–4 cm long
Fruit Type: Capsule c. 5 mm long, obconical, slightly 4-lobed. Seeds black, glabrous or puberulent
Flowering and Fruiting: October–January
IUCN Status: Not evaluated
Notes: Mostly found in degraded deciduous forests
Reference: Bot. Jahrb. Syst. 18: 118 (1893)

Merremia hirta (L.) Merr.
Convolvulaceae
Synonyms: *Convolvulus hirtus* L.
Climbing Mechanism: Stem Twiner
Distribution (Global): Assam, Bangladesh, Borneo, Cambodia, China South-Central, China Southeast, East Himalaya, India, Java, Laos, Lesser Sunda Is., Malaya, Maluku, Myanmar, New Guinea, Philippines, Queensland, Solomon Is., Sulawesi, Sumatera, Taiwan, Thailand, Vietnam
Distribution (India): Odisha
Leaf Type: Leaves alternate, simple, highly variable
Inflorescence: Flowers axillary, solitary to few flowered cymes
Fruit Type: Fruits capsular, globular to broad ovoid
Flowering and Fruiting: August–December
IUCN Status: Not evaluated
Reference: Philipp. J. Sci., C 7: 244 (1912)

Merremia kentrocaulos Rendle
Convolvulaceae
Synonyms: *Distimake kentrocaulos* (C.B. Clarke) A.R. Simões & Staples., *Convolvulus formosus* Roxb. ex C.B. Clarke, *Convolvulus kentrocaulos* Steud. ex Choisy, *Ipomoea kentrocaulos* C.B. Clarke, *Ipomoea tuberosa* A. Rich., *Operculina kentrocaulos* (C.B. Clarke) Hallier f.
Common Name: Spanish arborvine, Wood rose
Climbing Mechanism: Stem Twiner
Distribution (Global): Angola, Botswana, Cameroon, Eritrea, Ethiopia, India, Mozambique, Nigeria, Northern Provinces, Senegal, Sudan, Uganda, Zimbabwe
IUCN Status: Not evaluated
Reference: Bot. J. Linn. Soc. 183: 574 (2017)

Merremia mammosa (Lour.) Hallier f.
Convolvulaceae
Synonyms: *Convolvulus chrysorrhizus* Schrad. ex Steud., *Convolvulus mammosus* Lour., *Ipomoea gomezii* C.B. Clarke, *Ipomoea mammosa* (Lour.) Choisy
Climbing Mechanism: Stem Twiner
Distribution (Global): Andaman Is., Assam, Bangladesh, India, Myanmar, Thailand, Vietnam
Distribution (India): Arunachal Pradesh
Leaf Type: Leaves simple, entire
Inflorescence: Inflorescences paniculate or corymbiform, the lowermost bract often foliaceous
Fruit Type: Fruits usually valvate capsules and rarely dry indehiscent berries
IUCN Status: Not evaluated
Reference: Bot. J. Linn. Soc. 183: 571 (2017)

Merremia peltata (L.) Merr.
Convolvulaceae
Synonyms: *Decalobanthus peltatus* (L.) A.R. Simões & Staples., *Convolvulus nymphaeifolius* D. Dietr., *Convolvulus peltatus* L., *Ipomoea menispermacea* Domin, *Ipomoea nymphaeifolia* Blume, *Ipomoea peltata* (L.) Choisy, *Merremia nymphaeifolia* (Blume) Hallier f.
Climbing Mechanism: Stem Twiner
Distribution (Global): Andaman Is., Borneo, Christmas I., Comoros, Cook Is., Fiji, Java, Lesser Sunda Is., Madagascar, Malaya, Maluku, Mauritius, New Guinea, Nicobar Is., Niue, Philippines, Queensland, Rodrigues, Réunion, Samoa, Society Is., Solomon Is., Sulawesi, Sumatera, Tanzania, Thailand, Tonga, Tubuai Is., Vanuatu, Wallis-Futuna Is.
Distribution (India): Arunachal Pradesh, Great Nicobar Island
Leaf Type: Leaves peltate; blade ovate-orbicular, orbicular or oblate, 7–30 cm. long and broad, ± rounded at the base, acuminate or cuspidate at the apex, glabrous or slightly hairy below
Inflorescence: Inflorescence up to 40 cm. long, corymbose, several- to many-flowered, up to 2 per leaf axil
IUCN Status: Not evaluated
Reference: Bot. J. Linn. Soc. 183: 571 (2017)

Merremia quinata (R.Br.) Ooststr.
Convolvulaceae
Synonyms: *Distimake quinatus* (R.Br.) A.R. Simões & Staples., *Convolvulus brownii* Spreng., *Convolvulus pentadactylis* Wall., *Convolvulus quinatus* (R. Br.) Spreng., *Ipomoea hirsuta* R.Br., *Ipomoea pentadactylis* Choisy, *Ipomoea quinata* R.Br.
Common Name: Five-leaf morning glory, Five-leaf bindweed
Climbing Mechanism: Stem Twiner
Distribution (Global): Cambodia, China South-Central, China Southeast, Hainan, India, Laos, Myanmar, New Guinea, Northern Territory, Philippines, Queensland, Taiwan, Thailand, Vietnam, Western Australia
Distribution (India): Odisha, Tamil Nadu
Leaf Type: Leaf blade digitately compound with 5 leaflets which are linear, lance shaped, or oblong-elliptic
Inflorescence: Inflorescences 1–3-flowered, on peduncles 2.5–7 cm long
Fruit Type: Capsules are ovoid
Flowering and Fruiting: June–October
IUCN Status: Not evaluated
Notes: Usually found in degraded moist deciduous forests
Reference: Bot. J. Linn. Soc. 183: 574 (2017)

Merremia quinquefolia (L.) Hallier f.
Convolvulaceae

Synonyms: *Batatas quinquefolia* (L.) Choisy, *Batatas quinquefolia* var. *breviflora* Choisy, *Convolvulus ampelopsifolius* Cham. & Schltdl., *Convolvulus domingensis* Spreng., *Convolvulus hispaniolae* Spreng., *Convolvulus quinquefolius* (L.) L. + 10

Common Name: Five-fingered morning glory, Rock rosemary, Snakevine

Climbing Mechanism: Stem Twiner

Distribution (Global): Bahamas, Belize, Brazil North, Colombia, Costa Rica, Cuba, Dominican Republic, Ecuador, El Salvador, Florida, Guatemala, Guyana, Haiti, Honduras, Jamaica, Leeward Is., Mexico Central, Mexico Gulf, Mexico Northeast, Mexico Northwest, Mexico Southeast, Mexico Southwest, Nicaragua, Panamá, Peru, Puerto Rico, Trinidad-Tobago, Venezuela, Windward Is.

Distribution (India): Andhra Pradesh, Gujarat, Maharashtra, Rajasthan, Tamil Nadu, Telangana

Leaf Type: The leaves, which are palmately compound with 5 leaflets

Inflorescence: Flowers are borne in leaf axils, either singly or in several-flowered clusters

Fruit Type: Fruit is about 9 mm long, capsular, round, 4-valved; seeds, about 4.5 mm long, blackish

IUCN Status: Not evaluated

Reference: Bot. J. Linn. Soc. 183: 575 (2017)

Merremia rhynchorrhiza Hallier f.

Convolvulaceae

Climbing Mechanism: Stem Twiner

Distribution (India): Gujarat, Maharashtra

Leaf Type: Leaves Deeply palmate (5–7 segments)

Inflorescence: Peduncles l–2-flowered, 5–10 cm long, filiform, glabrous

Fruit Type: Capsules oblong with Persistent sepals

Flowering and Fruiting: July–September

IUCN Status: Not evaluated

Notes: Usually found in the crevices of lateritic rocky exposures in high rainfall zones

Merremia sibirica (L.) Hallier f.

Convolvulaceae

Synonyms: *Convolvuloides elongata* Moench.

Convolvulus sibiricus L., *Ipomoea sibirica* (L.) Pers.

Climbing Mechanism: Stem Twiner

Distribution (Global): Buryatiya, China North-Central, China South-Central, China Southeast, Chita, Khabarovsk, Manchuria, Mongolia, Primorye, West Himalaya

Leaf Type: Leaf blade ovate-cordate, 3–13 X 1.7–7.5 cm, base cordate, margin entire or undulate, apex attenuate-acuminate or caudate

Inflorescence: Inflorescences (1-)3-7(-20)-flowered

Fruit Type: Capsule ± globose, broadly ovoid, or conical-ovoid

IUCN Status: Not evaluated

Notes: Common across roadsides, open and exposed areas
Reference: Bot. Jahrb. Syst. 16: 552 (1893)

Merremia tridentata
Convolvulaceae
Climbing Mechanism: Stem Twiner
IUCN Status: Not evaluated

Merremia tuberosa (L.) Rendle
Convolvulaceae
Synonyms: *Distimake tuberos*us (L.) A.R. Simões & Staples., *Batatas tuberosa* (L.) Bojer, *Convolvulus gossypifolius* Kunth. +10
Common Name: Spanish arborvine, Hawaiian woodrose, Spanish woodbine, Yellow morning glory
Climbing Mechanism: Stem Twiner
Distribution (Global): Bahamas, Belize, Brazil Northeast, Brazil Southeast, Brazil West-Central, Colombia, Costa Rica, Cuba, Dominican Republic, Ecuador, Guatemala, Guyana, Haiti, Honduras, Mexico Gulf, Mexico Southeast, Mexico Southwest, Nicaragua, Panamá, Venezuela
Distribution (India): Maharashtra, Telangana
Leaf Type: Leaves are 6–20 cm long and wider than long, usually palmately 7-parted
Inflorescence: Inflorescence a 2 to 7-flowered raceme about 7–13 cm long
Fruit Type: Capsule are 2 cm long, 3 cm wide approximately, pale brown, thin-walled, enclosed in calyx
IUCN Status: Not evaluated
Reference: Bot. J. Linn. Soc. 183: 577 (2017)

Merremia umbellata (L.) Hallier f.
Convolvulaceae
Synonyms: *Camonea umbellata* (L.) A.R. Simões & Staples., *Camonea bifida* (Vahl) Raf., *Convolvulus aristolochiifolius* Mill., *Convolvulus bifidus* Vahl, *Convolvulus blandus* Roxb., *Convolvulus caliginosus* J. Koenig ex Choisy. +40
Common Name: Hogvine, Yellow Merremia, Yellow wood rose
Climbing Mechanism: Stem Twiner
Distribution (Global): Argentina Northeast, Argentina Northwest, Belize, Bolivia, Brazil North, Brazil Northeast, Brazil South, Brazil Southeast, Brazil West-Central, Cameroon, Cayman Is., Colombia, Costa Rica, Cuba, Dominican Republic, Ecuador, El Salvador, Florida, French Guiana, Gambia, Ghana, Guatemala, Guinea, Guyana, Haiti, Honduras, Ivory Coast, Jamaica, Leeward Is., Liberia, Mali, Mexico Central, Mexico Gulf, Mexico Northeast, Mexico Northwest, Mexico Southeast, Mexico Southwest, Nicaragua, Nigeria, Panamá, Paraguay, Peru, Puerto Rico, Sierra Leone, Suriname, Trinidad-Tobago, Venezuela, Venezuelan Antilles, Windward Is.

Distribution (India): Andaman and Nicobar Islands, Andhra Pradesh, Assam, Goa, Kerala, Madhya Pradesh, Maharashtra, Mizoram, Odisha, Tamil Nadu, Tripura, West Bengal
Flowering and Fruiting: February–August
IUCN Status: Not evaluated
Reference: Bot. J. Linn. Soc. 183: 583 (2017)

Merremia umbellata subsp. *orientalis* (Hallier f.) Ooststr.
Convolvulaceae
Synonyms: *Camonea pilosa* (Houtt.) A.R. Simões & Staples. *Amphione cymosa* (Desr.) Raf., *Convolvulus cymosus* Desr., *Ipomoea cymosa* Blume, *Ipomoea cymosa* (Desr.) Roem. & Schult., *Ipomoea modesta* Choisy, *Ipomoea pilosa* Houtt., *Ipomoea sepiaria* Zoll. & Moritzi.
Common Name: Hogvine
Climbing Mechanism: Stem Twiner
Distribution (Global): Andaman Is., Bangladesh, Borneo, Cambodia, Caroline Is., China South-Central, China Southeast, India, Java, Laos, Lesser Sunda Is., Malaya, Maluku, Myanmar, Nepal, New Guinea, Northern Territory, Philippines, Queensland, Seychelles, Sri Lanka, Sulawesi, Sumatera, Taiwan, Tanzania, Thailand, Vietnam, Western Australia
Leaf Type: Leaf blades about 6.5–9.5 x 4–4.5 cm, petioles about 1.5–2.5 cm long
Inflorescence: Flowers borne in tight umbel-like cyme
Fruit Type: Capsules 4-valved
IUCN Status: Not evaluated
Reference: Bot. J. Linn. Soc. 183: 583 (2017)

Merremia vitifolia (Burm.f.) Hallier f.
Convolvulaceae
Synonyms: *Camonea vitifolia* (Burm.f.) A.R. Simões & Staples., *Convolvulus angularis* Burm.f., *Convolvulus vitifolius* Burm.f., *Ipomoea angularis* (Burm.f.) Choisy, *Ipomoea vitifolia* Blume, *Ipomoea vitifolia* (Burm.f.) Sweet
Common Name: Grape-leaf wood rose
Climbing Mechanism: Stem Twiner
Distribution (Global): Andaman Is., Assam, Bangladesh, Borneo, Cambodia, China South-Central, China Southeast, East Himalaya, Hainan, India, Java, Laos, Lesser Sunda Is., Malaya, Maluku, Myanmar, Nepal, Philippines, Sri Lanka, Sulawesi, Sumatera, Taiwan, Thailand, Vietnam
Distribution (India): Andhra Pradesh, Assam, Goa, Gujarat, Kerala, Madhya Pradesh, Maharashtra, Mizoram, Odisha, Rajasthan, Tamil Nadu, Tripura, West Bengal, Arunachal Pradesh
Leaf Type: Leaf blade is circular in outline, 5–18 by 5–16 cm, cordate at the base, palmately 5–7-lobed
Inflorescence: Inflorescence axillary, 1–3-flowered cymes
Fruit Type: Capsule 1–1.2 cm across, globose, fruiting sepals much larger. Seeds 4, 6–8 mm long ovoid

Flowering and Fruiting: November–February
IUCN Status: Not evaluated
Notes: Often found in degraded moist deciduous forests
Reference: Bot. J. Linn. Soc. 183: 583 (2017)

Neuropeltis malabarica Ooststr.
Convolvulaceae
Climbing Mechanism: Stem Twiner
Distribution (Global): India
Distribution (India): Maharashtra
Leaf Type: Alternate
Inflorescence: Raceme
Fruit Type: Capsule globose
Flowering and Fruiting: November–March
IUCN Status: Not evaluated
Notes: Predominantly found in evergreen forests
Reference: Blumea 5: 272 (1942)

Neuropeltis racemosa Wall.
Convolvulaceae
Synonyms: *Erycibe integripetala* Merr. & Chun, *Neuropeltis bracteata* Griff., *Neuropeltis integripetala* (Merr. & Chun) C.Y. Wu, *Neuropeltis intermedia* Griff., *Neuropeltis ovata* Wall., *Sinomerrillia bracteata* Hu.
Climbing Mechanism: Stem Twiner
Distribution (Global): Borneo, Cambodia, China South-Central, Hainan, Laos, Malaya, Myanmar, Sumatera, Thailand, Vietnam
Distribution (India): Peninsular India
IUCN Status: Not evaluated
Reference: W. Roxburgh, Fl. Ind. 2: 44 (1824)

Neuropeltis racemosa Wall.
Convolvulaceae
Climbing Mechanism: Stem Twiner
IUCN Status: Not evaluated

Operculina riedeliana Ooststr.
Convolvulaceae
Synonyms: *Convolvulus bufalinus* Lour., Convolvulus *platypeltis* Zipp. ex Span. + 7
Climbing Mechanism: Stem Twiner
Distribution (Global): Andaman Islands to N. & NE. Australia
Distribution (India): Andaman and Nicobar Islands
IUCN Status: Not evaluated
Reference: Blumea 3: 366 (1939)

Operculina tansaensis Santapau & V. Patel
Convolvulaceae
Synonyms: *Operculina ventricosa* (Bertero) Peter
Climbing Mechanism: Stem Twiner
Distribution (India): Maharashtra
IUCN Status: Not evaluated
Reference: Trans. Bose Res. Inst. Calcutta 22: 33 (1958)

Operculina turpethum (L.) Silva Manso
Convolvulaceae
Synonyms: *Argyreia alata* Montrouz., Argyreia *alulata* Miq. + 20
Common Name: Indian jalap, Terpeth root
Climbing Mechanism: Stem Twiner
Distribution (Global): Somalia to S. Tropical Africa, W. Indian Ocean, Tropical & Subtropical Asia to Pacific
Distribution (India): Andaman and Nicobar Islands, Andhra Pradesh, Assam, Delhi, Gujarat, Madhya Pradesh, Maharashtra, Odisha, Rajasthan, Tamil Nadu, Telangana, Tripura, Uttar Pradesh, Western Ghats
Flowering and Fruiting: October–March
IUCN Status: Not evaluated
Notes: Often occur in degraded forest areas
Reference: Enum. Subst. Braz.: 16 (1836)

Porana malabarica C.B. Clarke
Convolvulaceae
Synonyms: *Dinetus malabaricus* (C.B. Clarke) Staples
Climbing Mechanism: Stem Twiner
Distribution (India): Maharashtra
Leaf Type: Simple, cordate at base
Inflorescence: Terminal, paiculate
Fruit Type: Dry, indehiscent
IUCN Status: Not evaluated
Notes: Mostly found in deciduous forests
Reference: J. D. Hooker, Fl. Brit. India 4: 223 (1883)

Porana paniculata Roxb.
Convolvulaceae
Synonyms: *Poranopsis paniculata* (Roxb.) Roberty
Common Name: Bridal creeper
Climbing Mechanism: Stem Twiner
Distribution (India): Arunachal Pradesh, Assam, Bihar, Odisha, Delhi, Himachal Pradesh, Madhya Pradesh, Maharashtra, Odisha, Puducherry, Rajasthan, Tamil Nadu, Uttar Pradesh
Leaf Type: Simple, cordate at base
Inflorescence: Terminal, paiculate

Fruit Type: Dry, indehiscent
Flowering and Fruiting: September–January
IUCN Status: Not evaluated
Reference: Pl. Coromandel 3: 31 (1819)

Porana spectabilis Kurz
Convolvulaceae
Synonyms: *Tridynamia spectabilis* (Kurz) Parmar
Climbing Mechanism: Stem Twiner
Distribution (India): Andaman and Nicobar Islands, Mizoram
Leaf Type: Simple, cordate at base
Inflorescence: Terminal, paiculate
Fruit Type: Dry, indehiscent
IUCN Status: Not evaluated
Reference: J. Bot. 11: 136 (1873)

Porana volubilis Burm.f.
Convolvulaceae
Common Name: Horse tail creeper
Climbing Mechanism: Stem Twiner
Distribution (Global): Borneo, Cambodia, Java, Laos, Lesser Sunda Is., Maluku, Sulawesi, Sumatera, Thailand, Vietnam
Distribution (India): Maharashtra, Tamil Nadu
Leaf Type: Simple, cordate at base
Inflorescence: Terminal, paiculate
Fruit Type: Dry, indehiscent
IUCN Status: Not evaluated
Reference: Fl. Indica: 51 (1768)

Poranopsis paniculata (Roxb.) Roberty
Convolvulaceae
Synonyms: *Dinetus paniculatus* (Roxb.) Sweet, *Porana tomentosa* Lesch. ex Choisy, *Porana paniculata* Roxb.
Common Name: Bridal creeper
Climbing Mechanism: Stem Twiner
Distribution (Global): Assam, Bangladesh, East Himalaya, India, Myanmar, Nepal, Pakistan, Tibet, West Himalaya
Distribution (India): Mizoram, Tamil Nadu
IUCN Status: Not evaluated
Reference: Candollea 14: 26 (1952)

Quamoclit longiflora (Humb. & Bonpl. ex Willd.) G. Don
Convolvulaceae
Climbing Mechanism: Stem Twiner
Leaf Type: Simple, ovate, cordate at base

Inflorescence: Solitary or few in cymes
Fruit Type: Dry, capsule, subglobose
IUCN Status: Not evaluated
Reference: Gen. Hist. 4: 259 (1837)

Rivea hypocrateriformis Choisy
Convolvulaceae
Synonyms: *Argyreia bona-nox* (Roxb.) Sweet, *Argyreia uniflora* (Roxb.) Sweet, *Convolvulus candicans* Roem. & Schult., *Convolvulus hypocrateriformis* Desr., *Convolvulus hypoleucus* Roem. & Schult., *Ipomoea hypocrateriformis* (Desr.) J.R.I. Wood & Scotland
Common Name: Midnapore creeper
Climbing Mechanism: Stem Twiner
Distribution (Global): Assam, Bangladesh, India, Nepal, Pakistan
Leaf Type: Simple, cordiform
Inflorescence: Cymes
Fruit Type: Dry, capsule
Flowering and Fruiting: August–March
IUCN Status: Not evaluated
Notes: Found on hedges and dry-deciduous forests and scrub jungles
Reference: Mém. Soc. Phys. Genève 6: 408 (1833 publ. 1834)

Rivea ornata Choisy
Convolvulaceae
Synonyms: *Argyreia ornata* (Roxb.) Sweet, *Ipomoea ornata* (Roxb.) J.R.I. Wood & Scotland, *Lettsomia ornata* Roxb., *Rivea clarkeana* Craib, *Rivea laotica* Ooststr., *Rivea roxburghii* Prain ex Brandis
Climbing Mechanism: Stem Twiner
Distribution (Global): Assam, East Himalaya, India, Laos, Myanmar, Nepal, Thailand
Distribution (India): Andhra Pradesh, Arunachal Pradesh, Eastern Ghats, Gujarat, Madhya Pradesh, Maharashtra, Meghalaya, Rajasthan, Tamil Nadu
Leaf Type: Simple, cordiform
Inflorescence: Cymes
Fruit Type: Dry, capsule
Flowering and Fruiting: August–February
IUCN Status: Not evaluated
Reference: Mém. Soc. Phys. Genève 6: 409 (1833 publ. 1834)

Stictocardia beraviensis (Vatke) Hallier f.
Convolvulaceae
Synonyms: *Argyreia bagshawei* Rendle., *Ipomoea beraviensis* Vatke., *Argyreia beraviensis* (Vatke) Baker., *Ipomoea hierniana* Rendle
Common Name: Hawaiian sunset vine, Hawaiian bell, Braveheart vine
Climbing Mechanism: Stem Twiner

Distribution (Global): Botswana, Burundi, Cameroon, Central African Repu, Chad, Congo, Ethiopia, Ghana, Guinea, Guinea-Bissau, Kenya, Liberia, Madagascar, Mali, Nigeria, Senegal, Sierra Leone, Sudan, Tanzania, Uganda, Zambia, Zaire
Distribution (India): Tamil Nadu
Leaf Type: Ovate, heart shaped
Inflorescence: Axillary, many-flowered
Fruit Type: Capsule, globose
IUCN Status: Not evaluated
Notes: Found in grasslands and in riverine forests
Reference: Bot. Jahrb. Syst. 18: 159 (1893)

Stictocardia sivarajanii Biju, Pushpangadan & P. Mathew
Convolvulaceae
Synonyms: *Ipomoea sivarajanii* (Biju, Pushp. & P. Mathew) J.R.I. Wood & Scotland
Climbing Mechanism: Stem Twiner
Distribution (Global): India.
Distribution (India): Peninsular India
Leaf Type: Leaves simple, alternate
Inflorescence: Axillary, solitary or in 2–3 flowered cymes
Fruit Type: Capsular, indehiscent, depressed globose
Flowering and Fruiting: October–February
IUCN Status: Not evaluated
Notes: Usually occur in semi-evergreen forests
Reference: Novon 9: 147 (1999)

Stictocardia tiliifolia (Desr.) Hallier f.
Convolvulaceae
Synonyms: *Amphione tiliifolia* (Desr.) Raf., *Argyreia campanulata* Alston., *Argyreia tiliifol*ia (Desr.) Wight +20
Climbing Mechanism: Stem Twiner
Distribution (Global): Bangladesh, Borneo, Caroline Is., Christmas I., Cook Is., Fiji, Hainan, Howland-Baker Is., India, Java, Kazan-retto, Lesser Sunda Is., Malaya, Maluku, Marianas, Myanmar, Nansei-shoto, New Caledonia, New Guinea, Northern Territory, Ogasawara-shoto, Philippines, Queensland, Samoa, Society Is., Sri Lanka, Sulawesi, Sumatera, Taiwan, Thailand, Tonga, Tubuai Is., Vanuatu, Vietnam
Distribution (India): Andaman and Nicobar Islands, Madhya Pradesh, Maharashtra, Odisha, Tamil Nadu
Leaf Type: Simple
Inflorescence: Cymes in peduncles
Fruit Type: Wet, globose
Flowering and Fruiting: November–February
IUCN Status: Least Concern
Reference: Bot. Jahrb. Syst. 18: 159 (1893)

Tridynamia megalantha (Merr.) Staples
Convolvulaceae
Synonyms: *Porana megalantha* Merr., *Porana spectabilis* var. *megalantha* (Merr.) F.C. How, *Porana sutepensis* Kerr, *Tridynamia eberhardtii* Gagnep.
Climbing Mechanism: Stem Twiner
Distribution (Global): Andaman Is., Assam, Cambodia, China South-Central, China Southeast, Hainan, Laos, Malaya, Myanmar, Thailand, Vietnam
Distribution (India): Andaman and Nicobar Islands
Leaf Type: Simple
Inflorescence: Single axillary flowers
Fruit Type: Dry, follicles
IUCN Status: Not evaluated
Reference: Novon 3: 201 (1993)

Turbina corymbosa (L.) Raf.
Convolvulaceae
Synonyms: *Ipomoea corymbosa* (L.) Roth., *Turbina corymbosa* f. mollissima (Webb & Berthel.) Stearn. +20
Climbing Mechanism: Stem Twiner
Distribution (Global): Bahamas, Belize, Bolivia, Brazil North, Brazil Northeast, Brazil South, Brazil Southeast, Brazil West-Central, Colombia, Costa Rica, Cuba, Dominican Republic, Ecuador, El Salvador, Guatemala, Haiti, Honduras, Jamaica, Leeward Is., Mexico Central, Mexico Gulf, Mexico Northeast, Mexico Northwest, Mexico Southeast, Mexico Southwest, Nicaragua, Panamá, Paraguay, Peru, Puerto Rico, Trinidad-Tobago, Venezuela, Windward Is.
Distribution (India): Maharashtra
Leaf Type: Simple, lobed
Inflorescence: Single axillary flowers
Fruit Type: Dry, capsule
IUCN Status: Not evaluated
Reference: J.J. Roemer & J.A.Schultes, Syst. Veg., ed. 15 bis 4: 232 (1819)

Xenostegia filiformis
Convolvulaceae
Climbing Mechanism: Stem Twiner
IUCN Status: Not evaluated

Alangium salviifolium subsp. *hexapetalum* (Lam.) Wangerin
Cornaceae
Climbing Mechanism: Hook Climber
IUCN Status: Not evaluated

Actinostemma tenerum Griff.
Cucurbitaceae

Synonyms: *Actinostemma chaffanjonii* Lév., *Actinostemma japonicum* Miq., *Actinostemma lobatum* (Maxim.) Maxim., *Actinostemma lobatum* var. *palmatum* Makino, *Actinostemma lobatum* f. *subintegrum* (Komarov) M. Kitagawa, *Actinostemma lobatum* var. *subintegrum* Kom.
Climbing Mechanism: Tendril Climber
Distribution (Global): Assam, Bangladesh, Cambodia, China North-Central, China South-Central, China Southeast, East Himalaya, India, Korea, Laos, Taiwan, Thailand, Tibet, Vietnam, West Himalaya
Distribution (India): Arunachal Pradesh, Manipur
Leaf Type: Simple, alternate
Inflorescence: Racemose sometimes paniculate
Flowering and Fruiting: July–November
IUCN Status: Not evaluated
Reference: Account Bot. Coll. Cantor: 25 (1845)

Alsomitra sarcophylla (Wall.) M.Roem.
Cucurbitaceae
Climbing Mechanism: Tendril Climber
IUCN Status: Not evaluated
Reference: Fam. Nat. Syn. Monogr. 2: 118 (1846)

Benincasa hispida (Thunb.) Cogn.
Cucurbitaceae
Synonyms: *Benincasa cerifera* Savi, *Benincasa cylindrica* Ser., *Benincasa pruriens* (Parkinson) W.J.de Wilde & Duyfjes, *Benincasa pruriens* f. *hispida* (Thunb.) W. J.de Wilde & Duyfjes, *Benincasa vacua* (F. Muell.) F. Muell. + 12
Common Name: Ash gourd, Wax gourd, White gourd melon
Climbing Mechanism: Tendril Climber
Distribution (Global): Bismarck Archipelago, Borneo, Java, Lesser Sunda Is., Maluku, New Caledonia, New Guinea, Queensland, Solomon Is., Sulawesi
Distribution (India): Arunachal Pradesh, Delhi, Madhya Pradesh, Manipur, Meghalaya, Odisha, Tamil Nadu, Telangana, Tripura, Uttar Pradesh, West Bengal
Leaf Type: Simple, lobed, base deeply heart shaped
Inflorescence: Female flowers epi-perigynous
Fruit Type: Wet, pepo
Flowering and Fruiting: October–January
IUCN Status: Not evaluated
Reference: A.L.P.P.de Candolle & A.C.P.de Candolle, Monogr. Phan. 3: 513 (1881)

Biswarea tonglensis (C.B. Clarke) Cogn.
Cucurbitaceae
Synonyms: *Herpetospermum tonglense* (C.B. Clarke) H. Schaef. & S.S. Renner
Climbing Mechanism: Tendril Climber
Distribution (Global): China, Myanmar, Nepal, India

.
Leaf Type: ovate-cordate
Inflorescence: peduncle
IUCN Status: Not evaluated
Reference: Bull. Soc. Roy. Bot. Belgique, Compt.-Rend. 21: 16 (1882)

Blastania fimbristipula Kotschy & Peyr.
Cucurbitaceae
Synonyms: *Blastania cerasiformis* (Stocks) A. Meeuse
Climbing Mechanism: Tendril Climber
Distribution (Global): Kenya and tropical East Africa
Distribution (India): Rajasthan
IUCN Status: Not evaluated
Reference: Pl. Tinn.: 15 (1867)

Bryonia dioica Sessé & Moc.
Cucurbitaceae
Common Name: White bryony
Climbing Mechanism: Tendril Climber
Distribution (India): Himachal Pradesh, Maharashtra
IUCN Status: Not evaluated
Reference: Fl. Mexic., ed. 2: 228 (1894)

Cayaponia laciniosa (L.) C. Jeffrey
Cucurbitaceae
Synonyms: *Bryonia laciniosa* L. *Bryonopsis laciniosa* (L.) Naudin., *Bryonopsis laciniosa* var. *typica* Domin.
Climbing Mechanism: Tendril Climber
Distribution (Global): Jamaica
Leaf Type: Ovate-Oblong
Inflorescence: Male-Axillary racemes, Female-Axillary solitary
Fruit Type: Berry, globose
Flowering and Fruiting: February–October
IUCN Status: Not evaluated
Notes: Found occasionally along the edges and bushes
Reference: Kew Bull. 15: 346 (1962)

Cerasiocarpum zeylanicum (Thwaites) C.B. Clarke
Cucurbitaceae
Synonyms: *Kedrostis courtallensis* (Arn.) C.Jeffrey., *Aechmandra zeylanica* Thwaites., *Bryonopsis bennettii* Miq., *Bryonopsis courtallensis* Arn., *Cerasiocarpum bennettii* (Miq.) Cogn., *Cerasiocarpum bennettii* var. *deshmukhii* M.R. Almeida., *Kedrostis bennettii* (Miq.) W.J
Climbing Mechanism: Tendril Climber

Distribution (Global): Borneo, India, Java, Lesser Sunda Is., Myanmar, Sri Lanka, Sulawesi, Sumatera
IUCN Status: Not evaluated
Notes: Found in evergreen forests of the Western Ghats
Reference: Kew Bull. 15: 353 (1961 publ. 1962)

Ceratosanthes palmata (L.) Urb.
Cucurbitaceae
Synonyms: Ceratosanthes corniculata (Lam.) Cogn., Ceratosanthes gracilis Cogn., Ceratosanthes latiloba Cogn., Ceratosanthes tuberosa J.F. Gmel., Trichosanthes corniculata Lam., Trichosanthes palmata L., Trichosanthes tuberosa (J.F. Gmel.) Willd.
Climbing Mechanism: Tendril Climber
Distribution (Global): Aruba, Bolivia, Brazil Northeast, Brazil South, Brazil West-Central, Colombia, French Guiana, Guyana, Netherlands Antilles, Suriname, Trinidad-Tobago, Venezuela, Venezuelan Antilles, Windward Is.
IUCN Status: Not evaluated
Reference: Repert. Spec. Nov. Regni Veg. 15: 323 (1918)

Citrullus colocynthis (L.) Schrad.
Cucurbitaceae
Synonyms: *Citrullus colocynthis* subsp. *insipidus* (Pangalo) Fursa., *Citrullus colocynthis* subsp. *stenotomus* (Pangalo) Fursa., *Citrullus colocynthoides* Pangalo., *Citrullus pseudocolocynthis* M. Roem., *Colocynthis officinalis* Schrad., *Colocynthis vulgaris* Schrad., Cucum
Common Name: Colocynth, Indian wild gourd, Citron, Bitter apple, Vine of sodom
Climbing Mechanism: Tendril Climber
Distribution (Global): Pan tropical
Distribution (India): Andhra Pradesh, Delhi, Rajasthan, Tamil Nadu, Telangana
Leaf Type: Alternate, pinnately lobulate
Fruit Type: Globose, smooth, longitudinally green striped
Flowering and Fruiting: May–October
IUCN Status: Not evaluated
Notes: Medicnally valuable in treating various ailments
Reference: Linnaea 12: 414 (1838)

Citrullus lanatus (Thunb.) Matsum. & Nakai
Cucurbitaceae
Synonyms: *Anguria citrullus* Mill., *Citrullus anguria* (Duchesne) H. Hara. +40
Common Name: Bitter apple, Bittermelon, Citron, Citron melon, Edible seed melon, Keme, Monkey apple, Tsama melon, Tsamma, Watermelon, Wild watermelon
Climbing Mechanism: Tendril Climber
Distribution (Global): Egypt, Ethiopia, Libya, Sudan
Distribution (India): Gujarat, Himachal Pradesh, Odisha, Rajasthan, Tamil Nadu, Telangana, West Bengal

Leaf Type: Deeply palmately lobed with 3–5 lobes
Fruit Type: Subglobose, greenish mottled with darker green
Flowering and Fruiting: Throughout the year
IUCN Status: Not evaluated
Notes: Cultivated, but often found in wild
Reference: Index Seminum (TI, Tokyo) 1915–1916: 30 (1916)

Coccinia grandis (L.) Voigt
Cucurbitaceae
Synonyms: *Bryonia acerifolia* D. Dietr., *Bryonia alceifolia* Willd., *Bryonia barbata* Buch. -Ham. ex Cogn., *Bryonia grandis* L., *Bryonia moimoi* Ser., *Bryonia sinuosa* Wall. +20
Common Name: Ivy gourd, Scarlet gourd, Tindora
Climbing Mechanism: Tendril Climber
Distribution (Global): Pan tropical.
Distribution (India): Andaman and Nicobar Islands, Andhra Pradesh, Assam, Gujarat, Jharkhand, Kerala, Madhya Pradesh, Maharashtra, Manipur, Odisha, Puducherry, Rajasthan, Tamil Nadu, Telangana, Tripura, Uttar Pradesh, West Bengal
Leaf Type: Palmately 3–5 lobed or angled
Inflorescence: Axillary, solitary
Fruit Type: Ovoid-oblong; pulp red
Flowering and Fruiting: Throughout the year
IUCN Status: Not evaluated
Notes: Common in wasteland and thickets. Fruits are edible and used in traditional cooking
Reference: Hort. Suburb. Calcutt.: 59 (1845)

Corallocarpus conocarpus (Dalzell & A. Gibson) Hook.f.
Cucurbitaceae
Synonyms: *Aechmandra conocarpa* Dalzell & A. Gibson
Climbing Mechanism: Tendril Climber
Distribution (Global): India.
Distribution (India): Gujarat, Madhya Pradesh, Maharashtra
Leaf Type: Simple, Deeply lobed
Inflorescence: Racemes, axillary peduncles
Fruit Type: Wet, berry
IUCN Status: Not evaluated
Reference: Fl. Brit. India 2: 628 (1879)

Corallocarpus epigaeus (Rottler) Hook.f.
Cucurbitaceae
Synonyms: *Aechmandra epigaea* (Rottler) Arn., *Bryonia epigaea* Rottler, *Bryonia sinuata* Wall., *Coniandra corallina* Fenzl ex Naudin. + 10
Common Name: Redfruit creeper

Climbing Mechanism: Tendril Climber
Distribution (Global): Assam, Burkina, Djibouti, Eritrea, Ethiopia, India, Kenya, Niger, Nigeria, Oman, Pakistan, Rwanda, Saudi Arabia, Senegal, Somalia, Sri Lanka, Sudan, Tanzania, Zaire
Distribution (India): Andhra Pradesh, Gujarat, Madhya Pradesh, Maharashtra, Rajasthan, Tamil Nadu, Telangana, West Bengal
Leaf Type: Palmately lobed
Inflorescence: Racemes
Fruit Type: Ellipsoid, beaked
Flowering and Fruiting: August–March
IUCN Status: Not evaluated
Notes: Tubers are believed to cleanse the wounds, help deal with obesity, skin disease, tumours, cough, bronchitis
Reference: Fl. Brit. India 2: 628 (1879)

Corallocarpus schimperi
Cucurbitaceae
Climbing Mechanism: Tendril Climber
IUCN Status: Not evaluated

Ctenolepis cerasiformis (Stocks) C.B. Clarke
Cucurbitaceae
Synonyms: *Blastania cerasiformis* (Stocks) A. Meeuse. *Blastania fimbristipula* Kotschy & Peyr., *Ctenolepis cerasiformis* (Stocks) Naudin, *Melothria fimbristipula* (Kotschy & Peyr.) Roberty, *Pilogyne cerasiformis* Stocks ex Aitch., *Zehneria cerasiformis* Stocks
Common Name: Cherry vine
Climbing Mechanism: Tendril Climber
Distribution (Global): Benin, Botswana, Burkina, Cameroon, Central African Repu, Chad, Eritrea, Ethiopia, India, Kenya, Malawi, Mali, Mauritania, Mozambique, Niger, Nigeria, Northern Provinces, Pakistan, Saudi Arabia, Senegal, Somalia, Sudan, Tanzania, Yemen, Zambia, Zimbabwe
Distribution (India): Gujarat, Maharashtra, Telangana
Leaf Type: Broadly ovate
Inflorescence: Racemes
Fruit Type: Spherical, cherry-like red
Flowering and Fruiting: August–October
IUCN Status: Not evaluated
Notes: Fruits are used in treating calculi, strangury and diabetes
Reference: Bothalia 8: 12 (1962)

Ctenolepis garcini (L.) C.B. Clarke
Cucurbitaceae
Common Name: Garcen's bur cucumber
Climbing Mechanism: Tendril Climber

Distribution (India): Andhra Pradesh, Delhi, Gujarat, Maharashtra, Puducherry, Tamil Nadu, Telangana
Leaf Type: Palmately 5 lobed
Inflorescence: Cymes
Fruit Type: Bright red, kidney shaped or hammer shaped
Flowering and Fruiting: August–March
IUCN Status: Not evaluated
Notes: Mostly found in dry deciduous forests and scrub jungles.
Reference: Naudin. In: Ann. Sc. Nat. Ser. V. 6: 13. (1866).

Cucumis dipsaceus Ehrenb. ex Spach
Cucurbitaceae
Synonyms: *Momordica dasycarpa* Hochst. ex A. Rich.
Common Name: Hedgehog cucumber, Teasel gourd
Climbing Mechanism: Tendril Climber
Distribution (Global): Central African Repu, Eritrea, Ethiopia, Kenya, Somalia, Sudan, Tanzania, Uganda
Distribution (India): Western Ghats
Leaf Type: Alternate
Inflorescence: Solitary, sometimes in fasicles
Fruit Type: Ellipsoid or spherical, densely covered with spines
Flowering and Fruiting: November–February
IUCN Status: Not evaluated
Notes: Leaves and roots are pounded and used as a poultice to treat wounds
Reference: Hist. Nat. Vég. 6: 211 (1838)

Cucumis hystrix Chakrav.
Cucurbitaceae
Synonyms: *Cucumis hystrix* var. *mizoramensis* Sutar & S.R. Yadav
Cucumis muriculatus Chakrav.
Climbing Mechanism: Tendril Climber
Distribution (Global): Assam, Bangladesh, China South-Central, East Himalaya, Laos, Myanmar, Thailand
Distribution (India): Peninsular India
Flowering and Fruiting: May–September
IUCN Status: Not evaluated
Notes: Mostly found in dry decifduous forests
Reference: J. Bombay Nat. Hist. Soc. 50: 896 (1952)

Cucumis indicus Ghebret. & Thulin
Cucurbitaceae
Synonyms: *Cucumella ritchiei* (Chakrav.) C.Jeffrey, *Melothria ritchiei* Chakrav.
Climbing Mechanism: Tendril Climber
Distribution (Global): India
Flowering and Fruiting: June–August

IUCN Status: Not evaluated
Reference: Novon 17: 177 (2007)

Cucumis leiospermus (Wight & Arn.) Ghebret. & Thulin
Cucurbitaceae
Synonyms: *Bryonia leiosperma* Wight & Arn., *Bryonia wightiana* Wall., *Melothria leiosperma* (Wight & Arn.) Cogn., *Mukia leiosperma* (Wight & Arn.) Arn.
Climbing Mechanism: Tendril Climber
Distribution (Global): Assam, East Himalaya, India
Flowering and Fruiting: Throughout the year
IUCN Status: Not evaluated
Notes: Mostly found in evergreen and moist deciduous forests
Reference: Novon 17: 177 (2007)

Cucumis melo L.
Cucurbitaceae
Synonyms: *Bryonia collosa* Rottler, *Cucumis acidus* Jacq. + 100
Climbing Mechanism: Tendril Climber
Distribution (Global): Afghanistan, Angola, Chad, Eritrea, Ethiopia, Gulf States, India, Iran, Iraq, Kenya, Lebanon-Syria, Malawi, Mozambique, New South Wales, Northern Provinces, Northern Territory, Oman, Pakistan, Palestine, Queensland, Saudi Arabia, Somalia, South Australia, Sri Lanka, Sudan, Tanzania, Uganda, Western Australia, Yemen, Zambia, Zimbabwe
Distribution (India): Andaman and Nicobar Islands, Arunachal Pradesh, Gujarat, Himachal Pradesh, Maharashtra, Odisha, Puducherry, Telangana, West Bengal
Leaf Type: Triangular, ovate
Inflorescence: Solitary or rarely in pairs or threes
Fruit Type: Pepo/Berry
Flowering and Fruiting: May–September
IUCN Status: Not evaluated
Notes: Often cultivated
Reference: Sp. Pl.: 1011 (1753)

Cucumis prophetarum L.
Cucurbitaceae
Synonyms: *Cucumis prophetarum* subsp. *dissectus* (Naudin) C. Jeffrey, *Cucumis prophetarum* subsp. *Prophetarum*
Common Name: Wild gourd, Wild cucumber
Climbing Mechanism: Tendril Climber
Distribution (Global): Algeria, Benin, Burkina, Burundi, Cameroon, Chad, Djibouti, Egypt, Eritrea, Ethiopia, Gulf States, India, Iran, Iraq, Kenya, Lebanon-Syria, Mali, Mauritania, Niger, Nigeria, Oman, Pakistan, Palestine, Rwanda, Saudi Arabia, Senegal, Sinai, Socotra, Somalia, Sudan, Tanzania, Uganda, Yemen, Zaire
Distribution (India): Karnataka, Maharashtra, Rajasthan, Tamil Nadu

Leaf Type: Ovate-round
Inflorescence: Fascicles, rarely solitary
Fruit Type: Ovoid or nearly spherical, yellow when ripe
Flowering and Fruiting: August–September
IUCN Status: Not evaluated
Reference: Cent. Pl. I: 32 (1755)

Cucumis ritchiei (C.B. Clarke) Ghebret. & Thulin
Cucurbitaceae
Climbing Mechanism: Tendril Climber
Distribution (India): Peninsular India
IUCN Status: Not evaluated
Reference: Ghebret., Thulin. In: Novon 17(2): 178. (2007)

Cucumis sativus L.
Cucurbitaceae
Synonyms: *Cucumis esculentus* Salisb., *Cucumis hardwickii* Royle, *Cucumis muricatus* Willd., *Cucumis rumphii* Hassk., *Cucumis sativus* f. *albus* M. Hiroe, *Cucumis sativus* f. *hardwickii* (Royle) W.J.de Wilde & Duyfjes, *Cucumis sativus* subsp. *hardwickii* (Royle) Banfi.
Common Name: Cucumber
Climbing Mechanism: Tendril Climber
Distribution (Global): Assam, Bangladesh, China South-Central, China Southeast, East Himalaya, Myanmar, Nepal, Thailand, West Himalaya
Distribution (India): Andhra Pradesh, Arunachal Pradesh, Bihar, Odisha, Delhi, Gujarat, Himachal Pradesh, Madhya Pradesh, Maharashtra, Odisha, Rajasthan, Tamil Nadu, Telangana, West Bengal
Flowering and Fruiting: September–May
IUCN Status: Not evaluated
Notes: Often cultivated
Reference: Sp. Pl.: 1012 (1753)

Cucumis silentvalleyi (Manilal, T. Sabu & P. Mathew) Ghebret. & Thulin
Cucurbitaceae
Synonyms: *Cucumella silentvalleyi* Manilal, T. Sabu & P. Mathew
Climbing Mechanism: Tendril Climber
Distribution (Global): India
Distribution (India): Tamil Nadu
Leaf Type: Obtusely 5-angular
Inflorescence: Axillary, solitary
Fruit Type: Hispid, ellipsoid
Flowering and Fruiting: July–December
IUCN Status: Not evaluated
Notes: Usually found in moist deciduous and semi-evergreen forests
Reference: Novon 17: 178 (2007)

Cucurbita maxima Duchesne
Cucurbitaceae
Synonyms: *Cucumis rapallito* (Carrière) Carrière, *Cucurbita farinae* Mozz. ex Naudin. + 10
Common Name: Squash gourd, Squash melon, Red gourd pumpkin
Climbing Mechanism: Tendril Climber
Distribution (Global): Argentina Northeast, Argentina Northwest, Bolivia
Distribution (India): Gujarat, Himachal Pradesh, Maharashtra, Odisha, Tamil Nadu, West Bengal
Leaf Type: Orbicular or more or less reniform
Inflorescence: Axillary, solitary, large
Fruit Type: Large fruits
Flowering and Fruiting: September–May
IUCN Status: Not evaluated
Notes: Cultivated for its edible fruits
Reference: Ess. Hist. Nat. Courges: 7 (1786)

Cucurbita moschata Duchesne
Cucurbitaceae
Synonyms: *Cucurbita colombiana* (Zhit.) Bukasov, *Cucurbita hippopera* Ser., *Cucurbita macrocarpa* Gasp., *Cucurbita meloniformis* Carrière. + 10
Common Name: Crooknecked squash, musk melon
Climbing Mechanism: Tendril Climber
Distribution (Global): Belize, Guatemala, Mexico Central, Mexico Gulf, Mexico Northeast, Mexico Southeast, Mexico Southwest.
Distribution (India): Gujarat, Maharashtra, Odisha, Tamil Nadu, Tripura
Leaf Type: Leaves 5–7-lobed
Fruit Type: Pepo
Flowering and Fruiting: September–June
IUCN Status: Not evaluated
Notes: Cultivated for its edible fruits
Reference: Ess. Hist. Nat. Courges: 7 (1786)

Cucurbita pepo L.
Cucurbitaceae
Synonyms: *Citrullus variegatus* Schrad. ex M. Roem., *Cucumis pepo* (L.) Dumort. +40
Common Name: Pumpkin
Climbing Mechanism: Tendril Climber
Distribution (Global): Mexico Central, Mexico Gulf, Mexico Northeast, Mexico Northwest, Mexico Southeast, Mexico Southwest
Distribution (India): Arunachal Pradesh, Himachal Pradesh, Maharashtra, Odisha, Tamil Nadu, Telangana
Leaf Type: Alternate, petioled
Inflorescence: Solitary, yellow colored flowers

Fruit Type: Pepo
Flowering and Fruiting: July–October
IUCN Status: Least Concern
Notes: Cultivated for its edible fruits.
Reference: Sp. Pl.: 1010 (1753)

Cucurbita pepo subsp. *ovifera* (L.) D.S. Decker
Cucurbitaceae
Synonyms: *Cucurbita melopepo* subsp. *Melopepo.*, *Cucurbita ovifera* L., *Cucurbita pepo* convar. *ovifera* (L.) Alef., *Cucurbita subverrucosa* Willd., *Cucurbita verrucosa* L., *Pepo ovifer* (L.) Peterm.
Climbing Mechanism: Tendril Climber
Distribution (Global): Arkansas, Kentucky, Louisiana, Mississippi, Missouri
Distribution (India): Tamil Nadu
IUCN Status: Not evaluated
Reference: D. S. Decker. In: Econ. Bot., 42(1): 11. (1988)

Dactyliandra welwitschii Hook.f.
Cucurbitaceae
Synonyms: *Ctenolepis welwitschii* (Hook.f.) Jafri, *Dactyliandra luederitziana* (Cogn.) Cogn.
Climbing Mechanism: Tendril Climber
Distribution (Global): Angola, Namibia
Distribution (India): Gujarat, Rajasthan
IUCN Status: Not evaluated
Reference: D. Oliver & auct. suc. (eds.), Fl. Trop. Afr. 2: 557 (1871)

Diplocyclos palmatus (L.) C. Jeffrey
Cucurbitaceae
Synonyms: *Bryonia palmata* L., *Coccinia palmata* (L.) M. Roem.
Common Name: Lollipop climber, Marble vine, Native bryony, Striped cucumber, Boiled sweet plant
Climbing Mechanism: Tendril Climber
Distribution (Global): Assam, Bangladesh, Burundi, Cambodia, Chad, China Southeast, Congo, East Himalaya, Ethiopia, Gulf of Guinea Is., Hainan, India, Java, Kenya, KwaZulu-Natal, Laos, Lesser Sunda Is., Malaya, Maluku, Mozambique, Myanmar, Namibia, Nansei-shoto, Nepal, New Caledonia, New Guinea, New South Wales, Norfolk Is., Northern Territory, Oman, Pakistan, Philippines, Queensland, Rwanda, Solomon Is., Sri Lanka, Sudan, Sulawesi, Sumatera, Taiwan, Tanzania, Thailand, Uganda, Vanuatu, Vietnam, West Himalaya, Yemen, Zaire, Zimbabwe
Distribution (India): Andhra Pradesh, Assam, Gujarat, Himachal Pradesh, Madhya Pradesh, Maharashtra, Odisha, Puducherry, Tamil Nadu, Telangana, Tripura, Uttar Pradesh, West Bengal
Leaf Type: Simple, Palmately 5-lobed

Inflorescence: Axillary clusters
Fruit Type: A globose berry
Flowering and Fruiting: November–January
IUCN Status: Not evaluated
Reference: Kew Bull. 15: 352 (1962)

Edgaria darjeelingensis C.B. Clarke
Cucurbitaceae
Climbing Mechanism: Tendril Climber
Distribution (India): Arunachal Pradesh, West Bengal
IUCN Status: Not evaluated
Reference: J. Linn. Soc., Bot. 15: 114 (1876)

Gomphogyne cissiformis Griff.
Cucurbitaceae
Synonyms: *Gomphogyne alleizettei* Gagnep., *Gomphogyne bonii* Gagnep., *Gomphogyne macrocarpa* Cogn.
Common Name: Ivy Gomphogyne
Climbing Mechanism: Tendril Climber
Distribution (Global): Assam, China South-Central, East Himalaya, Nepal, Vietnam, West Himalaya
Distribution (India): Arunachal Pradesh, Mizoram, West Bengal
Flowering and Fruiting: September–November
IUCN Status: Not evaluated
Reference: Account Bot. Coll. Cantor: 26 (1845)

Gomphogyne heterosperma (Wall.) Kurz
Cucurbitaceae
Synonyms: *Alsomitra heterosperma* (Wall.) M.Roem., *Gomphogyne heterosperma* f. vittata W.J.de Wilde & Duyfjes, *Hemsleya heterosperma* (Wall.) C.Jeffrey, *Zanonia heterosperma* Wall.
Climbing Mechanism: Tendril Climber
Distribution (Global): Bangladesh, China South-Central, Myanmar, Thailand
IUCN Status: Not evaluated
Reference: J. Asiat. Soc. Bengal, Pt. 2, Nat. Hist. 40: 58 (1871)

Gomphogyne macrocarpa Cogn.
Cucurbitaceae
Synonyms: *Hemsleya macrocarpa* (Cogn.) C.Y. Wu ex C. Jeffrey
Climbing Mechanism: Tendril Climber
Distribution (Global): H.G.A. Engler (ed.), Pflanzenr., IV, 275 I: 40 (1916)
Distribution (India): Arunachal Pradesh
IUCN Status: Not evaluated
Reference: H.G.A. Engler (ed.), Pflanzenr., IV, 275 I: 40 (1916)

Gymnopetalum chinense (Lour.) Merr.
Cucurbitaceae
Synonyms: *Trichosanthes costata* Blume
Common Name: Ribbed orange gourd
Climbing Mechanism: Tendril Climber
Distribution (India): West Bengal
Leaf Type: Simple, alternate, spiral
Inflorescence: Solitary
Fruit Type: Fruit is orange oblong-ovoid
IUCN Status: Not evaluated
Reference: Philipp. J. Sci. 15: 256 (1919)

Gymnopetalum pedata Bl.
Cucurbitaceae
Climbing Mechanism: Tendril Climber
Distribution (India): Arunachal Pradesh
IUCN Status: Not evaluated

Gymnopetalum scabrum (Lour.) W.J.de Wilde & Duyfjes
Cucurbitaceae
Synonyms: *Trichosanthes scabra* Lour.
Climbing Mechanism: Tendril Climber
IUCN Status: Not evaluated
Reference: Reinwardtia 12: 268 (2008)

Gymnopetalum tubiflorum (Wight & Arn.) Cogn.
Cucurbitaceae
Synonyms: *Trichosanthes tubiflora* (Wight & Arn.) H.J.de Boer
Climbing Mechanism: Tendril Climber
Flowering and Fruiting: October–November
IUCN Status: Not evaluated
Reference: A.L.P.P.de Candolle & A.C.P.de Candolle, Monogr. Phan. 3: 388 (1881)

*Gynostemma pentaphy*llum (Thunb.) Makino
Cucurbitaceae
Synonyms: *Gynostemma blumei* Hassk., *Gynostemma burmanicum* King ex Chakrav., *Gynostemma burmanicum* var. molle C.Y. Wu, *Gynostemma cissoides* (Wall.) Benth. & Hook.f. ex Franch. & Sav., *Gynostemma crenulatum* Ridl., *Gynostemma laxum* (Wall.) Cogn.
Common Name: Sweet tea vine
Climbing Mechanism: Tendril Climber
Distribution (Global): Assam, Bangladesh, Borneo, China North-Central, China South-Central, China Southeast, East Himalaya, Hainan, Japan, Java, Korea, Kuril Is., Laos, Lesser Sunda Is., Malaya, Maluku, Myanmar, Nansei-shoto,

Nepal, New Guinea, Philippines, Sri Lanka, Sulawesi, Sumatera, Thailand, Vietnam, West Himalaya
Distribution (India): Tamil Nadu, West Bengal
Leaf Type: 3–9 foliolate
Inflorescence: Panicle
IUCN Status: Not evaluated
Reference: Bot. Mag. (Tokyo) 16: 179 (1902)

Hemsleya graciliflora (Harms) Cogn.
Cucurbitaceae
Synonyms: *Alsomitra graciliflora* Harms, *Hemsleya graciliflora* var. *tianmuensis* X.J. Xue & H. Yao
Climbing Mechanism: Tendril Climber
Distribution (Global): China South-Central, China Southeast
Distribution (India): Andhra Pradesh
IUCN Status: Not evaluated
Reference: H.G.A. Engler (ed.), Pflanzenr., IV, 275 I: 24 (1916)

Herpetospermum pedunculosum (Ser.) C.B. Clarke
Cucurbitaceae
Synonyms: *Herpetospermum grandiflorum* Cogn., *Herpetospermum pedunculosum* (Ser.) Baill., *Rampinia herpetospermoides* C.B. Clarke, *Bryonia pedunculosa* Ser., *Herpetospermum caudigerum* Wall. ex C.B. Clarke
Common Name: Himalayan bitter gourd
Climbing Mechanism: Tendril Climber
Distribution (Global): Assam, China South-Central, East Himalaya, Nepal, Tibet, West Himalaya
Distribution (India): North Eastern India, West Bengal
IUCN Status: Not evaluated
Reference: J. Linn. Soc., Bot. 15: 115 (1876)

Hodgsonia heteroclita (Roxb.) Hook.f. & Thomson
Cucurbitaceae
Synonyms: *Trichosanthes heteroclita* Roxb., *Trichosanthes listeri* Chakrav., *Trichosanthes theba* Buch. -Ham. ex Wall.
Climbing Mechanism: Tendril Climber
Distribution (Global): Assam, Bangladesh, Cambodia, China South-Central, East Himalaya, Laos, Myanmar, Thailand, Tibet, Vietnam
Distribution (India): Arunachal Pradesh, Assam
Leaf Type: Simple, 3–5 lobed
Inflorescence: Simple
Fruit Type: Wet, globose
IUCN Status: Not evaluated
Reference: Proc. Linn. Soc. London 2: 257 (1853)

Hodgsonia macrocarpa (Blume) Cogn.
Cucurbitaceae
Synonyms: *Hodgsonia capniocarpa* Ridl., *Hodgsonia heteroclita* subsp. *indochinensis* W.J.de Wilde & Duyfjes, *Hodgsonia macrocarpa* var. *capniocarpa* (Ridl.) H.T. Tsai ex A.M. Lu & Zhi Y. Zhang, *Trichosanthes hexasperma* Blume, *Trichosanthes kadam* Miq.
Common Name: Chinese lardfruit, Pork fat nut
Climbing Mechanism: Tendril Climber
Distribution (Global): Borneo, Cambodia, Java, Malaya, Myanmar, Sumatera, Thailand, Vietnam
Distribution (India): Arunachal Pradesh, Assam, Madhya Pradesh, Manipur, Meghalaya, Mizoram, Tripura, West Bengal
Leaf Type: Leaves are 3–5 palmately lobed
Fruit Type: Tomentose
Flowering and Fruiting: February–November
IUCN Status: Not evaluated
Reference: A.L.P.P.de Candolle & A.C.P.de Candolle, Monogr. Phan. 3: 349 (1881)

Indofevillea khasiana Chatterjee
Cucurbitaceae
Climbing Mechanism: Tendril Climber
Distribution (Global): Assam, East Himalaya, Tibet
Distribution (India): Arunachal Pradesh
Leaf Type: Simple, alternate
Inflorescence: Axillary, corymbose
Flowering and Fruiting: May–July
IUCN Status: Not evaluated
Reference: Nature 158: 345 (1946)

Kedrostis foetidissima (Jacq.) Cogn.
Cucurbitaceae
Climbing Mechanism: Tendril Climber
IUCN Status: Not evaluated

Lagenaria siceraria (Molina) Standl.
Cucurbitaceae
Synonyms: *Cucumis bicirrha* J.R. Forst. ex Guill., *Cucumis lagenaria* (L.) Dumort., *Cucumis mairei* H. Lév., *Cucurbita ciceraria* Molina, *Cucurbita idololatrica* Willd., *Cucurbita lagenaria* L., *Cucurbita leucantha* Duchesne, *Cucurbita longa* W.M. Fletcher
Common Name: Bottle gourd, Monkey apple
Climbing Mechanism: Tendril Climber
Distribution (Global): Cameroon, Central African Repu, Chad, Congo, Ethiopia, Gabon, Gambia, Ghana, Guinea-Bissau, Gulf of Guinea Is., Ivory Coast, Kenya,

Liberia, Niger, Nigeria, Senegal, Sierra Leone, Somalia, Sudan, Tanzania, Togo, Uganda, Zaire

Distribution (India): Himachal Pradesh, Madhya Pradesh, Maharashtra, Manipur, Odisha, Rajasthan, Tamil Nadu, Telangana, Tripura, West Bengal

Leaf Type: Leaves palmate, broadly ovate, reniform or sub-orbicular, cordate, obscurely 3–5-lobed, lobes rounded

Inflorescence: Both male and female flowers solitary; male peduncle as long as petiole, female somewhat shorter

Fruit Type: Fruit large, subglobose, ellipsoid, lageniform, sometimes biventricose, green or greenish-yellow, hairy, indehiscent

Flowering and Fruiting: June–February

IUCN Status: Not evaluated

Reference: Publ. Field Mus. Nat. Hist., Bot. Ser. 3: 435 (1930)

Luffa acutangula (L.) Roxb.

Cucurbitaceae

Synonyms: *Cucumis acutangulus* L., *Cucumis lineatus* Bosc, *Cucumis megacarpus* G. Don, *Cucumis operculatus* Roxb. ex Wight & Arn. +20

Common Name: Loofah, Ridge gourd, Bitter Luffa, Wild luffa, Wild ribbed gourd, Wild ridge gourd

Climbing Mechanism: Tendril Climber

Distribution (Global): Assam, Bangladesh, East Himalaya, India, Pakistan, Sri Lanka, West Himalaya

Distribution (India): Andhra Pradesh, Arunachal Pradesh, Bihar, Odisha, Delhi, Gujarat, Himachal Pradesh, Jharkhand, Madhya Pradesh, Maharashtra, Manipur, Odisha, Rajasthan, Tamil Nadu, Telangana, Tripura, West Bengal

Leaf Type: Leaves alternate, simple, shallowly palmately 5–7-lobed with broadly triangular to broadly rounded lobes

Inflorescence: Male inflorescence racemose, female flowers solitary

Fruit Type: Fruit a club-shaped, dry, and fibrous capsule

Flowering and Fruiting: June–September

IUCN Status: Not evaluated

Reference: Fl. Ind. ed. 1832, 3: 713 (1832)

Luffa cylindrica (L.) M. Roem.

Cucurbitaceae

Synonyms: *Bryonia collosa* Rottler, *Cucumis acidus* Jacq. +100

Common Name: Cantaloupe melon, Dishcloth gourd, Snake melon, Sweet melon, Sugar melon

Climbing Mechanism: Tendril Climber

Distribution (Global): Pantropical

Distribution (India): Andaman and Nicobar Islands, Andhra Pradesh, Arunachal Pradesh, Delhi, Gujarat, Himachal Pradesh, Jharkhand, Madhya Pradesh, Maharashtra, Manipur, Mizoram, Rajasthan, Tamil Nadu, West Bengal

Leaf Type: Leaves rough, Simple, alternate, elliptic, lobed

Inflorescence: Golden yellow flowers with solitary cyme inflorescence
Fruit Type: Musky-scented, spherical to oblong berry with a rind (pepo), often furrowed with yellow
Flowering and Fruiting: July and August
IUCN Status: Not evaluated
Reference: Sp. Pl.: 1011 (1753)

Luffa echinata Roxb.
Cucurbitaceae
Synonyms: *Luffa bendaul* Roxb., *Luffa bindaal* Roxb., *Luffa bondel* Buch. -Ham. ex Steud., *Luffa echinata* var. *longistyla* (Müll.Stuttg.) C.B. Clarke, *Luffa longistyla* Müll.Stuttg., *Momordica echinocarpa* Fenzl ex Hook.f., *Momordica erinocarpa* Fenzl ex Naudin
Common Name: Bitter sponge gourd, Rag gourd
Climbing Mechanism: Tendril Climber
Distribution (Global): Assam, Bangladesh, Cameroon, Chad, Ethiopia, India, Mali, Mauritania, Niger, Nigeria, Pakistan, Somalia, Sudan, West Himalaya
Distribution (India): Andhra Pradesh, Bihar, Odisha, Gujarat, Himachal Pradesh, Maharashtra, Rajasthan
Leaf Type: Leaves are kidney shaped, round, shallowly or deeply 5-lobed. Tip is rounded or rarely pointed
Inflorescence: Male flowers are borne in 5–12-flowered, up to 15 cm long raceme
Fruit Type: Fruit is ashy, oblong, ovoid
Flowering and Fruiting: August–September
IUCN Status: Least Concern
Notes: Mostly found on hedges
Reference: Fl. Ind. ed. 1832, 3: 716 (1832)

Luffa graveolens Roxb.
Cucurbitaceae
Climbing Mechanism: Tendril Climber
Distribution (Global): Bangladesh, East Himalaya, India, Nepal
Distribution (India): Bihar, Odisha, Maharashtra
IUCN Status: Not evaluated
Reference: Fl. Ind. ed. 1832, 3: 716 (1832)

Luffa hermaphrodita N.B. Singh & U.C. Bhattach.
Cucurbitaceae
Climbing Mechanism: Tendril Climber
Distribution (India): West Bengal
IUCN Status: Not evaluated

Luffa kleinii Wight & Arn.
Cucurbitaceae

Climbing Mechanism: Tendril Climber
IUCN Status: Not evaluated

Luffa tuberosa Roxb.
Cucurbitaceae
Climbing Mechanism: Tendril Climber
IUCN Status: Not evaluated

Luffa umbellata M. Roem.
Cucurbitaceae
Synonyms: Synonym
Climbing Mechanism: Tendril Climber
Distribution (India): Rajasthan
IUCN Status: Not evaluated

Melothria angulata Chakrav.
Cucurbitaceae
Synonyms: *Solena umbellata* (J.G. Klein ex Willd.) W.J.de Wilde & Duyfjes., *Bryonia teedonda* Roxb. ex Willd., *Bryonia umbellata* J.G. Klein ex Willd., *Karivia umbellata* (J.G. Klein ex Willd.) Arn., *Momordica umbellata* (J.G. Klein ex Willd.) Roxb., *Solena angulata* (Chakrav.) Raizada
Climbing Mechanism: Tendril Climber
Distribution (Global): Assam, Bangladesh, India, Myanmar, Sri Lanka
Distribution (India): Peninsular India
Leaf Type: Leaves subsessile, heteromorphic, cordate or sagittate at base
Inflorescence: Male flowers subumbellate, axillary, female solitary and pedicellate
IUCN Status: Not evaluated
Reference: Blumea 49: 77 (2004)

Melothria perpusilla (Blume) Cogn.
Cucurbitaceae
Synonyms: *Zehneria scabra* subsp. *Scabra*., *Bryonia angulata* Thunb., *Bryonia cordata* Thunb., *Bryonia dentata* E. Mey., *Bryonia maderaspatana* P.J. Bergius, *Bryonia perpusilla* (Blume) Blume, *Bryonia punctata* Thunb., *Bryonia scrobiculata* Hochst. ex A. Rich.
Climbing Mechanism: Tendril Climber
Distribution (Global): Angola, Assam, Bangladesh, Benin, Burkina, Burundi, Cameroon, Cape Provinces, Congo, East Himalaya, Eritrea, Ethiopia, Free State, Gulf of Guinea Is., India, Ivory Coast, Java, Kenya, KwaZulu-Natal, Lesotho, Lesser Sunda Is., Liberia, Malawi, Mozambique, Northern Provinces, Philippines, Rwanda, Saudi Arabia, Somalia, Sudan, Sulawesi, Swaziland, Tanzania, Uganda, Yemen, Zambia, Zaire, Zimbabwe
Distribution (India): Arunachal Pradesh, Madhya Pradesh, Mizoram, Rajasthan
Leaf Type: Leaves to 7 x 7 cm, triangular, cordate at base, 3-lobed or entire, glandular toothed

Inflorescence: Umbel 3–5 flowered; peduncle 1.5 cm long
Fruit Type: Berry 7 mm across, globose; seeds 5 x 4 mm, obovate, biconvex, smooth, not marginate
Flowering and Fruiting: February–June
IUCN Status: Not evaluated
Notes: Usually found along the forest margins

Melothria zehnerioides Haines
Cucurbitaceae
Synonyms: *Zehneria odorata* (Hook.f. & Thomson ex Benth.) M.D. Dwivedi, A.K. Pandey & H. Schaef.
Aechmandra indica (Lour.) Wight ex Miq. +10
Climbing Mechanism: Tendril Climber
Distribution (Global): Assam, Bangladesh, East Himalaya, India, Myanmar, Nepal, West Himalaya
IUCN Status: Not evaluated
Reference: Taxon 67: 62 (2018)

Melothria zeylanica C.B. Clarke
Cucurbitaceae
Synonyms: *Zehneria thwaitesii* (Schweinf.) C.Jeffrey., *Aechmandra deltoidea* Arn., *Bryonia deltoidea* Arn., *Cucumella thwaitesii* (Schweinf.) M.R. Almeida, *Melothria deltoidea* (Arn.) Thwaites, *Melothria thwaitesii* Schweinf., *Neoachmandra deltoidea* (Arn.) W.J.de Wi
Climbing Mechanism: Tendril Climber
Distribution (Global): India, Sri Lanka
Leaf Type: Leaf-blade triangular, hastate or sagittate, ± scabrid-punctate above, smooth and glabrous except for a few hairs on nerves beneath, subentire or ± sinuate-dentate or rarely lobulate
Inflorescence: Female flowers solitary on slender 7–40 mm long pedicels, Male flowers 1–3, axillary, or at nodes on 30–95 mm long leafless shoots with shortened internodes and then appearing racemose
Fruit Type: Fruit berry, pendulous on a 5–37 mm long stalk, fusiform, smooth, glabrous, red when ripe
Flowering and Fruiting: August–December
IUCN Status: Not evaluated
Reference: Kew Bull. 15: 371 (1961 publ. 1962)

Momordica balsamina L.
Cucurbitaceae
Climbing Mechanism: Tendril Climber
Distribution (India): Gujarat, Maharashtra, Rajasthan, Tamil Nadu
Leaf Type: Simple, 3–7 lobed
Inflorescence: Flowers on peduncle, orbicular-reniform

Fruit Type: Wet, subglobose, terete
IUCN Status: Not evaluated

Momordica charantia L.
Cucurbitaceae
Climbing Mechanism: Tendril Climber
Distribution (India): Andaman and Nicobar Islands, Andhra Pradesh, Arunachal Pradesh, Assam, Delhi, Great, Gujarat, Himachal Pradesh, Jharkhand, Maharashtra, Manipur, Mizoram, Odisha, Rajasthan, Tamil Nadu, Telangana, Tripura, Uttar Pradesh, West Bengal
Leaf Type: Simple, 3–7 lobed
Inflorescence: Flowers on peduncle, orbicular-reniform
Fruit Type: Wet, subglobose, terete
IUCN Status: Not evaluated

Momordica cochinchinensis (Lour.) Spreng.
Cucurbitaceae
Climbing Mechanism: Tendril Climber
Distribution (India): Andaman and Nicobar Islands, Arunachal Pradesh, Gujarat, Karnataka, Maharashtra, Manipur, Odisha, Rajasthan, Tamil Nadu, Tripura, West Bengal
Leaf Type: Simple, 3–7 lobed
Inflorescence: Flowers on peduncle, orbicular-reniform
Fruit Type: Wet, subglobose, terete
IUCN Status: Not evaluated

Momordica cymbalaria Hook.f.
Cucurbitaceae
Climbing Mechanism: Tendril Climber
Distribution (India): Karnataka, Maharashtra, Tamil Nadu
Leaf Type: Simple, 3–7 lobed
Inflorescence: Flowers on peduncle, orbicular-reniform
Fruit Type: Wet, subglobose, terete
IUCN Status: Not evaluated

Momordica denudata C.B. Clarke
Cucurbitaceae
Climbing Mechanism: Tendril Climber
Distribution (India): Gujarat
Leaf Type: Simple, 3–7 lobed
Inflorescence: Flowers on peduncle, orbicular-reniform
Fruit Type: Wet, subglobose, terete
IUCN Status: Not evaluated

Momordica dioica Roxb. ex Willd.
Cucurbitaceae
Climbing Mechanism: Tendril Climber
Distribution (India): Andhra Pradesh, Arunachal Pradesh, Assam, Delhi, Gujarat, Himachal Pradesh, Jharkhand, Madhya Pradesh, Maharashtra, Manipur, Mizoram, Odisha, Rajasthan, Tamil Nadu, Telangana, Tripura, Uttar Pradesh
Leaf Type: Simple, 3–7 lobed
Inflorescence: Flowers on peduncle, orbicular-reniform
Fruit Type: Wet, subglobose, terete
IUCN Status: Not Evaluated

Momordica sahyadrica Kattuk. & V.T. Antony
Cucurbitaceae
Climbing Mechanism: Tendril Climber
Distribution (India): Peninsular India
Leaf Type: Simple, 3–7 lobed
Inflorescence: Flowers on peduncle, orbicular-reniform
Fruit Type: Wet, subglobose, terete
IUCN Status: Not evaluated

Momordica subangulata Blume
Cucurbitaceae
Climbing Mechanism: Tendril Climber
Distribution (India): Karnataka, Tamil Nadu
Leaf Type: Simple, 3–7 lobed
Inflorescence: Flowers on peduncle, orbicular-reniform
Fruit Type: Wet, subglobose, terete
IUCN Status: Not evaluated

Mukia maderaspatana (L.) M. Roem.
Cucurbitaceae
Climbing Mechanism: Tendril Climber
Distribution (India): Andaman and Nicobar Islands, Andhra Pradesh, Assam, Bihar, Odisha, Gujarat, Kerala, Maharashtra, Odisha, Puducherry, Tamil Nadu, Telangana, Uttar Pradesh, West Bengal
IUCN Status: Not evaluated

Neoalsomitra clavigera (M. Roem.) Hutch.
Cucurbitaceae
Synonyms: *Alsomitra beccariana* Cogn.
Climbing Mechanism: Tendril Climber
Distribution (Global): Assam, Bangladesh, Bismarck Archipelago, Cambodia, China South-Central, East Himalaya, Fiji, Hainan, India, Laos, Malaya, Maluku, Myanmar, Nepal, New Guinea, Philippines, Queensland, Solomon Is., Sulawesi, Sumatera, Taiwan, Thailand, Tibet, Vietnam, West Himalaya

Distribution (India): Arunachal Pradesh, Manipur, West Bengal
IUCN Status: Not evaluated
Reference: Ann. Bot. (Oxford), n.s., 6: 101 (1942)

Schizopepon bicirrhosa (C.B. Clarke) C. Jeffrey
Cucurbitaceae
Climbing Mechanism: Tendril Climber
Leaf Type: Simple, cordate at base
Inflorescence: Single axillary flowers
Fruit Type: Dry
IUCN Status: Not evaluated

Schizopepon dioicus Cogn. ex Oliv.
Cucurbitaceae
Climbing Mechanism: Tendril Climber
Distribution (Global): China North-Central, China South-Central, China Southeast
Distribution (India): West Bengal
Leaf Type: Simple, cordate at base
Inflorescence: Single axillary flowers
Fruit Type: Dry
IUCN Status: Not evaluated
Reference: Hooker's Icon. Pl. 23: t. 2224 (1892)

Sechium edule (Jacq.) Sw.
Cucurbitaceae
Synonyms: *Sicyos edulis* Jacq., *Chayota edulis* (Jacq.) Jacq., *Sechium americanum* Poir., *Sechium edule* subsp., *sylvestre* Lira & Castrejón., *Sicyos laciniatus* Descourt
Common Name: Chayote, Chow-chow, Pipinola
Climbing Mechanism: Tendril Climber
Distribution (Global): Belize, Mexico Central, Mexico Gulf, Mexico Northeast, Mexico Northwest, Mexico Southeast, Mexico Southwest
Distribution (India): Himachal Pradesh, Maharashtra, Manipur, Odisha, Tamil Nadu, West Bengal
Flowering and Fruiting: July–January
IUCN Status: Not evaluated
Reference: Enum. Syst. Pl.: 32 (1760)

Siraitia sikkimensis (Chakrav.) C.Jeffrey
Cucurbitaceae
Synonyms: *Neoluffa sikkimensis* Chakrav.
Climbing Mechanism: Tendril Climber
Distribution (Global): China South-Central, East Himalaya
IUCN Status: Not evaluated
Reference: Kew Bull. 36: 737 (1982)

Solena amplexicaulis (Lam.) Gandhi
Cucurbitaceae
Synonyms: *Bryonia amplexicaulis* Lam., *Bryonia heterophylla* Raeusch. +10
Common Name: Creeping cucumber, Diversely leaved melothria
Climbing Mechanism: Tendril Climber
Distribution (Global): Assam, Bangladesh, East Himalaya, India, Myanmar, Pakistan
Distribution (India): Andhra Pradesh, Assam, Kerala, Maharashtra, Manipur, Odisha, Tamil Nadu, Telangana, West Bengal, Western Ghats
Leaf Type: Simple, polymorphic
Inflorescence: Umbellate, clusters
Fruit Type: Wet
Flowering and Fruiting: April–January
IUCN Status: Not evaluated
Notes: Often found in deciduous forests
Reference: C.J. Saldanha & D.H. Nicolson, Fl. Hassan Distr.: 179 (1976)

Solena angulata (Chakrav.) Babu
Cucurbitaceae
Synonyms: *Solena umbellata* (J.G. Klein ex Willd.) W.J.de Wilde & Duyfjes., *Bryonia teedonda* Roxb. ex Willd., *Bryonia umbellata* J.G. Klein ex Willd., Karivia umbellata (J.G. Klein ex Willd.) Arn., Melothria angulata Chakrav., *Momordica umbellata* (J.G. Klein ex Willd.) R
Climbing Mechanism: Tendril Climber
Distribution (Global): Assam, Bangladesh, India, Myanmar, Sri Lanka
Distribution (India): Tamil Nadu
Leaf Type: Leaves plymorphic
Inflorescence: Racemes
Fruit Type: Pepo, ribbed, apically beaked
Flowering and Fruiting: July–January
IUCN Status: Not evaluated
Notes: Often found in deciduous forests
Reference: Blumea 49: 77 (2004)

Solena heterophylla Lour.
Cucurbitaceae
Synonyms: *Melothria heterophylla* (Lour.) Cogn., *Melothria marginata* var. *heterophylla* (Lour.) Boerl., Zehneria heterophylla (Lour.) Druce.
Common Name: Himalayan creeping cucumber
Climbing Mechanism: Tendril Climber
Distribution (Global): Afghanistan, Assam, Bangladesh, Cambodia, China South-Central, China Southeast, East Himalaya, India, Java, Laos, Malaya, Myanmar, Nepal, Taiwan, Thailand, Tibet, Vietnam, West Himalaya
Distribution (India): Gujarat, Uttar Pradesh
Leaf Type: Variable, undivided

Inflorescence: Solitary
Fruit Type: Pepo, reddish brown
Flowering and Fruiting: April–August
IUCN Status: Not evaluated
Reference: Fl. Cochinch.: 514 (1790)

Solena umbellata (Klein ex Willd.) W.J.de Wilde & Duyfjes
Cucurbitaceae
Synonyms: *Bryonia teedonda* Roxb. ex Willd., *Bryonia umbellata* J.G. Klein ex Willd., *Karivia umbellata* (J.G. Klein ex Willd.) Arn., *Melothria angulata* Chakrav., *Momordica umbellata* (J.G. Klein ex Willd.) Roxb., *Solena angulata* (Chakrav.) Raizada
Climbing Mechanism: Tendril Climber
Distribution (India): Tamil Nadu
Leaf Type: Simple, polymorphic
Inflorescence: Umbellate, clusters
Fruit Type: Wet
IUCN Status: Not evaluated
Reference: Blumea 49: 77 (2004)

Thladiantha cordifolia (Blume) Cogn.
Cucurbitaceae
Synonyms: *Gymnopetalum horsfleldii* Miq., *Thladiantha cordifolia* var. *tonkinensis* (Cogn.) A.M. Lu & Zhi Y. Zhang. +10
Common Name: Himalayan goldencreeper, Himalayan tuber-gourd
Climbing Mechanism: Tendril Climber
Distribution (Global): Assam, Bangladesh, China South-Central, China Southeast, East Himalaya, India, Java, Laos, Lesser Sunda Is., Malaya, Myanmar, Nepal, Sumatera, Thailand, Vietnam
Distribution (India): Andhra Pradesh, Arunachal Pradesh, Manipur, Mizoram, West Bengal
Leaf Type: Simple, cordate at base
Inflorescence: Rachis into pseudoraceme
Fruit Type: Wet, drupe
Flowering and Fruiting: May–October
IUCN Status: Not evaluated
Reference: Thladiantha cordifolia var. tonkinensis (Cogn.) A.M. Lu & Zhi Y. Zhang

Thladiantha dubia Bunge
Cucurbitaceae
Synonyms: *Thladiantha sparsiflora* E.H.L. Krause
Climbing Mechanism: Tendril Climber
Distribution (Global): China North-Central, China South-Central, Inner Mongolia, Khabarovsk, Korea, Manchuria, Primorye, Sakhalin
Leaf Type: Simple, cordate at base

Inflorescence: Rachis into pseudoraceme
Fruit Type: Wet, drupe
IUCN Status: Not evaluated
Reference: Enum. Pl. China Bor.: 29 (1833)

Thladiantha hookeri C.B. Clarke
Cucurbitaceae
Synonyms: *Hemsleya tonkinensis* Cogn., *Hemsleya tonkinensis* Cogn., *Hemsleya yunnanensis* Cogn., *Thladiantha digitata* H. Lév., *Thladiantha heptadactyla* Cogn., *Thladiantha hookeri* var. *heptadactyla* (Cogn.) A.M. Lu & Zhi Y. Zhang., *Thladiantha hookeri* var. *irregularis* Chak.
Climbing Mechanism: Tendril Climber
Distribution (Global): Assam, Bangladesh, China South-Central, East Himalaya, Laos, Myanmar, Thailand, Vietnam
Distribution (India): Manipur, Mizoram
Leaf Type: Simple, cordate at base
Inflorescence: Rachis into pseudoraceme
Fruit Type: Wet, drupe
IUCN Status: Not evaluated
Reference: J. D. Hooker, Fl. Brit. India 2: 631 (1879)

Trichosanthes anaimalaiensis Bedd.
Cucurbitaceae
Synonyms: *Anguina anamalayna* (Bedd.) Kuntze., *Trichosanthes anamalayana* Bedd., *Trichosanthes tricuspidata* var. *tomentosa* (B. Heyne ex C.B. Clarke) Kumari
Climbing Mechanism: Tendril Climber
Distribution (Global): Andaman Is., Assam, East Himalaya, India, Sri Lanka
Distribution (India): Andhra Pradesh, Tamil Nadu
Leaf Type: Simple, palmately compound
Inflorescence: Racemes
Fruit Type: Wet, globose
Flowering and Fruiting: April–September
IUCN Status: Not evaluated
Notes: Usually found in grasslands and margins of shola forests
Reference: Madras J. Lit. Sci., ser. 3, 1: 47 (1864)

Trichosanthes cordata Roxb.
Cucurbitaceae
Synonyms: *Anguina cordata* (Roxb.) Kuntze., *Trichosanthes macrosiphon* Kurz., *Trichosanthes tuberosa* Roxb. ex Wight & Arn.
Climbing Mechanism: Tendril Climber
Distribution (Global): Assam, Bangladesh, East Himalaya, India, Laos, Myanmar, Nepal, Tibet, West Himalaya

Distribution (India): Andhra Pradesh, Arunachal Pradesh, Manipur, Mizoram, Rajasthan, Tripura, West Bengal
Leaf Type: Leaves simple, alternate
Inflorescence: Solitary, racemes
Fruit Type: White banded, globose
Flowering and Fruiting: June–December
IUCN Status: Not evaluated
Reference: Fl. Ind. ed. 1832, 3: 703 (1832)

Trichosanthes cucumerina L.
Cucurbitaceae
Synonyms: *Anguina cucumerina* (L.) Kuntze
Common Name: Wild snake gourd
Climbing Mechanism: Tendril Climber
Distribution (Global): Assam, Bangladesh, Cambodia, China South-Central, China Southeast, East Himalaya, India, Java, Laccadive Is., Lesser Sunda Is., Maldives, Myanmar, Nepal, Northern Territory, Pakistan, Philippines, Queensland, Sri Lanka, Sulawesi, Thailand, Vietnam, West Himalaya, Western Australia
Distribution (India): Andhra Pradesh, Arunachal Pradesh, Bihar, Odisha, Delhi, Gujarat, Himachal Pradesh, Karnataka, Madhya Pradesh, Maharashtra, Odisha, Rajasthan, Tamil Nadu, Telangana, Uttar Pradesh, West Bengal
Leaf Type: Leaves simple, alternate
Inflorescence: Solitary, raceme
Flowering and Fruiting: June–May
IUCN Status: Not evaluated
Notes: Often found in margins of semi-evergreen and moist deciduous forests, also in the plains
Reference: Sp. Pl.: 1008 (1753)

Trichosanthes cuspidata Lam.
Cucurbitaceae
Synonyms: *Trichosanthes nervifolia* L., *Anguina cuspidata* (Lam.) Kuntze, *Anguina nervifolia* (L.) Kuntze., *Trichosanthes caudata* Willd.
Climbing Mechanism: Tendril Climber
Distribution (Global): Assam, East Himalaya, India, Sri Lanka
Leaf Type: Oavte-lanceolate
Inflorescence: Racemes
Flowering and Fruiting: September–January
IUCN Status: Not evaluated
Notes: Often distributed in moist deciduous forests of the Western Ghats
Reference: Sp. Pl.: 1008 (1753)

Trichosanthes dioica Roxb.
Cucurbitaceae
Synonyms: *Anguina dioica* (Roxb.) Kuntze

Common Name: Pointed gourd
Climbing Mechanism: Tendril Climber
Distribution (Global): Assam, Bangladesh, East Himalaya, India, Myanmar, Nepal, Pakistan, Sri Lanka, West Himalaya
Distribution (India): Arunachal Pradesh, Bihar, Odisha, Delhi, Gujarat, Himachal Pradesh, Madhya Pradesh, Maharashtra, Odisha, West Bengal
Leaf Type: Simple, palmately compound
Inflorescence: Racemes
Fruit Type: Wet, globose
Flowering and Fruiting: April–September
IUCN Status: Not evaluated
Reference: Fl. Ind. ed. 1832, 3: 701 (1832)

Trichosanthes hylonoma Hand. -Mazz.
Cucurbitaceae
Synonyms: *Trichosanthes leishanensis* C.Y. Cheng & C.H. Yueh, *Trichosanthes parviflora* C.Y. Wu ex S.K. Chen, *Trichosanthes zhejiangensis* C.H. Yueh & Lu Q. Huang
Climbing Mechanism: Tendril Climber
Distribution (Global): China South-Central, China Southeast
Leaf Type: Simple, palmately compound
Inflorescence: Racemes
Fruit Type: Wet, globose
IUCN Status: Not evaluated
Reference: Symb. Sin. 7: 1066 (1936)

Trichosanthes kerrii W.G. Craib
Cucurbitaceae
Synonyms: *Trichosanthes tomentosa* Chakrav.
Climbing Mechanism: Tendril Climber
Distribution (Global): Assam, China South-Central, China Southeast, Laos, Thailand, Vietnam
Leaf Type: Simple, palmately compound
Inflorescence: Racemes
Fruit Type: Wet, globose
IUCN Status: Not evaluated
Reference: Bull. Misc. Inform. Kew 1914: 7 (1914)

Trichosanthes lepiniana (Naudin) Cogn.
Cucurbitaceae
Synonyms: *Anguina lepiniana* (Cogn.) Kuntze., Involucraria lepiniana Naudin.
Climbing Mechanism: Tendril Climber
Distribution (Global): India.
Distribution (India): Karnataka, West Bengal
Leaf Type: Leaves alternate

Inflorescence: Male flowers, racemes, Females-Raceme
Fruit Type: Red, ovoid
Flowering and Fruiting: April–November
IUCN Status: Not evaluated
Reference: A.L.P.P.de Candolle & A.C.P.de Candolle, Monogr. Phan. 3: 377 (1881)

Trichosanthes lobata Roxb.
Cucurbitaceae
Synonyms: *Trichosanthes cucumerina* subsp. *Cucumerina*., *Trichosanthes cucumerina* var. *anguina* (L.) Haines. + 20
Common Name: Wild snake gourd, Bitter snake gourd
Climbing Mechanism: Tendril Climber
Distribution (Global): Assam, Bangladesh, Cambodia, China South-Central, China Southeast, East Himalaya, India, Java, Laccadive Is., Lesser Sunda Is., Maldives, Myanmar, Nepal, Northern Territory, Pakistan, Philippines, Queensland, Sri Lanka, Sulawesi, Thailand, Vietnam, West Himalaya, Western Australia
Distribution (India): Andhra Pradesh, Karnataka, Tamil Nadu
Leaf Type: 5-lobed
Inflorescence: Male flowers, racemes, Females-Raceme
Fruit Type: Berry
Flowering and Fruiting: Spetember–December
IUCN Status: Not evaluated
Notes: Usually found in evergreen forests
Reference: Roxb. In: Hort. Beng. 70 (1814), nomen, Fl. Ind. 3: 703. (1832)

Trichosanthes multiloba C.B. Clarke
Cucurbitaceae
Synonyms: *Trichosanthes wallichiana* (Ser.) Wight., *Anguina wallichiana* (Ser.) Kuntze, *Involucraria wallichiana* Ser., *Trichosanthes grandibracteata* Kurz.
Climbing Mechanism: Tendril Climber
Distribution (Global): Assam, China South-Central, China Southeast, East Himalaya, India, Malaya, Myanmar, Nepal, Tibet, West Himalaya
Leaf Type: Simple, palmately compound
Inflorescence: Racemes
Fruit Type: Wet, globose
IUCN Status: Not evaluated
Reference: Ann. Mag. Nat. Hist. 8: 270 (1842)

Trichosanthes nervifolia L.
Cucurbitaceae
Climbing Mechanism: Tendril Climber
Distribution (India): Tamil Nadu
Leaf Type: Simple, palmately compound
Inflorescence: Racemes
Fruit Type: Wet, globose

IUCN Status: Not evaluated
Reference: Published in: L. In: Sp. Pl. 1008. (1753)

Trichosanthes perrottetiana Cogn.
Cucurbitaceae
Synonyms: *Trichosanthes cucumerina* subsp. *villosula* (Cogn.) K. Pradheep, Pani & K.C. Bhatt., *Anguina perrottetiana* (Cogn.) Kuntze, *Anguina villosula* (Cogn.) Kuntze, *Trichosanthes villosula* Cogn.
Climbing Mechanism: Tendril Climber
Leaf Type: Simple, palmately compound
Inflorescence: Racemes
Fruit Type: Wet, globose
IUCN Status: Not evaluated
Notes: Often found in moist deciduous forests of the Western Ghats
Reference: Novon 24: 41 (2015)

Trichosanthes pilosa Lour.
Cucurbitaceae
Synonyms: *Anguina pilosa* (Lour.) Kuntze., *Trichosanthes horsfieldii* Miq.
Climbing Mechanism: Tendril Climber
Distribution (Global): Assam, Borneo, China South-Central, China Southeast, East Himalaya, Hainan, Japan, Java, Kazan-retto, Lesser Sunda Is., Malaya, Maluku, Myanmar, Nansei-shoto, New Guinea, Northern Territory, Ogasawara-shoto, Philippines, Queensland, Sulawesi, Sumatera, Taiwan, Thailand, Tibet, Vietnam, Western Australia
Distribution (India): West Bengal
Leaf Type: Simple, palmately compound
Inflorescence: Racemes
Fruit Type: Wet, globose
IUCN Status: Not evaluated
Reference: Fl. Cochinch.: 588 (1790)

Trichosanthes rosthornii Harms
Cucurbitaceae
Climbing Mechanism: Tendril Climber
Distribution (Global): China South-Central, China Southeast
Leaf Type: Simple, palmately compound
Inflorescence: Racemes
Fruit Type: Wet, globose
IUCN Status: Not evaluated
Reference: Bot. Jahrb. Syst. 29: 603 (1901)

Trichosanthes tricuspidata Lour.
Cucurbitaceae
Synonyms: *Anguina tricuspidata* (Lour.) Kuntze

Climbing Mechanism: Tendril Climber
Distribution (Global): Andaman Is., Borneo, Cambodia, Java, Laos, Lesser Sunda Is., Malaya, Maluku, Myanmar, Nicobar Is., Sulawesi, Sumatera, Thailand, Vietnam
Distribution (India): Andhra Pradesh, Arunachal Pradesh, Karnataka, Kerala, Maharashtra, Manipur, Meghalaya, Mizoram, Odisha, Tamil Nadu, Telangana
Leaf Type: Orbicular-cordate
Inflorescence: Male- axillary racemes
Fruit Type: Berries, globose red
Flowering and Fruiting: November–May
IUCN Status: Not evaluated
Notes: Found mostly in deciduous forests and also in the plains
Reference: Fl. Cochinch.: 529 (1790)

Trichosanthes tricuspidata var. *strigosa*
Cucurbitaceae
Synonyms: *Trichosanthes tricuspidata* var. *strigosa.*, *Anguina bracteata* (Lam.) Kuntze, *Trichosanthes hupehensis* C.Y. Cheng & Yueh, *Trichosanthes palmata* Roxb. *Modecca bracteata* Lam.
Climbing Mechanism: Tendril Climber
Distribution (Global): Assam, Bangladesh, China South-Central, East Himalaya, India, Myanmar, Nepal, Sri Lanka, West Himalaya
Distribution (India): West Bengal
Leaf Type: Simple, palmately compound
Inflorescence: Racemes
Fruit Type: Wet, globose
IUCN Status: Not evaluated
Reference: Hort. Suburb. Calcutt.: 58 (1845)

Trichosanthes tricuspidata var. *tomentosa*
Cucurbitaceae
Climbing Mechanism: Tendril Climber
Distribution (India): Tamil Nadu.
Leaf Type: Simple, palmately compound
Inflorescence: Racemes
Fruit Type: Wet, globose
IUCN Status: Not evaluated
Reference: Kumari. In: N. C. Nair & A. N Henry, Fl. Tamil Nadu, India, Ser. 1, l: 174. (1983)

Trichosanthes truncata C.B. Clarke
Cucurbitaceae
Synonyms: *Anguina ovata* (Cogn.) Kuntze, *Anguina truncata* (C.B. Clarke) Kuntze, *Trichosanthes crispisepala* C.Y. Wu ex S.K. Chen, *Trichosanthes ovata* Cogn.
Climbing Mechanism: Tendril Climber

Distribution (Global): Assam, Bangladesh, East Himalaya, India, Myanmar, Thailand, Vietnam
Distribution (India): Andhra Pradesh, Arunachal Pradesh, West Bengal
Leaf Type: Simple, palmately compound
Inflorescence: Racemes
Fruit Type: Wet, globose
IUCN Status: Not evaluated
Reference: J. D. Hooker, Fl. Brit. India 2: 608 (1879)

Trichosanthes villosula Cogn.
Cucurbitaceae
Climbing Mechanism: Tendril Climber
Distribution (India): Tamil Nadu.
Leaf Type: Simple, palmately compound
Inflorescence: Racemes
Fruit Type: Wet, globose
IUCN Status: Not evaluated
Reference: Cogn. In: DC. Monog. Phan. 3: 362. (1881)

Trichosanthes wallichiana (Ser.) Wight
Cucurbitaceae
Climbing Mechanism: Tendril Climber
Distribution (India): Arunachal Pradesh, Himachal Pradesh, Karnataka, Maharashtra, Manipur, Meghalaya, Tripura, West Bengal
Leaf Type: Simple, palmately compound
Inflorescence: Racemes
Fruit Type: Wet, globose
IUCN Status: Not evaluated
Reference: Wight. In: Madras J. Lit. Sci. 12: 52. (1840)

Zanonia indica L.
Cucurbitaceae
Climbing Mechanism: Tendril Climber
Distribution (Global): Andaman Is., Assam, Bangladesh, Borneo, Cambodia, China South-Central, China Southeast, East Himalaya, India, Java, Laos, Malaya, Maluku, Myanmar, New Guinea, Philippines, Sri Lanka, Sulawesi, Sumatera, Thailand, Vietnam
Distribution (India): Maharashtra, Tamil Nadu, West Bengal
Leaf Type: Ovate, oblong
Inflorescence: Raceme
Fruit Type: pepo
Flowering and Fruiting: October–November
IUCN Status: Not evaluated
Reference: Syst. Nat. ed. 10, 2: 1292 (1759)

Zehneria angulata (Chakrav.) J.L. Ellis
Cucurbitaceae
Synonyms: *Solena umbellata* (J.G. Klein ex Willd.) W.J.de Wilde & Duyfjes., *Bryonia teedonda* Roxb. ex Willd., *Bryonia umbellata* J.G. Klein ex Willd., *Karivia umbellata* (J.G. Klein ex Willd.) Arn., *Melothria angulata* Chakrav., *Momordica umbellata* (J.G. Klein ex Willd.)
Climbing Mechanism: Tendril Climber
Distribution (Global): Assam, Bangladesh, India, Myanmar, Sri Lanka
Distribution (India): Peninsular India
Leaf Type: Alternate
Inflorescence: Subumbellate
Flowering and Fruiting: January–May
IUCN Status: Not evaluated
Notes: Often found in deciduous forests
Reference: Blumea 49: 77 (2004)

Zehneria baueriana C.B. Clarke
Cucurbitaceae
Synonyms: *Melothria baueriana* (Endl.) F. Muell., *Pilogyne zehneria* Steud.
Climbing Mechanism: Tendril Climber
Distribution (Global): New Caledonia, Norfolk Is., Vanuatu
Leaf Type: Simple, broadly ovate-subdeltoid
Inflorescence: Umbellate racemes in clusters
Fruit Type: Wet, globose
IUCN Status: Not evaluated
Reference: Prodr. Fl. Norfolk.: 69 (1833)

Zehneria bodinieri (H. Lév.) W.J.de Wilde & Duyfjes
Cucurbitaceae
Synonyms: *Bryonia cissoides* Wall., *Bryonia oxyphylla* Wall., *Melothria bodinieri* H. Lév., *Pilogyne bodinieri* (H. Lév.) W.J.de Wilde & Duyfjes.
Common Name: Asian Zehneria, Emile bodinieri Zehneria, Wild cucumber
Climbing Mechanism: Tendril Climber
Distribution (Global): Borneo, Cambodia, China South-Central, China Southeast, Hainan, India, Laos, Malaya, Myanmar, Philippines, Sri Lanka, Sumatera, Thailand, Vietnam
Distribution (India): Tamil Nadu
Leaf Type: Simple, broadly ovate-subdeltoid
Inflorescence: Umbellate racemes in clusters
Fruit Type: Wet, globose
IUCN Status: Not evaluated
Reference: Thai Forest Bull., Bot. 32: 17 (2004)

Zehneria hookeriana Arn.
Cucurbitaceae

Synonyms: *Bryonia hookeriana* Wight & Arn., *Pilogyne hookeriana* (Wight & Arn.) W.J.de Wilde & Duyfjes.
Climbing Mechanism: Tendril Climber
Distribution (Global): India
Distribution (India): Tamil Nadu
Leaf Type: Simple, broadly ovate-subdeltoid
Inflorescence: Umbellate racemes in clusters
Fruit Type: Wet, globose
IUCN Status: Not evaluated
Reference: J. Bot. (Hooker) 3: 275 (1841)

Zehneria japonica (Thunb.) H.Y. Liu
Cucurbitaceae
Synonyms: *Bryonia japonica* Thunb., *Cucumis luzonicus* Blanco., *Melothria argyi* H. Lév., *Melothria formosana* Hayata., *Melothria japonica* (Thunb.) Maxim. ex Cogn., *Melothria regelii* Naudin., *Neoachmandra japonica* (Thunb.) W.J.de Wilde & Duyfjes., *Zehneria formosana*
Climbing Mechanism: Tendril Climber
Distribution (Global): Assam, Bangladesh, Borneo, China South-Central, China Southeast, East Himalaya, India, Japan, Java, Laos, Malaya, Myanmar, Philippines, Taiwan, Thailand, Vietnam, West Himalaya
Distribution (India): West Bengal
Leaf Type: Simple, broadly ovate-subdeltoid
Inflorescence: Umbellate racemes in clusters
Fruit Type: Wet, globose
IUCN Status: Not evaluated
Reference: Bull. Natl. Mus. Nat. Sci., Taichung 1: 40 (1989)

Zehneria maysorensis (Wight & Arn.) Arn.
Cucurbitaceae
Synonyms: *Bryonia maysorensis* Wight & Arn., *Melothria lucida* (Naudin ex C. Huber) Cogn., *Melothria maysorensis* (Wight & Arn.) C.C. Chang, *Pilogyne lucida* Naudin ex C. Huber, *Pilogyne maysorensis* (Wight & Arn.) W.J.de Wilde & Duyfjes, *Zehneria lucida* (Naudin ex C. Huber)
Climbing Mechanism: Tendril Climber
Distribution (Global): Assam, East Himalaya, India, Laos, Myanmar, Nepal, Sri Lanka, Vietnam
Distribution (India): Andhra Pradesh, Karnataka, Kerala, Odisha, Tamil Nadu, Telangana, West Bengal
Leaf Type: Simple, broadly ovate-subdeltoid
Inflorescence: Umbellate racemes in clusters
Fruit Type: Wet, globose
Flowering and Fruiting: June–November
IUCN Status: Not evaluated
Reference: J. Bot. (Hooker) 3: 275 (1841)

Zehneria maysorensis var. *oblonga*
Cucurbitaceae
Climbing Mechanism: Tendril Climber
Distribution (India): Peninsular India
Leaf Type: Simple, broadly ovate-subdeltoid
Inflorescence: Umbellate racemes in clusters
Fruit Type: Wet, globose
IUCN Status: Not evaluated
Reference: Arn. In: J. Econ. Taxon. Bot., 17(2): 471. (1993)

Zehneria maysorensis var. *umbellata*
Cucurbitaceae
Climbing Mechanism: Tendril Climber
Distribution (India): Tamil Nadu
Leaf Type: Simple, broadly ovate-subdeltoid
Inflorescence: Umbellate racemes in clusters
Fruit Type: Wet, globose
IUCN Status: Not evaluated
Reference: Arn. In: Hook., J. B. 3: 275. (1841)

Zehneria mucronata (Blume) Miq.
Cucurbitaceae
Synonyms: *Bryonia arguta* Zipp. ex Span., *Bryonia johnstonii* Seem., *Bryonia mucronata* Blume, *Karivia samoensis* A. Gray, *Melothria carnosula* Cogn., *Melothria grayana* Cogn. +10
Climbing Mechanism: Tendril Climber
Distribution (Global): Bismarck Archipelago, Caroline Is., Christmas I., Fiji, Java, Lesser Sunda Is., Maluku, Marquesas, New Guinea, Niue, Northern Territory, Philippines, Queensland, Samoa, Society Is., Solomon Is., South Australia, Sulawesi, Sumatera, Tonga, Western Australia
Distribution (India): Maharashtra
Leaf Type: Simple, broadly ovate-subdeltoid
Inflorescence: Umbellate racemes in clusters
Fruit Type: Wet, globose
Flowering and Fruiting: December–February
IUCN Status: Not evaluated
Reference: Fl. Ned. Ind. 1(1): 656 (1856)

Zehneria scabra Sond.
Cucurbitaceae
Synonyms: *Bryonia scabra* L.f., *Bryonia scabrata* Blume, *Cucurbita scabra* (L.f.) Blume, *Pilogyne scabra* (L.f.) W.J.de Wilde & Duyfjes.
Climbing Mechanism: Tendril Climber
Distribution (Global): Angola, Assam, Bangladesh, Benin, Burkina, Burundi, Cameroon, Cape Provinces, Congo, East Himalaya, Eritrea, Ethiopia, Free

State, Gulf of Guinea Is., India, Ivory Coast, Java, Kenya, KwaZulu-Natal, Lesotho, Lesser Sunda Is., Liberia, Malawi, Mozambique, Northern Provinces, Philippines, Rwanda, Saudi Arabia, Somalia, Sudan, Sulawesi, Swaziland, Tanzania, Uganda, Yemen, Zambia, Zaire, Zimbabwe
Distribution (India): Andhra Pradesh, Maharashtra, Tamil Nadu, Telangana
Leaf Type: Ovat, cordate at base
Inflorescence: Subumbelliform or shortly racemiform sessile or pedunculate axillary clusters
Fruit Type: Berry, bright red, globose, ellipsoid
Flowering and Fruiting: Throughout the year
IUCN Status: Not evaluated
Notes: Frequently found in semi evergreen and evergreen forests
Reference: W.H. Harvey & auct. suc. (eds.), Fl. Cap. 2: 486 (1862)

Zehneria thwaitesii (Schweinf.) C.Jeffrey
Cucurbitaceae
Synonyms: *Aechmandra deltoidea* Arn., *Bryonia deltoidea* Arn., *Cucumella thwaitesii* (Schweinf.) M.R. Almeida, *Melothria deltoidea* (Arn.) Thwaites., *Melothria thwaitesii* Schweinf., *Melothria zeylanica* C.B. Clarke., *Neoachmandra deltoidea* (Arn.) W.J.de Wilde & Duyfjes.
Climbing Mechanism: Tendril Climber
Distribution (Global): India, Sri Lanka
Distribution (India): Maharashtra, Tamil Nadu
Leaf Type: Variable, ovate to triangular
Inflorescence: Racemose, male flowers-axillary, female flowers-solitary
Fruit Type: Berry
Flowering and Fruiting: August–December
IUCN Status: Not evaluated
Notes: Found across the Western Ghats and the Eastern Ghats
Reference: Kew Bull. 15: 371 (1961 publ. 1962)

Zehneria wallichii (C.B. Clarke) C. Jeffrey
Cucurbitaceae
Synonyms: *Melothria wallichii* C.B. Clarke., *Neoachmandra wallichii* (C.B. Clarke) W.J.de Wilde & Duyfjes.
Climbing Mechanism: Tendril Climber
Distribution (Global): China South-Central, Myanmar, Thailand
Leaf Type: Simple, broadly ovate-subdeltoid
Inflorescence: Umbellate racemes in clusters
Fruit Type: Wet, globose
IUCN Status: Not evaluated
Reference: Kew Bull. 34: 802 (1980)

Dichapetalum laurocerasus (Hook.f.) Engl.
Dichapetalaceae

Synonyms: *Chailletia laurocerasus* Planch. ex Hook.f.
Climbing Mechanism: Scrambler-Unarmed
Distribution (Global): Malaya.
IUCN Status: Not evaluated
Reference: H.G.A. Engler & K.A.E. Prantl, Nat. Pflanzenfam. 3(4): 348 (1896)

Dichapetalum longipetalum (Turcz.) Engl.
Dichapetalaceae
Synonyms: *Chailletia hainanensis* Hance, *Chailletia longipetala* Turcz., *Dichapetalum baillonii* Pierre, *Dichapetalum hainanense* (Hance) Engl., *Dichapetalum tonkinense* Engl.
Climbing Mechanism: Scrambler-Unarmed
Distribution (Global): Cambodia, China Southeast, Hainan, Laos, Malaya, Myanmar, Thailand, Vietnam
IUCN Status: Not evaluated
Reference: H.G.A. Engler & K.A.E. Prantl, Nat. Pflanzenfam. 3(4): 348 (1896)

Dichapetalum platyphyllum Merr.
Dichapetalaceae
Climbing Mechanism: Scrambler-Unarmed
Distribution (Global): Philippines.
Distribution (India): Great Nicobar Island
IUCN Status: Not evaluated
Reference: Philipp. J. Sci. 30: 401 (1926)

Dichapetalum timoriense (DC.) Boerl.
Dichapetalaceae
Synonyms: *Chailletia benthamiana* Turcz., *Chailletia deflexifolia* Turcz., *Chailletia hookeri* King, *Chailletia tesselata* King, *Chailletia timoriensis* DC. + 20
Climbing Mechanism: Scrambler-Unarmed
Distribution (Global): Bismarck Archipelago, Borneo, Java, Lesser Sunda Is., Malaya, Maluku, New Guinea, Nicobar Is., Philippines, Solomon Is., Sulawesi, Vanuatu
Distribution (India): Great Nicobar Island
IUCN Status: Least Concern
Reference: Handl. Fl. Ned. Ind. 1: 199 (1890)

Delima sarmentosa L.
Dilleniaceae
Synonyms: *Tetracera sarmentosa* (L.) Vahl., *Tetracera delima*
Climbing Mechanism: Scrambler-Unarmed
Distribution (Global): Andaman Is., Assam, Bangladesh, Borneo, Cambodia, China South-Central, China Southeast, East Himalaya, Hainan, Java, Laos, Malaya, Maluku, Myanmar, New Caledonia, New Guinea, Nicobar Is., Philippines, Sri Lanka, Sulawesi, Sumatera, Thailand, Vietnam

Distribution (India): Andaman and Nicobar Islands, Assam, Maharashtra
Flowering and Fruiting: February–April
IUCN Status: Not evaluated
Reference: Symb. Bot. 3: 70 (1794)

Eleiastis laevis Raf.
Dilleniaceae
Climbing Mechanism: Scrambler-Unarmed
IUCN Status: Not evaluated
Reference: Sylva Tellur.: 165 (1838)

Tetracera akara Merr.
Dilleniaceae
Synonyms: *Calophyllum akara* Burm.f., *Tetracera axillaris* Martelli., *Tetracera axillaris* Martelli., *Tetracera sericea* Blume., *Tetracera sylvestris* Ridl.
Climbing Mechanism: Scrambler-Unarmed
Distribution (Global): Borneo, Cambodia, India, Java, Malaya, Philippines, Sri Lanka, Sulawesi, Sumatera, Thailand, Vietnam
Distribution (India): Kerala, Tamil Nadu
Leaf Type: Leaves simple, usually scabrid
Inflorescence: Terminal or lateral panicles
Fruit Type: Pseudocarps, carpels follicular, 1–4 globose
Flowering and Fruiting: March–May
IUCN Status: Not evaluated
Notes: Found mostly in semi-evergreen forests and also in sacred groves
Reference: Philipp. J. Sci. 19: 366 (1921)

Tetracera indica (Christm. & Panz.) Merr.
Dilleniaceae
Synonyms: *Wahlbomia indica* (Christm. & Panz.) Thunb., *Roehlingia suaveolens* Dennst., *Tetracera trigyna* Roxb. + 10
Climbing Mechanism: Scrambler-Unarmed
Distribution (Global): Bangladesh, Cambodia, Java, Laos, Lesser Sunda Is., Malaya, Myanmar, Sumatera, Thailand, Vietnam
Distribution (India): Peninsular India
Leaf Type: Leaves simple, usually scabrid
Inflorescence: Terminal or lateral panicles
Fruit Type: Dry dehiscent follicles globose
Flowering and Fruiting: March–May
IUCN Status: Not evaluated
Notes: Found in open and exposed areas, wastelands and in roadsides
Reference: Interpr. Herb. Amboin.: 367 (1917)

Tetracera sarmentosa (L.) Vahl
Dilleniaceae

Synonyms: *Delima sarmentosa* L., *Tetracera delima*
Climbing Mechanism: Scrambler-Unarmed
Distribution (Global): Andaman Is., Assam, Bangladesh, Borneo, Cambodia, China South-Central, China Southeast, East Himalaya, Hainan, Java, Laos, Malaya, Maluku, Myanmar, New Caledonia, New Guinea, Nicobar Is., Philippines, Sri Lanka, Sulawesi, Sumatera, Thailand, Vietnam
Distribution (India): Arunachal Pradesh, Manipur, Meghalaya, Tamil Nadu, Tripura, West Bengal
Flowering and Fruiting: February–April
IUCN Status: Not evaluated
Reference: Symb. Bot. 3: 70 (1794)

Tetracera scandens (L.) Merr.
Dilleniaceae
Synonyms: *Delima aspera* (Lour.) Blanco., *Delima frangulifolia* C. Presl., *Tetracera hebecarpa* (DC.) Boerl., *Traxilisa aspera* (Lour.) Raf. + 14
Climbing Mechanism: Scrambler-Unarmed
Distribution (Global): Bangladesh, Borneo, Cambodia, Java, Laos, Lesser Sunda Is., Malaya, Myanmar, Nicobar Is., Philippines, Sulawesi, Sumatera, Thailand, Vietnam
Distribution (India): Arunachal Pradesh
IUCN Status: Not evaluated
Reference: Interpr. Herb. Amboin.: 365 (1917)

Dioscorea alata L.
Dioscoreaceae
Synonyms: *Dioscorea atropurpurea* Roxb., *Dioscorea colocasiifolia* Pax. + 10
Common Name: Greater yam, Asiatic yam, Yam, Water yam, White yam
Climbing Mechanism: Stem Twiner
Distribution (Global): Assam, Bangladesh, Bismarck Archipelago, Borneo, Cambodia, Christmas I., East Himalaya, India, Java, Lesser Sunda Is., Malaya, Myanmar, Nepal, New Guinea, Philippines, Sulawesi, Sumatera, Thailand, Vietnam
Distribution (India): Andaman and Nicobar Islands, Andhra Pradesh, Arunachal Pradesh, Assam, Bihar, Odisha, Gujarat, Kerala, Madhya Pradesh, Odisha, Sikkim, Tamil Nadu, Telangana, Tripura, West Bengal
Leaf Type: Opposite or rarely subopposite
Inflorescence: Spike on axillary branchlet
Fruit Type: Capsule
Flowering and Fruiting: Throughout the year
IUCN Status: Not evaluated
Notes: Often cultivated for its edible tubers, also naturalized
Reference: Sp. Pl.: 1033 (1753)

Dioscorea belophylla (Prain) Voigt ex Haines
Dioscoreaceae
Synonyms: *Dioscorea nummularia* var. *belophylla* Prain, *Dioscorea sagittata* Royle
Climbing Mechanism: Stem Twiner
Distribution (Global): Assam, Bangladesh, East Himalaya, India, Nepal, West Himalaya
Distribution (India): Andaman and Nicobar Islands, Andhra Pradesh, Arunachal Pradesh, Gujarat, Himachal Pradesh, Jharkhand, Madhya Pradesh, Odisha, Tamil Nadu
Leaf Type: Simple, opposite
Inflorescence: Panicles
Fruit Type: Capsule
Flowering and Fruiting: September–March
IUCN Status: Not evaluated
Notes: Often found in moist deciduous forests
Reference: Forest Fl. Chota Nagpur: 530 (1910)

Dioscorea bulbifera L.
Dioscoreaceae
Synonyms: *Dioscorea anthropophagorum* A. Chev., *Dioscorea bulbifera* var. *albotuberosa* Y.F. Zhou, Z.L. Xu & Y.Y. Hang, *Dioscorea bulbifera* var. *brachybotryum* Y.Y. Hang & Y.F. Zhou. +18
Common Name: Potato yam, Air potato, Bulb bearing yam
Climbing Mechanism: Stem Twiner
Distribution (Global): Pan tropical
Distribution (India): Andaman and Nicobar Islands, Andhra Pradesh, Arunachal Pradesh, Assam, Delhi, Gujarat, Himachal Pradesh, Jharkhand, Kerala, Madhya Pradesh, Odisha, Rajasthan, Tamil Nadu, Telangana, Tripura, Uttar Pradesh, West Bengal
Leaf Type: Ovate-suborbicular
Inflorescence: Axillary, panicled spikes
Fruit Type: Capsule
Flowering and Fruiting: July–March
IUCN Status: Not evaluated
Notes: Often found on hills, climbing on thickets along roadsides, on bushes of scrub and deciduous forests
Reference: Sp. Pl.: 1033 (1753)

Dioscorea composita Hemsl.
Dioscoreaceae
Synonyms: *Dioscorea tepinapensis* Uline ex R. Knuth, *Dioscorea tepinapensis* var. *aggregata* Uline ex R. Knuth.
Climbing Mechanism: Stem Twiner
Distribution (Global): Belize, Costa Rica, El Salvador, Guatemala, Honduras, Mexico Central, Mexico Gulf, Mexico Southeast, Mexico Southwest

Distribution (India): Peninsular India
Leaf Type: Simple/compound, 3–5 foliolate
Inflorescence: Racemes/panicles, spikes
Fruit Type: Dry, capsule
Flowering and Fruiting: September–December
IUCN Status: Not evaluated
Reference: Biol. Cent. -Amer., Bot. 3: 354 (1884)

Dioscorea cumingii Prain & Burkill
Dioscoreaceae
Synonyms: *Dioscorea echinata* R. Knuth, *Dioscorea elmeri* Prain & Burkill, *Dioscorea heptaphylla* Sasaki, *Dioscorea inaequifolia* Elmer ex Prain & Burkill, *Dioscorea polyphylla* R. Knuth.
Climbing Mechanism: Stem Twiner
Distribution (Global): Philippines, Taiwan
Leaf Type: Simple/compound, 3–5 foliolate
Inflorescence: Racemes/panicles, spikes
Fruit Type: Dry, capsule
IUCN Status: Not evaluated
Reference: J. Proc. Asiat. Soc. Bengal 4: 449 (1908)

Dioscorea deltoidea Wall. ex Griseb.
Dioscoreaceae
Synonyms: *Dioscorea nepalensis* (Jacquem. ex Prain & Burkill) Sweet ex Bernardi, *Tamus nepalensis* Jacquem. ex Prain & Burkill.
Common Name: Nepal yam
Climbing Mechanism: Stem Twiner
Distribution (Global): Assam, Bangladesh, China South-Central, East Himalaya, India, Myanmar, Nepal, Thailand, Tibet, Vietnam, West Himalaya
Distribution (India): North Eastern India, Uttar Pradesh
Leaf Type: Simple, alternate
Inflorescence: Solitary
Fruit Type: Capsule
Flowering and Fruiting: May–July
IUCN Status: Not evaluated
Notes: Medicinally valuable: Used to alleviate constipation
Reference: C.F.P.von Martius & auct. suc. (eds.), Fl. Bras. 3(1): 43 (1842)

Dioscorea esculenta (Lour.) Burkill
Dioscoreaceae
Climbing Mechanism: Stem Twiner
Distribution (India): Andaman and Nicobar Islands, Andhra Pradesh, Arunachal Pradesh, Madhya Pradesh, Odisha, Rajasthan, Sikkim, Tamil Nadu
Leaf Type: Simple/compound, 3–5 foliolate
Inflorescence: Racemes/panicles, spikes

Fruit Type: Dry, capsule
IUCN Status: Not evaluated
Reference: Burkill. In: Gard. Bull. Straits Settlem. 1: 396. (1917)

Dioscorea floribunda M. Martens & Galeotti
Dioscoreaceae
Climbing Mechanism: Stem Twiner
Distribution (India): Odisha, Peninsular India
Leaf Type: Simple/compound, 3–5 foliolate
Inflorescence: Racemes/panicles, spikes
Fruit Type: Dry, capsule
Flowering and Fruiting: December–January
IUCN Status: Not evaluated
Notes: Often cultivated and grown as medicinal plants
Reference: M. Martens & Galeotti. In: Bull. Acad. Roy. Sci. Bruxelles 9(2): 391. (1842)

Dioscorea glabra Roxb.
Dioscoreaceae
Synonyms: *Dioscorea hongkongensis* Uline ex R. Knuth, *Dioscorea oryzetorum* var. *latifolia* Prain & Burkill, *Dioscorea oryzetorum* var. *mediifolia* Prain & Burkill, *Dioscorea siamensis* R. Knuth.
Climbing Mechanism: Stem Twiner
Distribution (Global): Andaman Is., Assam, Bangladesh, Cambodia, China South-Central, China Southeast, East Himalaya, India, Laos, Malaya, Myanmar, Nepal, Nicobar Is., Thailand, Vietnam, West Himalaya
Distribution (India): Andhra Pradesh, Arunachal Pradesh, Great Nicobar Island, Madhya Pradesh, Odisha, Sikkim, Tripura, West Bengal
Leaf Type: Simple, usually ovate
Fruit Type: Capsule not reflexed, oblate, seeds near middle of capsule
Flowering and Fruiting: September–January
IUCN Status: Not evaluated
Notes: Fairly common in forests and the tubers are edible
Reference: Fl. Ind. ed. 1832, 3: 804 (1832)

Dioscorea hamiltonii Hook.f.
Dioscoreaceae
Synonyms: *Dioscorea persimilis* Prain & Burkill, *Dioscorea persimilis* var. *pubescens* C.T. Ting & M.C. Chang, *Dioscorea raishaensis* Hayata
Climbing Mechanism: Stem Twiner
Distribution (Global): Assam, Bangladesh, China South-Central, China Southeast, East Himalaya, Hainan, India, Myanmar, Nepal, Taiwan, Thailand, Vietnam
Distribution (India): Andaman and Nicobar Islands, Andhra Pradesh, Arunachal Pradesh, Assam, Bihar, Odisha, Madhya Pradesh, Odisha, Sikkim, Tamil Nadu, Tripura

Leaf Type: Simple/compound, 3–5 foliolate
Inflorescence: Racemes/panicles, spikes
Fruit Type: Dry, capsule
Flowering and Fruiting: September–February
IUCN Status: Not evaluated
Notes: Occasionally found in the forests of Northern Eastern Ghats
Reference: Fl. Brit. India 6: 294 (1892)

Dioscorea hispida Dennst.
Dioscoreaceae
Synonyms: *Dioscorea daemona* Roxb., *Dioscorea daemona* var. *reticulata* Hook.f., *Dioscorea hirsuta* Blume, *Dioscorea hispida* var. *neoscaphoides* Prain & Burkill, *Dioscorea hispida* var. *reticulata* (Hook.f.) Prain & Burkill, *Dioscorea hispida* var. *scaphoides* Prain & Burkill
Common Name: Asiatic bitter yam, Intoxicating yam
Climbing Mechanism: Stem Twiner
Distribution (Global): Andaman Is., Assam, Bangladesh, Bismarck Archipelago, Borneo, Cambodia, China Southeast, East Himalaya, Hainan, India, Java, Lesser Sunda Is., Malaya, Maluku, Myanmar, Nepal, New Guinea, Philippines, Queensland, Sulawesi, Sumatera, Taiwan, Thailand, Tibet, Vietnam, West Himalaya
Distribution (India): Andaman and Nicobar Islands, Andhra Pradesh, Gujarat, Madhya Pradesh, Odisha, Rajasthan, Sikkim, Tamil Nadu, Telangana, West Bengal
Leaf Type: Alternate, trifoliate
Inflorescence: Racemes, fascicled
Flowering and Fruiting: June–January
IUCN Status: Not evaluated
Reference: Schlüssel Hortus Malab.: 33 (1818)

Dioscorea intermedia Thwaites
Dioscoreaceae
Climbing Mechanism: Stem Twiner
Distribution (Global): Sri Lanka
Distribution (India): Peninsular India, Madhya Pradesh
Leaf Type: Alternate
Inflorescence: Spikes simple, solitary, or paired
Fruit Type: Capsule
IUCN Status: Not evaluated
Reference: Enum. Pl. Zeyl.: 326 (1864)

Dioscorea japonica Thunb.
Dioscoreaceae
Climbing Mechanism: Stem Twiner

Distribution (Global): Assam, China South-Central, China Southeast, Japan, Korea, Taiwan
Distribution (India): Uttar Pradesh
Leaf Type: Simple/compound, 3–5 foliolate
Inflorescence: Racemes/panicles, spikes
Fruit Type: Dry, capsule
IUCN Status: Not evaluated
Reference: J.A. Murray, Syst. Veg. ed. 14: 889 (1784)

Dioscorea kalkapershadii Prain & Burkill
Dioscoreaceae
Climbing Mechanism: Stem Twiner
Distribution (Global): Bangladesh, India
Distribution (India): Tamil Nadu
Leaf Type: Alternate, compound
Inflorescence: Spikes
Fruit Type: Capsules 3-winged
IUCN Status: Not evaluated
Notes: Mostly found in evergreen forests
Reference: J. Proc. Asiat. Soc. Bengal 10: 24 (1914)

Dioscorea kamoonensis Kunth
Dioscoreaceae
Synonyms: *Dioscorea fargesii* Franch., *Dioscorea firma* R. Knuth, *Dioscorea kamoonensis* var. *brevifolia* Prain & Burkill, *Dioscorea kamoonensis* var. *fargesii* (Franch.) Prain & Burkill, *Dioscorea mairei* R. Knuth, *Dioscorea mengtzeana* R. Knuth, *Dioscorea ochroleuca*
Climbing Mechanism: Stem Twiner
Distribution (Global): Assam, Bangladesh, China South-Central, China Southeast, East Himalaya, Laos, Myanmar, Nepal, Thailand, Tibet, Vietnam, West Himalaya
Distribution (India): Sikkim
Leaf Type: Simple/compound, 3–5 foliolate
Inflorescence: Racemes/panicles, spikes
Fruit Type: Dry, capsule
IUCN Status: Not evaluated
Reference: Enum. Pl. 5: 395 (1850)

Dioscorea laurifolia Wall. ex Hook.f.
Dioscoreaceae
Climbing Mechanism: Stem Twiner
Distribution (Global): Borneo, Malaya, Thailand, Vietnam
Distribution (India): Arunachal Pradesh
Leaf Type: Simple/compound, 3–5 foliolate
Inflorescence: Racemes/panicles, spikes

Fruit Type: Dry, capsule
IUCN Status: Not evaluated
Reference: Fl. Brit. India 6: 293 (1892)

Dioscorea lepcharum Prain & Burkill
Dioscoreaceae
Synonyms: *Dioscorea lepcharum* var. *bhamoica* Prain & Burkill
Climbing Mechanism: Stem Twiner
Distribution (Global): Assam, Bangladesh, East Himalaya, Myanmar
Distribution (India): Arunachal Pradesh
Leaf Type: Simple/compound, 3–5 foliolate
Inflorescence: Racemes/panicles, spikes
Fruit Type: Dry, capsule
IUCN Status: Not evaluated
Reference: J. Proc. Asiat. Soc. Bengal 10: 36 (1914)

Dioscorea listeri Prain & Burkill
Dioscoreaceae
Climbing Mechanism: Stem Twiner
Distribution (Global): Assam
Distribution (India): Arunachal Pradesh
Leaf Type: Simple/compound, 3–5 foliolate
Inflorescence: Racemes/panicles, spikes
Fruit Type: Dry, capsule
IUCN Status: Not evaluated
Reference: J. Proc. Asiat. Soc. Bengal 4: 452 (1908)

Dioscorea melanophyma Prain & Burkill
Dioscoreaceae
Synonyms: *Dioscorea tenii* R. Knuth
Common Name: Black-bulb yam.
Climbing Mechanism: Stem Twiner
Distribution (Global): Assam, Bangladesh, China South-Central, East Himalaya, Nepal, Tibet, West Himalaya
Distribution (India): Arunachal Pradesh, Himachal Pradesh
Leaf Type: Simple/compound, 3–5 foliolate
Inflorescence: Racemes/panicles, spikes
Fruit Type: Dry, capsule
IUCN Status: Not evaluated
Reference: J. Proc. Asiat. Soc. Bengal 4: 452 (1908)

Dioscorea oppositifolia L.
Dioscoreaceae

Synonyms: *Dioscorea opposita* Thunb., *Dioscorea oppositifolia* var. *dukhunensis* Prain & Burkill, *Dioscorea oppositifolia* var. *linnaei* Prain & Burkill, *Dioscorea oppositifolia* var. *thwaitesii* Prain & Burkill.
Common Name: Chinese yam
Climbing Mechanism: Stem Twiner
Distribution (Global): Assam, Bangladesh, East Himalaya, India, Myanmar, Sri Lanka
Distribution (India): Andaman and Nicobar Islands, Andhra Pradesh, Arunachal Pradesh, Gujarat, Himachal Pradesh, Jharkhand, Kerala, Madhya Pradesh, Maharashtra, Odisha, Tamil Nadu, Telangana, Uttar Pradesh
Leaf Type: Opposite
Inflorescence: Spike
Fruit Type: Capsule
Flowering and Fruiting: August–November
IUCN Status: Not evaluated
Notes: Occurs in moist deciduous, semi-evergreen and evergreen forests
Reference: Sp. Pl.: 1033 (1753)

Dioscorea orbiculata Hook.f.
Dioscoreaceae
Climbing Mechanism: Stem Twiner
Distribution (Global): Malaya, Sumatera
Distribution (India): Arunachal Pradesh
Leaf Type: Simple/compound, 3–5 foliolate
Inflorescence: Racemes/panicles, spikes
Fruit Type: Dry, capsule
IUCN Status: Least Concern
Reference: Fl. Brit. India 6: 292 (1892)

Dioscorea pentaphylla L.
Dioscoreaceae
Synonyms: *Dioscorea changjiangensis* F.W. Xing & Z.X.Li, *Botryosicyos pentaphyllus* (L.) Hochst., *Dioscorea spinosa* Burm., *Ubium scandens* J.St.-Hil., *Dioscorea sumbawensis* R. Knuth, *Ubium quadrifarium* J.F. Gmel., *Dioscorea triphylla* L., *Hamatris triphylla* (L.) S
Common Name: Fiji yam, Five leaf yam, Kawan yam
Climbing Mechanism: Stem Twiner
Distribution (Global): Assam, Bangladesh, Borneo, Cambodia, Caroline Is., China South-Central, China Southeast, East Himalaya, Hainan, India, Java, Laos, Lesser Sunda Is., Malaya, Maldives, Maluku, Myanmar, Nansei-shoto, Nepal, New Guinea, Philippines, Queensland, Sri Lanka, Sulawesi, Sumatera, Taiwan, Thailand, Tibet, Vietnam, West Himalaya
Distribution (India): Andaman and Nicobar Islands, Andhra Pradesh, Arunachal Pradesh, Assam, Bihar, Odisha, Himachal Pradesh, Jharkhand, Kerala, Madhya

Pradesh, Maharashtra, Odisha, Rajasthan, Tamil Nadu, Telangana, Tripura, Uttar Pradesh, West Bengal
Leaf Type: 3–5 foliolate
Inflorescence: Spikes on axillary or terminal panicles
Fruit Type: Capsule
Flowering and Fruiting: September–December
IUCN Status: Not evaluated
Notes: Usually found in degraded deciduous forests and in wastelands
Reference: Sp. Pl.: 1032 (1753)

Dioscorea prazeri Prain & Burkill
Dioscoreaceae
Synonyms: *Dioscorea clarkei* Prain & Burkill, *Dioscorea sikkimensis* Prain & Burkill.
Climbing Mechanism: Stem Twiner
Distribution (Global): Assam, Bangladesh, East Himalaya, Malaya, Myanmar, Nepal, Thailand, Vietnam
Distribution (India): Arunachal Pradesh, Madhya Pradesh, Sikkim
Leaf Type: Simple/compound, 3–5 foliolate
Inflorescence: Racemes/panicles, spikes
Fruit Type: Dry, capsule
IUCN Status: Not evaluated
Reference: J. Asiat. Soc. Bengal, Pt. 2, Nat. Hist. 73(Suppl.): 2 (1904)

Dioscorea pubera Blume
Dioscoreaceae
Synonyms: *Dioscorea anguina* Roxb., *Dioscorea combilium* Buch. -Ham. ex Wall., *Dioscorea cornifolia* Kunth.
Climbing Mechanism: Stem Twiner
Distribution (Global): Assam, Bangladesh, East Himalaya, India, Java, Myanmar, Nepal, Sumatera
Distribution (India): Andhra Pradesh, Arunachal Pradesh, Gujarat, Madhya Pradesh, Odisha, Tripura, Uttar Pradesh
Leaf Type: Leaves very rarely alternate
Inflorescence: Spikes, solitary
IUCN Status: Not evaluated
Notes: Occurs in semi-evergreen forests and scrub jungles
Reference: Enum. Pl. Javae 1: 21 (1827)

Dioscorea pyrifolia Kunth
Dioscoreaceae
Synonyms: *Dioscorea diepenhorstiana* Miq., *Dioscorea ferruginea* Thunb. ex Prain & Burkill, *Dioscorea preangeriana* Uline ex R. Knuth, *Dioscorea sandakanensis* R. Knuth, *Dioscorea zollingeriana* Kunth.
Climbing Mechanism: Stem Twiner

Distribution (Global): Assam, Borneo, Cambodia, East Himalaya, Java, Malaya, Sumatera, Thailand, Vietnam
Leaf Type: Simple/compound, 3–5 foliolate
Inflorescence: Racemes/panicles, spikes
Fruit Type: Dry, capsule
IUCN Status: Not evaluated
Notes: It is commonly found in the lowlands, among hedges and thickets
Reference: Enum. Pl. 5: 384 (1850)

Dioscorea scortechinii Prain & Burkill
Dioscoreaceae
Climbing Mechanism: Stem Twiner
Distribution (Global): Bangladesh, East Himalaya, Hainan, Malaya, Sumatera, Thailand, Vietnam
Distribution (India): Arunachal Pradesh
Leaf Type: Simple/compound, 3–5 foliolate
Inflorescence: Racemes/panicles, spikes
Fruit Type: Dry, capsule
IUCN Status: Not evaluated
Notes: Usually found in moist evergreen and lower montane forests
Reference: J. Proc. Asiat. Soc. Bengal 4: 455 (1908)

Dioscorea spicata Roth
Dioscoreaceae
Synonyms: *Dioscorea nummularia* var. *glauca* Prain & Burkill, *Dioscorea spicata* var. *anamallayana* Prain & Burkill, *Dioscorea spicata* var. *parvifolia* Prain & Burkill.
Climbing Mechanism: Stem Twiner
Distribution (Global): Bangladesh, India, Sri Lanka
Distribution (India): Madhya Pradesh, Tamil Nadu
Leaf Type: Leaves opposite or subopposite, rarely alternate
Inflorescence: Axillary spikes
Flowering and Fruiting: August–December
IUCN Status: Not evaluated
Notes: Mostly found in shola and semi-evergreen forests
Reference: Nov. Pl. Sp.: 371 (1821)

Dioscorea tomentosa J. Koenig ex Spreng.
Dioscoreaceae
Synonyms: *Helmia tomentosa* (J. Koenig ex Spreng.) Kunth
Climbing Mechanism: Stem Twiner
Distribution (Global): Bangladesh, India, Sri Lanka
Distribution (India): Andaman and Nicobar Islands, Andhra Pradesh, Arunachal Pradesh, Madhya Pradesh, Odisha, Tamil Nadu
Leaf Type: Trifoliate

Inflorescence: Axillary Spikes
Fruit Type: Capsule
Flowering and Fruiting: July–December
IUCN Status: Not evaluated
Notes: Usually found in semi-evergreen, moist, and dry deciduous forests
Reference: Pl. Min. Cogn. Pug. 2: 92 (1815)

Dioscorea trifida L.f.
Dioscoreaceae
Synonyms: *Dioscorea affinis* Kunth, *Dioscorea angustifolia* Rusby, *Dioscorea articulata* Steud., *Dioscorea brasiliensis* Willd., *Dioscorea goyazensis* Griseb., *Dioscorea quinquelobata* Vell., *Dioscorea ruiziana* Klotzsch ex Kunth, *Dioscorea triloba* Lam.
Climbing Mechanism: Stem Twiner
Distribution (Global): Bolivia, Brazil North, Brazil Northeast, Brazil Southeast, Brazil West-Central, Colombia, Costa Rica, Ecuador, French Guiana, Guatemala, Guyana, Honduras, Leeward Is., Nicaragua, Panamá, Peru, Suriname, Trinidad-Tobago, Venezuela, Windward Is.
Leaf Type: Simple/compound, 3—5 foliolate
Inflorescence: Racemes/panicles, spikes
Fruit Type: Dry, capsule
IUCN Status: Not evaluated
Reference: Suppl. Pl.: 427 (1782)

Dioscorea trinervia Roxb. ex Prain & Burkill
Dioscoreaceae
Climbing Mechanism: Stem Twiner
Distribution (Global): Assam, Bangladesh, Myanmar
Distribution (India): Arunachal Pradesh
Leaf Type: Simple/compound, 3–5 foliolate
Inflorescence: Racemes/panicles, spikes
Fruit Type: Dry, capsule
IUCN Status: Not evaluated
Reference: J. Proc. Asiat. Soc. Bengal 10: 32 (1914)

Dioscorea vexans Prain & Burkill
Dioscoreaceae
Climbing Mechanism: Stem Twiner
Distribution (Global): India
Distribution (India): Andaman and Nicobar Islands
Leaf Type: Simple/compound, 3–5 foliolate
Inflorescence: Racemes/panicles, spikes
Fruit Type: Dry, capsule
IUCN Status: Not evaluated
Reference: J. Proc. Asiat. Soc. Bengal 4: 456 (1908)

Dioscorea villosa L.
Dioscoreaceae
Synonyms: *Dioscorea cliffortiana* Lam., *Dioscorea glauca* Muhl. ex L.C. Beck, *Dioscorea hexaphylla* Raf., *Dioscorea hirticaulis* Bartlett, *Dioscorea lloydiana* E.H.L. Krause, *Dioscorea longifolia* Raf., *Dioscorea megaptera* Raf., *Dioscorea paniculata* Michx.
Common Name: Wild Yam
Climbing Mechanism: Stem Twiner
Distribution (Global): Alabama, Arkansas, Connecticut, Delaware, District of Columbia, Florida, Georgia, Illinois, Indiana, Iowa, Kansas, Kentucky, Louisiana, Maryland, Massachusetts, Michigan, Minnesota, Mississippi, Missouri, Nebraska, New Jersey, New York, North Carolina, Ohio, Oklahoma, Ontario, Pennsylvania, Rhode I., South Carolina, Tennessee, Texas, Vermont, Virginia, West Virginia, Wisconsin
Leaf Type: Simple/compound, 3–5 foliolate
Inflorescence: Racemes/panicles, spikes
Fruit Type: Dry, capsule
IUCN Status: Not evaluated
Reference: Sp. Pl.: 1033 (1753)

Dioscorea wallichii Hook.f.
Dioscoreaceae
Synonyms: *Dioscorea wallichii* var. *christiei* Prain & Burkill, *Dioscorea wallichii* var. *vera* Prain & Burkill
Climbing Mechanism: Stem Twiner
Distribution (Global): Assam, Bangladesh, China South-Central, East Himalaya, India, Malaya, Myanmar, Thailand
Distribution (India): Andaman and Nicobar Islands, Andhra Pradesh, Gujarat, Madhya Pradesh, Odisha, Tamil Nadu, West Bengal
Leaf Type: Simple/compound, 3–5 foliolate
Inflorescence: Racemes/panicles, spikes
Fruit Type: Dry, capsule
Flowering and Fruiting: October–November
IUCN Status: Least Concern
Reference: Fl. Brit. India 6: 295 (1892)

Dioscorea wightii Hook.f.
Dioscoreaceae
Climbing Mechanism: Stem Twiner
Distribution (Global): Bangladesh, India
Distribution (India): Arunachal Pradesh, Madhya Pradesh, Tamil Nadu
Leaf Type: Simple
Inflorescence: Spikes, axillary
Flowering and Fruiting: July–November

IUCN Status: Data Deficient
Reference: Fl. Brit. India 6: 291 (1892)

Elaeagnus caudata Schltdl. ex Momiy.
Elaeagnaceae
Common Name: Bastard oleaster
Climbing Mechanism: Scrambler-Armed
Distribution (India): Andhra Pradesh, Arunachal Pradesh
Flowering and Fruiting: November–April
IUCN Status: Not evaluated
Reference: H. Hara, Fl. E. Himal., 2nd. Rep.: 85 (1971)

Elaeagnus conferta Roxb.
Elaeagnaceae
Synonyms: *Elaeagnus arborea* var. *dendroidea* Schlechtend. in DC., *Elaeagnus conferta* var. *calcuttensis* Serv., *Elaeagnus conferta* subsp. *euconferta* Servettaz, *Elaeagnus conferta* subsp. *javanica* (Bl.) Serv., *Elaeagnus conferta* var. *malaccensis* Serv.
Climbing Mechanism: Scrambler-Unarmed
Distribution (Global): Andaman Is., Assam, Bangladesh, China South-Central, China Southeast, East Himalaya, Java, Laos, Malaya, Myanmar, Nepal, Sri Lanka, Sumatera, Thailand, Vietnam, West Himalaya
Distribution (India): Goa, Madhya Pradesh, Maharashtra, Meghalaya, Tamil Nadu, Tripura, Western Ghats
Flowering and Fruiting: December–March
IUCN Status: Least Concern
Reference: Fl. Ind. 1: 460 (1820)

Elaeagnus indica Servett.
Elaeagnaceae
Climbing Mechanism: Scrambler-Unarmed
Distribution (Global): India
Distribution (India): Andhra Pradesh, Tamil Nadu
Flowering and Fruiting: December–April
IUCN Status: Not evaluated
Reference: Bull. Herb. Boissier, sér. 2, 8: 393 (1908)

Elaeagnus latifolia Lour.
Elaeagnaceae
Synonyms: *Elaeagnus kologa*, *Elaeagnus arborea*, *Elaeagnus elliptica*
Common Name: South Indian wild olive, Wild olive
Climbing Mechanism: Scrambler-Unarmed
Distribution (India): Andaman and Nicobar Islands, Himachal Pradesh, Karnataka, Mizoram, Tamil Nadu

IUCN Status: Not evaluated
Reference: Fl. Cochinch.: 89 (1790)

Ephedra foliata Boiss. ex C.A. Mey.
Ephedraceae
Synonyms: *Ephedra aitchisonii* (Stapf) V.A. Nikitin, *Ephedra alte* Brandis, *Ephedra asparagoides* Griff., *Ephedra ciliata* Aitch., nom. illeg., *Ephedra ciliata* var. *polylepis* (Boiss. & Hausskn.) Riedl, *Ephedra foliata* var. *aitchisonii* Stapf, *Ephedra foliata*
Climbing Mechanism: Scrambler-Unarmed
IUCN Status: Not evaluated
Reference: Mém. Acad. Imp. Sci. Saint-Pétersbourg, Sér. 6, Sci. Math., Seconde Pt. Sci. Nat. 7(2): 297 (1846)

Agapetes anonyma Airy Shaw
Ericaceae
Climbing Mechanism: Scrambler-Unarmed
Distribution (Global): East Himalaya
Distribution (India): Arunachal Pradesh
Leaf Type: Scattered, glabrous, elliptic-ovate
Inflorescence: Fasciculate, 1 or 2 flowered
Fruit Type: Wet, berry, globose
IUCN Status: Not evaluated
Reference: Kew Bull. 3: 95 (1948)

Agapetes burmanica W.E. Evans
Ericaceae
Synonyms: *Agapetes megacarpa* var. *megacarpa*
Climbing Mechanism: Scrambler-Unarmed
Distribution (India): Arunachal Pradesh
Leaf Type: Scattered, glabrous, elliptic-ovate
Inflorescence: Fasciculate, 1 or 2 flowered
Fruit Type: Wet, berry, globose
IUCN Status: Not evaluated
Reference: Notes Roy. Bot. Gard. Edinburgh 15: 199 (1927)

Agapetes interdicta (Hand. -Mazz.) Sleumer
Ericaceae
Synonyms: *Pentapterygium interdictum* Hand. -Mazz.
Climbing Mechanism: Scrambler-Unarmed
Distribution (Global): China South-Central, East Himalaya, Myanmar, Tibet
Distribution (India): Arunachal Pradesh
Leaf Type: Scattered, glabrous, elliptic-ovate
Inflorescence: Fasciculate, 1 or 2 flowered
Fruit Type: Wet, berry, globose

IUCN Status: Not evaluated
Reference: Bot. Jahrb. Syst. 70: 106 (1939)

Rhododendron pendulum Hook.f.
Ericaceae
Synonyms: *Azalea pendula* (Hook.f.) Kuntze, Gm, *Rhododendron modestum* Hook. f.
Climbing Mechanism: Scrambler-Unarmed
Distribution (Global): East Himalaya, Nepal, Tibet
Distribution (India): Arunachal Pradesh
Leaf Type: Simple, oblong
Inflorescence: 1–3 flowered
Fruit Type: Dry, capsule
IUCN Status: Not evaluated
Reference: Rhododendr. Sikkim-Himalaya 2: t. 13 (1851)

Chrozophora caudatus Geisel.
Euphorbiaceae
Climbing Mechanism: Scrambler-Unarmed
Distribution (India): Bihar, Odisha
IUCN Status: Not evaluated

Cnesmone javanica Blume
Euphorbiaceae
Climbing Mechanism: Stem Twiner
Distribution (Global): Andaman Is., Assam, Bangladesh, Borneo, East Himalaya, Java, Lesser Sunda Is., Malaya, Myanmar, Nicobar Is., Sumatera, Thailand, Vietnam
Distribution (India): Mizoram, Tripura
Leaf Type: Oblong or ovate
Inflorescence: Green flowers
IUCN Status: Not evaluated
Notes: Scattered in evergreen forest or bamboo forest, sometimes in clearings
Reference: Bijdr. Fl. Ned. Ind.: 630 (1826)

Cnesmone Javanica var. *glabriuscula*
Euphorbiaceae
Climbing Mechanism: Stem Twiner
Distribution (Global): Andaman Is.
IUCN Status: Not evaluated
Reference: Gard. Bull. Singapore 31: 49 (1978)

Croton caudatus Geiseler
Euphorbiaceae

Synonyms: *Croton aromaticus* Gaertn., *Croton caudatus* var. *denticulatus* Müll. Arg., *Croton caudatus* var. *globosus* Hook.f., *Croton caudatus* var. *harmandii* Gagnep., *Croton caudatus* var. *hispidus* Hook.f., *Croton caudatus* var. *malaccanus* Hook.f. + 10

Common Name: Climbing Croton

Climbing Mechanism: Scrambler-Unarmed

Distribution (Global): Assam, Bangladesh, Borneo, Cambodia, China South-Central, Christmas I., East Himalaya, India, Java, Laos, Lesser Sunda Is., Malaya, Myanmar, Nepal, Pakistan, Philippines, Queensland, Sri Lanka, Sulawesi, Sumatera, Thailand, Vietnam

Distribution (India): Arunachal Pradesh, Meghalaya, Mizoram, Odisha, Tamil Nadu, Western Ghats

Leaf Type: Ovate, acuminate

Inflorescence: Raceme

Fruit Type: Capsule

Flowering and Fruiting: April–May

IUCN Status: Not evaluated

Notes: Usually found in evergreen, semi-evergreen, scrub and open forests

Reference: Croton. Monogr.: 73 (1807)

Dalechampia ficifolia Lam.

Euphorbiaceae

Synonyms: *Dalechampia scandens* Vell., *Dalechampia tiliifolia* var. *ficifolia* (Lam.) Kuntze

Climbing Mechanism: Stem Twiner

Distribution (Global): Brazil Northeast, Brazil Southeast

Distribution (India): Andhra Pradesh, Eastern Ghats, Maharashtra, Tamil Nadu

IUCN Status: Not evaluated

Reference: Encycl. 2: 258 (1786)

Dalechampia indica Wight

Euphorbiaceae

Synonyms: *Dalechampia bidentata* Thwaites, *Dalechampia coromandeliana* Benth., Dalechampia ternata Müll.Arg.

Common Name: Indian spurge-creeper

Climbing Mechanism: Stem Twiner

Distribution (Global): India

Leaf Type: 3-foliolate, elliptic-ovate

Fruit Type: Depressed, nearly spherical

Flowering and Fruiting: November–February

IUCN Status: Not evaluated

Reference: Icon. Pl. Ind. Orient. 5: t. 1882 (1852)

Dalechampia scandens var. *cordofana* (Hochst. ex Webb) Müll.Arg.

Euphorbiaceae

Synonyms: *Dalechampia cordofana* Hochst. ex Webb, *Dalechampia parvifolia* Lam., *Dalechampia senegalensis* A. Juss. ex Webb, *Dalechampia tripartita* R. Br.
Common Name: Gujarat spurge-creeper
Climbing Mechanism: Stem Twiner
Distribution (Global): Angola, Burkina, Cape Verde, Chad, Congo, Djibouti, Eritrea, Ethiopia, Gabon, Guinea, Gulf States, India, Kenya, Mali, Mauritania, Mozambique, Namibia, Oman, Pakistan, Saudi Arabia, Senegal, Somalia, Sudan, Tanzania, Uganda, Yemen, Zaire
Distribution (India): Gujarat
Leaf Type: Cut to 3-lobed
Flowering and Fruiting: September–October
IUCN Status: Not evaluated
Reference: A.P.de Candolle, Prodr. 15(2): 1245 (1866)

Dalechampia stenoloba Raghavan & B.G.P. Kulk.
Euphorbiaceae
Climbing Mechanism: Stem Twiner
Distribution (Global): India
Distribution (India): Karnataka
IUCN Status: Not evaluated
Reference: Kew Bull. 35: 325 (1980)

Dalechampia tamifolia Lam.
Euphorbiaceae
Synonyms: *Dalechampia anisophylla* Müll.Arg., *Dalechampia boiviniana* Baill., *Dalechampia longipes* Müll.Arg., *Dalechampia semitriloba* Tausch ex Pax & K. Hoffm., *Dalechampia sieberi* Klotzsch ex Pax & K. Hoffm., *Dalechampia trilobata* Sieber ex Baker.
Climbing Mechanism: Stem Twiner
Distribution (Global): Comoros, India, Madagascar, Mauritius
Distribution (India): Peninsular India
IUCN Status: Not evaluated
Reference: Encycl. 2: 256 (1786)

Dalechampia velutina Wight
Euphorbiaceae
Common Name: Velvety spurge-creeper
Climbing Mechanism: Stem Twiner
Distribution (Global): India
Distribution (India): Karnataka, Tamil Nadu
Leaf Type: Tri-lobed
Inflorescence: Stalked clusters
Flowering and Fruiting: December–February
IUCN Status: Not evaluated

Notes: Mostly found in evergreen forests
Reference: Icon. Pl. Ind. Orient. 5: 1881 (1852)

Dimorphocalyx balakrishnanii Chakrab. & Premanath
Euphorbiaceae
Synonyms: *Tritaxis balakrishnanii* (Chakrab. & Premanath) R.Y. Yu & Welzen., *Dimorphocalyx dilipianus* N.P. Balakr. & Chakr.
Climbing Mechanism: Stem Twiner
Distribution (Global): Andaman Is.
IUCN Status: Endangered
Reference: Taxon 68: 932 (2019)

Dimorphocalyx beddomei (Benth.) Airy Shaw
Euphorbiaceae
Climbing Mechanism: Stem Twiner
IUCN Status: Endangered

Mallotus repandus (Willd.) Müll.Arg.
Euphorbiaceae
Synonyms: *Adisca timoriana* Span., *Croton baccifer* Benth., *Croton repandus* Rottler. + 20
Climbing Mechanism: Scrambler-Unarmed
Distribution (Global): Andaman Is., Assam, Bangladesh, Borneo, Cambodia, China North-Central, China South-Central, China Southeast, East Himalaya, Hainan, India, Java, Laos, Lesser Sunda Is., Malaya, Maluku, Myanmar, Nepal, New Caledonia, New Guinea, Nicobar Is., Philippines, Queensland, Sri Lanka, Sulawesi, Sumatera, Taiwan, Thailand, Vietnam
Distribution (India): Andhra Pradesh, Bihar, Odisha, Eastern Ghats, Goa, Karnataka, Kerala, Maharashtra
Leaf Type: Leaves, ovate or ovate-deltoid, base subcordate or truncate, margins entire or somewhat sinuate-toothed
Inflorescence: Male flowers in terminal panicled racemes
Fruit Type: Capsule, 2-valved, densely fulvous-tomentose and glandular without
Flowering and Fruiting: November–January
IUCN Status: Not evaluated
Notes: Mostly occur in dry evergreen and dry deciduous forests
Reference: Linnaea 34: 197 (1865)

Pachystylidium hirsutum (Blume) Pax & K. Hoffm.
Euphorbiaceae
Climbing Mechanism: Stem Twiner
Distribution (Global): Bangladesh, Cambodia, China South-Central, India, Java, Laos, Lesser Sunda Is., Philippines, Sulawesi, Thailand, Vietnam
Distribution (India): Peninsular India

IUCN Status: Not evaluated
Reference: H.G.A. Engler (ed.), Pflanzenr., IV, 147, IX: 108 (1919)

Plukenetia corniculata Sm.
Euphorbiaceae
Synonyms: *Hedraiostylus corniculatus* (Sm.) Hassk., *Hedraiostylus glaberrimus* Hassk., *Pterococcus corniculatus* (Sm.) Pax & K. Hoffm., *Pterococcus glaberrimus* Hassk., *Sajorium corniculatum* (Sm.) Baill., *Sajorium corniculatum* (Sm.) D. Dietr.
Climbing Mechanism: Stem Twiner
Distribution (Global): Assam, Bangladesh, Borneo, East Himalaya, India, Java, Lesser Sunda Is., Malaya, Maluku, Myanmar, Philippines, Sulawesi, Sumatera, Thailand
Flowering and Fruiting: July–August
IUCN Status: Not evaluated
Reference: Nova Acta Regiae Soc. Sci. Upsal. 6: 4 (1799)

Tragia bicolor Miq.
Euphorbiaceae
Synonyms: *Tragia miqueliana* Müll.Arg.
Climbing Mechanism: Stem Twiner
Distribution (Global): India
Distribution (India): Odisha
Leaf Type: Ovate, acuminate
Inflorescence: Racemes
Flowering and Fruiting: November–December
IUCN Status: Not evaluated
Reference: Linnaea 26: 222 (1854)

Tragia hildebrandtii Müll.Arg.
Euphorbiaceae
Synonyms: *Tragia hildebrandtii* subsp. *glaucescens* Pax, *Tragia mombassana* Vatke ex Prain.
Climbing Mechanism: Stem Twiner
Distribution (Global): Ethiopia, Kenya, Malawi, Tanzania
Distribution (India): Gujarat
Flowering and Fruiting: June–November
IUCN Status: Not evaluated
Reference: Bremen Abh. 7: 26 (1880)

Tragia hispida Willd.
Euphorbiaceae
Common Name: Bristly climbing nettle
Climbing Mechanism: Stem Twiner
Distribution (Global): Assam, Bangladesh, India, Sri Lanka, West Himalaya

Distribution (India): Karnataka, Maharashtra, West Bengal
Leaf Type: Lance shaped to oblong-lance shaped
Inflorescence: Racemes
Flowering and Fruiting: August–October
IUCN Status: Not evaluated
Reference: Sp. Pl., ed. 4, 4: 323 (1805)

Tragia involucrata L.
Euphorbiaceae
Synonyms: *Croton urens* L., *Tragia cordata* B. Heyne ex Benth., *Tragia trifida* Benth.
Climbing Mechanism: Stem Twiner
Distribution (Global): Bangladesh, India, Laccadive Is., Sri Lanka
Distribution (India): Andhra Pradesh, Jharkhand, Karnataka, Kerala, Madhya Pradesh, Maharashtra, Mizoram, Odisha, Puducherry, Tamil Nadu, Telangana, Tripura, Uttar Pradesh, West Bengal
Leaf Type: Ovate or elliptic
Inflorescence: Spikes axillary
Fruit Type: Capsule
Flowering and Fruiting: Throughout the year
IUCN Status: Not evaluated
Notes: Usually found in Scrub forests and wastelands
Reference: Sp. Pl.: 980 (1753)

Tragia montana (Thwaites) Müll.Arg.
Euphorbiaceae
Synonyms: *Tragia montana* var. *dioica* Chakrab. & N.P. Balakr., *Tragia muelleriana* Pax & K. Hoffm.
Climbing Mechanism: Stem Twiner
Distribution (Global): India, Sri Lanka
Distribution (India): Tamil Nadu
Flowering and Fruiting: August–January
IUCN Status: Not evaluated
Notes: Often found in evergreen and moist deciduous forests
Reference: Linnaea 34: 183 (1865)

Tragia plukenetii Radcl. -Sm.
Euphorbiaceae
Synonyms: *Croton hastatus* L., *Tragia brouniana* Prain, *Tragia cannabina* L.f., *Tragia gallabatensis* Prain, *Tragia tripartita* Beille
Common Name: Cannabis leaf nettle, Noseburn
Climbing Mechanism: Stem Twiner
Distribution (Global): Cameroon, Central African Repu, Chad, Djibouti, Eritrea, Ethiopia, India, Kenya, Mozambique, Nigeria, Somalia, Sri Lanka, Sudan, Tanzania, Uganda, Zambia, Zaire, Zimbabwe

Distribution (India): Andhra Pradesh, Madhya Pradesh, Maharashtra, Odisha, Tamil Nadu, Telangana
Leaf Type: Leaves are alternate, palmately 3-lobed
Inflorescence: Racemes
Flowering and Fruiting: September–February
IUCN Status: Not evaluated
Notes: Found occasionally on plains to foothills of scrub jungles, in wastelands, paddy fields and stream banks
Reference: Kew Bull. 37: 688 (1983)

Tragia praetervisa Chakrab. & N.P. Balakr.
Euphorbiaceae
Synonyms: *Tragia miqueliana* var. *unicolor* Müll.Arg., *Tragia praetervisa* var. *unicolor* (Müll.Arg.) Chakrab. & N.P. Balakr.
Climbing Mechanism: Stem Twiner
Distribution (Global): India, Sri Lanka
Distribution (India): Tamil Nadu
Leaf Type: Ovate, acuminate
Inflorescence: Spike axillary
Fruit Type: Capsule, 3-lobed
Flowering and Fruiting: January–June
IUCN Status: Not evaluated
Reference: Rheedea 16: 23 (2006)

Tragia sanjappae Chakrab. & N.P. Balakr.
Euphorbiaceae
Common Name: Narrow-leaf climbing nettle
Climbing Mechanism: Stem Twiner
Distribution (Global): India
Distribution (India): Tamil Nadu
Leaf Type: Leaves alternate
Inflorescence: Racemes, axillary, terminal or leaf-opposed
Fruit Type: Capsules subsessile, depressed
IUCN Status: Not evaluated
Notes: Often distributed in semi-evergreen forests
Reference: Rheedea 16: 25 (2006)

Abrus fruticulosus Wight & Arn.
Fabaceae
Synonyms: *Abrus acutifolius* Blume ex Miq., *Hoepfneria africana* Vatke, *Abrus schimperi* Baker.
Common Name: Woody rosary pea
Climbing Mechanism: Stem Twiner
Distribution (Global): Andaman Is., Assam, Benin, Cameroon, Central African Repu, Chad, China Southeast, Congo, Eritrea, Ethiopia, Gabon, Guinea-Bissau,

India, Ivory Coast, Kenya, Liberia, Madagascar, Malawi, Mozambique, Sudan, Tanzania, Uganda, Zambia, Zimbabwe
Distribution (India): Telengana, Tamil Nadu, Andhra Pradesh, Maharashtra, Tripura, Rajasthan, Eastern Ghats, Western Ghats
Leaf Type: Compound with 12–34 oblong, obovate-oblong or ovate leaflets, flat to broadly rounded and mucronulate at tip
Inflorescence: Slender, usually straight, flowers in clusters on cushion-like reduced branchlets
Fruit Type: Pod is oblong to linear-oblong, compressed, 4–12 seeded
Flowering and Fruiting: October–January
IUCN Status: Data Deficient
Notes: Used in local ornamental works.
Reference: Prodr. Fl. Ind. Orient. 1: 236 (1834)

Abrus precatorius L.
Fabaceae
Synonyms: *Glycine abrus* L *Abrus abrus* (L.) Wright, *Abrus cyaneus* R. Vig., *Abrus maculatus* Noronha, *Abrus minor* Desv., *Abrus pauciflorus* Desv., *Abrus precatorius* var. *novo-guineensis* Miq., *Abrus squamulosus* E. Mey., *Abrus tunguensis* Lima.
Common Name: Crab's eye, Indian liquorice, Jeguirity, Bead vine, Black-eyed susan, Buddhist rosary bead, Rosary pea, Wild liquorice
Climbing Mechanism: Stem Twiner
Distribution (Global): Aldabra, Andaman Is., Angola, Assam, Bangladesh, Benin, Botswana, Burkina, Burundi, Cambodia, Cameroon, Cape Provinces, Cape Verde, Caprivi Strip, Central African Repu, Chad, China South-Central, China Southeast, Comoros, Congo, East Himalaya, Equatorial Guinea, Eritrea, Ethiopia, Gabon, Gambia, Guinea, Guinea-Bissau, Gulf of Guinea Is., India, Ivory Coast, Kenya, KwaZulu-Natal, Laos, Liberia, Madagascar, Malawi, Malaya, Mali, Mauritania, Mozambique, Myanmar, Namibia, Nepal, New South Wales, Nicobar Is., Niger, Northern Provinces, Northern Territory, Pakistan, Queensland, Rwanda, Senegal, Seychelles, Somalia, South China Sea, Sri Lanka, Sudan, Swaziland, Taiwan, Tanzania, Thailand, Togo, Uganda, Vietnam, West Himalaya, Western Australia, Yemen, Zambia, Zaire, Zimbabwe
Distribution (India): Kerala, Arunachal Pradesh, Tamil Nadu, Puducherry, Andaman Islands, West Bengal, Goa, Gujarat, Madhya Pradesh, Himachal Pradesh, Delhi, Maharashtra, Manipur, Tripura, Rajasthan, Bihar, Odisha, Uttar Pradesh, Jharkhand
Leaf Type: Paripinnate
Inflorescence: Compound
Fruit Type: Pod
Flowering and Fruiting: Throughout the year
IUCN Status: Not evaluated
Notes: Seeds are used as beads and in percussion instruments.
Reference: Syst. Nat. ed. 12, 2: 472 (1767)

Abrus pulchellus Thwaites
Fabaceae
Synonyms: *Abrus fruticulosus* Auct. non Wight & Arn., *Abrus laevigatus* Sensu Breteler, *Abrus stictosperma* Berhaut
Common Name: Showy-rosary pea
Climbing Mechanism: Stem Twiner
Distribution (Global): Brazil
Distribution (India): Kerala, Tamil Nadu, Andhra Pradesh, Andaman Islands, West Bengal, Madhya Pradesh, Manipur, Bihar, Odisha, Mizoram, Nicobar Islands
Leaf Type: Paripinnate
Inflorescence: Axillary raceme
Fruit Type: Pod
Flowering and Fruiting: August–March
IUCN Status: Not evaluated
Notes: The leaves and roots of this plant are used to cure diseases like asthma, cough, tuberculosis, bronchitis, and chest pain
Reference: Enum. Pl. Zeyl. [Thwaites] 91

Acacia andamanica I.C. Nielsen
Fabaceae
Synonyms: *Acacia pennata* Sensu auct., *Acacia pseudo-intsia* Sensu auct., *Acacia pseudo-intsia* var. *ambigua* Prain, *Acacia pseudointsia* var. *ambigua* Prain
Climbing Mechanism: Scrambler-Armed
Distribution (Global): Andaman Is., China South-Central, China Southeast, Thailand, Vietnam
Distribution (India): Andaman and Nicobar Islands
Leaf Type: bipinnate leaves and cauline prickles
Inflorescence: spicate inflorescences
Fruit Type: pod
IUCN Status: Not evaluated
Notes: A black pigment in the bark and fruit can be used for dye or ink
Reference: Adansonia, n.s., 16: 354 (1980)

Acacia caesia (L.) Willd.
Fabaceae
Synonyms: *Acacia columnaris* Craib, *Acacia intsia* Sensu auct., *Acacia intsia* var. *caesia* (L.) Baker, *Mimosa caesia* (L.) Willd., *Mimosa intsia* Auct. non L.
Common Name: Soap bark, Black Catechu
Climbing Mechanism: Scrambler-Armed
Distribution (India): Kerala, Andrapradesh, Tamil Nadu, Puducherry, Telangana, West Bengal, Goa, Karnataka, Madhya Pradesh, Maharashtra, Bihar, Odisha, Mizoram, Manipur, Sikkim, North Eastern India
Leaf Type: Compound
Fruit Type: A stipitate pod, seeds more than 10
Flowering and Fruiting: October–December

IUCN Status: Least Concern
Notes: Frequently found in scrub jungles and along hill slopes, foothills of evergreen and semi-evergreen forests
Reference: Sp. Pl., ed. 4, 4: 1090 (1806)

Acacia caesia var. *subnuda* (Craib) I.C. Nielsen
Fabaceae
Climbing Mechanism: Scrambler-Armed
Leaf Type: Compound, Pinnae
Inflorescence: Heads terminal, panicled
Fruit Type: Dry, Pods
IUCN Status: Not evaluated
Reference: Adansonia, n.s., 19: 348 (1980)

Acacia concinna (Willd.) DC.
Fabaceae
Synonyms: *Acacia hooperiana*, *Mimosa concinna*
Common Name: Shikakai, Soap-pod, Soap pod wattle
Climbing Mechanism: Scrambler-Armed
Distribution (India): Maharashtra, Tripura, Bihar, Odisha, Uttar Pradesh, Andhra Pradesh, NorthWestern India
Leaf Type: Leaf alternate, 2-pinnate
Inflorescence: Paniculate
Fruit Type: Pod
Flowering and Fruiting: October–January
IUCN Status: Not evaluated
Notes: Commonly used to remove dandruff
Reference: Prodr. 2: 464 (1825)

Acacia dealbata Link
Fabaceae
Synonyms: *Acacia decurrens* var. *dealbata* (Link) Maiden, *Mimosa dealbata* (Link) Page, *Racosperma dealbatum* (Link) Pedley
Common Name: Silver wattle
Climbing Mechanism: Scrambler-Armed
Distribution (Global): New South Wales, Tasmania, Victoria
Distribution (India): Tamil Nadu, Western Ghats
Leaf Type: Bipinnate
Inflorescence: Axillary or terminal panicles
Fruit Type: Pod
Flowering and Fruiting: September–March
IUCN Status: Not evaluated
Notes: Introduced and planted on hills for its bark
Reference: Enum. Hort. Berol. Alt. 1: 445 (1821)

Acacia diadenia R. Parker
Fabaceae
Climbing Mechanism: Scrambler-Armed
Leaf Type: Compound, Pinnae
Inflorescence: Heads terminal, panicled
Fruit Type: Dry, Pods
IUCN Status: Least Concern
Reference: Indian Forester 55: 332 (1929)

Acacia gageana Craib.
Fabaceae
Climbing Mechanism: Scrambler-Armed
Distribution (India): Madhya Pradesh, Himachal Pradesh
Leaf Type: Compound, Pinnae
Inflorescence: Heads terminal, panicled
Fruit Type: Dry, Pods
IUCN Status: Not evaluated
Notes: Warm and subtropical forests, 200– 1100 m altitude
Reference: Bull. Misc. Inform. Kew 1915: 409 (1915)

Acacia glaucoptera Benth.
Fabaceae
Synonyms: *Acacia bossiaeoides* Seem., *Racosperma glaucopterum* (Benth.) Pedley, *Acacia sinuata* Jacques
Common Name: Clay wattle
Climbing Mechanism: Scrambler-Armed
Distribution (Global): Western Australia
Leaf Type: Compound, Pinnae
Inflorescence: Heads terminal, panicled
Fruit Type: Dry, Pods
IUCN Status: Not evaluated
Reference: Linnaea 36: 604 (1855)

Acacia grahamii Vajr.
Fabaceae
Climbing Mechanism: Scrambler-Armed
Distribution (India): Kerala
Leaf Type: Compound, Pinnae
Inflorescence: Heads terminal, panicled
Fruit Type: Dry, Pods
IUCN Status: Not evaluated
Notes: Usually found in scrub jungle
Reference: N.C. Nair & A.N. Henry, Fl. Tamil Nadu 1(Add.): ii (1983)

Acacia hohenackeri Craib
Fabaceae
Climbing Mechanism: Scrambler-Armed
Distribution (India): Peninsular India
Leaf Type: Compound, Pinnae
Inflorescence: Heads terminal, panicled
Fruit Type: Dry, Pods
Flowering and Fruiting: April–June
IUCN Status: Not evaluated
Notes: Commonly occur in dry deciduous forests
Reference: Bull. Misc. Inform. Kew 1915: 408 (1915)

Acacia intsia (L.) Willd.
Fabaceae
Synonyms: *Mimosa rubicaulis* Lam
Climbing Mechanism: Scrambler-Armed
Distribution (India): Madhya Pradesh, Himachal Pradesh, Tamil Nadu, Western Ghats, Eastern Ghats
Leaf Type: Compound, Pinnae
Inflorescence: Heads terminal, panicled
Fruit Type: Dry, Pods
IUCN Status: Least Concern
Reference: Sp. Pl., ed. 4, 4: 1091 (1806)

Acacia megaladena Desv.
Fabaceae
Synonyms: *Acacia arrophula* D. Don, *Acacia pennata* Auct. non (L.) Willd., *Acacia pennata* var. *arrophula* (D. Don) Baker
Common Name: Large-gland Acacia
Climbing Mechanism: Scrambler-Armed
Distribution (India): Karnataka, Rajasthan, Mizoram
Leaf Type: Compound
Flowering and Fruiting: August–September
IUCN Status: Not evaluated
Reference: J. Bot. Agric. 3: 69 (1814)

Acacia pennata (L.) Willd.
Fabaceae
Synonyms: *Acacia canescens* Graham ex Gamble, *Acacia grahamii* Vajr.
Common Name: Rusty Mimosa
Climbing Mechanism: Scrambler-Armed
Distribution (India): Kerala, Arunachal Pradesh, Tamil Nadu, Andhra Pradesh, Andaman Islands, West Bengal, Goa, Karnataka, North Eastern India, Bihar, Orissa, Jharkhand, Madhya Pradesh
Leaf Type: Bipinnate

Inflorescence: Umbelled panicles
Fruit Type: Pod
Flowering and Fruiting: May–February
IUCN Status: Least Concern
Notes: In foothills to 1200 m. Frequently occur in moist deciduous forests
Reference: Sp. Pl., ed. 4, 4: 1090 (1806)

Acacia pruinescens Kurz
Fabaceae
Climbing Mechanism: Scrambler-Armed
Distribution (India): Arunachal Pradesh, Manipur, Tripura, Mizoram
Leaf Type: Compound
Inflorescence: Axillary
Flowering and Fruiting: June–October
IUCN Status: Not evaluated
Notes: Frequently occur in moist deciduous forests
Reference: J. Asiat. Soc. Bengal, Pt. 2, Nat. Hist. 45: 296 (1876)

Acacia pseudo-intsia Miq.
Fabaceae
Climbing Mechanism: Scrambler-Armed
Distribution (India): Andaman Islands
Leaf Type: Compound, Pinnae
Inflorescence: Heads terminal, panicled
Fruit Type: Dry, Pods
IUCN Status: Not evaluated
Reference: Fl. Ned. Ind. 1(1): 12 (1855)

Acacia purpusii Brandegee
Fabaceae
Climbing Mechanism: Scrambler-Armed
Leaf Type: Compound, Pinnae
Inflorescence: Heads terminal, panicled
Fruit Type: Dry, Pods
IUCN Status: Not evaluated
Reference: Univ. Calif. Publ. Bot. 3: 380 (1909)

Acacia rugata Buch. -Ham. ex Benth.
Fabaceae
Climbing Mechanism: Scrambler-Armed
Distribution (India): Arunachal Pradesh, Meghalaya
Leaf Type: Compound, Pinnae
Inflorescence: Heads terminal, panicled
Fruit Type: Dry, Pods

IUCN Status: Not evaluated
Reference: London J. Bot. 1: 514 (1842)

Acacia tenuifolia (L.) Willd.
Fabaceae
Climbing Mechanism: Scrambler-Armed
Leaf Type: Compound, Pinnae
Inflorescence: Heads terminal, panicled
Fruit Type: Dry, Pods
IUCN Status: Not evaluated
Reference: Sp. Pl., ed. 4, 4: 1091 (1806)

Acacia torta (Roxb.) Craib
Fabaceae
Synonyms: *Acacia pennata* Sensu Baker, p.p. non Willd., *Mimosa torta* Roxb.
Common Name: Twisted Acacia
Climbing Mechanism: Scrambler-Armed
Distribution (India): Peninsular India, Madhya Pradesh, Himachal Pradesh, Arunachal Pradesh, Bihar, Odisha, Maharashtra, Rajasthan, Gujarat
Leaf Type: Compound, Pinnae
Inflorescence: Heads terminal, panicled
Fruit Type: Dry, Pods
Flowering and Fruiting: Throughout the year
IUCN Status: Not evaluated
Notes: Occurs in mixed and deciduous forests
Reference: Bull. Misc. Inform. Kew 1915: 410 (1915)

Aeschynomene indica L.
Fabaceae
Synonyms: *Aeschynomene burmanni* Raeusch., *Aeschynomene cachemiriana* Cambess., *Aeschynomene diffusa* J.G. Klein ex Willd., *Aeschynomene glaberrima* Poir., *Aeschynomene hirsuta* Vitman, *Aeschynomene kashmiriana* Baker, *Aeschynomene macropoda* DC., *Aeschynomene monta*
Common Name: Indian joint vetch.
Climbing Mechanism: Stem Twiner
Distribution (Global): Afghanistan, Andaman Is., Angola, Assam, Bangladesh, Benin, Borneo, Botswana, Burkina, Burundi, Cambodia, Cameroon, Central African Repu, Chad, China North-Central, China South-Central, China Southeast, Comoros, East Himalaya, Eritrea, Ethiopia, Florida, Gabon, Gambia, Ghana, Guinea-Bissau, Gulf of Guinea Is., Hainan, India, Iran, Ivory Coast, Japan, Kenya, Korea, KwaZulu-Natal, Laos, Madagascar, Malawi, Mali, Manchuria, Mauritania, Mexico Northeast, Mexico Southeast, Mozambique, Myanmar, Namibia, Nansei-shoto, Nepal, New South Wales, Nicobar Is., Niger, Nigeria, North Carolina, Northern Provinces, Northern Territory, Pakistan, Queensland, Rwanda, Senegal, Somalia, South Australia, Sri Lanka, Sudan, Taiwan,

Tanzania, Tennessee, Texas, Thailand, Togo, Uganda, Vietnam, West Himalaya, Western Australia, Zambia, Zaire, Zimbabwe
Leaf Type: Compound
Inflorescence: Axillary raceme
Fruit Type: linear-oblong
Flowering and Fruiting: September-November
IUCN Status: Least Concern
Reference: Sp. Pl.: 713 (1753)

Aganope agastyamalayana M.B. Viswan. Manik. & Tangav.
Fabaceae
Climbing Mechanism: Stem Twiner
Distribution (Global): India.
Distribution (India): Tamil Nadu
Leaf Type: Compound, imparipinnate
Inflorescence: Terminal flowers
Fruit Type: Dry, Pods
IUCN Status: Not evaluated
Reference: Adansonia, sér. 3, 25: 206 (2003)

Aganope heptaphylla (L.) Polhill
Fabaceae
Synonyms: *Aspalathus arborea* Lour., *Deguelia heptaphylla* (L.) S.F. Blake, *Deguelia macrol*oba (Miq.) Taub., *Deguelia sinu*ata (Benth. ex Thwaites) Taub., *Derris diadelpha* (Blanco) Merr., *Derris floribunda* Náves ex Fern. -Vill., *Derris heptaphylla* (L.) Merr.
Climbing Mechanism: Stem Twiner
Distribution (Global): Andaman Is., Assam, Bangladesh, Bismarck Archipelago, Borneo, Caroline Is., India, Java, Malaya, Maluku, Myanmar, New Guinea, Nicobar Is., Philippines, Solomon Is., Sri Lanka, Sumatera, Thailand
Distribution (India): West Bengal
Leaf Type: Compound, imparipinnate
Inflorescence: Terminal flowers
Fruit Type: Dry, Pods
IUCN Status: Not evaluated
Reference: Kew Bull. 25: 268 (1971)

Aganope polystachya (Benth.) Thoth. & D.N. Das
Fabaceae
Synonyms: *Deguelia polystachya* (Benth.) Taub., *Derris polystachya* Benth., *Pterocarpus polystachyus* (Benth.) Kuntze
Climbing Mechanism: Stem Twiner
Distribution (Global): Assam, East Himalaya
Leaf Type: Compound, imparipinnate
Inflorescence: Terminal flowers

Fruit Type: Dry, Pods
IUCN Status: Not evaluated
Reference: Rheedea 2: 63 (1992)

Aganope thyrsiflora (Benth.) Polhill
Fabaceae
Synonyms: *Aganope latifolia* (Prain) T.C. Chen & Pedley, *Aganope macrophylla* Miq., *Aganope subavenis* Miq., *Aganope thyrsiflo*ra var. *eualata* (Bedd.) Thoth. & D.N. Das, *Aganope thyrsiflora* var. *wallichii* (Prain) Thoth. & D.N. Das, *Amerimnon obovatum* Buch. -Ham. ex Ba
Climbing Mechanism: Stem Twiner
Distribution (Global): Andaman Is., Assam, Bangladesh, Borneo, Cambodia, China South-Central, China Southeast, East Himalaya, Hainan, India, Java, Laos, Malaya, Myanmar, New Guinea, Nicobar Is., Philippines, Sulawesi, Sumatera, Thailand, Vietnam
Distribution (India): Kerala, Tamil Nadu, Mizoram, Western Ghats
Leaf Type: 5–9 foliate
Flowering and Fruiting: December–August
IUCN Status: Not evaluated
Notes: Frequently found in Semi-evergreen forests
Reference: Kew Bull. 25: 268 (1971)

Albizia myriophylla Benth.
Fabaceae
Synonyms: *Acacia foliolosa* Graham, *Acacia microphylla* Roxb. ex Steud., *Acacia myriophylla* Steud., *Acacia roxburghii* Kostel., *Albizia microphylla* (Kuntze) J.F. Macbr., *Albizia thorelii* Pierre, *Mimosa microphylla* Roxb.
Climbing Mechanism: Scrambler-Armed
Distribution (Global): Assam, Bangladesh, Cambodia, East Himalaya, India, Laos, Malaya, Myanmar, Thailand, Vietnam
Distribution (India): Manipur
IUCN Status: Not evaluated
Reference: London J. Bot. 3: 90 (1844)

Apios carnea (Wall.) Benth.
Fabaceae
Synonyms: *Apios bodinieri* H. Lév., *Cyrtotropis carnea* Wall., *Pueraria stracheyi* Baker
Climbing Mechanism: Stem Twiner
Distribution (Global): Assam, China North-Central, China South-Central, China Southeast, East Himalaya, Laos, Myanmar, Nepal, Thailand, Tibet, Vietnam, West Himalaya
Distribution (India): Arunachal Pradesh, Mizoram
IUCN Status: Not evaluated
Reference: J. D. Hooker, Fl. Brit. India 2: 188 (1876)

Astragalus tenuicaulis Bunge
Fabaceae
Synonyms: *Astragalus inconspicuus* Baker, *Tragacantha tenuicaulis* (Benth. ex Bunge) Kuntze
Climbing Mechanism: Scrambler-Unarmed
Distribution (Global): East Himalaya, West Himalaya
IUCN Status: Not evaluated
Reference: Mém. Acad. Imp. Sci. Saint Pétersbourg, Sér. 7, 11(16): 23 (1868)

Bauhinia diphylla Buch. -Ham.
Fabaceae
Synonyms: *Bauhinia buchananii* Desv., *Phanera diphylla* (Buch. -Ham.) Benth.
Climbing Mechanism: Scrambler-Unarmed
Distribution (India): Maharashtra, Tamil Nadu
Leaf Type: Simple, lobed
Inflorescence: Panicle or racemes
Fruit Type: Dry, Pods
IUCN Status: Near Threatened
Reference: Account Embassy Kingd. Ava [Symes] 476, t. 24. 1800

Bauhinia divergens Baker
Fabaceae
Climbing Mechanism: Scrambler-Unarmed
Distribution (India): Arunachal Pradesh
Leaf Type: Simple, lobed
Inflorescence: Panicle or racemes
Fruit Type: Dry, Pods
IUCN Status: Not evaluated
Reference: J. D. Hooker, Fl. Brit. India 2: 282 (1878)

Bauhinia galpinii N.E.Br.
Fabaceae
Synonyms: *Bauhinia punctata* Bolle, *Perlebia galpinii* (N.E.Br.) A. Schmitz.
Common Name: Pride of The Cape, Red Bauhinia, African Plume, Nasturtium Bauhinia
Climbing Mechanism: Scrambler-Unarmed
Distribution (Global): Angola, KwaZulu-Natal, Mozambique, Northern Provinces, Swaziland, Zimbabwe
Distribution (India): Maharashtra
Leaf Type: Simple, lobed
Inflorescence: Panicle or racemes
Fruit Type: Dry, Pods
IUCN Status: Least Concern
Reference: Gard. Chron., ser. 3, 9: 728 (1891)

Bauhinia glauca (Benth.) Benth.
Fabaceae
Synonyms: *Bauhinia paraglauca* T. Tang & Wang, *Bauhinia viridiflora* Miq., *Phanera glauca* Benth.
Common Name: Glaucous climbing Bauhinia
Climbing Mechanism: Stem Twiner
Distribution (India): Arunachal Pradesh, Manipur, Mizoram
Leaf Type: Simple, lobed
Inflorescence: Panicle or racemes
Fruit Type: Dry, Pods
IUCN Status: Least Concern
Reference: Fl. Hongk.: 99 (1861)

Bauhinia glauca subsp. *tenuiflora* (C.B. Clarke) K. Larsen & S.S. Lar
Fabaceae
Climbing Mechanism: Scrambler-Unarmed
Distribution (India): Mizoram
Leaf Type: Simple, lobed
Inflorescence: Panicle or racemes
Fruit Type: Dry, Pods
IUCN Status: Not evaluated
Reference: Fl. Reipubl. Popularis Sin. 39: 196 (1988)

Bauhinia integrifolia Roxb.
Fabaceae
Common Name: Flame vine Bauhinia
Climbing Mechanism: Scrambler-Unarmed
Leaf Type: Simple, lobed
Inflorescence: Panicle or racemes
Fruit Type: Dry, Pods
IUCN Status: Not evaluated
Reference: Fl. Ind. ed. 1832, 2: 331 (1832)

Bauhinia involucellata Kurz
Fabaceae
Climbing Mechanism: Stem Twiner
Leaf Type: Simple, lobed
Inflorescence: Panicle or racemes
Fruit Type: Dry, Pods
IUCN Status: Not evaluated
Reference: J. Asiat. Soc. Bengal, Pt. 2, Nat. Hist. 42: 72 (1873)

Bauhinia khasiana Baker
Fabaceae

Synonyms: *Bauhinia howii* Merr. & Chun, *Bauhinia pierrei* Gagnep., *Bauhinia polystachya* Gagnep., *Phanera khasiana* (Baker)Thoth.
Common Name: Khasi Bauhinia.
Climbing Mechanism: Stem Twiner
Distribution (India): Arunachal Pradesh, Mizoram
Leaf Type: Simple, lobed
Inflorescence: Panicle or racemes
Fruit Type: Dry, Pods
IUCN Status: Not evaluated
Reference: J. D. Hooker, Fl. Brit. India 2: 281 (1878)

Bauhinia kunthiana Vogel
Fabaceae
Synonyms: *Schnella kunthiana* (Vogel) Wunderlin
Climbing Mechanism: Stem Twiner
Leaf Type: Simple, lobed
Inflorescence: Panicle or racemes
Fruit Type: Dry, Pods
IUCN Status: Not evaluated
Reference: Linnaea 13: 312 (1839)

Bauhinia nervosa (Benth.) Baker
Fabaceae
Synonyms: *Phanera nervosa* Benth.
Climbing Mechanism: Tendril Climber
Leaf Type: Simple, lobed
Inflorescence: Panicle or racemes
Fruit Type: Dry, Pods
IUCN Status: Not evaluated
Reference: J. D. Hooker, Fl. Brit. India 2: 283 (1878)

Bauhinia ornata Kurz
Fabaceae
Synonyms: *Bauhinia integrifolia* Auct. non Roxb.
Climbing Mechanism: Tendril Climber
Distribution (India): Arunachal Pradesh, Mizoram
Leaf Type: Simple, alternate
Inflorescence: Corymbose raceme
IUCN Status: Least Concern
Reference: J. Asiat. Soc. Bengal, Pt. 2, Nat. Hist. 42: 72 (1873)

Bauhinia ovatifolia T. Chen
Fabaceae
Climbing Mechanism: Tendril Climber
Distribution (India): Arunachal Pradesh

Leaf Type: Simple, lobed
Inflorescence: Panicle or racemes
Fruit Type: Dry, Pods
IUCN Status: Not evaluated
Reference: Guihaia 8: 50 (1988)

Bauhinia phoenicea Wight & Arn.
Fabaceae
Synonyms: *Bauhinia benthamii* Bedd., *Bauhinia ruficarpa* Desv., *Perlebia phoenicea* (B. Heyne ex Wight & Arn.) A. Schmitz, *Phanera phoenicea* (B. Heyne ex Wight & Arn.) Benth.
Common Name: Crimson mountain ebony
Climbing Mechanism: Scrambler-Unarmed
Distribution (Global): India
Distribution (India): Karnataka, Kerala, Maharashtra, Tamil Nadu, Western Ghats
Leaf Type: Simple, lobed
Inflorescence: Panicle or racemes
Fruit Type: Dry, Pods
Flowering and Fruiting: September–April
IUCN Status: Not evaluated
Notes: Common in evergreen and semi-evergreen forests of the Western Ghats
Reference: Prodr. Fl. Ind. Orient. 1: 296 (1832)

Bauhinia roxburghiana Voigt
Fabaceae
Synonyms: *Bauhinia emarginata* G. Don, *Bauhinia retusa* Roxb.
Common Name: Roxburgh's Bauhinia
Climbing Mechanism: Stem Twiner
Leaf Type: Simple, lobed
Inflorescence: Panicle or racemes
Fruit Type: Dry, Pods
IUCN Status: Not evaluated
Reference: Hort. Suburb. Calcutt.: 254 (1845)

Bauhinia scandens L.
Fabaceae
Synonyms: *Bauhinia anguina* Roxb., *Bauhinia anguina* var. *horsfieldii* (Miq.) Prain, *Bauhinia debilis* Hassk., *Bauhinia horsfieldii* (Miq.) J.F. Macbr., *Bauhinia piperifolia* Roxb. + 8
Common Name: Snake climber
Climbing Mechanism: Stem Twiner
Distribution (India): Kerala, Maharashtra
Leaf Type: Simple, lobed
Inflorescence: Panicle or racemes
Fruit Type: Dry, Pods

IUCN Status: Not evaluated
Reference: Sp. Pl.: 374 (1753)

Bauhinia stipularis Korth.
Fabaceae
Synonyms: *Phanera stipularis* (Korth.) Benth.
Common Name: Nicobar creeper
Climbing Mechanism: Scrambler-Unarmed
Distribution (India): Andaman and Nicobar Islands
Leaf Type: Simple, lobed
Inflorescence: Panicle or racemes
Fruit Type: Dry, Pods
IUCN Status: Least Concern
Reference: Nat. Gesch. Bot.: 92 (1839-42)

Bauhinia touranensis Gagnep.
Fabaceae
Synonyms: *Cheniella touranensis* (Gagnep.) R. Clark & Mackinder
Climbing Mechanism: Stem Twiner
Distribution (India): Arunachal Pradesh
Leaf Type: Simple, lobed
Inflorescence: Panicle or racemes
Fruit Type: Dry, Pods
IUCN Status: Not evaluated
Reference: Notul. Syst. (Paris) 2: 181 (1912)

Bauhinia vahlii Wight & Arn.
Fabaceae
Synonyms: *Bauhinia racemosa* Vahl, *Phanera vahlii* (Wight & Arn.) Benth.
Common Name: Maloo creeper
Climbing Mechanism: Stem Twiner
Distribution (India): Andhra Pradesh, Bihar, Odisha, Delhi, Eastern Ghats, Goa, Himachal Pradesh, Jharkhand, Madhya Pradesh, Maharashtra, Odisha, Tamil Nadu, Telangana, Uttar Pradesh, Western Ghats
Leaf Type: Simple, lobed
Inflorescence: Panicle or racemes
Fruit Type: Dry, Pods
Flowering and Fruiting: Throughout the year
IUCN Status: Not evaluated
Notes: Common in sal forests
Reference: R. Wight, Cat. Ind. Pl.: 38 (1833)

Bauhinia wallichii J.F. Macbr.
Fabaceae

Synonyms: *Bauhinia macrostachya* (Benth.) Baker, *Bauhinia melanophylla* Merr., *Phanera macrostachya* Benth., *Phanera wallichii* (J.F. Macbr.) Thoth.
Climbing Mechanism: Stem Twiner
Distribution (India): Arunachal Pradesh
Leaf Type: Simple, lobed
Inflorescence: Panicle or racemes
Fruit Type: Dry, Pods
Flowering and Fruiting: April–June
IUCN Status: Least Concern
Reference: Contr. Gray Herb. 59: 23 (1919)

Butea buteiformis (Voigt) Mabb.
Fabaceae
Synonyms: *Megalotropis buteiformis* (Voigt) Griff., *Meizotropis buteiformis* Voigt.
Common Name: Shrub Butea.
Climbing Mechanism: Scrambler-Unarmed
Distribution (Global): Assam, Bangladesh, East Himalaya, India, Myanmar, Nepal, Thailand, Tibet, West Himalaya
Distribution (India): Arunachal Pradesh
Leaf Type: Trifoliate
Inflorescence: Racemes or panicles
Fruit Type: Pods
IUCN Status: Not evaluated
Reference: Notes Roy. Bot. Gard. Edinburgh 37: 346 (1979)

Butea superba Roxb.
Fabaceae
Synonyms: *Rudolphia superba* (Roxb. ex Willd.) Poir.
Common Name: Climbing Palas
Climbing Mechanism: Scrambler-Unarmed
Distribution (Global): Bangladesh, Cambodia, India, Laos, Myanmar, Thailand, Vietnam
Distribution (India): Andhra Pradesh, Bihar, Odisha, Eastern Ghats, Goa, Gujarat, Jharkhand, Madhya Pradesh, Maharashtra, Odisha, Rajasthan, Telangana, West Bengal
Leaf Type: Trifoliate
Inflorescence: Racemes
Fruit Type: Pods
Flowering and Fruiting: February–July
IUCN Status: Not evaluated
Notes: Mostly found in dry-deciduous forests
Reference: Sp. Pl., ed. 4, 3: 917 (1802)

Caesalpinia andamanica (Prain) Hattink
Fabaceae

Synonyms: *Mezoneuron andamanicum* Prain
Climbing Mechanism: Scrambler-Armed
Distribution (India): Andaman and Nicobar Islands
Leaf Type: Compound, 2-pinnate
Inflorescence: Racemes/panicles
Fruit Type: Dry, Pods
IUCN Status: Not evaluated
Reference: Reinwardtia 9: 15 (1974)

Caesalpinia bonduc (L.) Roxb.
Fabaceae
Synonyms: *Bonduc minus* Medik., *Caesalpinia bonducella* (L.) Fleming, *Caesalpinia crista* L., p.p.A, *Caesalpinia cristata* Prowazek, *Caesalpinia grisebachiana* Kuntze, *Caesalpinia sepiaria* Auct. non Roxb., *Caesalpinia sogerensis* Baker f., *Guilandina bonduc* Griseb.
Common Name: Bonduc nut, Fever nut, Physic nut
Climbing Mechanism: Scrambler-Armed
Distribution (India): Andaman and Nicobar Islands, Andhra Pradesh, Arunachal Pradesh, Eastern Ghats, Himachal Pradesh, Jharkhand, Kerala, Madhya Pradesh, Maharashtra, Manipur, Odisha, Rajasthan, Telangana, Tripura, Uttar Pradesh, West Bengal
Leaf Type: Bipinnate
Inflorescence: Axillary or terminal
Fruit Type: Pods
Flowering and Fruiting: March–May
IUCN Status: Not evaluated
Reference: Fl. Ind. 2: 362 (1824)

Caesalpinia crista L.
Fabaceae
Synonyms: *Butea loureiroi* Spreng., *Caesalpinia axillaris* (Lam.) DC., *Caesalpinia crista* var. *parvistipula* Urb., *Caesalpinia laevigata* Perr., *Caesalpinia nuga* (L.) W.T. Aiton, *Caesalpinia paniculata* (Lam.) Roxb., *Caesalpinia scandens* B. Heyne ex Roth.
Common Name: Crested fever nut
Climbing Mechanism: Scrambler-Armed
Distribution (Global): Andaman Is., Assam, Bangladesh, Bismarck Archipelago, Borneo, Cambodia, Caroline Is., China South-Central, China Southeast, Christmas I., East Himalaya, El Salvador, Guatemala, Hainan, India, Japan, Java, Laccadive Is., Lesser Sunda Is., Malaya, Maluku, Myanmar, Nansei-shoto, New Caledonia, New Guinea, Nicobar Is., Panamá, Philippines, Queensland, Santa Cruz Is., Solomon Is., South China Sea, Sri Lanka, Sulawesi, Sumatera, Taiwan, Thailand, Tonga, Vanuatu, Vietnam, West Himalaya

Distribution (India): Andaman and Nicobar Islands, Andhra Pradesh, Bihar, Odisha, Delhi, Goa, Gujarat, Karnataka, Kerala, Maharashtra, Odisha, Tamil Nadu, West Bengal
Leaf Type: Bipinnate, alternate
Inflorescence: Axillary or terminal racemes
Fruit Type: Pods
Flowering and Fruiting: February–June
IUCN Status: Not evaluated
Notes: Occur in deciduous forests and mangrove forests
Reference: Sp. Pl.: 380 (1753)

Caesalpinia cucullata Roxb.
Fabaceae
Synonyms: *Mezoneurum cucullatum* (Roxb.) Wight & Arn., *Mezoneuron cucullatum* (Roxb.) Wight & Arn., *Mezoneuron cucullatum* var. *grandis* Baker, *Mezoneuron cucullatum* var. *robustum* Craib, *Mezoneuron macrophyllum* Miq.
Common Name: Sahyadri thorn, Hooded-flowered brasiletto
Climbing Mechanism: Scrambler-Armed
Distribution (India): Andaman and Nicobar Islands, Arunachal Pradesh, Assam, Eastern Ghats, Kerala, Maharashtra, Mizoram, Odisha, Tamil Nadu, Uttar Pradesh, West Bengal, Western Ghats
Leaf Type: Simple, alternate
Inflorescence: Raceme or panicle
Flowering and Fruiting: September–April
IUCN Status: Not evaluated
Reference: Fl. Ind. ed. 1832, 2: 358 (1832)

Caesalpinia decapetala (Roth) Alston
Fabaceae
Synonyms: *Biancaea scandens* Tod., *Biancaea sepiaria* (Roxb.) Tod., *Caesalpinia benguetensis* Elmer, *Caesalpinia crista* Auct. non L., *Caesalpinia decapetala* var. *japonica* (Siebold & Zucc.) H. Ohashi, *Caesalpinia ferox* Hassk. + 6
Common Name: Black bonduc, Mysore thorn
Climbing Mechanism: Scrambler-Armed
Distribution (India): Andhra Pradesh, Goa, Gujarat, Himachal Pradesh, Jharkhand, Kerala, Madhya Pradesh, Maharashtra, Manipur, Odisha, Rajasthan, Tamil Nadu, Uttar Pradesh, West Bengal
Leaf Type: Double compound
Inflorescence: Racemes terminal
Fruit Type: Pods
Flowering and Fruiting: October–May
IUCN Status: Not evaluated
Notes: On hills up to 1400 m, frequently occur in degraded forests
Reference: H. Trimen, Handb. Fl. Ceylon 6(Suppl.): 89 (1931)

Caesalpinia digyna Rottler
Fabaceae
Synonyms: *Caesalpinia gracilis* Miq., *Caesalpinia oleosperma* Roxb., *Caesalpinia olesperma* Roxb.
Common Name: Teri pod
Climbing Mechanism: Scrambler-Armed
Distribution (India): Andhra Pradesh, Bihar, Odisha, Eastern Ghats, Kerala, Madhya Pradesh, Manipur, Odisha, Tripura
Leaf Type: Compound, opposite
Inflorescence: Terminal or axillary racemes
Fruit Type: Legume
Flowering and Fruiting: August–March
IUCN Status: Not evaluated
Reference: Neue Schriften Ges. Naturf. Freunde Berlin 4: 200 (1803)

Caesalpinia enneaphylla Roxb.
Fabaceae
Synonyms: *Caesalpinia sepiaria* Auct. non Roxb., *Mezoneuron enneaphyllum* (Roxb.) Benth., *Mezoneuron glabrum* var. *enneaphyllum* (Roxb.) Kurz, *Mezoneurum enneaphyllum* (Roxb.) Wight & Arn.
Climbing Mechanism: Scrambler-Armed
Distribution (India): Andaman and Nicobar Islands, Arunachal Pradesh, Manipur, Mizoram
Leaf Type: Compound, bipinnate
Inflorescence: Racemes, axillary, terminal
Fruit Type: Pods
Flowering and Fruiting: September–February
IUCN Status: Not evaluated
Reference: Fl. Ind. 2: 363 (1824)

Caesalpinia gilliesii (Hook.) D.Dietr.
Fabaceae
Synonyms: *Erythrostemon gilliesii* (Hook.) Klotzsch
Climbing Mechanism: Scrambler-Armed
Distribution (India): Maharashtra
Leaf Type: Compound, 2-pinnate
Inflorescence: Racemes/panicles
Fruit Type: Dry, Pods
Flowering and Fruiting: January–February
IUCN Status: Not evaluated
Reference: Syn. Plant. 2: 1495 (1840)

Caesalpinia globulorum Bakh.f. & P. Royen
Fabaceae
Synonyms: *Caesalpinia jayabo* Maza (nom. illeg)

Climbing Mechanism: Scrambler-Armed
Leaf Type: Bipinnate
Inflorescence: Axillary or terminal racemes or panicles
Fruit Type: Pods
IUCN Status: Not evaluated
Reference: Blumea 12: 62 (1963)

Caesalpinia hymenocarpa (Prain) Hattink
Fabaceae
Synonyms: *Mezoneuron hymenocarpum* Prain, *Mezoneuron laoticum* Gagnep., *Mezoneuron pubescens* Sensu Baker, *Mezoneurum hymenocarpum* Wight & Arn.
Climbing Mechanism: Scrambler-Armed
Distribution (India): Andaman and Nicobar Islands, Mizoram
Leaf Type: Compound, 2-pinnate
Inflorescence: Racemes/panicles
Fruit Type: Dry, Pods
Flowering and Fruiting: September–January
IUCN Status: Not evaluated
Reference: Reinwardtia 9: 35 (1974)

*Caesalpinia jayab*o M. Gomez
Fabaceae
Climbing Mechanism: Scrambler-Armed
Leaf Type: Compound, 2-pinnate
Inflorescence: Racemes/panicles
Fruit Type: Dry, Pods
IUCN Status: Not Evaluated
Reference: Anales Soc. Esp. Hist. Nat. 19: 234 (1890)

Caesalpinia mimosoides Lam.
Fabaceae
Synonyms: *Caesalpinia resupinata* Roxb., *Caesalpinia simora* Roxb.
Climbing Mechanism: Scrambler-Armed
Distribution (India): Goa, Kerala, Maharashtra, Tamil Nadu
Leaf Type: Compound, 2-pinnate
Inflorescence: Racemes/panicles
Fruit Type: Dry, Pods
Flowering and Fruiting: October–March
IUCN Status: Not evaluated
Notes: Often found in secondary forests and farmlands
Reference: Encycl. 1: 462 (1785)

Caesalpinia pubescens (Desf.) Hattink
Fabaceae
Climbing Mechanism: Scrambler-Armed

Leaf Type: Compound, 2-pinnate
Inflorescence: Racemes/panicles
Fruit Type: Dry, Pods
IUCN Status: Not evaluated
Reference: Numer. List: n.° 5834 (1831)

Caesalpinia tortuosa Roxb.
Fabaceae
Synonyms: *Caesalpinia acanthobotrya* Miq., *Caesalpinia cinclidocarpa* Miq., *Caesalpinia microphylla* Prain, *Caesalpinia tortuosa* var. *grandifolia* Fedde, *Cinclidocarpus nitidus* Zoll.
Climbing Mechanism: Scrambler-Armed
Leaf Type: Compound, bipinnate
Inflorescence: Racemes or panicles terminal or axillary
Fruit Type: Legume
Flowering and Fruiting: June–August
IUCN Status: Data Deficient
Reference: Fl. Ind. 2: 365 (1824)

Cajanus albicans (Wight & Arn.) Maesen
Fabaceae
Synonyms: *Atylosia Atylosia* (Wight & Arn.) Benth., *Cajanus wightianus* Graham, *Cantharospermum albicans* Wight & Arn.
Common Name: Whitish pigeon pea
Climbing Mechanism: Stem Twiner
Distribution (Global): India, Sri Lanka
Distribution (India): Kerala, Madhya Pradesh, Maharashtra, Tamil Nadu
Leaf Type: Compound, pinnately digitate
Inflorescence: Terminal/pedunculate Racemes
Fruit Type: Dry, Pods
Flowering and Fruiting: September–April
IUCN Status: Not evaluated
Notes: occasionally found in deciduous, scrub forests and on the roadsides
Reference: Agric.Univ. Wageningen Pap.85(4);55(1985 publ.1986)

Cajanus crassus (King) Maesen
Fabaceae
Synonyms: Atylosia crassa Prain
Common Name: Thick-stemmed Cajanus
Climbing Mechanism: Stem Twiner
Distribution (Global): Andaman Is., Assam, Bangladesh, China South-Central, China Southeast, East Himalaya, Hainan, India, Java, Laos, Lesser Sunda Is., Malaya, Myanmar, Nepal, New Guinea, Pakistan, Philippines, Sulawesi, Thailand, Vietnam, West Himalaya
Distribution (India): Andaman and Nicobar Islands, Maharashtra, West Bengal

Leaf Type: Trifoliate
Inflorescence: Racemes
Flowering and Fruiting: November–April
IUCN Status: Not evaluated
Notes: Mostly found in scrub jungles
Reference: Agric. Univ. Wageningen Pap. 85(4): 105 (1985 publ. 1986)

Cajanus crassus var. *burmanicus* (Collett & Hemsl.) Maesen
Fabaceae
Climbing Mechanism: Stem Twiner
Leaf Type: Compound, pinnately digitate
Inflorescence: Terminal/pedunculate Racemes
Fruit Type: Dry, Pods
IUCN Status: Not evaluated
Reference: Agric. Univ. Wageningen Pap. 85(4): 109 (1985 publ. 1986)

Cajanus elongatus (Benth.) Maesen
Fabaceae
Synonyms: *Atylosia elongata* Benth., *Cantharospermum elongatum* (Benth.) Raizada, *Dolichos elongatus* Graham.
Climbing Mechanism: Stem Twiner
Distribution (Global): Assam, East Himalaya, India, Myanmar, Nepal, Thailand, Vietnam, West Himalaya
Distribution (India): West Bengal
Leaf Type: Compound, pinnately digitate
Inflorescence: Terminal/pedunculate Racemes
Fruit Type: Dry, Pods
IUCN Status: Not evaluated
Reference: Agric. Univ. Wageningen Pap. 85(4): 115 (1985 publ. 1986)

Cajanus goensis Dalzell
Fabaceae
Synonyms: *Atylosia barbata* (Benth.) Baker, *Atylosia calycina* (Miq.) Kurz, *Atylosia goensis* (Dalzell) Dalzell, *Atylosia siamensis* Craib, *Cantharospermum barbatum* (Benth.) Koord. ex Keuchenius, *Dolichos barbatus* Wall., *Dolichos ornatus* Wall., *Dunbaria barbata*
Climbing Mechanism: Stem Twiner
Distribution (Global): Assam, Bangladesh, China South-Central, East Himalaya, India, Java, Laos, Malaya, Myanmar, Thailand, Vietnam
Distribution (India): Kerala, Maharashtra, Manipur, Mizoram, Tamil Nadu
Leaf Type: Compound, pinnately digitate
Inflorescence: Terminal/pedunculate Racemes
Fruit Type: Dry, Pods
Flowering and Fruiting: September–February
IUCN Status: Not evaluated

Notes: Occurs in evergreen and moist deciduous forests
Reference: Hooker's J. Bot. Kew Gard. Misc. 2: 264 (1850)

Cajanus grandiflorus (Baker) Maesen
Fabaceae
Synonyms: *Atylosia grandiflora* Benth. ex Baker, *Dunbaria pulchra* Benth. ex Baker, *Pueraria seguinii* H. Lév. & Vaniot.
Climbing Mechanism: Stem Twiner
Distribution (Global): Assam, China South-Central, China Southeast, East Himalaya, Myanmar, West Himalaya
Leaf Type: Compound, pinnately digitate
Inflorescence: Terminal/pedunculate Racemes
Fruit Type: Dry, Pods
IUCN Status: Not evaluated
Reference: Agric. Univ. Wageningen Pap. 85(4): 125 (1985 publ. 1986)

Cajanus heynei (Wight & Arn.) Maesen
Fabaceae
Synonyms: *Atylosia kulnensis* (Dalzell) Dalzell, *Cajanus kulnensis* Dalzell, *Collaea gibba* Graham, *Dunbaria heynei* Wight & Arn., *Dunbaria oblonga* Wight ex Arn.
Common Name: Heyne's pigeon pea.
Climbing Mechanism: Stem Twiner
Distribution (Global): India, Sri Lanka
Distribution (India): Kerala, Madhya Pradesh, Maharashtra, Tamil Nadu
Leaf Type: Compound, pinnately digitate
Inflorescence: Terminal/pedunculate Racemes
Fruit Type: Dry, Pods
Flowering and Fruiting: October–February
IUCN Status: Not evaluated
Reference: Agric. Univ. Wageningen Pap. 85(4): 129 (1985 publ. 1986)

Cajanus mollis (Benth.) Maesen
Fabaceae
Synonyms: *Atylosia glandulosa* Dalzell, *Atylosia mollis* Benth., *Cantharospermum molle* (Benth.) Taub., *Collaea cinerascens* Graham, *Collaea mollis* Graham, *Dolichos blandus* Graham, *Dolichos crassus* Graham.
Climbing Mechanism: Stem Twiner
Distribution (Global): Assam, China South-Central, East Himalaya, India, Nepal, Pakistan, West Himalaya
Leaf Type: Compound, pinnately digitate
Inflorescence: Terminal/pedunculate Racemes
Fruit Type: Dry, Pods
IUCN Status: Not evaluated
Reference: Agric. Univ. Wageningen Pap. 85(4): 154 (1985 publ. 1986)

Cajanus platycarpus (Benth.) Maesen
Fabaceae
Synonyms: *Atylosia geminiflora* Dalzell, *Atylosia platycarpa* Benth., *Cantharospermum distans* Royle ex Baker, *Cantharospermum geminiflorum* (Dalzell) Raizada, *Cantharospermum platycarpum* (Benth.) Raizada.
Climbing Mechanism: Stem Twiner
Distribution (Global): India, Java, Pakistan, West Himalaya
Distribution (India): Maharashtra
Leaf Type: Compound, pinnately digitate
Inflorescence: Terminal/pedunculate Racemes
Fruit Type: Dry, Pods
IUCN Status: Not evaluated
Reference: Agric. Univ. Wageningen Pap. 85(4): 160 (1985 publ. 1986)

*Cajanus rugo*sus (Wight & Arn.) Maesen
Fabaceae
Synonyms: *Atylosia rugosa* Wight & Arn., *Rhynchosia velutina* Graham, *Cantharospermum rugosum* (Wight & Arn.) Alston.
Climbing Mechanism: Stem Twiner
Distribution (Global): India, Sri Lanka
Distribution (India): Kerala, Madhya Pradesh
Leaf Type: Compound, pinnately digitate
Inflorescence: Terminal/pedunculate Racemes
Fruit Type: Dry, Pods
Flowering and Fruiting: October–February
IUCN Status: Not evaluated
Notes: Mostly found on hills up to 1600m and in grasslands
Reference: Agric. Univ. Wageningen Pap. 85(4): 179 (1985 publ. 1986)

Cajanus scarabaeoides (L.) Thouars
Fabaceae
Synonyms: *Atylosia scarabaeoides* (L.) Benth., *Atylosia scarabaeoides* var. *typica* Domin, *Cantharospermum scarabaeoides* (L.) Koord., *Cantharospermum scarabaeoideum* Baill., *Dolichos scarabaeoides* L., *Rhynchosia scarabaeoides* (L.) DC.
Common Name: Showy pigeon pea, peanut grass, Tropical clover
Climbing Mechanism: Stem Twiner
Distribution (Global): Assam, Bangladesh, Cambodia, China South-Central, China Southeast, Comoros, East Himalaya, Hainan, India, Laos, Lesser Sunda Is., Madagascar, Malaya, Marianas, Myanmar, Nansei-shoto, Nepal, New Guinea, Nicobar Is., Northern Territory, Pakistan, Philippines, Queensland, South China Sea, Sri Lanka, Sumatera, Taiwan, Thailand, Vietnam, West Himalaya, Western Australia
Distribution (India): Jharkhand, Kerala, Maharashtra, Manipur, Tamil Nadu, Telangana, Uttar Pradesh, West Bengal

Leaf Type: Trifoliate
Inflorescence: Racemose
Fruit Type: Pods
Flowering and Fruiting: August–March
IUCN Status: Least Concern
Notes: Distributed mostly in the scrub jungle, moist- and dry-deciduous forests
Reference: G.-F. Cuvier, Dict. Sci. Nat., ed. 2. 6: 167 (1816)

Cajanus villosus (Baker) Maesen
Fabaceae
Synonyms: *Atylosia villosa* Benth. ex Baker
Climbing Mechanism: Stem Twiner
Distribution (Global): East Himalaya, India
Distribution (India): West Bengal
Leaf Type: Compound, pinnately digitate
Inflorescence: Terminal/pedunculate Racemes
Fruit Type: Dry, Pods
IUCN Status: Not evaluated
Reference: Agric. Univ. Wageningen Pap. 85(4): 205 (1985 publ. 1986)

Callerya cinerea (Benth.) Schot
Fabaceae
Synonyms: *Callerya dielsiana* var. *heterocarpa* (Chun ex T.C. Chen) X.Y. Zhu ex Z. Wei & Pedley, *Callerya dielsiana* var. *solida* (T.C. Chen ex Z. Wei) X.Y. Zhu ex Z. Wei & Pedley, *Millettia bockii* Harms ex Diels, *Millettia bracteosa* Gagnep., *Millettia cinerea* Benth.
Climbing Mechanism: Scrambler-Unarmed
Distribution (Global): Assam, Bangladesh, Cambodia, China South-Central, China Southeast, East Himalaya, Laos, Malaya, Myanmar, Nepal, Thailand, Tibet, Vietnam
Distribution (India): West Bengal
IUCN Status: Not evaluated
Reference: Blumea 39: 17 (1994)

Callerya eriantha (Benth.) Schot
Fabaceae
Climbing Mechanism: Scrambler-Unarmed
IUCN Status: Not evaluated
Reference: Blumea 39: 21 (1994)

Calopogonium mucunoides Desv.
Fabaceae
Synonyms: *Calopogonium brachycarpum* (Benth.) Benth. ex Hemsl., *Calopogonium flavidum* Brandegee, *Calopogonium orthocarpum* Urb.,

Stenolobium brachycarpum Benth., *Stenolobium brachycarpum* var. *brachystachyum* Benth.
Climbing Mechanism: Stem Twiner
Distribution (Global): Argentina Northeast, Belize, Bolivia, Brazil North, Brazil Northeast, Brazil South, Brazil Southeast, Brazil West-Central, Colombia, Costa Rica, Cuba, Dominican Republic, Ecuador, El Salvador, French Guiana, Guatemala, Guyana, Honduras, Jamaica, Leeward Is., Mexico Central, Mexico Gulf, Mexico Northeast, Mexico Northwest, Mexico Southeast, Mexico Southwest, Nicaragua, Panamá, Peru, Puerto Rico, Suriname, Venezuela, Windward Is.
Distribution (India): Karnataka, Kerala, Tamil Nadu
Leaf Type: Alternate, trifoliate
Inflorescence: Axillary racemes
Fruit Type: Pods
Flowering and Fruiting: August–February
IUCN Status: Not evaluated
Notes: Sometimes grown as a cover crop in rubber plantations
Reference: Ann. Sci. Nat. (Paris) 9: 423 (1826)

Canavalia africana Dunn
Fabaceae
Synonyms: *Canavalia ferruginea* Piper
Climbing Mechanism: Stem Twiner
Distribution (Global): Tropical Africa and India
Distribution (India): Kerala, Maharashtra, Telangana, West Bengal
Leaf Type: Compound, 3-foliate
Inflorescence: Axillary racemes
Fruit Type: Pods
Flowering and Fruiting: August–March
IUCN Status: Not evaluated
Notes: Usually found along stream banks in the plains and shola forests
Reference: Bull. Misc. Inform. Kew 1922: 135 (1922)

Canavalia cathartica Thouars
Fabaceae
Synonyms: *Canavalia bouquetii* Montrouz., *Canavalia microcarpa* (DC.) Piper, *Canavalia polystachya* Schweinf., *Canavalia turgida* Graham ex Miq., *Canavalia virosa* (Roxb.) Wight & Arn., *Canavalia virosa* Náves ex Fern.-Vill., *Dolichos virosus* Roxb.
Common Name: Maunaloa
Climbing Mechanism: Stem Twiner
Distribution (Global): Andaman Is., Bangladesh, Bismarck Archipelago, Borneo, Cambodia, Caroline Is., Chagos Archipelago, China Southeast, Christmas I., Cocos (Keeling) Is., Cook Is., Fiji, Hainan, India, Java, Kenya, KwaZulu-Natal, Laccadive Is., Lesser Sunda Is., Line Is., Malaya, Maldives, Maluku, Marianas, Marshall Is., Myanmar, Namibia, Nansei-shoto, Nauru, New Caledonia, New

Guinea, Nicobar Is., Northern Provinces, Northern Territory, Philippines, Queensland, Samoa, Saudi Arabia, Seychelles, Society Is., Solomon Is., Somalia, South China Sea, Sri Lanka, Sulawesi, Sumatera, Swaziland, Taiwan, Tanzania, Thailand, Tokelau-Manihiki, Tonga, Tubuai Is., Tuvalu, Vanuatu, Vietnam, Yemen

Distribution (India): Andaman and Nicobar Islands, Andhra Pradesh, Gujarat, Kerala, Maharashtra, Tamil Nadu, West Bengal

Leaf Type: Pinnately compound, trifoliate, alternate

Inflorescence: Axillary racemes

Fruit Type: Pubescent Pods

Flowering and Fruiting: June–December

IUCN Status: Not evaluated

Notes: Frequent along banks of rivers, canals, and backwaters

Reference: J. Bot. Agric. 1: 81 (1813)

Canavalia ensiformis (L.) DC.

Fabaceae

Synonyms: *Canavalia ensiformis* var. *albida* DC., *Canavalia ensiformis* var. *coriacea* Domin, *Canavalia ensiformis* var. *normalis* Kuntze, *Canavalia ensiformis* var. *truncata* Ricker. + 10

Common Name: Sword bean, Jack bean, Broad bean, Patagonian bean

Climbing Mechanism: Stem Twiner

Distribution (Global): Pan tropical

Distribution (India): Arunachal Pradesh, Bihar, Odisha, Gujarat, Maharashtra, Mizoram, Rajasthan, Tamil Nadu, Telangana, West Bengal

Leaf Type: Trifoliate leaves

Fruit Type: Pods

Flowering and Fruiting: August–March

IUCN Status: Not evaluated

Reference: Prodr. 2: 404 (1825)

Canavalia gladiata (Jacq.) DC.

Fabaceae

Synonyms: *Canavalia ensiformis* var. *gladiata* (Jacq.) Kuntze, *Canavalia gladiata* f. *alba* (Makino) H. Ohashi. +10

Common Name: Horse bean, Jack bean, Sword bean, Broad bean, Patagonian bean

Climbing Mechanism: Stem Twiner

Distribution (Global): Assam, Bangladesh, Cambodia, China South-Central, China Southeast, East Himalaya, Hainan, India, Japan, Laos, Myanmar, Nicobar Is., Panamá, Taiwan, Thailand, Vietnam, West Himalaya

Distribution (India): Andhra Pradesh, Gujarat, Jharkhand, Madhya Pradesh, Maharashtra, Odisha, Tamil Nadu, Telangana, West Bengal

Leaf Type: Pinnately compound, trifoliate

Inflorescence: Racemes

Fruit Type: Pods

Flowering and Fruiting: July–April
IUCN Status: Not evaluated
Notes: Cultivated. Wild populations are often found in deciduous forests
Reference: Prodr. 2: 404 (1825)

Canavalia lineata (Thunb.) DC.
Fabaceae
Synonyms: *Dolichos lineatus* Thunb.
Climbing Mechanism: Stem Twiner
Distribution (Global): Bangladesh, Cambodia, China Southeast, Hainan, Japan, Java, Kazan-retto, Korea, Malaya, Nansei-shoto, Ogasawara-shoto, Philippines, Sumatera, Taiwan, Thailand, Vietnam
Distribution (India): Andhra Pradesh, Maharashtra, Tamil Nadu, West Bengal
IUCN Status: Not evaluated
Notes: Occasionally found in sandy sea shores.
Reference: Prodr. 2: 404 (1825)

Canavalia mollis Wight & Arn.
Fabaceae
Climbing Mechanism: Stem Twiner
Distribution (Global): India, Java, Lesser Sunda Is., Sri Lanka, Sumatera
Distribution (India): Andhra Pradesh, Eastern Ghats, Tamil Nadu
Leaf Type: Pinnately compound, trifoliate
Inflorescence: Racemes
Fruit Type: Pods
Flowering and Fruiting: October–April
IUCN Status: Not evaluated
Notes: Mostly found on hills and forest boundaries above 1000 m
Reference: Prodr. Fl. Ind. Orient. 1: 253 (1834)

Canavalia rosea (Sw.) DC.
Fabaceae
Synonyms: *Canavalia apiculata* Piper, *Canavalia arenicola* Piper, *Canavalia baueriana* Endl., *Canavalia emarginata* (Jacq.) G. Don, *Canavalia maritima* Thouars. +20
Common Name: Beach bean
Climbing Mechanism: Stem Twiner
Distribution (Global): Aldabra, Andaman Is., Angola, Aruba, Bahamas, Bangladesh, Belize, Benin, Bermuda, Bismarck Archipelago, Borneo, Brazil North, Brazil Northeast, Brazil South, Brazil Southeast, Burkina, Cambodia, Cameroon, Cape Provinces, Caroline Is., Cayman Is., Central American Pac, China Southeast, Colombia, Comoros, Congo, Costa Rica, Cuba, Dominican Republic, Ecuador, El Salvador, Equatorial Guinea, Fiji, Florida, French Guiana, Gabon, Galápagos, Gambia, Ghana, Guatemala, Guinea, Guinea-Bissau, Gulf of Guinea Is., Guyana, Hainan, Haiti, Honduras, India, Ivory Coast, Jamaica, Java, Kazan-retto, Kenya,

Kermadec Is., KwaZulu-Natal, Laos, Leeward Is., Lesser Sunda Is., Liberia, Madagascar, Malaya, Maldives, Mali, Maluku, Marianas, Marquesas, Marshall Is., Mauritius, Mexican Pacific Is., Mexico Central, Mexico Gulf, Mexico Northeast, Mexico Northwest, Mexico Southeast, Mexico Southwest, Mozambique, Mozambique Channel I, Myanmar, Nansei-shoto, Nauru, Netherlands Antilles, New Caledonia, New Guinea, New South Wales, Nicaragua, Nicobar Is., Nigeria, Norfolk Is., Northern Territory, Panamá, Peru, Philippines, Pitcairn Is., Puerto Rico, Queensland, Rodrigues, Réunion, Samoa, Santa Cruz Is., Senegal, Seychelles, Sierra Leone, Society Is., Solomon Is., Somalia, South China Sea, Southwest Caribbean, Sri Lanka, Sulawesi, Sumatera, Suriname, Taiwan, Tanzania, Texas, Thailand, Togo, Tonga, Trinidad-Tobago, Tubuai Is., Turks-Caicos Is., Vanuatu, Venezuela, Venezuelan Antilles, Vietnam, Wallis-Futuna Is., Western Australia, Windward Is., Zaire
Distribution (India): Gujarat, Maharashtra, Odisha, Tamil Nadu, Karnataka, West Bengal
Leaf Type: Compound leaves
Inflorescence: Erect spikes
Fruit Type: Pods
Flowering and Fruiting: January–March
IUCN Status: Not evaluated
Notes: Common on coastal sands. Tender pods are used as vegetable
Reference: Prodr. 2: 404 (1825)

Centrosema molle Benth.
Fabaceae
Synonyms: *Centrosema pubescens* "sensu auct., non Benth."; *Centrosema virginianum* auct. non (L.) Benth.; *Clitoria virginiana* L.
Common Name: Spurred butterfly pea, Large butterfly pea, Climbing Centrosema
Climbing Mechanism: Stem Twiner
Distribution (Global): Belize, Bolivia, Brazil North, Brazil Northeast, Brazil Southeast, Brazil West-Central, Colombia, Costa Rica, Cuba, Dominican Republic, Ecuador, El Salvador, French Guiana, Guatemala, Guyana, Haiti, Honduras, Jamaica, Leeward Is., Mexico Gulf, Mexico Southwest, Nicaragua, Panamá, Peru, Puerto Rico, Suriname, Trinidad-Tobago, Venezuela, Windward Is.
Distribution (India): Kerala, Tamil Nadu
Leaf Type: 3-Foliate, ovate
Inflorescence: Racemes axillary
Fruit Type: Legumes/Pods
Flowering and Fruiting: September–January
IUCN Status: Not evaluated
Notes: Non-native, cultivated, and naturalized widely in India.
Reference: Comm. Legum. Gen.: 55 (1837)

Centrosema pubescens Benth.
Fabaceae

Synonyms: *Bradburya ferruginea* (A. Rich.) Kuntze., *Bradburya pubescens* (Benth.) Kuntze., *Bradburya schiedeana* (Schltdl.) Rose. +10.
Common Name: Bluebell, Butterfly pea
Climbing Mechanism: Stem Twiner
Distribution (Global): Pan tropical
Distribution (India): Karnataka, Kerala, Maharashtra, Odisha, Tamil Nadu
IUCN Status: Not evaluated
Reference: Comm. Legum. Gen.: 55 (1837)

Centrosema virginianum (L.) Benth.
Fabaceae
Climbing Mechanism: Stem Twiner
Distribution (India): Maharashtra
IUCN Status: Not evaluated
Reference: Comm. Legum. Gen.: 56 (1837)

Cheniella corymbosa (Roxb. ex DC.) R. Clark & Mackinder
Fabaceae
Synonyms: *Bauhinia corymbosa* Roxb. ex DC., *Bauhinia corymbosa* var. *longipes* Hosok., *Bauhinia parvifolia* (Teijsm. & Binn.) Teijsm. & Binn., *Bauhinia scandens* Burm.f., *Phanera corymbosa* (Roxb. ex DC.) Benth., *Phanera corymbosa* var. *longipes* (Hosok.) X.Y. Zhu., Phaner
Climbing Mechanism: Scrambler-Unarmed
Distribution (Global): China Southeast, Hainan, Vietnam
Distribution (India): Tamil Nadu
IUCN Status: Not evaluated
Reference: Eur. J. Taxon. 360: 19 (2017)

Clitoria arborescens R.Br.
Fabaceae
Synonyms: *Clitoria amoena* Miq., *Clitoria poitaei* Benth., *Dolichos spurius* G. Mey., *Mucuna spuria* (G. Mey.) Steud., *Stizolobium meyeri* Spreng., *Stizolobium spurium* (G. Mey.) Kuntze., *Ternatea arborescens* Kuntze
Climbing Mechanism: Scrambler-Unarmed
Distribution (Global): Colombia, French Guiana, Guyana, Suriname, Trinidad-Tobago, Venezuela, Windward Is.
Leaf Type: Compound, 3-foliolate
Inflorescence: Racemes
Fruit Type: Dry, Pods
IUCN Status: Least Concern
Reference: W.T. Aiton, Hortus Kew. 4: 302 (1812)

Clitoria mexicana Link
Fabaceae
Climbing Mechanism: Stem Twiner

Distribution (Global): El Salvador, Guatemala, Honduras, Mexico Central, Mexico Gulf, Mexico Northeast, Mexico Southeast, Mexico Southwest, Nicaragua, Panamá
Leaf Type: Compound, 3-foliolate
Inflorescence: Racemes
Fruit Type: Dry, Pods
IUCN Status: Not evaluated
Reference: Enum. Hort. Berol. Alt. 2: 235 (1822)

Clitoria ternatea L.
Fabaceae
Synonyms: *Clitoria albiflora* Mattei., *Clitoria bracteata* Poir., *Clitoria mearnsii* De Wild., *Clitoria parviflora* Raf., *Clitoria philippensis* Perr., *Clitoria spectabilis* Salisb. + 20
Common Name: Butterfly pea, Clitoria
Climbing Mechanism: Stem Twiner
Distribution (Global): Pan tropical
Distribution (India): Andaman and Nicobar Islands, Andhra Pradesh, Assam, Delhi, Gujarat, Himachal Pradesh, Jharkhand, Karnataka, Kerala, Madhya Pradesh, Manipur, Mizoram, Odisha, Puducherry, Rajasthan, Tamil Nadu, Telangana
Leaf Type: Alternate distichous, Imparipinnate
Inflorescence: Solitary/Clusters
Fruit Type: Pods/Legumes
Flowering and Fruiting: March-October.
IUCN Status: Not evaluated
Notes: Usually found in deciduous forests also cultivated as ornamental plant
Reference: Sp. Pl.: 753 (1753)

Clitoria ternatea var. *pleniflora*
Fabaceae
Climbing Mechanism: Stem Twiner
Distribution (India): Peninsular India
Leaf Type: Compound, 3-foliolate
Inflorescence: Racemes
Fruit Type: Dry, Pods
IUCN Status: Not evaluated
Reference: Moscosoa 6: 164 (1990)

Cochlianthus gracilis Benth.
Fabaceae
Synonyms: *Cochlianthus gracilis* var. *brevipes* C.F. Wei, *Mucuna gracilis* Graham.
Climbing Mechanism: Stem Twiner
Distribution (Global): China South-Central, East Himalaya, Nepal, Tibet
IUCN Status: Not evaluated
Reference: F.A.W. Miquel, Pl. Jungh.: 234 (1852)

Codariocalyx motorius (Houtt.) H. Ohashi
Fabaceae
Synonyms: *Codariocalyx gyrans* (L.f.) Hassk. + 10
Common Name: Telegraph plant, Indian telegraph plant
Climbing Mechanism: Stem Twiner
Distribution (Global): Assam, Bangladesh, Borneo, Cambodia, China South-Central, China Southeast, East Himalaya, India, Java, Laos, Lesser Sunda Is., Malaya, Myanmar, Nepal, Pakistan, Philippines, Sri Lanka, Sulawesi, Sumatera, Taiwan, Thailand, Tibet, Vietnam, West Himalaya
Leaf Type: 3 foliate
Inflorescence: Racemes, axillary/terminal
Fruit Type: Pods/Legumes
Flowering and Fruiting: August–February
IUCN Status: Not evaluated
Notes: Occasional in hills as an undergrowth along stream banks up to 1000 m
Reference: J. Jap. Bot. 40: 367 (1965)

Dalbergia assamica Benth.
Fabaceae
Synonyms: *Amerimnon assamicum* (Benth.) Kuntze, *Dalbergia assamica* var. *laccifera* (Eberh. & Dubard) Niyomdham, *Dalbergia balansae* Prain, *Dalbergia hupeana* var. *laccifera* Eberh. & Dubard, *Dalbergia lanceolaria* var. *assamica* (Benth.) Thoth.
Climbing Mechanism: Scrambler-Unarmed
Distribution (Global): Assam, Bangladesh, Cambodia, China South-Central, China Southeast, East Himalaya, Hainan, Laos, Myanmar, Thailand, Vietnam
Leaf Type: Compound, imparipinnate, alternate
Inflorescence: Axillary panicles
Fruit Type: Pods
Flowering and Fruiting: April–January
IUCN Status: Least Concern
Notes: Usually found in forest edges and near hilly slopes
Reference: F.A.W. Miquel, Pl. Jungh.: 256 (1852)

Dalbergia beddomei Thoth.
Fabaceae
Synonyms: *Dalbergia rubiginosa* Roxb., *Amerimnon rubiginosum* (Roxb.) Kuntze
Climbing Mechanism: Scrambler-Unarmed
Distribution (Global): Bangladesh, India
Distribution (India): Kerala
Leaf Type: Compound, imparipinnate
Inflorescence: Panicles, axillary
Fruit Type: Pods
Flowering and Fruiting: October–March
IUCN Status: Not evaluated

Notes: Mostly found in evergreen forests along stream banks
Reference: Pl. Coromandel 2: 9 (1799)

Dalbergia benthamii Prain
Fabaceae
Climbing Mechanism: Scrambler-Unarmed
Distribution (Global): China South-Central, China Southeast, Hainan, Taiwan
Distribution (India): Kerala
Leaf Type: Compound, imparipinnate
Inflorescence: Cyme, panicles
Fruit Type: Dry, pods
Flowering and Fruiting: January–May
IUCN Status: Not evaluated
Notes: Usually found in deciduous forests.
Reference: J. Asiat. Soc. Bengal, Pt. 2, Nat. Hist. 67: 289 (1898)

Dalbergia bracteolata Baker
Fabaceae
Synonyms: *Amerimnon bracteolatum* (Baker) Kuntze, *Dalbergia goetzei* Harms, *Dalbergia grandidieri* Baill., *Dalbergia richardii* Baill.
Climbing Mechanism: Scrambler-Unarmed
Distribution (Global): Kenya, Madagascar, Mozambique, Tanzania
Leaf Type: Pinnately compound
Inflorescence: Axillary and terminal panicles
Fruit Type: Pods
IUCN Status: Least Concern
Reference: D. Oliver & auct. suc. (eds.), Fl. Trop. Afr. 2: 234 (1871)

Dalbergia candenatensis (Dennst.) Prain
Fabaceae
Synonyms: *Amerimnon tortum* (Graham ex A. Gray) Kuntze, *Cassia candenatensis* Dennst., *Dalbergia monosperma* Dalzell, *Dalbergia torta* Graham ex King, *Dalbergia torta* Graham ex A. Gray, *Drepanocarpus monospermus* (Dalzell) Kurz.
Climbing Mechanism: Scrambler-Unarmed
Distribution (Global): Andaman Is., Bangladesh, Bismarck Archipelago, Borneo, Cambodia, Caroline Is., China Southeast, Fiji, India, Java, Malaya, Maluku, Marianas, Myanmar, Nansei-shoto, New Caledonia, New Guinea, Nicobar Is., Northern Territory, Philippines, Queensland, Solomon Is., Sri Lanka, Sulawesi, Sumatera, Thailand, Tonga, Vanuatu, Vietnam
Distribution (India): Andaman and Nicobar Islands, Bihar, Odisha, Eastern Ghats, Kerala, Maharashtra, West Bengal
Leaf Type: Compound, imparipinnate
Inflorescence: Branched raceme
Fruit Type: Pods

Flowering and Fruiting: May–June
IUCN Status: Not evaluated
Reference: J. Asiat. Soc. Bengal, Pt. 2, Nat. Hist. 70: 49 (1901)

Dalbergia clarkei Thoth.
Fabaceae
Climbing Mechanism: Scrambler-Unarmed
Distribution (Global): Assam, East Himalaya
Distribution (India): Arunachal Pradesh.
Leaf Type: Compound, imparipinnate
Inflorescence: Cyme, panicles
Fruit Type: Dry, pods
IUCN Status: Not evaluated
Reference: J. Jap. Bot. 46: 73 (1971)

Dalbergia confertiflora Benth.
Fabaceae
Synonyms: *Amerimnon confertiflorum* (Benth.) Kuntze, *Dalbergia confertiflora* var. *listeri* Thoth.
Climbing Mechanism: Scrambler-Unarmed
Distribution (Global): Andaman Is., Assam, Bangladesh, East Himalaya, Myanmar
Distribution (India): Andaman and Nicobar Islands.
Leaf Type: Compound, imparipinnate
Inflorescence: Cyme, panicles
Fruit Type: Dry, pods
IUCN Status: Not evaluated
Reference: F.A.W. Miquel, Pl. Jungh.: 255 (1852)

Dalbergia congesta Wight & Arn.
Fabaceae
Synonyms: *Amerimnon congestum* (Graham ex Wight & Arn.) Kuntze
Climbing Mechanism: Stem Twiner
Distribution (Global): India
Distribution (India): Kerala, Tamil Nadu
Leaf Type: Compound, imparipinnate
Inflorescence: Panicles, axillary
Fruit Type: Pods
Flowering and Fruiting: December–April
IUCN Status: Endangered
Notes: Usually distributed in semi-evergreen forests
Reference: Prodr. Fl. Ind. Orient. 1: 265 (1834)

Dalbergia coromandeliana Prain
Fabaceae
Climbing Mechanism: Scrambler-Armed

Distribution (Global): India.
Leaf Type: Compound, imparipinnate
Inflorescence: Cyme, panicles
Fruit Type: Dry, pods
IUCN Status: Not evaluated
Notes: Mostly found in evergreen forests of the Western Ghats
Reference: J. Asiat. Soc. Bengal, Pt. 2, Nat. Hist. 70: 60 (1901)

Dalbergia gardneriana Benth.
Fabaceae
Climbing Mechanism: Scrambler-Armed
Distribution (India): Tamil Nadu
Leaf Type: Compound, imparipinnate
Inflorescence: Cyme, panicles
Fruit Type: Dry, pods
IUCN Status: Not evaluated
Reference: J. Proc. Linn. Soc., Bot. 4(Suppl.): 42 (1860)

Dalbergia horrida (Dennst.) Mabb.
Fabaceae
Synonyms: *Amerimnon horridum* Dennst.
Common Name: Prickly Dalbergia
Climbing Mechanism: Scrambler-Armed
Distribution (Global): Andaman Is., Bangladesh, Cambodia, India, Laos, Malaya, Myanmar, Thailand, Vietnam
Distribution (India): Andhra Pradesh, Goa, Karnataka, Kerala, Maharashtra, Tamil Nadu, Western Ghats
Leaf Type: Unequal pinnate
Inflorescence: Dense Cymes
Fruit Type: Pods/Legumes
Flowering and Fruiting: September–February
IUCN Status: Not evaluated
Notes: Usually found in semi-evergreen and moist deciduous forests
Reference: Taxon 26: 538 (1977)

Dalbergia horrida var. *concanensis*
Fabaceae
Synonyms: *Dalbergia horrida* var. *horrida*., *Amerimnon spinosum* (Roxb.) Kuntze, *Amerimnon sympatheticum* (Nimmo) Kuntze, *Dalbergia ferruginea* Hochst. ex Benth., *Dalbergia spinosa* Roxb., *Dalbergia sympathetica* Nimmo, *Drepanocarpus spinosus* (Roxb.) Kurz.
Climbing Mechanism: Scrambler-Armed
Distribution (Global): Andaman Is., Bangladesh, India, Malaya, Myanmar
Distribution (India): Maharashtra
Leaf Type: Pinnately compound

Fruit Type: Pods
IUCN Status: Not evaluated
Reference: Taxon. Revis. Dalbergieae Ind.: 110 (1987)

Dalbergia horrida var. *glabrescens* (Prain) Thoth. & Nair
Fabaceae
Synonyms: *Dalbergia multiflora* var. *glabrescens* Prain
Climbing Mechanism: Scrambler-Armed
Distribution (Global): Cambodia, India, Laos, Thailand, Vietnam
Distribution (India): Peninsular India
Leaf Type: Pinnately compound
Fruit Type: Pods
Flowering and Fruiting: November–April
IUCN Status: Not evaluated
Reference: Taxon 30: 46 (1981)

Dalbergia junghuhnii Benth.
Fabaceae
Synonyms: *Amerimnon junghuhnii* (Benth.) Kuntze, *Dalbergia curtisii* Prain, *Dalbergia sennoides* Blume ex Miq., *Dalbergia stercoracea* Maingay ex Prain, *Dalbergia subsympathetica* Prain
Climbing Mechanism: Scrambler-Unarmed
Distribution (Global): Borneo, Java, Laos, Malaya, Maluku, Myanmar, Sulawesi, Sumatera, Thailand, Vietnam
Distribution (India): Andaman and Nicobar Islands
Leaf Type: Pinnately compound
Fruit Type: Pods
IUCN Status: Not evaluated
Reference: F.A.W. Miquel, Pl. Jungh.: 254 (1852)

Dalbergia malabarica Prain
Fabaceae
Climbing Mechanism: Scrambler-Unarmed
Distribution (Global): India.
Distribution (India): Karnataka, Kerala, Maharashtra, Tamil Nadu
Leaf Type: Pinnately compound, Imparipinnate, alternate
Inflorescence: Panicles
Fruit Type: Pod, one-seeded
Flowering and Fruiting: December–February
IUCN Status: Not evaluated
Reference: J. Asiat. Soc. Bengal, Pt. 2, Nat. Hist. 70: 48 (1901)

Dalbergia matthewii Soosairaj, P. Raja & Britto
Fabaceae
Climbing Mechanism: Scrambler-Unarmed

Distribution (Global): India
Distribution (India): Tamil Nadu
Leaf Type: Pinnately compound
Fruit Type: Pods
IUCN Status: Not evaluated
Reference: Phytotaxa 360: 283 (2018)

Dalbergia millettii var. *oldhamii*
Fabaceae
Climbing Mechanism: Scrambler-Unarmed
Distribution (Global): Assam.
Leaf Type: Compound, imparipinnate
Inflorescence: Cyme, panicles
Fruit Type: Dry, pods
IUCN Status: Not evaluated
Reference: J. Jap. Bot. 46: 75 (1971)

Dalbergia mimosoides Franch.
Fabaceae
Synonyms: *Dalbergia millettii* var. *mimosoides* (Franch.) Thoth.
Climbing Mechanism: Scrambler-Unarmed
Distribution (Global): Assam, China North-Central, China South-Central, East Himalaya, Myanmar, Tibet
Distribution (India): North Eastern India
Leaf Type: Compound, imparipinnate
Inflorescence: Cyme, panicles
Fruit Type: Dry, pods
IUCN Status: Not evaluated
Reference: Bull. Bot. Surv. India 25: 170 (1983 publ. 1985)

Dalbergia pinnata (Lour.) Prain
Fabaceae
Synonyms: *Amerimnon pinnatum* (Lour.) Kuntze, *Derris pinnata* Lour.
Climbing Mechanism: Scrambler-Unarmed
Distribution (Global): Andaman Is., Assam, Bangladesh, Borneo, China South-Central, China Southeast, East Himalaya, Hainan, India, Java, Laos, Lesser Sunda Is., Malaya, Maluku, Myanmar, Nepal, Philippines, Sulawesi, Sumatera, Thailand, Tibet, Vietnam
Distribution (India): Andaman and Nicobar Islands, Arunachal Pradesh, Kerala, Manipur, Meghalaya, Mizoram, Odisha, West Bengal
Leaf Type: Pinnate leaves
Fruit Type: Pods
Flowering and Fruiting: February–May
IUCN Status: Least Concern

Notes: Rarely found in hill forests along streams of the northern Eastern Ghats
Reference: Ann. Roy. Bot. Gard. (Calcutta) 10: 48 (1909)

Dalbergia pinnata var. *acaciifolia* (Dalzell) Thoth.
Fabaceae
Synonyms: *Dalbergia acaciifolia* Dalzell
Climbing Mechanism: Scrambler-Unarmed
Distribution (Global): India
Distribution (India): Maharashtra
Leaf Type: Alternate
Inflorescence: Axillary panicles with corymbose
Fruit Type: Pods, samaroid
Flowering and Fruiting: March–May
IUCN Status: Not evaluated
Reference: Bull. Bot. Surv. India 25: 170 (1983 publ. 1985)

Dalbergia pseudo-sissoo Miq.
Fabaceae
Synonyms: *Dalbergia championii* Thwaites, *Dalbergia rostrata* Graham ex Prain.
Climbing Mechanism: Stem Twiner
Distribution (India): Kerala, Tamil Nadu
Leaf Type: Compound, imparipinnate
Inflorescence: Cyme, panicles
Fruit Type: Dry, pods
Flowering and Fruiting: April–September
IUCN Status: Not evaluated

Dalbergia rimosa Roxb.
Fabaceae
Synonyms: *Amerimnon discolor* (Blume ex Miq.) Kuntze, *Amerimnon rimosum* (Roxb.) Kuntze, *Dalbergia discolor* Blume ex Miq., *Dalbergia mengsuoensis* Y.Y. Qian, *Dalbergia rimosa* var. *griffithii* Thoth., *Dalbergia rimosa* var. *laevis* Thoth., *Dalbergia ximengensis* Y.Y.
Climbing Mechanism: Scrambler-Unarmed
Distribution (Global): Assam, Bangladesh, Borneo, Cambodia, China South-Central, East Himalaya, Java, Laos, Myanmar, Philippines, Sulawesi, Thailand, Vietnam
Distribution (India): Arunachal Pradesh, Meghalaya, Mizoram, Tripura, West Bengal
Leaf Type: compound, imparipinnate, alternate
Inflorescence: Axillary and terminal, panicles with corymbose branches
Fruit Type: Pods, star shaped
Flowering and Fruiting: April–December
IUCN Status: Least Concern
Reference: Fl. Ind. ed. 1832, 3: 233 (1832)

Dalbergia rostrata Hassk.
Fabaceae
Synonyms: *Amerimnon championii* (Thwaites) Kuntze, *Amerimnon pseudosissoo* (Miq.) Kuntze, *Dalbergia championii* Thwaites. + 10
Climbing Mechanism: Scrambler-Unarmed
Distribution (Global): Bangladesh, Borneo, India, Java, Malaya, Maluku, New Guinea, Nicobar Is., Philippines, Sri Lanka, Sulawesi, Thailand
Leaf Type: Compound, imparipinnate
Inflorescence: Cyme, panicles
Fruit Type: Dry, pods
IUCN Status: Not evaluated
Reference: Flora 25(2, Beibl. 1): 53 (1842)

Dalbergia rubiginosa Roxb.
Fabaceae
Synonyms: *Amerimnon rubiginosum* (Roxb.) Kuntze, *Dalbergia beddomei* Thoth.
Common Name: Rusty Dalbergia
Climbing Mechanism: Stem Twiner
Distribution (Global): Bangladesh, India
Distribution (India): Andhra Pradesh, Bihar, Odisha, Eastern Ghats, Goa, Karnataka, Kerala, Maharashtra, Odisha, Tamil Nadu, Western Ghats
Leaf Type: Alternately arranged, compound
Inflorescence: Compound spikes
Fruit Type: Pods
Flowering and Fruiting: January–June
IUCN Status: Not evaluated
Notes: Usually found on hills up to 1250 m, dry slopes and in shola borders
Reference: Pl. Coromandel 2: 9 (1799)

Dalbergia stipulacea Roxb.
Fabaceae
Synonyms: *Dalbergia penduliflora* Blume ex Miq., *Dalbergia stipulacea* var. *kurzii* Thoth., *Dalbergia stipulacea* var. *mogkokensis* Thoth., *Dalbergia stipulacea* f. *puberula* Thoth., *Endespermum penduliflorum* Zipp. ex Miq.
Common Name: East Himalayan Dalbergia
Climbing Mechanism: Scrambler-Unarmed
Distribution (Global): Assam, Bangladesh, Cambodia, China South-Central, East Himalaya, Laos, Myanmar, Nepal, Thailand, Vietnam
Distribution (India): Manipur, Mizoram, Tripura, West Bengal
Leaf Type: Compound, Stipules
Inflorescence: Panicles, axillary
Fruit Type: Pods
Flowering and Fruiting: April–May
IUCN Status: Least Concern

Notes: Used in folk medicine
Reference: Fl. Ind. ed. 1832, 3: 233 (1832)

Dalbergia thomsonii Benth.
Fabaceae
Synonyms: *Amerimnon thomsonii* (Benth.) Kuntze
Climbing Mechanism: Scrambler-Unarmed
Distribution (Global): Assam, East Himalaya
Distribution (India): Tripura.
Leaf Type: Compound, imparipinnate
Inflorescence: Cyme, panicles
Fruit Type: Dry, pods
IUCN Status: Not evaluated
Reference: J. Proc. Linn. Soc., Bot. 4(Suppl.): 33 (1860)

Dalbergia tinnevelliensis Thoth.
Fabaceae
Climbing Mechanism: Scrambler-Unarmed
Distribution (Global): India
Distribution (India): Tamil Nadu
Leaf Type: Compound, imparipinnate
Inflorescence: Cyme, panicles
Fruit Type: Dry, pods
IUCN Status: Not evaluated
Notes: Commonly found in dry deciduous forests
Reference: Ceylon J. Sci., Biol. Sci. 12: 47 (1976)

Dalbergia travancorica Thoth.
Fabaceae
Climbing Mechanism: Scrambler-Unarmed
Distribution (Global): India
Distribution (India): Kerala
Leaf Type: Imparipinnate, alternate
Inflorescence: Short, axillary, panicles
Fruit Type: Pods
Flowering and Fruiting: January–November
IUCN Status: Not evaluated
Notes: Endemic to Southern Western Ghats
Reference: Reinwardtia 8: 329 (1972)

Dalbergia volubilis Roxb.
Fabaceae
Synonyms: *Amerimnon volubile* (Roxb.) Kuntze, *Dalbergia volubilis* var. *assamica* Thoth.
Common Name: Climbing Dalbergia, Bastard rosewood climber

Climbing Mechanism: Stem Twiner
Distribution (Global): Andaman Is., Assam, Bangladesh, Cambodia, China South-Central, East Himalaya, India, Laos, Myanmar, Nepal, Thailand, Vietnam
Distribution (India): Andaman and Nicobar Islands, Andhra Pradesh, Bihar, Odisha, Eastern Ghats, Goa, Gujarat, Karnataka, Kerala, Madhya Pradesh, Maharashtra, Meghalaya, Mizoram, Odisha, Rajasthan, Tamil Nadu, Telangana
Leaf Type: Compound, paripinnate
Inflorescence: Copious panicles
Fruit Type: Pods
Flowering and Fruiting: February–June
IUCN Status: Not evaluated
Notes: Medicinally valuable; used in treating various ailments
Reference: Pl. Coromandel 2: 48 (1805)

Dalhousiea bracteata (Roxb.) Benth.
Fabaceae
Synonyms: *Dalhousiea paucisperma* Griff., *Podalyria bracteata* Roxb.
Climbing Mechanism: Scrambler-Unarmed
Distribution (Global): Assam, Bangladesh, East Himalaya, Myanmar
Distribution (India): Arunachal Pradesh, Manipur, Meghalaya, Mizoram, Tripura
Leaf Type: Simple
Inflorescence: Branch-end panicles, Corymb like
Fruit Type: Pods, leathery
IUCN Status: Not evaluated
Notes: Medicinally valuable climber found mostly in evergreen forests
Reference: Comm. Legum. Gen.: 5 (1837)

Derris acuminata Benth.
Fabaceae
Synonyms: *Deguelia acu*minata (Benth.) Taub., *Kraunhia acuminata* (Benth.) Prain, *Millettia acuminata* (Benth.) Prain, *Pongamia acuminata* Graham, *Pongamia elongata* Graham, *Pterocarpus acuminatus* (Benth.) Kuntze
Climbing Mechanism: Stem Twiner
Distribution (Global): East Himalaya, Java, Laos, Nepal, Thailand, Vietnam
Distribution (India): Arunachal Pradesh, West Bengal
Leaf Type: Compound, imparipinnate
Fruit Type: Pods
IUCN Status: Least Concern
Reference: F.A.W. Miquel, Pl. Jungh.: 252 (1852)

Derris amoena Benth.
Fabaceae
Synonyms: *Deguelia amoena* (Benth.) Taub., *Deguelia maingayana* (Baker) Taub., *Derris maingayana* Baker, *Pterocarpus amoenus* (Benth.) Kuntze, *Pterocarpus maingayanus* (Baker) Kuntze

Climbing Mechanism: Stem Twiner
Distribution (Global): Borneo, Malaya, Myanmar, Sumatera, Thailand, Vietnam
Leaf Type: Compound, imparipinnate
Inflorescence: Racemes on peduncles
Fruit Type: Dry, pods
IUCN Status: Least Concern
Reference: F.A.W. Miquel, Pl. Jungh.: 252 (1852)

Derris andamanica Prain
Fabaceae
Synonyms: *Derris sinuata* Prain
Climbing Mechanism: Stem Twiner
Distribution (Global): Andaman Is., Nicobar Is.
Distribution (India): Andaman and Nicobar Islands
Leaf Type: Compound, imparipinnate
Inflorescence: Racemes on peduncles
Fruit Type: Dry, pods
IUCN Status: Not evaluated
Reference: J. Asiat. Soc. Bengal, Pt. 2, Nat. Hist. 66: 104 (1897)

Derris benthamii (Thwaites) Thwaites
Fabaceae
Synonyms: *Brachypterum benthamii* Thwaites, *Deguelia benthamii* (Thwaites) S.F. Blake
Climbing Mechanism: Stem Twiner
Distribution (Global): India, Sri Lanka
Distribution (India): Kerala, Tamil Nadu, Western Ghats
Leaf Type: Compound, imparipinnate
Inflorescence: Racemes on peduncles
Fruit Type: Dry, pods
Flowering and Fruiting: December–May
IUCN Status: Not evaluated
Notes: Mostly found in evergreen forests
Reference: Enum. Pl. Zeyl.: 413 (1864)

Derris benthamii var. *wightii* (Baker) Thoth.
Fabaceae
Synonyms: *Deguelia wightii* (Baker) Taub., *Derris wightii* Baker, *Pterocarpus wightii* (Baker) Kuntze
Climbing Mechanism: Stem Twiner
Distribution (Global): India
Distribution (India): Tamil Nadu
Leaf Type: Compound, imparipinnate
Inflorescence: Racemes on peduncles
Fruit Type: Dry, pods

IUCN Status: Not evaluated
Reference: J. Jap. Bot. 51: 145 (1976)

Derris brevipes (Benth.) Baker
Fabaceae
Synonyms: *Deguelia brevipes* (Baker) Taub., *Pterocarpus brevipes* (Benth.) Kuntze
Climbing Mechanism: Stem Twiner
Distribution (Global): India.
Distribution (India): Kerala, Maharashtra, Tamil Nadu, Western Ghats
Leaf Type: Compound, imparipinnate
Inflorescence: Panicles, axillary
Fruit Type: Pods
Flowering and Fruiting: February–December
IUCN Status: Not evaluated
Notes: Endemic to Southern Western Ghats
Reference: J. D. Hooker, Fl. Brit. India 2: 244 (1878)

Derris brevipes var. *coriacea* (Benth.) Baker
Fabaceae
Climbing Mechanism: Stem Twiner
Distribution (Global): India
Distribution (India): Tamil Nadu
Leaf Type: Alternate
Inflorescence: Axillary/terminal panicles
Fruit Type: Pods
Flowering and Fruiting: August–December
IUCN Status: Not evaluated
Notes: Usually found in Evergreen and semi-Evergreen forests
Reference: J. D. Hooker, Fl. Brit. India 2: 244 (1878)

Derris brevipes var. *travancorica*
Fabaceae
Climbing Mechanism: Stem Twiner
IUCN Status: Not evaluated

Derris canarensis (Dalzell) Baker
Fabaceae
Synonyms: *Brachypterum canarense* (Dalzell) Dalzell & A. Gibson, *Deguelia canarensis* (Baker) Taub., *Deguelia oblonga* (Benth.) Taub., *Derris oblonga* Benth., *Paraderris canarensis* (Dalzell) Adema, *Pongamia canarensis* Dalzell, *Pterocarpus canarensis* (Dalzell) Kun
Common Name: Kanara Derris
Climbing Mechanism: Stem Twiner
Distribution (Global): India, Sri Lanka, Taiwan
Distribution (India): Goa, Kerala, Maharashtra, Tamil Nadu, Western Ghats

Leaf Type: Compound, imparipinnate
Inflorescence: Panicles
Fruit Type: Pods
Flowering and Fruiting: September–January
IUCN Status: Not evaluated
Notes: Often found in moist deciduous forests
Reference: J. D. Hooker, Fl. Brit. India 2: 246 (1878)

Derris cuneifolia Benth.
Fabaceae
Synonyms: *Deguelia cuneifolia* (Benth.) Taub., *Deguelia discolor* (Benth.) Taub., *Deguelia hanceana* Taub., *Derris discolor* Benth., *Derris glauca* Merr. & Chun., *Derris hancei* Hemsl., *Derris oblonga* Hance., *Derris truncata* Craib., *Galedupa marginata* Roxb. + 10
Climbing Mechanism: Stem Twiner
Distribution (Global): Assam, Bangladesh, Cambodia, China Southeast, East Himalaya, Hainan, India, Laos, Malaya, Myanmar, Nepal, Sulawesi, Thailand, Vietnam
Distribution (India): Arunachal Pradesh, Assam, Bihar, Odisha, Manipur, Meghalaya, West Bengal
Leaf Type: Compound, imparipinnate
Inflorescence: Racemes on peduncles
Fruit Type: Dry, pods
IUCN Status: Least Concern
Reference: F.A.W. Miquel, Pl. Jungh.: 253 (1852)

Derris elegans Benth.
Fabaceae
Synonyms: *Deguelia elegans* (Benth.) Taub., *Pterocarpus elegans* (Benth.) Kuntze
Climbing Mechanism: Stem Twiner
Distribution (Global): Andaman Is., Assam, Bangladesh, Bismarck Archipelago, Borneo, Cambodia, East Himalaya, Java, Laos, Malaya, Maluku, Myanmar, New Guinea, Nicobar Is., Philippines, Solomon Is., Sulawesi, Sumatera, Thailand, Vanuatu, Vietnam
Distribution (India): Andaman and Nicobar Islands
Leaf Type: Compound, imparipinnate
Inflorescence: Raccemes axillary
Fruit Type: Pods
Flowering and Fruiting: June–December
IUCN Status: Least Concern
Notes: Used in folk and Siddha medicine
Reference: F.A.W. Miquel, Pl. Jungh.: 252 (1852)

Derris elegans var. *vestita* Prain
Fabaceae

Climbing Mechanism: Stem Twiner
Distribution (India): Andaman and Nicobar Islands
Leaf Type: Compound, imparipinnate
Inflorescence: Racemes on peduncles
Fruit Type: Dry, pods
IUCN Status: Not evaluated
Reference: F.A.W. Miquel, Pl. Jungh.: 252 (1852)

Derris elliptica (Wall.) Benth.
Fabaceae
Synonyms: *Dalbergia elliptica* Span.,
Dalbergia glaucescens Zipp. ex Miq., *Deguelia elliptica* (Wall.) Taub., *Derris elliptica* var. *chittagongensis* Thoth., *Derris elliptica* var. *glaucophylla* (Miq.) Kaneh. & Hatus., *Galedupa elliptica* (Wall.) Roxb. + 10
Climbing Mechanism: Stem Twiner
Distribution (Global): Andaman Is., Assam, Bangladesh, Borneo, Cambodia, Christmas I., India, Java, Laos, Malaya, Maluku, Myanmar, Nepal, New Guinea, Nicobar Is., Philippines, Sri Lanka, Sulawesi, Sumatera, Thailand, Vietnam
Distribution (India): Andaman and Nicobar Islands, Assam, Kerala, Tamil Nadu
Leaf Type: Alternate, pinnate
Inflorescence: Racemes, axillary/terminal
Fruit Type: Pods
Flowering and Fruiting: March–June
IUCN Status: Not evaluated
Reference: J. Proc. Linn. Soc., Bot. 4(Suppl.): 111 (1860)

Derris ferruginea (Roxb.) Benth.
Fabaceae
Synonyms: *Deguelia ferruginea* (Wall. ex Voigt) Taub., *Pongamia ferruginea* Graham, *Pterocarpus ferrugineus* (Wall. ex Voigt) Kuntze, *Robinia ferruginea* Roxb., *Tephrosia ferruginea* Wall. ex Voigt.
Climbing Mechanism: Stem Twiner
Distribution (Global): Assam, Bangladesh, Cambodia, China South-Central, China Southeast, Hainan, India, Laos, Myanmar, Thailand, Vietnam
Distribution (India): Arunachal Pradesh, Assam, Meghalaya, Tamil Nadu, West Bengal
Leaf Type: Compound, imparipinnate
Inflorescence: Racemes on peduncles
Fruit Type: Dry, pods
IUCN Status: Not evaluated
Reference: F.A.W. Miquel, Pl. Jungh.: 252 (1852)

Derris gamblei S. Soosairaj, P. Raja & N. Dhatchanamoorthy
Fabaceae

Climbing Mechanism: Stem Twiner
IUCN Status: Not evaluated

Derris laxiflora Benth.
Fabaceae
Synonyms: *Deguelia laxiflora* (Benth.) Taub., *Pterocarpus laxiflorus* (Benth.) Kuntze
Climbing Mechanism: Stem Twiner
Distribution (Global): Taiwan
Distribution (India): Bihar, Odisha
Leaf Type: Compound, imparipinnate
Inflorescence: Racemes on peduncles
Fruit Type: Dry, pods
IUCN Status: Not evaluated
Reference: J. Proc. Linn. Soc., Bot. 4(Suppl.): 105 (1860)

Derris lushaiensis Thoth.
Fabaceae
Synonyms: *Paraderris lushaiensis* (Thoth.) Adema
Climbing Mechanism: Stem Twiner
Distribution (Global): Assam
Distribution (India): Mizoram
Leaf Type: Compound, imparipinnate
Inflorescence: Racemes on peduncles
Fruit Type: Dry, pods
IUCN Status: Not evaluated
Reference: Bull. Bot. Surv. India 12: 104 (1970 publ. 1972)

Derris macrocarpa Thoth.
Fabaceae
Climbing Mechanism: Stem Twiner
Distribution (Global): India, Laos
Distribution (India): Maharashtra
Leaf Type: Compound, imparipinnate
Inflorescence: Racemes on peduncles
Fruit Type: Dry, pods
Flowering and Fruiting: April–May
IUCN Status: Not evaluated
Reference: Bull. Bot. Surv. India 3: 192 (1962)

Derris marginata (Roxb.) Benth.
Fabaceae
Synonyms: *Aganope marginata* (Roxb.) Miq., *Dalbergia marginata* Roxb., *Deguelia marginata* (Benth.) Taub., *Pongamia reflexa* Graham, *Pterocarpus marginatus* (Roxb.) Kuntze

Climbing Mechanism: Stem Twiner
Distribution (Global): Assam, Bangladesh, China South-Central, China Southeast, East Himalaya, Laos, Myanmar, Nepal, Thailand, Vietnam
Distribution (India): North Eastern India
Leaf Type: Compound, imparipinnate
Inflorescence: Racemes on peduncles
Fruit Type: Dry, pods
IUCN Status: Not evaluated
Reference: F.A.W. Miquel, Pl. Jungh.: 252 (1852)

Derris matthewii Kottaim
Fabaceae
Climbing Mechanism: Stem Twiner
Distribution (Global): India
Distribution (India): Tamil Nadu
Leaf Type: Compound, imparipinnate
Inflorescence: Racemes on peduncles
Fruit Type: Dry, pods
IUCN Status: Not evaluated
Reference: Webbia 72: 97 (2016)

Derris monticola (Kurz) Prain
Fabaceae
Synonyms: *Deguelia microptera* (Benth.) Taub., *Derris acuminata* var. *sikkimensis* Thoth., *Derris microptera* Benth., *Millettia monticola* Kurz, *Phaseoloides monticola* (Kurz) Kuntze, *Pterocarpus micropterus* (Benth.) Kuntze
Climbing Mechanism: Stem Twiner
Distribution (Global): Assam, Bangladesh, East Himalaya, Myanmar, Nepal, Thailand
Distribution (India): Mizoram, West Bengal
Leaf Type: Compound, imparipinnate
Inflorescence: Racemes on peduncles
Fruit Type: Dry, pods
IUCN Status: Not evaluated
Reference: J. Asiat. Soc. Bengal, Pt. 2, Nat. Hist. 66: 361 (1897)

Derris oblonga Benth.
Fabaceae
Climbing Mechanism: Stem Twiner
Distribution (India): Andhra Pradesh
Leaf Type: Compound, imparipinnate
Inflorescence: Racemes on peduncles
Fruit Type: Dry, pods
IUCN Status: Not evaluated
Reference: J. Proc. Linn. Soc., Bot. 4(Suppl.): 112 (1860)

Derris ovalifolia (Wight & Arn.) Benth.
Fabaceae
Synonyms: *Millettia ovalifolia* (Wight & Arn.) Kurz, *Phaseoloides ovalifolium* (Wight & Arn.) Kuntze, *Pongamia ovalifolia* Wight & Arn., *Pterocarpus ovalifolius* (Benth.) Kuntze
Climbing Mechanism: Stem Twiner
Distribution (Global): India
Distribution (India): Maharashtra, Puducherry, Tamil Nadu
Leaf Type: Compound, imparipinnate
Inflorescence: Racemes copious, axillary
Fruit Type: Pods
IUCN Status: Not evaluated
Reference: F.A.W. Miquel, Pl. Jungh.: 252 (1852)

Derris parviflora Benth.
Fabaceae
Synonyms: *Brachypterum elegans* Thwaites, *Deguelia parviflora* (Benth.) Taub., *Pterocarpus parviflorus* (Benth.) Kuntze
Climbing Mechanism: Stem Twiner
Distribution (Global): Sri Lanka
Leaf Type: Compound, imparipinnate
Inflorescence: Racemes on peduncles
Fruit Type: Dry, pods
IUCN Status: Not evaluated
Reference: J. Proc. Linn. Soc., Bot. 4(Suppl.): 105 (1860)

Derris pseudorobusta Thoth.
Fabaceae
Synonyms: *Brachypterum pseudorobustum* (Thoth.) Adema & Sirich., *Solori pseudorobusta* (Thoth.) Sirich. & Adema.
Climbing Mechanism: Stem Twiner
Distribution (Global): Assam, East Himalaya
Distribution (India): Arunachal Pradesh, Meghalaya
Leaf Type: Compound, imparipinnate
Inflorescence: Racemes on peduncles
Fruit Type: Dry, pods
IUCN Status: Not evaluated
Reference: Thai Forest Bull., Bot. 48: 58 (2020)

Derris robusta (DC.) Benth.
Fabaceae
Synonyms: *Brachypterum robustum* (Roxb. ex DC.) Dalzell & A. Gibson., *Dalbergia krowee* Roxb. + 10
Climbing Mechanism: Stem Twiner

Distribution (Global): Assam, Bangladesh, China South-Central, China Southeast, East Himalaya, India, Java, Laos, Myanmar, Sulawesi, Sumatera, Thailand, Vietnam
Distribution (India): Assam
Leaf Type: Pinnately compound
Inflorescence: Axillary Racemes
Fruit Type: Pods, linear
IUCN Status: Not evaluated
Notes: Mostly occur in deciduous forests, especially in teak forests
Reference: Bombay Fl.: 77 (1861)

Derris scandens (Roxb.) Benth.
Fabaceae
Synonyms: *Brachypterum scandens* (Roxb.) Miq., *Brachypterum timorense* (DC.) Benth., *Dalbergia robusta* Miq. ex Benth., *Dalbergia scanden*s Roxb., *Dalbergia timoriensis* DC., *Dalbergia venusta* Zipp. ex Span., *Deguelia timoriensis* (DC.) Taub., *Derris scandens* (Roxb.)
Common Name: Hog creeper
Climbing Mechanism: Stem Twiner
Distribution (Global): Andaman Is., Assam, Bangladesh, Borneo, Cambodia, China South-Central, India, Java, Laos, Lesser Sunda Is., Malaya, Maluku, Myanmar, Nepal, New Guinea, New South Wales, Nicobar Is., Philippines, Queensland, Sri Lanka, Sulawesi, Sumatera, Thailand, Vietnam, West Himalaya
Distribution (India): Andaman and Nicobar Islands, Andhra Pradesh, Arunachal Pradesh, Bihar, Odisha, Delhi, Eastern Ghats, Goa, Gujarat, Kerala, Madhya Pradesh, Maharashtra, Manipur, Odisha, Puducherry, Tamil Nadu, Telangana, Uttar Pradesh, West Bengal, Western Ghats
Leaf Type: Compound, alternate
Inflorescence: Racemes
Fruit Type: Pods, narrowly winged
Flowering and Fruiting: July–February
IUCN Status: Least Concern
Notes: Usually found in scrub jungle and dry deciduous forests. Also, grow as mangrove associates
Reference: Fl. Ned. Ind. 1: 138 (1855)

Derris scandens var. *saharanpurensis* (Thoth.) Thoth.
Fabaceae
Climbing Mechanism: Stem Twiner
Leaf Type: Compound, imparipinnate
Inflorescence: Racemes on peduncles
Fruit Type: Dry, pods
IUCN Status: Not evaluated
Reference: Bull. Bot. Surv. India 13: 164 (1971 publ. 1973)

Derris secunda Baker
Fabaceae
Synonyms: *Amerimnon secundum* Buch. -Ham. ex Baker, *Pongamia secunda* Graham, *Pterocarpus secundus* (Baker) Kuntze
Climbing Mechanism: Stem Twiner
Distribution (Global): Assam, East Himalaya
Distribution (India): Arunachal Pradesh
Leaf Type: Compound, imparipinnate
Inflorescence: Racemes on peduncles
Fruit Type: Dry, pods
IUCN Status: Not evaluated
Reference: J. D. Hooker, Fl. Brit. India 2: 247 (1878)

Derris thothathrii Bennet
Fabaceae
Climbing Mechanism: Stem Twiner
Distribution (India): Kerala, Maharashtra, Tamil Nadu
Leaf Type: Compound, imparipinnate
Inflorescence: Panicles, axillary/terminal
Fruit Type: Pods
Flowering and Fruiting: December–March
IUCN Status: Not evaluated
Notes: Endemic to Western Ghats
Reference: Indian J. Forest. 1: 23 (1978)

Derris trifoliata Lour
Fabaceae
Synonyms: *Brachypterum floribundum* Miq., *Dalbergia acuminata* Hassk., *Dalbergia heterophylla* Willd. +28
Climbing Mechanism: Stem Twiner
Distribution (Global): Andaman Is., Assam, Bangladesh, Bismarck Archipelago, Borneo, Cambodia, Caroline Is., China Southeast, Comoros, East Himalaya, Fiji, Hainan, India, Java, Kenya, KwaZulu-Natal, Lesser Sunda Is., Madagascar, Malaya, Maluku, Marianas, Mozambique, Myanmar, Nansei-shoto, Nauru, New Caledonia, New Guinea, Nicobar Is., Northern Territory, Philippines, Queensland, Réunion, Samoa, Seychelles, Solomon Is., Somalia, Sri Lanka, Sulawesi, Sumatera, Swaziland, Taiwan, Tanzania, Thailand, Tonga, Vanuatu, Vietnam, Wallis-Futuna Is.
Distribution (India): Andaman and Nicobar Islands, Andhra Pradesh, Eastern Ghats, Goa, Gujarat, Kerala, Maharashtra, Meghalaya, Odisha, Tamil Nadu, Tripura, West Bengal, Western Ghats
Leaf Type: Compound, alternate
Inflorescence: Pseudoracemes terminal or axillary
Fruit Type: Pods
Flowering and Fruiting: January–October

IUCN Status: Not evaluated
Notes: Usually grow along the stream banks, canals and sea coasts
Reference: Fl. Cochinch.: 433 (1790)

Desmodium benthamii N.P. Balakr.
Fabaceae
Synonyms: *Grona brachystachya* (Graham ex Benth.) H. Ohashi & K. Ohashi., *Desmodium brachystachyum* Graham ex Benth., *Meibomia brachystachya* (Graham ex Benth.) Kuntze, *Nicolsonia brachystachya* (Graham ex Benth.) Schindl.
Climbing Mechanism: Stem Twiner
Distribution (Global): India
Distribution (India): Maharashtra
IUCN Status: Not evaluated
Reference: J. Jap. Bot. 93: 110 (2018)

Desmodium dichotomum (Willd.) DC.
Fabaceae
Synonyms: *Bouffordia dichotoma* (Willd.) H. Ohashi & K. Ohashi., *Desmodium amplexicaule* Zoll. & Moritzi, *Desmodium quinquangulatum* (Roxb.) G. Don, *Desmodium sennaarense* Schweinf., *Desmodium willdenowii* G. Don, *Hedysarum dichotomum* Willd., *Hedysarum diffusum* Willd.
Climbing Mechanism: Stem Twiner
Distribution (Global): Bangladesh, Borneo, China South-Central, India, Java, Myanmar, Oman, Sri Lanka, Sulawesi
Distribution (India): Andhra Pradesh, Gujarat, Puducherry
Leaf Type: Alternate
Inflorescence: Racemes
Fruit Type: Pods
Flowering and Fruiting: December–February
IUCN Status: Not evaluated
Reference: J. Jap. Bot. 93: 180 (2018)

Desmodium gangeticum (L.) DC.
Fabaceae
Synonyms: *Pleurolobus gangeticus* (L.) J.St.-Hil. ex H. Ohashi & K. Ohashi., *Aeschynomene gangetica* (L.) Poir. + 18
Climbing Mechanism: Stem Twiner
Distribution (Global): Andaman Is., Angola, Assam, Bangladesh, Benin, Bismarck Archipelago, Borneo, Burkina, Cambodia, Cameroon, Caroline Is., Central African Repu, Chad, China South-Central, China Southeast, Congo, East Himalaya, Equatorial Guinea, Eritrea, Ethiopia, Gabon, Gambia, Ghana, Guinea, Guinea-Bissau, Gulf of Guinea Is., Hainan, India, Ivory Coast, Java, Kenya, KwaZulu-Natal, Laccadive Is., Laos, Lesser Sunda Is., Liberia, Malawi, Malaya, Maldives, Mali, Maluku, Mozambique, Myanmar, Nansei-shoto, Nepal, New Guinea, New

South Wales, Nicobar Is., Nigeria, Northern Provinces, Northern Territory, Pakistan, Philippines, Queensland, Rwanda, Senegal, Sierra Leone, Sri Lanka, Sudan, Sumatera, Taiwan, Tanzania, Thailand, Togo, Uganda, Vietnam, West Himalaya, Western Australia, Yemen, Zambia, Zaire, Zimbabwe

Distribution (India): Madhya Pradesh, Maharashtra

Leaf Type: Unifoliate

Inflorescence: Pseudoracemose or paniculate, terminal and axillary

Fruit Type: Pods

Flowering and Fruiting: October–January

IUCN Status: Not evaluated

Notes: Mostly found in moist and dry deciduous forests and also in forest plantations

Reference: J. Jap. Bot. 93: 184 (2018)

Desmodium heterocarpon (L.) DC.

Fabaceae

Synonyms: *Hedysarum heterocarpon* L., *Meibomia heterocarpon* (L.) Kuntze, *Pleurolobus heterocarpon* (L.) J.St.-Hil.

Common Name: Asian tick trefoil

Climbing Mechanism: Stem Twiner

Distribution (Global): Andaman Is., Assam, Bangladesh, Borneo, Cambodia, Caroline Is., China South-Central, China Southeast, East Himalaya, Fiji, Hainan, India, Japan, Java, Korea, Laos, Malaya, Myanmar, Nansei-shoto, Nepal, New Caledonia, New Guinea, New South Wales, Nicobar Is., Northern Territory, Philippines, Queensland, Samoa, Sri Lanka, Sulawesi, Sumatera, Taiwan, Tanzania, Thailand, Tonga, Vanuatu, Vietnam, Wallis-Futuna Is., West Himalaya, Western Australia

Distribution (India): Andhra Pradesh, Goa, Karnataka

Leaf Type: 1-3-foliolate, leaflets usually elliptic

Inflorescence: Racemes

Fruit Type: Pods

Flowering and Fruiting: August–April

IUCN Status: Not evaluated

Notes: Mostly grow on hills above 900m, in moist places among grasses and under partial shade

Reference: J. Jap. Bot. 93: 112 (2018)

Desmodium heterophyllum (Willd.) DC.

Fabaceae

Synonyms: *Hedysarum heterophyllum* Willd., *Hedysarum reptans* Roxb., *Meibomia heterophylla* (Willd.) Kuntze

Climbing Mechanism: Stem Twiner

Distribution (Global): Andaman Is., Assam, Bangladesh, Borneo, Cambodia, China South-Central, China Southeast, East Himalaya, Hainan, India, Java, Madagascar, Malaya, Myanmar, Nansei-shoto, Nepal, New Guinea, Nicobar

Is., Ogasawara-shoto, Philippines, South China Sea, Sri Lanka, Sumatera, Taiwan, Thailand, Vietnam
Distribution (India): Kerala, Madhya Pradesh
Leaf Type: 3-foliolate
Inflorescence: Racemes
Fruit Type: Pods
Flowering and Fruiting: May–December
IUCN Status: Not evaluated
Reference: J. Jap. Bot. 93: 114 (2018)

Desmodium laxiflorum DC.
Fabaceae
Synonyms: *Sohmaea laxiflora* (DC.) H. Ohashi & K. Ohashi., *Desmodium kibalanka* Blume ex Miq. + 20
Climbing Mechanism: Stem Twiner
Distribution (Global): Andaman Is., Assam, Bangladesh, Borneo, China South-Central, China Southeast, East Himalaya, Hainan, India, Java, Laos, Lesser Sunda Is., Malaya, Maluku, Myanmar, Nepal, New Guinea, Nicobar Is., Pakistan, Philippines, Sulawesi, Sumatera, Taiwan, Thailand, Vietnam, West Himalaya
Distribution (India): Puducherry
Leaf Type: Compound trifoliate
Inflorescence: Racemes
Fruit Type: Pods
Flowering and Fruiting: September–April
IUCN Status: Not evaluated
Notes: Used in Chinese medicine to treat dysentery, rheumatism, and fever
Reference: J. Jap. Bot. 93: 162 (2018)

Desmodium renifolium (L.) Schindl.
Fabaceae
Synonyms: *Huangtcia renifolia* (L.) H. Ohashi & K. Ohashi. *Desmodium reniforme* DC., *Hedysarum nummulariifolium* var. *renifolium* L., *Hedysarum renifolium* L., *Hedysarum reniforme* L., *Meibomia reniformis* Kuntze
Climbing Mechanism: Stem Twiner
Distribution (Global): Assam, Bangladesh, Cambodia, China South-Central, East Himalaya, Hainan, India, Java, Laos, Myanmar, Nepal, Taiwan, Thailand
Distribution (India): West Bengal
IUCN Status: Not evaluated
Reference: J. Jap. Bot. 93: 183 (2018)

Desmodium scorpiurus (Sw.) Desv.
Fabaceae
Synonyms: *Desmodium akoense* Hayata, *Desmodium arenarium* Kunth, *Desmodium cinereum* Poepp. ex Griseb. +10

Common Name: Scorpion ticktrefoil
Climbing Mechanism: Stem Twiner
Distribution (Global): Arizona, Aruba, Belize, Bolivia, Brazil North, Brazil Northeast, Brazil Southeast, Brazil West-Central, Colombia, Costa Rica, Cuba, Dominican Republic, Ecuador, El Salvador, Florida, French Guiana, Guatemala, Guyana, Haiti, Honduras, Jamaica, Leeward Is., Mexico Central, Mexico Gulf, Mexico Northeast, Mexico Northwest, Mexico Southeast, Mexico Southwest, Nicaragua, Panamá, Peru, Puerto Rico, Trinidad-Tobago, Venezuela, Venezuelan Antilles, Windward Is.
Distribution (India): Maharashtra, West Bengal
Leaf Type: Trifoliate, alternate
Inflorescence: Racemes
Fruit Type: Pods
Flowering and Fruiting: December–July
IUCN Status: Not evaluated
Notes: An introduced species, now occur as a weed in wastelands
Reference: Prodr. 2: 333 (1825)

Desmodium styracifolium (Osbeck) Merr.
Fabaceae
Synonyms: *Grona styracifolia* (Osbeck) H. Ohashi & K. Ohashi., *Desmodium strigillosum* subsp. *celebicum* (Schindl.) H. Ohashi. + 10
Climbing Mechanism: Stem Twiner
Distribution (Global): Assam, Bangladesh, Cambodia, Caroline Is., China South-Central, China Southeast, East Himalaya, Hainan, India, Java, Lesser Sunda Is., Malaya, Maluku, Myanmar, Philippines, Sri Lanka, Sulawesi, Sumatera, Thailand, Vietnam
Distribution (India): Kerala
Leaf Type: Unifoliate
Inflorescence: Racemes
Fruit Type: Pods
Flowering and Fruiting: March–October
IUCN Status: Not evaluated
Notes: Usually found in moist deciduous forests
Reference: J. Jap. Bot. 93: 116 (2018)

Dioclea hexandra (Ralph) Mabb.
Fabaceae
Synonyms: *Macropsychanthus hexander* (Ralph) L.P. Queiroz & Snak., *Dioclea panamensis* Seem. ex Hemsl., *Dioclea reflexa* var. *australiensis* Domin, *Dolichos coriaceus* Graham, *Macropsychanthus novoguineensis* Pulle, *Mucuna hexandra* Ralph, *Mucuna wertheimii* Burck.
Climbing Mechanism: Stem Twiner
Distribution (Global): Andaman Is., Angola, Bangladesh, Benin, Bismarck Archipelago, Borneo, Cameroon, Central African Repu, Equatorial Guinea, Gabon,

Ghana, Guinea, Guinea-Bissau, Gulf of Guinea Is., Ivory Coast, Java, Liberia, Madagascar, Malaya, Maluku, Myanmar, New Guinea, Nicobar Is., Nigeria, Philippines, Queensland, Senegal, Sierra Leone, Solomon Is., Sri Lanka, Sulawesi, Sumatera, Thailand, Togo, Zaire
Distribution (India): Andaman and Nicobar Islands
IUCN Status: Not evaluated
Reference: PhytoKeys 164: 97 (2020)

Dioclea reflexa Hook.f.
Fabaceae
Synonyms: *Macropsychanthus comosus* (G. Mey.) L.P. Queiroz & Snak., *Dioclea comosa* (G. Mey.) Kuntze
Dolichos comosus G. Mey., *Mucuna como*sa (G. Mey.) DC.
Climbing Mechanism: Stem Twiner
Distribution (Global): Belize, Benin, Bolivia, Brazil North, Cameroon, Colombia, Costa Rica, Cuba, Ecuador, French Guiana, Gabon, Ghana, Guatemala, Guinea-Bissau, Gulf of Guinea Is., Guyana, Haiti, Honduras, Ivory Coast, Liberia, Madagascar, Mexico Southeast, Nicaragua, Nigeria, Panamá, Peru, Sierra Leone, Suriname, Togo, Trinidad-Tobago, Venezuela, Zaire
Distribution (India): Andaman and Nicobar Islands
IUCN Status: Not evaluated
Reference: PhytoKeys 164: 94 (2020)

Dioclea virgata (Rich.) Amshoff
Fabaceae
Synonyms: *Dolichos virgatus* Rich
Climbing Mechanism: Stem Twiner
Distribution (Global): Argentina Northeast, Belize, Bolivia, Brazil North, Brazil Northeast, Brazil South, Brazil Southeast, Brazil West-Central, Colombia, Costa Rica, Ecuador, French Guiana, Guatemala, Guyana, Honduras, Mexico Southeast, Nicaragua, Panamá, Peru, Suriname, Venezuela
IUCN Status: Not evaluated
Reference: Meded. Bot. Mus. Herb. Rijks Univ. Utrecht 52: 69 (1939)

Dolichos capensis L.
Fabaceae
Climbing Mechanism: Stem Twiner
Distribution (India): Peninsular India
IUCN Status: Not evaluated
Reference: Pl. Rar. Afr.: 17 (1760)

Dolichos trilobus L.
Fabaceae
Synonyms: *Dolichos cristatus*, *Dolichos falcatus*, *Dolichos tuberosus*
Climbing Mechanism: Stem Twiner

Distribution (Global): Angola, Assam, Bangladesh, Benin, Cameroon, China South-Central, China Southeast, East Himalaya, Ethiopia, Ghana, Hainan, India, Java, Kenya, KwaZulu-Natal, Laos, Lesser Sunda Is., Liberia, Malawi, Mozambique, Myanmar, Namibia, Nepal, Nigeria, Northern Provinces, Philippines, Rwanda, Senegal, Sierra Leone, Somalia, Sri Lanka, Sudan, Swaziland, Taiwan, Tanzania, Togo, Uganda, Vietnam, West Himalaya, Yemen, Zambia, Zimbabwe
Distribution (India): Andhra Pradesh, Gujarat, Karnataka, Kerala, Madhya Pradesh, Maharashtra, Odisha, Tamil Nadu, Telangana, West Bengal
Leaf Type: Trifolioate
Inflorescence: Racemes
Fruit Type: Pods
Flowering and Fruiting: September–January
IUCN Status: Not evaluated
Reference: Sp. Pl.: 726 (1753)

Dumasia cordifolia Baker
Fabaceae
Climbing Mechanism: Stem Twiner
Distribution (Global): Assam, China South-Central, China Southeast, Tibet
IUCN Status: Not evaluated
Reference: J. D. Hooker, Fl. Brit. India 2: 183 (1876)

Dumasia villosa DC.
Fabaceae
Climbing Mechanism: Stem Twiner
Distribution (Global): Assam, Bangladesh, Burundi, Cape Provinces, China North-Central, China South-Central, China Southeast, East Himalaya, Ethiopia, India, Java, Kenya, KwaZulu-Natal, Laos, Lesser Sunda Is., Madagascar, Malawi, Malaya, Mozambique, Myanmar, Nepal, New Guinea, Northern Provinces, Pakistan, Philippines, Sri Lanka, Sulawesi, Sumatera, Swaziland, Taiwan, Tanzania, Thailand, Tibet, Uganda, Vietnam, West Himalaya, Zambia, Zaire, Zimbabwe

Distribution (India): Andhra Pradesh, Arunachal Pradesh, Bihar, Odisha, Jharkhand, Karnataka, Kerala, Madhya Pradesh, Maharashtra, Manipur, Rajasthan, Sikkim, Tamil Nadu, Uttar Pradesh, West Bengal
IUCN Status: Least Concern
Reference: Mém. Légum.: 257 (1826)

Dunbaria circinalis (Benth.) Baker
Fabaceae
Synonyms: *Atylosia circinnalis* Benth., *Dunbaria thorelii* Gagnep., *Phaseolus circinalis* Buch. -Ham.
Climbing Mechanism: Stem Twiner

Distribution (Global): Assam, China South-Central, East Himalaya, Laos, Lesser Sunda Is., Myanmar, Sumatera, Thailand, Vietnam
Distribution (India): West Bengal
IUCN Status: Not evaluated
Reference: J. D. Hooker, Fl. Brit. India 2: 219 (1876)

Dunbaria debilis Baker
Fabaceae
Synonyms: *Dunbaria parvifolia* X.X. Chen, *Dunbaria singuliflora* F. Muell.
Climbing Mechanism: Stem Twiner
Distribution (Global): Assam, China Southeast, East Himalaya, New Guinea, Northern Territory, Thailand
Distribution (India): Manipur
IUCN Status: Not evaluated
Reference: J. D. Hooker, Fl. Brit. India 2: 218 (1876)

Dunbaria ferruginea Wight & Arn.
Fabaceae
Synonyms: *Collaea venosa* Graham, *Cylista ferruginea* Wight & Arn.
Common Name: Rusty Dunbaria
Climbing Mechanism: Stem Twiner
Distribution (Global): India, Sri Lanka, Vietnam
Distribution (India): Andhra Pradesh, Eastern Ghats, Karnataka, Puducherry, Tamil Nadu
Leaf Type: Trifoliate
Inflorescence: Racemes
IUCN Status: Not evaluated
Reference: Prodr. Fl. Ind. Orient. 1: 258 (1834)

Dunbaria fusca (Wall.) Kurz
Fabaceae
Synonyms: *Atylosia crinita* Dunn, *Dunbaria crinita* (Dunn) Maesen, *Dunbaria flavescens* Thuan, *Phaseolus fuscus* Wall., *Vigna fusca* (Wall.) A.S. Chauhan
Climbing Mechanism: Stem Twiner
Distribution (Global): China South-Central, China Southeast, Hainan, Laos, Myanmar, Thailand, Vietnam
IUCN Status: Least Concern
Reference: J. Asiat. Soc. Bengal, Pt. 2, Nat. Hist. 45: 255 (1876 publ. 1877)

Dunbaria glandulosa (Dalzell & A. Gibson) Prain
Fabaceae
Synonyms: *Atylosia rostrata* Baker, *Cajanus glandulosus* Dalzell & A. Gibson
Climbing Mechanism: Stem Twiner
Distribution (Global): Assam, Bangladesh, India, Myanmar, Nepal, Thailand

Distribution (India): Gujarat, Madhya Pradesh, Maharashtra, Tamil Nadu, West Bengal
IUCN Status: Not evaluated
Reference: J. Asiat. Soc. Bengal, Pt. 2, Nat. Hist. 66: 433 (1897)

Dunbaria podocarpa Kurz
Fabaceae
Climbing Mechanism: Stem Twiner
Distribution (Global): Assam, Cambodia, China Southeast, East Himalaya, Hainan, Java, Laos, Malaya, Myanmar, Sumatera, Thailand, Vietnam
Distribution (India): Mizoram
IUCN Status: Not evaluated
Reference: J. Asiat. Soc. Bengal, Pt. 2, Nat. Hist. 43: 185 (1874)

Dunbaria punctata (Wight & Arn.) Benth.
Fabaceae
Synonyms: *Atylosia punc*tata (Wight & Arn.) Dalzell, *Dolichos conspersus* Graham, *Dolichos finlaysonianus* Graham, *Dolichos punctatus* Wight & Arn., *Dolichos rhynchosioides* Miq., *Dunbaria conspersa* Benth., *Dunbaria rotundifolia* (Lour.) Merr.
Common Name: Dotted Dunbaria
Climbing Mechanism: Stem Twiner
Distribution (Global): Assam, Bangladesh, Cambodia, China South-Central, China Southeast, East Himalaya, Hainan, India, Java, Laos, Myanmar, Nepal, Philippines, Queensland, Sulawesi, Taiwan, Thailand, Vietnam
Distribution (India): Tamil Nadu
Leaf Type: Trifoliate
Fruit Type: Pod
IUCN Status: Not evaluated
Reference: F.A.W. Miquel, Pl. Jungh.: 242 (1852)

Dysolobium dolichoides (Roxb.) Prain
Fabaceae
Synonyms: *Canavalia dolichoides* Kurz, *Dolichos dasycarpus* Miq., *Dolichos schomburgkii* Gagnep., *Dysolobium dolichoides* var. *schomburgkii* (Gagnep.) Maréchal, *Mucuna recta* Buch. -Ham., *Phaseolus dolichoides* Roxb., *Vigna dolichoides* (Kurz) Baker
Climbing Mechanism: Stem Twiner
Distribution (Global): Assam, Bangladesh, Cambodia, India, Java, Thailand, Vietnam
Distribution (India): Kerala
Leaf Type: Trifoliate
Inflorescence: Axillary racemes
Fruit Type: Pod
Flowering and Fruiting: November–January

IUCN Status: Not evaluated
Reference: J. Asiat. Soc. Bengal, Pt. 2, Nat. Hist. 66: 427 (1897)

Dysolobium grande (Benth.) Prain
Fabaceae
Synonyms: *Canavalia grandis* Kurz, *Dolichos grandis* (Wall. ex Benth.) Gagnep., *Mucuna chienkweiensis* G.Z.Li, *Phaseolus velutinus* Graham ex Baker
Climbing Mechanism: Stem Twiner
Distribution (Global): Assam, Bangladesh, China South-Central, China Southeast, East Himalaya, India, Myanmar, Nepal, Thailand
Distribution (India): Arunachal Pradesh, Madhya Pradesh, Mizoram, West Bengal
IUCN Status: Not evaluated
Reference: J. Asiat. Soc. Bengal, Pt. 2, Nat. Hist. 66: 427 (1897)

Dysolobium lucens (Benth.) Prain
Fabaceae
Synonyms: *Canavalia lucens* (Benth.) Kurz, *Vigna lucens* (Benth.) Baker
Climbing Mechanism: Stem Twiner
Distribution (Global): Assam, Bangladesh, East Himalaya, Myanmar
IUCN Status: Not evaluated
Reference: J. Asiat. Soc. Bengal, Pt. 2, Nat. Hist. 66: 427 (1897)

Dysolobium pilosum (Willd.) Marechal
Fabaceae
Synonyms: *Dolichos apioides* Gagnep., *Dolichos formosanus* (Hayata) S.S. Ying, *Dolichos formosanus* var. *rhombifolius* (Hayata) S.S. Ying, *Dolichos pilosus* J.G. Klein ex Willd., *Dolichos rhombifolius* (Hayata) Hosok., *Dolichovigna formosana* Hayata
Climbing Mechanism: Stem Twiner
Distribution (Global): Andaman Is., Assam, Bangladesh, Cambodia, East Himalaya, India, Java, Laos, Malaya, Myanmar, Nepal, Philippines, Taiwan, Thailand, Vietnam
Distribution (India): Telangana
IUCN Status: Not evaluated
Reference: Bull. Jard. Bot. Natl. Belg. 47: 483 (1977)

Dysolobium tetragonum Prain
Fabaceae
Climbing Mechanism: Stem Twiner
Distribution (India): West Bengal
IUCN Status: Not evaluated
Reference: J. Asiat. Soc. Bengal, Pt. 2, Nat. Hist. 66: 427. 1897 [1898 publ. 1897]

Eleiotis monophyllos (Burm.f.) DC.
Fabaceae

Synonyms: *Eleiotis sororia* (L.) DC., *Glycine monophylla* Burm.f., *Glycine monophyllos* Burm.f., *Hallia sororia* (L.) Willd., *Hedysarum sororium* L.
Common Name: One-leaf glycine, One-leaved Eleiotis
Climbing Mechanism: Scrambler-Unarmed
Distribution (India): Puducherry
Leaf Type: Compound
Inflorescence: Racemes
Fruit Type: Pods
IUCN Status: Not evaluated
Reference: Prodr. 2: 348. 1825 "monophylla"

Eleiotis rottleri Wight & Arn.
Fabaceae
Synonyms: *Desmodium bakeri* N.P. Balakr., *Desmodium rottleri* (Wight & Arn.) Baker, *Eleiotis trifoliolata* T. Cooke
Climbing Mechanism: Scrambler-Unarmed
Distribution (Global): India, Myanmar, Sri Lanka
IUCN Status: Vulnerable
Reference: Prodr. Fl. Ind. Orient. 1: 231 (1834)

Endosamara racemosa (Roxb.) R. Geesink
Fabaceae
Synonyms: *Endosamara racemosa* var. *pallida* (Dalzell & A. Gibson) J. Compton & Schrire, *Millettia leiogyna* Kurz, *Millettia orissae* Panigrahi & S.C. Mishra, *Millettia pallida* (Dalzell & A. Gibson) Dalzell, *Millettia racemosa* (Roxb.) Benth., *Phaseoloides leiogynum*
Climbing Mechanism: Scrambler-Unarmed
Distribution (Global): Assam, Bangladesh, India, Laos, Malaya, Myanmar, Philippines, Thailand, Vietnam
Distribution (India): Tamil Nadu, West Bengal
IUCN Status: Not evaluated
Reference: Leiden Bot. Ser. 8: 93 (1984)

Entada gigas (L.) Fawc. & Rendle
Fabaceae
Synonyms: *Entada gigalobium* DC., *Entada planosemi*nata (De Wild.) G.C.C. Gilbert & Boutique, *Entada umbonata* (De Wild.) G.C.C. Gilbert & Boutique, *Mimosa gigas* L.
Climbing Mechanism: Scrambler-Unarmed
Distribution (Global): Angola, Belize, Benin, Cameroon, Central African Repu, Central American Pac, Colombia, Congo, Costa Rica, Cuba, Dominican Republic, Ecuador, Equatorial Guinea, Gabon, Ghana, Haiti, Honduras, Ivory Coast, Jamaica, Kenya, Leeward Is., Liberia, Mexico Central, Mexico Gulf, Mexico Southeast, Mexico Southwest, Nicaragua, Panamá, Sierra Leone, Sudan, Togo, Uganda, Venezuela, Windward Is., Zambia, Zaire

Distribution (India): Assam
IUCN Status: Not evaluated
Reference: Fl. Jamaica 4: 124 (1920)

Entada phaseoloides (L.) Merr.
Fabaceae
Synonyms: *Acacia scandens* (L.) Willd., *Adenanthera scandens* (L.) G. Forst., *Entada adenanthera* DC., *Entada gandu* Hoffmanns., *Entada parrana* Spreng., *Entada rumphii* Scheff., *Entada scand*ens (L.) Benth., *Entada scandens* var. *aequilatera* Domin.
Climbing Mechanism: Scrambler-Unarmed
Distribution (Global): Bangladesh, Bismarck Archipelago, Borneo, Caroline Is., China South-Central, China Southeast, Cook Is., East Himalaya, Fiji, Hainan, Hawaii, Japan, Java, Lesser Sunda Is., Maluku, Marianas, Marshall Is., Nansei-shoto, New Caledonia, New Guinea, Northern Territory, Philippines, Phoenix Is., Queensland, Samoa, Santa Cruz Is., Solomon Is., South China Sea, Sulawesi, Sumatera, Taiwan, Tibet, Tonga, Tubuai Is., Vanuatu, Vietnam, Wallis-Futuna Is.
Distribution (India): Arunachal Pradesh, Madhya Pradesh, Tripura
IUCN Status: Not evaluated
Reference: Philipp. J. Sci., C 9: 86 (1914)

Entada rheedii Spreng.
Fabaceae
Synonyms: *Entada rheedei* Spreng., *Entada phaseoloides* Sensu auct., *Entada pursaetha* DC., *Adenanthera gogo* Blanco, *Entada gigalobium* Sensu auct., p. p., *Entada gigas* G.C.C. Gilbert & Boutique, *Entada gogo* (Blanco)I.M. Johnst., *Entada pusaetha* DC.
Common Name: Sea bean, African dream herb
Climbing Mechanism: Scrambler-Unarmed
Distribution (India): Andaman and Nicobar Islands, Andhra Pradesh, Goa, Kerala, Maharashtra, Manipur, Odisha, Tamil Nadu, Telangana, West Bengal, Western Ghats
Leaf Type: Bipinnate
Inflorescence: Axillary pendulous spikes
Fruit Type: Pod
IUCN Status: Not evaluated
Reference: Syst. Veg. 2: 325 (1825)

Flemingia tuberosa Dalzell
Fabaceae
Synonyms: *Maughania tuberosa* (Dalzell) Kuntze
Common Name: Tuberous Flemingia
Climbing Mechanism: Stem Twiner
Distribution (Global): India
Distribution (India): Kerala

Leaf Type: Trifolioate
Inflorescence: Solitary, axillary racemes
Fruit Type: Pod
Flowering and Fruiting: September–October
IUCN Status: Not evaluated
Reference: Hooker's J. Bot. Kew Gard. Misc. 2: 34 (1850)

Galactia filiformis
Fabaceae
Climbing Mechanism: Stem Twiner
IUCN Status: Not evaluated

Galactia latifolia (Baker) Thuan
Fabaceae
Climbing Mechanism: Stem Twiner
Distribution (Global): India, Vietnam
IUCN Status: Not evaluated
Reference: Fl. Camb., Laos & Vietn. 17: 90 (1979)

Galactia longiflora Arn.
Fabaceae
Synonyms: *Collaea longiflora* (Arn.) Benth.
Climbing Mechanism: Stem Twiner
Distribution (Global): Honduras, Leeward Is., Windward Is.
Distribution (India): Andhra Pradesh
IUCN Status: Not evaluated
Reference: R. Wight & G.A.W. Arnott, Prodr. Fl. Ind. Orient. 1: 206 (1834)

Galactia longifolia (Jacq.) Benth.
Fabaceae
Climbing Mechanism: Stem Twiner
Distribution (India): Tamil Nadu, West Bengal
IUCN Status: Not evaluated
Reference: Comm. Legum. Gen.: 63 (1837)

Galactia regularis (L.) Britton & al.
Fabaceae
Synonyms: *Clitoria glabella* (Michx.) Desf., *Clitoria lactescens* L., *Dolichos regularis* L., *Ervum volubile* Walter, *Galactia angustifolia* var. *retusa* C. Wright ex Griseb., *Galactia glabella* Michx., *Hedysarum volubile* L.
Climbing Mechanism: Stem Twiner
Distribution (Global): Alabama, Arkansas, Cuba, Delaware, Florida, Georgia, Illinois, Kansas, Kentucky, Louisiana, Maryland, Mississippi, Missouri, New Jersey, New York, North Carolina, Oklahoma, Pennsylvania, South Carolina, Texas, Virginia

IUCN Status: Not evaluated
Reference: Prelim. Cat.: 14 (1888)

Galactia striata (Jacq.) Urb.
Fabaceae
Synonyms: *Glycine striata* Jacq., *Odonia retusa* Rose, *Phaseolus tomentosus* Andersson, *Teramnus tenuiflorus* (Willd.) Spreng., *Galactia striata* var. *tomentosa* (Bertol.) Urb., *Galactia tenuiflora* var. *minor* Baker, *Leucodictyon malvensis* Dalzell & A. Gibson
Climbing Mechanism: Stem Twiner
Distribution (Global): Andaman Is., Angola, Argentina Northeast, Argentina Northwest, Assam, Bahamas, Bangladesh, Belize, Benin, Bolivia, Brazil North, Brazil Northeast, Brazil South, Brazil Southeast, Brazil West-Central, Cayman Is., Central African Repu, China South-Central, China Southeast, Christmas I., Colombia, Comoros, Congo, Costa Rica, Cuba, Dominican Republic, East Himalaya, Ecuador, El Salvador, Eritrea, Florida, Gabon, Galápagos, Ghana, Guatemala, Guyana, Hainan, Haiti, Honduras, India, Ivory Coast, Jamaica, Java, Kenya, KwaZulu-Natal, Leeward Is., Lesser Sunda Is., Madagascar, Mauritius, Mexico Central, Mexico Gulf, Mexico Northeast, Mexico Northwest, Mexico Southeast, Mexico Southwest, Mozambique, Myanmar, Nepal, Netherlands Antilles, New Guinea, New South Wales, Nicaragua, Nigeria, Northern Provinces, Northern Territory, Panamá, Paraguay, Peru, Philippines, Puerto Rico, Queensland, Réunion, South Australia, Southwest Caribbean, Sri Lanka, Sudan, Sulawesi, Swaziland, Taiwan, Tanzania, Thailand, Togo, Trinidad-Tobago, Turks-Caicos Is., Uganda, Uruguay, Venezuela, Vietnam, West Himalaya, Western Australia, Yemen, Zaire, Zimbabwe
IUCN Status: Not evaluated
Reference: Symb. Antill. 2: 320 (1900)

Galactia striata var. *villosa* (Wight & Arn.) Verdc.
Fabaceae
Synonyms: *Clitoria phyrne* Steud., *Copisma subsericeum* Sond., *Galactia diversifolia* Bojer, *Galactia dumetorum* Benth., *Galactia graminea* B. Heyne ex Baker, *Galactia lanceolata* Hayata, *Galactia mucronata* Klotzsch, *Galactia obcordata* (Baill.) Verdc.
Climbing Mechanism: Stem Twiner
Distribution (Global): Andaman Is., Angola, Argentina Northeast, Argentina Northwest, Bangladesh, Benin, Brazil Northeast, Brazil Southeast, Central African Repu, China South-Central, China Southeast, Christmas I., Comoros, Congo, East Himalaya, Eritrea, Gabon, Ghana, Hainan, India, Ivory Coast, Java, Kenya, KwaZulu-Natal, Lesser Sunda Is., Madagascar, Mauritius, Mozambique, Nepal, New Guinea, New South Wales, Nigeria, Northern Provinces, Northern Territory, Paraguay, Philippines, Queensland, Réunion, South Australia, Sri Lanka, Sudan, Sulawesi, Swaziland, Taiwan, Tanzania, Thailand, Togo, Uganda, Venezuela, Vietnam, West Himalaya, Western Australia, Yemen, Zaire, Zimbabwe

Distribution (India): Tamil Nadu
IUCN Status: Not evaluated
Reference: Fl. Zambes. 3(5): 42 (2001)

Galactia tenuiflora (Willd.) Wight & Arn.
Fabaceae
Synonyms: *Galactia obcordata* (Baill.) Verdc., *Galactia sericea* var. *phrynoides* DC., *Glycine tenuiflora* Willd., *Teramnus obcordatus* Baill.
Common Name: Slender flowered milkpea
Climbing Mechanism: Stem Twiner
Distribution (India): Andhra Pradesh, Bihar, Odisha, Madhya Pradesh, Odisha, Rajasthan, Tamil Nadu, West Bengal
Leaf Type: Trifoliate
Inflorescence: Racemes
Fruit Type: Pods
IUCN Status: Least Concern
Reference: Prodr. Fl. Ind. Orient. 1: 206 (1834)

Geissaspis cristata Wight & Arn.
Fabaceae
Synonyms: *Zornia disperma* Graham
Climbing Mechanism: Scrambler-Unarmed
Distribution (Global): Bangladesh, Cambodia, China Southeast, East Himalaya, India, Laos, Malaya, Myanmar, Nepal, Nicobar Is., Sri Lanka, Thailand, Vietnam
Distribution (India): Gujarat, Kerala
Leaf Type: Compound
Inflorescence: Spikes
Fruit Type: Pods
Flowering and Fruiting: July–September
IUCN Status: Least Concern
Reference: Prodr. Fl. Ind. Orient. 1: 218 (1834)

Geissaspis tenella Benth.
Fabaceae
Synonyms: *Geissaspis cristata* var. *tenella* (Benth.) M.R. Almeida
Climbing Mechanism: Scrambler-Unarmed
Distribution (India): Karnataka
Leaf Type: Alternate
Inflorescence: Axillary racemes
Fruit Type: Pods
Flowering and Fruiting: August–November
IUCN Status: Least Concern
Reference: Flora 32: 559 (1849)

Hardenbergia comptoniana (Andrews) Benth.
Fabaceae
Synonyms: *Caulinia comptoniana* (Andrews) F. Muell., *Glycine comptoniana* Andrews, *Hardenbergia digitata* Lindl., *Hardenbergia huegelii* Benth., *Hardenbergia lindleyi* Meisn., *Hardenbergia lindleyi* var. *digitata* (Lindl.) Meisn., *Hardenbergia lindleyi* var. *trifolia*
Common Name: Wild sarsaparilla
Climbing Mechanism: Scrambler-Unarmed
Distribution (Global): Western Australia
Distribution (India): Tamil Nadu
IUCN Status: Not evaluated
Reference: Comm. Legum. Gen.: 60 (1837)

Hegnera obcordata (Miq.) Schindl.
Fabaceae
Synonyms: *Desmodium obcordatum* (Miq.) Kurz, *Meibomia obcordata* (Miq.) Kuntze, *Uraria obcordata* Miq.
Climbing Mechanism: Stem Twiner
Distribution (Global): Cambodia, Java, Laos, Malaya, Myanmar, Sumatera, Thailand, Vietnam
IUCN Status: Not evaluated
Reference: Report. Spec. Nov. Regni Veg. 20: 285 (1924)

Indigofera trifoliata L.
Fabaceae
Synonyms: *Anil trifoliata* (L.) Kuntze, *Indigofera moluccana*, *Indigofera barberi*, *Indigofera vestita*
Common Name: Three-leaf indigo
Climbing Mechanism: Scrambler-Unarmed
Distribution (Global): Assam, Bangladesh, Bismarck Archipelago, China South-Central, China Southeast, East Himalaya, Hainan, India, Java, Laos, Lesser Sunda Is., Maluku, Myanmar, Nansei-shoto, Nepal, New Guinea, Northern Territory, Pakistan, Philippines, Queensland, Sri Lanka, Sulawesi, Taiwan, Thailand, Vietnam, West Himalaya, Western Australia
Distribution (India): Arunachal Pradesh, Madhya Pradesh
Leaf Type: Trifoliate
Inflorescence: Axillary fascicles
Fruit Type: Pods
Flowering and Fruiting: September–December
IUCN Status: Not evaluated
Reference: Cent. Pl. II: 29 (1756)

Indigofera trita L.f.
Fabaceae
Synonyms: *Indigofera cinerea*, *Indigofera timoriensis*, *Indigofera rigida*

Common Name: Asian indigo, Three-leaved indigo
Climbing Mechanism: Scrambler-Unarmed
Distribution (Global): Assam, Bangladesh, Botswana, Burundi, Cameroon, Congo, Costa Rica, Djibouti, East Himalaya, Eritrea, Ethiopia, Gabon, Ghana, Gulf of Guinea Is., India, Ivory Coast, Java, Kenya, Lesser Sunda Is., Liberia, Madagascar, Malawi, Mozambique, Myanmar, Namibia, Nigeria, Oman, Pakistan, Philippines, Queensland, Rwanda, Saudi Arabia, Senegal, Sierra Leone, Somalia, Sri Lanka, Sudan, Swaziland, Tanzania, Togo, Uganda, West Himalaya, Western Australia, Yemen, Zambia, Zaire, Zimbabwe
Distribution (India): Puducherry
Leaf Type: Trifoliate
IUCN Status: Least Concern
Reference: Suppl. Pl.: 335 (1782)

Indigofera wightii Wight & Arn.
Fabaceae
Synonyms: *Indigofera foliolosa* Graham, *Indigofera hainanensis* H.T. Tsai & T.T. Yu, *Indigofera inamoena* Thwaites, *Indigofera pallida* Craib
Common Name: Wight's indigo
Climbing Mechanism: Scrambler-Unarmed
Distribution (Global): Cambodia, Hainan, India, Laos, Myanmar, Sri Lanka, Thailand, Vietnam
Distribution (India): Rajasthan
Leaf Type: Compound
Inflorescence: Racemes
Fruit Type: Pods
Flowering and Fruiting: August–July
IUCN Status: Not evaluated
Reference: Prodr. Fl. Ind. Orient. 1: 202 (1834)

Kunstleria keralensis C.N. Mohanan & N.C. Nair
Fabaceae
Common Name: Kerala Kunstleria
Climbing Mechanism: Scrambler-Armed
Distribution (Global): India
Distribution (India): Karnataka, Kerala, Tamil Nadu, Western Ghats
Leaf Type: Leaves are imparipinnate; leaflets opposite, ovate-oblong, tapering
Inflorescence: Flowers are borne in panicles at branch ends and in leaf-axils, branches slender, rusty woolly
Fruit Type: Pods
Flowering and Fruiting: April–June
IUCN Status: Not evaluated
Notes: Usually found in evergreen and semi-evergreen forests
Reference: Proc. Indian Acad. Sci., Pl. Sci. 90: 208 (1981)

Lablab purpureus (L.) Sweet
Fabaceae
Synonyms: *Dolichos purpureus* L.
Common Name: Lablab bean, Hyacinth bean, Bonavista bean, Egyptian bean
Climbing Mechanism: Stem Twiner
Distribution (Global): Pantropical
Distribution (India): Andaman and Nicobar Islands, Andhra Pradesh, Arunachal Pradesh, Gujarat, Himachal Pradesh, Jharkhand, Kerala, Madhya Pradesh, Maharashtra, Manipur, Mizoram, Odisha, Rajasthan, Tamil Nadu, Telangana
Leaf Type: Pinnately divided leaves with 3 large leaflets, Compound (Pinnately, Bipinnately, Palmately)
Inflorescence: Pink Purple/Lavender White colored fragrant flowers on axillary inflorescence
Fruit Type: Pods
Flowering and Fruiting: September–November
IUCN Status: Not evaluated
Notes: Often cultivated
Reference: Hort. Brit.: 481 (1826)

Lathyrus aphaca L.
Fabaceae
Synonyms: *Aphaca disperma* Alef., *Aphaca marmorata* Alef., *Aphaca pseudophaca* (Boiss.) Alef., *Aphaca vulgaris* C. Presl. + 10
Common Name: Yellow Pea, Yellow vetchling
Climbing Mechanism: Scrambler-Armed
Distribution (Global): Afghanistan, Albania, Algeria, Assam, Baleares, Bangladesh, Bulgaria, Canary Is., Corse, Cyprus, East Aegean Is., East Himalaya, Egypt, France, Greece, India, Iran, Iraq, Italy, Kazakhstan, Kirgizstan, Kriti, Krym, Lebanon-Syria, Libya, Madeira, Morocco, Nepal, North Caucasus, Pakistan, Palestine, Portugal, Romania, Sardegna, Sicilia, Sinai, South European Russi, Spain, Tadzhikistan, Transcaucasus, Tunisia, Turkey, Turkey-in-Europe, Turkmenistan, Ukraine, Uzbekistan, West Himalaya, Yugoslavia
Distribution (India): Gujarat, Madhya Pradesh, Rajasthan, Uttar Pradesh, West Benga
Flowering and Fruiting: February–April
IUCN Status: Least Concern
Notes: Distributed in high altitudes of the Western Ghats
Reference: Sp. Pl.: 729 (1753)

Lathyrus odoratus L.
Fabaceae
Synonyms: *Lathyrus cyprius* Rech.f., *Lathyrus maccaguenii* Tod. ex Nyman, *Lathyrus odoratus-zeylanicus* Burm.f., *Pisum odoratum* (L.) E.H.L. Krause
Common Name: Sweet pea
Climbing Mechanism: Scrambler-Unarmed

Distribution (Global): Italy, Sicilia
Distribution (India): Gujarat, Madhya Pradesh, Manipur, Tripura, Uttar Pradesh
Leaf Type: Leaves with branched ten-dril at apex, leaflets 1-paired, ovate-oblong or elliptic
Inflorescence: Raceme longer than leaf, 1-3(or 4)-flowered
Fruit Type: Legume brown-yellow
Flowering and Fruiting: May–July
IUCN Status: Critically endangered
Reference: Sp. Pl.: 732 (1753)

Lathyrus purpureus C. Presl
Fabaceae
Synonyms: *Lathyrus clymenum* L., *Clymenum alatum* (Ten.) Link, *Clymenum tenuifolium* (Desf.) Link, *Clymenum uncinatum* Moench, *Lathyros clymenus* (L.) St.-Lag., *Lathyrus alatus* Sm., *Lathyrus alatus* Ten., *Lathyrus articulatus* Loisel., *Lathyrus auriculatus* Bertol
Climbing Mechanism: Scrambler-Unarmed
Distribution (Global): Albania, Algeria, Baleares, Canary Is., Corse, East Aegean Is., France, Greece, Italy, Kriti, Libya, Madeira, Morocco, Palestine, Portugal, Sicilia, Spain, Tunisia, Turkey, Turkey-in-Europe, Yugoslavia
Distribution (India): Tripura
Leaf Type: With broad leaf-like petiole and rachis, the lower linear-lanceolate, without leaflets, the upper with 2-4(-5) pairs of leaflets
Inflorescence: Racemes 1- to 5-flowered
Fruit Type: Pods/Legumes
Flowering and Fruiting: April–May
IUCN Status: Least Concern
Notes: Usually found in open and dry shrubby vegetation and along the margins of cultivated fields
Reference: Sp. Pl.: 732 (1753)

Lathyrus sativus L.
Fabaceae
Synonyms: *Cicercula alata* Moench, *Cicercula alba* Medik., *Cicercula caerulea* Medik., *Cicercula sativa* (L.) Medik., *Lathyrus asiaticus* (Zalkind) Kudrj., *Lathyrus azureus* Booth ex Bosse, *Lathyrus sativus* subsp. *albus* Smekalova. +14
Common Name: Grass pea, Chickling pea, Indian vetch
Climbing Mechanism: Scrambler-Unarmed
Distribution (Global): Bulgaria, Yugoslavia.
Distribution (India): Gujarat, Rajasthan, Uttar Pradesh, West Bengal.
Leaf Type: Leaves have tendril at tip, axis winged. Stipules are semiarrow shaped, leaflets,1-paired
Inflorescence: Flowers are borne 1 or 2 together
Fruit Type: Pods, nearly elliptic
Flowering and Fruiting: June–July

IUCN Status: Not evaluated
Reference: Sp. Pl.: 730 (1753)

Macroptilium atropurpureum (DC.) Urb.
Fabaceae
Synonyms: *Dolichos rhynchosioides* Desv., *Phaseolus affinis* Piper, *Phaseolus atropurpureus* Torr., *Phaseolus atropurpureus* DC., *Phaseolus atropurpureus* var. *canescens* (M. Martens & Galeotti) Hassl. +10
Common Name: Purple bush bean
Climbing Mechanism: Stem Twiner
Distribution (Global): Belize, Bolivia, Colombia, Costa Rica, Ecuador, El Salvador, French Guiana, Galápagos, Honduras, Mexico Central, Mexico Gulf, Mexico Northeast, Mexico Northwest, Mexico Southeast, Mexico Southwest, Nicaragua, Panamá, Peru, Trinidad-Tobago, Venezuela
Distribution (India): Maharashtra, Telangana
Leaf Type: Leaves trifoliolate; leaflet blades, dark green and finely hairy on the upper surface
Inflorescence: Inflorescence is a raceme comprising 6–12 often paired flowers on a short rachis
Fruit Type: Pods dehisce violently (shatter) along both sutures when ripe
IUCN Status: Not evaluated
Reference: Symb. Antill. 9: 457 (1928)

Macroptilium lathyroides (L.) Urb.
Fabaceae
Synonyms: *Lotus maritimus* Vell., *Macroptilium lathyroides* var. *bustarretianum* Stehlé & Quentin, *Macroptilium lathyroides* var. *semierectum* (L.) Urb., *Phasellus lathyroides* (L.) Moench, *Phaseolus candidus* f. *maritimus* (Benth.) Hassl., *Phaseolus cinnabarinus* Sal
Common Name: Phasey bean, Wild bean, Wild bush bean, Wild pea bean, One leaf clover, Quail bean, Wild dolly bean.
Climbing Mechanism: Stem Twiner
Distribution (Global): Argentina Northeast, Bahamas, Belize, Bolivia, Brazil North, Brazil Northeast, Brazil South, Brazil Southeast, Brazil West-Central, Cayman Is., Colombia, Costa Rica, Cuba, Dominican Republic, Ecuador, Florida, French Guiana, Galápagos, Georgia, Guyana, Haiti, Honduras, Jamaica, Leeward Is., Louisiana, Mexico Gulf, Mexico Northeast, Mexico Northwest, Mexico Southeast, Mexico Southwest, Netherlands Antilles, Nicaragua, Panamá, Paraguay, Peru, Puerto Rico, Southwest Caribbean, Suriname, Trinidad-Tobago, Turks-Caicos Is., Uruguay, Venezuela, Windward Is.
Distribution (India): Puducherry
Leaf Type: leaves are trifoliate long, leaflets
Inflorescence: The inflorescences are long racemes borne on 30 cm long flower-cluster-stalks, and bear red to red-purple (occasionally white or pink) pea-like flowers

Fruit Type: The fruits are velvet-hairy, linear, splitting pods
IUCN Status: Not evaluated
Reference: Symb. Antill. 9: 457 (1928)

Macrotyloma ciliatum (Willd.) Verdc.
Fabaceae
Synonyms: *Dolichos argenteus* Roxb. ex Wight & Arn., *Dolichos ciliatus* J.G. Klein ex Willd., *Dolichos prostratus* Buch. -Ham. & K.D. Koenig ex Roxb.
Climbing Mechanism: Stem Twiner
Distribution (Global): India, Sri Lanka
Distribution (India): Andhra Pradesh, Madhya Pradesh, Tamil Nadu, West Bengal
Flowering and Fruiting: September–December
IUCN Status: Not evaluated
Reference: Kew Bull. 24: 404 (1970)

Macrotyloma uniflorum (Lam.) Verdc.
Fabaceae
Synonyms: *Dolichos uniflorus* Lam., *Glycine uniflora* (Lam.) Dalzell, *Kerstingiella uniflora* (Lam.) J.A. Lackey
Climbing Mechanism: Stem Twiner
Distribution (Global): Angola, Assam, Bangladesh, Botswana, Cameroon, East Himalaya, Ethiopia, Guinea, India, Kenya, Mozambique, Myanmar, Namibia, Pakistan, Rwanda, Senegal, Somalia, Sri Lanka, Sudan, Tanzania, Uganda, West Himalaya, Zaire, Zimbabwe
Distribution (India): Andhra Pradesh, Madhya Pradesh, Maharashtra, Manipur, Mizoram, Telangana
Leaf Type: Leaflets ovate-rhomboid, obovate or elliptic, rounded or subacute at the tip, glabrescent to pubescent
Inflorescence: Flowers in axillary (1–)2–3(–5)-flowered clusters
Fruit Type: Pod, upcurved towards the apex, pods pilose with an understorey of very fine hairs, or glabrous
Flowering and Fruiting: October–December
IUCN Status: Least Concern
Notes: Usually occur in open and exposed areas
Reference: Kew Bull. 24: 322 (1970)

Mastersia assamica Benth.
Fabaceae
Synonyms: *Mastersia cleistocarpa* Baker
Climbing Mechanism: Stem Twiner
Distribution (Global): Assam, East Himalaya, Myanmar, Tibet
Distribution (India): Arunachal Pradesh
Leaf Type: Leaf rachis, stipels linear, leaflets subrhombic, elliptic, or ovate
Inflorescence: Racemes
Fruit Type: Legumes blackish

IUCN Status: Not evaluated
Reference: Trans. Linn. Soc. London 25: 300 (1865)

Meizotropis buteiformis Voigt
Fabaceae
Synonyms: *Butea buteiformis* (Voigt) Grierson., *Megalotropis buteiformis* (Voigt) Griff.
Common Name: Shrub Butea
Climbing Mechanism: Scrambler-Unarmed
Distribution (Global): Assam, Bangladesh, East Himalaya, India, Myanmar, Nepal, Thailand, Tibet, West Himalaya.
Distribution (India): West Bengal
Leaf Type: Leaves are trifoliate, with stalks 10–20 cm long. Leaflets are broadly ovate-elliptic
Inflorescence: Flowers are borne in many flowered racemes or panicles
Fruit Type: Pods, velvety with brownish hairs
Flowering and Fruiting: April–August
IUCN Status: Not evaluated
Notes: Found mostly in dry valley slopes and open grasslands
Reference: Notes Roy. Bot. Gard. Edinburgh 37: 346 (1979)

Melodinus cochinchinensis (Lour.) Merr.
Fabaceae
Synonyms: *Echaltium piscidium* (Roxb.) Wight, *Lycimnia suaveolens* Hance, *Melodinus annamensis* Pit., *Melodinus crassipetalus* Kerr. + 18
Common Name: Mountain orange
Climbing Mechanism: Stem Twiner
Distribution (Global): Assam, Bangladesh, Cambodia, China South-Central, China Southeast, East Himalaya, Hainan, India, Laos, Malaya, Myanmar, Thailand, Vietnam
Distribution (India): North Eastern India
Leaf Type: Leaf blade elliptic or narrowly so, 6–19 X 2.2–6.5 cm, papery, base cuneate, apex acute or acuminate
Inflorescence: Flowers are borne in panicle-like cymes, at branch-ends, 3-branched, 4–5.5 cm, minutely hairy
Fruit Type: Berries are narrowly ellipsoid, about 9 x 5 cm
Flowering and Fruiting: April–May
IUCN Status: Not evaluated
Notes: Occurs mostly in humid forests and valleys.
Reference: Trans. Amer. Philos. Soc., n.s., 24: 310 (1935)

Mezoneurum glabrum Desf.
Fabaceae
Climbing Mechanism: Scrambler-Armed
IUCN Status: Not evaluated

Mezoneurum sumatranum Wight & Arn.
Fabaceae
Climbing Mechanism: Scrambler-Armed
IUCN Status: Not evaluated

Millettia caerulea Baker
Fabaceae
Climbing Mechanism: Scrambler-Unarmed
Leaf Type: Compound, imparipinnate, opposite
Inflorescence: Fasciled racemes in terminal panicles
Fruit Type: Dry, pods
IUCN Status: Not evaluated

Millettia caudata (Benth.) Baker
Fabaceae
Climbing Mechanism: Scrambler-Unarmed
Distribution (India): Meghalaya, Tripura
Leaf Type: Compound, imparipinnate, opposite
Inflorescence: Fasciled racemes in terminal panicles
Fruit Type: Dry, pods
IUCN Status: Not evaluated

Millettia dorwardi Collett & Hemsl.
Fabaceae
Climbing Mechanism: Scrambler-Unarmed
Distribution (India): Mizoram
Leaf Type: Compound, imparipinnate, opposite
Inflorescence: Fasciled racemes in terminal panicles
Fruit Type: Dry, pods
IUCN Status: Not evaluated

Millettia extensa (Benth.) Baker
Fabaceae
Climbing Mechanism: Scrambler-Unarmed
Distribution (India): Arunachal Pradesh, Jharkhand, Madhya Pradesh, Maharashtra, Manipur, Odisha, Rajasthan, West Bengal
Leaf Type: Compound, imparipinnate, opposite
Inflorescence: Fasciled racemes in terminal panicles
Fruit Type: Dry, pods
IUCN Status: Not evaluated

Millettia fruticosa (DC.) Baker
Fabaceae
Synonyms: *Amerimnon fasciculatum* Buch. -Ham. ex Baker, *Cracca fruticosa* (DC.) Kuntze, *Otosema fruticosa* (DC.) Benth., *Phaseoloides fruticosum* (DC.) Kuntze,

Pongamia fruticosa Graham, *Pongamia fruticosa* (DC.) Ralph, *Robinia fruticosa* (DC.) Roxb.
Climbing Mechanism: Scrambler-Unarmed
Distribution (Global): Assam, Bangladesh, Myanmar, Nepal
Distribution (India): Manipur
Leaf Type: Compound, imparipinnate, opposite
Inflorescence: Fasciled racemes in terminal panicles
Fruit Type: Dry, pods
IUCN Status: Data Deficient
Notes: Frequently found in open and exposed areas and also in forest edges
Reference: J. D. Hooker, Fl. Brit. India 2: 109 (1876)

Millettia glaucescens Kurz
Fabaceae
Synonyms: *Millettia prainii* Dunn, *Phaseoloides glaucescens* Kuntze
Climbing Mechanism: Scrambler-Unarmed
Distribution (Global): Assam, East Himalaya, Myanmar, Nepal, Thailand.
Leaf Type: Leaves imparipinnate; stipels present or absent; leaflet blades opposite or rarely subalternate
Inflorescence: Inflorescence as a pseudoraceme, unbranched
Fruit Type: Legume 2-valved, flat, sometimes cylindric or inflated only around ellipsoid seeds
IUCN Status: Least Concern
Reference: J. Asiat. Soc. Bengal, Pt. 2, Nat. Hist. 42: 67 (1873)

Millettia orissae Panigrahi & S.C. Mishra
Fabaceae
Climbing Mechanism: Scrambler-Unarmed
Distribution (India): Maharashtra
Leaf Type: Compound, imparipinnate, opposite
Inflorescence: Fasciled racemes in terminal panicles
Fruit Type: Dry, pods
IUCN Status: Not evaluated

Millettia pachycarpa Benth.
Fabaceae
Synonyms: *Millettia dunnii* Merr., *Millettia fooningensis* Hu, *Millettia taiwaniana* (Hayata) Hayata, *Phaseoloides pachycarpum* (Benth.) Kuntze, *Pongamia taiwaniana* Hayata, *Whitfordiodendron taiwanianum* (Hayata) Ohwi
Climbing Mechanism: Scrambler-Unarmed
Distribution (Global): Assam, Bangladesh, China South-Central, China Southeast, East Himalaya, Laos, Myanmar, Nepal, Taiwan, Thailand, Tibet, Vietnam
Distribution (India): Arunachal Pradesh, Assam, Manipur, Meghalaya, Mizoram, Tripura, West Bengal
Leaf Type: Compound, imparipinnate, opposite

Inflorescence: Fasciled racemes in terminal panicles
Fruit Type: Dry, pods
IUCN Status: Least Concern
Notes: Frequently found in open and exposed areas and also in forest edges
Reference: F.A.W. Miquel, Pl. Jungh.: 250 (1852)

Millettia piscidia (Roxb.) Wight
Fabaceae
Synonyms: *Galedupa piscidia* Roxb., *Phaseoloides piscidia* Kuntze, *Pongamia piscidia* (Roxb.) Sweet
Climbing Mechanism: Scrambler-Unarmed
Distribution (Global): Assam, Bangladesh, East Himalaya, Myanmar, Thailand
Leaf Type: Compound, imparipinnate, opposite
Inflorescence: Fasciled racemes in terminal panicles
Fruit Type: Dry, pods
IUCN Status: Not evaluated
Reference: Prodr. Fl. Ind. Orient. 1: 263 (1834)

Millettia rubiginosa Wight & Arn.
Fabaceae
Synonyms: *Phaseoloides rubiginosum* Kuntze
Climbing Mechanism: Scrambler-Unarmed
Distribution (Global): India
Distribution (India): Kerala, Tamil Nadu, Western Ghats
Leaf Type: Leaves pinnately 3–5-foliolate; leaflets to 20 x 8 cm, ovate-oblong, acuminate at apex; minutely golden brown hairy
Inflorescence: Flowers in axillary simple or branched racemes
Fruit Type: Pods, obovate, compressed, woody, densely hairy
Flowering and Fruiting: October–November
IUCN Status: Not evaluated
Notes: Mostly occur in evergreen forests
Reference: Prodr. Fl. Ind. Orient. 1: 263 (1834)

Millettia sanjappae
Fabaceae
Climbing Mechanism: Scrambler-Unarmed
IUCN Status: Not evaluated

Millettia sericea Wight & Arn.
Fabaceae
Synonyms: *Dalbergia angustifolia* Hassk., *Dalbergia sericea* (Vent.) Spreng., *Millettia argentea* Miq., *Millettia heterophylla* Blume ex Miq., *Millettia obtusa* Blume ex Miq., *Millettia subcordata* Blume, *Millettia turgida* Miq., *Millettia zollingeriana* Miq.
Climbing Mechanism: Scrambler-Unarmed

Distribution (Global): Bangladesh, Borneo, Java, Malaya, Myanmar, Sumatera, Thailand, Vietnam
Leaf Type: Compound, imparipinnate, opposite
Inflorescence: Fasciled racemes in terminal panicles
Fruit Type: Dry, pods
IUCN Status: Not evaluated
Reference: F.A.W. Miquel, Pl. Jungh.: 250 (1852)

Millettia splendens Wight & Arn.
Fabaceae
Synonyms: *Galactia terminiflora* Blanco
Phaseoloides splendens Kuntze
Climbing Mechanism: Scrambler-Unarmed
Distribution (Global): India
Distribution (India): Kerala, Tamil Nadu, Western Ghats
Leaf Type: Leaves alternate, 7–9 pinnate, leaflets up to 17 x 6 cm, obovate, bluntly acuminate, cuneate, glabrous above
Inflorescence: Flowers pink, fascicled on long terminal racemes
Fruit Type: Pod flat, velvety, pubescent, up to 10 x 2 cm; endocarp not separable
Flowering and Fruiting: April–September
IUCN Status: Not evaluated
Notes: Mostly occur in evergreen forests
Reference: Prodr. Fl. Ind. Orient. 1: 263 (1834)

Mimosa depauperata Benth.
Fabaceae
Synonyms: *Acacia canescens* M. Martens & Galeotti
Climbing Mechanism: Scrambler-Armed
Distribution (Global): Mexico Central, Mexico Gulf, Mexico Northeast
Leaf Type: Compound, even-pinnate, alternate
Inflorescence: Flowers heads in racemes or cluster
Fruit Type: Dry, pods
IUCN Status: Not evaluated
Reference: Pl. Hartw.: 13 (1839)

Mimosa diplotricha C. Wright
Fabaceae
Common Name: Giant sensitive plant, Giant false sensitive plant, Nila grass
Climbing Mechanism: Scrambler-Armed
Distribution (Global): Argentina Northeast, Bolivia, Brazil North, Brazil Northeast, Brazil South, Brazil Southeast, Brazil West-Central, Colombia, Costa Rica, Cuba, Ecuador, Haiti, Honduras, Jamaica, Mexico Central, Mexico Gulf, Mexico Northeast, Mexico Northwest, Mexico Southeast, Mexico Southwest, Nicaragua, Panamá, Paraguay, Peru, Puerto Rico, Trinidad-Tobago, Venezuela
Distribution (India): Kerala, Tamil Nadu

Leaf Type: Leaves are up to 10 cm long; sidestalks opposite
Inflorescence: Flower-head are up to 1.5 cm across, in branch-end racemes, stalked
Fruit Type: Pods 2–5 cm long, bristly, black when mature
Flowering and Fruiting: November–March
IUCN Status: Not evaluated
Notes: Often cultivated
Reference: Anales Acad. Ci. Méd. Habana 5: 405 (1869)

Mimosa diplotricha var. *inermis* (Adelb.) M.K. Alam & M. Yusof
Fabaceae
Synonyms: *Mimosa diplotricha* var. *diplotricha*., *Mimosa diplotricha* var. *inermis* (Adelb.) Veldkamp, *Mimosa invisa* Mart., *Mimosa ixiamensis* Rusby, *Mimosa longisiliqua* Vell., *Morongia pilosa* Standl., *Schrankia brachycarpa* Benth., *Schrankia heterocarpa* Standl.
Climbing Mechanism: Scrambler-Armed
Distribution (Global): Argentina Northeast, Bolivia, Brazil North, Brazil Northeast, Brazil South, Brazil Southeast, Brazil West-Central, Colombia, Costa Rica, Cuba, Ecuador, Haiti, Honduras, Jamaica, Mexico Central, Mexico Gulf, Mexico Northeast, Mexico Northwest, Mexico Southeast, Mexico Southwest, Nicaragua, Panamá, Paraguay, Peru, Puerto Rico, Trinidad-Tobago, Venezuela
Distribution (India): Peninsular India
Leaf Type: Leaves to 10 cm long; pinnae opposite, to 5 cm long; leaflets to 35 pairs
Inflorescence: Head to 1.5 cm across, in terminal racemes, peduncled
Fruit Type: Pods 2–5 cm long, bristly, black when mature
Flowering and Fruiting: November–March
IUCN Status: Not evaluated
Notes: Found as a weed in degraded forests

Mimosa hamata Willd.
Fabaceae
Synonyms: *Mimosa armata* Rottler ex Spreng.
Common Name: Hooked Mimosa
Climbing Mechanism: Scrambler-Armed
Distribution (Global): Andaman Is., India, Pakistan
Distribution (India): Delhi
Leaf Type: Leaf bipinnate. Leaflets 6–10 pairs, more or less stalkles
Inflorescence: Flowers in spherical heads, solitary or paired in upper axils, peduncle 1.5–3.5 cm long
Fruit Type: Pods/Legumes
Flowering and Fruiting: April–August
IUCN Status: Not evaluated
Reference: Sp. Pl., ed. 4, 4: 1033 (1806)

Mimosa himalayana Gamble
Fabaceae

Synonyms: *Mimosa rubicaulis* subsp. *himalayana* (Gamble) H. Ohashi
Common Name: Himalayan Mimosa
Climbing Mechanism: Scrambler-Armed
Distribution (Global): Assam, Bangladesh, East Himalaya, India, Nepal, Pakistan, West Himalaya
Distribution (India): Madhya Pradesh, Rajasthan, West Bengal
Leaf Type: Leaf double compound, axis 10–23 cm long, prickly, prickles mostly hooked, hairy, ribbed; pinnae 5–12 pairs
Inflorescence: Inflorescence spherical pedunculate head forming at branch-ends panicle
Fruit Type: Pods
Flowering and Fruiting: June–August
IUCN Status: Not evaluated
Reference: Enum. Fl. Pl. Nepal 2: 126 (1979)

Mimosa polyancistra Benth.
Fabaceae
Synonyms: *Mimosa adenantheroides* (M. Martens & Galeotti) Benth., *Acacia adenantheroides* M. Martens & Galeotti, *Acacia cylindriflora* M. Martens & Galeotti, *Mimosa gomezii* Britton & Rose, *Mimosa hystricosa* Brandegee, *Mimosa remota* Benth.
Climbing Mechanism: Scrambler-Armed
Distribution (Global): Honduras, Mexico Central, Mexico Southeast, Mexico Southwest
Distribution (India): Andhra Pradesh, Eastern Ghats, Tamil Nadu
Leaf Type: Compound, even-pinnate, alternate
Inflorescence: Flowers heads in racemes or cluster
Fruit Type: Dry, pods
IUCN Status: Least Concern
Reference: London J. Bot. 5: 88 (1846)

Mimosa prainiana Gamble
Fabaceae
Climbing Mechanism: Scrambler-Armed
Distribution (India): Andhra Pradesh, Maharashtra
Leaf Type: Compound, even-pinnate, alternate
Inflorescence: Flowers heads in racemes or cluster
Fruit Type: Dry, pods
Flowering and Fruiting: July–October
IUCN Status: Not evaluated
Notes: Mostly found in deciduous forests
Reference: Fl. Madras: 421 (1919)

Mimosa rubicaulis Lam.
Fabaceae

Climbing Mechanism: Scrambler-Armed
Distribution (India): Delhi, Himachal Pradesh, Maharashtra
Leaf Type: Compound, even-pinnate, alternate
Inflorescence: Flowers heads in racemes or cluster
Fruit Type: Dry, pods
IUCN Status: Not evaluated

Moullava spicata (Dalzell) Nicolson
Fabaceae
Climbing Mechanism: Scrambler-Armed
Distribution (India): Goa, Gujarat, Karnataka, Kerala, Maharashtra, Rajasthan, Tamil Nadu, Western Ghats
Leaf Type: Compound, leaflets 5–6 pairs
Inflorescence: Racemes panicled
Fruit Type: Dry, pods
IUCN Status: Not evaluated

Mucuna biplicata Kurz
Fabaceae
Climbing Mechanism: Stem Twiner
Leaf Type: Compound, 3-foliolate
Inflorescence: Racemes, flowers clustered on nodes
Fruit Type: Dry, pods
IUCN Status: Not evaluated

Mucuna bracteata DC.
Fabaceae
Climbing Mechanism: Stem Twiner
Distribution (India): Manipur, Tripura
Leaf Type: Compound, 3-foliolate
Inflorescence: Racemes, flowers clustered on nodes
Fruit Type: Dry, pods
IUCN Status: Least Concern

Mucuna gigantea (Willd.) DC.
Fabaceae
Climbing Mechanism: Stem Twiner
Distribution (India): Andaman and Nicobar Islands, Andhra Pradesh, Eastern Ghats, Kerala, Maharashtra, Odisha, Tamil Nadu, West Bengal
Leaf Type: Compound, 3-foliolate
Inflorescence: Racemes, flowers clustered on nodes
Fruit Type: Dry, pods
IUCN Status: Not evaluated

Mucuna imbricata Baker
Fabaceae
Climbing Mechanism: Stem Twiner
Distribution (India): Andaman and Nicobar Islands, Assam, Bihar, Odisha, Meghalaya, Mizoram
Leaf Type: Compound, 3-foliolate
Inflorescence: Racemes, flowers clustered on nodes
Fruit Type: Dry, pods
IUCN Status: Least Concern

Mucuna laticifera Ingalhalikar, N. V. Page & Gaikwad
Fabaceae
Climbing Mechanism: Stem Twiner
Distribution (India): Sikkim
Leaf Type: Compound, 3-foliolate
Inflorescence: Racemes, flowers clustered on nodes
Fruit Type: Dry, pods
IUCN Status: Not evaluated

Mucuna macrocarpa Wall.
Fabaceae
Climbing Mechanism: Stem Twiner
Distribution (India): Arunachal Pradesh, Meghalaya, Sikkim, West Bengal
Leaf Type: Compound, 3-foliolate
Inflorescence: Racemes, flowers clustered on nodes
Fruit Type: Dry, pods
IUCN Status: Not evaluated

Mucuna monosperma Wight
Fabaceae
Climbing Mechanism: Stem Twiner
Distribution (India): Andaman and Nicobar Islands, Andhra Pradesh, Bihar, Odisha, Eastern Ghats, Goa, Karnataka, Kerala, Madhya Pradesh, Maharashtra, Mizoram, Odisha, Tamil Nadu
Leaf Type: Compound, 3-foliolate
Inflorescence: Racemes, flowers clustered on nodes
Fruit Type: Dry, pods
IUCN Status: Not evaluated

Mucuna nigricans (Lour.) Steud.
Fabaceae
Climbing Mechanism: Stem Twiner
Distribution (India): Andhra Pradesh, Arunachal Pradesh, Eastern Ghats, Himachal Pradesh, Madhya Pradesh, Odisha, West Bengal
Leaf Type: Compound, 3-foliolate

Inflorescence: Racemes, flowers clustered on nodes
Fruit Type: Dry, pods
IUCN Status: Not evaluated

Mucuna pruriens (L.) DC.
Fabaceae
Climbing Mechanism: Stem Twiner
Distribution (India): Andaman and Nicobar Islands, Andhra Pradesh, Arunachal Pradesh, Assam, Bihar, Odisha, Delhi, Goa, Gujarat, Himachal Pradesh, Jharkhand, Karnataka, Kerala, Madhya Pradesh, Maharashtra, Meghalaya, Mizoram, Odisha
Leaf Type: Compound, 3-foliolate
Inflorescence: Racemes, flowers clustered on nodes
Fruit Type: Dry, pods
IUCN Status: Least Concern

Mucuna pruriens var. *hirsuta* (Wight & Arn.) Wilmot-Dear
Fabaceae
Climbing Mechanism: Stem Twiner
Distribution (India): Tamil Nadu, Telangana
Leaf Type: Compound, 3-foliolate
Inflorescence: Racemes, flowers clustered on nodes
Fruit Type: Dry, pods
IUCN Status: Not evaluated

Mucuna pruriens var. *thekkadiensis*
Fabaceae
Climbing Mechanism: Stem Twiner
Distribution (India): Peninsular India
Leaf Type: Compound, 3-foliolate
Inflorescence: Racemes, flowers clustered on nodes
Fruit Type: Dry, pods
IUCN Status: Not evaluated

Mucuna pruriens var. *utilis* (Wall. ex Wight) L.H. Bailey
Fabaceae
Climbing Mechanism: Stem Twiner
Distribution (India): Telangana
Leaf Type: Compound, 3-foliolate
Inflorescence: Racemes, flowers clustered on nodes
Fruit Type: Dry, pods
IUCN Status: Not evaluated

Mucuna sanjappae Aitawade & S.R. Yadav
Fabaceae

Climbing Mechanism: Stem Twiner
Distribution (India): Maharashtra
Leaf Type: Compound, 3-foliolate
Inflorescence: Racemes, flowers clustered on nodes
Fruit Type: Dry, pods
IUCN Status: Not evaluated

Mucuna sempervirens Hemsl.
Fabaceae
Climbing Mechanism: Stem Twiner
Distribution (India): West Bengal
Leaf Type: Compound, 3-foliolate
Inflorescence: Racemes, flowers clustered on nodes
Fruit Type: Dry, pods
IUCN Status: Least Concern

Mucuna yadaviana Gaikwad, Lawand & Gurav
Fabaceae
Climbing Mechanism: Stem Twiner
Distribution (India): Andaman and Nicobar Islands
Leaf Type: Compound, 3-foliolate
Inflorescence: Racemes, flowers clustered on nodes
Fruit Type: Dry, pods
IUCN Status: Least Concern

Neonotonia wightii (Wight & Arn.) J.A. Lackey
Fabaceae
Synonyms: Glycine wightii (Wight & Arn.) Verdc.
Common Name: Perennial soyabean
Climbing Mechanism: Stem Twiner
Distribution (Global): Tropical & S. Africa, SW. Arabian Peninsula, India, Sri Lanka
Distribution (India): Kerala, Madhya Pradesh, Maharashtra, Tamil Nadu.
Leaf Type: Leaves 3 foliate
Inflorescence: Raceme
Fruit Type: Pods
Flowering and Fruiting: November–December
IUCN Status: Least Concern
Notes: Distributed mostly in semi-evergreen and moist deciduous forests
Reference: Phytologia 37: 210 (1977)

Neptunia oleracea Lour.
Fabaceae
Synonyms: *Neptunia prostrata* [Lam]Baill, *Desmanthus natans* Wild, *Mimosa natans* Roxb.
Common Name: Sensitive water plant

Climbing Mechanism: Stem Twiner
Flowering and Fruiting: January–November
IUCN Status: Least Concern
Reference: Fl. Cochinch.: 654 (1790)

Nesphostylis bracteata (Baker) D. Potter & J.J. Doyle
Fabaceae
Synonyms: *Dolichos ghaticus* Sant and Panthaki
Climbing Mechanism: Stem Twiner
Leaf Type: Trifoliate
Inflorescence: Flowers are solitary in leaf axils
Fruit Type: Pods
IUCN Status: Not evaluated
Reference: Syst. Bot. 19: 401 (1994)

Nesphostylis lanceolata (Baker) H. Ohashi & Tateishi
Fabaceae
Synonyms: *Phaseolus repens* Graham, *Vigna repens* Baker.
Climbing Mechanism: Stem Twiner
Distribution (Global): Myanmar
IUCN Status: Not evaluated
Reference: Bot. Mag. (Tokyo) 90: 150 (1977)

Nogra dalzellii (Baker) Merr.
Fabaceae
Synonyms: *Galactia simplicifolia* Dalzell, *Grona dalzellii* Baker, *Grona simplicifolia* Raizada, *Nogra simplicifolia* (Raizada) Raizada
Climbing Mechanism: Stem Twiner
Distribution (Global): India
Distribution (India): Madhya Pradesh, Maharashtra
IUCN Status: Not evaluated
Reference: Trans. Amer. Philos. Soc., n.s., 24(2): 201 (1935)

Nogra filicaulis
Fabaceae
Climbing Mechanism: Stem Twiner
IUCN Status: Not evaluated

Nogra grahamii (Benth.) Merr.
Fabaceae
Synonyms: *Apios esquirolii* (H. Lév.) H. Lév., *Glycine grahamii* Wall. ex Benth., *Grona grahamii* (Wall. ex Benth.) Benth., *Uraria esquirolii* H. Lév.
Climbing Mechanism: Stem Twiner
Distribution (Global): China South-Central, Laos, Thailand
Distribution (India): Madhya Pradesh, Odisha

Flowering and Fruiting: August–September
IUCN Status: Not evaluated
Reference: Trans. Amer. Philos. Soc., n.s., 24(2): 201 (1935)

Ophrestia pentaphylla (Dalzell) Verdc.
Fabaceae
Synonyms: *Glycine pentaphylla* Dalzell, *Paraglycine pentaphylla* (Dalzell) F.J. Herm.
Climbing Mechanism: Stem Twiner
Distribution (Global): India
Distribution (India): Kerala, Maharashtra
Flowering and Fruiting: September–February
IUCN Status: Not evaluated
Reference: Kew Bull. 24: 259 (1970)

Pachyrhizus erosus (L.) Urb.
Fabaceae
Synonyms: *Cacara erosa* (L.) Kuntze, *Cacara palmatiloba* (DC.) Kuntze, *Dolichos articulatus* Lam.+19
Common Name: Yam bean
Climbing Mechanism: Stem Twiner
Distribution (Global): Mexico to Central America
Distribution (India): Madhya Pradesh, Maharashtra, Odisha, West Bengal
IUCN Status: Not evaluated
Reference: Symb. Antill. 4: 311 (1905)

Paracalyx scariosus (Roxb.) Ali
Fabaceae
Synonyms: *Cylista scariosa* Roxb.
Common Name: Ran ghevada
Climbing Mechanism: Stem Twiner
Distribution (Global): Bangladesh, India, Myanmar, Nepal, Thailand.
Distribution (India): Andhra Pradesh, Eastern Ghats, Kerala, Madhya Pradesh, Maharashtra, Odisha, Telangana, West Bengal
Leaf Type: Leaves 3-foliate
Inflorescence: Raceme
Flowering and Fruiting: January–May
IUCN Status: Not evaluated
Notes: Usually found in moist and dry deciduous forests
Reference: Univ. Stud. (Karachi) 5: 95 (1968)

Parochetus communis D. Don
Fabaceae
Synonyms: *Cosmiusa repens* Alef., *Parochetus major* Buch. -Ham. ex D. Don, *Parochetus maculata* Benn., *Parochetus oxalidifolius* Royle.

Common Name: Blue Oxalis
Climbing Mechanism: Stem Twiner
Distribution (Global): Assam, Bangladesh, China South-Central, East Himalaya, India, Java, Lesser Sunda Is., Myanmar, Nepal, Pakistan, Sri Lanka, Thailand, Tibet, Vietnam, West Himalaya
Distribution (India): Manipur, Tamil Nadu
Flowering and Fruiting: December–March
IUCN Status: Least Concern
Notes: Often found in grasslands
Reference: Prodr. Fl. Nepal.: 240 (1825)

Phanera murthyi Vadhyar & J.H.F. Benj
Fabaceae
Climbing Mechanism: Hook Climber
Distribution (India): Kerala
IUCN Status: Not evaluated
Reference: Phytotaxa 401: 140 (2019)

Phaseolodes canum (Benth.) Kuntze
Fabaceae
Climbing Mechanism: Scrambler-Unarmed
IUCN Status: Not evaluated

Phaseolus coccineus L.
Fabaceae
Synonyms: *Lipusa formosa* (Kunth) Alef., *Lipusa multiflora* Alef., *Phaseolus bicolor* Vell., *Phaseolus coccineus* Moc. & Sessé ex G. Don. + 31
Common Name: Scarlet runner
Climbing Mechanism: Stem Twiner
Distribution (Global): Guatemala, Honduras, Mexico Central, Mexico Gulf, Mexico Northeast, Mexico Northwest, Mexico Southeast, Mexico Southwest, Nicaragua, Panamá
Distribution (India): Himachal Pradesh, Tamil Nadu
Leaf Type: Trifoliate
Inflorescence: Raceme
Fruit Type: Pod
IUCN Status: Least Concern
Reference: Sp. Pl.: 724 (1753)

Phaseolus lunatus L.
Fabaceae
Synonyms: *Dolichos tonkinensis* Bui-quang-Chieu, *Phaseolus amazonicus* Benth., *Phaseolus bipunctatus* Jacq., *Phaseolus compressus* Zoll. & Moritzi +27
Common Name: Lima bean
Climbing Mechanism: Stem Twiner

Distribution (Global): Belize, Colombia, Costa Rica, El Salvador, Guatemala, Honduras, Mexican Pacific Is., Mexico Central, Mexico Gulf, Mexico Northeast, Mexico Northwest, Mexico Southeast, Mexico Southwest, Nicaragua, Panamá
Distribution (India): Bihar, Odisha, Delhi, Himachal Pradesh, Tamil Nadu, Telangana
IUCN Status: Least Concern
Reference: Sp. Pl.: 724 (1753)

Phaseolus pauciflorus Sessé & Moc. ex G. Don
Fabaceae
Synonyms: *Minkelersia biflora* Hemsl., *Minkelersia galactioides* M. Martens & Galeotti, *Phaseolus galactioides* (M. Martens & Galeotti) Maréchal, Mascherpa & Stainier, *Phaseolus lambertianus* D. Dietr.
Climbing Mechanism: Stem Twiner
Distribution (Global): Guatemala, Mexico Central, Mexico Northeast, Mexico Northwest, Mexico Southeast, Mexico Southwest
IUCN Status: Least Concern
Reference: Gen. Hist. 2: 356 (1832)

Phaseolus vulgaris L.
Fabaceae
Synonyms: *Phaseolus aborigineus* Burkart, *Phaseolus albus* Hoffmanns., *Phaseolus angulosus* Schübl. & G. Martens, *Phaseolus asparagioides* Schur. + 53
Common Name: Common bean, Kidney bean, French bean, Haricate bean
Climbing Mechanism: Stem Twiner
Distribution (Global): Costa Rica, Guatemala, Honduras, Mexico Central, Mexico Gulf, Mexico Northeast, Mexico Northwest, Mexico Southeast, Mexico Southwest, Nicaragua, Panamá
Distribution (India): Bihar, Odisha, Delhi, Himachal Pradesh, Madhya Pradesh, Rajasthan, Tamil Nadu
Flowering and Fruiting: October–February
IUCN Status: Least Concern
Reference: Sp. Pl.: 723 (1753)

Piptanthus nepalensis (Hook.) D.Don
Fabaceae
Synonyms: *Anagyris indica* Lindl., *Anagyris nepalensis* (Hook.) Graham, *Baptisia nepalensis* Hook., *Piptanthus bicolor* Craib, *Piptanthus bombycinus* C. Marquand. + 12
Common Name: Evergreen Laburnum
Climbing Mechanism: Scrambler-Unarmed
Distribution (Global): Assam, China North-Central, China South-Central, East Himalaya, India, Myanmar, Nepal, Tibet, West Himalaya
Leaf Type: Trifoliate

IUCN Status: Not evaluated
Reference: Brit. Fl. Gard. 3: t. 264 (1828)

Pisum sativum L.
Fabaceae
Synonyms: *Lathyrus oleraceus* Lam., *Pisum arvens* L., *Pisum vulgare* Jundz.
Common Name: Garden pea, Common pea, Pea
Climbing Mechanism: Stem Twiner
Distribution (India): Gujarat, Madhya Pradesh, Maharashtra, Odisha, Rajasthan, Telangana
Flowering and Fruiting: November–February
IUCN Status: Least Concern
Reference: Sp. Pl.: 727 (1753)

Pseudarthria viscida (L.) Wight & Arn.
Fabaceae
Synonyms: *Desmodium leschenaultii* DC., *Desmodium peripterum* Zipp. ex Miq., *Desmodium timoriense* DC., *Desmodium viscidum* (L.) DC., *Glycine viscida* (L.) Willd., *Hedysarum prostratum* Roxb. ex Wight & Arn., *Hedysarum timoriense* (DC.) Spreng. + 6
Common Name: Sticky pod weed
Climbing Mechanism: Stem Twiner
Distribution (Global): Bangladesh, India, Java, Lesser Sunda Is., Maluku, Myanmar, Philippines, Sri Lanka, Sulawesi
Distribution (India): Karnataka, Madhya Pradesh, Maharashtra, Puducherry, Tamil Nadu, Telangana
Leaf Type: Trifoliate
Inflorescence: Axillary or terminal
Fruit Type: Pods
Flowering and Fruiting: August–February
IUCN Status: Not evaluated
Notes: Occurs mostly in moist deciduous forests
Reference: Prodr. Fl. Ind. Orient. 1: 209 (1834)

Psophocarpus scandens (Endl.) Verdc.
Fabaceae
Synonyms: *Diesingia scandens* Endl., *Dolichos subaequalis* Graham, *Dolichos suffultus* Graham, *Mucuna comorensis* Vatke, *Psophocarpus comorensis* (Vatke) Baill. + 4
Climbing Mechanism: Stem Twiner
Distribution (Global): Angola, Burundi, Cameroon, Central African Repu, Comoros, Congo, Gabon, Gulf of Guinea Is., Kenya, Madagascar, Malawi, Mauritius, Mozambique, Sudan, Tanzania, Uganda, Zambia, Zaire
Distribution (India): Peninsular India

IUCN Status: Not evaluated
Reference: Taxon 17: 539 (1968)

Psophocarpus tetragonolobus (L.) DC.
Fabaceae
Synonyms: *Botor tetragonoloba* (L.) Kuntze, *Dolichos ovatus* Graham, *Dolichos tetragonolobus* L.
Common Name: Malabar kino, Goa bean, Asparagus bean, Winged bean
Climbing Mechanism: Stem Twiner
Distribution (Global): Tanzania
Distribution (India): Madhya Pradesh, Maharashtra, Manipur, Tamil Nadu, Tripura
Flowering and Fruiting: February–October
IUCN Status: Not evaluated
Reference: Prodr. 2: 403 (1825)

Pterolobium hexapetalum (Roth) Santapau & Wagh
Fabaceae
Synonyms: *Caesalpinia lacerans* Roxb., *Cantuffa hexapetala* (Roth) Kuntze, *Cantuffa indica* (A. Rich.) Kuntze, *Cantuffa lacerans* (Roxb.) Taub., *Pterolobium indicum* A. Rich., *Pterolobium insigne* Blume ex Miq., *Pterolobium lacerans* (Roxb.) Wight & Arn.
Common Name: White brasiletto climber, Climber siege
Climbing Mechanism: Scrambler-Armed
Distribution (Global): Bangladesh, East Himalaya, India, Myanmar
Distribution (India): Andhra Pradesh, Karnataka, Kerala, Puducherry, Tamil Nadu, Telangana, Western Ghats
Leaf Type: Bipinnate
Inflorescence: Racemes axillary or terminal
Fruit Type: Samaroid
Flowering and Fruiting: March–November
IUCN Status: Not evaluated
Notes: Mostly found in foothills of scrub jungles and deciduous forests
Reference: Bull. Bot. Surv. India 5: 108 (1964)

Pterolobium macropterum Kurz
Fabaceae
Synonyms: *Pterolobium sinense* J.E. Vidal
Climbing Mechanism: Scrambler-Armed
Distribution (Global): Andaman Is., China South-Central, China Southeast, East Himalaya, Hainan, Java, Laos, Myanmar, Thailand, Vietnam
Distribution (India): Andaman and Nicobar Islands
IUCN Status: Not evaluated
Reference: J. Asiat. Soc. Bengal, Pt. 2, Nat. Hist. 42: 71 (1873)

Pterolobium punctatum Hemsl.
Fabaceae
Synonyms: *Caesalpinia aestivalis* Chun & F.C. How, *Prosopis esquirolii* H. Lév., *Cantuffa punctata* (Hemsl.) Kuntze
Climbing Mechanism: Scrambler-Armed
Distribution (Global): China South-Central, China Southeast, Hainan
IUCN Status: Not evaluated
Reference: J. Linn. Soc., Bot. 23: 207 (1887)

Pueraria bella Prain
Fabaceae
Climbing Mechanism: Stem Twiner
Distribution (Global): East Himalaya, Myanmar
Distribution (India): Arunachal Pradesh
Leaf Type: Compound, 3 lobed, entire
Inflorescence: Racemes, simple, or branched
Fruit Type: Dry, pods
IUCN Status: Not evaluated
Reference: J. Asiat. Soc. Bengal, Pt. 2, Nat. Hist. 67: 288 (1898)

Pueraria edulis Pamp.
Fabaceae
Synonyms: *Pueraria bicalcarata* Gagnep., *Pueraria edulis* var. *likiangensis* P.C. Li, *Pueraria quadristipellata* C.B. Clarke ex W.W.Sm.
Climbing Mechanism: Stem Twiner
Distribution (Global): Assam, China South-Central, East Himalaya
Distribution (India): West Bengal
Leaf Type: Compound, 3 lobed, entire
Inflorescence: Racemes, simple, or branched
Fruit Type: Dry, pods
IUCN Status: Not evaluated
Reference: Nuovo Giorn. Bot. Ital., n.s., 17: 28 (1910)

Pueraria montana (Lour.) Merr.
Fabaceae
Synonyms: *Dolichos montanus* Lour., *Pueraria lobata* var. *montana* (Lour.) Maesen.
Climbing Mechanism: Stem Twiner
Distribution (Global): Assam, Bangladesh, Cambodia, Caroline Is., China North-Central, China South-Central, China Southeast, East Himalaya, Hainan, India, Inner Mongolia, Japan, Java, Korea, Laos, Malaya, Maluku, Manchuria, Myanmar, Nansei-shoto, Nepal, New Guinea, Northern Territory, Ogasawara-shoto, Philippines, Primorye, Solomon Is., Taiwan, Thailand, Tibet, Vietnam
Distribution (India): Assam, Tamil Nadu, West Bengal
Leaf Type: Compound, 3 lobed, entire

Inflorescence: Racemes, simple, or branched
Fruit Type: Dry, pods
IUCN Status: Not evaluated
Reference: Trans. Amer. Philos. Soc., n.s., 24(2): 10, 210 (1935)

Pueraria montana var. *chinensis* (Ohwi) Sanjappa & Pradeep
Fabaceae
Synonyms: *Pueraria montana* var. *lobata* (Willd.) Maesen & S.M. Almeida ex Sanjappa & Predeep, *Dolichos grandiflorus* Wall., *Dolichos grandifolius* Wall, *Dolichos trilobus* Lour., *Pachyrhizus trilobus* DC., *Pueraria chinensis* Ohwi. + 6
Climbing Mechanism: Stem Twiner
Distribution (India): Tripura
Leaf Type: Compound, 3 lobed, entire
Inflorescence: Racemes, simple, or branched
Fruit Type: Dry, pods
IUCN Status: Not evaluated
Reference: M. Sanjappa, Legumes India: 288 (1992)

Pueraria montana var. *lobata* (Willd.) Sanjappa & Pradeep
Fabaceae
Climbing Mechanism: Stem Twiner
Distribution (India): Maharastra
Leaf Type: Compound, 3 lobed, entire
Inflorescence: Racemes, simple or branched
Fruit Type: Dry, pods
IUCN Status: Not evaluated

Pueraria peduncularis (Benth.) Benth.
Fabaceae
Synonyms: *Toxicopueraria peduncularis* (Benth.) A.N. Egan & B. Pan, *Derris bonatiana* Pamp., *Neustanthus peduncularis* Benth., *Pueraria peduncularis* var. *violacea* Franch., *Pueraria yunnanensis* Franch.
Climbing Mechanism: Stem Twiner
Distribution (India): Arunachal Pradesh, West Bengal
Leaf Type: Compound, 3 lobed, entire
Inflorescence: Racemes, simple, or branched
Fruit Type: Dry, pods
IUCN Status: Not evaluated
Reference: J. Linn. Soc., Bot. 9: 124 (1867)

Pueraria phaseoloides (Roxb.) Benth.
Fabaceae
Synonyms: *Dolichos phaseoloides* Roxb., *Neustanthus phaseoloides* Benth.
Climbing Mechanism: Stem Twiner

Distribution (India): Arunachal Pradesh, Kerala, Madhya Pradesh, Tamil Nadu, Tripura, West Bengal
Leaf Type: Compound, 3 lobed, entire
Inflorescence: Racemes, simple, or branched
Fruit Type: Dry, pods
Flowering and Fruiting: October–March
IUCN Status: Not evaluated
Notes: Frequently occurs along the margins of cultivated lands
Reference: J. Linn. Soc., Bot. 9: 125 (1867)

Pueraria phaseoloides var. *subspicata* (Benth.) Maesen
Fabaceae
Climbing Mechanism: Stem Twiner
Distribution (India): West Bengal
Leaf Type: Compound, 3 lobed, entire
Inflorescence: Racemes, simple, or branched
Fruit Type: Dry, pods
IUCN Status: Not evaluated

Pueraria sikkimensis Prain
Fabaceae
Climbing Mechanism: Stem Twiner
Distribution (Global): Assam, East Himalaya, Nepal, West Himalaya
Distribution (India): North Eastern India, West Bengal
Leaf Type: Compound, 3 lobed, entire
Inflorescence: Racemes, simple, or branched
Fruit Type: Dry, pods
IUCN Status: Not evaluated
Reference: J. Asiat. Soc. Bengal, Pt. 2, Nat. Hist. 66: 419 (1897)

Pueraria tuberosa (Willd.) DC.
Fabaceae
Synonyms: *Desmodium tuberosum* (Roxb. ex Willd.) G. Don, *Hedysarum tuberosum* Roxb. ex Willd.
Common Name: Tuberous Honey Suckle, Indian kudzu
Climbing Mechanism: Stem Twiner
Distribution (Global): Assam, Bangladesh, East Himalaya, India, Nepal, Pakistan, West Himalaya
Distribution (India): Andaman and Nicobar Islands, Andhra Pradesh, Bihar, Odisha, Eastern Ghats, Goa, Gujarat, Himachal Pradesh, Jharkhand, Karnataka, Kerala, Madhya Pradesh, Maharashtra, Odisha, Rajasthan, Tamil Nadu, Telangana
Leaf Type: Trifoliate
Inflorescence: Racemes axillary
Fruit Type: Pods
Flowering and Fruiting: December–June

IUCN Status: Not evaluated
Notes: Found mostly in dry and moist deciduous forests
Reference: Ann. Sci. Nat. (Paris) 4: 97 (1825)

Pueraria wallichii DC.
Fabaceae
Synonyms: *Dolichos frutescens* D. Don
Climbing Mechanism: Stem Twiner
Distribution (India): Meghalaya, Mizoram, West Bengal.
Leaf Type: Compound, 3 lobed, entire
Inflorescence: Racemes, simple, or branched
Fruit Type: Dry, pods
IUCN Status: Not evaluated
Reference: Ann. Sci. Nat. (Paris) 4: 97 (1825)

Rhynchosia acutissima Thwaites
Fabaceae
Climbing Mechanism: Stem Twiner
Distribution (Global): Assam, East Himalaya, India
Distribution (India): Kerala
Leaf Type: Compound, 3-foliolate
Inflorescence: Racemes, flowers 1 or in pairs
Fruit Type: Dry, pods
Flowering and Fruiting: November–April
IUCN Status: Not evaluated
Notes: Usually found in evergreen and semi-evergreen forests
Reference: Enum. Pl. Zeyl.: 413 (1864)

Rhynchosia aurea (Willd.) DC.
Fabaceae
Synonyms: *Dolichos rufescens* Graham, *Glycine aurea* Willd., *Glycine elongata* Roth, *Hedysarum aureum* (Willd.) Rottler, *Nomismia aurea* (Willd.) Wight & Arn., *Rhynchosia densiflora* Wall., *Rhynchosia elongata* DC.
Common Name: Golden-flowered snoutbean
Climbing Mechanism: Stem Twiner
Distribution (Global): Assam, Bangladesh, India, Sri Lanka, West Himalaya
Distribution (India): Andhra Pradesh, Himachal Pradesh, Tamil Nadu
Leaf Type: Compound, 3-foliolate
Inflorescence: Racemes, flowers 1 or in pairs
Fruit Type: Dry, pods
Flowering and Fruiting: February–March
IUCN Status: Not evaluated
Reference: Prodr. 2: 386 (1825)

Rhynchosia bracteata Baker
Fabaceae
Synonyms: *Atylosia candicans* Kurz, *Dolichos bracteatus* Wall., *Cajanus candicans* Wall., *Rhynchosia mollissima* Dalzell.
Climbing Mechanism: Stem Twiner
Distribution (Global): Bangladesh, Cambodia, India, Laos, Myanmar, Thailand
Distribution (India): Andhra Pradesh, Bihar, Odisha, Eastern Ghats, Gujarat, Madhya Pradesh, Maharashtra, Odisha, Rajasthan, West Bengal
Leaf Type: Compound, 3-foliolate
Inflorescence: Racemes, flowers 1 or in pairs
Fruit Type: Dry, pods
Flowering and Fruiting: August–January
IUCN Status: Not evaluated
Notes: Occurs mostly in evergreen forests
Reference: J. D. Hooker, Fl. Brit. India 2: 225 (1876)

Rhynchosia capitata (Roth) DC.
Fabaceae
Synonyms: *Glycine capitata* B. Heyne ex Roth, *Nomismia capitata* (B. Heyne ex Roth) Wight & Arn.
Common Name: Clustered-flower snoutbean
Climbing Mechanism: Stem Twiner
Distribution (Global): Assam, East Himalaya, India, Pakistan, Sri Lanka, West Himalaya
Distribution (India): Delhi, Kerala, Madhya Pradesh, Odisha, Rajasthan, Tamil Nadu
Leaf Type: Compound, 3-foliolate
Inflorescence: Racemes, flowers 1 or in pairs
Fruit Type: Dry, pods
Flowering and Fruiting: August–February
IUCN Status: Not evaluated
Notes: Frequently occur in open deciduous forests and scrub forests
Reference: Prodr. 2: 386 (1825)

Rhynchosia courtallensis Maesen
Fabaceae
Synonyms: *Dunbaria latifolia* Wight & Arn., *Dunbaria ferruginea* Sensu Baker.
Common Name: Courtallam snoutbean
Climbing Mechanism: Stem Twiner
Distribution (Global): India
Distribution (India): Kerala, Tamil Nadu
Leaf Type: Compound, 3-foliolate
Inflorescence: Racemes, flowers 1 or in pairs
Fruit Type: Dry, pods
Flowering and Fruiting: January–April
IUCN Status: Not evaluated

Notes: Mostly found in moist deciduous forests
Reference: Rheedea 5: 56 (1995)

Rhynchosia densiflora (Roth) DC.
Fabaceae
Synonyms: *Glycine densiflora* B. Heyne ex Roth
Common Name: Dense-flowered snoutbean
Climbing Mechanism: Stem Twiner
Distribution (Global): Angola, Benin, Botswana, Burkina, Cameroon, Central African Repu, Eritrea, Ethiopia, Gabon, Ghana, Guinea-Bissau, Gulf of Guinea Is., India, Ivory Coast, Kenya, KwaZulu-Natal, Liberia, Malawi, Mozambique, Myanmar, Namibia, Nigeria, Northern Provinces, Rwanda, Sierra Leone, Somalia, Sudan, Swaziland, Tanzania, Togo, Uganda, Zambia, Zaire, Zimbabwe
Distribution (India): Andhra Pradesh, Kerala, Tamil Nadu, West Bengal
Leaf Type: Compound, 3-foliolate
Inflorescence: Racemes, flowers 1 or in pairs
Fruit Type: Dry, pods
Flowering and Fruiting: December–March
IUCN Status: Not evaluated
Notes: Mostly found in dry and moist deciduous forests
Reference: Prodr. 2: 386 (1825)

Rhynchosia falconeri Baker
Fabaceae
Common Name: Falconer's snoutbean
Climbing Mechanism: Stem Twiner
Distribution (Global): India, Nepal, West Himalaya
Distribution (India): Himachal Pradesh
Leaf Type: Compound, 3-foliolate
Inflorescence: Racemes, flowers 1 or in pairs
Fruit Type: Dry, pods
IUCN Status: Least Concern
Reference: J. D. Hooker, Fl. Brit. India 2: 224 (1876)

Rhynchosia filipes Benth.
Fabaceae
Climbing Mechanism: Stem Twiner
Distribution (Global): India.
Distribution (India): Tamil Nadu
Leaf Type: Compound, 3-foliolate
Inflorescence: Racemes, flowers 1 or in pairs
Fruit Type: Dry, pods
Flowering and Fruiting: January–March
IUCN Status: Not evaluated
Reference: Linnaea 24: 644 (1852)

Rhynchosia heynei Wight & Arn.
Fabaceae
Synonyms: *Cylista reticulata* B. Heyne, *Rhynchosia codoorensis* Bedd.
Climbing Mechanism: Stem Twiner
Distribution (Global): India
Distribution (India): Tamil Nadu
Leaf Type: Compound, 3-foliolate
Inflorescence: Racemes, flowers 1 or in pairs
Fruit Type: Dry, pods
Flowering and Fruiting: November–April
IUCN Status: Vulnerable
Reference: Prodr. Fl. Ind. Orient. 1: 240 (1834)

Rhynchosia himalensis Baker
Fabaceae
Common Name: Himalayan snoutbean
Climbing Mechanism: Stem Twiner
Distribution (Global): Assam, China South-Central, Myanmar, Nepal, Pakistan, Tibet, West Himalaya
Distribution (India): Himachal Pradesh
Leaf Type: Compound, 3-foliolate
Inflorescence: Racemes, flowers 1 or in pairs
Fruit Type: Dry, pods
IUCN Status: Least Concern
Reference: J. D. Hooker, Fl. Brit. India 2: 225 (1876)

Rhynchosia hirta (Andrews) Meikle & Verdc.
Fabaceae
Synonyms: *Cyanospermum albiflorum* (Sims) Wight & Arn., *Cyanospermum tomentosum* (Roxb.) Wight & Arn., *Cylista albiflora* Sims, *Cylista tomentosa* Roxb., *Cylista villosa* W.T. Aiton, *Dolichos hirtus* Andrews, *Rhynchosia albiflora* (Sims) Alston, *Rhynchosia cyanosper*
Common Name: Hairy snoutbean
Climbing Mechanism: Stem Twiner
Distribution (Global): Angola, Benin, Burkina, Burundi, Cameroon, Central African Repu, Chad, Ethiopia, Gabon, Gulf of Guinea Is., India, Ivory Coast, Kenya, Malawi, Mozambique, Nigeria, Northern Provinces, Rwanda, Sri Lanka, Sudan, Tanzania, Togo, Uganda, Zambia, Zaire, Zimbabwe
Distribution (India): Andhra Pradesh, Eastern Ghats, Gujarat, Karnataka, Kerala, Maharashtra, Tamil Nadu
Leaf Type: Trifoliate
Inflorescence: Raceme
Fruit Type: Pods
Flowering and Fruiting: December–March
IUCN Status: Not evaluated

Notes: Often found in mixed deciduous forests
Reference: Taxon 16: 462 (1967)

Rhynchosia jacobii Chandrab. & B.V. Shetty
Fabaceae
Climbing Mechanism: Stem Twiner
Distribution (Global): India
Distribution (India): Tamil Nadu
Leaf Type: Compound, 3-foliolate
Inflorescence: Racemes, flowers 1 or in pairs
Fruit Type: Dry, pods
IUCN Status: Not evaluated
Notes: Usually found in evergreen forests
Reference: Bull. Bot. Surv. India 15: 139 (1973 publ. 1976)

Rhynchosia meeboldii P. Satyanar. & Thoth
Fabaceae
Climbing Mechanism: Stem Twiner
Distribution (Global): India
Leaf Type: Compound, 3-foliolate
Inflorescence: Racemes, flowers 1 or in pairs
Fruit Type: Dry, pods
IUCN Status: Not evaluated
Reference: Bull. Bot. Surv. India 28: 243 (1986 publ. 1988)

Rhynchosia minima (L.) DC.
Fabaceae
Synonyms: *Dolicholus minimus* (L.) Medik., *Dolichos minimus* L., *Rhynchosia minima* var. *normalis* Kuntze
Climbing Mechanism: Stem Twiner
Distribution (Global): Afghanistan, Algeria, Angola, Argentina Northeast, Argentina Northwest, Aruba, Assam, Bahamas, Bangladesh, Belize, Benin, Bolivia, Botswana, Brazil North, Brazil Northeast, Brazil South, Brazil Southeast, Brazil West-Central, Burkina, Burundi, Cameroon, Cape Provinces, Cape Verde, Cayman Is., Central African Repu, Chad, Chile North, China South-Central, Colombia, Costa Rica, Cuba, Dominican Republic, East Himalaya, Ecuador, Egypt, El Salvador, Eritrea, Ethiopia, Florida, Free State, French Guiana, Galápagos, Georgia, Ghana, Guatemala, Gulf of Guinea Is., Gulf States, Guyana, Haiti, Honduras, India, Ivory Coast, Jamaica, Java, Kazan-retto, Kenya, KwaZulu-Natal, Leeward Is., Lesser Sunda Is., Liberia, Louisiana, Madagascar, Malawi, Mali, Mauritania, Mexican Pacific Is., Mexico Central, Mexico Gulf, Mexico Northeast, Mexico Northwest, Mexico Southeast, Mexico Southwest, Morocco, Mozambique, Namibia, Nansei-shoto, Nepal, Netherlands Antilles, New Guinea, New South Wales, Nicaragua, Niger, Nigeria, Northern Provinces, Northern Territory, Oman, Pakistan, Palestine, Panamá, Paraguay, Peru,

Philippines, Puerto Rico, Queensland, Rodrigues, Rwanda, Saudi Arabia, Senegal, Sierra Leone, Sinai, Socotra, Somalia, South Australia, South China Sea, Southwest Caribbean, Sri Lanka, Sudan, Suriname, Swaziland, Taiwan, Tanzania, Texas, Thailand, Togo, Trinidad-Tobago, Turks-Caicos Is., Uganda, Venezuela, Venezuelan Antilles, Vietnam, West Himalaya, Western Australia, Windward Is., Yemen, Zambia, Zaire, Zimbabwe
Distribution (India): Andhra Pradesh, Bihar, Odisha, Delhi, Himachal Pradesh, Karnataka, Madhya Pradesh, Maharashtra, Rajasthan, Tamil Nadu, Uttar Pradesh, West Bengal
Leaf Type: Trifoliate
Flowering and Fruiting: November–February
IUCN Status: Least Concern
Notes: Mostly found in scrub jungles and dry deciduous forests
Reference: Prodr. 2: 385 (1825)

Rhynchosia nummularia (L.) DC.
Fabaceae
Synonyms: *Glycine nummularia* L., *Nomismia nummularia* Wight & Arn.
Climbing Mechanism: Stem Twiner
Distribution (Global): India, Sri Lanka, Vietnam
Leaf Type: Compound, 3-foliolate
Inflorescence: Racemes, flowers 1 or in pairs
Fruit Type: Dry, pods
IUCN Status: Not evaluated
Reference: Prodr. 2: 386 (1825)

Rhynchosia pilosa Wall.
Fabaceae
Synonyms: *Pseudarthria viscida* (L.) Wight & Arn.
Climbing Mechanism: Stem Twiner
Leaf Type: Compound, 3-foliolate
Inflorescence: Racemes, flowers 1 or in pairs
Fruit Type: Dry, pods
IUCN Status: Not evaluated
Reference: Numer. List: 190, n.° 5499 (1831)

Rhynchosia pulverulenta Stocks
Fabaceae
Synonyms: *Rhynchosia elachistantha* Chiov., *Rhynchosia rhombifolia* Blatt. & Hallb.
Climbing Mechanism: Stem Twiner
Distribution (Global): Eritrea, Ethiopia, India, Iran, Kenya, Oman, Pakistan, Saudi Arabia, Socotra, Somalia, Sudan, Tanzania, Yemen
Distribution (India): Rajasthan
Leaf Type: Compound, 3-foliolate

Inflorescence: Racemes, flowers 1 or in pairs
Fruit Type: Dry, pods
IUCN Status: Not evaluated
Reference: Hooker's J. Bot. Kew Gard. Misc. 4: 147 (1852)

Rhynchosia rothii Benth. ex Aitchison
Fabaceae
Synonyms: *Dolichos macrodon* Graham, *Dolichos tomentosus* Roth, *Rhynchosia mollissima* Zoll. & Moritzi, *Rhynchosia sericea* Span., *Rhynchosia tomentosa* Kurz.
Common Name: Silky snoutbean
Climbing Mechanism: Stem Twiner
Distribution (Global): Bangladesh, China Southeast, India, Java, Lesser Sunda Is., Myanmar, Nepal, Pakistan, Taiwan, Thailand, West Himalaya
Distribution (India): Andhra Pradesh, Eastern Ghats, Gujarat, Himachal Pradesh, Karnataka, Kerala, Madhya Pradesh, Maharashtra, Mizoram, Rajasthan, Tamil Nadu
Leaf Type: Alternate, trifoliate
Inflorescence: Raceme
Fruit Type: Pod
Flowering and Fruiting: October–February
IUCN Status: Not evaluated
Notes: Often found in open and exposed areas in semi-evergreen forests
Reference: Cat. Pl. Punjab Sindh: 50 (1869)

Rhynchosia rufescens (Willd.) DC.
Fabaceae
Synonyms: *Cyanospermum javanicum* Miq., *Cylista suaveolens* Graham, *Flemingia rothiana* DC., *Glycine monosperma* Willd. ex Spreng., *Glycine pondicheriensis* Spreng., *Glycine rufescens* Willd., *Hallia trifoliata* Roth., *Hedysarum rufescens* Willd. ex Rottl.
Common Name: Rusty snoutbean
Climbing Mechanism: Stem Twiner
Distribution (Global): Assam, Bangladesh, Cambodia, China South-Central, China Southeast, India, Java, Lesser Sunda Is., Myanmar, Sri Lanka, Thailand
Distribution (India): Bihar, Odisha, Karnataka, Madhya Pradesh, Maharashtra, Odisha, Tripura
Leaf Type: Trifoliate
Inflorescence: Raceme
Fruit Type: Pod
Flowering and Fruiting: October–February
IUCN Status: Not evaluated
Notes: Usually found in moist and dry-deciduous forests
Reference: Prodr. 2: 387 (1825)

Rhynchosia schimperi Boiss.
Fabaceae
Synonyms: *Rhynchosia arenaria* Blatt. & Hallb., *Rhynchosia cliffordii* Hutch. & E.A. Bruce.
Climbing Mechanism: Stem Twiner
Distribution (Global): Ethiopia, Gulf States, India, Iran, Oman, Pakistan, Somalia, Yemen
Distribution (India): Rajasthan
Leaf Type: Compound, 3-foliolate
Inflorescence: Racemes, flowers 1 or in pairs
Fruit Type: Dry, pods
IUCN Status: Not evaluated
Reference: Fl. Orient. 2: 626 (1872)

Rhynchosia suaveolens (L.f.) DC.
Fabaceae
Synonyms: *Cajanus suaveolens* (L.f.) Graham, *Hedysarum venosum* Rottler, *Glycine suaveolens* L.f.
Common Name: Sweet-scented snoutbean
Climbing Mechanism: Stem Twiner
Distribution (Global): India
Distribution (India): Telangana
Leaf Type: Trifoliate
Inflorescence: Axillary
Fruit Type: Pod
Flowering and Fruiting: August–March
IUCN Status: Not evaluated
Notes: Usually found in moist and dry-deciduous forests
Reference: Prodr. 2: 387 (1825)

Rhynchosia velutina Wight & Arn.
Fabaceae
Climbing Mechanism: Stem Twiner
Distribution (Global): Comoros, India, Kenya, Madagascar, Mozambique, Somalia, Sri Lanka, Tanzania
Distribution (India): Tamil Nadu
Leaf Type: Compound, 3-foliolate
Inflorescence: Racemes, flowers 1 or in pairs
Fruit Type: Dry, pods
IUCN Status: Not evaluated
Notes: Usually found in moist and dry-deciduous forests
Reference: Prodr. Fl. Ind. Orient. 1: 238 (1834)

Rhynchosia viscosa (Roth) DC.
Fabaceae

Synonyms: *Glycine viscosa* Roth
Climbing Mechanism: Stem Twiner
Distribution (Global): Angola, Assam, Benin, Cameroon, Central African Repu, China South-Central, Comoros, East Himalaya, Ethiopia, Ghana, Guinea-Bissau, India, Kenya, Madagascar, Malawi, Mauritius, Mozambique, Nepal, Nigeria, Sierra Leone, Somalia, Sri Lanka, Sudan, Tanzania, Togo, Uganda, West Himalaya, Zambia, Zaire
Distribution (India): Andhra Pradesh, Assam, Eastern Ghats, Gujarat, Himachal Pradesh, Karnataka, Kerala, Madhya Pradesh, Odisha, Tamil Nadu, Telangana, West Bengal
Leaf Type: Trifoliate
Fruit Type: Pod
Flowering and Fruiting: August–January
IUCN Status: Not evaluated
Notes: Usually found in moist and dry-deciduous forests
Reference: Prodr. 2: 387 (1825)

Shuteria hirsuta Baker
Fabaceae
Synonyms: *Harashuteria hirsuta* (Baker) K. Ohashi & H. Ohashi. , *Amphicarpaea linearis* Chun & H.Y.Chen., *Pueraria anabaptista* Kurz., *Shuteria anabaptista* C.Y.Wu., *Shuteria lancangensis* Y.Y.Qian
Climbing Mechanism: Stem Twiner
Distribution (Global): Assam, China South-Central, East Himalaya, Hainan, Laos, Myanmar, Nepal, Thailand, Vietnam
Distribution (India): Manipur, West Bengal
IUCN Status: Not evaluated
Reference: J. Jap. Bot. 92: 40 (2017)

Shuteria involucrata (Wall.) Wight & Arn.
Fabaceae
Synonyms: *Glycine involucrat*a Wall., *Shuteria suffulta* var. *sinensis* (Hemsl.) Niyomdham., *Shuteria sinensis* Hemsl.
Common Name: Clothed shuteria
Climbing Mechanism: Stem Twiner
Distribution (Global): Assam, Cambodia, China South-Central, China Southeast, East Himalaya, India, Myanmar, Nepal, Pakistan, Thailand, Vietnam, West Himalaya
Distribution (India): Andhra Pradesh, Arunachal Pradesh, Kerala, Manipur, Odisha, Tamil Nadu
Leaf Type: Leaves are pinnately 3-foliolate
Inflorescence: Raceme, axillary, red, purple or light purple colored
Fruit Type: Legumes/pods
Flowering and Fruiting: November-April.
IUCN Status: Not evaluated

Notes: Usually found in evergreen and semi-evergreen forests.
Reference: Repert. Bot. Syst. 1: 761 (1842)

Shuteria involucrata var. *glabrata* (Wight & Arn.) H. Ohashi
Fabaceae
Synonyms: *Shuteria vestita* Wight & Arn., *Shuteria involucrata* var. *vestita* (Wight & Arn.) H. Ohashi., *Shuteria involucrata* var. *villosa* (Pamp.) H. Ohashi. + 10
Climbing Mechanism: Stem Twiner
Distribution (Global): Assam, China South-Central, China Southeast, East Himalaya, Hainan, India, Lesser Sunda Is., Myanmar, Nepal, New Guinea, Sri Lanka, Sulawesi, Sumatera, Thailand, Vietnam, West Himalaya
Distribution (India): Peninsular India
IUCN Status: Not evaluated
Reference: Prodr. Fl. Ind. Orient. 1: 207 (1834)

Shuteria suffulta Benth.
Fabaceae
Synonyms: *Glycine suffulta* Wall., *Shuteria involucrata* f. suffulta (Benth.) H. Ohashi & K. Ohashi
Climbing Mechanism: Stem Twiner
Distribution (Global): Myanmar, Thailand
IUCN Status: Not evaluated
Reference: F.A.W. Miquel, Pl. Jungh.: 232 (1852)

Spatholobus acuminatus Benth.
Fabaceae
Synonyms: *Butea listeri* (Prain) Blatt., *Butea riparia* (Prain) Blatt. + 10
Climbing Mechanism: Stem Twiner
Distribution (Global): Andaman Is., Bangladesh, Cambodia, East Himalaya, India, Laos, Malaya, Myanmar, Thailand, Vietnam.
Distribution (India): Andaman and Nicobar Islands
Leaf Type: Compound, 7–9 pairs
Inflorescence: Panicles, axillary in clusters
Fruit Type: Dry, pods
IUCN Status: Not evaluated
Reference: F.A.W. Miquel, Pl. Jungh.: 238 (1852)

Spatholobus gyrocarpus Benth.
Fabaceae
Climbing Mechanism: Stem Twiner
Distribution (Global): Borneo, Malaya, Philippines
Leaf Type: Compound, 7–9 pairs
Inflorescence: Panicles, axillary in clusters
Fruit Type: Dry, pods

IUCN Status: Least Concern
Reference: F.A.W. Miquel, Pl. Jungh.: 238 (1852)

Spatholobus parviflorus (DC.) Kuntze
Fabaceae
Synonyms: *Butea parviflora* Roxb. ex G. Don., *Flemingia hallaya* Buch. -Ham. ex Wall., *Butea sericophylla* Wall., *Spatholobus roxburghii* Benth.
Common Name: Flame of the forest, Palas climber
Climbing Mechanism: Stem Twiner
Distribution (Global): Assam, Bangladesh, Cambodia, China South-Central, East Himalaya, India, Laos, Myanmar, Nepal, Sri Lanka, Thailand, Vietnam, West Himalaya
Distribution (India): Assam, Goa, Karnataka, Kerala, Maharashtra, Manipur, Meghalaya, Mizoram, Tripura, Uttar Pradesh, West Bengal, Western Ghats
Leaf Type: Compound, 7–9 pairs
Inflorescence: Panicles, axillary in clusters
Fruit Type: Dry, pods
Flowering and Fruiting: August–April
IUCN Status: Least Concern
Notes: Mostly occur in semi-evergreen and moist deciduous forests
Reference: Revis. Gen. Pl. 1: 205 (1891)

Spatholobus purpureus Prain
Fabaceae
Synonyms: *Butea purpurea* (Benth. ex Baker) Blatt.
Climbing Mechanism: Scrambler-Unarmed
Distribution (Global): India, Myanmar
Distribution (India): Kerala, Maharashtra, Tamil Nadu
Leaf Type: Compound, 7–9 pairs
Inflorescence: Panicles, axillary in clusters
Fruit Type: Dry, pods
Flowering and Fruiting: December–May
IUCN Status: Not evaluated
Notes: Found mostly in evergreen and semi-evergreen forests
Reference: J. D. Hooker, Fl. Brit. India 2: 194 (1876)

Stenochlaena palustris (DC.) Kuntze
Fabaceae
Synonyms: *Lomaria haenkeana* C. Presl., *Lomaria juglandifolia* C. Presl + 15
Climbing Mechanism: Stem Twiner
Distribution (Global): Assam, Bangladesh, Bismarck Archipelago, Borneo, Cambodia, Caroline Is., China South-Central, China Southeast, Fiji, Hainan, India, Java, Laos, Lesser Sunda Is., Malaya, Maluku, Nepal, New Guinea, Nicobar Is., Northern Territory, Philippines, Queensland, Samoa, Solomon Is.,

Sri Lanka, Sulawesi, Sumatera, Thailand, Tonga, Vietnam, Wallis-Futuna Is., Western Australia
Distribution (India): Assam, Bihar, Odisha, Odisha, Tripura
Leaf Type: Bipinnate leaves
Fruit Type: Sporangial capsule, globose
Flowering and Fruiting: January–February
IUCN Status: Not evaluated
Notes: Not common, found growing along canals in partially shaded places
Reference: Ferns Brit. India, Suppl.: 26 (1876)

Strongylodon lucidus (G. Forst.) Seem.
Fabaceae
Synonyms: *Arthroclianthus vieillardii* Schindl., *Glycine lucida* G. Forst.
Climbing Mechanism: Stem Twiner
Distribution (Global): Andaman Is., Bismarck Archipelago, Caroline Is., Christmas I., Fiji, Marianas, New Caledonia, New Guinea, Philippines, Queensland, Samoa, Society Is., Solomon Is., Tonga, Vanuatu
Distribution (India): Andaman and Nicobar Islands
Leaf Type: Compound, 3-foliolate
Inflorescence: Racemes
Fruit Type: Dry, pods
IUCN Status: Not evaluated
Reference: Fl. Vit.: 61 (1865)

Strongylodon macrobotrys A. Gray
Fabaceae
Synonyms: *Strongylodon megaphyllus* Merr., *Strongylodon warburgii* Perkins.
Common Name: Jade vine
Climbing Mechanism: Stem Twiner
Distribution (Global): Philippines
Leaf Type: Compound, 3-foliolate
Inflorescence: Racemes
Fruit Type: Dry, pods
IUCN Status: Not evaluated
Notes: Often occur in damp forests, along streams, or in ravines
Reference: U.S. Expl. Exped., Phan. 1: 448 (1854)

Strongylodon ruber Vogel
Fabaceae
Synonyms: *Mucuna altissima* Hook. & Arn., *Rhynchosia lucida* DC
Climbing Mechanism: Stem Twiner
Distribution (Global): Hawaii
Distribution (India): Andaman and Nicobar Islands
Leaf Type: Compound, 3-foliolate
Inflorescence: Racemes

Fruit Type: Dry, pods
IUCN Status: Not evaluated
Reference: Linnaea 10: 585 (1836)

Tephrosia maxima (L.) Pers.
Fabaceae
Synonyms: *Galega maxima* L.
Climbing Mechanism: Stem Twiner
Distribution (Global): India, Sri Lanka, Tanzania
Distribution (India): Puducherry
Leaf Type: 13–18-foliolate, lanceolate
Inflorescence: short pseudoracemes, purple and white flowers
Flowering and Fruiting: June–January
IUCN Status: Not evaluated
Notes: Usually occur in rocky areas, waste places and sea coasts, also in foothills of deciduous forests
Reference: Syn. Pl. 2: 329 (1807)

Teramnus flexilis Benth.
Fabaceae
Synonyms: *Glycine flexilis* Graham
Climbing Mechanism: Stem Twiner
Distribution (Global): Assam, Bangladesh, East Himalaya, Myanmar
Distribution (India): West Bengal
Leaf Type: ovate-elliptic
Inflorescence: Racemes
Fruit Type: Pods/Legumes
IUCN Status: Not evaluated
Reference: J. Linn. Soc., Bot. 8: 265 (1865)

Teramnus hookerianus A. Sen
Fabaceae
Synonyms: *Teramnus labialis* subsp. *Labialis*., *Bujacia gampsonychia* E. Mey., *Hedysarum hamatum* Reinw. ex Miq., *Kennedia arabica* Hochst. ex Benth. + 20
Climbing Mechanism: Stem Twiner
Distribution (Global): Angola, Assam, Bahamas, Bangladesh, Benin, Bismarck Archipelago, Burundi, Cambodia, Cameroon, Cape Provinces, Cayman Is., Central African Repu, Chad, Comoros, Cuba, Dominican Republic, East Himalaya, Gabon, Ghana, Guatemala, Guinea-Bissau, Gulf of Guinea Is., Hainan, Haiti, India, Jamaica, Kenya, KwaZulu-Natal, Laccadive Is., Laos, Leeward Is., Liberia, Madagascar, Mali, Mexico Gulf, Mexico Southeast, Mexico Southwest, Mozambique, Myanmar, Nepal, New Guinea, Nicaragua, Nigeria, Northern Provinces, Pakistan, Panamá, Puerto Rico, Rwanda, Senegal, Sierra Leone, Southwest Caribbean, Sri Lanka, Swaziland, Taiwan, Tanzania, Thailand,

Togo, Trinidad-Tobago, Uganda, Vietnam, West Himalaya, Windward Is., Yemen, Zambia, Zaire, Zimbabwe
Distribution (India): West Bengal
IUCN Status: Not evaluated
Reference: J. Bombay Nat. Hist. Soc. 73: 187 (1976)

Teramnus labialis (L.f.) Spreng.
Fabaceae
Synonyms: *Glycine labialis* L.f.
Climbing Mechanism: Stem Twiner
Distribution (Global): Angola, Assam, Bahamas, Bangladesh, Benin, Bismarck Archipelago, Burkina, Burundi, Cambodia, Cameroon, Cape Provinces, Cape Verde, Cayman Is., Central African Repu, Chad, Comoros, Cuba, Dominican Republic, East Himalaya, Eritrea, Ethiopia, Gabon, Ghana, Guatemala, Guinea-Bissau, Gulf of Guinea Is., Hainan, Haiti, India, Ivory Coast, Jamaica, Kenya, KwaZulu-Natal, Laccadive Is., Laos, Leeward Is., Liberia, Madagascar, Malawi, Mali, Mauritius, Mexico Gulf, Mexico Southeast, Mexico Southwest, Mozambique, Myanmar, Nepal, New Guinea, Nicaragua, Nigeria, Northern Provinces, Pakistan, Panamá, Puerto Rico, Rodrigues, Rwanda, Réunion, Senegal, Sierra Leone, Somalia, Southwest Caribbean, Sri Lanka, Sudan, Swaziland, Taiwan, Tanzania, Thailand, Togo, Trinidad-Tobago, Uganda, Vietnam, West Himalaya, Windward Is., Yemen, Zambia, Zaire, Zimbabwe
Distribution (India): Andhra Pradesh, Bihar, Odisha, Delhi, Gujarat, Himachal Pradesh, Karnataka, Kerala, Madhya Pradesh, Maharashtra, Odisha, Rajasthan, Tamil Nadu, Telangana, Uttar Pradesh, West Bengal
Leaf Type: Pinnately compound, trifoliate
Inflorescence: Racemes, axillary
Fruit Type: Pods/Legumes
Flowering and Fruiting: July–January
IUCN Status: Not evaluated
Notes: Frequently found in dry deciduous forests and in the plains
Reference: Syst. Veg. ed. 16, 3: 235 (1826)

Teramnus mollis Benth.
Fabaceae
Synonyms: *Teramnus labialis* var. *mollis* Baker
Common Name: Blue wiss, Rabbit vine, Horse vine
Climbing Mechanism: Stem Twiner
Distribution (Global): Assam, Bangladesh, East Himalaya, India, Myanmar, Pakistan, Sri Lanka, West Himalaya
Distribution (India): Andhra Pradesh, Gujarat, Karnataka, Kerala, Madhya Pradesh, Maharashtra, Odisha, Tamil Nadu, West Bengal
Flowering and Fruiting: September–March
IUCN Status: Not evaluated

Notes: Mostly occurs in semi-evergreen and moist deciduous forests
Reference: J. Linn. Soc., Bot. 8: 265 (1865)

Tetramnus hookerianus A. Sen
Fabaceae
Climbing Mechanism: Stem Twiner
IUCN Status: Not evaluated

Vicia pallida Hook. & Arn.
Fabaceae
Synonyms: *Abacosa pallida* (Hook. & Arn.) Alef., *Ervum chilense* Fisch. & C.A. Mey. ex Regel, *Ervum pallidum* (Hook. & Arn.) Stank., *Ervum vestitum* Stank., *Vicia chilensis* D. Dietr., *Vicia paposana* Phil., *Vicia vestita* Clos.
Climbing Mechanism: Scrambler-Unarmed
Distribution (Global): Chile Central
Distribution (India): Himachal Pradesh
Leaf Type: Compound, 3–8 paired
Inflorescence: Solitary in leaf axile
Fruit Type: Dry, pods
IUCN Status: Least Concern
Reference: Bot. Misc. 3: 196 (1833)

Vicia sativa L.
Fabaceae
Common Name: Common vetch, Spring vetch.
Climbing Mechanism: Scrambler-Unarmed
Distribution (Global): Pan tropical
Distribution (India): Gujarat, Madhya Pradesh, Maharashtra, Manipur, Rajasthan, Tamil Nadu
Leaf Type: Pinnately compound
Fruit Type: Pods
Flowering and Fruiting: July–August
IUCN Status: Least Concern
Reference: Sp. Pl.: 736 (1753)

Vicia sativa subsp. *nigra* (L.) Ehrh.
Fabaceae
Synonyms: *Cracca timbaliana* Debeaux, *Vicia abyssinica* Alef., *Vicia angustifolia* Roth. + 40
Common Name: Black-pod vetch, Common vetch.
Climbing Mechanism: Scrambler-Unarmed
Distribution (Global): Pan tropical
Leaf Type: Leaves are compound with 5–6 pairs of opposite leaflets
Fruit Type: Pods
Flowering and Fruiting: June–July

IUCN Status: Least Concern
Reference: Hannover. Mag. 18: 229 (1780)

Vicia tenera Benth.
Fabaceae
Common Name: Delicate Himalayan vetch.
Climbing Mechanism: Scrambler-Unarmed
Distribution (Global): Bangladesh, India, West Himalaya
Distribution (India): Himachal Pradesh
Leaf Type: Compound, 3–8 paired
Inflorescence: Solitary in leaf axile
Fruit Type: Dry, pods
IUCN Status: Not evaluated
Reference: J.F. Royle, Ill. Bot. Himal. Mts.: 200 (1835)

Vicia tenuifolia Roth
Fabaceae
Synonyms: *Cracca tenuifolia* (Roth) Opiz, *Ervum tenuifolium* (Roth) Trautv., *Vicia cracca* subsp. *tenuifolia* (Roth) Gaudin
Climbing Mechanism: Scrambler-Unarmed
Distribution (Global): Afghanistan, Albania, Algeria, Altay, Austria, Baltic States, Belarus, Belgium, Bulgaria, Central European Rus, China South-Central, Corse, Czechoslovakia, Denmark, East Aegean Is., East European Russia, France, Germany, Greece, Hungary, Iran, Iraq, Italy, Kazakhstan, Kirgizstan, Korea, Krasnoyarsk, Kriti, Krym, Lebanon-Syria, Mongolia, Morocco, Nepal, North Caucasus, North European Russi, Northwest European R, Pakistan, Palestine, Poland, Portugal, Romania, Sardegna, Sicilia, South European Russi, Spain, Sweden, Switzerland, Tadzhikistan, Transcaucasus, Turkey, Turkey-in-Europe, Tuva, Ukraine, Uzbekistan, West Himalaya, West Siberia, Xinjiang, Yugoslavia
Leaf Type: Compound leaflets
Fruit Type: Pods
IUCN Status: Least Concern
Reference: Tent. Fl. Germ. 1: 309 (1788)

Vicia tetrasperma (L.) Schreb.
Fabaceae
Synonyms: *Ervilia tetrasperma* (L.) Opiz, *Ervum glabrum* Gilib., *Ervum monanthos* Georgi, *Ervum soloniense* Thuill., *Ervum tetraspermum* L., *Lathyrus monanthos* Baumg., *Vicia agrestis* Scheele, *Vicia gemella* Crantz, *Vicia pusilla* Muhl. ex Willd., *Vicia tetrasperm*
Common Name: Lentil vetch, Smooth tare, Smooth vetch, Sparrow vetch.
Climbing Mechanism: Scrambler-Unarmed
Distribution (Global): Pan tropical
Distribution (India): Madhya Pradesh, Rajasthan, West Bengal
Leaf Type: Compound, alternate

Inflorescence: Stalked raceme
Fruit Type: Pods
Flowering and Fruiting: February–August
IUCN Status: Not evaluated
Reference: Spic. Fl. Lips.: 26 (1771)

Vigna aconitifolia (Jacq.) Marechal
Fabaceae
Synonyms: *Dolichos dissectus* Lam., *Glycine triloba* Burm.f., *Phaseolus aconitifolius* Jacq., *Phaseolus palmatus* Forssk., *Vigna aridicola* N. Tomooka & Maxted
Common Name: Moth bean, Mat bean, Turkish gram
Climbing Mechanism: Stem Twiner
Distribution (Global): Assam, Bangladesh, China South-Central, East Himalaya, India, Myanmar, Pakistan, Sri Lanka, West Himalaya
Distribution (India): Gujarat, Karnataka, Tamil Nadu, West Bengal
Leaf Type: 3-Foliate
Inflorescence: Racemes capitate
Fruit Type: Pods
Flowering and Fruiting: August–January
IUCN Status: Data Deficient
Notes: Found mostly in scrub jungles and plains
Reference: Bull. Jard. Bot. Natl. Belg. 39: 160 (1969)

Vigna adenantha (G. Mey.) Marechal & al.
Fabaceae
Synonyms: *Leptospron adenanthum* (G. Mey.) A. Delgado., *Dolichos oleraceus* Schumach. & Thonn., *Phaseolus adenanthus* G. Mey., *Phaseolus adenanthus* f. bahiensis Hassl. + 35
Climbing Mechanism: Stem Twiner
Distribution (Global): Pan tropical
Distribution (India): Andaman and Nicobar Islands, Andhra Pradesh, Kerala, Madhya Pradesh, West Bengal
Leaf Type: Pinnately compound, Oavte-elliptic
Inflorescence: Racemes
Fruit Type: Pods
Flowering and Fruiting: August–November
IUCN Status: Least Concern
Notes: Often found in coastal areas along the banks of backwaters
Reference: Amer. J. Bot. 98: 1710 (2011)

Vigna bournaea Gamble
Fabaceae
Synonyms: *Vigna trinervia* var. *bourneae* (Gamble) Tateishi & Maxted
Climbing Mechanism: Stem Twiner

Distribution (Global): India
Distribution (India): Tamil Nadu
Leaf Type: 3-Foliate
Inflorescence: Racemes axillary
Fruit Type: Pods
IUCN Status: Not evaluated
Notes: Endemic to Southern Western Ghats
Reference: Fl. Madras: 365 (1918)

Vigna clarkei Prain
Fabaceae
Climbing Mechanism: Stem Twiner
Distribution (Global): Assam, East Himalaya
Distribution (India): Arunachal Pradesh, West Bengal
Leaf Type: Compound, 3-foliolate
Inflorescence: Racemes, subumbellate in clusters
Fruit Type: Dry, pods
IUCN Status: Not evaluated
Reference: J. Asiat. Soc. Bengal, Pt. 2, Nat. Hist. 66: 429 (1897)

Vigna dalzelliana (Kuntze) Verdc.
Fabaceae
Synonyms: *Phaseolus dalzellianus* Kuntze., *Phaseolus pauciflorus* Dalzell.
Climbing Mechanism: Stem Twiner
Distribution (Global): Cambodia, India, Laos, Pakistan, Philippines, Vietnam
Distribution (India): Andhra Pradesh, Gujarat, Karnataka, Kerala, Madhya Pradesh, Maharashtra, Rajasthan, Tamil Nadu
Leaf Type: Compound, 3-foliolate
Inflorescence: Racemes, subumbellate in clusters
Fruit Type: Dry, pods
Flowering and Fruiting: August–March
IUCN Status: Least Concern
Reference: Kew Bull. 24: 558 (1970)

Vigna fusca (Wall.) A.S. Chauhan
Fabaceae
Synonyms: *Dunbaria fusca* (Wall.) Kurz., *Atylosia crinita* Dunn., *Dunbaria crinita* (Dunn) Maesen., *Dunbaria flavescens* Thuan., *Phaseolus fuscus* Wall.
Climbing Mechanism: Stem Twiner
Distribution (Global): China South-Central, China Southeast, Hainan, Laos, Myanmar, Thailand, Vietnam
Distribution (India): Manipur
Leaf Type: Compound, 3-foliolate
Inflorescence: Racemes, subumbellate in clusters
Fruit Type: Dry, pods

IUCN Status: Not evaluated
Reference: J. Asiat. Soc. Bengal, Pt. 2, Nat. Hist. 45: 255 (1876 publ. 1877)

Vigna grahamiana (Wight & Arn.) Verdc.
Fabaceae
Synonyms: *Wajira grahamiana* (Wight & Arn.) Thulin & Lavin., *Dolichos grahamianus* (Wight & Arn.) Niyomdham., *Dolichos subcarnosus* Wight & Arn., *Phaseolus grahamianus* Wight & Arn., *Phaseolus macrorhynchus* Harms., *Phaseolus stenocarpus* Harms., *Vigna grahamiana* (Wight & Arn.) Verdc.
Common Name: Graham's wild pea
Climbing Mechanism: Stem Twiner
Distribution (Global): Bangladesh, Burundi, Ethiopia, India, Kenya, Malawi, Mozambique, Nigeria, Rwanda, Somalia, Sri Lanka, Sudan, Tanzania, Thailand, Uganda, Zambia, Zaire, Zimbabwe
Distribution (India): Kerala, Maharashtra, Tamil Nadu
Leaf Type: Compound, 3-foliolate
Inflorescence: Racemes, subumbellate in clusters
Fruit Type: Dry, pods
Flowering and Fruiting: September–December
IUCN Status: Not evaluated
Notes: Often found in evergreen forests
Reference: Syst. Bot. 29: 908 (2004)

Vigna hainiana Babu, Gopin. & S.K. Sharma
Fabaceae
Synonyms: *Phaseolus wightii* Graham ex Wight & Arn.
Climbing Mechanism: Stem Twiner
Distribution (Global): Assam, India
Distribution (India): Peninsular India
Leaf Type: Compound, 3-foliolate
Inflorescence: Racemes, subumbellate in clusters
Fruit Type: Dry, pods
Flowering and Fruiting: August–December
IUCN Status: Not evaluated
Notes: Usually found in dry deciduous forests
Reference: Bull. Bot. Surv. India 27: 15 (1985 publ. 1987)

Vigna hosei (Craib) Backer
Fabaceae
Synonyms: *Dolichos hosei* Craib., *Vigna hosei* var. *pubescens* Maréchal, Mascherpa & Stainier
Climbing Mechanism: Stem Twiner
Distribution (Global): Borneo, Java, Malaya, Taiwan
Distribution (India): Kerala, Tamil Nadu

Leaf Type: Compound, 3-foliolate
Inflorescence: Racemes, subumbellate in clusters
Fruit Type: Dry, pods
IUCN Status: Data Deficient
Reference: Geill. Handb. Jav. Theeonkr.: 153 (1924)

Vigna khandalensis (Santapau) Raghavan & Wadhwa
Fabaceae
Synonyms: *Phaseolus grandis* Dalzell & A. Gibson., *Phaseolus khandallensis* Santapau., *Vigna grandis* Verdc.
Common Name: Khandala wild pea
Climbing Mechanism: Stem Twiner
Distribution (Global): India
Distribution (India): Tamil Nadu
Leaf Type: Compound, 3-foliolate
Inflorescence: Racemes, subumbellate in clusters
Fruit Type: Dry, pods
IUCN Status: Near Threatened
Notes: Mostly found in evergreen forests
Reference: Curr. Sci. 41: 429 (1972)

Vigna konkanensis
Fabaceae
Climbing Mechanism: Stem Twiner
IUCN Status: Not evaluated

Vigna luteola (Jacq.) Benth.
Fabaceae
Synonyms: *Calopogonium pedunculatum* Standl., *Dolichos abyssinicus* Steud., *Dolichos gangeticus* Roxb., *Dolichos glycinoides* Kunth, *Dolichos helicopus* (E. Mey.) Steud., *Dolichos luteolus* Jacq., *Dolichos mexicanus* Schltdl. + 30
Common Name: Yellow water pea
Climbing Mechanism: Stem Twiner
Distribution (Global): Pan tropical
Distribution (India): West Bengal
Leaf Type: Compound, 3-foliolate
Inflorescence: Racemes, subumbellate in clusters
Fruit Type: Dry, pods
IUCN Status: Least Concern
Reference: C.F.P.von Martius & auct. suc. (eds.), Fl. Bras. 15(1): 194 (1859)

Vigna marina (Burm.) Merr.
Fabaceae
Synonyms: *Callicysthus volubilis* Endl., *Dolichos luteus* Sw., *Dolichos retusus* E. Mey., *Dolichos savii* Montrouz., Phaseolus marinus Burm. + 10

Common Name: Beach pea, Notched cowpea
Climbing Mechanism: Stem Twiner
Distribution (Global): Aldabra, Andaman Is., Bangladesh, Bismarck Archipelago, Borneo, Cameroon, Caroline Is., Central African Repu, Chagos Archipelago, China Southeast, Cocos (Keeling) Is., Congo, Cook Is., Equatorial Guinea, Fiji, Gabon, Ghana, Gilbert Is., Gulf of Guinea Is., Hainan, Hawaii, India, Ivory Coast, Japan, Java, Kazan-retto, KwaZulu-Natal, Laccadive Is., Lesser Sunda Is., Liberia, Madagascar, Malaya, Maldives, Maluku, Marianas, Marshall Is., Mauritius, Mozambique, Nansei-shoto, Nauru, New Guinea, New South Wales, Nicobar Is., Nigeria, Niue, Norfolk Is., Northern Territory, Ogasawara-shoto, Philippines, Queensland, Réunion, Samoa, Seychelles, Society Is., Solomon Is., South China Sea, Sri Lanka, Sulawesi, Sumatera, Taiwan, Tanzania, Thailand, Tonga, Tuamotu, Tubuai Is., Tuvalu, Vietnam, Wallis-Futuna Is.
Distribution (India): Andaman and Nicobar Islands
Leaf Type: Leaves are 3-foliolate
Inflorescence: Racemes
Fruit Type: Legume
IUCN Status: Least Concern
Reference: Interpr. Herb. Amboin.: 285 (1917)

Vigna mungo (L.) Hepper
Fabaceae
Synonyms: *Azukia muogo* (L.) Masam., *Phaseolus aureus* Zuccagni., *Phaseolus chlorospermus* Ten., *Phaseolus gibbosus* Pi. Savi., *Phaseolus glabrescens* Steud. + 12
Common Name: Black gram
Climbing Mechanism: Stem Twiner
Distribution (Global): Assam, East Himalaya, India, Pakistan, West Himalaya
Distribution (India): Karnataka, Madhya Pradesh, Manipur, Puducherry, Tripura
Leaf Type: Compound, 3-foliolate
Inflorescence: Racemes, subumbellate in clusters
Fruit Type: Dry, pods
Flowering and Fruiting: August–December
IUCN Status: Not evaluated
Reference: Kew Bull. 11: 128 (1956)

Vigna pandeyana R.D. Gore, S.P. Gaikwad & S.D. Randive
Fabaceae
Climbing Mechanism: Stem Twiner
Distribution (Global): India
Distribution (India): Maharashtra
Leaf Type: Compound, 3-foliolate
Inflorescence: Racemes, subumbellate in clusters
Fruit Type: Dry, pods

IUCN Status: Not evaluated
Reference: Biodivers. Data J. 3(e4606): 2 (2015)

Vigna pilosa (Klein ex Willd.) Baker
Fabaceae
Synonyms: *Dysolobium pilosum* (J.G. Klein ex Willd.) Maréchal. +10
Climbing Mechanism: Stem Twiner
Distribution (Global): Andaman Is., Assam, Bangladesh, Cambodia, East Himalaya, India, Java, Laos, Malaya, Myanmar, Nepal, Philippines, Taiwan, Thailand, Vietnam
Distribution (India): Andaman and Nicobar Islands, Bihar, Odisha, Karnataka, Kerala, Madhya Pradesh, Mizoram, Tripura
Leaf Type: Compound, 3-foliolate
Inflorescence: Racemes, subumbellate in clusters
Fruit Type: Dry, pods
Flowering and Fruiting: October–March
IUCN Status: Not evaluated
Notes: Mostly found in semi-evergreen forests
Reference: Bull. Jard. Bot. Natl. Belg. 47: 483 (1977)

Vigna radiata (L.) R. Wilczek
Fabaceae
Synonyms: *Azukia radiata* (L.) Ohwi. *Phaseolus radiatus* L.
Common Name: Green gram, Mung, Golden gram
Climbing Mechanism: Stem Twiner
Distribution (Global): Assam, Bangladesh, India, Laos, Lesser Sunda Is., Myanmar, New South Wales, Northern Territory, Pakistan, Queensland, Sri Lanka, Taiwan, Thailand, Vietnam, West Himalaya, Western Australia, Yemen
Distribution (India): Madhya Pradesh, Maharashtra, Manipur, Puducherry, Tamil Nadu
Leaf Type: Alterntate, trilobate
Fruit Type: Legume
Flowering and Fruiting: May–December
IUCN Status: Least Concern
Reference: Fl. Congo Belge 6: 286 (1954)

Vigna radiata var. setulosa (Dalzell) Ohwi & Ohashi
Fabaceae
Synonyms: *Phaseolus novoguineensis* Baker f., *Phaseolus scaberulus* Miq., *Phaseolus setulosus* Dalzell., *Vigna opisotricha* A. Rich.
Climbing Mechanism: Stem Twiner
Distribution (Global): Assam, India, Pakistan, Sri Lanka
Distribution (India): Telangana, West Bengal
Leaf Type: Compound, 3-foliolate
Inflorescence: Racemes, subumbellate in clusters

Fruit Type: Dry, pods
IUCN Status: Not evaluated
Reference: J. Jap. Bot. 44: 31 (1969)

Vigna radiata var. *sublobata* (Roxb.) Verdc.
Fabaceae
Synonyms: *Phaseolus sublobatus* Roxb., *Vigna mungo* var. *sublobata* (Roxb.) M. Sharma, *Vigna stipulata* Hayata, *Vigna sublobata* (Roxb.) Bairig. & al.
Common Name: Wild moong, Mung bean, Wild black gram
Climbing Mechanism: Stem Twiner
Distribution (Global): Bangladesh, India, Laos, Lesser Sunda Is., New South Wales, Northern Territory, Queensland, Taiwan, Thailand, Vietnam, Western Australia
Distribution (India): Andhra Pradesh, Gujarat, Karnataka, Tamil Nadu
Leaf Type: Compound, 3-foliolate
Inflorescence: Racemes, subumbellate in clusters
Fruit Type: Dry, pods
Flowering and Fruiting: March–April
IUCN Status: Not evaluated
Notes: Mostly found in evergreen forests of the Western Ghats
Reference: Kew Bull. 24: 559 (1970)

Vigna ricciardiana (Ten.) Babu & S.K. Sharma
Fabaceae
Synonyms: *Vigna umbellata* (Thunb.) Ohwi & H. Ohashi., *Azukia umbellata* (Thunb.) Ohwi., *Dolichos trilobus* Blanco, *Dolichos umbellatus* Thunb.
Climbing Mechanism: Stem Twiner
Distribution (Global): Assam, Bangladesh, China South-Central, China Southeast, East Himalaya, Hainan, India, Japan, Java, Laccadive Is., Laos, Lesser Sunda Is., Malaya, New Guinea, Pakistan, Philippines, Solomon Is., Sri Lanka, Taiwan, Thailand, Vietnam, West Himalaya
Distribution (India): Peninsular India
Leaf Type: Compound, 3-foliolate
Inflorescence: Racemes, subumbellate in clusters
Fruit Type: Dry, pods
Flowering and Fruiting: September–November
IUCN Status: Not evaluated
Reference: J. Jap. Bot. 44: 31 (1969)

Vigna sahyadriana M.M. Aitawade, S.P. Sutar, S.R. Rao, S.K. Malik, S.R. Yadav & K.V. Bhat
Fabaceae
Climbing Mechanism: Stem Twiner
Distribution (Global): India
Distribution (India): Peninsular India
Leaf Type: Compound, 3-foliolate

Inflorescence: Racemes, subumbellate in clusters
Fruit Type: Dry, pods
Flowering and Fruiting: August–November
IUCN Status: Not evaluated
Notes: Often found in evergreen forests
Reference: Rheedea 22: 20 (2012)

Vigna subramaniana (Babu ex Raizada) M. Sharma
Fabaceae
Synonyms: *Phaseolus subramanianus* Babu ex Raizada
Climbing Mechanism: Stem Twiner
Distribution (Global): India
Distribution (India): Tamil Nadu
Leaf Type: Compound, 3-foliolate
Inflorescence: Racemes, subumbellate in clusters
Fruit Type: Dry, pods
IUCN Status: Not evaluated
Notes: Often found in evergreen forests

Vigna subramanianus
Fabaceae
Climbing Mechanism: Stem Twiner
IUCN Status: Not evaluated

Vigna trilobata (L.) Verdc.
Fabaceae
Synonyms: *Dolichos trilobatus* L., *Phaseolus trilobatus* (L.) Schreb.
Climbing Mechanism: Stem Twiner
Distribution (Global): Assam, Bangladesh, China South-Central, East Himalaya, India, Myanmar, Nepal, Pakistan, Sri Lanka, Taiwan, Vietnam, West Himalaya
Distribution (India): Andhra Pradesh, Gujarat, Karnataka, Madhya Pradesh, Odisha, Puducherry, Rajasthan, Tamil Nadu, Telangana, West Bengal
Leaf Type: Compound, 3-foliolate
Inflorescence: Racemes, subumbellate in clusters
Fruit Type: Dry, pods
Flowering and Fruiting: January–February
IUCN Status: Not evaluated
Notes: Found usually in rocky areas in dry and moist deciduous forests
Reference: Taxon 17: 172 (1968)

Vigna umbellata (Thunb.) Ohwi & H. Ohashi
Fabaceae
Climbing Mechanism: Stem Twiner

Distribution (India): Andhra Pradesh, Arunachal Pradesh, Gujarat, Karnataka, Kerala, Madhya Pradesh, Maharashtra, Manipur, Rajasthan, Tamil Nadu, West Bengal
Leaf Type: Compound, 3-foliolate
Inflorescence: Racemes, subumbellate in clusters
Fruit Type: Dry, pods
IUCN Status: Not evaluated

Vigna unguiculata (L.) Walp.
Fabaceae
Synonyms: *Dolichos unguiculatus* L., *Phaseolus unguiculatus* (L.) Piper., *Vigna unguiculata* subvar. *typica* Bertoni
Common Name: Cowpea, Asparagus bean, Barbati, Black eye pea, Blue goose southern pea, Common cowpea, Southern pea
Climbing Mechanism: Stem Twiner
Distribution (Global): Pan tropical
Distribution (India): Andaman and Nicobar Islands, Arunachal Pradesh, Himachal Pradesh, Kerala, Madhya Pradesh, Manipur, Rajasthan, Tamil Nadu, Telangana, Tripura
Leaf Type: Alternate
Inflorescence: Ra-cemes axillary
Fruit Type: Pods
Flowering and Fruiting: Throughout the year
IUCN Status: Not evaluated
Reference: Repert. Bot. Syst. 1: 779 (1843)

Vigna unguiculata subsp. *cylindrica* (L.) Verdc.
Fabaceae
Synonyms: *Vigna unguiculata* subsp. *unguiculata*., *Vigna catjang* (Burm.f.) Walp., *Vigna cylindrica* (L.) Skeels. + 35
Climbing Mechanism: Stem Twiner
Distribution (Global): Pan tropical
Distribution (India): Andaman, Nicobar, Islands, Gujarat
Leaf Type: Compound, 3-foliolate
Inflorescence: Racemes, subumbellate in clusters
Fruit Type: Dry, pods
Flowering and Fruiting: September–April
IUCN Status: Not evaluated

Vigna vexillata (L.) A. Rich.
Fabaceae
Synonyms: *Dolichos vexillatus* (L.) Kunth., *Phaseolus vexillatus* L., *Vigna vexillata* f. typica Domin.
Common Name: Zombi pea, Wild mung, Wild cowpea
Climbing Mechanism: Stem Twiner

Distribution (Global): Pan tropical
Distribution (India): Bihar, Odisha, Gujarat, Himachal Pradesh, Karnataka, Kerala, Madhya Pradesh, Manipur, Odisha, Rajasthan, Tamil Nadu, Telangana, West Bengal
Leaf Type: Compound, 3-foliate
Fruit Type: Pods
Flowering and Fruiting: March–April
IUCN Status: Not evaluated
Notes: Found mostly in moist deciduous forests
Reference: R.de la Sagra, Hist. Fis. Cuba, Bot. 10: 191 (1845)

Vigna vexillata var. *angustifolia* (Schum. & Thonn.) Baker
Fabaceae
Synonyms: *Dolichos angustifolius* (Schumach. & Thonn.) Vahl ex Guill. & Perr., *Dolichos stenophyllus* Gagnep., *Phaseolus novoguineensis* Warb., *Plectrotropis angustifolia* Schumach. & Thonn., *Vigna angustifolia* (Schumach. & Thonn.) Hook.f., *Vigna valetonii* Pulle, Vigna
Common Name: Narrow-leaved Zombi pea
Climbing Mechanism: Stem Twiner
Distribution (Global): Angola, Assam, Benin, Cambodia, Cape Provinces, Congo, East Himalaya, Ethiopia, Free State, India, Kenya, KwaZulu-Natal, Madagascar, Malawi, Mozambique, New South Wales, Northern Provinces, Northern Territory, Queensland, Senegal, Sudan, Tanzania, Togo, West Himalaya, Western Australia, Yemen, Zambia, Zaire, Zimbabwe
Distribution (India): Tamil Nadu
Leaf Type: Compound, 3-foliolate
Inflorescence: Racemes, subumbellate in clusters
Fruit Type: Dry, pods
IUCN Status: Not evaluated
Reference: Fl. Trop. Afr. 2: 200 (1871)

Vigna vexillata var. *sepiaria*
Fabaceae
Synonyms: *Phaseolus sepiarius* Dalzell
Climbing Mechanism: Stem Twiner
Distribution (Global): India
Distribution (India): Maharashtra, Tamil Nadu
Leaf Type: Compound, 3-foliolate
Inflorescence: Racemes, subumbellate in clusters
Fruit Type: Dry, pods
Flowering and Fruiting: November-January
IUCN Status: Not evaluated
Notes: Usually found in semi-evergreen forests
Reference: Bull. Bot. Surv. India 27: 27 (1985 publ. 1987)

Vigna vexillata var. *stocksii*
Fabaceae
Climbing Mechanism: Stem Twiner
Distribution (India): Tamil Nadu
Leaf Type: Compound, 3-foliolate
Inflorescence: Racemes, subumbellate in clusters
Fruit Type: Dry, pods
Flowering and Fruiting: July–December
IUCN Status: Not evaluated

Vigna vexillata var. *wightii* (Bedd.) Babu & S.K. Sharma
Fabaceae
Synonyms: *Vigna wightii* Benth. ex Bedd.
Climbing Mechanism: Stem Twiner
Distribution (Global): India
Distribution (India): Tamil Nadu
Leaf Type: Compound, 3-foliolate
Inflorescence: Racemes, subumbellate in clusters
Fruit Type: Dry, pods
Flowering and Fruiting: November-December.
IUCN Status: Not evaluated
Notes: Mostly occur in moist-deciduous forests
Reference: Bull. Bot. Surv. India 27: 28 (1985 publ. 1987)

Vigna yadavii S.P. Gaikwad, R.D. Gore, S.D.Randive & K.U.Garad
Fabaceae
Climbing Mechanism: Stem Twiner
Distribution (Global): India
Distribution (India): Maharashtra
Leaf Type: Elliptic/Oblanceolate
Inflorescence: Raceme axillary, solitary, umbellate
Fruit Type: Pods
Flowering and Fruiting: May-January.
IUCN Status: Not evaluated
Reference: Biodivers. Data J. 2-e4281: 2 (2014)

Wagatea spicata Dalzell
Fabaceae
Synonyms: *Moullava spicata* (Dalzell) Nicolson., *Almeloveenia spinosa* Dennst., *Caesalpinia ferox* Hohen. ex Baker, *Caesalpinia heyneana* Steud., *Caesalpinia mimosoides* B. Heyne, *Caesalpinia spicata* Dalzel
Common Name: Orange brasiletto climber, Candy corn plant
Climbing Mechanism: Scrambler-Armed
Distribution (Global): India
Distribution (India): Gujarat, Madhya Pradesh, Uttar Pradesh

Leaf Type: Double pinnate
Fruit Type: Pod
Flowering and Fruiting: October–March
IUCN Status: Not evaluated
Notes: Usually found in moist deciduous forests and in open plains
Reference: K.S. Manilal, Bot. Hist. Hort. Malabaricus: 184 (1980)

Wajira grahamiana (Wight & Arn.) Thulin & Lavin
Fabaceae
Synonyms: *Dolichos grahamianus* (Wight & Arn.) Niyomdham., *Dolichos subcarnosus* Wight & Arn., *Phaseolus grahamianus* Wight & Arn., *Phaseolus macrorhynchus* Harms., *Phaseolus stenocarpus* Harms., *Vigna grahamiana* (Wight & Arn.) Verdc., *Vigna macrorhyncha* (Harms) Milne-R
Common Name: Graham's wild pea
Climbing Mechanism: Stem Twiner
Distribution (Global): Bangladesh, Burundi, Ethiopia, India, Kenya, Malawi, Mozambique, Nigeria, Rwanda, Somalia, Sri Lanka, Sudan, Tanzania, Thailand, Uganda, Zambia, Zaire, Zimbabwe
Distribution (India): Tamil Nadu
Leaf Type: Linear—lance shaped
Fruit Type: Pods
IUCN Status: Not evaluated
Reference: Syst. Bot. 29: 908 (2004)

Wisteria polystachya
Fabaceae
Climbing Mechanism: Scrambler-Unarmed
IUCN Status: Not evaluated

Wisteria sinensis (Sims) Sweet
Fabaceae
Climbing Mechanism: Stem Twiner
Distribution (India): Tamil Nadu
IUCN Status: Not evaluated
Reference: Hort. Brit. [Sweet] 121. 1826

Flagellaria indica L.
Flagellariaceae
Synonyms: *Flagellaria catenata* Lour. ex B.A. Gomes, *Flagellaria indica* var. *australiensis* Wepfer & H.P. Linder, *Flagellaria indica* var. *bifurcata* Wepfer & H.P. Linder, *Flagellaria indica* var. *borneensis* Wepfer & H.P. Linder, *Flagellaria loureiroi* Steud
Climbing Mechanism: Scrambler-Unarmed
Distribution (Global): Andaman Is., Assam, Bangladesh, Bismarck Archipelago, Borneo, Cambodia, Caroline Is., China Southeast, Comoros, Fiji, Hainan, India,

Japan, Java, Lesser Sunda Is., Madagascar, Malaya, Maldives, Maluku, Marianas, Mauritius, Mozambique, Myanmar, Nansei-shoto, New Caledonia, New Guinea, New South Wales, Nicobar Is., Norfolk Is., Northern Territory, Philippines, Queensland, Rodrigues, Réunion, Santa Cruz Is., Seychelles, Solomon Is., Sri Lanka, Sulawesi, Sumatera, Taiwan, Tanzania, Thailand, Vanuatu, Vietnam, Wallis-Futuna Is., Western Australia

Distribution (India): Andaman and Nicobar Islands, Andhra Pradesh, Bihar, Odisha, Madhya Pradesh, Tamil Nadu, Telangana, Tripura

Flowering and Fruiting: August–October

IUCN Status: Not evaluated

Reference: Sp. Pl.: 333 (1753)

Gelsemium elegans (Gardner & Chapm.) Benth.

Gelsemiaceae

Synonyms: *Gelsemium sumatranum* (Blume) Boerl., *Leptopteris sumatrana* Blume, *Medicia elegans* Gardner & Champ.

Climbing Mechanism: Scrambler-Unarmed

Distribution (Global): Assam, Borneo, China South-Central, China Southeast, Hainan, Java, Laos, Malaya, Myanmar, Sumatera, Taiwan, Thailand, Vietnam

Distribution (India): Meghalaya, Mizoram

IUCN Status: Not evaluated

Reference: J. Proc. Linn. Soc., Bot. 1: 90 (1856)

Crawfurdia affinis Wall. ex C.B. Clarke

Gentianaceae

Synonyms: *Crawfurdia affinis* Wall. ex C.B. Clarke, *Crawfurdia fasciculata* Wall., *Gentiana confusa* C. Marquand, *Gentiana fascicularis* (Wall.) C.Marquand, *Tripterospermum affine* (Wall. ex C.B. Clarke) Harry Sm.

Climbing Mechanism: Stem Twiner

Distribution (Global): Assam, Bangladesh, East Himalaya, Myanmar, Nepal, Vietnam

IUCN Status: Not evaluated

Reference: Enum. Fl. Pl. Nepal 3: 98 (1982)

Crawfurdia angustata C.B. Clarke

Gentianaceae

Synonyms: *Crawfurdia trailiana* Forrest, *Gentiana angustata* (C.B. Clarke) C. Marquand, *Gentiana helenii* C. Marquand, *Tripterospermum angustatum* (C.B. Clarke) Raizada, *Tripterospermum trailianum* (Forrest) Raizada

Common Name: Pink gentian vine

Climbing Mechanism: Stem Twiner

Distribution (Global): Assam, China South-Central, East Himalaya, Myanmar, Tibet

Leaf Type: Elliptic to ovate

Inflorescence: Cymes

Flowering and Fruiting: October–December

IUCN Status: Not evaluated
Reference: J. D. Hooker, Fl. Brit. India 4: 106 (1883)

Crawfurdia arunachalensis S.S. Dash, Gogoi & A.A. Mao
Gentianaceae
Climbing Mechanism: Stem Twiner
Distribution (Global): East Himalaya
Distribution (India): Arunachal Pradesh
IUCN Status: Not evaluated
Reference: J. Jap. Bot. 86: 127 (2011)

Crawfurdia campanulacea Wall. & Griff. ex C.B. Clarke
Gentianaceae
Synonyms: *Crawfurdia bulleyana* Forrest, *Crawfurdia speciosa* C.B. Clarke, *Gentiana bulleyana* (Forrest) C.Marquand, *Gentiana campanulacea* (Wall. & Griff. ex C.B. Clarke) C.Marquand, *Tripterospermum bulleyanum* (Forrest) Raizada, *Tripterospermum campanulaceum* (Wall. & Griff. ex C. B. Clarke) Raizada.
Climbing Mechanism: Stem Twiner
Distribution (Global): Assam, China South-Central, East Himalaya, Myanmar, Vietnam
Leaf Type: Elliptic, ovate
Inflorescence: Cymes
Fruit Type: Capsule, ellipsoid
IUCN Status: Not evaluated
Reference: J. D. Hooker, Fl. Brit. India 4: 106 (1883)

Crawfurdia puberula C.B. Clarke
Gentianaceae
Synonyms: *Tripterospermum puberulum* (C.B. Clarke) Raizada
Climbing Mechanism: Stem Twiner
Distribution (Global): East Himalaya, Tibet
IUCN Status: Not evaluated
Reference: J. Linn. Soc., Bot. 14: 442 (1875)

Crawfurdia speciosa C.B. Clarke
Gentianaceae
Climbing Mechanism: Stem Twiner
Distribution (India): West Bengal
IUCN Status: Not evaluated
Reference: C. B. Clarke. In: Journ. L. Soc. 14: 442 p. p. (1875)

Gentiana ornata (D. Don) Wall. ex Griseb.
Gentianaceae

Synonyms: *Gentianodes ornata* (D. Don) Á.Löve & D.Löve, *Pneumonanthe ornata* G.Don
Common Name: Decorated gentian, Showy gentian
Climbing Mechanism: Stem Twiner
Distribution (Global): East Himalaya, Nepal, Tibet
Distribution (India): West Bengal
IUCN Status: Not evaluated
Reference: Gen. Sp. Gent.: 277 (1838)

Halenia elliptica D. Don
Gentianaceae
Synonyms: *Halenia vaniotii* H. Lév., *Swertia centrostemma* Wall., *Swertia peloria* Griff., *Tetragonanthus centrostemma* (Wall.) Kuntze
Common Name: Spurred Gentian
Climbing Mechanism: Stem Twiner
Distribution (Global): Assam, China North-Central, China South-Central, China Southeast, East Himalaya, Inner Mongolia, Kazakhstan, Kirgizstan, Manchuria, Myanmar, Nepal, Qinghai, Tibet, West Himalaya, Xinjiang
Distribution (India): Arunachal Pradesh
IUCN Status: Not evaluated
Reference: Trans. Linn. Soc. London 17: 529 (1837)

Tripterospermum fasciculatum (Wall.) Chater
Gentianaceae
Synonyms: *Crawfurdia affinis* Wall. ex C.B. Clarke, *Crawfurdia fasciculata* Wall., *Gentiana confusa* C. Marquand, *Gentiana fascicularis* (Wall.) C.Marquand, *Tripterospermum affine* (Wall. ex C.B. Clarke) Harry Sm.
Climbing Mechanism: Stem Twiner
Distribution (Global): Assam, Bangladesh, East Himalaya, Myanmar, Nepal, Vietnam
Distribution (India): Arunachal Pradesh
Leaf Type: Simple, opposite
Inflorescence: Single axillary flowers
Fruit Type: Dry, capsule
IUCN Status: Not evaluated
Reference: Enum. Fl. Pl. Nepal 3: 98 (1982)

Tripterospermum japonicum Maxim.
Gentianaceae
Synonyms: *Tripterospermum trinervium* (Thunb.) H. Ohashi & H. Nakai., *Tripterospermum involubile* N. Yonez. + 15
Climbing Mechanism: Stem Twiner
Distribution (Global): Japan, Korea, Kuril Is., Sakhalin
Leaf Type: Simple, opposite
Inflorescence: Single axillary flowers

Fruit Type: Dry, capsule
IUCN Status: Not evaluated
Reference: J. Jap. Bot. 71: 112 (1996)

Tripterospermum nigrobaccatum H. Hara
Gentianaceae
Climbing Mechanism: Stem Twiner
Distribution (Global): East Himalaya, Nepal
Distribution (India): West Bengal
Leaf Type: Simple, opposite
Inflorescence: Single axillary flowers
Fruit Type: Dry, capsule
IUCN Status: Not evaluated
Reference: J. Jap. Bot. 40: 19 (1965)

Tripterospermum volubile (D. Don) H. Hara
Gentianaceae
Climbing Mechanism: Stem Twiner
Distribution (India): Mizoram
Leaf Type: Simple, opposite
Inflorescence: Single axillary flowers
Fruit Type: Dry, capsule
IUCN Status: Not evaluated
Reference: J. Jap. Bot. 40: 21 (1965)

Pelargonium peltatum (L.) L'Hér.
Geraniaceae
Synonyms: *Dibrachya clypeata* Eckl. & Zeyh., *Dibrachya peltata* (L.) Eckl. & Zeyh., *Dibrachya scutata* (Sweet) Eckl. & Zeyh., *Geraniospermum peltatum* (L.) Kuntze, *Geranium peltatum* L., *Pelargonium bachmannii* R. Knuth, *Pelargonium clypeatum* (Eckl. & Zeyh.) Steud.
Common Name: Ivy-leaf geranium
Climbing Mechanism: Scrambler-Unarmed
Distribution (Global): Cape Provinces
Distribution (India): Peninsular India
Leaf Type: Ivy shaped
Inflorescence: Umbel
Flowering and Fruiting: December–February
IUCN Status: Not evaluated
Reference: W. Aiton, Hort. Kew. 2: 427 (1789)

Aeschynanthus zeylanica var. *pinguis*
Gesneriaceae
Climbing Mechanism: Scrambler-Unarmed
Leaf Type: Simple, opposite

Inflorescence: Clusters at tips
Fruit Type: Dry, capsule
IUCN Status: Not evaluated
Reference: Ann. Bot. Syst. 3: 95 (1852)

Gnetum contractum Markgr.
Gnetaceae
Synonyms: *Thoa contracta* (Markgr.) Doweld
Climbing Mechanism: Stem Twiner
Distribution (Global): India
Distribution (India): Tamil Nadu
IUCN Status: Not evaluated
Reference: Bull. Jard. Bot. Buitenzorg, sér. 3, 10: 470 (1930)

Gnetum gnemonoides Brongn.
Gnetaceae
Synonyms: *Gnetum kerstingii* K. Schum. & Lauterb., *Gnetum macrocarpum* Becc., *Gnetum moluccense* H. Karst. ex Markgr., *Gnetum ovalifolium* H. Karst., *Gnetum rumphianum* Becc., *Gnetum verrucosum* H. Karst., *Gnetum wrayi* Gamble, *Thoa gnemonoides* (Brongn.) Doweld.
Climbing Mechanism: Stem Twiner
Distribution (Global): Bismarck Archipelago, Borneo, Malaya, Maluku, New Guinea, Philippines, Sulawesi, Sumatera
IUCN Status: Not evaluated
Reference: L.I. Duperrey, Voy. Monde, Phan.: 12 (1829)

Gnetum latiifolium var. *macropodum* (Kurz) Markgr.
Gnetaceae
Climbing Mechanism: Stem Twiner
IUCN Status: Not evaluated
Reference: Markgr. In: Bull. Jard. Bot. Buitenzorg, sér. 3, 10: 466. (1930)

Gnetum montanum Markgr.
Gnetaceae
Synonyms: *Thoa montana* (Markgr.) Doweld
Climbing Mechanism: Stem Twiner
Distribution (Global): Andaman Is., Assam, Bangladesh, Cambodia, China South-Central, China Southeast, East Himalaya, Laos, Myanmar, Nepal, Thailand, Vietnam
Distribution (India): North Eastern India, Odisha
Leaf Type: Opposite
IUCN Status: Not evaluated
Reference: Bull. Jard. Bot. Buitenzorg, sér. 3, 10: 466 (1930)

Gnetum neglectum Blume
Gnetaceae
Synonyms: *Gnemon neglecta* (Blume) Kuntze
Thoa neglecta (Blume) Doweld
Climbing Mechanism: Stem Twiner
Distribution (Global): Borneo
IUCN Status: Not evaluated
Reference: Rumphia 4: 6 (1849)

Gnetum oblongum Markgr.
Gnetaceae
Synonyms: *Thoa oblonga* (Markgr.) Doweld
Climbing Mechanism: Stem Twiner
Distribution (Global): Bangladesh, Myanmar
Distribution (India): Tripura
IUCN Status: Not evaluated
Reference: Bull. Jard. Bot. Buitenzorg, sér. 3, 10: 471 (1930)

Gnetum scandens Roxb.
Gnetaceae
Common Name: Joint Fir.
Climbing Mechanism: Stem Twiner
Distribution (India): Andaman and Nicobar Islands, Bihar, Odisha
IUCN Status: Not evaluated
Reference: Hort. Bengal.: 66 (1814)

Gnetum ula Brongn.
Gnetaceae
Synonyms: *Gnetum edule* (Willd.) Blume
Common Name: India jointfir
Climbing Mechanism: Stem Twiner
Distribution (India): Andhra Pradesh, Eastern Ghats, Karnataka, Odisha, Tamil Nadu, Western Ghats
IUCN Status: Least Concern
Reference: L.I. Duperrey, Voy. Monde, Phan.: 12 (1829)

Illigera appendiculata Blume
Hernandiaceae
Climbing Mechanism: Stem Twiner
Distribution (Global): Java, Lesser Sunda Is., Sumatera
Distribution (India): Andaman and Nicobar Islands
IUCN Status: Not evaluated
Reference: Bijdr. Fl. Ned. Ind.: 1153 (1827)

Illigera khasiana C.B. Clarke
Hernandiaceae
Climbing Mechanism: Stem Twiner
Distribution (Global): Assam, Bangladesh, Myanmar
Distribution (India): Arunachal Pradesh, Assam, Meghalaya, Mizoram
Leaf Type: Compound, alternate, trifoliate
Inflorescence: Panicles terminal or axillary, cymose
Flowering and Fruiting: August–April
IUCN Status: Not evaluated
Reference: J. D. Hooker, Fl. Brit. India 2: 461 (1878)

Illigera trifoliata (Griff.) Dunn
Hernandiaceae
Synonyms: *Coryzadenia trifoliata* Griff.
Climbing Mechanism: Stem Twiner
Distribution (Global): Andaman Is., Assam, Bangladesh, China South-Central, Laos, Malaya, Myanmar, Sumatera, Thailand, Vietnam
IUCN Status: Not evaluated
Reference: J. Linn. Soc., Bot. 38: 294 (1908)

Illigera villosa C.B. Clarke
Hernandiaceae
Synonyms: *Illigera villosa* C.B. Clarke ex C.E.C. Fisch.
Climbing Mechanism: Stem Twiner
Distribution (Global): Assam, Myanmar
IUCN Status: Not evaluated
Reference: J. Linn. Soc., Bot. 25: 22 (1889)

Hydrangea anomala D. Don
Hydrangeaceae
Synonyms: *Hydrangea altissima* Wall., *Hydrangea anomala* var. *sericea* C.C. Yang, *Hydrangea glabra* Hayata, *Hydrangea glaucophylla* C.C. Yang, *Hydrangea glaucophylla* var. *sericea* (C.C.Yang) C.F.Wei.
Common Name: Climbing Hydrangea, Japanese climbing-hydrangea, Himalayan climbing-hydrangea
Climbing Mechanism: Stem Twiner
Distribution (Global): Assam, China North-Central, China South-Central, China Southeast, East Himalaya, Myanmar, Nepal, Taiwan, Tibet
Distribution (India): North Eastern India
IUCN Status: Not evaluated
Reference: Prodr. Fl. Nepal.: 211 (1825)

Philadelphus tomentosus Wall. ex G. Don
Hydrangeaceae
Synonyms: *Philadelphus coronarius* var. *tomentosus*

Common Name: Fuzzy mock orange
Climbing Mechanism: Scrambler-Unarmed
Distribution (Global): Assam, China South-Central, East Himalaya, Nepal, Tibet, West Himalaya
Distribution (India): North Eastern India
IUCN Status: Not evaluated
Reference: Gen. Hist. 2: 807 (1832)

Iodes cirrhosa Turcz.
Icacinaceae
Synonyms: *Iodes brandisii* Kurz, *Iodes floribunda* Merr., *Iodes tomentella* Miq., *Natsiatum oppositifolium* Planch
Climbing Mechanism: Tendril Climber
Distribution (Global): Andaman Is., Assam, Borneo, China South-Central, China Southeast, Java, Laos, Malaya, Myanmar, Nicobar Is., Philippines, Sulawesi, Sumatera, Thailand, Vietnam
Distribution (India): Great Nicobar Island
Leaf Type: Simple, opposite
Inflorescence: Cymes
Fruit Type: Drupe
Flowering and Fruiting: January–October
IUCN Status: Not evaluated
Reference: Bull. Soc. Imp. Naturalistes Moscou 27(II): 281 (1854 publ. 1855)

Iodes ovalis Blume
Icacinaceae
Synonyms: *Iodes horsfieldii* Baill., *Iodes oblonga* Planch. ex Mast., *Polyporandra junghuhnii* Koord.
Climbing Mechanism: Scrambler-Unarmed
Distribution (Global): Java, Malaya, Myanmar, Sumatera
IUCN Status: Not evaluated
Reference: Bijdr. Fl. Ned. Ind.: 30 (1825)

Mappianthus hookerianus (Baill.) Sleumer
Icacinaceae
Synonyms: *Iodes hookeriana* Baill., *Iodes thomsoniana* Baill., *Mappianthus borneensis* Merr.
Climbing Mechanism: Scrambler-Unarmed
Distribution (Global): Assam, Bangladesh, Borneo, East Himalaya, Malaya, Myanmar, Sumatera.
Leaf Type: Leaves simple, opposite
Inflorescence: Inflorescences in extra-axillary cymes
Fruit Type: Drupes, red or orange-red, ellipsoid, ovoid, compressed
Flowering and Fruiting: April–October
IUCN Status: Not evaluated

Notes: Occurs usually in open woodlands
Reference: Blumea 17: 225 (1969)

Miquelia dentata Bedd.
Icacinaceae
Climbing Mechanism: Stem Twiner
Distribution (India): Karnataka, Kerala, Tamil Nadu
Leaf Type: Simple, base rounded or truncate, dentate
Inflorescence: Peduncled globose spike
Fruit Type: Wet, drupe
IUCN Status: Not evaluated

Miquelia gibba Baill.
Icacinaceae
Climbing Mechanism: Stem Twiner
Leaf Type: Simple, base rounded or truncate, dentate
Inflorescence: Peduncled globose spike
Fruit Type: Wet, drupe
IUCN Status: Not evaluated

Miquelia kleinii Meisn.
Icacinaceae
Climbing Mechanism: Stem Twiner
Distribution (India): Arunachal Pradesh, Assam
Leaf Type: Simple, base rounded or truncate, dentate
Inflorescence: Peduncled globose spike
Fruit Type: Wet, drupe
IUCN Status: Not evaluated

Natsiatum herpeticum Buch. -Ham. ex Arn.
Icacinaceae
Synonyms: *Sicyos pentandrus* Wall. ex G. Don, *Natsiatum tonkinense* Gagnep.
Climbing Mechanism: Stem Twiner
Distribution (Global): Assam, Bangladesh, China South-Central, East Himalaya, India, Laos, Myanmar, Nepal, Thailand, Vietnam.
Distribution (India): Andhra Pradesh, Arunachal Pradesh, Assam, Bihar, Odisha, Eastern Ghats, Odisha, West Bengal.
Leaf Type: Simple, alternate
Inflorescence: Spikes like racemes
Fruit Type: Ovate drupe
Flowering and Fruiting: December–April
IUCN Status: Not evaluated
Notes: Commonly found in scrub jungles
Reference: Edinburgh New Philos. J. 16: 314 (1834)

Phytocrene bracteata Wall.
Icacinaceae
Synonyms: *Phytocrene macrocarpa* Griff., *Gynocephalum bracteatum* (Wall.) Trécul.
Climbing Mechanism: Scrambler-Unarmed
Distribution (Global): Malaya, Myanmar, Sumatera, Thailand
IUCN Status: Not evaluated
Reference: Pl. Asiat. Rar. 3: 12 (1832)

Phytocrene macrophylla (Blume) Blume
Icacinaceae
Synonyms: *Gynocephalum macrophyllum* Blume
Climbing Mechanism: Scrambler-Unarmed
Distribution (Global): Bangladesh, Borneo, Java, Myanmar, Philippines, Sulawesi, Sumatera
IUCN Status: Not evaluated
Reference: Rumphia 4: 36 (1849)

Phytocrene oblonga Wall.
Icacinaceae
Synonyms: *Miquelia cancellata* Kurz, *Gynocephalum oblongum* (Wall.) Trécul.
Climbing Mechanism: Scrambler-Unarmed
Distribution (Global): Malaya, Myanmar, Thailand, Vietnam
IUCN Status: Not evaluated
Reference: Pl. Asiat. Rar. 3: 12 (1832)

Phytocrene palmata Wall.
Icacinaceae
Synonyms: *Phytocrene stylocarpa* Griff., *Gynocephalum palmatum* (Wall.) Trécul.
Climbing Mechanism: Scrambler-Unarmed
Distribution (Global): Malaya, Sumatera, Thailand
IUCN Status: Not evaluated
Reference: Pl. Asiat. Rar. 3: 12 (1832)

Pyrenacantha volubilis Hook.
Icacinaceae
Synonyms: *Cavanilla volubilis* Kuntze
Climbing Mechanism: Stem Twiner
Distribution (Global): Cambodia, Hainan, India, Laos, Sri Lanka, Thailand, Vietnam
Distribution (India): Andhra Pradesh, Eastern Ghats, Kerala, Puducherry, Tamil Nadu
Flowering and Fruiting: June–November
IUCN Status: Not evaluated
Reference: Bot. Misc. 2: 107 (1830)

Sarcostigma kleinii Wight & Arn.
Icacinaceae
Synonyms: *Chailletia edulis* Kurz., *Sarcostigma edule* Kurz., *Sarcostigma horsfieldii* R.Br., *Sarcostigma roxburghii* Wall. ex Griff., *Sarcostigma wallichii* Baill.
Common Name: Odal oil plant
Climbing Mechanism: Stem Twiner
Distribution (Global): Andaman Is., Borneo, India, Java, Malaya, Myanmar, Vietnam
Distribution (India): Andaman and Nicobar Islands, Goa, Karnataka, Maharashtra, Tamil Nadu, Western Ghats
Leaf Type: Simple
Inflorescence: Spikes, axillary
Fruit Type: Wet, drupe
Flowering and Fruiting: October–February
IUCN Status: Not evaluated
Notes: Evergreen and semi-Evergreen forests, along the ghats, also in the sacred groves
Reference: Edinburgh New Philos. J. 14: 299 (1833)

Clerodendrum capitatum
Lamiaceae
Climbing Mechanism: Scrambler-Unarmed
IUCN Status: Least Concern

Clerodendrum japonicum
Lamiaceae
Climbing Mechanism: Scrambler-Unarmed
IUCN Status: Least Concern

Clerodendrum splendens G. Don
Lamiaceae
Synonyms: *Clerodendrum aurantium* G. Don, *Clerodendrum giletii* De Wild. & T. Durand, *Clerodendrum rollisonii* Rollisson, *Clerodendrum rollissonii* Rollisson, *Siphonanthus splendens* (G. Don) Hiern.
Common Name: Flaming glorybower, Flaming glorybower, Pagoda flower, Bleeding heart vine
Climbing Mechanism: Stem Twiner
Distribution (Global): Angola, Benin, Burundi, Cabinda, Cameroon, Central African Repu, Congo, Equatorial Guinea, Gabon, Ghana, Guinea, Guinea-Bissau, Ivory Coast, Liberia, Nigeria, Senegal, Sierra Leone, Zaire
Distribution (India): Delhi, Gujarat, Madhya Pradesh, Maharashtra, Odisha, Uttar Pradesh
Leaf Type: Simple, decussate-opposite or whorled
Inflorescence: Axillary cyme, terminal thyrse or rarely corymb

Fruit Type: Fruit drupaceous with 4 pyrenes or schizocarp
Flowering and Fruiting: September–January
IUCN Status: Not evaluated
Notes: Cultivated in gardens
Reference: Edinburgh Philos. J. 11: 349 (1824)

Clerodendrum thomsoniae Balf.f.
Lamiaceae
Synonyms: *Clerodendrum balfourii* (B.D. Jacks. ex Dombrain) Dombrain, *Clerodendrum delectum* Pynaert, *Clerodendrum thomsoniae* var. *balfourii* B.D. Jacks. ex Dombrain
Common Name: Bleeding heart vine, Bleeding glory bower, Bag flower
Climbing Mechanism: Scrambler-Unarmed
Distribution (Global): Cameroon, Gambia, Ghana, Gulf of Guinea Is., Nigeria, Senegal, Sierra Leone, Zaire
Distribution (India): Maharashtra, Tamil Nadu, Tripura
Leaf Type: Simple, decussate-opposite
Inflorescence: Axillary cyme, terminal thyrse, or rarely corymb
Fruit Type: Schizocarp
Flowering and Fruiting: Throughout the year
IUCN Status: Not evaluated
Notes: Often cultivated in gardens
Reference: Gard. Chron. 1862: 70 (1862)

Clerodendrum volubile
Lamiaceae
Climbing Mechanism: Scrambler-Unarmed
IUCN Status: Not evaluated

Congea tomentosa Roxb.
Lamiaceae
Synonyms: *Calochlamys capitata* C. Presl., *Congea azurea* Wall., *Congea oblonga* Pierre ex Dop., *Congea petelotii* Moldenke., *Congea villosa* (Roxb.) Voigt., *Roscoea villosa* Roxb.
Common Name: Wooly Congea, Shower orchid, White Congea
Climbing Mechanism: Scrambler-Unarmed
Distribution (Global): Assam, Bangladesh, Cambodia, China South-Central, India, Laos, Myanmar, Thailand, Vietnam
Distribution (India): Maharashtra, Mizoram, Tamil Nadu
Leaf Type: Opposite pairs
Inflorescence: Terminal or axillary, of 3–9 flowered capitate cymes
Fruit Type: Drupes
IUCN Status: Not evaluated
Reference: Pl. Coromandel 3: 90 (1820)

Congea tomentosa var. *azurea*
Lamiaceae
Climbing Mechanism: Scrambler-Unarmed
Distribution (India): Peninsular India
Leaf Type: Simple
Inflorescence: Panicles, flowers pink
Fruit Type: Wet, drupe
IUCN Status: Not evaluated
Reference: Roxb. In: Pl. Coromandel 3: 90. (1820)

Congea velutina Wight
Lamiaceae
Climbing Mechanism: Scrambler-Unarmed
Distribution (Global): Malaya, Myanmar, Sumatera, Thailand
Distribution (India): Peninsular India
Leaf Type: Simple
Inflorescence: Panicles, flowers pink
Fruit Type: Wet, drupe
IUCN Status: Not evaluated
Reference: Icon. Pl. Ind. Orient.: t. 1497/3 (1849)

Congea vestita Griff.
Lamiaceae
Climbing Mechanism: Scrambler-Unarmed
Distribution (Global): Malaya, Myanmar, Thailand, Vietnam
Leaf Type: Simple
Inflorescence: Panicles, flowers pink
Fruit Type: Wet, drupe
IUCN Status: Not evaluated
Reference: Not. Pl. Asiat. 4: 174 (1854)

Glossocarya mollis Wall. ex Griff.
Lamiaceae
Synonyms: *Glossocarya mollis* var. *maxwellii* Moldenke
Climbing Mechanism: Scrambler-Unarmed
Distribution (Global): Malaya, Myanmar, New Guinea, Thailand, Vietnam
IUCN Status: Not evaluated
Reference: Calcutta J. Nat. Hist. 3: 366 (1843)

Glossocarya scandens (L.f.) Trimen
Lamiaceae
Synonyms: *Clerodendrum linnaei* Thwaites, *Clerodendrum scandens* (L.f.) Druce, *Glossocarya linnaei* (Thwaites) Maxim., *Glossocarya scandens* var. *pubescens* Moldenke, *Glossocarya scandens* f. *pubescens* (Moldenke) Moldenke, *Volkameria scandens* L.f.

Climbing Mechanism: Scrambler-Unarmed
Distribution (Global): New Guinea, Sri Lanka
IUCN Status: Not evaluated
Reference: Syst. Cat. Fl. Pl. Ceylon: 69 (1885)

Holmskioldia sanguinea Retz.
Lamiaceae
Synonyms: *Hastingia coccinea* Sm., *Hastingia scandens* Roxb., *Holmskioldia rubra* Pers., *Holmskioldia sanguinea* f. *aurantiaca* Yin Yin Kyi & DeFilipps, *Holmskioldia scandens* Sweet, *Platunum rubrum* A. Juss.
Common Name: Chinese hat plant, Cup and saucer plant, Mandarin's hat, Parasol flower
Climbing Mechanism: Scrambler-Unarmed
Distribution (Global): Assam, East Himalaya, India, Myanmar, Nepal, West Himalaya
Distribution (India): Arunachal Pradesh, Bihar, Odisha, Delhi, Eastern Ghats, Gujarat, Maharashtra, Meghalaya, Odisha, Rajasthan, Tamil Nadu, Telangana, Tripura
Leaf Type: Leaves decussate opposite
Inflorescence: Terminal thyrse of lax cymes or racemes
Fruit Type: Drupe
Flowering and Fruiting: October–December
IUCN Status: Not evaluated
Reference: Observ. Bot. 6: 31 (1791)

Hosea lobbii (C.B. Clarke) Ridl.
Lamiaceae
Synonyms: *Clerodendrum discolor* Becc., *Clerodendrum lobbii* C.B. Clarke, *Hoseanthus lobbii* (C.B. Clarke) Merr.
Climbing Mechanism: Scrambler-Unarmed
Distribution (Global): Borneo.
Leaf Type: Simple, ovate
Inflorescence: Axillary cymes
Fruit Type: Wet, fusiform
IUCN Status: Not evaluated
Reference: J. Straits Branch Roy. Asiat. Soc. 50: 125 (1908)

Hymenopyramis brachiata
Lamiaceae
Climbing Mechanism: Scrambler-Unarmed
IUCN Status: Not evaluated

Petraeovitex bambusetorum King & Gamble
Lamiaceae
Climbing Mechanism: Scrambler-Unarmed

Distribution (Global): Borneo, Malaya
Distribution (India): Tamil Nadu
IUCN Status: Not evaluated
Reference: Bull. Misc. Inform. Kew 1908: 113 (1908)

Pogostemon hedgei V.S. Kumar & B.D. Sharma
Lamiaceae
Climbing Mechanism: Scrambler-Unarmed
Distribution (Global): India.
IUCN Status: Not evaluated
Notes: Usually occur in evergreen forests
Reference: Nordic J. Bot. 15: 163 (1995)

Premna coriacea C.B. Clarke
Lamiaceae
Synonyms: *Gumira coriacea* (C.B. Clarke) Kuntze
Climbing Mechanism: Scrambler-Unarmed
Distribution (Global): Andaman Is., Assam, Bangladesh, Cambodia, East Himalaya, India, Laos, Nepal, Thailand, Vietnam
Distribution (India): Bihar, Odisha, Goa, Gujarat, Maharashtra, Odisha, Tamil Nadu, Tripura, Western Ghats
Leaf Type: Simple
Inflorescence: Axillary or terminal, cyme panicle, spike, thyrse or corymb
Fruit Type: Drupe
Flowering and Fruiting: April–June
IUCN Status: Not evaluated
Notes: Mostly found in evergreen and semi-evergreen forests
Reference: J. D. Hooker, Fl. Brit. India 5: 573 (1888)

Premna coriacea var. *villosa* (C.B. Clarke) A. Rajendran & P. Daniel
Lamiaceae
Climbing Mechanism: Scrambler-Unarmed
Distribution (India): Kerala, Tamil Nadu
Flowering and Fruiting: March–October
IUCN Status: Not evaluated
Notes: Usually occur in moist deciduous forests and semi-evergreen forests

Premna corymbosa Rottler & Willd.
Lamiaceae
Synonyms: *Gumira corymbosa* (Rottler) Kuntze, *Premna alstonii* Moldenke, *Premna alstonii* var. *mollis* Moldenke, *Premna alstonii* var. *subcrenata* Moldenke
Climbing Mechanism: Scrambler-Unarmed
Distribution (Global): Andaman Is., India, Nicobar Is., Sri Lanka
Distribution (India): Puducherry, Tamil Nadu

Leaf Type: Simple
Inflorescence: Axillary or terminal, cyme panicle, spike, thyrse or corymb
Fruit Type: Drupe
Flowering and Fruiting: December–September
IUCN Status: Not evaluated
Notes: Usually found in the foothills of the ghats and also in dry forests
Reference: Neue Schriften Ges. Naturf. Freunde Berlin 4: 187 (1803)

Premna divaricata Wall. ex Schauer
Lamiaceae
Synonyms: *Premna serratifolia* L.
Climbing Mechanism: Scrambler-Unarmed
Leaf Type: Simple
Inflorescence: Axillary or terminal, cyme panicle, spike, thyrse or corymb
Fruit Type: Drupe
IUCN Status: Not evaluated
Reference: A.P.de Candolle, Prodr. 11: 631 (1847)

Premna interrupta Wall. ex Schauer
Lamiaceae
Synonyms: *Gumira interrupta* (Wall. ex Schauer) Kuntze, *Gumira racemosa* (Wall. ex Schauer) Kuntze, *Premna derryana* King & Gamble, *Premna racemosa* Wall. ex Schauer, *Surfacea derryana* (King & Gamble) Moldenke + 4
Climbing Mechanism: Scrambler-Unarmed
Distribution (Global): Assam, Bangladesh, Cambodia, China South-Central, China Southeast, East Himalaya, Laos, Malaya, Myanmar, Nepal, Thailand, Tibet, Vietnam, West Himalaya
Distribution (India): Mizoram
Leaf Type: Simple
Inflorescence: Axillary or terminal, cyme panicle, spike, thyrse or corymb
Fruit Type: Drupe
IUCN Status: Not evaluated
Reference: A.P.de Candolle, Prodr. 11: 633 (1847)

Premna khasiana C.B. Clarke
Lamiaceae
Synonyms: *Gumira khasiana* (C.B. Clarke) Kuntze
Climbing Mechanism: Scrambler-Unarmed
Distribution (Global): Assam, Bangladesh, Thailand
Leaf Type: Simple
Inflorescence: Axillary or terminal, cyme panicle, spike, thyrse or corymb
Fruit Type: Drupe
IUCN Status: Not evaluated
Reference: J. D. Hooker, Fl. Brit. India 4: 575 (1885)

Premna paucinervis (C.B. Clarke) Gamble
Lamiaceae
Climbing Mechanism: Scrambler-Unarmed
Distribution (Global): India.
Distribution (India): Kerala, Tamil Nadu
Leaf Type: Simple
Inflorescence: Axillary or terminal, cyme panicle, spike, thyrse or corymb
Fruit Type: Drupe
Flowering and Fruiting: December–May
IUCN Status: Not evaluated
Notes: Usually found in semi-evergreen forests
Reference: Fl. Madras: 1095 (1924)

Premna purpurascens Thwaites
Lamiaceae
Synonyms: *Gumira purpurascens* (Thwaites) Kuntze, *Phaenicanthus zeylanicus* Thwaites
Climbing Mechanism: Scrambler-Unarmed
Distribution (Global): Sri Lanka
IUCN Status: Not evaluated
Reference: Enum. Pl. Zeyl.: 242 (1861)

Premna scandens Roxb.
Lamiaceae
Synonyms: *Gumira scandens* (Roxb.) Kuntze, *Premna cordifolia* Brandis, *Premna ramosa* Wall., *Premna cordifolia* Brandis, nom. illeg., *Premna coriacea* var. *cuneata* C.B. Clarke, *Premna coriacea* var. *oblonga* C.B. Clarke, *Premna ramosa* Wall., nom. nud.
Climbing Mechanism: Scrambler-Unarmed
Distribution (Global): Andaman Is., Assam, Bangladesh, China South-Central, East Himalaya, India, Malaya, Myanmar, Nepal, Thailand, Vietnam
Distribution (India): Arunachal Pradesh, Bihar, Odisha, Meghalaya
Leaf Type: Simple
Inflorescence: Axillary or terminal, cyme panicle, spike, thyrse or corymb
Fruit Type: Drupe
IUCN Status: Not evaluated
Reference: Fl. Ind. ed. 1832, 3: 82 (1832)

Premna tomentosa Willd.
Lamiaceae
Synonyms: *Gumira adenosticta* (Schauer) Kuntze, *Gumira cardiophylla* (Schauer) Kuntze, *Gumira cumingiana* (Schauer) Kuntze, *Gumira pyramidata* (Wall. ex Schauer) Kuntze, *Premna adenosticta* Schauer, *Premna cardiophylla* Schauer +12
Common Name: Wooly leaved fire brand teak, Bastard teak

Climbing Mechanism: Scrambler-Unarmed
Distribution (Global): Andaman Is., Assam, Bangladesh, Borneo, Cambodia, East Himalaya, India, Java, Malaya, Myanmar, Nepal, Nicobar Is., Philippines, Queensland, Solomon Is., Sri Lanka, Sulawesi, Sumatera, Thailand, Vietnam
Distribution (India): Tamil Nadu
Leaf Type: Simple
Inflorescence: Terminal or axillary corymb
Fruit Type: Drupe
Flowering and Fruiting: January–May
IUCN Status: Least Concern
Notes: Usually found in dry-deciduous forests
Reference: Sp. Pl., ed. 4, 3: 314 (1800)

Pseudocaryopteris foetida (D. Don) P.D. Cantino
Lamiaceae
Synonyms: *Callicarpa esquirolii* H. Lév., *Callicarpa martini* H. Lév., *Callicarpa vansteenisii* Moldenke, *Caryopteris foetida* (D. Don) Thell., *Caryopteris grata* (Wall. ex Walp.) Benth. & Hook.f. ex C.B. Clarke + 6
Common Name: Stinking bluebeard
Climbing Mechanism: Scrambler-Unarmed
Distribution (Global): Assam, Bangladesh, China South-Central, China Southeast, East Himalaya, India, Myanmar, Nepal, Pakistan, Sumatera, Thailand, West Himalaya
Leaf Type: Simple
Inflorescence: Axillary, cyme, corymb
Fruit Type: Globose, capsule
IUCN Status: Not evaluated
Reference: Syst. Bot. 23: 381 (1998 publ. 1999)

Salvia dorrii var. *pilosa* (A. Gray) Strachan & Reveal
Lamiaceae
Synonyms: *Audibertia incana* var. *pilosa* A. Gray., *Salvia carnosa* var. *pilosa* (A. Gray) H.M.Hall., *Ramona pilosa* (A.Gray) Abrams., *Salvia carnosa* subsp. *pilosa* (A. Gray) Epling
Climbing Mechanism: Stem Twiner
Distribution (Global): Arizona, California, Nevada
Leaf Type: Simple
Inflorescence: Flowers in clusters
Fruit Type: Wet, drupe
IUCN Status: Not evaluated
Reference: Great Basin Naturalist 41: 199 (1981)

Sphenodesme eryciboides Kurz
Lamiaceae
Synonyms: *Sphenodesme grossa* Kurz., *Symphorema grossum* Kurz.

Climbing Mechanism: Stem Twiner
Distribution (Global): Myanmar, Thailand
Leaf Type: Compound, 7–9 pairs
Inflorescence: Panicles, axillary in clusters
Fruit Type: Dry, pods
IUCN Status: Not evaluated
Reference: J. Asiat. Soc. Bengal, Pt. 2, Nat. Hist. 40: 76 (1871)

Sphenodesme griffithiana Wight
Lamiaceae
Synonyms: *Decadontia caerulescens* Griff., *Sphenodesme robinsonii* Dop.
Climbing Mechanism: Stem Twiner
Distribution (Global): Myanmar, Vietnam
Leaf Type: Compound, 7–9 pairs
Inflorescence: Panicles, axillary in clusters
Fruit Type: Dry, pods
IUCN Status: Not evaluated
Reference: Icon. Pl. Ind. Orient. 4(3): 14 (1849)

Sphenodesme involucrata (C. Presl) B.L. Rob.
Lamiaceae
Synonyms: *Vitex involucratus* C. Presl
Climbing Mechanism: Stem Twiner
Distribution (Global): Andaman Is., Assam, Bangladesh, China Southeast, Hainan, India, Myanmar, Nicobar Is., Sumatera, Taiwan, Thailand, Vietnam
Leaf Type: Compound, 7–9 pairs
Inflorescence: Panicles, axillary in clusters
Fruit Type: Dry, pods
IUCN Status: Not evaluated
Reference: Proc. Amer. Acad. Arts 51: 531 (1916)

Sphenodesme involucrata var. *paniculata* (C.B. Clarke) Munir
Lamiaceae
Synonyms: *Sphenodesme paniculata* C.B. Clarke
Common Name: Panicled varshiki
Climbing Mechanism: Stem Twiner
Distribution (Global): India, Myanmar
Distribution (India): Andaman and Nicobar Islands, Eastern Ghats, Tamil Nadu
Leaf Type: Simple, Opposite
Inflorescence: Panicles composed of 6–12 opposite dichotomously branched cymes
Fruit Type: Carcerulus
Flowering and Fruiting: December–August
IUCN Status: Not evaluated
Notes: Often occur in semi-evergreen and evergreen forests
Reference: Gard. Bull. Singapore 21: 338 (1966)

Sphenodesme pentandra var. *wallichiana* (Schauer) Munir
Lamiaceae
Synonyms: *Congea pentandra* (Roxb.) Steud., *Roscoea pentandra* Roxb., *Sphenodesme wallichiana* Schauer.
Climbing Mechanism: Stem Twiner
Distribution (Global): Assam, Bangladesh, Borneo, Cambodia, China South-Central, China Southeast, Hainan, Laos, Malaya, Myanmar, Nicobar Is., Sri Lanka, Thailand, Vietnam
Distribution (India): Arunachal Pradesh, Meghalaya
Leaf Type: Leaves simple, opposite, petiolate, exstipulate
Inflorescence: Axillary or terminal paniculate thyrses, cyme 3–7 flowered
Fruit Type: Indehiscent, carcerulus, obovoid, Yellow
Flowering and Fruiting: February–May
IUCN Status: Not evaluated
Notes: Found mostly in evergreen forests
Reference: Gard. Bull. Singapore 21: 360 (1966)

Symphorema involucratum Roxb.
Lamiaceae
Synonyms: *Congea involucratum* (Roxb.) Wall., *Analectis speciosa* Vahl., *Congea paniculata* Wall.
Climbing Mechanism: Scrambler-Unarmed
Distribution (Global): Andaman Is., Bangladesh, China South-Central, India, Myanmar, Sri Lanka, Thailand
Distribution (India): Andhra Pradesh, Eastern Ghats, Maharashtra, Odisha, Tamil Nadu, Telangana
Leaf Type: Elliptic, oblong
Inflorescence: Capitate cymes
Fruit Type: 1-seeded. enclosed by the calyx
Flowering and Fruiting: February–May
IUCN Status: Not evaluated
Notes: Found in a wide range of habitats: Evergreen, semi-evergreen and dry forests
Reference: Pl. Coromandel 2: 46 (1805)

Symphorema polyandrum Wight
Lamiaceae
Common Name: South Indian Symphorema, South Indian star-vervain climber
Climbing Mechanism: Scrambler-Unarmed
Distribution (Global): India
Distribution (India): Andhra Pradesh, Bihar, Odisha, Eastern Ghats, Madhya Pradesh, Maharashtra, Odisha, Tamil Nadu, Telangana.
Leaf Type: Simple, oppositely arranged
Inflorescence: Head-like cymes
Flowering and Fruiting: March–May
IUCN Status: Not evaluated

Notes: Found mostly in dry deciduous forests
Reference: Icon. Pl. Ind. Orient.: t. 363 (1840)

Volkameria inermis L.
Lamiaceae
Synonyms: *Clerodendrum buxifolium* (Willd.) Spreng., *Clerodendrum capsulare* Blanco, *Clerodendrum commersonii* (Poir.) Spreng., *Clerodendrum coriaceum* Poir., *Clerodendrum coromandelianum* Spreng. + 14
Common Name: Common hedge bower, Glory bower, Garden quinine, Sorcerers bush, Wild jasmine
Climbing Mechanism: Scrambler-Unarmed
Distribution (Global): Andaman Is., Bangladesh, Bismarck Archipelago, Borneo, Cambodia, Caroline Is., China Southeast, Fiji, Gilbert Is., India, Japan, Java, Laccadive Is., Lesser Sunda Is., Malaya, Maldives, Maluku, Marianas, Marshall Is., Myanmar, Nansei-shoto, Nauru, New Caledonia, New Guinea, New South Wales, Nicobar Is., Niue, Northern Territory, Pakistan, Philippines, Queensland, Samoa, Solomon Is., South China Sea, Sri Lanka, Sulawesi, Sumatera, Taiwan, Thailand, Tonga, Tuvalu, Vanuatu, Vietnam, Wallis-Futuna Is., West Himalaya
Distribution (India): Tamil Nadu
Leaf Type: Leaves opposite, simple
Inflorescence: Cymes or Umbells
Flowering and Fruiting: Throughout the year
IUCN Status: Not evaluated
Notes: Found in plains, scrub jungles, and wastelands
Reference: Sp. Pl.: 637 (1753)

Holboellia angustifolia Wall.
Lardizabalaceae
Synonyms: *Holboellia fargesii*, *Holboellia acuminata.*
Common Name: Narrow-leaf sausage vine, Farge's Holboellia
Climbing Mechanism: Scrambler-Unarmed
Leaf Type: Leaves are trifoliate to digitate
Inflorescence: Racemose corymbs or fascicled, axillary
Fruit Type: Several seeded berries
Flowering and Fruiting: April–May/June–July
IUCN Status: Not evaluated
Reference: Tent. Fl. Napal.: 25 (1824)

Holboellia latifolia Wall.
Lardizabalaceae
Synonyms: *Stauntonia latifolia* (Wall.) R.Br. ex Wall.
Common Name: Sausage vine, Holboellia
Climbing Mechanism: Scrambler-Unarmed
Distribution (India): Arunachal Pradesh, Himachal Pradesh, Manipur, Meghalaya, Sikkim

Leaf Type: Leaves are palmately compound
Inflorescence: Racemes
IUCN Status: Not evaluated
Reference: Tent. Fl. Napal.: 24 (1824)

Stauntonia brunoniana (Decne.) Hemsl.
Lardizabalaceae
Synonyms: *Parvatia brunoniana* Decne., *Parvatia brunoniana* subsp. *elliptica* (Hemsl.) H.N. Qin., *Parvatia elliptica* (Hemsl.) M.P. Nayar & T.K. Paul., *Stauntonia elliptica* Hemsl., *Stauntonia trifoliata* Griff.
Climbing Mechanism: Scrambler-Unarmed
Distribution (Global): Assam, Bangladesh, China South-Central, China Southeast, Nepal, Thailand, Vietnam
Leaf Type: Leaves alternate, pinnately compound 3–9 foliate
Inflorescence: Racemose, fascicled, axillary
Fruit Type: Berries, fruitlet 1–3, indehiscent
IUCN Status: Not evaluated
Reference: Hooker's Icon. Pl. 29: t. 2843 (1907)

Stauntonia elliptica Hemsl.
Lardizabalaceae
Synonyms: *Stauntonia brunoniana* (Decne.) Wall. ex Hemsl., *Parvatia brunoniana* Decne., *Parvatia brunoniana* Decne., *Parvatia elliptica* (Hemsl.) M.P. Nayar & T.K. Paul., *Parvatia elliptica* (Hemsl.) M.P. Nayar & T.K. Paul.
Climbing Mechanism: Scrambler-Unarmed
Distribution (Global): Assam, Bangladesh, China South-Central, China Southeast, Nepal, Thailand, Vietnam
Leaf Type: Leaves alternate, pinnately 3–9 foliate
Inflorescence: Racemose, fascicled, axillary
Fruit Type: Berries, greyish brown when ripe
Flowering and Fruiting: August–November
IUCN Status: Not evaluated
Reference: Hooker's Icon. Pl. 29: t. 2843 (1907)

Stauntonia filamentosa Griff.
Lardizabalaceae
Synonyms: *Holboellia filamentosa* (Griff.) H.N. Qin., *Holboellia khasiana* T.K. Paul & M.P. Nayar., *Parvatia filamentosa* (Griff.) Gagnep., *Stauntonia racemosa* Chatterjee
Climbing Mechanism: Scrambler-Unarmed
Distribution (Global): Assam, Myanmar
Leaf Type: Compound, alternate, palmate
Inflorescence: racemes
Fruit Type: Dry, pods

IUCN Status: Not evaluated
Reference: Not. Pl. Asiat. 4: 384 (1854)

Cassytha capillaris
Lauraceae
Climbing Mechanism: Stem Twiner
IUCN Status: Not evaluated

Cassytha glabella R.Br.
Lauraceae
Common Name: Lone vine
Climbing Mechanism: Stem Twiner
Distribution (Global): New South Wales, Queensland, South Australia, Tasmania, Victoria, Western Australia
Leaf Type: Parasitic twiner
Inflorescence: Spikes, sub-sessile
Fruit Type: Wet, drupe
IUCN Status: Not evaluated
Reference: Prodr. Fl. Nov. Holland.: 404 (1810)

Hugonia belli Sedgw.
Linaceae
Climbing Mechanism: Hook Climber
Distribution (Global): India
Distribution (India): Karnataka
IUCN Status: Not evaluated
Reference: Indian Forester 46: 424 (1920)

Hugonia ferruginea Wight & Arn.
Linaceae
Climbing Mechanism: Hook Climber
Distribution (Global): India, Sri Lanka
Distribution (India): Karnataka, Tamil Nadu, Western Ghats
IUCN Status: Not evaluated
Reference: Prodr. Fl. Ind. Orient. 1: 72 (1834)

Hugonia mystacina St.-Lag.
Linaceae
Synonyms: *Hugonia mystax* L.
Climbing Mechanism: Hook Climber
Distribution (India): Andhra Pradesh, Bihar, Odisha, Eastern Ghats, Maharashtra, Odisha, Puducherry, Tamil Nadu, Telangana
Leaf Type: Simple
Inflorescence: Axillary
Fruit Type: Drupe

Flowering and Fruiting: August–October
IUCN Status: Not evaluated
Reference: Ann. Soc. Linn. Lyon 7: 127 (1880)

Indorouchera griffithiana (Planch.) Hallier f.
Linaceae
Climbing Mechanism: Hook Climber
Distribution (India): Great Nicobar Island
IUCN Status: Not evaluated
Reference: Meded. Rijks-Herb. 35: 50 (1918)

Gardneria angustifolia Wall.
Loganiaceae
Synonyms: *Gardneria distincta* P.T.Li, *Gardneria glabra* Wall. ex D. Don, *Pseudogardneria angustifolia* (Wall.) Racib.
Climbing Mechanism: Scrambler-Unarmed
Distribution (Global): Assam, Bangladesh, China South-Central, East Himalaya, Nepal, West Himalaya
Distribution (India): Arunachal Pradesh, Meghalaya, Mizoram, West Bengal
IUCN Status: Not evaluated
Reference: W. Roxburgh, Fl. Ind. 1: 318 (1820)

Gardneria ovata Wall.
Loganiaceae
Synonyms: *Gardneria wallichii* Wight ex Wall.
Common Name: Oval leaf Gardneria
Climbing Mechanism: Scrambler-Unarmed
Distribution (Global): Assam, Bangladesh, China South-Central, China Southeast, India, Java, Laos, Malaya, Myanmar, Sri Lanka, Sumatera, Thailand, Tibet, Vietnam
Distribution (India): Arunachal Pradesh, Meghalaya, Tamil Nadu, Western Ghats
Flowering and Fruiting: February–May
IUCN Status: Not evaluated
Reference: W. Roxburgh, Fl. Ind. 1: 400 (1820)

Strychnos acuminata Wall.
Loganiaceae
Synonyms: *Strychnos rufa* C.B. Clarke
Climbing Mechanism: Hook Climber
Distribution (Global): Malaya, Myanmar
Distribution (India): Andaman and Nicobar Islands
Leaf Type: Simple
Inflorescence: Cymes in solitary
Fruit Type: Wet, berry

IUCN Status: Not evaluated
Reference: J. D. Hooker, Fl. Brit. India 4: 89 (1883)

Strychnos aenea A.W. Hill
Loganiaceae
Climbing Mechanism: Hook Climber
Distribution (Global): Assam, India
Distribution (India): Western Ghats
Leaf Type: Simple
Inflorescence: Cymes in solitary
Fruit Type: Wet, berry
IUCN Status: Not evaluated
Reference: Bull. Misc. Inform. Kew 1917: 138 (1917)

Strychnos andamanensis A.W. Hill
Loganiaceae
Climbing Mechanism: Hook Climber
Distribution (Global): Andaman Is.
Distribution (India): Andaman and Nicobar Islands
Leaf Type: Simple
Inflorescence: Cymes in solitary
Fruit Type: Wet, berry
IUCN Status: Not evaluated
Reference: Bull. Misc. Inform. Kew 1917: 146 (1917)

Strychnos axillaris Colebr.
Loganiaceae
Synonyms: *Strychnos arborea* A.W. Hill., *Strychnos armata* A.W. Hill., *Strychnos palembanica* Miq., *Strychnos penicillata* A.W. Hill. + 20
Climbing Mechanism: Hook Climber
Distribution (Global): Assam, Bangladesh, Borneo, Cambodia, China South-Central, Java, Laos, Lesser Sunda Is., Malaya, Maluku, Myanmar, New Guinea, New South Wales, Nicobar Is., Philippines, Queensland, Sri Lanka, Sulawesi, Sumatera, Thailand, Vietnam
Distribution (India): Andaman and Nicobar Islands, Arunachal Pradesh
Leaf Type: Simple
Inflorescence: Cymes in solitary
Fruit Type: Wet, berry
IUCN Status: Not evaluated
Reference: Trans. Linn. Soc. London 12: 356 (1819)

Strychnos beddomei C.B. Clarke
Loganiaceae
Synonyms: *Strychnos minor* var. *minor.*, *Strychnos bancroftiana* F.M. Bailey., *Strychnos barbata* A.W. Hill., *Strychnos leuconeura* Gilg & Gilg-Ben. + 20

Climbing Mechanism: Hook Climber
Distribution (Global): Bangladesh, Bismarck Archipelago, Borneo, Cambodia, India, Lesser Sunda Is., Malaya, Maluku, Myanmar, New Guinea, Nicobar Is., Philippines, Queensland, Santa Cruz Is., Solomon Is., Sri Lanka, Sulawesi, Sumatera, Thailand, Vietnam
Leaf Type: Simple
Inflorescence: Cymes in solitary
Fruit Type: Wet, berry
IUCN Status: Not evaluated
Reference: C. B. Clarke. In: Hook. fil. Fl. Brit. Ind. 4: 88. (1883)

Strychnos benthami C.B. Clarke
Loganiaceae
Climbing Mechanism: Hook Climber
Leaf Type: Simple
Inflorescence: Cymes in solitary
Fruit Type: Wet, berry
IUCN Status: Vulnerable
Reference: C. B. Clarke. In: Hook. fil. Fl. Brit. Ind. 4: 87. (1883)

Strychnos bicirrhosa Lesch. ex Roxb.
Loganiaceae
Synonyms: *Strychnos minor* var. *minor*., *Strychnos bancroftiana* F.M. Bailey., *Strychnos barbata* A.W.Hill., *Strychnos leuconeura* Gilg & Gilg-Ben. + 20
Climbing Mechanism: Hook Climber
Distribution (Global): Bangladesh, Bismarck Archipelago, Borneo, Cambodia, India, Lesser Sunda Is., Malaya, Maluku, Myanmar, New Guinea, Nicobar Is., Philippines, Queensland, Santa Cruz Is., Solomon Is., Sri Lanka, Sulawesi, Sumatera, Thailand, Vietnam
Leaf Type: Simple
Inflorescence: Cymes in solitary
Fruit Type: Wet, berry
IUCN Status: Not evaluated
Reference: Lesch. ex Roxb. In: Fl. Ind. ed. Carey, 2: 267. (1824)

Strychnos colubrina Stokes
Loganiaceae
Climbing Mechanism: Hook Climber
Distribution (India): Andhra Pradesh, Eastern Ghats, Goa, Kerala, Western Ghats
Leaf Type: Simple
Inflorescence: Cymes in solitary
Fruit Type: Wet, berry
IUCN Status: Not evaluated
Reference: Strychnos benthamii C.B. Clarke

Strychnos dalzellii C.B. Clarke
Loganiaceae
Synonyms: *Strychnos axillaris* Dalzell & A. Gibson
Climbing Mechanism: Hook Climber
Distribution (Global): India
Distribution (India): Karnataka, Maharashtra, Tamil Nadu, Western Ghats
Leaf Type: Leaves opposite, ovate
Inflorescence: Axillary cyme
Fruit Type: Berries
Flowering and Fruiting: November–December
IUCN Status: Not evaluated
Notes: Usually found in evergreen forests
Reference: J. D. Hooker, Fl. Brit. India 4: 87 (1883)

Strychnos dalzellii var. *lanceolaris*
Loganiaceae
Climbing Mechanism: Hook Climber
Distribution (India): Peninsular India
Leaf Type: Leaves opposite, lanceolate
Inflorescence: Axillary congested cymes
Fruit Type: Berry, globose, indehiscent
Flowering and Fruiting: November–April
IUCN Status: Not evaluated
Notes: Usually found in evergreen forests

Strychnos hypogyna C.B. Clarke
Loganiaceae
Synonyms: *Strychnos minor* var. *minor.*, *Strychnos bancroftiana* F.M. Bailey., *Strychnos leuconeura* Gilg & Gilg-Ben., *Strychnos ligustrina* Zipp. ex Span. + 20
Climbing Mechanism: Hook Climber
Distribution (Global): Bangladesh, Bismarck Archipelago, Borneo, Cambodia, India, Lesser Sunda Is., Malaya, Maluku, Myanmar, New Guinea, Nicobar Is., Philippines, Queensland, Santa Cruz Is., Solomon Is., Sri Lanka, Sulawesi, Sumatera, Thailand, Vietnam
Leaf Type: Simple
Inflorescence: Cymes in solitary
Fruit Type: Wet, berry
IUCN Status: Not evaluated
Reference: C. B. Clarke. In: Hook. fil. Fl. Brit. Ind. 4: 86. (1883)

Strychnos laurina Thwaites
Loganiaceae
Synonyms: *Strychnos minor* var. *minor.*, *Strychnos bancroftiana* F.M. Bailey., *Strychnos leuconeura* Gilg & Gilg-Ben., *Strychnos ligustrina* Zipp. ex Span. + 20
Climbing Mechanism: Hook Climber

Distribution (Global): Bangladesh, Bismarck Archipelago, Borneo, Cambodia, India, Lesser Sunda Is., Malaya, Maluku, Myanmar, New Guinea, Nicobar Is., Philippines, Queensland, Santa Cruz Is., Solomon Is., Sri Lanka, Sulawesi, Sumatera, Thailand, Vietnam
Distribution (India): Arunachal Pradesh
Leaf Type: Simple
Inflorescence: Cymes in solitary
Fruit Type: Wet, berry
IUCN Status: Not evaluated
Reference: Thwaites. In: Enum. Pl. Zeyl. 201. (1859)

Strychnos lenticellata A.W. Hill
Loganiaceae
Synonyms: *Strychnos minor* var. *minor*., *Strychnos bancroftiana* F.M. Bailey., *Strychnos leuconeura* Gilg & Gilg-Ben., *Strychnos ligustrina* Zipp. ex Span. +20
Climbing Mechanism: Stem Twiner
Distribution (Global): Bangladesh, Bismarck Archipelago, Borneo, Cambodia, India, Lesser Sunda Is., Malaya, Maluku, Myanmar, New Guinea, Nicobar Is., Philippines, Queensland, Santa Cruz Is., Solomon Is., Sri Lanka, Sulawesi, Sumatera, Thailand, Vietnam
Distribution (India): Andhra Pradesh, Eastern Ghats, Tamil Nadu
Leaf Type: Simple
Inflorescence: Cymes in solitary
Fruit Type: Wet, berry
IUCN Status: Not evaluated
Reference: A. W. Hill. In: Kew Bull., 159. (1917)

Strychnos maingayi C.B. Clarke
Loganiaceae
Synonyms: *Strychnos laurina* Wall. ex A. DC.
Climbing Mechanism: Hook Climber
Distribution (Global): Assam, Malaya, Myanmar
Leaf Type: Simple
Inflorescence: Cymes in solitary
Fruit Type: Wet, berry
IUCN Status: Not evaluated
Reference: J. D. Hooker, Fl. Brit. India 4: 88 (1883)

Strychnos malaccensis Benth.
Loganiaceae
Synonyms: *Strychnos arborea* A.W. Hill., *Strychnos axillaris* Colebr., *Strychnos armata* A.W. Hill., *Strychnos gaultheriana* Pierre ex C.B. Clarke
Climbing Mechanism: Hook Climber
Distribution (Global): Assam, Bangladesh, Borneo, Cambodia, China South-Central, Java, Laos, Lesser Sunda Is., Malaya, Maluku, Myanmar, New Guinea,

New South Wales, Nicobar Is., Philippines, Queensland, Sri Lanka, Sulawesi, Sumatera, Thailand, Vietnam
Leaf Type: Simple
Inflorescence: Cymes in solitary
Fruit Type: Wet, berry
IUCN Status: Not evaluated
Reference: Trans. Linn. Soc. London 12: 356 (1819)

Strychnos micrantha Thwaites
Loganiaceae
Climbing Mechanism: Hook Climber
Leaf Type: Simple
Inflorescence: Cymes in solitary
Fruit Type: Wet, berry
IUCN Status: Not evaluated
Reference: Thwaites. In: Enum. Pl. Zeyl. 425. (1859)

Strychnos minor Dennst.
Loganiaceae
Synonyms: *Strychnos minor* var. *minor*., *Strychnos minor* var. *thorelii* (Hill) Tirel
Climbing Mechanism: Stem Twiner
Distribution (Global): Bangladesh, Bismarck Archipelago, Borneo, Cambodia, India, Lesser Sunda Is., Malaya, Maluku, Myanmar, New Guinea, Nicobar Is., Philippines, Queensland, Santa Cruz Is., Solomon Is., Sri Lanka, Sulawesi, Sumatera, Thailand, Vietnam
Distribution (India): Andaman and Nicobar Islands, Puducherry, Tamil Nadu
Leaf Type: Ovate, elliptic
Inflorescence: Cymes, white in color
Fruit Type: Berry, woody
Flowering and Fruiting: September–October
IUCN Status: Not evaluated
Notes: Usually found in evergreen forests
Reference: Schlüssel Hortus Malab.: 33 (1818)

Strychnos narcondamensis A.W. Hill
Loganiaceae
Climbing Mechanism: Hook Climber
Distribution (Global): Andaman Is.
Leaf Type: Simple
Inflorescence: Cymes in solitary
Fruit Type: Wet, berry
IUCN Status: Not evaluated
Reference: Bull. Misc. Inform. Kew 1917: 203 (1917)

Strychnos pubescens C.B. Clarke
Loganiaceae
Synonyms: *Strychnos acuminata* Wall., *Strychnos arborea* A. W. Hill., *Strychnos axillaris* Colebr. + 23
Climbing Mechanism: Hook Climber
Distribution (Global): Borneo, Malaya
Leaf Type: Simple
Inflorescence: Cymes in solitary
Fruit Type: Wet, berry
IUCN Status: Not evaluated
Reference: J. D. Hooker, Fl. Brit. India 4: 89 (1883)

Strychnos quintuplinervis A.W. Hill
Loganiaceae
Synonyms: *Strychnos axillaris* Colebr., *Strychnos polytoma* Gilg & Gilg-Ben., *Strychnos scortechinii* A.W. Hill., *Strychnos gaultheriana* Pierre ex C.B. Clarke. + 20
Climbing Mechanism: Hook Climber
Distribution (Global): Assam, Bangladesh, Borneo, Cambodia, China South-Central, Java, Laos, Lesser Sunda Is., Malaya, Maluku, Myanmar, New Guinea, New South Wales, Nicobar Is., Philippines, Queensland, Sri Lanka, Sulawesi, Sumatera, Thailand, Vietnam
Distribution (India): Great Nicobar Island, Meghalaya
Leaf Type: Simple
Inflorescence: Cymes in solitary
Fruit Type: Wet, berry
IUCN Status: Not evaluated
Reference: Trans. Linn. Soc. London 12: 356 (1819)

Strychnos rufa C.B. Clarke
Loganiaceae
Synonyms: *Strychnos acuminata* Wall.
Climbing Mechanism: Hook Climber
Distribution (Global): Malaya, Myanmar
Leaf Type: Simple
Inflorescence: Cymes in solitary
Fruit Type: Wet, berry
IUCN Status: Not evaluated
Reference: J. D. Hooker, Fl. Brit. India 4: 89 (1883)

Strychnos septemnervis C.B. Clarke
Loganiaceae
Synonyms: *Strychnos minor* var. *minor.*, *Strychnos bancroftiana* F.M. Bailey., *Strychnos leuconeura* Gilg & Gilg-Ben., *Strychnos ligustrina* Zipp. ex Span. + 20
Climbing Mechanism: Hook Climber

Distribution (Global): Bangladesh, Bismarck Archipelago, Borneo, Cambodia, India, Lesser Sunda Is., Malaya, Maluku, Myanmar, New Guinea, Nicobar Is., Philippines, Queensland, Santa Cruz Is., Solomon Is., Sri Lanka, Sulawesi, Sumatera, Thailand, Vietnam
Leaf Type: Simple
Inflorescence: Cymes in solitary
Fruit Type: Wet, berry
IUCN Status: Not evaluated
Reference: C. B. Clarke. In: Hook. fil. Fl. Brit. Ind. 4: 88. (1883)

Strychnos vanprukii Craib
Loganiaceae
Synonyms: *Strychnos nitida* Gagnep., *Strychnos vanprukii* var. *acuminata* (A.W. Hill)., *Strychnos quadrangularis* A.W. Hill. H.B. Naithani & S. Biswas
Climbing Mechanism: Hook Climber
Distribution (Global): Assam, Borneo, Laos, Malaya, Thailand, Vietnam
Distribution (India): Great Nicobar Island, Meghalaya
Leaf Type: Elliptic, acute
Inflorescence: Axillary
Flowering and Fruiting: October–March
IUCN Status: Not evaluated
Notes: Usually found in evergreen forests
Reference: Bull. Misc. Inform. Kew 1911: 421 (1911)

Strychnos wallichiana Steud. ex A. DC.
Loganiaceae
Synonyms: *Strychnos bourdilloni* Brandis., *Strychnos lucida* Wall., *Strychnos cinnamomifolia* Thwaites., *Strychnos pierreana* A.W. Hill., *Strychnos cirrhosa* Stokes., *Strychnos cirrhosa* Stokes., *Strychnos colubrina* L., *Strychnos tubiflora* A.W. Hill
Climbing Mechanism: Hook Climber
Distribution (Global): Andaman Is., Assam, Bangladesh, China South-Central, Myanmar, Nicobar Is., Sri Lanka, Vietnam
Distribution (India): Andaman and Nicobar Islands, Andhra Pradesh, Arunachal Pradesh, Eastern Ghats, Maharashtra, Meghalaya, Tamil Nadu, Western Ghats
Leaf Type: Simple
Inflorescence: Cymes in solitary
Fruit Type: Wet, berry
Flowering and Fruiting: October–December
IUCN Status: Not evaluated
Reference: A.P.de Candolle, Prodr. 9: 13 (1845)

Lophopyxis maingayi Hook.f.
Lophopyxidaceae

Synonyms: *Combretum perakense* M.G. Gangop. & Chakrab., *Lophopyxis combretocarpa* Engl. ex Pax, *Lophopyxis pentaptera* Engl., *Lophopyxis pierrei* Boerl., *Combretopsis pentaptera* (Engl.) K. Schum., *Homalium gilgianum* Lauterb.
Climbing Mechanism: Scrambler-Unarmed
Distribution (Global): Bismarck Archipelago, Borneo, Caroline Is., Malaya, Maluku, New Guinea, Solomon Is., Sulawesi, Thailand
Leaf Type: Leaves narrowly
ovate, ovate to oblong, apex attenuate to acuminate, base rounded to obtuse, chartaceous to subcoriaceous, hairy
Inflorescence: Inflorescences axillary in distal, leaves and terminal, spike-like racemes
Fruit Type: Capsule obovoid-ellipsoid in outline, 5-winged, green, later dark brown, glabrescent
IUCN Status: Not evaluated
Reference: Hooker's Icon. Pl. 18: t. 1714 (1887)

Aspidopterys balakrishnanii R.C. Srivast.
Malpighiaceae
Climbing Mechanism: Stem Twiner
Distribution (Global): India, West Himalaya
Distribution (India): West Bengal
Leaf Type: Single, opposite, ovate to heart shaped
Inflorescence: Panicles in leaf axils
Fruit Type: Dry, samara
IUCN Status: Not evaluated
Reference: J. Econ. Taxon. Bot. 4: 1003 (1983)

Aspidopterys canarensis Dalzell
Malpighiaceae
Synonyms: *Aspidopterys glomerata* Wight
Climbing Mechanism: Scrambler-Unarmed
Distribution (Global): India
Distribution (India): Maharashtra, Tamil Nadu.
Leaf Type: Single, opposite, ovate to heart shaped
Inflorescence: Panicles in leaf axils
Fruit Type: Dry, samara
Flowering and Fruiting: February–May
IUCN Status: Not evaluated
Notes: Distributed usually in evergreen and semi-evergreen forests
Reference: Hooker's J. Bot. Kew Gard. Misc. 3: 37 (1851)

Aspidopterys cordata (B. Heyne ex Wall.) A. Juss.
Malpighiaceae

Synonyms: *Aspidopterys cordata* var. *vermae* R.C. Srivast. & N.P. Balakr., Hiraea cordata B. Heyne ex Wall.
Climbing Mechanism: Stem Twiner
Distribution (Global): East Himalaya, India
Distribution (India): Andhra Pradesh, Eastern Ghats, Gujarat, Madhya Pradesh, Maharashtra, Rajasthan, Tamil Nadu, Telangana
Leaf Type: Single, opposite, ovate to heart shaped
Inflorescence: Panicles in leaf axils
Fruit Type: Dry, samara
Flowering and Fruiting: August–January
IUCN Status: Not evaluated
Notes: Occurs in dry and moist deciduous forests
Reference: Ann. Sci. Nat., Bot., sér. 2, 13: 267 (1840)

Aspidopterys elliptica (Blume) A. Juss.
Malpighiaceae
Synonyms: *Aspidopterys macrocarpa* Dop, *Aspidopterys ovata* (Turcz.) Merr. & Rolfe, *Hiraea elliptica* Blume, *Ryssopterys ovata* Turcz.
Climbing Mechanism: Stem Twiner
Distribution (Global): Andaman Is., Assam, Borneo, Java, Lesser Sunda Is., Philippines, Sumatera, Vietnam
Distribution (India): Andaman and Nicobar Islands, Arunachal Pradesh, Meghalaya, Tripura
Leaf Type: Simple
Fruit Type: Samara
Flowering and Fruiting: February–June
IUCN Status: Not evaluated
Reference: Ann. Sci. Nat., Bot., sér. 2, 13: 266 (1840)

Aspidopterys glabriuscula A. Juss.
Malpighiaceae
Synonyms: *Aspidopterys heterocarpa* Arènes, *Hiraea glabriuscula* Wall.
Climbing Mechanism: Stem Twiner
Distribution (Global): Assam, Bangladesh, China South-Central, China Southeast, East Himalaya, Hainan, Vietnam
Distribution (India): Arunachal Pradesh, Madhya Pradesh, Mizoram
Leaf Type: Simple
Fruit Type: Samara
Flowering and Fruiting: May–November
IUCN Status: Not evaluated
Reference: Ann. Sci. Nat., Bot., sér. 2, 13: 267 (1840)

Aspidopterys indica (Willd.) W. Theob.
Malpighiaceae

Synonyms: *Aspidopterys roxburghiana* A. Juss., *Hiraea indica* (Willd.) Roxb., *Triopterys indica* Willd.
Climbing Mechanism: Scrambler-Unarmed
Distribution (Global): Assam, Bangladesh, East Himalaya, India, Myanmar
Distribution (India): Andhra Pradesh, Arunachal Pradesh, Bihar, Odisha, Madhya Pradesh, Maharashtra, Manipur, Meghalaya, Odisha, Tamil Nadu
Leaf Type: Simple, opposite
Fruit Type: Samara
Flowering and Fruiting: April–January
IUCN Status: Not evaluated
Reference: F. Mason, Burmah, ed. 3, 2: 599 (1883)

Aspidopterys jainii R.C. Srivast.
Malpighiaceae
Climbing Mechanism: Stem Twiner
Distribution (Global): Assam
Leaf Type: Single, opposite, ovate to heart shaped
Inflorescence: Panicles in leaf axils
Fruit Type: Dry, samara
IUCN Status: Not evaluated
Reference: J. Jap. Bot. 59: 329 (1984)

Aspidopterys lanuginosa (Wall.) A. Juss.
Malpighiaceae
Synonyms: *Aspidopterys nutans* (Roxb. ex DC.) A. Juss
Climbing Mechanism: Stem Twiner
Leaf Type: Single, opposite, ovate to heart shaped
Inflorescence: Panicles in leaf axils
Fruit Type: Dry, samara
IUCN Status: Not evaluated
Reference: Ann. Sci. Nat., Bot., sér. 2, 13: 267 (1840)

Aspidopterys nutans (Roxb. ex DC.) A. Juss.
Malpighiaceae
Climbing Mechanism: Stem Twiner
IUCN Status: Not evaluated

Aspidopterys oxyphylla A. Juss.
Malpighiaceae
Synonyms: *Hiraea oxyphylla* Wall.
Climbing Mechanism: Stem Twiner
Distribution (Global): Assam, Bangladesh.
Leaf Type: Single, opposite, ovate to heart shaped
Inflorescence: Panicles in leaf axils
Fruit Type: Dry, samara

IUCN Status: Not evaluated
Reference: Ann. Sci. Nat., Bot., sér. 2, 13: 267 (1840)

Aspidopterys tomentosa (Blume) A. Juss.
Malpighiaceae
Synonyms: *Hiraea tomentosa* Blume
Climbing Mechanism: Stem Twiner
Distribution (Global): Andaman Is., Bangladesh, Borneo, Cambodia, China South-Central, Hainan, India, Java, Laos, Myanmar, Philippines, Thailand, Vietnam
Distribution (India): Andaman and Nicobar Islands, Odisha
Leaf Type: Single, opposite, ovate to heart shaped
Inflorescence: Panicles in leaf axils
Fruit Type: Dry, samara
IUCN Status: Not evaluated
Reference: Ann. Sci. Nat., Bot., sér. 2, 13: 267 (1840)

Aspidopterys tomentosa var. *hutchinsonii* (Haines) R.C. Srivast.
Malpighiaceae
Climbing Mechanism: Stem Twiner
Distribution (India): Odisha
Leaf Type: Single, opposite, ovate to heart shaped
Inflorescence: Panicles in leaf axils
Fruit Type: Dry, samara
Flowering and Fruiting: December–June
IUCN Status: Not evaluated
Reference: R. C. Srivast. In: J. Bombay Nat. Hist. Soc., 81(3): 728 1984, with incorrect basionym page, et J. Econ. Taxon. Bot., 6(1): 70. (1985)

Aspidopterys wallichii Hook.f.
Malpighiaceae
Synonyms: *Aspidopterys wallichii* var. *dehradunensis* R.C. Srivast., Hiraea nutans Wall.
Climbing Mechanism: Stem Twiner
Distribution (Global): Assam, East Himalaya, India, Nepal, West Himalaya
Leaf Type: Single, opposite, ovate to heart shaped
Inflorescence: Panicles in leaf axils
Fruit Type: Dry, samara
IUCN Status: Not evaluated
Reference: Fl. Brit. India 1: 421 (1874)

Heteropterys laevifolia (A. Juss.) R.C. Srivast.
Malpighiaceae
Synonyms: *Banisteriopsis laevifolia* (A. Juss.) B. Gates
Climbing Mechanism: Scrambler-Unarmed
Distribution (India): Maharashtra

IUCN Status: Not evaluated
Reference: Nation. Acad. Sci. Lett. 7: 325 (1984)

Hipatge jacobsii R.C. Srivast.
Malpighiaceae
Climbing Mechanism: Scrambler-Unarmed
IUCN Status: Not evaluated

Hiptage acuminata Wall. ex A. Juss.
Malpighiaceae
Climbing Mechanism: Scrambler-Unarmed
Distribution (Global): Assam, Bangladesh, China South-Central, Myanmar, Thailand.
Distribution (India): Arunachal Pradesh, Assam, Meghalaya.
Leaf Type: Leaves opposite, lanceolate
Inflorescence: Axillary, racemes
Flowering and Fruiting: February–August
IUCN Status: Not evaluated
Reference: Ann. Sci. Nat., Bot., sér. 2, 13: 269 (1840)

Hiptage bengalensis Kuntze
Malpighiaceae
Synonyms: *Hiptage parvifolia* Wight & Arn., *Gaertnera racemosa* (Cav.) Roxb., *Hiptage obtusifolia* (Roxb.) DC., *Banisteria benghalensis* L., *Banisteria javanica* Thunb., *Banisteria tetraptera* Sonner
Common Name: Clustered hiptage, Helicopter flower
Climbing Mechanism: Scrambler-Unarmed
Distribution (Global): Andaman Is., Assam, Bangladesh, Borneo, Cambodia, China South-Central, China Southeast, East Himalaya, Hainan, India, Java, Laos, Lesser Sunda Is., Malaya, Myanmar, Nepal, Nicobar Is., Pakistan, Philippines, Sri Lanka, Sulawesi, Sumatera, Taiwan, Thailand, Vietnam
Distribution (India): Andaman and Nicobar Islands, Andhra Pradesh, Arunachal Pradesh, Delhi, Eastern Ghats, Gujarat, Himachal Pradesh, Karnataka, Madhya Pradesh, Maharashtra, Manipur, Meghalaya, Mizoram, Odisha, Rajasthan, Tamil Nadu, Telangana, Tripura, West Bengal, Western Ghats
Leaf Type: Simple
Inflorescence: Axillary or Terminal racemes
Fruit Type: 3-winged samara
Flowering and Fruiting: January–November
IUCN Status: Not evaluated
Reference: J. Asiat. Soc. Bengal, Pt. 2, Nat. Hist. 43: 136 (1874)

Hiptage benghalensis (L.) Kurz
Malpighiaceae
Synonyms: *Banisteria benghalensis* L.

Climbing Mechanism: Scrambler-Unarmed
Leaf Type: Simple, ovate, elliptic-oblong
Inflorescence: Flowers white
Fruit Type: Dry, samara with unequal wings
IUCN Status: Least Concern

Hiptage benghalensis var. *rothinii*
Malpighiaceae
Climbing Mechanism: Scrambler-Unarmed
Leaf Type: Simple, ovate, elliptic-oblong
Inflorescence: Flowers white
Fruit Type: Dry, samara with unequal wings
IUCN Status: Not evaluated

Hiptage jacobsii R.C. Srivast.
Malpighiaceae
Climbing Mechanism: Scrambler-Unarmed
Distribution (Global): Assam
Leaf Type: Simple, ovate, elliptic-oblong
Inflorescence: Flowers white
Fruit Type: Dry, samara with unequal wings
IUCN Status: Not evaluated
Reference: Indian Forester 110: 499 (1984)

Hiptage laxiflora Sujana & Vadhyar
Malpighiaceae
Climbing Mechanism: Scrambler-Unarmed
Leaf Type: Simple, ovate, elliptic-oblong
Inflorescence: Flowers white
Fruit Type: Dry, samara with unequal wings
IUCN Status: Not evaluated

Hiptage madablota Gaertn.
Malpighiaceae
Synonyms: *Hiptage benghalensis* var. *benghalensis*
Climbing Mechanism: Scrambler-Unarmed
Distribution (India): Andaman and Nicobar Islands, Assam, Bihar, Odisha
Leaf Type: Simple, ovate, elliptic-oblong
Inflorescence: Flowers white
Fruit Type: Dry, samara with unequal wings
IUCN Status: Not evaluated
Reference: Fruct. Sem. Pl. 2: 169 (1790)

Hiptage parviflora Wight ex Hook.f.
Malpighiaceae

Synonyms: *Hiptage sericea* Hook.f.
Climbing Mechanism: Scrambler-Unarmed
Distribution (India): Tamil Nadu
Leaf Type: Simple, ovate, elliptic-oblong
Inflorescence: Flowers white
Fruit Type: Dry, samara with unequal wings
IUCN Status: Not evaluated
Reference: Fl. Brit. India 1: 419 (1874)

Hiptage sericea Hook.f.
Malpighiaceae
Synonyms: *Hiptage parviflora* Wight ex Hook.f.
Climbing Mechanism: Scrambler-Unarmed
Distribution (Global): Malaya, Myanmar, Thailand
Distribution (India): Tamil Nadu
Leaf Type: Simple, ovate, elliptic-oblong
Inflorescence: Flowers white
Fruit Type: Dry, samara with unequal wings
IUCN Status: Not evaluated
Reference: Fl. Brit. India 1: 419 (1874)

Hiptage thothathrii N.P. Balakr. & R.C. Srivast.
Malpighiaceae
Climbing Mechanism: Scrambler-Unarmed
Distribution (Global): Andaman Is.
Distribution (India): Andaman and Nicobar Islands.
Leaf Type: Simple, ovate, elliptic-oblong
Inflorescence: Flowers white
Fruit Type: Dry, samara with unequal wings
IUCN Status: Not evaluated
Reference: J. Econ. Taxon. Bot. 4: 985 (1983)

Stigmaphyllon ciliatum (Lam.) A. Juss.
Malpighiaceae
Synonyms: *Stigmaphyllon ciliatum.*, *Banisteria glauca* Desf., *Banisteria nitida* Vell.
Climbing Mechanism: Stem Twiner
Distribution (Global): Belize, Brazil Northeast, Brazil South, Brazil Southeast, Colombia, Honduras, Nicaragua, Trinidad-Tobago, Uruguay, Venezuela
Distribution (India): Peninsular India
Leaf Type: Simple, heart shaped
Inflorescence: Clusters on long stalks
Fruit Type: Dry
Flowering and Fruiting: January-April
IUCN Status: Not evaluated
Reference: A.F.C.P.de Saint-Hilaire & al., Fl. Bras. Merid. 3: 49 (1833)

Stigmaphyllon emarginatum
Malpighiaceae
Climbing Mechanism: Scrambler-Unarmed
IUCN Status: Not evaluated

Tristellateia australasiae A. Rich.
Malpighiaceae
Synonyms: *Platynema laurifolium* Wight & Arn., *Tristellateia australasiae* f. *obtusiuscula* Nied., *Tristellateia australasiae* f. *typica* Nied., *Tristellateia australis* A. Rich., *Tristellateia malintana* Blanco, *Tristellateia novaeguineensis* Blume ex A. Juss.
Common Name: Shower of gold climber, Vining galphimia, Vining milkweed, Australian gold vine, Shower of gold climber
Climbing Mechanism: Stem Twiner
Distribution (Global): Bismarck Archipelago, Borneo, Cambodia, Caroline Is., Comoros, Java, Lesser Sunda Is., Malaya, Maluku, Marianas, Myanmar, Nansei-shoto, New Caledonia, New Guinea, Philippines, Queensland, Sulawesi, Sumatera, Taiwan, Thailand, Vietnam
Distribution (India): Maharashtra, Tamil Nadu
Leaf Type: Simple, opposite
Inflorescence: Racemose
Fruit Type: Dry, star shaped
Flowering and Fruiting: September–May
IUCN Status: Not evaluated
Reference: J.S.C. Dumont d'Urville, Voy. Astrolabe 2: 159 (1834)

Byttneria andamanensis Kurz
Malvaceae
Climbing Mechanism: Scrambler-Unarmed
Distribution (India): Andaman and Nicobar Islands
Leaf Type: 3–5 lobed, cordate ot orbicular
Inflorescence: Pedicles are slender
Fruit Type: Dry, capsule, globose
IUCN Status: Not evaluated
Reference: Kurz. In: Journ. As. Soc. Beng. 2: 47, et Flora, 54: 277. (1871)

Byttneria aspera Collebr. ex Wall.
Malvaceae
Climbing Mechanism: Scrambler-Unarmed
Distribution (India): Assam, Tripura
Leaf Type: 3–5 lobed, cordate to orbicular
Inflorescence: Pedicles are slender
Fruit Type: Dry, capsule, globose
IUCN Status: Not evaluated
Reference: Colebr. ex Wall. In: Roxb., Fl. Ind. 2: 383. (1824)

Byttneria herbacea Roxb.
Malvaceae
Synonyms: *Ayenia herbacea* (Roxb.) ined.
Common Name: Common hog weed, Herbaceous Byttneria
Climbing Mechanism: Scrambler-Unarmed
Distribution (India): Telangana
Leaf Type: Simple, alternate
Flowering and Fruiting: Throughout the year
IUCN Status: Not evaluated
Notes: Often found in scrub jungles and dry-deciduous forests
Reference: Pl. Coromandel 1: 28 (1795)

Byttneria jackiana Wall.
Malvaceae
Synonyms: *Ayenia jackiana* (Wall.) ined.
Climbing Mechanism: Scrambler-Unarmed
Leaf Type: 3-5 lobed, cordate ot orbicular
Inflorescence: Pedicles are slender
Fruit Type: Dry, capsule, globose
IUCN Status: Not evaluated
Reference: W. Roxburgh, Fl. Ind. 2: 386 (1824)

Byttneria pilosa Roxb.
Malvaceae
Synonyms: *Ayenia elegans* (Ridl.) ined., *Byttneria elegans* Ridl., *Byttneria velutina* Wall., *Chaetaea pilosa* (Roxb.) Adelb., *Commersonia pilosa* G. Don.
Common Name: Elegant Ayenia
Climbing Mechanism: Scrambler-Unarmed
Distribution (India): Assam, Mizoram, Tripura, West Bengal
Leaf Type: 3–5 lobed, cordate to orbicular
Inflorescence: Pedicles are slender
Fruit Type: Dry, capsule, globose
IUCN Status: Not evaluated
Reference: Fl. Ind. 2: 381 (1824)

Gossypium stocksii Mast.
Malvaceae
Climbing Mechanism: Scrambler-Unarmed
Distribution (Global): Oman, Pakistan, Somalia, Yemen
Distribution (India): Maharashtra
IUCN Status: Not evaluated
Reference: J. D. Hooker, Fl. Brit. India 1: 346 (1874)

Grewia abutilifolia Vent. ex Juss.
Malvaceae

Synonyms: *Grewia arbutifolia* Pers., *Grewia aspera* Roxb. ex Roth, *Grewia hirsutovelutina* Burret, *Grewia kainantensis* Masam., *Sterculia tiliacea* H. Lév.
Common Name: Mallow-leaved crossberry
Climbing Mechanism: Scrambler-Unarmed
Distribution (Global): Assam, Bangladesh, Borneo, Cambodia, China South-Central, China Southeast, Christmas I., East Himalaya, Hainan, India, Java, Laos, Malaya, Myanmar, New Guinea, Taiwan, Thailand, Vietnam, West Himalaya
Distribution (India): Andhra Pradesh, Karnataka
Leaf Type: Leaves are broadly ovate
Inflorescence: Cymes
Flowering and Fruiting: April–December
IUCN Status: Least Concern
Reference: Ann. Mus. Natl. Hist. Nat. 4: 92 (1804)

Grewia bracteata Roth
Malvaceae
Synonyms: *Grewia obtusa* Wall. ex Dunn, *Grewia wightiana* J.R. Drumm.
Common Name: Wight's crossberry
Climbing Mechanism: Scrambler-Unarmed
Distribution (Global): India, Sri Lanka
Distribution (India): Karnataka.
Leaf Type: Simple, caducous
Inflorescence: Axillary cymes, sometimes umbellate
Fruit Type: Wet, drupe
IUCN Status: Not evaluated
Reference: Nov. Pl. Sp.: 243 (1821)

Grewia carpinifolia Juss.
Malvaceae
Synonyms: *Grewia biflora* G. Don, *Grewia rautanenii* Schinz, *Grewia rhamnifolia* Roth, *Grewia rowlandii* K.Schum., *Vincentia carpinifolia* (Juss.) Burret, *Vinticena carpinifolia* (Juss.) Burret.
Common Name: Hornbeam-leaved crossberry
Climbing Mechanism: Scrambler-Unarmed
Distribution (Global): Angola, Benin, Burkina, Cameroon, Gambia, Ghana, Gulf of Guinea Is., Ivory Coast, Liberia, Mauritania, Nigeria, Sierra Leone, Togo, Zaire
Distribution (India): Western Ghats
Leaf Type: Simple, alternate distichous
Inflorescence: Axillary cymes
Fruit Type: Drupe
Flowering and Fruiting: November–March
IUCN Status: Not evaluated
Reference: Ann. Mus. Natl. Hist. Nat. 4: 91 (1804)

Grewia damine Gaertn.
Malvaceae
Synonyms: *Grewia salvifolia* Heyne
Common Name: Salvia leaved crossberry
Climbing Mechanism: Scrambler-Unarmed
Distribution (India): Karnataka
Leaf Type: Simple, caducous
Inflorescence: Axillary cymes, sometimes umbellate
Fruit Type: Wet, drupe
Flowering and Fruiting: July–September
IUCN Status: Not evaluated
Reference: Fruct. Sem. Pl. 2: 113 (1790)

Grewia denticulata Wall. ex Prain
Malvaceae
Synonyms: *Grewia nagensium* Prain
Climbing Mechanism: Scrambler-Unarmed
Distribution (India): Arunachal Pradesh, Assam
Leaf Type: Simple, alternate
Inflorescence: Axillary
Fruit Type: Drupe
Flowering and Fruiting: May–February
IUCN Status: Not evaluated
Reference: Wall. ex Prain. In: Ann. Roy. Bot. Gard. Calc. 9: 10. t. 12. (1901).

Grewia flavescens Juss.
Malvaceae
Synonyms: *Vincentia flavescens* (Juss.) Burret, *Vinticena flavescens* (Juss.) Burret
Common Name: Donkey berry, Rough-leaved raisin, Sandpaper raisin
Climbing Mechanism: Scrambler-Unarmed
Distribution (Global): Angola, Botswana, Burkina, Burundi, Cameroon, Caprivi Strip, Central African Repu, Chad, Eritrea, Ethiopia, Gabon, Gambia, Guinea, India, Ivory Coast, Kenya, KwaZulu-Natal, Malawi, Mali, Mauritania, Mozambique, Namibia, Niger, Northern Provinces, Rwanda, Saudi Arabia, Senegal, Sudan, Swaziland, Tanzania, Togo, Uganda, Yemen, Zambia, Zaire, Zimbabwe
Distribution (India): Bihar, Odisha, Gujarat, Karnataka, Madhya Pradesh, Tamil Nadu, Telangana
Leaf Type: Simple
Inflorescence: Axillary
Fruit Type: Fruit berry-like hard
Flowering and Fruiting: November–March
IUCN Status: Least Concern
Reference: Ann. Mus. Natl. Hist. Nat. 4: 91 (1804)

Grewia heterotricha Mast.
Malvaceae
Synonyms: *Microcos heterotricha* (Mast.) Burret
Climbing Mechanism: Scrambler-Unarmed
Distribution (India): Karnataka, Maharashtra, Tamil Nadu, Western Ghats, Andhra Pradesh
Leaf Type: Alternate
Inflorescence: Axillary or Umbelled cymes
Fruit Type: Drupe
Flowering and Fruiting: March–December
IUCN Status: Not evaluated
Reference: J. D. Hooker, Fl. Brit. India 1: 385 (1874)

Grewia hirsuta Vahl
Malvaceae
Synonyms: *Grewia montana* J. Koenig ex Wight & Arn., *Grewia obliqua* Juss., *Grewia pilosa* Roxb., *Grewia roxburghii* G.Don, *Grewia tomentosa* Roxb. ex Wight & Arn., *Grewia trichodes* Voigt.
Climbing Mechanism: Scrambler-Unarmed
Distribution (Global): Assam, Bangladesh, Cambodia, China Southeast, India, Laos, Malaya, Myanmar, Nepal, Sri Lanka, Thailand, Vietnam
Distribution (India): Gujarat
Leaf Type: Simple, caducous
Inflorescence: Axillary cymes, sometimes umbellate
Fruit Type: Wet, drupe
Flowering and Fruiting: June–January
IUCN Status: Not evaluated
Reference: Symb. Bot. 1: 84 (1790)

Grewia laevigata Vahl
Malvaceae
Synonyms: *Grewia mallococa* Blanco, *Grewia multiflora* Blanco, *Grewia odorata* Blume, *Grewia ovalifolia* Juss., *Grewia pedicellata* Roxb., *Grewia racemosa* Noronha, *Grewia umbellata* Roxb. ex DC., *Mallococca parva* Blanco.
Common Name: Two-lobed crossberry
Climbing Mechanism: Scrambler-Unarmed
Distribution (Global): Andaman Is., Assam, Bangladesh, Borneo, Cambodia, Christmas I., Java, Laos, Lesser Sunda Is., Malaya, Maluku, Myanmar, New Guinea, Nicobar Is., Philippines, Sulawesi, Sumatera, Thailand, Vietnam
Leaf Type: Simple, caducous
Inflorescence: Axillary cymes, sometimes umbellate
Fruit Type: Wet, drupe
IUCN Status: Least Concern
Reference: Symb. Bot. 1: 34 (1790)

Grewia nervosa (Lour.) Panigrahi
Malvaceae
Climbing Mechanism: Scrambler-Unarmed
Distribution (India): Maharashtra
Leaf Type: Simple, caducous
Inflorescence: Axillary cymes, sometimes umbellate
Fruit Type: Wet, drupe
Flowering and Fruiting: April–December
IUCN Status: Least Concern
Reference: Taxon 34: 702 (1985)

Grewia oppositifolia Roxb. ex DC.
Malvaceae
Synonyms: *Grewia emarginata* Wight & Arn., *Grewia orientalis* Roxb. ex Wight & Arn., *Grewia rhamnifolia* B. Heyne ex Dunn.
Climbing Mechanism: Scrambler-Unarmed
Distribution (Global): India, Nepal, Sri Lanka, West Himalaya
Distribution (India): Karnataka, Tamil Nadu
Leaf Type: Alternate
Inflorescence: Cymes axillary
Flowering and Fruiting: October-December
IUCN Status: Not evaluated
Reference: Prodr. 1: 509 (1824)

Grewia orbiculata Rottler
Malvaceae
Synonyms: *Grewia rotundifolia* Juss., *Grewia orbicularis* G. Don.
Common Name: Round-leaved Grewia
Climbing Mechanism: Scrambler-Unarmed
Distribution (Global): India
Distribution (India): Andhra Prades
Leaf Type: Simple
Flowering and Fruiting: March–October
IUCN Status: Not evaluated
Reference: Neue Schriften Ges. Naturf. Freunde Berlin 4: 205 (1803)

Grewia palodensis E.S.S. Kumar & al.
Malvaceae
Climbing Mechanism: Scrambler-Unarmed
Distribution (Global): India
Distribution (India): Peninsular India
Leaf Type: Alternate
Inflorescence: Axillary, terminal, umbellate cymes
Fruit Type: Drupe
Flowering and Fruiting: May–October

IUCN Status: Not evaluated
Reference: Rheedea 11: 41 (2001)

Grewia rhamnifolia Burret
Malvaceae
Climbing Mechanism: Scrambler-Unarmed
Distribution (India): Andhra Pradesh, Bihar, Odisha, Eastern Ghats, Karnataka, Puducherry, Tamil Nadu
Leaf Type: Simple, caducous
Inflorescence: Axillary cymes, sometimes umbellate
Fruit Type: Wet, drupe
IUCN Status: Not evaluated
Reference: Burret M (1926) Beiträge zur Kenntnis der Tiliaceen. Notizblatt Des Königl. Botanischen Gartens Und Museums Zu Berlin 9(35): 592-880. doi: 10.2307/3994327

Grewia tenax (Forssk.) Fiori
Malvaceae
Synonyms: *Grewia chadara* Lam., *Grewia populifolia* Vahl, *Chadara tenax* Forssk.
Common Name: White crossberry, Phalsa cherry, Raisin bush
Climbing Mechanism: Scrambler-Unarmed
Distribution (Global): Afghanistan, Algeria, Angola, Assam, Bangladesh, Botswana, Burkina, Caprivi Strip, Chad, Djibouti, Egypt, Eritrea, Ethiopia, Gulf States, India, Iran, Kenya, Mali, Mauritania, Mauritius, Morocco, Namibia, Niger, Nigeria, Oman, Pakistan, Saudi Arabia, Senegal, Sinai, Socotra, Somalia, Sri Lanka, Sudan, Tanzania, Uganda, West Himalaya, Western Sahara, Yemen, Zimbabwe
Distribution (India): Maharashtra, Rajasthan
Leaf Type: Alternate
IUCN Status: Least Concern
Reference: Agric. Colon. 5(Suppl.): 23 (1911 publ. 1912)

Grewia umbellifera Bedd.
Malvaceae
Common Name: Ghat crossberry
Climbing Mechanism: Scrambler-Unarmed
Distribution (Global): India
Distribution (India): Goa, Maharashtra, Western Ghats
Leaf Type: Simple, caducous
Inflorescence: Axillary cymes, sometimes umbellate
Fruit Type: Wet, drupe
Flowering and Fruiting: March–May
IUCN Status: Not evaluated
Reference: Fl. Sylv. S. India: xxxvii (1870)

Hibiscus hispidissimus Griff.
Malvaceae
Synonyms: *Furcaria furcellata* Ulbr., *Furcaria roxburghii* Kostel., *Hibiscus aculeatus* Roxb., *Hibiscus hamatus* E. Mey. ex Harv.
Common Name: Wild Hibiscus, Hill hemp bendy
Climbing Mechanism: Scrambler-Unarmed
Distribution (Global): Assam, Bangladesh, Cambodia, China South-Central, India, Myanmar, Sri Lanka, Thailand, Vietnam
Distribution (India): Maharashtra
Leaf Type: Simple, palmilobed or palmiparted
Inflorescence: Axillary or terminal, solitary or racemes
Fruit Type: Capsule
Flowering and Fruiting: September–March
IUCN Status: Not evaluated
Reference: Not. Pl. Asiat. 4: 521 (1854)

Hibiscus rostellatus Guill. & Perr.
Malvaceae
Synonyms: *Abelmoschus rostellatus* (Guill. & Perr.) Walp., *Hibiscus furcellatoides* Hochr.
Climbing Mechanism: Scrambler-Armed
Distribution (Global): Angola, Benin, Burkina, Cameroon, Central African Repu, Chad, Congo, Ethiopia, Gabon, Gambia, Ghana, Guinea, Guinea-Bissau, Ivory Coast, Kenya, Liberia, Mali, Mozambique, Nigeria, Rwanda, Senegal, Sierra Leone, Sudan, Tanzania, Togo, Uganda, Zambia, Zaire
IUCN Status: Not evaluated
Reference: Fl. Seneg. Tent.: 55 (1831)

Hibiscus surattensis L.
Malvaceae
Synonyms: *Abelmoschus aculeatus* Walp., *Furcaria surattensis* (L.) Kostel., *Hibiscus aculeatus* G. Don, *Hibiscus appendiculatus* Stokes, *Hibiscus bifurcatus* Blanco, *Hibiscus hypoglossus* E.Mey. ex Harv., *Hibiscus involucratus* Salisb., *Hibiscus trinitarius* Noron
Common Name: Bush sorrel, Wild sour, Bush Althea
Climbing Mechanism: Scrambler-Armed
Distribution (Global): Angola, Assam, Bangladesh, Benin, Borneo, Burundi, Cambodia, Cameroon, Cape Verde, Central African Repu, China South-Central, China Southeast, Congo, East Himalaya, Ethiopia, Gabon, Gambia, Ghana, Guinea, Guinea-Bissau, Gulf of Guinea Is., Hainan, India, Ivory Coast, Kenya, Laos, Liberia, Malawi, Malaya, Mozambique, Myanmar, Nigeria, Northern Provinces, Philippines, Senegal, Sierra Leone, Sri Lanka, Sudan, Tanzania, Thailand, Togo, Uganda, Vietnam, Zambia, Zaire, Zimbabwe
Distribution (India): Arunachal Pradesh, Gujarat, Karnataka, Maharashtra, Tripura
Leaf Type: Simple, alternate, palmilobed or palmiparted

Inflorescence: Axillary or terminal, racemes
Fruit Type: Capsule
Flowering and Fruiting: October–January
IUCN Status: Not evaluated
Reference: Sp. Pl.: 696 (1753)

Melochia corchorifolia L.
Malvaceae
Synonyms: *Waltheria indica* L., *Melochia corchorifolia* Wall., *Waltheria africana* Schumach. & Thonn., *Waltheria americana* L., *Waltheria americana var. elliptica* K. Schum., *Waltheria americana var. glandulosa* R.E.Fr. + 15
Common Name: Sleepy morning, Velvet leaf, Marsh-mallow, Monkey bush, Boater bush, Leather coat, Buff coat
Climbing Mechanism: Scrambler-Unarmed
Distribution (Global): Pantropical
Distribution (India): Puducherry
Leaf Type: The leaves ate simple, alternate leaves are narrowly ovate or oblong with a rounded to heart-shaped base
Inflorescence: Inflorescences are usually dense clusters in leaf axils
Fruit Type: Capsule holds one tiny, black, obovoid seed
Flowering and Fruiting: October–January
IUCN Status: Least Concern
Notes: Usually found in degraded moist deciduous forests
Reference: Sp. Pl.: 673 (1753)

Melochia nodiflora Sw.
Malvaceae
Synonyms: *Melochia borbonica* Cav., *Melochia carpinifolia* J.C. Wendl., *Melochia conglobata* Sessé & Moc., *Melochia nodiflora var. longipetiolata* Kitan., *Mougeotia nodiflora* (Sw.) Kunth, *Riedlea borbonica* (Cav.) DC., *Riedlea nodiflora* (Sw.) DC., *Visenia nodiflo.*
Common Name: Hornbeam leaved Melochia
Climbing Mechanism: Scrambler-Unarmed
Distribution (Global): Bahamas, Belize, Brazil Northeast, Brazil Southeast, Cayman Is., Colombia, Costa Rica, Cuba, Dominican Republic, El Salvador, Florida, Guatemala, Haiti, Honduras, Jamaica, Leeward Is., Mexico Central, Mexico Gulf, Mexico Northeast, Mexico Northwest, Mexico Southeast, Mexico Southwest, Nicaragua, Panamá, Puerto Rico, Trinidad-Tobago, Venezuela, Windward Is.
Distribution (India): Puducherry
Leaf Type: Leaves are very much like those of Hornbeam trees. Flowers are borne in a cluster of 3–15, in leaf-axils
Inflorescence: Flowers are borne in a cluster of 3-15, axillary in leaf-axils
Fruit Type: Capsules subspherical, longitudinally 5-grooved, bristly, reddish when young

Flowering and Fruiting: November–December
IUCN Status: Least Concern
Reference: Prodr. Veg. Ind. Occ.: 97 (1788)

Microcos heterotricha Burret
Malvaceae
Climbing Mechanism: Scrambler-Unarmed
Distribution (India): Tamil Nadu
IUCN Status: Not evaluated

Pterospermum heyneanum G. Don
Malvaceae
Synonyms: *Pterospermum xylocarpum* (Gaertn.) Oken
Common Name: Lance wood tree.
Climbing Mechanism: Scrambler-Unarmed
Distribution (India): Arunachal Pradesh
Flowering and Fruiting: January–March
IUCN Status: Not evaluated
Reference: Gen. Hist. 1: 538 (1831)

Pterygota alata (Roxb.) R.Br.
Malvaceae
Synonyms: *Clompanus alata* (Roxb.) Kuntze., *Pterygota roxburghii* Schott & Endle, *Pterygota alata var. irregularis* (W.W.Sm.) Deb & S.K. Basu, *Sterculia alata* Roxb., *Sterculia coccinea* Wall., *Sterculia heynei* Bedd.
Common Name: Buddha's coconut tree
Climbing Mechanism: Scrambler-Unarmed
Distribution (Global): Andaman Is., Assam, Bangladesh, Cambodia, China South-Central, East Himalaya, Hainan, India, Laos, Malaya, Myanmar, Nicobar Is., Sumatera, Thailand, Vietnam
Distribution (India): Arunachal Pradesh
Leaf Type: Simple
Inflorescence: Axillary panicle
Fruit Type: Follicles
Flowering and Fruiting: April–May
IUCN Status: Not evaluated
Notes: Often found in evergreen and semi-evergreen forests
Reference: Pterocymbium: 234 (1844)

Senra incana Cav.
Malvaceae
Synonyms: *Dumreichera arabica* Hochst. & Steud., *Gossypium bakeri* G.Watt., *Senra arabica* Webb., *Senra bakeri* (G.Watt) Prokh., *Senra nubica* Webb., *Senra zoes* Volkens & Schweinf., *Serraea arabica* (Hochst. & Steud.) Mattei., *Serraea incana* (Cav.) Spreng.

Climbing Mechanism: Scrambler-Unarmed
Distribution (Global): Djibouti, Eritrea, Ethiopia, India, Kenya, Oman, Pakistan, Saudi Arabia, Socotra, Somalia, Sudan, Yemen.
Distribution (India): Gujarat
Leaf Type: Leaves alternate, simple, 3 palmately lobed
Inflorescence: Axillary solitary, rarely terminal racemes
Fruit Type: dehiscent, capsule, dehiscing loculicidally
Flowering and Fruiting: November–April
IUCN Status: Not evaluated
Reference: Diss. 2: 83 (1786)

Waltheria indica L.
Malvaceae
Synonyms: *Melochia corchorifolia* Wall., *Waltheria africana* Schumach. & Thonn. +20
Common Name: Boater bush, Sleepy morning, Velvet leaf
Climbing Mechanism: Scrambler-Unarmed
Distribution (Global): Pan tropical
Distribution (India): Puducherry
Leaf Type: Simple, alternate-distichous
Inflorescence: Axillary clusters
Flowering and Fruiting: Throughout the year
IUCN Status: Not evaluated
Notes: Usually occur in degraded moist deciduous forests
Reference: Sp. Pl.: 673 (1753)

Anplectrum annulatum Triana
Melastomataceae
Climbing Mechanism: Scrambler-Unarmed
IUCN Status: Not evaluated
Reference: Trans. Linn. Soc. London 28: 84 (1871 publ. 1872)

Anplectrum divaricatum Triana
Melastomataceae
Climbing Mechanism: Scrambler-Unarmed
IUCN Status: Not evaluated
Reference: Trans. Linn. Soc. London 28: 84 (1871 publ. 1872)

Anplectrum glaucum Triana
Melastomataceae
Climbing Mechanism: Scrambler-Unarmed
IUCN Status: Not evaluated
Reference: Trans. Linn. Soc. London 28: 84 (1871 publ. 1872)

Anplectrum pallens Blume
Melastomataceae
Climbing Mechanism: Scrambler-Unarmed
IUCN Status: Not evaluated
Reference: Blume. In: Mus. Bot. Lugd. Bat. 1: 38. (1849).

Heterotis rotundifolia (Sm.) Jacq. -Fél.
Melastomataceae
Synonyms: *Asterostoma rotundifolia* Blume, *Dissotis plumosa* Hook.f., *Dissotis rotundifolia* (Sm.) Triana, *Heterotis plumosa* Benth., *Kadalia rotundifolia* Raf., *Melastoma plumosum* D. Don, *Osbeckia rotundifolia* Sm.
Common Name: Spanish shawl, Pink lady, Trailing dissotis
Climbing Mechanism: Stem Twiner
Distribution (Global): Angola, Benin, Cameroon, Central African Repu, Congo, Gabon, Guinea-Bissau, Gulf of Guinea Is., Mozambique, Sierra Leone, Sudan, Togo, Zambia, Zaire, Zimbabwe
Distribution (India): Peninsular India
IUCN Status: Least Concern
Reference: Adansonia, n.s., 20: 417 (1981)

Kendrickia walkeri Hook.f.
Melastomataceae
Synonyms: *Kendrickia walkeri* (Thwaites) Hook.f. ex Triana, *Medinilla walkeri* Gardner, *Pachycentria walkeri* Thwaites
Climbing Mechanism: Stem Twiner
Distribution (Global): India, Sri Lanka
Distribution (India): Tamil Nadu
Leaf Type: Leaves on sterile branches more or less appressed to the substrate, elliptic-ovate
Inflorescence: Flowers solitary or few together in the leaf axils and terminally on the branchlets
Fruit Type: Fruit subglobose, crowned by the 4 thick sepals
IUCN Status: Not evaluated
Reference: G. Bentham & J. D. Hooker, Gen. Pl. 1: 752 (1867)

Marumia nemorosa Blume
Melastomataceae
Synonyms: *Macrolenes nemorosa* (Jack) Bakh.f., *Dissochaeta affinis* (Korth.) Clausing, *Dissochaeta nemorosa* G. Don,*Marumia affinis* Korth., *Marumia bancana* Scheff., *Marumia impressa* Craib, *Marumia leprosa* Korth., *Marumia verrucosa* Cogn., *Melastoma nemorosum* Ja.
Climbing Mechanism: Scrambler-Unarmed
Distribution (Global): Borneo, Malaya, Sumatera, Thailand
Leaf Type: Leaves are opposite and arranged in 2 rows, base of the leaf blade Abaxially with a pair of hair

cushion domatia
Inflorescence: The inflorescences of Macrolenes are axillary and consist of few-flowered cymes
with 1-15 flowers. The main axes of the cymes are usually terete, angular
Fruit Type: Berries,
ovoid to urceolate and colorful when mature, with four persistent remnants of the calyx lobes, dark blue to purple
Flowering and Fruiting: Throughout the year
IUCN Status: Not evaluated
Notes: Usually found in tropical evergreen and perpetually wet forest types
Reference: Contr. Melastom.: 206 (1943)

Marumia reticulata Blume
Melastomataceae
Synonyms: *Macrolenes pachygyna* (Korth.) M.P. Nayar., *Anplectrum reformatum* (Blume) Triana, *Diplectria reformata* (Blume) Kuntze, *Dissochaeta cyanocarpa* Korth. ex Blume, *Dissochaeta pachygyna* (Korth.) I.M. Turner, *Dissochaeta reformata* Blume, *Macrolenes reticulata*
Climbing Mechanism: Scrambler-Unarmed
Distribution (Global): Borneo, Sumatera.
Leaf Type: Leaves are opposite and arranged in 2 rows, base of leaf blade Abaxially with a pair of hair cushion domatia
Inflorescence: The inflorescences of Macrolenes are axillary and consist of few-flowered cymes
with 1-15 flowers. The main axes of the cymes are usually terete, angular
Fruit Type: Berries, dark blue to purple
Flowering and Fruiting: Throughout the year.
IUCN Status: Not evaluated
Reference: J. Jap. Bot. 55: 49 (1980)

Marumia zeylanica Blume
Melastomataceae
Synonyms: *Macrolenes annulata* (Vent.) Naudin., *Huberia annulata* (Vent.) DC., *Macrolenes zeylanica* (Blume) Bakh.f., *Maieta annulata* Vent., *Marumia annulata* (Vent.) Triana, *Marumia horsfieldii* Miq., *Melastoma annulatum* Poir. ex Steud.
Climbing Mechanism: Scrambler-Unarmed
Distribution (Global): Java, Sumatera
Leaf Type: Leaves are opposite and are arranged in 2 rows
Inflorescence: Axillary flowered Cymes
Fruit Type: Berries, ovoid, initially green and dark blue to purple when ripe
Flowering and Fruiting: Throughout the year
IUCN Status: Not evaluated
Notes: Usually occur in tropical evergreen and wet forests
Reference: Ann. Sci. Nat., Bot., sér. 3, 15: 311 (1851)

Medinilla malabarica Bedd. & C.E.C. Fisch.
Melastomataceae
Climbing Mechanism: Scrambler-Unarmed
Distribution (Global): India
Distribution (India): Tamil Nadu
Leaf Type: Leaves to 7 x 4 cm, ovate, acute, 5-nerved from base, glabrous, thick; ribs reddish brown
Inflorescence: Cymes axillary, 1–3-flowered. Flowers light pink
Fruit Type: Berries
Flowering and Fruiting: September–December
IUCN Status: Not evaluated
Notes: Occurs mostly in evergreen and shola forests
Reference: Bull. Misc. Inform. Kew 1938: 124 (1938)

Neodissochaeta celebica Bakh.f.
Melastomataceae
Climbing Mechanism: Scrambler-Unarmed
Distribution (India): Great Nicobar Island
IUCN Status: Not evaluated

Otanthera nicobarensis Teijsm. & Binn.
Melastomataceae
Climbing Mechanism: Scrambler-Unarmed
Distribution (India): Great Nicobar Island
IUCN Status: Not evaluated
Reference: Tijdschr. Ned. -Indië 25: 427 (1863)

Oxyspora cernua (Roxb.) Hook.f. & Thomson ex Triana
Melastomataceae
Synonyms: *Allozygia cernua* Naudin, *Melastoma cernuum* Roxb.
Climbing Mechanism: Scrambler-Unarmed
Distribution (Global): Assam, Bangladesh, East Himalaya, Tibet
Distribution (India): Manipur, Meghalaya
IUCN Status: Not evaluated
Reference: Trans. Linn. Soc. London 28: 73 (1871 publ. 1872)

Albertisia mecistophylla (Miers) Forman
Menispermaceae
Synonyms: *Pycnarrhena mecistophylla* Miers
Climbing Mechanism: Stem Twiner
Distribution (Global): Assam
Leaf Type: Simple, alternate
Inflorescence: Axillary, cymes
Fruit Type: Drupe

IUCN Status: Not evaluated
Reference: Kew Bull. 30: 84 (1975)

Anamirta cocculus (L.) Wight & Arn.
Menispermaceae
Synonyms: *Anamirta baueriana* Endl., *Anamirta jucunda* Miers, *Anamirta paniculata* Colebr., *Anamirta racemosa* Colebr. ex Steud., *Anamirta toxifera* Miers, *Cocculus indicus* Royle, *Cocculus lacunosus* DC., *Cocculus populifolius* DC., *Cocculus suberosus* DC., Menisp.
Common Name: Fish Bbrry, Indian berry, Levant nut
Climbing Mechanism: Stem Twiner
Distribution (Global): Andaman Is., Assam, Bangladesh, Cambodia, India, Java, Laos, Lesser Sunda Is., Maluku, Myanmar, New Guinea, Nicobar Is., Philippines, Sri Lanka, Sumatera, Thailand, Vietnam
Distribution (India): Peninsular India, Goa, Maharashtra, Bihar, Odisha, North Eastern India, Andaman and Nicobar Islands
Leaf Type: Simple, alternate
Inflorescence: Drooping panicles
Fruit Type: Wet, drupe
Flowering and Fruiting: August–May
IUCN Status: Not evaluated
Notes: Common in Moist deciduous and evergreen forests
Reference: Prodr. Fl. Ind. Orient. 1: 446 (1834)

Antitaxis calocarpa Kurz
Menispermaceae
Synonyms: *Antitaxis fasciculata* Miers, *Antitaxis lucida* Miers, *Antitaxis nodiflora* Gagnep., *Cocculus lucidus* Teijsm. & Binn., *Pycnarrhena calocarpa* Diels, *Pycnarrhena fasciculata* (Miers) Diels, *Telotia nodiflora* Pierre.
Climbing Mechanism: Stem Twiner
Distribution (India): Andaman Islands
Leaf Type: Simple, alternate
Inflorescence: Axillary, cymose
Fruit Type: Drupe
IUCN Status: Not evaluated
Reference: J. Bot. 13: 324 (1875)

Aspidocarya uvifera Hook.f. & Thomson
Menispermaceae
Climbing Mechanism: Stem Twiner
Distribution (Global): Assam, China South-Central, East Himalaya, Myanmar, Thailand
Distribution (India): Arunachal Pradesh, Assam, Manipur, West Bengal
Leaf Type: Simple, alternate
Inflorescence: Axillary

Fruit Type: Drupe
IUCN Status: Not evaluated
Reference: Fl. Ind. 1: 180 (1855)

Cissampelos glaberrima A.St.-Hil.
Menispermaceae
Synonyms: *Cissampelos clematidea* C. Presl, *Cissampelos errabunda* Miers, *Cissampelos galapagensis* A.Stewart,*Cissampelos glaberrima var. orbicularis* Chodat & Hassl., *Cissampelos parmata* Miers ex Diels, *Cissampelos parriera* Vell.
Climbing Mechanism: Stem Twiner
Distribution (Global): Bolivia, Brazil North, Brazil Northeast, Brazil South, Brazil Southeast, Brazil West-Central, Colombia, Ecuador, Galápagos, Paraguay, Peru, Venezuela
Distribution (India): Andhra Pradesh
Leaf Type: Simple, orbicular/cordate
Inflorescence: Corymbose cyme
Fruit Type: Wet, drupe
IUCN Status: Not evaluated
Reference: Fl. Bras. Merid. 1: 57 (1825)

Cissampelos pareira L.
Menispermaceae
Synonyms: *Cissampelos acuminata* Benth., *Cissampelos argentea* Kunth, *Cissampelos auriculata* Miers, *Cissampelos australis* A.St.-Hil., *Cissampelos benthamiana* Miers. +74
Common Name: Velvet leaf, False pareira brava, Abuta, Pereira root, Barbasco
Climbing Mechanism: Stem Twiner
Distribution (Global): Pan tropical
Distribution (India): Andaman and Nicobar Islands, Andhra Pradesh, Arunachal Pradesh, Assam, Delhi, Goa, Gujarat, Himachal Pradesh, Jharkhand, Karnataka, Kerala, Madhya Pradesh, Maharashtra, Manipur, Meghalaya, Mizoram, Odisha, Rajasthan, Tamil Nadu, Telangana
Leaf Type: Ovate to nearly round
Inflorescence: Long Cymes
Flowering and Fruiting: August–October
IUCN Status: Not evaluated
Notes: Velvet Leaf is known as Midwife's herb as mainly used in women's ailments.
Reference: Sp. Pl.: 1031 (1753)

Cocculus hirsutus (L.) W. Theob.
Menispermaceae
Synonyms: *Cebatha hirsuta* (L.) Kuntze., *Cebatha villosa* C.Chr., *Cocculus aristolochiae* DC., *Cocculus hastatus* DC. +15.
Common Name: Broom creeper, Ink berry

Climbing Mechanism: Stem Twiner
Distribution (Global): Angola, Bangladesh, Botswana, Caprivi Strip, China Southeast, Eritrea, Ethiopia, Gulf States, India, Kenya, KwaZulu-Natal, Malawi, Mozambique, Myanmar, Namibia, Nepal, Northern Provinces, Pakistan, Saudi Arabia, Sri Lanka, Sudan, Swaziland, Tanzania, Thailand, West Himalaya, Yemen, Zambia, Zimbabwe
Distribution (India): Andaman and Nicobar Islands, Andhra Pradesh, Bihar, Odisha, Delhi, Eastern Ghats, Gujarat, Karnataka, Madhya Pradesh, Maharashtra, Odisha, Puducherry, Rajasthan, Tamil Nadu, Telangana, Tripura, Uttar Pradesh, West Bengal
Leaf Type: Densely velvety, Ovate
Inflorescence: Axillary fascicles or racemes
Fruit Type: Drupe, globose, purple
Flowering and Fruiting: Throughout the year
IUCN Status: Not evaluated
Notes: Medicinally valuable: Roots are laxative and also used in treating Chronic rheumatism, mineral diseases, leaves are used externally for eczema. Plains to 750 m.
Reference: F. Mason, Burmah, ed. 3, 2: 657 (1860)

Cocculus laurifolius DC.
Menispermaceae
Synonyms: *Cinnamomum esquirolii* H. Lév., *Cebatha laurifolia* (DC.) Kuntze., *Cocculus angustifolius* Hassk., *Cocculus bariensis* Pierre ex Gagnep., *Galloa trinervis* Hassk., *Holopeira australis* Miers., *Holopeira fusiformis* Miers., *Holopeira laurifolia* (DC.) Miers., Menis
Climbing Mechanism: Stem Twiner
Distribution (Global): Assam, Bangladesh, China South-Central, China Southeast, East Himalaya, India, Japan, Java, Laos, Myanmar, Nansei-shoto, Nepal, Philippines, Sumatera, Taiwan, Thailand, Vietnam, West Himalaya
Leaf Type: Simple, alternate, non-peltate
Inflorescence: Cymose or thyrsoid
Fruit Type: Drupes curved, obovoid
Flowering and Fruiting: April–October
IUCN Status: Not evaluated
Notes: Alkaloid extracted from the bark is used as muscle-relaxant
Reference: Syst. Nat. 1: 520 (1817)

Cocculus orbiculatus (L.) DC.
Menispermaceae
Synonyms: *Cebatha cuneifolia* (Miers) Kuntze, *Cebatha ferrandiana* (Gaudich.) Kuntze, *Cebatha integra* (Hillebr.) Kuntze, *Cebatha lonchophylla* (Miers) Kuntze, *Cebatha mollis* (Miers) Kuntze, *Cebatha orbiculata* (L.) Kuntze. +20
Climbing Mechanism: Stem Twiner

Distribution (Global): Assam, China North-Central, China South-Central, China Southeast, Cook Is., East Himalaya, Hainan, Hawaii, Japan, Java, Korea, Laos, Lesser Sunda Is., Malaya, Nepal, Philippines, Pitcairn Is., Sumatera, Taiwan, Thailand, Tubuai Is., Vietnam
Distribution (India): Arunachal Pradesh, Meghalaya
Leaf Type: Simple, alternate
Inflorescence: Cymose or thyrsoid, terminal or axillary
Fruit Type: Drupes
Flowering and Fruiting: May–August
IUCN Status: Not evaluated
Reference: Syst. Nat. 1: 523 (1817)

Cocculus pendulus (J.R. Forst. & G. Forst.) Diels
Menispermaceae
Synonyms: *Adenocheton phyllanthoides* Fenzl., *Bricchettia somalensis* Pax., *Cebatha pendula* (J.R.Forst. & G. Forst.) Kuntze., *Cocculus cebatha* DC., *Cocculus ellipticus* DC., *Cocculus epibaterium* DC., *Cocculus glaber* Wight & Arn. +10
Common Name: Pilwan
Climbing Mechanism: Stem Twiner
Distribution (Global): Angola, Cape Verde, Chad, Djibouti, Egypt, Eritrea, Ethiopia, Gambia, Guinea, Gulf States, India, Iran, Kenya, Libya, Mali, Mauritania, Morocco, Niger, Nigeria, Oman, Pakistan, Palestine, Saudi Arabia, Senegal, Sinai, Socotra, Somalia, Sudan, West Himalaya, Western Sahara, Yemen
Distribution (India): Andaman and Nicobar Islands, Gujarat, Himachal Pradesh, Madhya Pradesh, Maharashtra, Rajasthan, Tamil Nadu
Leaf Type: Simple, alternate
Inflorescence: Cymose or thyrsoid, terminal or axillary
Fruit Type: Drupes
Flowering and Fruiting: October–January
IUCN Status: Not evaluated
Notes: Usually found in dry deciduous forests of the Western Ghats.
Reference: H.G.A. Engler (ed.), Pflanzenr., IV, 94: 237 (1910)

Coscinium blumeanum Miers ex Hook.f. & Thomson
Menispermaceae
Climbing Mechanism: Stem Twiner
Distribution (Global): Malaya, Thailand, Vietnam
IUCN Status: Not evaluated
Reference: Fl. Ind. 1: 178 (1855)

Coscinium fenestratum (Goetgh.) Colebr.
Menispermaceae
Synonyms: *Coscinium maingayi* Pierre, *Coscinium miosepalum* Diels, *Coscinium peltatum* Merr., *Coscinium usitatum* Pierre, *Coscinium wallichianum* Miers,

Coscinium wightianum Miers ex Diels, *Menispermum fenestratum* Gaertn., *Pereiria medica* Lindl.
Common Name: False columba, Tree turmeric
Climbing Mechanism: Stem Twiner
Distribution (Global): Borneo, Cambodia, India, Java, Laos, Malaya, Sri Lanka, Sumatera, Thailand, Vietnam
Distribution (India): Karnataka, Tamil Nadu, Western Ghats
Leaf Type: Ovate
Inflorescence: Supra-axillary or cauliflorous
Fruit Type: Drupes
Flowering and Fruiting: August–October
IUCN Status: Data Deficient
Notes: Usually found in semi-evergreen and evergreen forests of Western Ghats
Reference: Trans. Linn. Soc. London 13: 65 (1821)

Cyclea barbata Miers
Menispermaceae
Synonyms: *Cyclea ciliata* Craib, *Cyclea wallichii* Diels
Climbing Mechanism: Stem Twiner
Distribution (Global): Andaman Is., Assam, Bangladesh, Cambodia, China Southeast, East Himalaya, Hainan, Java, Laos, Lesser Sunda Is., Myanmar, Nicobar Is., South China Sea, Sumatera, Thailand, Vietnam
Distribution (India): Kerala, Tripura
Leaf Type: Alternate, ovate
Inflorescence: axillary or cauliflorous, Cymes, flowers sessile
Fruit Type: Drupes obovoid to globose
Flowering and Fruiting: July–May
IUCN Status: Not evaluated
Notes: Usually found in evergreen and semi-evergreen forests
Reference: Contr. Bot. 3: 237 (1871)

Cyclea bicristata (Griff.) Diels
Menispermaceae
Synonyms: *Natsiatum gamosepalum* Griff., *Cyclea populifolia* Hook.f. & Thomson, *Lophophyllum bicristata* Griff., *Peraphora robusta* Miers.
Climbing Mechanism: Stem Twiner
Distribution (Global): Assam, East Himalaya, Myanmar, Vietnam
Distribution (India): Arunachal Pradesh, Meghalaya, Mizoram
Leaf Type: Simple, alternate
Inflorescence: Cauliflorous, racemose or thyrsoid paniculated, axillary, extra-axillary
Fruit Type: Drupes, obovoid
Flowering and Fruiting: August–May
IUCN Status: Not evaluated
Reference: H.G.A. Engler (ed.), Pflanzenr., IV, 94: 317 (1910)

Cyclea fissicalyx Dunn
Menispermaceae
Climbing Mechanism: Stem Twiner
Distribution (Global): India
Distribution (India): Kerala
Leaf Type: Ovate, apex sharply acuminate
Inflorescence: Panicles
Flowering and Fruiting: December–April
IUCN Status: Not evaluated
Notes: Rarely found in evergreen forests
Reference: J.S. Gamble, Fl. Madras: 31 (1915)

Cyclea meeboldii Diels
Menispermaceae
Climbing Mechanism: Stem Twiner
Distribution (Global): Assam, China South-Central
Distribution (India): Mizoram
Leaf Type: Simple, alternate, non-peltate
Inflorescence: Racemose or thyrsoid paniculated, axillary, extra-axillary
Fruit Type: Drupes, obovoid
Flowering and Fruiting: December–March
IUCN Status: Not evaluated
Notes: Usually found in subtropical evergreen forests
Reference: H.G.A. Engler (ed.), Pflanzenr., IV, 94: 315 (1910)

Cyclea peltata (Lam.) Hook.f. & Thomson
Menispermaceae
Synonyms: *Clypea burmanni* (DC.) Wight & Arn., *Cocculus burmanni* DC., *Cocculus peltatus* DC., *Cyclea arnottii* Miers, *Cyclea burmanni* Miers, *Cyclea burmanni* (DC.) Arn. ex Wight, *Cyclea discolor* Miers, *Cyclea peltata var. arnottii* (Miers) Ridl., *Cyclea versicol*
Common Name: Pata root, Indian moon-seed, buckler-leaved moon-seed
Climbing Mechanism: Stem Twiner
Distribution (Global): Andaman Is., Assam, Bangladesh, Cambodia, India, Myanmar, Nicobar Is., Sri Lanka, Vietnam
Distribution (India): Andaman and Nicobar Islands, Andhra Pradesh, Arunachal Pradesh, Assam, Goa, Great, Nicobar, Island, Gujarat, Karnataka, Kerala, Maharashtra, Puducherry, Tamil Nadu, Telangana
Leaf Type: Alternate
Inflorescence: Congested Cymes
Fruit Type: Drupes, spherical
Flowering and Fruiting: November–June
IUCN Status: Not evaluated
Notes: Usually found in moist deciduous, semi-evergreen, and Evergreen forests
Reference: Fl. Ind. 1: 201 (1855)

Cyclea pendulina Miers
Menispermaceae
Climbing Mechanism: Stem Twiner
Distribution (Global): Andaman Is., Nicobar Is.
Distribution (India): Andaman and Nicobar Islands
Leaf Type: Simple, alternate, non-peltate
Inflorescence: Thyrsoid paniculated, axillary, extra-axillary
Fruit Type: Drupes, ovoid
Flowering and Fruiting: November–February
IUCN Status: Not evaluated
Reference: Contr. Bot. 3: 248 (1871)

Cyclea wattii Diels
Menispermaceae
Synonyms: *Paracyclea wattii* (Diels) Yamam.
Climbing Mechanism: Scrambler-Unarmed
Distribution (Global): Assam, China South-Central
Leaf Type: Simple, alternate
Inflorescence: Racemose or thyrsoid paniculated, axillary, extra-axillary, terminal
Fruit Type: Drupes, ovoid
Flowering and Fruiting: May–June
IUCN Status: Not evaluated
Notes: Usually found in forest edges
Reference: Assam, China South-Central

Diploclisia glaucescens (Blume) Diels
Menispermaceae
Synonyms: *Cebatha macrocarpa* (Wight & Arn.) Kuntze, *Cocculus glaucescens* Blume, *Menispermum glaucescens* (Blume) Spreng., *Cocculus macrocarpa* Wight, *Diploclisia inclyta* Miers, *Quinio cocculoides* Schlecht.
Common Name: Glaucous Diploclisia
Climbing Mechanism: Stem Twiner
Distribution (Global): Andaman Is., Assam, Bangladesh, Borneo, Cambodia, China South-Central, China Southeast, Hainan, India, Java, Malaya, Maluku, Myanmar, New Guinea, Nicobar Is., Philippines, Sri Lanka, Sulawesi, Sumatera, Thailand, Vietnam
Distribution (India): Andaman and Nicobar Islands, Andhra Pradesh, Goa, Gujarat, Karnataka, Kerala, Maharashtra, Tamil Nadu, Western Ghats
Flowering and Fruiting: March–August
IUCN Status: Not evaluated
Reference: H.G.A. Engler (ed.), Pflanzenr., IV, 94: 225 (1910)

Fibraurea darshanii Udayan & K. Ravik.
Menispermaceae
Climbing Mechanism: Stem Twiner

Distribution (Global): India
Distribution (India): Peninsular India
Flowering and Fruiting: January–April
IUCN Status: Not evaluated
Reference: Rheedea 17: 9 (2007)

Fibraurea tinctoria Lour.
Menispermaceae
Synonyms: *Cocculus fibraurea* DC., *Fibraurea chloroleuca* Miers, *Fibraurea fasciculata* Miers, *Fibraurea laxa* Miers, *Fibraurea manipurensis* Brace ex Diels, *Fibraurea trotteri* Watt ex Diels, *Menispermum tinctorium* Spreng.
Climbing Mechanism: Stem Twiner
Distribution (Global): Assam, Borneo, Cambodia, India, Java, Laos, Malaya, Myanmar, Nicobar Is., Philippines, Sulawesi, Sumatera, Thailand, Vietnam
Leaf Type: Simple, alternate
Inflorescence: Axillary, extra axillary or terminal, ramiflorous
Fruit Type: Drupe
Flowering and Fruiting: May–June
IUCN Status: Not evaluated
Reference: Fl. Cochinch.: 626 (1790)

Haematocarpus thompsonii Miers
Menispermaceae
Synonyms: *Baterium validum* Miers, *Haematocarpus comptus* Miers, *Haematocarpus thomsonii* Miers.
Climbing Mechanism: Stem Twiner
Distribution (India): Assam
Leaf Type: Simple, alternate
Inflorescence: cauliflorous, axillary, extra-axillary, terminal panicle, or raceme
Fruit Type: Drupe
Flowering and Fruiting: April–December
IUCN Status: Not evaluated

Haematocarpus validus (Miers) Bakh.f. ex Forman
Menispermaceae
Synonyms: *Baterium validum* Miers, *Haematocarpus comptus* Miers, *Haematocarpus incusus* Miers, *Haematocarpus thomsonii* Miers
Climbing Mechanism: Stem Twiner
Distribution (India): Arunachal Pradesh, Great Nicobar Island
IUCN Status: Not evaluated

Hypserpa nitida Miers ex Benth.
Menispermaceae
Synonyms: *Cocculus cuspidatus* Wall., *Hypserpa borneensis* Becc., *Hypserpa cuspidata* (Hook.f. & Thomson) Miers, *Hypserpa heteromera* Miers, *Hypserpa*

jagorii Diels, *Hypserpa laevifolia* Diels, *Hypserpa nandinifolia* Yamam., *Hypserpa pauciflora* Miers
Climbing Mechanism: Stem Twiner
Distribution (Global): Andaman Is., Assam, Bangladesh, Borneo, Cambodia, China South-Central, China Southeast, Hainan, Laos, Malaya, Myanmar, Philippines, Sri Lanka, Sulawesi, Sumatera, Thailand
Distribution (India): Andhra Pradesh
Leaf Type: Simple, alternate
Inflorescence: Cymose
Fruit Type: Drupe
IUCN Status: Not evaluated
Reference: Hooker's J. Bot. Kew Gard. Misc. 3: 258 (1851)

Limacia oblonga Miers
Menispermaceae
Synonyms: *Cocculus oblongus* Wall., *Limacia distincta* Miers, *Limacia inornata* Miers
Climbing Mechanism: Stem Twiner
Distribution (Global): Borneo, Malaya, Sumatera, Thailand
IUCN Status: Not evaluated
Reference: Fl. Ind. 1: 189 (1855)

Limacia triandra Miers
Menispermaceae
Synonyms: *Tiliacora triandra* (Colebr.) Diels., *Aristega laevifolia* Miers, *Cocculus triandrus* Colebr., *Limacia amherstiana* Miers, *Limacia triandra* (Colebr.) Hook. f. & Thomson, *Limacia wallichiana* Miers, *Menispermum triandrum* (Colebr.) Roxb., *Sebicea stipulari*
Climbing Mechanism: Stem Twiner
Distribution (Global): Andaman Is., Assam, Cambodia, Laos, Malaya, Myanmar, Thailand, Vietnam
Leaf Type: Leaves simple, alternate, lanceolate, broadly ovate, pinnately 3–5 veined, coriaceous
Inflorescence: Inflorescences axillary, panicle, pseudoracemes or cauliflorous, few-flowered, peduncled.
Fruit Type: Fruits drupes, subglobose or obovoid, stipitate shortly, condyle linear, endocarp hard
Flowering and Fruiting: May–September
IUCN Status: Not evaluated
Reference: H.G.A. Engler (ed.), Pflanzenr., IV, 94: 62 (1910)

Limacia velutina Miers
Menispermaceae
Synonyms: *Limacia scandens* Lour., *Cocculus limacia* DC., *Cocculus velutinus* Wall., *Limacia cerasifera* Becc., *Menispermum limacia* Spreng.

Climbing Mechanism: Stem Twiner
Distribution (Global): Borneo, Laos, Malaya, Myanmar, Philippines, Sumatera, Thailand, Vietnam
IUCN Status: Not evaluated
Reference: Fl. Cochinch.: 621 (1790)

Lophophyllum bicristata Griff.
Menispermaceae
Synonyms: *Cyclea bicristata* (Griff.) Diels. *Natsiatum gamosepalum* Griff., *Cyclea populifolia* Hook.f. & Thomson, *Peraphora robusta* Miers
Climbing Mechanism: Stem Twiner
Distribution (Global): Assam, East Himalaya, Myanmar, Vietnam
Distribution (India): Assam
Leaf Type: Leaves simple, alternate, subpeltate, deltoid-elongate to ovate, 9-16 x 5-8 cm across, base cordate or broadly cordate, margin entire
Inflorescence: Inflorescence axillary, extra axillary, leaf opposed or on old leafless stems, fascicled, compound panicle of corymb cymes
Fruit Type: Fruits drupes, subobovoid, compressed, pilose, endocarp subovoid
Flowering and Fruiting: August–May
IUCN Status: Not evaluated
Reference: H.G.A.Engler (ed.), Pflanzenr., IV, 94: 317 (1910)

Pachygone ovata (Poir.) Diels
Menispermaceae
Synonyms: *Cebatha pubescens* (F.Muell.) Kuntze, *Cissampelos ovata* Poir.+22.
Climbing Mechanism: Stem Twiner
Distribution (Global): Borneo, Christmas I., India, Java, Lesser Sunda Is., Maluku, Myanmar, New Guinea, Northern Territory, Queensland, Sri Lanka, Sulawesi, Western Australia
Distribution (India): Andhra Pradesh, Eastern Ghats, Karnataka, Puducherry, Tamil Nadu, Telangana
Leaf Type: Simple
Inflorescence: Racemes axillary
Fruit Type: Drupe
Flowering and Fruiting: December–October
IUCN Status: Not evaluate
Notes: Often found in dry deciduous forests and in foothills of scrub jungles
Reference: Fl. Ind. 1: 203 (1855)

Pachygone zeylanica Santapau & Wagh
Menispermaceae
Synonyms: *Pachygone ovata* (Poir.) Miers ex Hook.f. & Thomson
Climbing Mechanism: Stem Twiner
Flowering and Fruiting: June–November

IUCN Status: Not evaluated
Reference: Bull. Bot. Surv. India 5: 107 (1964)

Parabaena sagittata Miers
Menispermaceae
Synonyms: *Parabaena ferruginea* Miers, *Parabaena oleracea* Miers, *Parabaena heterophylla* Miers, *Parabaena racemosa* Gagnep
Climbing Mechanism: Stem Twiner
Distribution (Global): Andaman Is., Assam, Bangladesh, China South-Central, China Southeast, East Himalaya, Laos, Myanmar, Nepal, Nicobar Is., Thailand, Tibet, Vietnam
Distribution (India): Andaman and Nicobar Islands, Andhra Pradesh, Assam, Eastern Ghats, Manipur, Mizoram, Odisha, Tripura, West Bengal
Leaf Type: Alternate
Inflorescence: Thyrsoid or Cymose
Fruit Type: Drupe
Flowering and Fruiting: April–December
IUCN Status: Not evaluated
Reference: Fl. Ind. 1: 181 (1855)

Pericampylus aduncus Miers
Menispermaceae
Climbing Mechanism: Stem Twiner
Distribution (India): Arunachal Pradesh, Great Nicobar Island, Meghalaya, Mizoram, Sikkim, Tripura, West Bengal
IUCN Status: Not evaluated
Reference: Contr. Bot. 3: 119 (1871)

Pericampylus glaucus (Lam.) Merr.
Menispermaceae
Synonyms: *Cebatha diffusa* (Miers) Kuntze, *Clypea tomentosa* Blume, *Cocculus cinereus* Zoll. & Moritzi, *Cocculus corymbosus* Blume+24
Climbing Mechanism: Stem Twiner
Distribution (Global): Assam, Bangladesh, Borneo, Cambodia, China South-Central, China Southeast, East Himalaya, Hainan, Java, Laos, Lesser Sunda Is., Malaya, Maluku, Myanmar, Nepal, New Guinea, Nicobar Is., Philippines, Sulawesi, Sumatera, Taiwan, Thailand, Vietnam
Distribution (India): Kerala
Leaf Type: Simple, alternate
Inflorescence: Cymose or sub-umbellate, axillary, solitary, or fascicled cymes
Fruit Type: Drupe
Flowering and Fruiting: April–August
IUCN Status: Not evaluated
Reference: Interpr. Herb. Amboin.: 219 (1917)

Pericampylus incanus (Colebr.) Miers ex Hook.f. & Thomson
Menispermaceae
Climbing Mechanism: Stem Twiner
IUCN Status: Not evaluated
Reference: Fl. Ind. 1: 194 (1855)

Pycnarrhena longifolia (Decne. ex Miq.) Becc.
Menispermaceae
Synonyms: *Antitaxis cauliflora* Miers, *Antitaxis longifolia* (Decne. ex Miq.) Miers, *Cocculus longifolius* Decne. ex Miq., *Cocculus timorensis* Decne. ex Diels, *Gabila longifolia* (Decne. ex Miq.) Baill., *Limacia longifolia* (Decne. ex Miq.) Miers, *Pycnarrhena caul*
Climbing Mechanism: Stem Twiner
Distribution (Global): Andaman Is., Java, Lesser Sunda Is., Nicobar Is., Sumatera
Distribution (India): Andaman and Nicobar Islands
IUCN Status: Not evaluated
Reference: Malesia 1: 160 (1877)

Pycnarrhena lucida (Teijsm. & Binn.) Miq.
Menispermaceae
Synonyms: *Antitaxis calocarpa* Kurz, *Antitaxis fasciculata* Miers, *Antitaxis lucida* Miers, *Antitaxis nodiflora* (Pierre) Gagnep., *Cocculus lucidus* Teijsm. & Binn., *Pycnarrhena calocarpa* (Kurz) Diels, *Pycnarrhena fasciculata* (Miers) Diels, *Telotia nodiflora* Pier
Climbing Mechanism: Stem Twiner
Distribution (Global): Andaman Is., Bangladesh, Cambodia, Hainan, Java, Laos, Malaya, Nicobar Is., Sumatera, Thailand
Distribution (India): Andaman and Nicobar Islands
Leaf Type: Simple, alternate
Inflorescence: Axillary, Cymose
Fruit Type: Drupe
IUCN Status: Not evaluated
Reference: Ann. Mus. Bot. Lugduno-Batavi 4: 87 (1868)

Pycnarrhena pleniflora Miers
Menispermaceae
Synonyms: *Cocculus planiflorus* Wall.
Climbing Mechanism: Stem Twiner
Distribution (Global): Assam, Bangladesh, East Himalaya, Myanmar
Distribution (India): Arunachal Pradesh, Assam, Mizoram, Tripura
Leaf Type: Simple, alternate
Inflorescence: Axillary, Cymose
Fruit Type: Drupe
IUCN Status: Not evaluated
Reference: Fl. Ind. 1: 206 (1855)

Stephania andamanica Diels
Menispermaceae
Climbing Mechanism: Stem Twiner
Distribution (Global): Andaman Is.
Distribution (India): Andaman and Nicobar Islands
Leaf Type: Peltate/non-peltate, Palmately compound
Inflorescence: Axillary or on old leafless stems, umbelliform cymes or disciform capitula/Corymbose cyme
Fruit Type: Drupes, ovoid to obovoid
Flowering and Fruiting: June
IUCN Status: Not evaluated
Reference: H.G.A.Engler (ed.), Pflanzenr., IV, 94: 266 (1910)

Stephania elegans Hook.f. & Thomson
Menispermaceae
Common Name: Elegant tape vine
Climbing Mechanism: Stem Twiner
Distribution (Global): Assam, Bangladesh, China South-Central, East Himalaya, Myanmar, Nepal, Thailand
Distribution (India): Andaman and Nicobar Islands, Arunachal Pradesh, Assam, Goa, Himachal Pradesh, Mizoram, West Bengal.
Leaf Type: Palmately viened
Inflorescence: Simple (or compound) umbelliform cymes
Fruit Type: Red, broadly obovate-globose
Flowering and Fruiting: March–September
IUCN Status: Not evaluated
Reference: Fl. Ind. 1: 195 (1855)

Stephania glabra (Roxb.) Miers
Menispermaceae
Synonyms: *Stephania rotunda* Lour., *Cissampelos glabra* Roxb., *Clypea glabra* (Roxb.) Wight & Arn. ex Voigt., *Clypea rotunda* Steud., *Menispermum japonicum* Roxb. ex DC., *Menispermum roxburghii* Spreng., *Stephania hexandra* Miers.
Common Name: Hairless tape vine.
Climbing Mechanism: Stem Twiner
Distribution (Global): Assam, Bangladesh, Cambodia, East Himalaya, India, Laos, Myanmar, Nepal, Thailand, Tibet, Vietnam.
Distribution (India): Arunachal Pradesh, Himachal Pradesh, Madhya Pradesh, Mizoram, West Bengal
Leaf Type: Ovate/round
Inflorescence: Umbel-like cymes
Fruit Type: Stalked drupes
Flowering and Fruiting: May–April

IUCN Status: Not evaluated
Reference: Fl. Cochinch.: 608 (1790)

Stephania glandulifera Miers
Menispermaceae
Climbing Mechanism: Stem Twiner
Distribution (Global): Assam, East Himalaya, Myanmar, Nepal, Thailand
Distribution (India): Arunachal Pradesh, Assam, Jharkhand, Mizoram, Tripura
Leaf Type: Non-peltate or peltate, palmately 8–13 veined
Inflorescence: Axillary or on old leafless stems, umbelliform cymes, or disciform capitula
Fruit Type: Drupes, ovoid to obovoid
Flowering and Fruiting: November–May
IUCN Status: Not evaluated
Notes: Found in mixed tropical and subtropical forests
Reference: Ann. Mag. Nat. Hist., ser. 3, 18: 15 (1866)

Stephania hernandiifolia (Willd.) Walp.
Menispermaceae
Synonyms: *Stephania japonica var. discolor* (Blume) Forman., *Cissampelos hernandiifolia* Willd., *Cissampelos hexandra* Roxb. +10.
Climbing Mechanism: Stem Twiner
Distribution (Global): Andaman Is., Assam, Borneo, Cambodia, China South-Central, China Southeast, East Himalaya, Java, Laos, Lesser Sunda Is., Malaya, Maluku, Myanmar, Nepal, New Guinea, New South Wales, Nicobar Is., Queensland, Sulawesi, Sumatera, Thailand, Vietnam
Distribution (India): Andaman and Nicobar Islands, Assam, Bihar, Odisha
Leaf Type: Simple, alternate, suborbicular
Inflorescence: Cyme in peduncles
Fruit Type: Wet, drupe
IUCN Status: Not evaluated
Notes: Usually found in evergreen forests
Reference: Kew Bull. 11: 56 (1956)

Stephania japonica (Thunb.) Miers
Menispermaceae
Synonyms: *Cocculus japonicus* (Thunb.) DC., *Menispermum japonicum* Thunb.
Common Name: Tape vine
Climbing Mechanism: Stem Twiner
Distribution (Global): Andaman Is., Assam, Bangladesh, Borneo, Cambodia, China South-Central, China Southeast, East Himalaya, Hainan, India, Japan, Java, Korea, Laos, Lesser Sunda Is., Malaya, Maluku, Marquesas, Myanmar, Nepal, New Caledonia, New Guinea, New South Wales, Nicobar Is., Norfolk Is., Northern Territory, Philippines, Queensland, Samoa, Society Is., Solomon Is., Sri Lanka, Sulawesi, Sumatera, Thailand, Tonga, Vanuatu, Vietnam

Distribution (India): Andaman and Nicobar Islands, Andhra Pradesh, Arunachal Pradesh, Assam, Goa, Gujarat, Karnataka, Kerala, Maharashtra, Meghalaya, Mizoram, Odisha, Tamil Nadu, Telangana, Tripura, Uttar Pradesh, West Bengal.
Leaf Type: Leaves simple, alternate, Palmately viened
Inflorescence: Axillary,umbelliform cymes or disciform capitula
Fruit Type: Drupes, Red, obovate
Flowering and Fruiting: Throughout the year
IUCN Status: Not evaluated
Notes: Often found in evergreen and moist deciduous forests
Reference: Ann. Mag. Nat. Hist., ser. 3, 18: 14 (1866)

Stephania japonica var. discolor (Blume) Forman
Menispermaceae
Climbing Mechanism: Stem Twiner
Distribution (India): Andaman and Nicobar Islands
Leaf Type: Simple, alternate, suborbicular
Inflorescence: Cyme in peduncles
Fruit Type: Wet, drupe
IUCN Status: Not evaluated
Reference: In: Kew Bull. 11: 56. (1956)

Stephania japonica var. timoriensis (DC.) Forman
Menispermaceae
Synonyms: *Cissampelos glabra* Buch.-Ham. ex Wight & Arn., *Clypea glaucescens* Decne. *Clypea longa* G.Don., *Cocculus japonicus var. timoriensis* DC., *Cocculus roxburghianus* DC., *Menispermum peltatum* J.R.Forst. ex DC., *Stephania concinna* Miers., *Stephania exigua* Miers.
Climbing Mechanism: Stem Twiner
Distribution (Global): Bangladesh, Cambodia, China South-Central, China Southeast, Java, Lesser Sunda Is., Maluku, Marquesas, Myanmar, New Caledonia, New Guinea, Norfolk Is., Northern Territory, Philippines, Queensland, Samoa, Society Is., Sri Lanka, Sulawesi, Tonga, Vanuatu
Distribution (India): Peninsular India
Leaf Type: Simple, alternate, suborbicular
Inflorescence: Cyme in peduncles
Fruit Type: Wet, drupe
Flowering and Fruiting: May–December
IUCN Status: Not evaluated
Reference: Kew Bull. 11: 55 (1956)

Stephania rotunda Lour.
Menispermaceae
Synonyms: *Cissampelos glabra* Roxb., *Clypea glabra* (Roxb.) Wight & Arn. ex Voigt., *Clypea rotunda* Steud., *Menispermum japonicum* Roxb. ex DC.,

Menispermum japonicum Roxb. ex DC., *Stephania glabra* (Roxb.) Miers., *Stephania hexandra* Miers.
Climbing Mechanism: Stem Twiner
Distribution (Global): Assam, Bangladesh, Cambodia, East Himalaya, India, Laos, Myanmar, Nepal, Thailand, Tibet, Vietnam
Leaf Type: Simple, alternate, suborbicular
Inflorescence: Cyme in peduncles
Fruit Type: Wet, drupe
IUCN Status: Not evaluated
Reference: Fl. Cochinch.: 608 (1790)

Stephania wightii Dunn
Menispermaceae
Synonyms: *Clypea wightii* Arn. ex Wight
Climbing Mechanism: Stem Twiner
Distribution (Global): India.
Distribution (India): Kerala, amil Nadu.
Leaf Type: Leaves simple, alternate
Inflorescence: Axillary,umbelliform cymes or disciform capitula
Fruit Type: drupes, ovoid to obovoid
Flowering and Fruiting: April–June
IUCN Status: Not evaluated
Notes: Found mostly in evergreen and semi-evergreen forests.
Reference: Clypea wightii Arn. ex Wight

Tiliacora racemosa Colebr.
Menispermaceae
Synonyms: *Tiliacora acuminata* (Lam.) Miers. *Braunea menispermoides* Willd., *Cocculus acuminatus* (Lam.) DC., *Cocculus polycarpus* Wall., *Cocculus radiatus* DC., *Menispermum acuminatum* Lam., *Menispermum glabrum* J.Koenig ex Willd., *Menispermum polycarpon* Roxb.
Common Name: Tapering-leaf Tiliacora
Climbing Mechanism: Stem Twiner
Distribution (Global): Bangladesh, India, Myanmar, Nepal, Sri Lanka, Vietnam
Distribution (India): Telangana, Uttar Pradesh
Leaf Type: Ovate
Inflorescence: Stalked racemose, Cymes
Fruit Type: Drupes oblong-ovoid to obovoid
Flowering and Fruiting: Throughout the year
IUCN Status: Not evaluated
Notes: Often found in stream banks of deciduous and moist deciduous forests
Reference: Ann. Mag. Nat. Hist., ser. 2, 7: 39 (1851)

Tinomiscium micranthum Diels
Menispermaceae

Synonyms: *Tinomiscium petiolare* Hook.f. & Thomson.*Burasaia javanensis* Zipp. ex Diels, *Cocculus petiolaris* Wall., *Tinomiscium coriaceum* Miers, *Tinomiscium elasticum* Becc., *Tinomiscium javanicum* Miers. *Tinomiscium molle* Diels, *Tinomiscium nicobaricum* N.P.
Climbing Mechanism: Stem Twiner
Distribution (Global): Assam, Borneo, China South-Central, China Southeast, Java, Malaya, Myanmar, New Guinea, Nicobar Is., Philippines, Sumatera, Thailand, Vietnam
Distribution (India): Arunachal Pradesh, Assam
Leaf Type: Leaves simple, alternate
Inflorescence: Axillary,Racemose cymes
Fruit Type: Drupes, compressed, ellipsoid
Flowering and Fruiting: August–April
IUCN Status: Not evaluated
Reference: Fl. Ind. 1: 205 (1855)

Tinomiscium nicobaricum N.P.Balakr.
Menispermaceae
Climbing Mechanism: Stem Twiner
Leaf Type: Simple, alternate
Inflorescence: Axillary racemes
Fruit Type: Wet, drupe
IUCN Status: Not evaluated
Reference: N. P. Balakr. In: New Bot., et in Bull. Bot. Surv. India, 7: 7 1980 (1982), 24(1-4): 56 [1982]. (1983)

Tinomiscium petiolare Hook.f. & Thomson
Menispermaceae
Climbing Mechanism: Stem Twiner
Distribution (India): Arunachal Pradesh, Great Nicobar Island
Leaf Type: Simple, alternate
Inflorescence: Axillary racemes
Fruit Type: Wet, drupe
IUCN Status: Not evaluated

Tinospora cordifolia var. congesta
Menispermaceae
Climbing Mechanism: Stem Twiner
IUCN Status: Not evaluated

Tinospora crispa (L.) Hook.f. & Thomson
Menispermaceae
Synonyms: *Chasmanthera crispa* (L.) Baill., *Cocculus bantamensis* Blume, *Cocculus crispus* (L.) DC., *Cocculus rimosus* Blume, *Cocculus verrucosus* (Roxb.) Wall., *Menispermum bantamense* (Blume) Spreng. +10

Climbing Mechanism: Stem Twiner
Distribution (Global): Assam, Bangladesh, Cambodia, China South-Central, India, Java, Laos, Lesser Sunda Is., Malaya, Myanmar, Philippines, Thailand, Vietnam
Distribution (India): Arunachal Pradesh, Assam
Leaf Type: Leaves simple, alternate
Inflorescence: Axillary,pseudoracemes, pseudopaniculate
Fruit Type: drupes, subglobose or ellipsoid, columnar carpophore
Flowering and Fruiting: February–June
IUCN Status: Not evaluated
Notes: Occurs mostly in mixed deciduous forests
Reference: Fl. Ind. 1: 183 (1855)

Tinospora formanii Udayan & Pradeep
Menispermaceae
Climbing Mechanism: Stem Twiner
Distribution (Global): India
Distribution (India): Peninsular India
Leaf Type: Simple, cordate
Inflorescence: Pseudoracemes
Fruit Type: Wet, drupes globose
IUCN Status: Not evaluated
Reference: Edinburgh J. Bot. 66: 77 (2009)

Tinospora glabra (Burm.f.) Merr.
Menispermaceae
Synonyms: *Chasmanthera cordifolia* (DC.) Baill., *Cocculus* convolvulaceus Endl. ex Walp., *Cocculus cordifolius* DC., *Cocculus coriaceus* Blume, *Menispermum coriaceum* (Blume) Spreng., *Menispermum glabrum* Burm.f., *Tinospora andamanica* Diels, *Tinospora convolvulac.*
Climbing Mechanism: Stem Twiner
Distribution (Global): Andaman Is., Bismarck Archipelago, Borneo, India, Java, Lesser Sunda Is., Malaya, Maluku, New Guinea, Nicobar Is., Philippines, Solomon Is., Sumatera
Distribution (India): Andaman and Nicobar Islands
Leaf Type: Leaves simple, alternate
Inflorescence: pseudoracemes, pseudopaniculate, thyrsoid
Fruit Type: Drupes, subglobose or ellipsoid
Flowering and Fruiting: Almost throughout the year
IUCN Status: Not evaluated
Notes: Found mostly in dense riparian forests, open forests, and scrubs.
Reference: J. Arnold Arbor. 19: 340 (1938)

Tinospora mahajanii Mishra, Khristi & Solanki
Menispermaceae
Climbing Mechanism: Stem Twiner

Leaf Type: Simple, cordate
Inflorescence: Pseudoracemes
Fruit Type: Wet, drupes globose
IUCN Status: Not evaluated
Reference: Shakun Mishra, Khristi & Solanki. In: Rheedea 30(4): 450. (2020)

Tinospora maqsoodiana Mujaffar, Moinuddin, and Mustakim
Menispermaceae
Climbing Mechanism: Stem Twiner
Distribution (Global): India.
Distribution (India): Madhya Pradesh
Leaf Type: Simple, cordate
Inflorescence: Pseudoracemes
Fruit Type: Wet, drupes globose
IUCN Status: Not evaluated
Reference: Indian Forester 140: 528 (2014)

Tinospora sinensis (Lour.) Merr.
Menispermaceae
Synonyms: *Campylus sinensis* Lour., *Cocculus chondodendrum* DC., *Cocculus malabaricus* (Lam.) DC., *Cocculus tomentosus* Colebr., *Epibaterium tomentosum* (Colebr.) Pers., *Menispermum chondrodendron* (DC.) Spreng., *Menispermum malabaricum* Lam., *Menispermum tomentosu*
Common Name: Malabar gulbel, Chinese Tinospora
Climbing Mechanism: Stem Twiner
Distribution (Global): Assam, Bangladesh, Cambodia, China South-Central, China Southeast, East Himalaya, India, Myanmar, Nepal, Sri Lanka, Thailand, Vietnam
Distribution (India): Andhra Pradesh, Arunachal Pradesh, Eastern Ghats, Gujarat, Karnataka, Maharashtra, Manipur, Mizoram, Odisha, Tamil Nadu, Uttar Pradesh, West Bengal
Leaf Type: Leaves simple, alternate
Inflorescence: pseudoracemes, pseudopaniculate, thyrsoid
Fruit Type: Drupes, subglobose, red colored
Flowering and Fruiting: February–June
IUCN Status: Not evaluated
Notes: Found often in evergreen and moist deciduous forests
Reference: Sunyatsenia 1: 193 (1934)

Tinospora uliginosa Miers
Menispermaceae
Climbing Mechanism: Stem Twiner
Leaf Type: Simple, cordate
Inflorescence: Pseudoracemes
Fruit Type: Wet, drupes globose

IUCN Status: Not evaluated
Reference: Miers. In: Ann. & Mag. Nat. Hist. Ser. III. 13: 321. (1864)

Broussonetia kurzii (Hook.f.) Corner
Moraceae
Synonyms: *Allaeanthus kurzii* Hook.f.
Climbing Mechanism: Scrambler-Unarmed
Distribution (India): Mizoram
Flowering and Fruiting: May–July
IUCN Status: Not evaluated
Reference: Gard. Bull. Singapore 19: 234 (1962)

Cudrania pubescens Trécul
Moraceae
Synonyms: *Maclura cochinchinensis* (Lour.) Corner. *Cudrania acuminata* Miq., *Cudrania amboinensis* (Blume) Miq., *Cudrania cambodiana* Gagnep., *Cudrania cochinchinensis* (Lour.) Yakuro Kudo & Masam., *Cudrania fruticosa* (Roxb.) Wight ex Kurz. +30
Climbing Mechanism: Scrambler-Armed
Distribution (Global): Assam, Bangladesh, Borneo, Cambodia, China South-Central, China Southeast, Christmas I., East Himalaya, Hainan, Java, Laos, Lesser Sunda Is., Malaya, Maluku, Myanmar, Nepal, New Caledonia, New Guinea, New South Wales, Nicobar Is., Philippines, Queensland, Sri Lanka, Sulawesi, Sumatera, Taiwan, Thailand, Tibet, Vietnam, West Himalaya
IUCN Status: Not evaluated
Reference: Gard. Bull. Singapore 19: 239 (1962)

Ficus diversiformis Miq.
Moraceae
Synonyms: *Ficus disticha* Thwaites, *Ficus stipulata* Moon, *Ficus thwaitesii* Miq.
Climbing Mechanism: Root Climber
Distribution (Global): India, Sri Lanka
Leaf Type: Simple, alt and spiral, stipules single or paired
Inflorescence: Figs (syconia), globose
Fruit Type: Wet, achenes
IUCN Status: Not evaluated
Reference: London J. Bot. 7: 441 (1848)

Ficus excavata King
Moraceae
Synonyms: *Ficus abbreviata* Wall.
Climbing Mechanism: Root Climber
Distribution (Global): Borneo, Malaya, Sumatera
Leaf Type: Simple, alt and spiral, stipules single or paired
Inflorescence: Figs (syconia), globose

Fruit Type: Wet, achenes
IUCN Status: Not evaluated
Reference: Ann. Roy. Bot. Gard. (Calcutta) 1: 127 (1888)

Ficus glaberrima
Moraceae
Climbing Mechanism: Scrambler-Unarmed
IUCN Status: Least Concern

Ficus globosa Blume
Moraceae
Synonyms: *Ficus manok* (Miq.) Miq., *Urostigma globosum* (Blume) Miq., *Urostigma manok* Miq., *Urostigma onustum* Miq.
Climbing Mechanism: Scrambler-Unarmed
Distribution (Global): Andaman Is., Borneo, Cambodia, Java, Malaya, Myanmar, Sumatera, Thailand, Vietnam
Leaf Type: Simple, alt and spiral, stipules single or paired
Inflorescence: Figs (syconia), globose
Fruit Type: Wet, achenes
IUCN Status: Least Concern
Reference: Bijdr. Fl. Ned. Ind.: 449 (1825)

Ficus hederacea Roxb.
Moraceae
Synonyms: *Ficus anabatos* Voigt, *Ficus cantoniensis* Bodinier ex H.Lév., *Ficus fructicosa* Roxb., *Ficus longipes* Griff., *Ficus ludens* Wall.
Common Name: Ivy fig
Climbing Mechanism: Scrambler-Unarmed
Distribution (Global): Andaman Is., Assam, Bangladesh, China South-Central, China Southeast, East Himalaya, Hainan, Laos, Myanmar, Nepal, Nicobar Is., Thailand, Vietnam, West Himalaya
Distribution (India): Himachal Pradesh, Mizoram
Leaf Type: Distichous
Fruit Type: Syconus
IUCN Status: Not evaluated
Reference: Fl. Ind. ed. 1832, 3: 538 (1832)

Ficus heterophylla L.f.
Moraceae
Synonyms: *Ficus acutiloba* Miq., *Ficus aquatica* J.Koenig ex Willd., *Ficus biglandula* Blume, *Ficus cannabina* Lour., *Ficus denticulata* Vahl, *Ficus denticulata* Willd., *Ficus elongata* Miq., *Ficus panduriformis* Miq.+8
Common Name: Creeping fig.
Climbing Mechanism: Scrambler-Unarmed

Distribution (Global): Andaman Is., Assam, Bangladesh, Borneo, Cambodia, China South-Central, China Southeast, East Himalaya, Hainan, India, Java, Laos, Malaya, Myanmar, Nicobar Is., Sri Lanka, Thailand, Vietnam
Distribution (India): Arunachal Pradesh, Bihar, Odisha, Goa, Karnataka, Kerala, Meghalaya, Mizoram, Uttar Pradesh
Leaf Type: Simple, alt and spiral, stipules single or paired
Inflorescence: Figs (syconia), globose
Fruit Type: Wet, achenes
Flowering and Fruiting: September–December
IUCN Status: Not evaluated
Reference: Suppl. Pl.: 442 (1782)

Ficus laevis Blume
Moraceae
Synonyms: *Ficus ceylanica* (Miq.) Miq., *Ficus jamini* H.Lév. & Vaniot, *Ficus vagans* Roxb., *Ficus zeylanica* Trimen, *Pogonotrophe assamica* Miq., *Pogonotrophe ceylanica* Miq.+6
Climbing Mechanism: Scrambler-Unarmed
Distribution (Global): Andaman Is., Assam, Bangladesh, Borneo, Cambodia, China South-Central, China Southeast, East Himalaya, India, Java, Laos, Malaya, Myanmar, Nepal, Nicobar Is., Sri Lanka, Sumatera, Thailand, Vietnam
Distribution (India): Meghalaya, Mizoram
Leaf Type: Simple, alt and spiral, stipules single or paired
Inflorescence: Figs (syconia), globose
Fruit Type: Wet, achenes
IUCN Status: Not evaluated
Reference: Bijdr. Fl. Ned. Ind.: 437 (1825)

Ficus laevis var. macrocarpa (Miq.) Corner
Moraceae
Climbing Mechanism: Scrambler-Unarmed
Distribution (India): Peninsular India
Leaf Type: Simple, alt and spiral, stipules single or paired
Inflorescence: Figs (syconia), globose
Fruit Type: Wet, achenes
IUCN Status: Not evaluated
Reference: Corner. In: Fl. Brit. India 5: 534. (1888)

Ficus lawesii King
Moraceae
Synonyms: Ficus adamii Elmer, Ficus rigida var. bracteata (Corner) Bennet
Climbing Mechanism: Scrambler-Unarmed
Distribution (Global): Borneo, Maluku, New Guinea, Philippines
Leaf Type: Simple, alt and spiral, stipules single or paired
Inflorescence: Figs (syconia), globose

Fruit Type: Wet, achenes
IUCN Status: Not evaluated
Reference: J. Asiat. Soc. Bengal, Pt. 2, Nat. Hist. 55: 403 (1887)

Ficus pendens Corner
Moraceae
Climbing Mechanism: Root Climber
Distribution (Global): Andaman Is., Borneo, Java, Malaya, Nicobar Is., Sumatera
Distribution (India): Great Nicobar Island
Leaf Type: Simple, alt and spiral, stipules single or paired
Inflorescence: Figs (syconia), globose
Fruit Type: Wet, achenes
IUCN Status: Not evaluated
Reference: Gard. Bull. Singapore 18: 15 (1960)

Ficus pubigera (Wall. ex Miq.) Kurz
Moraceae
Synonyms: *Pogonotrophe pubigera* Wall. ex Miq.
Ficus pubigera (Wall. ex Miq.) Brandis, 1874
Common Name: Khasi fig.
Climbing Mechanism: Scrambler-Unarmed
Distribution (Global): Assam, Cambodia, China South-Central, China Southeast, East Himalaya, Laos, Malaya, Myanmar, Nepal, Thailand, Tibet, Vietnam, West Himalaya
Distribution (India): Arunachal Pradesh
Leaf Type: Simple, alt and spiral, stipules single or paired
Inflorescence: Figs (syconia), globose
Fruit Type: Wet, achenes
IUCN Status: Least Concern
Reference: Forest Fl. N.W. India: 424 (1874)

Ficus punctata Thunb.
Moraceae
Synonyms: *Ficus aurantiacea* Griff., *Ficus aurantiacea var. angustifolia* (Corner) Corner, *Ficus aurantiacea var. parvifolia* (Corner) Corner, *Ficus falcata* Thunb., *Ficus kallicarpa* Miq., *Ficus kallicarpa var. angustifolia* Corner, *Ficus kallicarpa var. parvifoli*
Climbing Mechanism: Root Climber
Distribution (Global): Andaman Is., Borneo, Cambodia, Java, Laos, Lesser Sunda Is., Malaya, Maluku, Myanmar, Nicobar Is., Philippines, Sulawesi, Sumatera, Taiwan, Thailand, Vietnam
Leaf Type: Simple, alt and spiral, stipules single or paired
Inflorescence: Figs (syconia), globose
Fruit Type: Wet, achenes

IUCN Status: Least Concern
Reference: Ficus: 9 (1786)

Ficus sagittata Vahl
Moraceae
Synonyms: *Ficus adhaerens* Miq., *Ficus bordenii* Merr., *Ficus carophylla* Miq., *Ficus coccinella* Zipp. ex Miq., *Ficus compressicaulis* Blume, *Ficus crininervia* Miq., *Ficus oligosperma* Miq., *Ficus radicans* Desf., *Ficus ramentacea* Roxb., *Ficus rigescens* Miq.
Climbing Mechanism: Root Climber
Distribution (Global): Andaman Is., Assam, Bangladesh, Borneo, Cambodia, Caroline Is., China South-Central, China Southeast, East Himalaya, Hainan, Java, Laos, Lesser Sunda Is., Malaya, Maluku, Myanmar, Philippines, Sulawesi, Sumatera, Thailand, Vietnam
Distribution (India): Andaman and Nicobar Islands
Leaf Type: Simple, alt and spiral, stipules single or paired
Inflorescence: Figs (syconia), globose
Fruit Type: Wet, achenes
IUCN Status: Least Concern
Reference: Symb. Bot. 1: 83 (1790)

Ficus sarmentosa Buch.-Ham. ex Sm.
Moraceae
Synonyms: *Ficus foveolata* Wall., *Ficus ludens* Wall., *Ficus nepalensis* Spreng., *Ficus ovata* D. Don, *Ficus reticulata* (Miq.) Miq., *Ficus sarmentosa var. oleiformis* (King) V. Singh & P. Singh +3
Common Name: Nepal fig.
Climbing Mechanism: Scrambler-Unarmed
Distribution (Global): Assam, Bangladesh, China North-Central, China South-Central, China Southeast, East Himalaya, Hainan, India, Japan, Korea, Myanmar, Nansei-shoto, Nepal, Pakistan, Taiwan, Tibet, Vietnam, West Himalaya
Distribution (India): Andaman and Nicobar Islands, Arunachal Pradesh, Himachal Pradesh, Meghalaya, Mizoram, Tripura
Leaf Type: Simple, alt and spiral, stipules single or paired
Inflorescence: Figs (syconia), globose
Fruit Type: Wet, achenes
IUCN Status: Not evaluated
Reference: A.Rees, Cycl. 14: n.° 45 (1810)

Ficus tinctoria
Moraceae
Climbing Mechanism: Scrambler-Unarmed
IUCN Status: Least Concern

Ficus tinctoria subsp. gibbosa (Blume) Corner
Moraceae
Synonyms: *Ficus gibbosa*, *Ficus parasitica*, *Ficus swinhoei*
Common Name: Dye fig, Humped fig.
Climbing Mechanism: Scrambler-Unarmed
Distribution (India): Odisha.
Leaf Type: Simple, alt and spiral, stipules single or paired
Inflorescence: Figs (syconia), globose
Fruit Type: Wet, achenes
IUCN Status: Not evaluated
Reference: Corner. In: Gard. Bull. Singapore 17: 476. (1959)

Ficus travancorica King
Moraceae
Climbing Mechanism: Scrambler-Unarmed
Distribution (Global): India
Distribution (India): Kerala, Tamil Nadu
Leaf Type: Simple, alt and spiral, stipules single or paired
Inflorescence: Figs (syconia), globose
Fruit Type: Wet, achenes
IUCN Status: Data Deficient
Reference: Ann. Roy. Bot. Gard. (Calcutta) 1: 28 (1887)

Maclura andamanica (King ex J. D. Hooker) C.C.Berg
Moraceae
Synonyms: *Cudrania poilanei* Gagnep., *Plecospermum andamanicum* King ex Hook.f.
Climbing Mechanism: Scrambler-Unarmed
Distribution (Global): Andaman Is., Cambodia, Laos, Myanmar, Thailand, Vietnam
IUCN Status: Not evaluated
Reference: Proc. Kon. Ned. Akad. Wetensch. C 89: 245 (1986)

Maclura cochinchinensis (Lour.) Corner
Moraceae
Synonyms: *Cudrania acuminata* Miq., *Cudrania amboinensis* (Blume) Miq., *Cudrania cambodiana* Gagnep., *Cudrania cochinchinensis* (Lour.) Yakuro Kudo & Masam., *Cudrania fruticosa* (Roxb.) Wight ex Kurz, *Cudrania grandifolia* Merr. +30
Common Name: Cockspur thorn
Climbing Mechanism: Scrambler-Armed
Distribution (Global): Assam, Bangladesh, Borneo, Cambodia, China South-Central, China Southeast, Christmas I., East Himalaya, Hainan, Java, Laos, Lesser Sunda Is., Malaya, Maluku, Myanmar, Nepal, New Caledonia, New Guinea, New South Wales, Nicobar Is., Philippines, Queensland, Sri Lanka, Sulawesi, Sumatera, Taiwan, Thailand, Tibet, Vietnam, West Himalay

Distribution (India): Tamil Nadu
Leaf Type: Leaves are elliptic-lance shaped to oblong, 3–8 cm long, wide, papery to leathery
Inflorescence: Male inflorescences is a head capitulum, Female inflorescences are velvety
Fruit Type: The fruit is a compound, irregularly shaped berry as large as a small custard apple
Flowering and Fruiting: April–May
IUCN Status: Not evaluated
Notes: Mostly found in evergreen to deciduous forests
Reference: Gard. Bull. Singapore 19: 239 (1962)

Maclura fruticosa (Roxb.) Corner
Moraceae
Climbing Mechanism: Scrambler-Armed
Distribution (Global): Assam, Bangladesh, China South-Central, Myanmar, Thailand, Vietnam
Distribution (India): Arunachal Pradesh
Leaf Type: Leaves oblong or elliptic-oblong, margin entire
Inflorescence: Inflorescence axillary in pairs on stout peduncles, puberulous. Male inflorescence with slender peduncle
Fruit Type: Achenes protruding
Flowering and Fruiting: April–August
IUCN Status: Not evaluated
Notes: Often occur in deciduous forests
Reference: Gard. Bull. Singapore 19: 239 (1962)

Maclura spinosa (Roxb. ex Willd.) C.C.Berg
Moraceae
Synonyms: *Cudrania spinosa* (Willd.) Hochr., *Plecospermum spinosum* (Willd.) Trécul, *Trophis aculeata* Roth, *Trophis spinosa* Willd.
Common Name: Spiny maclura
Climbing Mechanism: Scrambler-Armed
Distribution (Global): East Himalaya, India, Sri Lanka
Distribution (India): Peninsular India
Leaf Type: Leaves are simple, alternate distichous, elliptic-ovate to inverted-lance shaped, base pointed
Inflorescence: In axillary, pedunculate capitate clusters, sessile
Fruit Type: An angled syncarp, irregularly shaped, enclosing a few anthocarp
Flowering and Fruiting: March–September
IUCN Status: Not evaluated
Reference: Proc. Kon. Ned. Akad. Wetensch. C 89: 245 (1986)

Plecospermum spinosum Trécul
Moraceae

Synonyms: *Maclura spinosa* (Willd.) C.C.Berg
Climbing Mechanism: Scrambler-Armed
Distribution (India): Andhra Pradesh, Eastern Ghats, Karnataka, Madhya Pradesh, Maharashtra, Puducherry, Tamil Nadu
Flowering and Fruiting: February-September.
IUCN Status: Not evaluated
Notes: Mostly found in semi-evergreen and deciduous forests also in scrub and open forests
Reference: Ann. Sci. Nat., Bot., sér. 3, 8: 124 (1847)

Prainea scandens King
Moraceae
Synonyms: *Artocarpus frutescens* (Becc.) Renner, *Prainea frutescens* Becc., *Artocarpus scandens* (King) Renner.
Climbing Mechanism: Scrambler-Unarmed
Distribution (Global): Borneo, Malaya, Thailand
IUCN Status: Not evaluated
Reference: J. D. Hooker, Fl. Brit. India 5: 547 (1888)

Nepenthes distillatoria L.
Nepenthaceae
Synonyms: *Nepenthes chapmanii* N.P.Balakr.
Climbing Mechanism: Tendril Climber
Distribution (Global): Sri Lanka
IUCN Status: Vulnerable
Reference: Sp. Pl.: 955 (1753)

Nepenthes khasiana Hook.f.
Nepenthaceae
Synonyms: *Nepenthes rubra* Rafarin
Common Name: Indian pitcher plant
Climbing Mechanism: Tendril Climber
Distribution (Global): Assam, Bangladesh
Distribution (India): North Eastern India
Leaf Type: Alternate
Inflorescence: Racemes or panicles
Fruit Type: Leathery capsule
IUCN Status: Endangered
Notes: Often found in Hilly areas
Reference: A.P.de Candolle, Prodr. 17: 102 (1873)

Nepenthes mirabilis (Lour.) Druce
Nepenthaceae
Synonyms: *Nepenthes albolineata* F.M.Bailey
Nepenthes macrostachya Blume

Climbing Mechanism: Tendril Climber
IUCN Status: Least Concern
Reference: Rep. Bot. Soc. Exch. Club Brit. Isles 1916: 637 (1917)

Bougainvillea buttiana Holttum & Standl.
Nyctaginaceae
Climbing Mechanism: Scrambler-Armed
Distribution (India): Maharashtra, Tamil Nadu
IUCN Status: Not evaluated
Reference: Publ. Field Mus. Nat. Hist., Bot. Ser. 23: 44 (1944)

Bougainvillea glabra Choisy
Nyctaginaceae
Synonyms: *Bougainvillea glabra var. graciliflora* Heimerl, *Bougainvillea arborea* Glaz., *Bougainvillea brachycarpa* Heimerl, *Bougainvillea glabra var. acutibracteata* Heimerl, *Bougainvillea glabra var. alba* Mendes & Viégas, *Bougainvillea glabra var. brachycarpa* (Heimerl) Heimerl
Common Name: Paper flower, Bougainvillea
Climbing Mechanism: Scrambler-Armed
Distribution (Global): Brazil Northeast, Brazil South, Brazil Southeast
Distribution (India): Delhi, Gujarat, Madhya Pradesh, Maharashtra, Odisha, Rajasthan, Tamil Nadu
Flowering and Fruiting: Throughout the year
IUCN Status: Least Concern
Notes: Often cultivated and planted in gardens and hedges. Sometimes found growing near forest clearings and road cuttings across ghats
Reference: A.P.de Candolle, Prodr. 13(2): 437 (1849)

Bougainvillea spectabilis Willd.
Nyctaginaceae
Synonyms: *Bougainvillea bracteata* Pers., *Bougainvillea peruviana* Nees & Mart., *Bougainvillea speciosa* Schnizl., *Bougainvillea spectabilis var. hirsutissima* J.A. Schmidt, *Bougainvillea spectabilis var. parviflora* Mart. ex J.A.Schmidt.
Common Name: Great Bougainvillea
Climbing Mechanism: Scrambler-Armed
Distribution (Global): Brazil Northeast, Brazil South, Brazil Southeast
Distribution (India): Andaman and Nicobar Islands, Delhi, Gujarat, Kerala, Madhya Pradesh, Maharashtra, Mizoram, Odisha, Rajasthan, Tamil Nadu, Telangana, West Bengal
Leaf Type: Simple, alternate
Fruit Type: Achene
Flowering and Fruiting: November–June
IUCN Status: Not evaluated
Notes: Grown as a garden plant
Reference: Sp. Pl., ed. 4, 2: 348 (1799)

Pisonia aculeata L.
Nyctaginaceae
Synonyms: *Pallavia aculeata* (L.) Vell., *Pisonia aculeata var. guaranitica* Chodat, *Pisonia aculeata var. hirsutissima* J.A.Schmidt, *Pisonia aculeata var. typica* Heimerl, *Pisonia acuminata* Mart. ex J.A.Schmidt, *Pisonia anisophylla* Hassk., *Pisonia comosa* Choisy
Common Name: West Indian cock's spur, Hairy water lily
Climbing Mechanism: Scrambler-Armed
Distribution (Global): Aldabra, Andaman Is., Argentina Northeast, Argentina Northwest, Bahamas, Belize, Bolivia, Borneo, Brazil North, Brazil Northeast, Brazil South, Brazil Southeast, Brazil West-Central, Cambodia, Cayman Is., Colombia, Costa Rica, Cuba, Dominican Republic, Ecuador, El Salvador, Ethiopia, Fiji, Florida, Guatemala, Hainan, Haiti, Honduras, India, Jamaica, Java, Kenya, KwaZulu-Natal, Laos, Leeward Is., Lesser Sunda Is., Madagascar, Malaya, Mexico Central, Mexico Gulf, Mexico Northeast, Mexico Northwest, Mexico Southeast, Mexico Southwest, Mozambique, Myanmar, New Caledonia, New Guinea, New South Wales, Nicaragua, Nicobar Is., Northern Territory, Panamá, Paraguay, Peru, Philippines, Puerto Rico, Queensland, Saudi Arabia, Southwest Caribbean, Sri Lanka, Sulawesi, Sumatera, Taiwan, Tanzania, Texas, Thailand, Tonga, Uganda, Venezuela, Venezuelan Antilles, Vietnam, Western Australia, Windward Is., Yemen
Distribution (India): Andaman and Nicobar Islands, Andhra Pradesh, Bihar, Odisha, Karnataka, Maharashtra, Tamil Nadu
Leaf Type: Simple
Inflorescence: Umbels axillary
Flowering and Fruiting: March–April
IUCN Status: Least Concern
Reference: Sp. Pl.: 1026 (1753)

Pisonia mitis L.
Nyctaginaceae
Synonyms: *Pisonia aculeata* L.
Climbing Mechanism: Scrambler-Armed
Distribution (India): Gujarat
IUCN Status: Not evaluated
Reference: Sp. Pl.: 1026 (1753)

Erythropalum scandens Blume
Olacaceae
Synonyms: *Dactylium vagum* Griff., *Decastrophia inconspicua* Griff., *Erythropalum grandifolium* Elmer, *Erythropalum populifolium* (Arn.) Mast., *Erythropalum vagum* (Griff.) Mast., *Mackaya populifolia* Arn., *Modeccopsis vaga* Griff., *Passiflora heyneana* Wall.
Common Name: Redstake climber
Climbing Mechanism: Tendril Climber

Distribution (Global): Andaman Is., Assam, Bangladesh, Borneo, Cambodia, China South-Central, China Southeast, East Himalaya, Hainan, India, Java, Laos, Lesser Sunda Is., Malaya, Myanmar, Nepal, Philippines, Sulawesi, Sumatera, Thailand, Tibet, Vietnam
Distribution (India): Arunachal Pradesh, Assam, Karnataka, Kerala, Madhya Pradesh, Mizoram, Tripura, Western Ghats
Leaf Type: Alternate
Inflorescence: Axillary peduncled cymes
Flowering and Fruiting: December–March
IUCN Status: Least Concern
Reference: Bijdr. Fl. Ned. Ind.: 922 (1826)

Olax acuminata Wall. ex Benth.
Olacaceae
Synonyms: *Olax heyneana* Wall.
Climbing Mechanism: Scrambler-Unarmed
Distribution (Global): Assam, Bangladesh, China South-Central, East Himalaya, Myanmar
Leaf Type: Simple, alternate
Inflorescence: Solitary or fasciculate raceme
Fruit Type: Drupe
Flowering and Fruiting: March–September
IUCN Status: Not evaluated
Reference: Proc. Linn. Soc. London 1: 89 (1840)

Olax imbricata Roxb.
Olacaceae
Synonyms: *Loranthus mitchellii* Wall., *Fissilia psittacorum* Blanco +9.
Climbing Mechanism: Scrambler-Unarmed
Distribution (Global): Tropical & Subtropical Asia to Caroline Islands (Palau)
Distribution (India): Andaman and Nicobar Islands, Eastern Ghats, Goa, Karnataka, Kerala, Maharashtra
Flowering and Fruiting: December–June
IUCN Status: Least Concern
Notes: Distributed across a wide range of habitats ranging from evergreen, semi-evergreen, moist-deciduous, and shola forest
Reference: Fl. Ind. 1: 169 (1820)

Olax psittacorum (Lam.) Vahl
Olacaceae
Synonyms: *Turraea thouarsiana* (Baill.) Cavaco & Keraudren, *Fissilia psittacorum* Lam., *Olax breonii* Baill., *Olax thouarsiana* Baill., *Olax psittacorum* (Lam.) Vahl appears in other Key resources
Climbing Mechanism: Scrambler-Unarmed
Distribution (Global): Mauritius, Réunion

Distribution (India): Maharashtra
IUCN Status: Critically endangered
Reference: Enum. Pl. 2: 33 (1805)

Olax scandens Roxb.
Olacaceae
Synonyms: *Loranthus securidacoides* Warb., *Drebbelia subarborescens* Zoll., *Olax bador* Buch.-Ham., *Olax obtusa* Blume, *Roxburghia baccata* J.Koenig ex Roxb.
Common Name: Parrot Olax
Climbing Mechanism: Scrambler-Unarmed
Distribution (Global): India, Sri Lanka, Bangladesh to Peninsula Malaysia, Java to Lesser Sunda Islands (Bali)
Distribution (India): Andhra Pradesh, Arunachal Pradesh, Bihar, Odisha, Eastern Ghats, Karnataka, Madhya Pradesh, Odisha, Puducherry, Tamil Nadu, Telangana, West Bengal, Western Ghats
Flowering and Fruiting: March–October
IUCN Status: Not evaluated
Notes: Often occurs in deciduous forests
Reference: Pl. Coromandel 2: 2 (1800)

Jasminum adenophyllum Wall. ex C.B.Clarke
Oleaceae
Synonyms: *Jasminum malayanum* Kiew, *Jasminum trangense* Kerr.
Climbing Mechanism: Stem Twiner
Distribution (Global): Andaman Is., Assam, Laos, Malaya, Thailand, Vietnam
Leaf Type: Smooth, Simple/Unifoliate
Foliar Arrangement Along Stem Opposite
Inflorescence: Cluster/Inflorescence
Fruit Type: Berries, blue to black.
IUCN Status: Not evaluated
Reference: J. D. Hooker, Fl. Brit. India 3: 597 (1882)

Jasminum agastyamalayanum Sabeena, Asmitha, Mulani, E.S.S.Kumar & Sibin
Oleaceae
Climbing Mechanism: Stem Twiner
Distribution (Global): India
Leaf Type: Leaves trifoliolate, opposite, terminal leaflet ovate-elliptic
Inflorescence: Inflorescence is a short axillary cyme usually 1–3 together from the node, with 3–5 flowers
Fruit Type: Berry ovoid, single or often paired
Flowering and Fruiting: August–March
IUCN Status: Not evaluated
Reference: Indian J. Forest. 30: 123 (2007)

Jasminum angustifolium (L.) Willd.
Oleaceae
Synonyms: *Nyctanthes angustifolia* L.
Common Name: Wild jasmine
Climbing Mechanism: Stem Twiner
Distribution (Global): Andaman Is., India, Sri Lanka
Distribution (India): Andaman and Nicobar Islands, Andhra Pradesh, Eastern Ghats, Gujarat, Maharashtra, Puducherry, Tamil Nadu
Leaf Type: Oppositely arranged, simple leaves are very variable even on the same plant
Inflorescence: The flowers are either solitary or more usually in threes
Fruit Type: Fruit a paired berry, black when ripe
Flowering and Fruiting: June–August
IUCN Status: Not evaluated
Notes: Commonly found in dry evergreen to dry deciduous forests
Reference: Sp. Pl., ed. 4, 1: 36 (1797)

Jasminum angustifolium var. sessiliflorum (Vahl) P.S.Green
Oleaceae
Synonyms: *Jasminum sessiliflorum* Vahl, *Mogorium sessiliflorum* (Vahl) Poir.
Climbing Mechanism: Stem Twiner
Distribution (Global): India, Sri Lanka
Distribution (India): Andhra Pradesh, Puducherry, Tamil Nadu
Leaf Type: Leaves, lanceolate, acuminate at apex, base rounded, glabrous
Inflorescence: Cymes 3-flowered, peduncled or sessile
Fruit Type: glabrous. Berry, oblong, obtuse, glabrous
Flowering and Fruiting: July–January
IUCN Status: Not evaluated
Reference: Kew Bull. 40: 227 (1985)

Jasminum arborescens Roxb.
Oleaceae
Climbing Mechanism: Stem Twiner
IUCN Status: Not evaluated

Jasminum attenuatum Roxb. ex DC.
Oleaceae
Synonyms: *Jasminum banlanense* P.Y.Pai
Jasminum robustifolium Kobuski
Climbing Mechanism: Stem Twiner
Distribution (Global): Andaman Is., Assam, Bangladesh, China South-Central, Myanmar, Thailand, Vietnam
Distribution (India): Arunachal Pradesh, Mizoram
Leaf Type: Leaves opposite, simple

Inflorescence: Cymes axillary, or terminal, racemose or in few-branched panicles, 5-9 (-20) flowered
Fruit Type: Berry green-white becoming orange-red, ovoid
Flowering and Fruiting: July–February
IUCN Status: Not evaluated
Reference: Gen. Hist. 4: 62 (1837)

Jasminum auriculatum Vahl
Oleaceae
Climbing Mechanism: Stem Twiner
IUCN Status: Not evaluated

Jasminum azoricum L.
Oleaceae
Synonyms: *Jasminum suaveolens* Salisb., *Jasminum trifoliatum* Moench
Common Name: Azores jasmine
Climbing Mechanism: Stem Twiner
Distribution (Global): Madeira
Distribution (India): Andaman and Nicobar Islands, Andhra Pradesh, Arunachal Pradesh, Maharashtra, Mizoram, Rajasthan, Tamil Nadu, Western Ghats
Leaf Type: Oppositely arranged leaves are divided into three leaflets, i.e., trifoliate
Inflorescence: Flowers fragrant, in many-flowered terminal cymes or panicles
Fruit Type: Small green berries which ripen to black
Flowering and Fruiting: June-September
IUCN Status: Critically endangered
Notes: Often cultivated as ornamental plant
Reference: Sp. Pl.: 7 (1753)

Jasminum betchei F.Muell.
Oleaceae
Synonyms: *Jasminum pedicellatum* A.C.Sm., *Jasminum unifoliolatum* Gillespie
Climbing Mechanism: Scrambler-Unarmed
Distribution (Global): Fiji, Niue, Samoa, Tonga
Leaf Type: Simple, 3-foliolate/paripinnate
Inflorescence: Cymes corymboid/paniculate
Fruit Type: Wet, berry, globose
IUCN Status: Not evaluated
Reference: Australas. Chem. Druggist 1881(Suppl. 4): 29 (1881)

Jasminum brevilobum DC.
Oleaceae
Climbing Mechanism: Stem Twiner
IUCN Status: Not evaluated

Jasminum calophyllum Wall. & G.Don
Oleaceae
Synonyms: *Jasminum courtallense* Wight
Climbing Mechanism: Stem Twiner
Distribution (Global): India
Distribution (India): Gujarat, Maharashtra, Tamil Nadu
Leaf Type: Leaflets subequal,ovate, membranous, glabrous
Inflorescence: Cymes terminal, white flowers
Fruit Type: Berry
Flowering and Fruiting: Throughout the year
IUCN Status: Not evaluated
Notes: Found mostly in evergreen forests.
Reference: Gen. Hist. 4: 63 (1837)

Jasminum cardiomorphum P.S.Green
Oleaceae
Climbing Mechanism: Scrambler-Unarmed
Distribution (Global): Assam
Leaf Type: Simple, 3-foliolate/paripinnate
Inflorescence: Cymes corymboid/paniculate
Fruit Type: Wet, berry, globose
IUCN Status: Not evaluated
Reference: Kew Bull. 58: 292 (2003)

Jasminum caudatum Wall. ex Lindl.
Oleaceae
Synonyms: *Jasminum ovatum* Wall.
Climbing Mechanism: Stem Twiner
Distribution (Global): Andaman Is., Assam, Bangladesh, East Himalaya, India, Nepal
Distribution (India): Andaman and Nicobar Islands, Arunachal Pradesh, Bihar, Odisha, Maharashtra, Tripura
Leaf Type: Leaves opposite, trifoliolate, glabrous, subcoriaceous; rachis to 3 cm long, usually with a pulvinous base
Inflorescence: Inflorescence terminal or occasionally axillary, cymose-paniculate,3 to many-flowered
Fruit Type: Fruit, ovoid, ripening black, often paired
Flowering and Fruiting: August–February
IUCN Status: Not evaluated
Notes: Found mostly in evergreen and semi-evergreen forests
Reference: Edwards's Bot. Reg. 28: t. 26 (1842)

Jasminum coarctatum Roxb.
Oleaceae

Climbing Mechanism: Stem Twiner
IUCN Status: Not evaluated

Jasminum cordifolium Wall. & G.Don
Oleaceae
Synonyms: *Jasminum erectiflorum* A.DC., *Jasminum glabellum* Wall., *Jasminum malabaricum* var. lawii C.B.Clarke
Climbing Mechanism: Stem Twiner
Distribution (Global): India
Distribution (India): Andaman and Nicobar Islands, Tamil Nadu
Leaf Type: Leaves, broadly ovate, acuminate, truncate, or cordate at base
Inflorescence: Flowers many, in terminal or axillary cymes
Fruit Type: Berry, oblong, obtuse, glabrous
Flowering and Fruiting: January–April
IUCN Status: Not evaluated
Notes: Found mostly in moist deciduous and evergreen forests
Reference: Gen. Hist. 4: 60 (1837)

Jasminum crassifolium Blume
Oleaceae
Synonyms: *Jasminum aphanodon* Miq. ex Holthuis, *Jasminum rotundifolium* Elmer, *Jasminum suberosum* Holthuis, *Jasminum triplinervium* Merr.
Climbing Mechanism: Stem Twiner
Distribution (Global): Borneo, Java, Philippines, Sulawesi, Sumatera
Leaf Type: Simple, 3-foliolate/paripinnate
Inflorescence: Cymes corymboid/paniculate
Fruit Type: Wet, berry, globose
IUCN Status: Not evaluated
Reference: Bijdr. Fl. Ned. Ind.: 679 (1826)

Jasminum cuspidatum Rottl. & Willd.
Oleaceae
Synonyms: *Jasminum ligustrifolium* Wall. ex Bojer, *Jasminum myrtophyllum* Zenker, *Jasminum perrottetianum* A.DC., *Jasminum rigidum* Zenker, *Jasminum rigidum var. tetraphis* (Wight & Gardner) C.B.Clarke, *Jasminum tetraphis* Wight & Gardner, *Jasminum triflorum* B.Heyn
Common Name: Pointed-leaf jasmine
Climbing Mechanism: Stem Twiner
Distribution (Global): India, Myanmar
Distribution (India): Andhra Pradesh, Tamil Nadu
Leaf Type: Leaves, lance shaped, pointed at tip, round at base
Inflorescence: Flowers are borne in cymes terminal or axillary at branch ends or in leaf-axils
Fruit Type: Berry is ovoid, hairless
Flowering and Fruiting: January–April

IUCN Status: Not evaluated
Notes: Usually found in evergreen and shola forests
Reference: Neue Schriften Ges. Naturf. Freunde Berlin 4: 192 (1803)

Jasminum decussatum Wall. & G.Don
Oleaceae
Synonyms: *Jasminum puberulum* Ridl.
Climbing Mechanism: Stem Twiner
Distribution (Global): Malaya, Myanmar, Sumatera, Thailand
Leaf Type: Leaves unifoliolate, lamina broadly lanceolate to ovate, venation pinnate
Inflorescence: Inflorescences terminal on short side shoots, long with one or two pairs of leaves or from the axils of the upper leaf pair, many flowered (more than 30), lax thyrse (trichotomous)
Fruit Type: Fruits 2 lobed berry, ripening blackish, lobes ellipsoid
IUCN Status: Not evaluated
Notes: Found mostly in lowlands and in open thickets
Reference: Gen. Hist. 4: 62 (1837)

Jasminum dispermum Wall.
Oleaceae
Climbing Mechanism: Stem Twiner
IUCN Status: Not evaluated

Jasminum elongatum (P.J.Bergius) Willd.
Oleaceae
Synonyms: *Jasminum acuminatissimum* Blume, *Jasminum aemulum* R.Br., *Jasminum aemulum var. brassii* P.S.Green, *Jasminum aemulum var. genuinum* Domin, *Jasminum aemulum var. glaberrimum* Domin, *Jasminum aemulum* f. interstans Domin. +40
Common Name: Ear-leaf jasmine
Climbing Mechanism: Stem Twiner
Distribution (Global): Andaman Is., Assam, Borneo, Cambodia, China South-Central, China Southeast, East Himalaya, Hainan, Java, Laos, Lesser Sunda Is., Malaya, Myanmar, Nepal, New Guinea, Nicobar Is., Northern Territory, Philippines, Queensland, Sulawesi, Sumatera, Thailand, Vietnam, Western Australia
Distribution (India): Madhya Pradesh, Mizoram, West Bengal
Leaf Type: Leaves opposite, simple, ovate-lanceolate, acuminate
Inflorescence: Inflorescences short dichasial, flowers scentless, occurring 6–10 in terminal flat-topped clusters
Fruit Type: Berry simple or didymous, globose, black when ripe
IUCN Status: Not evaluated
Notes: Distributed mostly in forest margins, hedges, rubber plantations, and riverbanks
Reference: Sp. Pl., ed. 4, 1: 37 (1797)

Jasminum flexile Vahl
Oleaceae
Synonyms: *Jasminum acuminatum* B.Heyne ex Wall., *Jasminum azoricum var. travancorense* (Gamble) M.Mohanan, *Jasminum burmannianum* Blume, *Jasminum flexile var. hookerianum* Wall. ex C.B.Clarke, *Jasminum flexile var. ovatum* Wall. ex C.B.Clarke, *Jasminum flexile* var
Common Name: River jasmine
Climbing Mechanism: Stem Twiner
Distribution (Global): Assam, China South-Central, India, Myanmar, Sri Lanka, Thailand
Distribution (India): Arunachal Pradesh, Bihar, Odisha, Gujarat, Madhya Pradesh, Maharashtra, Odisha, Tamil Nadu, Telangana, West Bengal
Leaf Type: Leaves are trifoliate, with leaflets n, ovate-lance shaped
Inflorescence: White flowers are borne in cymes, at branch ends, corymb-like, stalked
Fruit Type: Berry is, oblique-obovoid, black, glabrous
Flowering and Fruiting: October–March
IUCN Status: Not evaluated
Notes: Often found in evergreen, semi-evergreen, and shola forests
Reference: Symb. Bot. 3: 1 (1794)

Jasminum fluminense subsp. fluminense
Oleaceae
Synonyms: *Jasminum azoricum var. bahiense* (DC.) Eichler, *Jasminum bahiense* DC., *Jasminum blandum* S.Moore, *Jasminum fluminense subsp. holstii* (Gilg) Turrill, *Jasminum fluminense subsp. mauritianum* (Bojer ex DC.) Turrill, *Jasminum fluminense subsp. nairobiense*
Climbing Mechanism: Stem Twiner
Distribution (Global): Angola, Botswana, Burundi, Cameroon, Caprivi Strip, Gulf of Guinea Is., Kenya, Malawi, Mozambique, Namibia, Nigeria, Northern Provinces, Rwanda, Sierra Leone, Swaziland, Tanzania, Uganda, Zambia, Zimbabwe
Leaf Type: Simple, 3-foliolate/paripinnate
Inflorescence: Cymes corymboid/paniculate
Fruit Type: Wet, berry, globose
IUCN Status: Not evaluated
Reference: Accepted by, Figueiredo, E. & Smith, G.F. (2008)

Jasminum grandiflorum
Oleaceae
Climbing Mechanism: Stem Twiner
IUCN Status: Not evaluated

Jasminum griffithii C.B.Clarke
Oleaceae
Climbing Mechanism: Stem Twiner

Distribution (Global): Malaya
Leaf Type: Leaves unifoliolate; petioles 0.5–1 cm long, densely tomentose, venation pinnate
Inflorescence: Inflorescences terminal or terminal on short side shoots or sometimes axillary
Fruit Type: Fruits ripening waxy white, lobes more or less globose, 7–9 mm in diameter
IUCN Status: Not evaluated
Notes: Usually found in primary lowland dipterocarp forest, sometimes in seasonally swampy areas, or in hill dipterocarp forest
Reference: J. D. Hooker, Fl. Brit. India 3: 593 (1882)

Jasminum humile L.
Oleaceae
Synonyms: *Chrysojasminum humile* (L.) Banfi
Climbing Mechanism: Stem Twiner
Distribution (Global): Afghanistan, Assam, China North-Central, China South-Central, East Himalaya, Iran, Myanmar, Nepal, Pakistan, Tadzhikistan, Tibet, West Himalaya
Distribution (India): Arunachal Pradesh
Leaf Type: Leaves are pinnate with 3–7 ovate to lancelike leathery leaflets
Inflorescence: Inflorescences are lax clusters of yellow tubular flowers at the end of branches
Fruit Type: Fruit is blackberry, with crimson juice
Flowering and Fruiting: April–June
IUCN Status: Not evaluated
Reference: Atti Soc. Ital. Sci. Nat. Mus. Civico Storia Nat. Milano 1(1): 5 (2014)

Jasminum lanceolaria Roxb.
Oleaceae
Climbing Mechanism: Stem Twiner
Distribution (Global): Andaman Is., Assam, Bangladesh, Cambodia, China North-Central, China South-Central, China Southeast, East Himalaya, Hainan, Malaya, Myanmar, Nansei-shoto, Taiwan, Thailand, Vietnam
Distribution (India): Andaman and Nicobar Islands, Arunachal Pradesh, Assam, Mizoram
Leaf Type: Leaves trifoliolate, below inflorescences on side branches occasionally unifoliate, glabrous
petioles, leaflets more or less isomorphic
Inflorescence: Inflorescences 1 or 2 axillary or terminal on side branches, thyrse lax, branches widely spaced, many flowered
Fruit Type: Fruits ripening blackish brown or reddish and slightly
rough, lobes ellipsoid, stalk conspicuously swollen
IUCN Status: Not evaluated
Reference: Fl. Ind. 1: 98 (1820)

Jasminum latipetalum C.B.Clarke
Oleaceae
Climbing Mechanism: Stem Twiner
Distribution (Global): Laos, Myanmar, Thailand
Leaf Type: Simple, 3-foliolate/paripinnate
Inflorescence: Cymes corymboid/paniculate
Fruit Type: Wet, berry, globose
IUCN Status: Not evaluated
Reference: J. D. Hooker, Fl. Brit. India 3: 599 (1882)

Jasminum laurifolium Roxb. ex Hornem.
Oleaceae
Synonyms: *Jasminum angustifolium var. laurifolium* (Roxb. ex Hornem.) Ker Gawl., *Jasminum laurifolium var. genuinum* Kurz, nom. inval.
Common Name: Angelwing jasmine, Royal jasmine, Shining jasmine, Star jasmine
Climbing Mechanism: Stem Twiner
Distribution (Global): Assam, Bangladesh, China South-Central, China Southeast, East Himalaya, Hainan, Myanmar, Nepal, Thailand, Tibet, West Himalaya
Distribution (India): Arunachal Pradesh, Mizoram
Leaf Type: Glossy, leathery leaves are elliptic to lance shaped, oppositely arranged
IUCN Status: Not evaluated
Reference: Hort. Bot. Hafn.: 112 (1819)

Jasminum listeri King ex Gage
Oleaceae
Climbing Mechanism: Stem Twiner
Distribution (Global): Assam, Bangladesh
Distribution (India): Arunachal Pradesh, Mizoram, Tripura
Leaf Type: Simple, 3-foliolate/paripinnate
Inflorescence: Cymes corymboid/paniculate
Fruit Type: Wet, berry, globose
IUCN Status: Not evaluated
Reference: Rec. Bot. Surv. India 1: 349 (1901)

Jasminum maingayi C.B.Clarke
Oleaceae
Climbing Mechanism: Stem Twiner
IUCN Status: Not evaluated

Jasminum malabaricum Wight
Oleaceae
Common Name: Malabar jasmine
Climbing Mechanism: Stem Twiner
Distribution (Global): India

Distribution (India): Andhra Pradesh, Goa, Gujarat, Kerala, Madhya Pradesh, Maharashtra, Tamil Nadu, Western Ghats
Leaf Type: Oppositely arranged membranous leaves, broad, ovate with a sharp tip
Inflorescence: Fragrant white flowers appear in branched terminal cymes at the end of branches, up to 50 in a single cyme
Fruit Type: Ovoid, drying black, single or paired
Flowering and Fruiting: March–November
IUCN Status: Not evaluated
Notes: Commonly found in semi-evergreen forests and scrub jungles
Reference: Icon. Pl. Ind. Orient. 4: t. 1250 (1848)

Jasminum matthewii P.S.Green
Oleaceae
Synonyms: *Jasminum angustifolium var. hirsutum* P.S.Green
Climbing Mechanism: Stem Twiner
Distribution (Global): India, Sri Lanka
Distribution (India): Peninsular India
Leaf Type: Leaves opposite, broadly ovate to elliptic-oblanceolate
Inflorescence: Inflorescence is a cyme, terminal, only on side shoots
Fruit Type: Berries
paired, rarely single, black and shiny when ripe
Flowering and Fruiting: Throughout the year
IUCN Status: Not evaluated
Notes: Distributed from the eastern coastal plains to 650 m on the eastern hillslopes of the Western Ghats
Reference: Kew Bull. 40: 228 (1985)

Jasminum mesnyi Hance
Oleaceae
Synonyms: *Jasminum primulinum* Hemsl. ex Baker
Common Name: Primrose jasmine, Japanese jasmine
Climbing Mechanism: Stem Twiner
Distribution (Global): China South-Central, Vietnam
Distribution (India): Maharashtra, Tamil Nadu
Leaf Type: Leaves opposite, 3-foliolate or simple at base of branchlets
Inflorescence: Flowers are usually solitary, in leaf-axils or rarely at branch-ends
Fruit Type: Berry
IUCN Status: Not evaluated
Reference: J. Bot. 20: 37 (1882)

Jasminum multiflorum (Burm.f.) Andrews
Oleaceae
Synonyms: *Jasminum congestum* Buch.-Ham. ex Wall., *Jasminum gracillimum* Hook.f., *Jasminum multiflorum var. nicobaricum* Thoth., *Jasminum pubescens*

(Retz.) Willd., *Mogorium multiflorum* (Burm.f.) Lam., *Mogorium pubescens* (Retz.) Lam., *Nyctanthes multiflora* Burm.f

Common Name: Downy jasmine, Indian jasmine, Musk jasmine, Star jasmine, Winter jasmine

Climbing Mechanism: Stem Twiner

Distribution (Global): Bangladesh, India, Laos, Myanmar, Nepal, Thailand, Vietnam, West Himalaya

Distribution (India): Andaman and Nicobar Islands, Andhra Pradesh, Arunachal Pradesh, Delhi, Gujarat, Himachal Pradesh, Kerala, Madhya Pradesh, Maharashtra, Odisha, Puducherry, Rajasthan, Tamil Nadu, Telangana, Tripura, West Bengal

Leaf Type: Leaves are opposite, simple, densely hairy, Leaf blade si ovate-heart shaped

Inflorescence: Flowers are borne in congested clusters at branch ends on small side shoots, many flowered

Fruit Type: Berry

Flowering and Fruiting: Almost throughout the year

IUCN Status: Not evaluated

Reference: Bot. Repos. 8: t. 496 (1807)

Jasminum nepalense Spreng.

Oleaceae

Synonyms: *Jasminum dichotomum* D.Don, *Jasminum glandulosum* Wall. ex G.Don, *Jasminum hamiltonii* G.Don.

Climbing Mechanism: Stem Twiner

Distribution (Global): Assam, East Himalaya, Myanmar, Nepal, West Himalaya

Distribution (India): Arunachal Pradesh, Mizoram, West Bengal

Leaf Type: Simple, 3-foliolate/paripinnate

Inflorescence: Cymes corymboid/paniculate

Fruit Type: Wet, berry, globose

IUCN Status: Not evaluated

Reference: Syst. Veg. ed. 16, 4(2): 12 (1827)

Jasminum nervosum Lour.

Oleaceae

Synonyms: *Jasminum anastomosans* Wall. ex DC., *Jasminum elegans* (Hemsl.) Yamam., *Jasminum finlaysonianum* Wall. ex G.Don, *Jasminum hemsleyi* Yamam., *Jasminum lindleyanum* Blume, *Jasminum silhetense* Blume, *Jasminum smalianum* Brandis, *Jasminum stenopetalum* Lindl.

Common Name: Wild kund

Climbing Mechanism: Stem Twiner

Distribution (Global): Assam, Bangladesh, Cambodia, China South-Central, China Southeast, East Himalaya, Hainan, Laos, Malaya, Myanmar, Nepal, Taiwan, Thailand, Tibet, Vietnam

Distribution (India): Arunachal Pradesh, Mizoram

Leaf Type: Leaves are opposite, simple, ovate to lance shaped, papery
Inflorescence: Flowers are borne at branch ends or in leaf-axils, solitary or in 3–5-flowered cymes
Fruit Type: Berry is red becoming black, spherical
Flowering and Fruiting: March–July
IUCN Status: Not evaluated
Reference: Fl. Cochinch.: 20 (1790)

Jasminum nobile C.B.Clarke
Oleaceae
Synonyms: *Jasminum nobile* subsp. *rex* (Dunn) P.S.Green, Jasminum rex Dunn
Climbing Mechanism: Stem Twiner
Distribution (Global): Cambodia, Laos, Myanmar, Thailand
Leaf Type: Leaves are simple/Unifoliate oppositely arranged along the stem, oval in shape
Inflorescence: Axillary, Cluster/Inflorescence
Fruit Type: Berry
IUCN Status: Not evaluated
Reference: J. D. Hooker, Fl. Brit. India 3: 597 (1882)

Jasminum odoratissimum L.
Oleaceae
Synonyms: *Chrysojasminum odoratissimum* (L.) Banfi., *Jasminum augeronii* A. Cabrera, *Jasminum barrelieri* Webb & Berthel., *Jasminum gomeraeum* Gand., *Jasminum odorum* Salisb.
Climbing Mechanism: Stem Twiner
Distribution (Global): Canary Is., Madeira
Distribution (India): Maharashtra, Rajasthan
Leaf Type: Simple, 3-foliolate/paripinnate
Inflorescence: Cymes corymboid/paniculate
Fruit Type: Wet, berry, globose
IUCN Status: Not evaluated
Reference: Atti Soc. Ital. Sci. Nat. Mus. Civico Storia Nat. Milano 1(1): 5 (2014)

Jasminum officinale L.
Oleaceae
Synonyms: *Jasminum affine Royle* ex Lindl., *Jasminum ochroleucum* Bosse, *Jasminum officinale var. piliferum* P.Y.Pai, *Jasminum officinale var. tibeticum* C.Y.Wu, *Jasminum viminale* Salisb., *Jasminum vulgatius* Lam.
Common Name: Common jasmine, True jasmine, Poet's jasmine
Climbing Mechanism: Stem Twiner
Distribution (Global): Afghanistan, Bangladesh, China South-Central, East Himalaya, Iran, Nepal, Pakistan, Tadzhikistan, Tibet, Transcaucasus, Turkey, West Himalaya

Distribution (India): Delhi, Gujarat, Himachal Pradesh, Madhya Pradesh, Maharashtra, Rajasthan, Tamil Nadu
Leaf Type: Leaves are opposite and pinnate; leaflets are ovate and pointed
Inflorescence: Terminal inflorescences with five to twelve flowers arranged in an umbellate fashion
Fruit Type: Berry
Flowering and Fruiting: June–August
IUCN Status: Not evaluated
Notes: Often cultivated
Reference: Sp. Pl.: 7 (1753)

Jasminum parkeri Dunn
Oleaceae
Synonyms: *Chrysojasminum parkeri*
Common Name: Dwarf jasmine, Himalayan jasmine, Parker jasmine
Climbing Mechanism: Scrambler-Unarmed
Distribution (Global): West Himalaya
Leaf Type: Leaves are alternate, estipulate, stalked, pinnately compound, 3–5 foliolate, imparipinnate
Inflorescence: Flowers are borne in cymes in leaf-axils and at branch-ends. Cymes are 1.9–2.2 cm long, stalked
Fruit Type: Fruit is a berry, ellipsoid
Flowering and Fruiting: May–July
IUCN Status: Not evaluated
Reference: Atti Soc. Ital. Sci. Nat. Mus. Civico Storia Nat. Milano 1(1): 5 (2014)

Jasminum ritchiei C.B.Clarke
Oleaceae
Synonyms: *Jasminum ritchiei var. purpureum* B.Heyne ex C.B.Clarke
Common Name: Ritchie's jasmine
Climbing Mechanism: Stem Twiner
Distribution (Global): Andaman Is., East Himalaya, India
Distribution (India): Goa, Maharashtra, Tamil Nadu.
Leaf Type: Leaves are opposite, elliptic tapering, wedge shaped at base, lateral nerves 4 or 5 pairs
Inflorescence: Flowers are borne in 3–9 flowered, somewhat panicled cymes
Fruit Type: Berry is solitary, spherical
Flowering and Fruiting: July–August
IUCN Status: Not evaluated
Notes: Occurs mostly in moist deciduous forests
Reference: J. D. Hooker, Fl. Brit. India 3: 598 (1882)

Jasminum sambac (L.) Aiton
Oleaceae

Synonyms: Jasminum bicorollatum Noronha, *Jasminum blancoi* Hassk., *Jasminum fragrans* Salisb., *Jasminum heyneanum* Wall. ex G.Don, *Jasminum odoratum* Noronha, *Jasminum pubescens* Buch.-Ham. ex Wall., *Jasminum quadrifolium* Buch.-Ham. ex Wall., *Jasminum quinqueflor*
Common Name: Arabian jasmine
Climbing Mechanism: Stem Twiner
Distribution (Global): East Himalaya, India
Distribution (India): Andaman and Nicobar Islands, Andhra Pradesh, Arunachal Pradesh, Bihar, Odisha, Delhi, Gujarat, Madhya Pradesh, Maharashtra, Odisha, Rajasthan, Tamil Nadu, Telangana, Tripura, West Bengal
Leaf Type: Simple, 3-foliolate/paripinnate
Inflorescence: Cymes corymboid/paniculate
Fruit Type: Wet, berry, globose
IUCN Status: Not evaluated
Reference: Hort. Kew. 1: 8 (1789)

Jasminum scandens (Retz.) Vahl
Oleaceae
Climbing Mechanism: Scrambler-Unarmed
Distribution (Global): Assam, Bangladesh, Cambodia, East Himalaya, India, Myanmar, Thailand
Distribution (India): Andhra Pradesh, Bihar, Odisha, Eastern Ghats, Madhya Pradesh, Mizoram, Odisha, Tamil Nadu, Tripura, West Bengal, Western Ghats
Leaf Type: Simple, 3-foliolate/paripinnate
Inflorescence: Cymes corymboid/paniculate
Fruit Type: Wet, berry, globose
IUCN Status: Not evaluated

Jasminum smilacifolium Griff. ex C.B.Clarke
Oleaceae
Synonyms: *Jasminum crassum* Merr.
Climbing Mechanism: Stem Twiner
Distribution (Global): Malaya, Sumatera
Leaf Type: Simple, 3-foliolate/paripinnate
Inflorescence: Cymes corymboid/paniculate
Fruit Type: Wet, berry, globose
IUCN Status: Not evaluated
Notes: Occurs mostly in lowland forests
Reference: J. D. Hooker, Fl. Brit. India 3: 600 (1882)

Jasminum subglandulosum Kurz
Oleaceae
Synonyms: *Jasminum sempervirens* Kerr, *Jasminum wangii* Kobuski
Climbing Mechanism: Stem Twiner

Distribution (Global): Assam, Bangladesh, China South-Central, Myanmar, Thailand, Vietnam
Distribution (India): Andaman and Nicobar Islands, Arunachal Pradesh, Mizoram, Tripura
Leaf Type: Leaves opposite, simple, glabrous
Inflorescence: Inflorescence terminal or sometimes axillary, glabrous
Fruit Type: Fruit
ellipsoid or globose, purple-black
Flowering and Fruiting: August–October
IUCN Status: Not evaluated
Reference: J. Bot. 13: 329 (1875)

Jasminum subhumile W.W.Sm.
Oleaceae
Synonyms: *Chrysojasminum subhumile var. glabricymosum* (W.W.Sm.) Banfi & Galasso, *Jasminum arboreum* Buch.-Ham. ex Loudon, *Jasminum diversifolium* Kobuski, *Jasminum heterophyllum* Roxb., *Jasminum heterophyllum var. glabricymosum* W.W.Sm., *Jasminum macrophyllum* DC.
Climbing Mechanism: Stem Twiner
Distribution (Global): Assam, China South-Central, Myanmar, Nepal
Leaf Type: Leaves alternate, compound, or simple
Inflorescence: Paniculate cymes terminal
Fruit Type: Berry black or red-black, globose or ellipsoid
Flowering and Fruiting: March–August
IUCN Status: Not evaluated
Reference: Atti Soc. Ital. Sci. Nat. Mus. Civico Storia Nat. Milano 1(1): 5 (2014)

Jasminum syringifolium Wall. & G.Don
Oleaceae
Climbing Mechanism: Stem Twiner
IUCN Status: Not evaluated

Jasminum trichotomum B. Heyne ex Roth
Oleaceae
Synonyms: *Jasminum gardnerianum* Wight, *Jasminum intermedium* Wight ex C.B. Clarke, *Jasminum wightii* C.B. Clarke
Climbing Mechanism: Stem Twiner
Distribution (Global): India
Distribution (India): Tamil Nadu
Leaf Type: Leaves simple, opposite, petioalte, ovate or lanceolate, rounded at base
Inflorescence: Cymes terminal and axillary; flowers white
Fruit Type: A obovoid-ellipsoid berry; dark purple when ripe
Flowering and Fruiting: March–June
IUCN Status: Not evaluated

Notes: Found mostly in dry deciduous forests
Reference: Nov. Pl. Sp.: 6 (1821)

Jasminum wengeri C.E.C. Fisch.
Oleaceae
Synonyms: *Jasminum anisophyllum* Kobuski
Climbing Mechanism: Stem Twiner
Distribution (Global): Assam, China South-Central, Myanmar
Distribution (India): Mizoram
Leaf Type: Leaves opposite, 3-foliolate, rarely simple
Inflorescence: Cymes terminal or axillary, many flowered, congested
Fruit Type: Fruit not seen
Flowering and Fruiting: August–September
IUCN Status: Not evaluated
Reference: Bull. Misc. Inform. Kew 1931: 283 (1931)

Myxopyrum nervosum Blume
Oleaceae
Climbing Mechanism: Stem Twiner
IUCN Status: Not evaluated

Myxopyrum smilacifolium (Wall.) Blume
Oleaceae
Climbing Mechanism: Stem Twiner
Distribution (India): Andaman and Nicobar Islands, Arunachal Pradesh, Assam, Kerala, Meghalaya, Tamil Nadu, West Bengal, Western Ghats.
IUCN Status: Not evaluated

Cansjera parvifolia Kurz
Opiliaceae
Synonyms: *Cansjera helferiana* Valeton, *Tsjeru-caniram helferianum* (Valeton) Kuntze, *Tsjeru-caniram parvifolium* (Kurz) Kuntze
Climbing Mechanism: Scrambler-Armed
Distribution (Global): Bangladesh, Myanmar
Leaf Type: Simple, ovate/oblong-lanceolate
Inflorescence: Spikes, sub-sessile
Fruit Type: Wet, drupe
IUCN Status: Not evaluated
Reference: J. Asiat. Soc. Bengal, Pt. 2, Nat. Hist. 41: 298 (1872)

Cansjera rheedei J.F.Gmel.
Opiliaceae
Synonyms: *Cansjera lanceolata* Benth., *Cansjera malabarica* Lam., *Cansjera monostachya* M.Roem., *Cansjera polystachya* M.Roem., *Cansjera scandens*

Roxb., *Cansjera zizyphifolia* Griff., *Octoplis polistachya* Raf., *Opilia amentacea* Wall., *Tsjeru-caniram lanceolatum* (Benth.)
Climbing Mechanism: Scrambler-Armed
Distribution (Global): Andaman Is., Bangladesh, Borneo, Cambodia, China South-Central, China Southeast, Hainan, India, Laos, Malaya, Myanmar, Nepal, Nicobar Is., Philippines, Sri Lanka, Sumatera, Thailand, Vietnam
Distribution (India): Andaman and Nicobar Islands, Andhra Pradesh, Bihar, Odisha, Eastern Ghats, Goa, Gujarat, Kerala, Madhya Pradesh, Maharashtra, Odisha, Puducherry, Tamil Nadu, Telangana, Western Ghats
Leaf Type: Alternate, 8–10
Inflorescence: Axillary spike
Fruit Type: Drupe, orange-red when ripe
Flowering and Fruiting: October–May
IUCN Status: Not evaluated
Notes: Commonly found in deciduous forests and scrub jungles
Reference: Syst. Nat. ed. 13[bis]: 280 (1791)

Vanilla andamanica Rolfe
Orchidaceae
Climbing Mechanism: Stem Twiner
Distribution (Global): Andaman Is., Nicobar Is.
Distribution (India): Andaman and Nicobar Islands
Leaf Type: Epiphytes, simple
Inflorescence: Axillary racemes
Fruit Type: Dry, pods
IUCN Status: Not evaluated
Reference: Bull. Misc. Inform. Kew 1918: 237 (1918)

Vanilla sanjappae Rasingam, R.P.Pandey, J.J.Wood & S.K.Srivast.
Orchidaceae
Climbing Mechanism: Stem Twiner
Distribution (Global): Andaman Is.
Leaf Type: Epiphytes, simple
Inflorescence: Axillary racemes
Fruit Type: Dry, pods
IUCN Status: Not evaluated
Reference: Orchid Rev. 115: 350 (2007)

Vanilla wightiana Lindl.
Orchidaceae
Climbing Mechanism: Stem Twiner
IUCN Status: Not evaluated

Brandisia discolor Hook.f. & Thomson
Orobanchaceae

Climbing Mechanism: Scrambler-Unarmed
Distribution (Global): Assam, China South-Central, Laos, Myanmar, Thailand, Vietnam
IUCN Status: Not evaluated
Reference: J. Linn. Soc., Bot. 8: 11 (1865)

Lindenbergia grandiflora (Buch.-Ham. ex D.Don) Benth.
Orobanchaceae
Synonyms: *Stemodia grandiflora* Buch.-Ham. ex D.Don
Common Name: Large-flower Lindenbergia
Climbing Mechanism: Stem Twiner
Distribution (Global): East Himalaya, Myanmar, Nepal, Tibet, West Himalaya
Distribution (India): Himachal Pradesh
Leaf Type: Leaves Opposite or the upper alternate, ovate, acute, serrate, glandular hairy
Inflorescence: Flowers are borne in lax spikes
Fruit Type: Capsules are ovoid, tip protruding from persistent calyx
Flowering and Fruiting: September–December
IUCN Status: Not evaluated
Reference: Scroph. Ind.: 22 (1835)

Lindenbergia hookeri C.B.Clarke ex Hook.f.
Orobanchaceae
Synonyms: *Lindenbergia bhutanica* T.Yamaz.
Common Name: Bhutan Lindenbergia
Climbing Mechanism: Stem Twiner
Distribution (Global): Assam, East Himalaya
Leaf Type: Leaves decussate, shortly stalked or nearly stalkless, elliptic, sawtoothed or minutely toothed, base wedge shaped
Inflorescence: Racemes, at branch ends, mostly covered with minute, somewhat club-shaped hair
Fruit Type: Capsules are ovoid
Flowering and Fruiting: September–December
IUCN Status: Not evaluated
Reference: Fl. Brit. India 4: 261 (1884)

Freycinetia insignis Blume
Pandanaceae
Climbing Mechanism: Root Climber
Distribution (Global): Java, Lesser Sunda Is., Sulawesi
Distribution (India): Andaman and Nicobar Islands
IUCN Status: Not evaluated
Reference: Rumphia 1: 158 (1837)

Dactylicapnos grandifoliolata Merr.
Papaveraceae
Synonyms: *Dactylicapnos ventii* (Khánh) Lidén, *Dicentra grandifoliolata* (Merr.) K. R.Stern, *Dicentra ventii* Khánh.
Climbing Mechanism: Scrambler-Unarmed
Distribution (Global): Assam, East Himalaya, Myanmar, Nepal, Tibet
IUCN Status: Not evaluated
Reference: Brittonia 4: 64 (1941)

Dactylicapnos roylei (Hook.f. & Thomson) Hutch.
Papaveraceae
Synonyms: *Capnorchis roylei* (Hook.f. & Thomson) Kuntze, *Dicentra roylei* Hook. f. & Thomson
Climbing Mechanism: Scrambler-Unarmed
Distribution (Global): Assam, China South-Central, East Himalaya, Myanmar, Nepal, Tibet, West Himalaya
Leaf Type: Alternate, ternate, or pinnatifid
Inflorescence: Corymbose or Umbellate
Fruit Type: Follicles, linear
Flowering and Fruiting: June–October
IUCN Status: Not evaluated
Notes: Used in traditional medicine
Reference: Bull. Misc. Inform. Kew 1921: 104 (1921)

Dactylicapnos scandens (D.Don) Hutch
Papaveraceae
Synonyms: *Capnorchis scandens* (D.Don) Kuntze., *Capnorchis thalictrifolia* (Wall.) Kuntze., *Corydalis scandens* Spreng., *Dactylicapnos asplenifolia* Wall. ex Steud., *Dactylicapnos multiflorus* Hu., *Dactylicapnos thalictrifolia* Wall., *Dicentra scandens* (D.Don) Walp., Dic
Climbing Mechanism: Scrambler-Unarmed
Distribution (Global): Assam, China South-Central, China Southeast, East Himalaya, Myanmar, Nepal, Thailand, Tibet, Vietnam
IUCN Status: Not evaluated
Notes: Cultivated in gardens
Reference: Bull. Misc. Inform. Kew 1921: 105 (1921)

Dactylicapnos torulosa (Hook.f. & Thomson) Hutch
Papaveraceae
Synonyms: *Capnorchis torulosa* (Hook.f. & Thomson) Kuntze, *Dicentra torulosa* Hook.f. & Thomson, *Dicentra wolfdietheri* Fedde
Climbing Mechanism: Scrambler-Unarmed
Distribution (Global): Assam, Bangladesh, China South-Central, East Himalaya, Myanmar, Tibet
Leaf Type: alternate, ternate, or pinnatifid

Inflorescence: Corymbose or umbellate
Fruit Type: Capsules, linear
Flowering and Fruiting: September-November
IUCN Status: Not evaluated
Notes: Usually found in subtropical to temperate forests
Reference: Bull. Misc. Inform. Kew 1921: 104 (1921)

Dactylicapnos ventii (Khanh) Lidén
Papaveraceae
Climbing Mechanism: Scrambler-Unarmed
IUCN Status: Not evaluated
Reference: Lidén. In: Nordic J. Bot. 25(1-2): 35. (2008)

Adenia cardiophylla (Mast.) Engl.
Passifloraceae
Synonyms: *Modecca cardiophylla* Mast., *Modecca parviflora* Blanco, *Passiflora parviflora* Blanco.
Climbing Mechanism: Tendril Climber
Distribution (Global): Assam, Bangladesh, Cambodia, China South-Central, East Himalaya, Laos, Myanmar, Thailand, Vietnam
Distribution (India): Andhra Pradesh, Manipur, Andaman and Nicobar Islands
Leaf Type: Simple & alternate
Inflorescence: Cymose
Flowering and Fruiting: July–September
IUCN Status: Not evaluated
Reference: Bot. Jahrb. Syst. 14: 376 (1891)

Adenia cordifolia (Blume) Engl.
Passifloraceae
Synonyms: *Adenia obtusa* (Blume) Engl., *Adenia quadrifida* Merr., *Adenia vespertilio* Hallier f., *Microblepharis cordifolia* (Blume) M.Roem., *Microblepharis obtusa* (Blume) M.Roem., *Modecca cordifolia* Blume, *Modecca obtusa* Blume
Climbing Mechanism: Tendril Climber
Distribution (Global): Andaman Is., Borneo, Java, Malaya, Nicobar Is., Philippines, Sumatera, Vietnam
Leaf Type: Simple
Fruit Type: capsule fruit
IUCN Status: Not evaluated
Notes: The stem juice is used in an eyewash in Sumatra to treat conjunctivitis
Reference: Bot. Jahrb. Syst. 14: 376 (1891)

Adenia heterophylla subsp. *andamanensis*
Passifloraceae

Synonyms: *Microblepharis heterophylla* (Blume) M.Roem., *Modecca heterophylla* Blume
Climbing Mechanism: Tendril Climber
Distribution (Global): Andaman Is., Bismarck Archipelago, Borneo, Cambodia, China Southeast, Hainan, Java, Laos, Lesser Sunda Is., Maluku, New Guinea, Nicobar Is., Northern Territory, Philippines, Queensland, Sulawesi, Sumatera, Taiwan, Thailand, Vietnam, Western Australia
Leaf Type: Simple, Deeply lobed
Inflorescence: Simple, white
Fruit Type: Wet, Capsule
IUCN Status: Not evaluated
Reference: Exkurs.-Fl. Java 2: 637 (1912)

Adenia hondala (Gaertn.) W.J.de Wilde
Passifloraceae
Synonyms: *Adenia palmata* Engl., *Granadilla hondala* Gaertn., *Modecca integrifolia* Lam., *Modecca palmata* Lam., *Modecca tuberosa* Roxb., *Passiflora hondala* Steud., *Passiflora itondala* Steud.
Climbing Mechanism: Tendril Climber
Distribution (Global): India, Sri Lanka, Vietnam
Distribution (India): Tamil Nadu, Puducherry, Karnataka, Kerala, Western Ghats
Leaf Type: Palmately compound
Inflorescence: Axillary cymes
Fruit Type: Capsule
Flowering and Fruiting: March–October
IUCN Status: Not evaluated
Notes: Commonly occur in semi-evergreen and moist deciduous forests.
Reference: Blumea 15: 265 (1967)

Adenia natalensis W.J.de Wilde
Passifloraceae
Climbing Mechanism: Tendril Climber
Distribution (Global): KwaZulu-Natal, South Africa
Leaf Type: Simple, Deeply lobed
Inflorescence: Simple, white
Fruit Type: Wet, Capsule
IUCN Status: Not evaluated
Reference: Meded. Landbouwhoogeschool 71(18): 161 (1971)

Adenia penangiana (Wall. ex G.Don) W.J.de Wilde
Passifloraceae
Synonyms: *Passiflora penangiana* Wall. ex G.Don, *Adenia catharinae* Merrill, *Anthactinia penangiana* M. Roem., *Disemma penangiana* Miq.
Climbing Mechanism: Tendril Climber

Distribution (Global): Andaman Is., China South-Central, Laos, Malaya, Nicobar Is., Sumatera, Thailand, Vietnam
Distribution (India): Nicobar Islands
Leaf Type: Simple, Deeply lobed
Inflorescence: Simple, white
Fruit Type: Wet, Capsule
IUCN Status: Not evaluated
Reference: Blumea 15: 266 (1967)

Adenia trilobata (Roxb.) Engl.
Passifloraceae
Synonyms: *Modecca saponaria* Blanco, *Modecca trilobata* Roxb., *Passiflora saponaria* Blanco.
Climbing Mechanism: Tendril Climber
Distribution (Global): Andaman Is., Assam, Bangladesh, East Himalaya, Myanmar, Pakistan, West Himalaya
Distribution (India): North Eastern India
Leaf Type: Simple, Deeply lobed
Inflorescence: Simple, white
Fruit Type: Wet, Capsule
IUCN Status: Not evaluated
Reference: Bot. Jahrb. Syst. 14: 375 (1891)

Adenia wightiana (Wall. ex Wight & Arn.) Engl.
Passifloraceae
Synonyms: *Microblepharis wightiana* (Wall. ex Wight & Arn.) M.Roem.
Climbing Mechanism: Tendril Climber
Distribution (Global): India, Kenya, Sri Lanka, Tanzania
Distribution (India): Tamil Nadu, Andhra Pradesh, Puducherry
Leaf Type: Simple, Deeply lobed
Inflorescence: Simple, white
Fruit Type: Wet, Capsule
Flowering and Fruiting: July–December
IUCN Status: Not evaluated
Reference: Bot. Jahrb. Syst. 14: 376 (1891)

Decaloba leschenaultii (DC.) M.Roem.
Passifloraceae
Synonyms: *Passiflora leschenaultii* DC., *Bryonia tricuspidata* Lesch. ex Steud.
Climbing Mechanism: Tendril Climber
Distribution (Global): Assam, Bangladesh, Myanmar
Distribution (India): Tamil Nadu, Western Ghats
Flowering and Fruiting: March–May
IUCN Status: Not evaluated

Notes: Endemic to Peninsular India. Often found in shola forests
Reference: Prodr. 3: 326 (1828)

Passiflora +alato-caerulea [hybrid formula]
Passifloraceae
Climbing Mechanism: Tendril Climber
Distribution (India): Peninsular India
Leaf Type: Alternate, palmately lobed
Inflorescence: Flowers bisexual
Fruit Type: Wet, berry, indehiscent
IUCN Status: Not evaluated

Passiflora antioquiensis H. Karst.
Passifloraceae
Climbing Mechanism: Tendril Climber
Distribution (Global): Colombia
Distribution (India): Tamil Nadu
Leaf Type: Alternate, palmately lobed
Inflorescence: Flowers bisexual
Fruit Type: Wet, berry, indehiscent
IUCN Status: Not evaluated
Reference: Linnaea 30: 162 (1859)

Passiflora biflora Lam.
Passifloraceae
Synonyms: *Cieca glabrata* M.Roem., *Decaloba biflora* (Lam.) M.Roem., *Lortetia biflora* Ser., *Passiflora brighamii* S.Watson ex Brigham +5
Climbing Mechanism: Tendril Climber
Distribution (Global): Bahamas, Belize, Colombia, Costa Rica, Ecuador, El Salvador, Guatemala, Honduras, Mexico Central, Mexico Gulf, Mexico Northeast, Mexico Northwest, Mexico Southeast, Mexico Southwest, Nicaragua, Panamá, Southwest Caribbean, Venezuela
Distribution (India): Maharashtra
Leaf Type: Alternate, palmately lobed
Inflorescence: Flowers bisexual
Fruit Type: Wet, berry, indehiscent
Flowering and Fruiting: Throughout the year
IUCN Status: Not evaluated
Reference: Encycl. 3: 36 (1789)

Passiflora caerulea L.
Passifloraceae
Synonyms: *Granadilla caerulea* (L.) Medik., *Passiflora caerulea var. angustifolia* G.Don, *Passiflora caerulea var. glauca* Mast., *Passiflora caerulea var. imbricata* Mast.+8

Common Name: Blue passion flower, Common passion flower
Climbing Mechanism: Tendril Climber
Distribution (Global): Argentina Northeast, Argentina Northwest, Bolivia, Brazil Northeast, Brazil South, Brazil Southeast, Paraguay
Distribution (India): Maharashtra, Manipur, Tamil Nadu
Leaf Type: Palmately compound
Flowering and Fruiting: Throughout the year
IUCN Status: Not evaluated
Notes: Usually cultivated at high altitudes
Reference: Sp. Pl.: 959 (1753)

Passiflora calcarata Mast.
Passifloraceae
Climbing Mechanism: Tendril Climber
Distribution (India): Tamil Nadu
Leaf Type: Alternate, palmately lobed
Inflorescence: Flowers bisexual
Fruit Type: Wet, berry, indehiscent
Flowering and Fruiting: December–May
IUCN Status: Not evaluated
Reference: Trans. Linn. Soc. London 27: 638 (1871)

Passiflora ciliata Aiton
Passifloraceae
Synonyms: *Dysosmia ciliata* (Aiton) M.Roem., *Passiflora foetida var. ciliata* (Aiton) Mast
Climbing Mechanism: Tendril Climber
Distribution (Global): Bahamas, Belize, Colombia, Costa Rica, Cuba, Dominican Republic, El Salvador, Florida, Guatemala, Haiti, Honduras, Jamaica, Mexico Central, Mexico Gulf, Mexico Southeast, Mexico Southwest, Nicaragua, Panamá, Venezuela
Distribution (India): Maharashtra, Tamil Nadu
Leaf Type: Alternate, palmately lobed
Inflorescence: Flowers bisexual
Fruit Type: Wet, berry, indehiscent
Flowering and Fruiting: June–September
IUCN Status: Not evaluated
Reference: Hort. Kew. 3: 310 (1789)

Passiflora coccinea Aubl.
Passifloraceae
Synonyms: *Passiflora coccinea var. minor* Mast., *Passiflora coccinea var. velutina* (DC.) Mast., *Passiflora fulgens* Wallis ex É.Morren, *Passiflora monnieri* Buc'hoz +4
Common Name: Scarlet passion flower, Red passion flower

Climbing Mechanism: Tendril Climber
Distribution (Global): Brazil North, Brazil Northeast, Brazil West-Central, Colombia, French Guiana, Guyana, Peru, Suriname, Venezuela
Distribution (India): Tamil Nadu
Leaf Type: Alternate, palmately lobed
Inflorescence: Flowers bisexual
Fruit Type: Wet, berry, indehiscent
Flowering and Fruiting: August–January
IUCN Status: Not evaluated
Notes: Grown as a garden plant
Reference: Hist. Pl. Guiane: 828 (1775)

Passiflora edulis Sims
Passifloraceae
Synonyms: *Passiflora cuneifolia Cav., Passiflora* diaden Vell., *Passiflora edulis var. pomifera* (M.Roem.) Mast., *Passiflora edulis var. rubricaulis* (J.Jacq.) Mast
Common Name: Passion fruit, The edible passion, Purple granadilla
Climbing Mechanism: Tendril Climber
Distribution (Global): Argentina Northeast, Brazil Northeast, Brazil South, Brazil Southeast, Brazil West-Central, Paraguay
Distribution (India): Assam, Gujarat, Kerala, Madhya Pradesh, Maharashtra, Manipur, Mizoram, Puducherry, Tamil Nadu, Telangana, Tripura, West Bengal
Leaf Type: Alternate, palmately lobed
Inflorescence: Flowers bisexual
Fruit Type: Wet, berry, indehiscent
Flowering and Fruiting: March–May
IUCN Status: Not evaluated
Reference: Bot. Mag. 45: t. 1989 (1818)

Passiflora foetida L.
Passifloraceae
Synonyms: *Dysosmia foetida* (L.) M.Roem., *Granadilla foetida* (L.) Gaertn., *Tripsilina foetida* (L.) Raf.
Common Name: Foetid passion flower, Pop vine, Common passion flower, Running pop, Stinking passion flower
Climbing Mechanism: Tendril Climber
Distribution (Global): Argentina Northeast, Argentina Northwest, Aruba, Bahamas, Belize, Bolivia, Brazil North, Brazil Northeast, Brazil South, Brazil Southeast, Brazil West-Central, Chile North, Colombia, Costa Rica, Cuba, Dominican Republic, Ecuador, El Salvador, French Guiana, Guatemala, Guyana, Haiti, Honduras, Jamaica, Leeward Is., Mexico Central, Mexico Gulf, Mexico Northeast, Mexico Southeast, Mexico Southwest, Netherlands Antilles, Nicaragua, Panamá, Paraguay, Peru, Puerto Rico, Suriname, Texas, Trinidad-Tobago, Uruguay, Venezuela, Venezuelan Antilles, Windward Is.

Distribution (India): Andaman and Nicobar Islands, Andhra Pradesh, Arunachal Pradesh, Assam, Gujarat, Kerala, Madhya Pradesh, Maharashtra, Manipur, Mizoram, Odisha, Puducherry, Rajasthan, Tamil Nadu, Telangana, Tripura, Uttar Pradesh, West Bengal, Western Ghats
Leaf Type: Simple
Fruit Type: Berry, Sub-globose berry
Flowering and Fruiting: June–December
IUCN Status: Not evaluated
Reference: Sp. Pl.: 959 (1753)

Passiflora foetida var. hispida (DC. ex Triana & Planch.) Killip
Passifloraceae
Climbing Mechanism: Tendril Climber
Distribution (India): Peninsular India
Leaf Type: Alternate, palmately lobed
Inflorescence: Flowers bisexual
Fruit Type: Wet, berry, indehiscent
Flowering and Fruiting: November–March
IUCN Status: Not evaluated
Notes: Commonly found in wastelands and dry deciduous forests

Passiflora gracilis J.Jacq. ex Link
Passifloraceae
Synonyms: *Cieca gracilis* (J.Jacq. ex Link) M.Roem., *Passiflora pediculata* Mast
Climbing Mechanism: Tendril Climber
Distribution (Global): Colombia, Costa Rica, Ecuador, Guatemala, Nicaragua, Panamá, Venezuela
Distribution (India): Maharashtra
Leaf Type: Alternate, palmately lobed
Inflorescence: Flowers bisexual
Fruit Type: Wet, berry, indehiscent
Flowering and Fruiting: August–September
IUCN Status: Not evaluated
Reference: Enum. Hort. Berol. Alt. 2: 182 (1822)

Passiflora holosericea L.
Passifloraceae
Synonyms: *Decaloba holosericea* (L.) M.Roem., *Lortetia holosericea* Ser., *Passiflora reticulata* C.Wright, *Passiflora tuxtlensis* Sessé & Moc.
Common Name: Silky passion flower
Climbing Mechanism: Tendril Climber
Distribution (Global): Colombia, Costa Rica, Cuba, El Salvador, Guatemala, Honduras, Mexico Gulf, Mexico Northwest, Mexico Southeast, Mexico Southwest, Nicaragua, Panamá, Venezuela
Distribution (India): Maharashtra, Tamil Nadu

Leaf Type: Alternate, palmately lobed
Inflorescence: Flowers bisexual
Fruit Type: Wet, berry, indehiscent
Flowering and Fruiting: May–August
IUCN Status: Not evaluated
Reference: Sp. Pl.: 958 (1753)

Passiflora kermesina Link & Otto
Passifloraceae
Synonyms: *Decaloba dentata* (Vell.) M.Roem., *Decaloba kermesina* (Link & Otto) M.Roem., *Passiflora dentata* Vell., *Passiflora raddiana* DC.
Climbing Mechanism: Tendril Climber
Distribution (Global): Brazil Southeast
Distribution (India): Tamil Nadu
Leaf Type: Alternate, palmately lobed
Inflorescence: Flowers bisexual
Fruit Type: Wet, berry, indehiscent
IUCN Status: Not evaluated
Reference: Verh. Vereins Beförd. Gartenbaues Königl. Preuss. Staaten 2: 403 (1826)

Passiflora laurifolia L.
Passifloraceae
Synonyms: *Granadilla laurifolia* (L.) Medik., *Passiflora oblongifolia* Pulle, *Passiflora laurifolia var. tinifolia* (Juss.) Bois.
Climbing Mechanism: Tendril Climber
Distribution (Global): Brazil North, Brazil Northeast, Colombia, Cuba, Dominican Republic, French Guiana, Guyana, Haiti, Jamaica, Leeward Is., Puerto Rico, Suriname, Trinidad-Tobago, Venezuela, Venezuelan Antilles, Vietnam, Windward Is.
Distribution (India): Maharashtra, Tamil Nadu
Leaf Type: Alternate, palmately lobed
Inflorescence: Flowers bisexual
Fruit Type: Wet, berry, indehiscent
Flowering and Fruiting: September–November
IUCN Status: Not evaluated
Reference: Sp. Pl.: 956 (1753)

Passiflora ligularis Juss.
Passifloraceae
Synonyms: *Passiflora lowei* Heer, *Passiflora serratistipula* (Moc. & Sesse, ex) DC., *Passiflora tiliifolia* L.
Common Name: Granadilla, Sweet granadilla
Climbing Mechanism: Tendril Climber
Distribution (Global): Bolivia, Colombia, Panamá, Peru, Venezuela
Bolivia, Colombia, Panamá, Peru, Venezuela

Distribution (India): Kerala, Tamil Nadu
Leaf Type: Alternate, palmately lobed
Inflorescence: Flowers bisexual
Fruit Type: Wet, berry, indehiscent
Flowering and Fruiting: January–June
IUCN Status: Not evaluated
Reference: Ann. Mus. Hist. Nat. 6: 113 (1805)

Passiflora mollissima (Kunth) L.H. Bailey
Passifloraceae
Synonyms: *Tacsonia mollissima* H.B.K., *Passiflora tripartita mollissima* (Kunth) L.B. Holm-Nielsen & P.M. Jørgensen, *Murucuia mollissima* (Kunth) Sprengel, *Passiflora tomentosa var. mollissima* (Kunth) Triana & Planch
Common Name: Banana poka
Climbing Mechanism: Tendril Climber
Distribution (Global): South America
Distribution (India): Tamil Nadu
Leaf Type: Alternate, palmately lobed
Inflorescence: Flowers bisexual
Fruit Type: Wet, berry, indehiscent
Flowering and Fruiting: September–May
IUCN Status: Not evaluated
Reference: Rhodora 18: 156 (1916)

Passiflora morifolia Mast.
Passifloraceae
Synonyms: *Passiflora dumetosa* Barb.Rodr., *Passiflora erosa* Rusby, *Passiflora heydei* Killip, *Passiflora warmingii* Mast., *Passiflora warmingii subsp. chacoensis* R.E.Fr.
Common Name: Woodland passion flower, Mulberry leaf passion flower
Climbing Mechanism: Tendril Climber
Distribution (Global): Argentina Northeast, Argentina Northwest, Bolivia, Brazil South, Brazil Southeast, Brazil West-Central, Colombia, Ecuador, Guatemala, Mexico Southeast, Mexico Southwest, Paraguay, Peru, Venezuela
Distribution (India): Maharashtra, Manipur
Leaf Type: Alternate, palmately lobed
Inflorescence: Flowers bisexual
Fruit Type: Wet, berry, indehiscent
Flowering and Fruiting: July–September
IUCN Status: Not evaluated
Reference: C.F.P.von Martius & auct. suc. (eds.), Fl. Bras. 13(1): 555 (1872)

Passiflora napalensis Wall.
Passifloraceae

Synonyms: *Anthactinia napalensis* (Wall.) M.Roem., *Passiflora geminiflora* D.Don, *Passiflora nipaulensis* G.Don.
Climbing Mechanism: Tendril Climber
Distribution (Global): Assam, East Himalaya, Nepal
Distribution (India): Arunachal Pradesh, Manipur, Mizoram
Leaf Type: Alternate, palmately lobed
Inflorescence: Flowers bisexual
Fruit Type: Wet, berry, indehiscent
IUCN Status: Not evaluated
Reference: Tent. Fl. Napal.: 20 (1826)

Passiflora quadrangularis L.
Passifloraceae
Synonyms: *Granadilla quadrangularis* (L.) Medik., *Passiflora grandiflora* Salisb., *Passiflora hexangularis* Raeusch., *Passiflora hullettii* Mast., *Passiflora macrocarpa* W.Thomps., *Passiflora tetragona* M.Roem.+ 2
Common Name: Giant granadilla
Climbing Mechanism: Tendril Climber
Distribution (Global): Brazil North, Brazil Northeast, Brazil South, Brazil Southeast, Brazil West-Central, Colombia
Distribution (India): Maharashtra, Tamil Nadu, West Bengal
Leaf Type: Alternate, palmately lobed
Inflorescence: Flowers bisexual
Fruit Type: Wet, berry, indehiscent
Flowering and Fruiting: November–April
IUCN Status: Not evaluated
Reference: Syst. Nat. ed. 10, 2: 1248 (1759)

Passiflora racemosa Brot.
Passifloraceae
Climbing Mechanism: Tendril Climber
Distribution (India): Maharashtra
Leaf Type: Alternate, palmately lobed
Inflorescence: Flowers bisexual
Fruit Type: Wet, berry, indehiscent
Flowering and Fruiting: April–August
IUCN Status: Not evaluated
Reference: Trans. Linn. Soc. London 12: 71 (1817)

Passiflora stipulata Aubl.
Passifloraceae
Climbing Mechanism: Tendril Climber
Distribution (Global): French Guiana
Distribution (India): Maharashtra, Tamil Nadu
Leaf Type: Leaves 3 lobed, cordate, glaucous

IUCN Status: Not evaluated
Reference: First published in Hist. Pl. Guiane: 830 (1775)

Passiflora suberosa L.
Passifloraceae
Synonyms: *Cieca suberosa* (L.) Moench, *Meioperis suberosa* (L.) Raf., *Granadilla suberosa* (L.) Gaertn., *Monactineirma suberosa* (L.) Bory.
Common Name: Small passion flower, Corky passion flower, Corkystem passion flower, Wildpassion fruit, Devil's pumpkin
Climbing Mechanism: Tendril Climber
Distribution (Global): Argentina Northeast, Argentina Northwest, Bahamas, Belize, Bolivia, Brazil North, Brazil Northeast, Brazil South, Brazil Southeast, Brazil West-Central, Colombia, Costa Rica, Cuba, Dominican Republic, Ecuador, El Salvador, Galápagos, Guatemala, Haiti, Honduras, Leeward Is., Mexico Central, Mexico Gulf, Mexico Northeast, Mexico Northwest, Mexico Southeast, Mexico Southwest, Nicaragua, Panamá, Paraguay, Peru, Puerto Rico, Southwest Caribbean, Trinidad-Tobago, Uruguay, Venezuela, Venezuelan Antilles, Windward Is.
Distribution (India): Gujarat, Maharashtra, Tamil Nadu, West Bengal
Leaf Type: Simple
Fruit Type: Berry
Flowering and Fruiting: July–February
IUCN Status: Not evaluated
Reference: Sp. Pl.: 958 (1753)

Passiflora subpeltata Ortega
Passifloraceae
Synonyms: *Passiflora adenophylla* Mast., *Passiflora alba* Link & Otto, *Passiflora atomaria* Planch. ex Mast., *Passiflora calcarata* Mast., *Passiflora holosericea* Ruiz & Pav. ex Mast., *Passiflora lutea* Ruiz & Pav. ex Mast., *Passiflora neillii* Regel
Common Name: White passion flower
Climbing Mechanism: Tendril Climber
Distribution (Global): Colombia, El Salvador, Guatemala, Mexico Central, Mexico Gulf, Mexico Northeast, Mexico Southeast, Mexico Southwest, Panamá, Venezuela
Distribution (India): Karnataka, Manipur, Tamil Nadu, West Bengal
Leaf Type: Alternate, palmately lobed
Inflorescence: Flowers bisexual
Fruit Type: Wet, berry, indehiscent
Flowering and Fruiting: September-May
IUCN Status: Not evaluated
Notes: Frequently occur in degraded forest areas
Reference: Nov. Rar. Pl. Descr. Dec.: 78 (1798)

Passiflora vitifolia Kunth
Passifloraceae
Synonyms: *Macrophora sanguinea* (Sm.) Raf., *Passiflora buchananii* (Lem.) Jacob-Makoy, *Passiflora sanguinea* Sm., *Passiflora servitensis* H.Karst., *Passiflora servitensis var. bracteosa* H.Karst., *Passiflora vitifolia var. bracteosa* (H. Karst.) Killip, *Tacsonia buc*
Common Name: Granadilla silvestre
Climbing Mechanism: Tendril Climber
Distribution (Global): Bolivia, Brazil North, Brazil West-Central, Colombia, Costa Rica, Ecuador, Honduras, Nicaragua, Panamá, Peru, Venezuela
Distribution (India): Tamil Nadu
Leaf Type: Leaves are 3 lobed
Flowering and Fruiting: August–May
IUCN Status: Not evaluated
Reference: F.W.H.von Humboldt, A.J.A.Bonpland & C.S.Kunth, Nov. Gen. Sp. 2: 138 (1817)

Passiflora wilsonii Hemsl.
Passifloraceae
Synonyms: *Passiflora assamica* Chakrav., *Passiflora celata* G.Cusset, *Passiflora perpera* Mast., *Passiflora rhombiformis* S.Y.Bao, *Passiflora spirei* G.Cusset.
Climbing Mechanism: Tendril Climber
Distribution (Global): Assam, Bangladesh, China South-Central, Laos, Myanmar, Thailand, Tibet, Vietnam
Leaf Type: Alternate, palmately lobed
Inflorescence: Flowers bisexual
Fruit Type: Wet, berry, indehiscent
IUCN Status: Not evaluated
Reference: Bull. Misc. Inform. Kew 1908: 17 (1908)

Turnera ulmifolia L.
Passifloraceae
Common Name: Yellow alder, Yellow buttercups, Sage rose, Cuban buttercup
Climbing Mechanism: Stem Twiner
Distribution (Global): Bahamas, Belize, Bermuda, Cayman Is., Central American Pac, Costa Rica, Cuba, Dominican Republic, El Salvador, Guatemala, Haiti, Honduras, Jamaica, Leeward Is., Mexico Central, Mexico Gulf, Mexico Northeast, Mexico Northwest, Mexico Southeast, Mexico Southwest, Netherlands Antilles, Nicaragua, Panamá, Southwest Caribbean, Turks-Caicos Is., Windward Is.
Distribution (India): Puducherry
Leaf Type: Simple, clustered at base
Inflorescence: Single, yellow
Fruit Type: Dry, capsule
Flowering and Fruiting: Throughout the year

IUCN Status: Not evaluated
Reference: Sp. Pl.: 271 (1753)

Chaetocarpus globosus subsp. oblongatus (Alain) Borhidi
Peraceae
Synonyms: *Chaetocarpus oblongatus* (Alain) Borhidi, *Chaetocarpus oblongatus var. monticola* Borhidi, *Chaetocarpus oblongatus var. subenervis* Borhidi, *Mettenia globosa subsp. oblongata* (Alain) Borhidi, *Mettenia oblongata* Alain
Climbing Mechanism: Stem Twiner
Distribution (Global): Cuba
IUCN Status: Not evaluated
Reference: Acta Bot. Acad. Sci. Hung. 25: 17 (1979)

Breynia racemosa (Blume) Müll.Arg.
Phyllanthaceae
Synonyms: *Breynia acuminata* (Müll.Arg.) Müll.Arg., *Breynia racemosa var. aromatica* Airy Shaw, *Breynia reclinata* (Roxb.) Hook.f., *Melanthesa acuminata* Müll.Arg., *Melanthesa racemosa* Blume, *Melanthesa reclinata* (Roxb.) Müll. Arg.+6
Climbing Mechanism: Scrambler-Unarmed
Distribution (Global): Bismarck Archipelago, Borneo, Java, Lesser Sunda Is., Malaya, Myanmar, New Guinea, Nicobar Is., Philippines, Sulawesi, Sumatera, Thailand
IUCN Status: Least Concern
Reference: A.P.de Candolle, Prodr. 15(2): 441 (1866)

Bridelia montana (Roxb.) Willd.
Phyllanthaceae
Synonyms: *Clutia montana* Roxb., *Andrachne elliptica* Roth, *Bridelia hamiltoniana* Wall. ex Müll.Arg.
Common Name: Mountain Bridelia, Hamilton's Bridelia
Climbing Mechanism: Scrambler-Unarmed
Distribution (Global): India
Flowering and Fruiting: August–December
IUCN Status: Not evaluated
Reference: Sp. Pl., ed. 4, 4: 978 (1806)

Bridelia stipularis (L.) Blume
Phyllanthaceae
Synonyms: *Clutia scandens* Roxb., *Clutia stipularis* L., *Bridelia alnifolia* Griff., *Bridelia dasycalyx* Kurz, *Bridelia montana* Woodrow ex J.J.Sm., *Bridelia scandens* (Roxb.) Willd., *Bridelia stipularis subsp. philippinensis* Jabl., *Bridelia zollingeri* Miq.
Common Name: Climbing Bridelia
Climbing Mechanism: Scrambler-Unarmed

Distribution (Global): Andaman Is., Assam, Bangladesh, Borneo, Cambodia, China South-Central, China Southeast, East Himalaya, Hainan, India, Java, Lesser Sunda Is., Malaya, Myanmar, Nepal, Philippines, Sumatera, Taiwan, Thailand, Vietnam, West Himalaya
Distribution (India): Meghalaya, Andaman and Nicobar Islands, Andhra Pradesh, Bihar, Odisha, Eastern Ghats, Gujarat, Himachal Pradesh, Kerala, Maharashtra, Mizoram, Odisha, Rajasthan, Tamil Nadu, Tripura, Uttar Pradesh, West Bengal
Flowering and Fruiting: May–March
IUCN Status: Least Concern
Notes: Often found in semi-evergreen forests
Reference: Bijdr. Fl. Ned. Ind.: 597 (1826)

Bridelia tomentosa Blume
Phyllanthaceae
Synonyms: *Amanoa tomentosa* Baill., *Bridelia glabrifolia* Merr., *Bridelia lancifolia* Roxb., *Bridelia nayarii* P.Basu, *Bridelia phyllanthoides* W.Fitzg., *Bridelia rhamnoides* Griff.+5
Climbing Mechanism: Scrambler-Unarmed
Distribution (Global): Andaman Is., Assam, Bangladesh, Borneo, Cambodia, China South-Central, China Southeast, East Himalaya, Hainan, India, Java, Lesser Sunda Is., Malaya, Maluku, Myanmar, Nepal, New Guinea, Nicobar Is., Northern Territory, Philippines, Queensland, Sulawesi, Sumatera, Taiwan, Thailand, Vietnam, West Himalaya, Western Australia
Distribution (India): North Eastern India
IUCN Status: Least Concern
Reference: Bijdr. Fl. Ned. Ind.: 597 (1826)

Flueggea leucopyrus Willd.
Phyllanthaceae
Synonyms: *Acidoton leucopyrus* (Willd.) Kuntze, *Cicca leucopyrus* (Willd.) Kurz, *Flueggea wallichiana* Baill., *Flueggea xerocarpa* A.Juss., *Phyllanthus albicans* Benth.+4
Common Name: Spinous fluggea, Bushweed, Cool pot, Indian snow berry, Thermacole plant, white honey shrub
Climbing Mechanism: Scrambler-Unarmed
Distribution (Global): Bangladesh, China South-Central, Ethiopia, India, Myanmar, Pakistan, Saudi Arabia, Socotra, Somalia, Sri Lanka
Distribution (India): Tamil Nadu
Leaf Type: Simple, alternate
Inflorescence: Axillary clusters
Fruit Type: Capsule
Flowering and Fruiting: June–September
IUCN Status: Least Concern
Reference: Sp. Pl., ed. 4, 4: 757 (1806)

Flueggea virosa (Roxb. ex Willd.) Royle
Phyllanthaceae
Synonyms: *Acidoton virosus* (Roxb. ex Willd.) Kuntze, *Phyllanthus virosus* Roxb. ex Willd., *Securinega virosa* (Roxb. ex Willd.) Baill.
Common Name: White-berry bush, Common bushweed
Climbing Mechanism: Scrambler-Unarmed
Distribution (Global): Andaman Is., Angola, Assam, Bangladesh, Benin, Borneo, Botswana, Burkina, Burundi, Cambodia, Cameroon, Caprivi Strip, Central African Repu, Chad, China North-Central, China South-Central, China Southeast, Comoros, Djibouti, East Himalaya, Egypt, Eritrea, Ethiopia, Gambia, Ghana, Guinea, Guinea-Bissau, India, Ivory Coast, Java, Kenya, KwaZulu-Natal, Lesser Sunda Is., Madagascar, Malawi, Malaya, Mali, Maluku, Mauritania, Mozambique, Mozambique Channel I, Myanmar, Namibia, Nepal, New Guinea, Nicobar Is., Nigeria, Northern Provinces, Northern Territory, Oman, Philippines, Queensland, Rwanda, Réunion, Saudi Arabia, Senegal, Sierra Leone, Socotra, Somalia, South China Sea, Sri Lanka, Sudan, Sulawesi, Sumatera, Swaziland, Taiwan, Tanzania, Thailand, Togo, Uganda, West Himalaya, Western Australia, Yemen, Zambia, Zaire, Zimbabwe
Leaf Type: Alternate
Inflorescence: Axillary clusters
Fruit Type: Fruit white, globose
Flowering and Fruiting: March–September
IUCN Status: Least Concern
Reference: Ill. Bot. Himal. Mts.: 328 (1836)

Phyllanthus chandrabosei Govaerts & Radcl.-Sm.
Phyllanthaceae
Synonyms: *Phyllanthus stipulaceus* (Gamble) Kumari & Chandrab., *Reidia stipulacea* Gamble
Climbing Mechanism: Scrambler-Unarmed
Distribution (Global): India
Flowering and Fruiting: March–April
IUCN Status: Not evaluated
Notes: Usually found along the margins of evergreen forests
Reference: Kew Bull. 51: 176 (1996)

Phyllanthus lawii J.Graham
Phyllanthaceae
Synonyms: *Diasperus lawii* (J.Graham) Kuntze, *Phyllanthus nemorum* Russell ex Müll.Arg., *Diasperus nemorum* (Russell ex Müll.Arg.) Kuntze, *Phyllanthus juniperinoides* Müll.Arg., *Phyllanthus polyphyllus* Dalzell & A.Gibson, *Phyllanthus spinulosus* B.Heyne ex Benth
Common Name: Wild gooseberry
Climbing Mechanism: Scrambler-Unarmed
Distribution (Global): India

Distribution (India): Goa, Tamil Nadu
IUCN Status: Not evaluated
Notes: Usually found along riverbeds and deciduous forests
Reference: Cat. Pl. Bombay: 181 (1839)

Phyllanthus reticulatus Poir.
Phyllanthaceae
Synonyms: *Anisonema dubium* Blume, *Anisonema intermedium* Decne., *Anisonema jamaicense* (Griseb.) Griseb., *Anisonema multiflorum* (Baill.) Wight +36
Common Name: Potato bush, Black honey shrub, Black featherfoil
Climbing Mechanism: Scrambler-Unarmed
Distribution (Global): Andaman Is., Assam, Bangladesh, Borneo, Cambodia, China South-Central, China Southeast, East Himalaya, Hainan, India, Java, Laos, Lesser Sunda Is., Malaya, Maluku, Myanmar, Nansei-shoto, New Guinea, Nicobar Is., Northern Territory, Pakistan, Philippines, Queensland, Santa Cruz Is., Solomon Is., Sri Lanka, Sulawesi, Sumatera, Taiwan, Thailand, Vietnam, West Himalaya, Western Australia
Distribution (India): Andaman and Nicobar Islands, Goa, Madhya Pradesh, Maharashtra, Mizoram, Tamil Nadu, Telangana, Tripura
Leaf Type: Simple
Inflorescence: Axillary cyme
Fruit Type: Berry
Flowering and Fruiting: June–March
IUCN Status: Least Concern
Reference: J.B.A.M.de Lamarck, Encycl. 5: 298 (1804)

Ercilla spicata (Vertero) Moq.
Phytolaccaceae
Synonyms: *Apodostachys densiflora* Turcz., *Bridgesia spicata* (Bertero) Hook. & Arn., *Ercilla volubilis* A.Juss., *Phytolacca volubilis* (A.Juss.) Heimerl, *Galvezia spicata* Bertero, *Pitavia spicata* Steud., *Suriana volubilis* Dombrain & Cav.
Climbing Mechanism: Stem Twiner
Distribution (Global): Chile Central, Chile South
Distribution (India): Tamil Nadu
IUCN Status: Not evaluated
Reference: A.P.de Candolle, Prodr. 13(2): 35 (1849)

Peperomia filiformis Ruiz & Pav.
Piperaceae
Synonyms: *Peperomia brachystachyon* (Vahl) A.Dietr., *Piper filiforme* (Ruiz & Pav.) Pers., *Piper brachystachyon* Vahl, *Troxirum filiforme* (Ruiz & Pav.) Raf.
Climbing Mechanism: Root Climber
Distribution (Global): Peru
IUCN Status: Not evaluated
Reference: Fl. Peruv. 1: 31 (1798)

Peperomia meeboldii C.DC.
Piperaceae
Climbing Mechanism: Root Climber
Distribution (Global): India
Distribution (India): Peninsular India
IUCN Status: Not evaluated
Reference: Repert. Spec. Nov. Regni Veg. 13: 297 (1914)

Peperomia serpens (Sw.) Loudon
Piperaceae
Synonyms: *Acrocarpidium guildingianum* (Spreng.) Miq., *Acrocarpidium pulicare* (Opiz) Miq., *Acrocarpidium repens* (Kunth) Miq., *Acrocarpidium repens f. minus* Miq.+20
Climbing Mechanism: Root Climber
Distribution (Global): Bolivia, Brazil North, Brazil Northeast, Brazil Southeast, Brazil West-Central, Colombia, Dominican Republic, Ecuador, French Guiana, Guyana, Haiti, Honduras, Jamaica, Leeward Is., Mexico Gulf, Mexico Southeast, Mexico Southwest, Nicaragua, Panamá, Peru, Puerto Rico, Suriname, Trinidad-Tobago, Venezuela, Venezuelan Antilles, Windward Is.
Distribution (India): Tamil Nadu
IUCN Status: Not evaluated
Reference: J.C.Loudon, Hort. Brit.: 13 (1830)

Peperomia tetraphylla (G.Forst.) Hook. & Arn.
Piperaceae
Synonyms: *Piper tetraphyllum* G.Forst.
Common Name: Fore-leaf Peperomia
Climbing Mechanism: Root Climber
Distribution (Global): Argentina Northeast, Argentina Northwest, Assam, Bangladesh, Belize, Bolivia, Brazil Northeast, Brazil South, Brazil Southeast, Brazil West-Central, Cameroon, Cape Provinces, China North-Central, China South-Central, China Southeast, Colombia, Costa Rica, Cuba, Dominican Republic, East Himalaya, Ecuador, El Salvador, Ethiopia, Free State, Gabon, Guatemala, Guinea, Gulf of Guinea Is., Guyana, Haiti, Honduras, Kenya, KwaZulu-Natal, Lesotho, Lesser Sunda Is., Liberia, Malawi, Mauritius, Mexican Pacific Is., Mexico Central, Mexico Gulf, Mexico Northeast, Mexico Northwest, Mexico Southeast, Mexico Southwest, Mozambique, Myanmar, Nepal, New South Wales, New Zealand North, Nicaragua, Norfolk Is., Northern Provinces, Panamá, Paraguay, Peru, Philippines, Puerto Rico, Queensland, Rwanda, Réunion, Sierra Leone, Society Is., Socotra, Sri Lanka, Sudan, Swaziland, Taiwan, Tanzania, Thailand, Tibet, Tubuai Is., Uganda, Uruguay, Venezuela, Vietnam, West Himalaya, Zambia, Zaire, Zimbabwe
Flowering and Fruiting: August-March
IUCN Status: Not evaluated

Notes: Mostly occur in evergreen and moist deciduous forests
Reference: Bot. Beechey Voy.: 97 (1832)

Peperomia urocarpa
Piperaceae
Climbing Mechanism: Root Climber
IUCN Status: Not evaluated

Piper anisotis Hook.f.
Piperaceae
Climbing Mechanism: Root Climber
Distribution (Global): Assam, Bangladesh, East Himalaya
Leaf Type: Simple, alternate, 5–9 nerved from base
Inflorescence: Axillary spikes, unisexual flowers
Fruit Type: Wet, berry, black or brown
IUCN Status: Not evaluated
Reference: Fl. Brit. India 5: 86 (1886)

Piper argyrophyllum Miq.
Piperaceae
Climbing Mechanism: Root Climber
Distribution (Global): India, Sri Lanka
Distribution (India): Tamil Nadu, Western Ghats
Leaf Type: Simple, alternate, 5–9 nerved from base
Inflorescence: Axillary spikes, unisexual flowers
Fruit Type: Wet, berry, black or brown
Flowering and Fruiting: July–February
IUCN Status: Not evaluated
Notes: Mostly found in semi-evergreen and evergreen forests
Reference: Syst. Piperac.: 330 (1844)

Piper arunachalense Gajurel, Rethy & Y.Kumar
Piperaceae
Climbing Mechanism: Root Climber
Distribution (Global): East Himalaya
Leaf Type: Simple, alternate, 5–9 nerved from base
Inflorescence: Axillary spikes, unisexual flowers
Fruit Type: Wet, berry, black or brown
IUCN Status: Not evaluated
Reference: Bot. J. Linn. Soc. 137: 418 (2001)

Piper attenuatum Buch.-Ham. ex Miq.
Piperaceae
Synonyms: *Chavica melamiris* Miq., *Chavica pseudomelamiris* Miq., *Piper karok* Blume, *Piper malamiris* Roxb., *Piper sirium* C.DC.

Climbing Mechanism: Root Climber
Distribution (Global): Assam, Bangladesh, East Himalaya, India, Java, Myanmar, Philippines
Distribution (India): Andhra Pradesh, Bihar, Odisha, Sikkim, West Bengal
Leaf Type: Simple, alternate, 5–9 nerved from base
Inflorescence: Axillary spikes, unisexual flowers
Fruit Type: Wet, berry, black or brown
IUCN Status: Not evaluated
Reference: Syst. Piperac.: 306 (1843)

Piper aurorubrum C.DC.
Piperaceae
Climbing Mechanism: Root Climber
Distribution (Global): Assam
Leaf Type: Simple, alternate, 5–9 nerved from base
Inflorescence: Axillary spikes, unisexual flowers
Fruit Type: Wet, berry, black or brown
IUCN Status: Not evaluated
Reference: Repert. Spec. Nov. Regni Veg. 10: 519 (1912)

Piper barberi Gamble
Piperaceae
Climbing Mechanism: Root Climber
Distribution (Global): India
Distribution (India): Tamil Nadu
Leaf Type: Simple, alternate, 5–9 nerved from base
Inflorescence: Axillary spikes, unisexual flowers
Fruit Type: Wet, berry, black or brown
Flowering and Fruiting: March–September
IUCN Status: Endangered
Notes: Usually found in evergreen forests
Reference: Bull. Misc. Inform. Kew 1924: 287 (1924)

Piper betle L.
Piperaceae
Synonyms: *Artanthe hexagyna* Miq., *Betela mastica* Raf., *Chavica auriculata* Miq., *Chavica betle* (L.) Miq.
Common Name: Beetle leaf
Climbing Mechanism: Root Climber
Distribution (Global): Cambodia, Laos, Lesser Sunda Is., Malaya, Myanmar, Philippines, Thailand, Vietnam
Distribution (India): Andaman and Nicobar Islands, Arunachal Pradesh, Assam, Bihar, Odisha, Great, Nicobar, Island, Gujarat, Jharkhand, Maharashtra, Mizoram, Tripura, West Bengal
Leaf Type: Heart shaped

Inflorescence: Catkin
Flowering and Fruiting: Throughout the year
IUCN Status: Not evaluated
Reference: Sp. Pl.: 28 (1753)

Piper betleoides Chaveer. & Tanomtong
Piperaceae
Climbing Mechanism: Root Climber
Distribution (India): Arunachal Pradesh, Mizoram
Leaf Type: Simple, alternate, 5–9 nerved from base
Inflorescence: Axillary spikes, unisexual flowers
Fruit Type: Wet, berry, black or brown
IUCN Status: Not evaluated

Piper boehmeriifolium (Miq.) Wall. ex C.DC.
Piperaceae
Synonyms: *Chavica boehmeriifolia* Miq.
Common Name: False-nettled leaved pepper
Climbing Mechanism: Root Climber
Distribution (Global): Assam, Bangladesh, Cambodia, China South-Central, China Southeast, East Himalaya, Laos, Malaya, Myanmar, Nepal, Philippines, Sumatera, Thailand, Vietnam
Distribution (India): Arunachal Pradesh, Mizoram, Tripura
Leaf Type: Simple, alternate, 5–9 nerved from base
Inflorescence: Axillary spikes, unisexual flowers
Fruit Type: Wet, berry, black or brown
IUCN Status: Not evaluated
Reference: A.P.de Candolle, Prodr. 16(1): 348 (1869)

Piper canescens
Piperaceae
Climbing Mechanism: Root Climber
IUCN Status: Not evaluated

Piper caninum Blume
Piperaceae
Climbing Mechanism: Root Climber
Distribution (India): Mizoram.
Leaf Type: Simple, alternate, 5–9 nerved from base
Inflorescence: Axillary spikes, unisexual flowers
Fruit Type: Wet, berry, black or brown
IUCN Status: Not evaluated
Reference: Verh. Batav. Genootsch. Kunsten 11: 214 (1826)

Piper caudilimbum C.DC.
Piperaceae
Climbing Mechanism: Root Climber
Leaf Type: Simple, alternate, 5–9 nerved from base
Inflorescence: Axillary spikes, unisexual flowers
Fruit Type: Wet, berry, black or brown
IUCN Status: Not evaluated
Reference: Candollea 1: 198 (1923)

Piper clypeatum Wall.
Piperaceae
Climbing Mechanism: Root Climber
Distribution (Global): Nicobar Is.
Leaf Type: Simple, alternate, 5–9 nerved from base
Inflorescence: Axillary spikes, unisexual flowers
Fruit Type: Wet, berry, black or brown
IUCN Status: Not evaluated
Reference: Fl. Brit. India 5: 95, 537 (1888)

Piper cornilimbum C.DC.
Piperaceae
Climbing Mechanism: Root Climber
Distribution (Global): Assam.
Leaf Type: Simple, alternate, 5–9 nerved from base
Inflorescence: Axillary spikes, unisexual flowers
Fruit Type: Wet, berry, black or brown
IUCN Status: Not evaluated
Reference: Candollea 1: 191 (1923)

Piper crenulatibracteum C.DC.
Piperaceae
Climbing Mechanism: Root Climber
Distribution (Global): India
Distribution (India): Peninsular India
Leaf Type: Simple, alternate, 5–9 nerved from base
Inflorescence: Axillary spikes, unisexual flowers
Fruit Type: Wet, berry, black or brown
IUCN Status: Not evaluated
Reference: Repert. Spec. Nov. Regni Veg. 10: 521 (1912)

Piper cubeba Bojer
Piperaceae
Synonyms: *Cubeba cubeba* (L.f.) H.Karst., *Cubeba segetum* Miq., *Cubeba officinalis* Raf.
Common Name: Tail pepper, Cubebs

Climbing Mechanism: Root Climber
Distribution (Global): Borneo, Cambodia, Lesser Sunda Is., Malaya, Maluku, Myanmar, Sulawesi, Sumatera, Vietnam
Distribution (India): Peninsular India
Leaf Type: Simple, alternate, 5–9 nerved from base
Inflorescence: Axillary spikes, unisexual flowers
Fruit Type: Wet, berry, black or brown
IUCN Status: Not evaluated
Reference: Suppl. Pl.: 90 (1782)

Piper curtistipes C.DC.
Piperaceae
Climbing Mechanism: Root Climber
Distribution (Global): Assam, East Himalaya
Distribution (India): Arunachal Pradesh
Leaf Type: Simple, alternate, 5–9 nerved from base
Inflorescence: Axillary spikes, unisexual flowers
Fruit Type: Wet, berry, black or brown
IUCN Status: Not evaluated
Reference: Candollea 1: 225 (1923)

Piper dekkoanum C.DC.
Piperaceae
Common Name: Mountain long pepper
Climbing Mechanism: Root Climber
Distribution (India): Arunachal Pradesh
Leaf Type: Simple, alternate, 5–9 nerved from base
Inflorescence: Axillary spikes, unisexual flowers
Fruit Type: Wet, berry, black or brown
IUCN Status: Not evaluated
Reference: Candollea 1: 199 (1923)

Piper diffusum Blume ex Miq.
Piperaceae
Synonyms: *Piper sarmentosum* Roxb.
Climbing Mechanism: Root Climber
Distribution (India): Arunachal Pradesh, Assam, Mizoram
Leaf Type: Simple, alternate, 5–9 nerved from base
Inflorescence: Axillary spikes, unisexual flowers
Fruit Type: Wet, berry, black or brown
IUCN Status: Not evaluated
Reference: Linnaea 20: 130 (1847)

Piper falconeri C.DC.
Piperaceae

Climbing Mechanism: Root Climber
Distribution (Global): Assam
Distribution (India): Mizoram, Tripura
Leaf Type: Simple, alternate, 5–9 nerved from base
Inflorescence: Axillary spikes, unisexual flowers
Fruit Type: Wet, berry, black or brown
IUCN Status: Not evaluated
Reference: Candollea 1: 187 (1923)

Piper filipedunculum C.DC.
Piperaceae
Climbing Mechanism: Root Climber
Distribution (Global): East Himalaya
Leaf Type: Simple, alternate, 5–9 nerved from base
Inflorescence: Axillary spikes, unisexual flowers
Fruit Type: Wet, berry, black or brown
IUCN Status: Not evaluated
Reference: Candollea 1: 186 (1923)

Piper galeatum (Miq.) C.DC.
Piperaceae
Synonyms: *Muldera galeata* Miq.
Climbing Mechanism: Root Climber
Distribution (Global): India
Distribution (India): Karnataka, Tamil Nadu
Leaf Type: Simple, alternate, 5–9 nerved from base
Inflorescence: Axillary spikes, unisexual flowers
Fruit Type: Wet, berry, black or brown
Flowering and Fruiting: December–March
IUCN Status: Not evaluated
Notes: Usually found in semi-evergreen and evergreen forests
Reference: A.P.de Candolle, Prodr. 16(1): 242 (1869)

Piper gamblei C.DC.
Piperaceae
Synonyms: *Piper suipigua* Buch.-Ham. ex D.Don
Climbing Mechanism: Root Climber
Distribution (India): Arunachal Pradesh
Leaf Type: Simple, alternate, 5–9 nerved from base
Inflorescence: Axillary spikes, unisexual flowers
Fruit Type: Wet, berry, black or brown
IUCN Status: Not evaluated
Reference: Candollea 1: 204 (1923)

Piper gibsonii
Piperaceae
Climbing Mechanism: Root Climber
IUCN Status: Not evaluated

Piper glabramentum C.DC.
Piperaceae
Synonyms: *Piper indicum* C.DC.
Climbing Mechanism: Root Climber
Distribution (India): Arunachal Pradesh
Leaf Type: Simple, alternate, 5–9 nerved from base
Inflorescence: Axillary spikes, unisexual flowers
Fruit Type: Wet, berry, black or brown
IUCN Status: Not evaluated
Reference: Candollea 1: 219 (1923)

Piper glabrirhahe C.DC.
Piperaceae
Climbing Mechanism: Root Climber
Leaf Type: Simple, alternate, 5–9 nerved from base
Inflorescence: Axillary spikes, unisexual flowers
Fruit Type: Wet, berry, black or brown
IUCN Status: Not evaluated

Piper griffithii C.DC.
Piperaceae
Synonyms: Piper longibracteum C.DC.
Climbing Mechanism: Root Climber
Distribution (Global): Assam, Bangladesh, East Himalaya, Malaya, Myanmar, Thailand, Vietnam
Distribution (India): Arunachal Pradesh, Assam, Meghalaya, Mizoram
Leaf Type: Simple, alternate, 5–9 nerved from base
Inflorescence: Axillary spikes, unisexual flowers
Fruit Type: Wet, berry, black or brown
IUCN Status: Not evaluated
Reference: J. Bot. 4: 166 (1866)

Piper hamiltonii C.DC.
Piperaceae
Synonyms: *Piper plantagineum* Buch.-Ham. ex Wall.
Climbing Mechanism: Root Climber
Distribution (Global): Assam, Bangladesh, East Himalaya, Myanmar, Nepal
Leaf Type: Simple, alternate, 5–9 nerved from base
Inflorescence: Axillary spikes, unisexual flowers
Fruit Type: Wet, berry, black or brown

IUCN Status: Not evaluated
Reference: A.P.de Candolle, Prodr. 16(1): 360 (1869)

Piper hapnium Buch.-Ham.
Piperaceae
Synonyms: *Piper retrofractum* Vahl, *Chavica arnottiana* Miq., *Piper siriboa* Heyne.
Climbing Mechanism: Root Climber
Distribution (India): Tamil Nadu
Leaf Type: Simple, alternate, 5–9 nerved from base
Inflorescence: Axillary spikes, unisexual flowers
Fruit Type: Wet, berry, black or brown
IUCN Status: Not evaluated
Reference: Fl. Brit. India 5: 86 (1886)

Piper hookeri Miq.
Piperaceae
Climbing Mechanism: Root Climber
Distribution (Global): India
Distribution (India): Maharashtra
Leaf Type: Simple, alternate, 5–9 nerved from base
Inflorescence: Axillary spikes, unisexual flowers
Fruit Type: Wet, berry, black or brown
Flowering and Fruiting: February–May
IUCN Status: Not evaluated
Reference: London J. Bot. 4: 437 (1845)

Piper hymenophyllum (Miq.) Wight
Piperaceae
Synonyms: *Piper lanatum* Wight ex Miq., *Piper nilghirianum* C.DC.
Climbing Mechanism: Root Climber
Distribution (Global): Assam, India, Laos, Sri Lanka, Vietnam
Distribution (India): Andhra Pradesh, Arunachal Pradesh, Mizoram, Tamil Nadu, Western Ghats
Leaf Type: Simple, alternate, 5–9 nerved from base
Inflorescence: Axillary spikes, unisexual flowers
Fruit Type: Wet, berry, black or brown
Flowering and Fruiting: September–April
IUCN Status: Not evaluated
Notes: Distributed mostly in semi-evergreen and evergreen forests
Reference: London J. Bot. 5: 554 (1846)

Piper isopleurum C.DC.
Piperaceae
Synonyms: *Piper peepuloides* Roxb.

Climbing Mechanism: Root Climber
Leaf Type: Simple, alternate, 5–9 nerved from base
Inflorescence: Axillary spikes, unisexual flowers
Fruit Type: Wet, berry, black or brown
IUCN Status: Not evaluated
Reference: Candollea 1: 193 (1923)

Piper karpuragandhum J. Mathew & Yohannan
Piperaceae
Climbing Mechanism: Root Climber
Distribution (Global): India
Distribution (India): Kerala
Leaf Type: Simple, alternate, 5–9 nerved from base
Inflorescence: Axillary spikes, unisexual flowers
Fruit Type: Wet, berry, black or brown
IUCN Status: Not evaluated
Reference: Ann. Bot. Fenn. 54: 270 (2017)

Piper khasianum C.DC.
Piperaceae
Synonyms: *Piper grandipedunculum* C.DC., *Piper curtistipes* C. DC., *Piper mannii* C. DC.
Climbing Mechanism: Root Climber
Distribution (Global): Assam, Bangladesh, East Himalaya, Malaya, Myanmar, Nepal, Vietnam
Distribution (India): Mizoram, Tripura
Leaf Type: Simple, alternate, 5–9 nerved from base
Inflorescence: Axillary spikes, unisexual flowers
Fruit Type: Wet, berry, black or brown
IUCN Status: Not evaluated
Reference: A.P.de Candolle, Prodr. 16(1): 349 (1869)

Piper lainatakanum C.DC.
Piperaceae
Climbing Mechanism: Root Climber
Distribution (Global): Assam
Leaf Type: Simple, alternate, 5–9 nerved from base
Inflorescence: Axillary spikes, unisexual flowers
Fruit Type: Wet, berry, black or brown
IUCN Status: Not evaluated
Reference: Repert. Spec. Nov. Regni Veg. 10: 519 (1912)

Piper laxivenum C.DC.
Piperaceae
Climbing Mechanism: Root Climber

Distribution (Global): East Himalaya
Leaf Type: Simple, alternate, 5–9 nerved from base
Inflorescence: Axillary spikes, unisexual flowers
Fruit Type: Wet, berry, black or brown
IUCN Status: Not evaluated
Reference: Candollea 1: 213 (1923)

Piper longum L.
Piperaceae
Synonyms: *Chavica longa* (L.) H.Karst., *Chavica roxburghii* Miq., *Piper latifolium* W.Hunter, *Piper roxburghii* (Miq.) J.Presl, *Piper sarmentosum* Wall., *Piper turbinatum* Noronha
Common Name: Long pepper, Indian long pepper
Climbing Mechanism: Root Climber
Distribution (Global): Assam, Bangladesh, Cambodia, China South-Central, East Himalaya, Laos, Myanmar, Nicobar Is., Thailand, Vietnam
Distribution (India): Andaman and Nicobar Islands, Andhra Pradesh, Arunachal Pradesh, Assam, Bihar, Odisha, Kerala, Madhya Pradesh, Maharashtra, Tamil Nadu, Tripura, Uttar Pradesh, Western Ghats
Leaf Type: Simple, alternate, 5–9 nerved from base
Inflorescence: Axillary spikes, unisexual flowers
Fruit Type: Wet, berry, black or brown
Flowering and Fruiting: August–January
IUCN Status: Not evaluated
Notes: Distributed mostly in semi-evergreen and evergreen forests
Reference: Sp. Pl.: 29 (1753)

Piper makruense C.DC.
Piperaceae
Synonyms: *Piper dekkoanum* C.DC.
Climbing Mechanism: Root Climber
Distribution (Global): Assam, East Himalaya
Leaf Type: Simple, alternate, 5–9 nerved from base
Inflorescence: Axillary spikes, unisexual flowers
Fruit Type: Wet, berry, black or brown
IUCN Status: Not evaluated
Reference: Repert. Spec. Nov. Regni Veg. 10: 521 (1912)

Piper meeboldii C.DC.
Piperaceae
Climbing Mechanism: Root Climber
Distribution (Global): Assam
Leaf Type: Simple, alternate, 5–9 nerved from base
Inflorescence: Axillary spikes, unisexual flowers
Fruit Type: Wet, berry, black or brown

IUCN Status: Not evaluated
Reference: Repert. Spec. Nov. Regni Veg. 10: 521 (1912)

Piper megacarpum J. Mathew
Piperaceae
Climbing Mechanism: Root Climber
Distribution (Global): India
Distribution (India): Kerala
Leaf Type: Simple, alternate, 5–9 nerved from base
Inflorescence: Axillary spikes, unisexual flowers
Fruit Type: Wet, berry, black or brown
IUCN Status: Not evaluated
Reference: Ann. Bot. Fenn. 54: 269 (2017)

Piper mullesua Buch.-Ham. ex D.Don
Piperaceae
Synonyms: *Piper peepuloides* Roxb., *Artanthe brachystachya* Miq. ex C. DC., *Chavica guigual* Miq., *Chavica mullesua* (Buch.-Ham. ex D. Don) Miq., *Chavica sphaerostachya* Wall. ex Miq., *Piper brachystachyum* Wall., *Piper guigual* Buch.-Ham. ex D. Don, *Piper vasculo.*
Common Name: Wild pepper, Hill pepper
Climbing Mechanism: Root Climber
Distribution (India): Arunachal Pradesh, Himachal Pradesh, Karnataka, Meghalaya, Tamil Nadu, Western Ghats
Leaf Type: Simple, alternate, 5–9 nerved from base
Inflorescence: Axillary spikes, unisexual flowers
Fruit Type: Wet, berry, black or brown
Flowering and Fruiting: January–April
IUCN Status: Not evaluated
Notes: Mostly found in evergreen and shola forests
Reference: Prodr. Fl. Nepal.: 20 (1825)

Piper muneyporense C.DC.
Piperaceae
Climbing Mechanism: Root Climber
Distribution (Global): Assam
Leaf Type: Simple, alternate, 5–9 nerved from base
Inflorescence: Axillary spikes, unisexual flowers
Fruit Type: Wet, berry, black or brown
IUCN Status: Not evaluated
Reference: Candollea 1: 222 (1923)

Piper mungpooanum C.DC.
Piperaceae
Synonyms: *Piper voigtii* C.DC.

Climbing Mechanism: Root Climber
Leaf Type: Simple, alternate, 5–9 nerved from base
Inflorescence: Axillary spikes, unisexual flowers
Fruit Type: Wet, berry, black or brown
IUCN Status: Not evaluated
Reference: Candollea 1: 185 (1923)

Piper nigrum L.
Piperaceae
Synonyms: *Muldera multinervis* Miq.,
Muldera wightiana Miq., *Piper aromaticum* Lam., *Piper baccatum* C.DC., *Piper colonum* C.Presl, *Piper denudatum* Opiz, *Piper glabrispica* C.DC., *Piper glyphicum* Hoffmanns. ex Kunth.
Common Name: Black pepper, Common pepper, Pepper
Climbing Mechanism: Root Climber
Distribution (Global): India
Distribution (India): Andhra Pradesh, Arunachal Pradesh, Assam, Bihar, Odisha, Goa, Karnataka, Kerala, Maharashtra, Mizoram, Odisha, Tamil Nadu, Western Ghats
Leaf Type: Simple, opposite
Inflorescence: Spike
Fruit Type: Drupe
Flowering and Fruiting: Throughout the year
IUCN Status: Not evaluated
Notes: Commercially cultivated. Found mostly in evergreen and semi-evergreen forests
Reference: Sp. Pl.: 28 (1753)

Piper nigrum var. hirtellosum
Piperaceae
Synonyms: *Piper nigrum var. Nigrum* L.
Climbing Mechanism: Root Climber
Leaf Type: Simple, alternate, 5–9 nerved from base
Inflorescence: Axillary spikes, unisexual flowers
Fruit Type: Wet, berry, black or brown
Flowering and Fruiting: April–May
IUCN Status: Not evaluated
Notes: Distributed mostly in the evergreen forests

Piper nilghirianum C.DC.
Piperaceae
Synonyms: *Piper hymenophyllum* Miq.
Climbing Mechanism: Root Climber
Distribution (India): Peninsular India
Leaf Type: Simple, alternate, 5–9 nerved from base

Inflorescence: Axillary spikes, unisexual flowers
Fruit Type: Wet, berry, black or brown
IUCN Status: Not evaluated
Reference: A.P.de Candolle, Prodr. 16(1): 364 (1869)

Piper nirjulianum Gajurel, Rethy & Y.Kumar
Piperaceae
Synonyms: Piper peepuloides Roxb.
Climbing Mechanism: Root Climber
Leaf Type: Simple, alternate, 5—9 nerved from base
Inflorescence: Axillary spikes, unisexual flowers
Fruit Type: Wet, berry, black or brown
IUCN Status: Not evaluated
Reference: Rheedea 17: 35 (2007)

Piper obtusistigmum C.DC.
Piperaceae
Climbing Mechanism: Root Climber
Distribution (Global): India
Distribution (India): Tamil Nadu
Leaf Type: Simple, alternate, 5–9 nerved from base
Inflorescence: Axillary spikes, unisexual flowers
Fruit Type: Wet, berry, black or brown
IUCN Status: Not evaluated
Notes: Distributed mostly in the evergreen forests
Reference: Candollea 1: 217 (1923)

Piper oldhamii C.DC.
Piperaceae
Climbing Mechanism: Root Climber
Distribution (Global): Assam
Leaf Type: Simple, alternate, 5–9 nerved from base
Inflorescence: Axillary spikes, unisexual flowers
Fruit Type: Wet, berry, black or brown
IUCN Status: Not evaluated
Reference: Candollea 1: 195 (1923)

Piper ootacamundense C.DC.
Piperaceae
Synonyms: *Piper schmidtii* Hook.f.
Climbing Mechanism: Root Climber
Leaf Type: Simple, alternate, 5–9 nerved from base
Inflorescence: Axillary spikes, unisexual flowers
Fruit Type: Wet, berry, black or brown
IUCN Status: Not evaluated

Notes: Distributed mostly in the evergreen forests
Reference: Candollea 1: 184 (1923)

Piper opacilimbum C.DC.
Piperaceae
Synonyms: *Piper schmidtii* Hook.f.
Climbing Mechanism: Root Climber
Leaf Type: Simple, alternate, 5–9 nerved from base
Inflorescence: Axillary spikes, unisexual flowers
Fruit Type: Wet, berry, black or brown
IUCN Status: Not evaluated
Notes: Distributed mostly in the evergreen forests
Reference: Candollea 1: 184 (1923)

Piper ovatistigmum C.DC.
Piperaceae
Climbing Mechanism: Root Climber
Distribution (Global): East Himalaya
Leaf Type: Simple, alternate, 5–9 nerved from base
Inflorescence: Axillary spikes, unisexual flowers
Fruit Type: Wet, berry, black or brown
IUCN Status: Not evaluated
Reference: Candollea 1: 188 (1923)

Piper parvilimbus C.DC.
Piperaceae
Climbing Mechanism: Root Climber
Leaf Type: Simple, alternate, 5–9 nerved from base
Inflorescence: Axillary spikes, unisexual flowers
Fruit Type: Wet, berry, black or brown
IUCN Status: Not evaluated

Piper pedicellatum C.DC.
Piperaceae
Synonyms: *Piper clarkei* C.DC., *Piper curtipedunculum* C.DC., *Piper nigramentum* C.DC., *Piper pedicellatum var. eglandulatum* Chaveer. & Mokkamul, *Piper subpedicellatum* Van Heurck & Müll.Arg.
Climbing Mechanism: Root Climber
Distribution (Global): Assam, Bangladesh, China South-Central, East Himalaya, Myanmar, Thailand, Vietnam
Distribution (India): Andaman and Nicobar Islands, Mizoram
Leaf Type: Simple, alternate, 5–9 nerved from base
Inflorescence: Axillary spikes, unisexual flowers
Fruit Type: Wet, berry, black or brown

IUCN Status: Vulnerable
Reference: J. Bot. 4: 164 (1866)

Piper pedicellosum Wall.
Piperaceae
Synonyms: *Piper crassipes* Korth. ex Miq.
Climbing Mechanism: Root Climber
Distribution (India): Great Nicobar Islands
Leaf Type: Simple, alternate, 5–9 nerved from base
Inflorescence: Axillary spikes, unisexual flowers
Fruit Type: Wet, berry, black or brown
IUCN Status: Not evaluated
Reference: A.P.de Candolle, Prodr. 16(1): 343 (1869)

Piper peepuloides Wall.
Piperaceae
Synonyms: *Chavica guigual* (Buch.-Ham. ex D.Don) Miq., *Chavica mullesua* (Buch.-Ham. ex D.Don) Miq., *Chavica neesiana* Miq., *Chavica peepuloides* (Roxb.) Miq.
Climbing Mechanism: Root Climber
Distribution (Global): Assam, Bangladesh, China South-Central, East Himalaya, Hainan, India, Myanmar, Nepal, Thailand, West Himalaya
Distribution (India): Arunachal Pradesh, Bihar, Odisha, Mizoram
Leaf Type: Simple, alternate, 5–9 nerved from base
Inflorescence: Axillary spikes, unisexual flowers
Fruit Type: Wet, berry, black or brown
IUCN Status: Not evaluated
Reference: Fl. Ind. 1: 159 (1820)

Piper petiolatum Hook.f.
Piperaceae
Synonyms: *Piper thomsonii* (C.DC.) Hook.f.
Climbing Mechanism: Root Climber
Distribution (India): Arunachal Pradesh
Leaf Type: Simple, alternate, 5–9 nerved from base
Inflorescence: Axillary spikes, unisexual flowers
Fruit Type: Wet, berry, black or brown
IUCN Status: Not evaluated
Reference: J. Bot. 4: 161 (1866)

Piper phalangense
Piperaceae
Climbing Mechanism: Root Climber
IUCN Status: Not evaluated

Piper pykarahense C.DC.
Piperaceae
Climbing Mechanism: Root Climber
Distribution (Global): India.
Distribution (India): Tamil Nadu
Leaf Type: Simple, alternate, 5–9 nerved from base
Inflorescence: Axillary spikes, unisexual flowers
Fruit Type: Wet, berry, black or brown
IUCN Status: Not evaluated
Notes: Distributed mostly in the evergreen forests
Reference: Repert. Spec. Nov. Regni Veg. 13: 300 (1914)

Piper redactum Trel.
Piperaceae
Climbing Mechanism: Root Climber
Distribution (Global): Peru
Leaf Type: Simple, alternate, 5–9 nerved from base
Inflorescence: Axillary spikes, unisexual flowers
Fruit Type: Wet, berry, black or brown
IUCN Status: Not evaluated
Reference: Publ. Field Mus. Nat. Hist., Bot. Ser. 13(2): 221 (1936)

Piper relictum
Piperaceae
Climbing Mechanism: Root Climber
IUCN Status: Not evaluated

Piper retrofractum Vahl
Piperaceae
Synonyms: *Amalago antillana* Raf., *Amalago malamiri* Raf., *Chavica arnottiana* Miq., *Chavica chaba* Miq., *Chavica labillardierei* Miq., *Chavica maritima* Miq. +14
Climbing Mechanism: Root Climber
Distribution (Global): Bangladesh, Cambodia, China South-Central, Laos, Lesser Sunda Is., Malaya, Philippines, Thailand, Vietnam
Distribution (India): Tamil Nadu
Leaf Type: Simple, alternate, 5–9 nerved from base
Inflorescence: Axillary spikes, unisexual flowers
Fruit Type: Wet, berry, black or brown
Flowering and Fruiting: April–September
IUCN Status: Not evaluated
Reference: Enum. Pl. 1: 314 (1804)

Piper ribesioides Wall.
Piperaceae

Synonyms: *Cubeba sumatrana* Miq., *Cubeba wallichii* Miq., *Piper pseudorubeba* Korth. ex Miq., *Piper sumatranum* (Miq.) C.DC.
Climbing Mechanism: Root Climber
Distribution (Global): Andaman Is., Assam, Cambodia, India, Laos, Malaya, Myanmar, Sri Lanka, Sumatera, Thailand, Vietnam
Leaf Type: Simple, alternate, 5–9 nerved from base
Inflorescence: Axillary spikes, unisexual flowers
Fruit Type: Wet, berry, black or brown
IUCN Status: Not evaluated
Reference: Pl. Asiat. Rar. 1: 79 (1830)

Piper schiedeanum Steud.
Piperaceae
Synonyms: *Artanthe benthamiana* Miq., *Artanthe pothomorpha* Miq., *Artanthe tiliifolia* Miq., *Piper benthamianum* (Miq.) C.DC., *Piper megalophyllum* C. DC., *Piper megalophyllum var. connivens* C.DC., *Piper pothomorphe* (Miq.) C.DC., *Piper tiliifolium* Schltdl. & Cham.
Climbing Mechanism: Root Climber
Distribution (Global): Mexico Central, Mexico Gulf, Mexico Northeast, Mexico Southeast, Mexico Southwest
Leaf Type: Simple, alternate, 5–9 nerved from base
Inflorescence: Axillary spikes, unisexual flowers
Fruit Type: Wet, berry, black or brown
IUCN Status: Least Concern
Reference: Nomencl. Bot., ed. 2, 2: 343 (1841)

Piper schmidtii Hook.f.
Piperaceae
Synonyms: *Piper gibsonii* C.DC., *Piper glabrirhache* C.DC., *Piper ootacamundense* C.DC., *Piper opacilimbum* C.DC., *Piper ovatostemon* C. DC.
Climbing Mechanism: Root Climber
Distribution (Global): India
Distribution (India): Karnataka, Tamil Nadu, Western Ghats
Leaf Type: Simple, alternate, 5–9 nerved from base
Inflorescence: Axillary spikes, unisexual flowers
Fruit Type: Wet, berry, black or brown
Flowering and Fruiting: November–April
IUCN Status: Not evaluated
Notes: Distributed mostly in evergreen and shola forests
Reference: Fl. Brit. India 5: 89 (1886)

Piper sikkimense C. DC.
Piperaceae
Synonyms: *Piper suipigua* Buch. -Ham. ex D. Don
Climbing Mechanism: Root Climber

Leaf Type: Simple, alternate, 5–9 nerved from base
Inflorescence: Axillary spikes, unisexual flowers
Fruit Type: Wet, berry, black or brown
IUCN Status: Not evaluated
Reference: Candollea 1: 192 (1923)

Piper silentvalleyense Ravindran, M.K.Nair & Asokan Nair
Piperaceae
Climbing Mechanism: Root Climber
Distribution (Global): East Himalaya
Distribution (India): Peninsular India
Leaf Type: Simple, alternate, 5–9 nerved from base
Inflorescence: Axillary spikes, unisexual flowers
Fruit Type: Wet, berry, black or brown
Flowering and Fruiting: March–May
IUCN Status: Not evaluated
Notes: Distributed mostly in evergreen forests
Reference: J. Econ. Taxon. Bot. 10: 167 (1987 publ. 1988)

Piper sonadense C. DC.
Piperaceae
Climbing Mechanism: Root Climber
Distribution (Global): East Himalaya
Leaf Type: Simple, alternate, 5–9 nerved from base
Inflorescence: Axillary spikes, unisexual flowers
Fruit Type: Wet, berry, black or brown
IUCN Status: Not evaluated
Reference: Repert. Spec. Nov. Regni Veg. 13: 298 (1914)

Piper suipigua Buch. -Ham. ex D. Don
Piperaceae
Synonyms: *Chavica suipigua* (Buch. -Ham. ex D.Don) Miq., *Piper gamblei* C.DC., *Piper gammiei* C.DC., *Piper nepalense* Miq., *Piper sikkimense* C.DC., *Piper subrigidilimbum* C.DC.
Climbing Mechanism: Root Climber
Distribution (Global): Assam, Bangladesh, China South-Central, East Himalaya, Myanmar, Nepal, Thailand
Leaf Type: Simple, alternate, 5–9 nerved from base
Inflorescence: Axillary spikes, unisexual flowers
Fruit Type: Wet, berry, black or brown
IUCN Status: Not evaluated
Reference: Prodr. Fl. Nepal.: 20 (1825)

Piper sumatranum (Miq.) C.DC.
Piperaceae

Synonyms: *Piper ribesioides* Wall.
Climbing Mechanism: Root Climber
Leaf Type: Simple, alternate, 5–9 nerved from base
Inflorescence: Axillary spikes, unisexual flowers
Fruit Type: Wet, berry, black or brown
IUCN Status: Not evaluated
Reference: A.P.de Candolle, Prodr. 16(1): 343 (1869)

Piper sylvaticum Roxb.
Piperaceae
Synonyms: *Chavica sylvatica* (Roxb.) Miq.
Common Name: Mountain long pepper
Climbing Mechanism: Root Climber
Distribution (Global): Bangladesh, East Himalaya
Distribution (India): Arunachal Pradesh, Assam, Mizoram
Leaf Type: Simple, alternate, 5–9 nerved from base
Inflorescence: Axillary spikes, unisexual flowers
Fruit Type: Wet, berry, black or brown
IUCN Status: Not evaluated
Reference: Fl. Ind. 1: 158 (1820)

Piper sylvestre Lam.
Piperaceae
Synonyms: *Piper fallax* Vahl
Climbing Mechanism: Root Climber
Distribution (India): Telangana
Leaf Type: Simple, alternate, 5–9 nerved from base
Inflorescence: Axillary spikes, unisexual flowers
Fruit Type: Wet, berry, black or brown
IUCN Status: Not evaluated
Reference: Tabl. Encycl. 1: 79 (1791)

Piper tenuibracteatum C.DC.
Piperaceae
Climbing Mechanism: Root Climber
Leaf Type: Simple, alternate, 5–9 nerved from base
Inflorescence: Axillary spikes, unisexual flowers
Fruit Type: Wet, berry, black or brown
IUCN Status: Not evaluated

Piper thomsonii (C.DC.) Hook.f.
Piperaceae
Synonyms: *Chavica petiolata* (C.DC.) C.DC., *Chavica thomsonii* C.DC., *Piper bavinum* C.DC., *Piper malmoris* Wall., *Piper petiolatum* C.DC., *Piper*

punctulivenum C.DC., *Piper thomsonii var. microphyllum* Y.C.Tseng, *Piper thomsonii var. trichostigma* Chaveer. & Sudmoo
Climbing Mechanism: Root Climber
Distribution (Global): Assam, Bangladesh, China South-Central, East Himalaya, Laos, Myanmar, Nepal, Sri Lanka, Thailand, Tibet, Vietnam
Distribution (India): Arunachal Pradesh
Leaf Type: Simple, alternate, 5–9 nerved from base
Inflorescence: Axillary spikes, unisexual flowers
Fruit Type: Wet, berry, black or brown
IUCN Status: Not evaluated
Reference: Fl. Brit. India 5: 87 (1886)

Piper trichostachyon (Miq.) C.DC.
Piperaceae
Synonyms: *Muldera trichostachya* Miq.
Climbing Mechanism: Root Climber
Distribution (Global): India
Distribution (India): Goa, Karnataka, Maharashtra, Tamil Nadu, Western Ghats
Leaf Type: Simple, alternate, 5–9 nerved from base
Inflorescence: Axillary spikes, unisexual flowers
Fruit Type: Wet, berry, black or brown
Flowering and Fruiting: January–May
IUCN Status: Not evaluated
Notes: Distributed mostly in semi-evergreen and moist deciduous forests
Reference: A.P.de Candolle, Prodr. 16(1): 242 (1869)

Piper trioicum Roxb.
Piperaceae
Synonyms: *Piper nigrum* L.,
Piper attenuatum Buch-Ham.
Climbing Mechanism: Root Climber
Distribution (India): Arunachal Pradesh, Assam, Odisha, Tamil Nadu, Tripura
Leaf Type: Simple, alternate, 5–9 nerved from base
Inflorescence: Axillary spikes, unisexual flowers
Fruit Type: Wet, berry, black or brown
Flowering and Fruiting: August–May
IUCN Status: Not evaluated
Notes: Distributed mostly in semi-evergreen and moist deciduous forests
Reference: Fl. Ind. 1: 153 (1820)

Piper umbellatum L.
Piperaceae
Synonyms: *Heckeria sidifolia var. subglabrata* Kunth, *Heckeria sidifolia* (Link & Otto) Kunth, *Heckeria subpeltata* (Willd.) Kunth, *Heckeria umbellata* (L.)

Kunth, *Lepianthes umbellata* (L.) Raf. ex Ramamoorthy, *Peperidia afzeliana* Kostel., *Peperidia sidifolia* (Link & Otto) Kostel.

Common Name: Cow-foot leaf

Climbing Mechanism: Root Climber

Distribution (Global): Belize, Bolivia, Brazil North, Brazil Northeast, Brazil South, Brazil Southeast, Brazil West-Central, Colombia, Costa Rica, Cuba, Dominican Republic, Ecuador, El Salvador, Guatemala, Haiti, Honduras, Jamaica, Leeward Is., Mexico Central, Mexico Gulf, Mexico Northeast, Mexico Southeast, Mexico Southwest, Nicaragua, Panamá, Peru, Puerto Rico, Trinidad-Tobago, Venezuela, Windward Is.

Leaf Type: Simple, alternate, 5–9 nerved from base

Inflorescence: Axillary spikes, unisexual flowers

Fruit Type: Wet, berry, black or brown

Flowering and Fruiting: June–February

IUCN Status: Not evaluated

Notes: Often occur in semi-evergreen and evergreen forests

Reference: Sp. Pl.: 30 (1753)

Piper velayudhanii

Piperaceae

Climbing Mechanism: Root Climber

IUCN Status: Not evaluated

Piper wallichii (Miq.) Hand.-Mazz.

Piperaceae

Synonyms: *Chavica chuvya* Miq., *Chavica wallichii* Miq., *Piper accrescens* Van Heurck & Müll.Arg. , *Piper aurantiacum* Wall., *Piper caudilimbum* C.DC., *Piper emeiense* Y.C.Tseng, *Piper flaviflorum* C.DC., *Piper henryei* C.DC., *Piper ichangense* C.DC., *Piper martinii*

Climbing Mechanism: Root Climber

Distribution (Global): Assam, Bangladesh, China North-Central, China South-Central, China Southeast, East Himalaya, Laos, Myanmar, Nepal, Thailand

Distribution (India): Andaman and Nicobar Islands

Leaf Type: Simple, alternate, 5–9 nerved from base

Inflorescence: Axillary spikes, unisexual flowers

Fruit Type: Wet, berry, black or brown

IUCN Status: Not evaluated

Reference: Symb. Sin. 7: 155 (1929)

Piper wightii Miq.

Piperaceae

Climbing Mechanism: Root Climber

Distribution (Global): India

Distribution (India): Tamil Nadu

Leaf Type: Simple, alternate, 5–9 nerved from base

Inflorescence: Axillary spikes, unisexual flowers
Fruit Type: Wet, berry, black or brown
Flowering and Fruiting: July–February
IUCN Status: Not evaluated
Notes: Found mostly in evergreen and shola forests.
Reference: London J. Bot. 5: 552 (1846)

Asarina antirrhiniflora
Plantaginaceae
Climbing Mechanism: Scrambler-Unarmed
IUCN Status: Not evaluated
Lophospermum scandens D.Don
Plantaginaceae
Synonyms: *Asarina lophospermum* (L.H.Bailey) Pennell, *Maurandya erubescens* var. glabrata I.M.Johnst., *Maurandya glabrata* (I.M.Johnst.) D.Ramírez, *Maurandya lophospermum* L.H.Bailey, *Maurandya scandens* (D.Don) A.Gray
Common Name: Mystic purple, Mystic rose
Climbing Mechanism: Stem Twiner
Distribution (Global): Mexico Central
Distribution (India): Tamil Nadu
Leaf Type: The leaves are narrowly heart shaped
Inflorescence: The flowers are borne singly on stems
Flowering and Fruiting: May–November
IUCN Status: Not evaluated
Notes: Found mostly in deciduous oak forests and scrub, on cliffs, canyon walls, and rocky outcrops
Reference: Trans. Linn. Soc. London 15: 353 (1827)

Maurandya barclayana Lindl.
Plantaginaceae
Synonyms: *Asarina barclayana* (Lindl.) Pennell, *Usteria barclayana* (Lindl.) H. Jacq.
Climbing Mechanism: Stem Twiner
Distribution (Global): Mexico Central, Mexico Gulf, Mexico Northeast, Mexico Southwest
Distribution (India): Maharashtra, Tamil Nadu
IUCN Status: Not evaluated
Reference: Bot. Reg. 13: t. 1108 (1827)

Maurandya lophospermum L.H.Bailey
Plantaginaceae
Synonyms: *Asarina lophospermum* (L.H.Bailey) Pennell, *Maurandya erubescens var. glabrata* I.M.Johnst., *Maurandya glabrata* (I.M.Johnst.) D.Ramírez, *Maurandya lophospermum* L.H.Bailey, *Maurandya scandens* (D.Don) A.Gray.
Climbing Mechanism: Stem Twiner

Distribution (Global): Mexico Central
Distribution (India): Maharashtra, Tamil Nadu
Leaf Type: The leaves are narrowly heart shaped, wide, with a pointed apex and toothed edges
Inflorescence: The flowers are borne singly on stems (peduncles)
Fruit Type: An unsymmetrical ovoid capsule forms, filled with brown seeds
Flowering and Fruiting: May-November.
IUCN Status: Not evaluated
Reference: Trans. Linn. Soc. London 15: 353 (1827)

Maurandya scandens (Cav.) Pers.
Plantaginaceae
Synonyms: Synonym
Climbing Mechanism: Stem Twiner
IUCN Status: Not evaluated

Arthraxon microphyllus (Trin.) Hochst.
Poaceae
Synonyms: *Andropogon microphyllus* Trin., *Andropogon muralis* Buch.-Ham. ex Wall., *Arthraxon sikkimensis* Bor, *Pleuroplitis microphylla* (Trin.) Regel.
Climbing Mechanism: Scrambler-Unarmed
Distribution (Global): Assam, China South-Central, East Himalaya, Myanmar, Nepal, Thailand, West Himalaya
Flowering and Fruiting: August–March
IUCN Status: Not evaluated
Notes: Usually found in grasslands and deciduous forests
Reference: Flora 39: 188 (1856)

Dinochloa andamanica Kurz
Poaceae
Climbing Mechanism: Scrambler-Unarmed
Distribution (Global): Andaman Is., Myanmar, Nicobar Is., Thailand
Distribution (India): Andaman and Nicobar Islands
IUCN Status: Not evaluated
Reference: J. Asiat. Soc. Bengal, Pt. 2, Nat. Hist. 42: 253 (1873)

Dinochloa macclellandii (Munro) Kurz
Poaceae
Synonyms: *Arundarbor macclellandii* (Munro) Kuntze, *Bambusa macclellandii* Munro, *Melocalamus macclellandii* (Munro) H.B.Naithani
Climbing Mechanism: Scrambler-Unarmed
Distribution (Global): Assam, Bangladesh, East Himalaya, Myanmar, Thailand, Vietnam
Distribution (India): Arunachal Pradesh

IUCN Status: Not evaluated
Reference: J. Asiat. Soc. Bengal, Pt. 2, Nat. Hist. 42: 253 (1873)

Dinochloa scandens (Blume ex Nees) Kuntze
Poaceae
Synonyms: *Bambusa scandens* Blume ex Nees, *Chusquea amplopaniculata* Steud., *Dinochloa macrocarpa* Elmer, *Dinochloa tjankorreh* Buse, Nastus tjankorreh Schult.f., Schizostachyum parviflorum Munro.
Climbing Mechanism: Scrambler-Unarmed
Distribution (Global): Andaman Is., Borneo, Java, Malaya, Nicobar Is., Thailand
Distribution (India): Andaman and Nicobar Islands
IUCN Status: Not evaluated
Reference: Revis. Gen. Pl. 2: 773 (1891)

Jansenella griffithiana (C. Muell.) Bor
Poaceae
Synonyms: *Arundinella avenacea* Munro ex Thwaites, *Arundinella campbelliana* Lisboa, *Arundinella griffithiana* (C. Muell.) Bor, *Avena malabarica* B. Heyne ex Hook.f., *Danthonia griffithiana* Müll.Hal., *Danthoniopsis griffithiana* (C. Muell.) Bor.
Common Name: Jansen grass
Climbing Mechanism: Stem Twiner
Distribution (Global): Assam, India, Myanmar, Sri Lanka
Leaf Type: Leaves are elliptic-lanceolate, base amplexicaul, apex acuminate; sheaths slightly keeled
Inflorescence: Inflorescence occurs at the end of stem, compact panicles. Spikelets are stalkless
Fruit Type: Caryopsis with adherent pericarp
Flowering and Fruiting: August–November
IUCN Status: Not evaluated
Notes: Often occur in moist rocky grasslands and swamps
Reference: Kew Bull. 10: 98 (1955)

Ochlandra setigera Gamble
Poaceae
Synonyms: *Neomicrocalamus setiger* (Gamble) Hsueh & T.P.Yi
Climbing Mechanism: Scrambler-Unarmed
Distribution (Global): India
IUCN Status: Not evaluated
Reference: Ann. Roy. Bot. Gard. (Calcutta) 7: 128 (1896)

Ochlandra wightii (Munro) C.E.C.Fisch.
Poaceae
Synonyms: *Bambusa wightii* Munro, *Ochlandra brandisii* Gamble, *Teinostachyum wightii* (Munro) Bedd.

Climbing Mechanism: Scrambler-Unarmed
Distribution (Global): India
Flowering and Fruiting: December–April
IUCN Status: Not evaluated
Notes: Usually found in semi-evergreen and evergreen forests
Reference: J.S.Gamble, Fl. Madras: 1864 (1934)

Pseudodichanthium serrafalcoides (Cooke & Stapf) Bor
Poaceae
Synonyms: *Andropogon cookei* Stapf ex T.Cooke, *Andropogon serrafalcoides* Cooke & Stapf, *Dichanthium serrafalcoides* (Cooke & Stapf) Blatt. & McCann, *Pseudodichanthium cookei* M.R.Almeida
Climbing Mechanism: Scrambler-Unarmed
Distribution (Global): India, Oman
Flowering and Fruiting: September–November
IUCN Status: Not evaluated
Reference: Indian Forester 66: 272 (1940)

Pseudoxytenanthera bourdillonii (Gamble) H.B.Naithani
Poaceae
Synonyms: *Oxytenanthera bourdillonii* Gamble, *Pseudotenanthera bourdillonii* (Gamble) R.B.Majumdar
Climbing Mechanism: Scrambler-Unarmed
Distribution (Global): India
Flowering and Fruiting: October–February
IUCN Status: Not evaluated
Reference: J. Bombay Nat. Hist. Soc. 87: 440 (1991)

Pseudoxytenanthera monadelpha (Thwaites) Soderstr. & R.P.Ellis
Poaceae
Synonyms: *Dendrocalamus monadelphus* Thwaites, *Oxytenanthera monadelpha* (Thwaites) Alston, *Oxytenanthera thwaitesii* Munro, *Pseudotenanthera monadelpha* (Thwaites) R.B.Majumdar
Climbing Mechanism: Scrambler-Unarmed
Distribution (Global): India, Laos, Myanmar, Sri Lanka, Vietnam
Distribution (India): Andhra Pradesh
Flowering and Fruiting: February–January
IUCN Status: Not evaluated
Notes: Mostly found in evergreen and shola forests
Reference: Smithsonian Contr. Bot. 72: 52 (1988)

Cobaea scandens Cav.
Polemoniaceae
Synonyms: *Cobaea lasseri* Pittier, *Rosenbergia scandens* (Cav.) House
Common Name: Cup & saucer vine, Mexican ivy, Monastery bell

Climbing Mechanism: Stem Twiner
Distribution (Global): Mexico Central, Mexico Gulf, Mexico Northeast, Mexico Southeast, Mexico Southwest
Distribution (India): Tamil Nadu
IUCN Status: Not evaluated
Notes: Native of tropical America, naturalized in the Western Ghats
Reference: Icon. 1: 11 (1791)

Polygala tricholopha Chodat
Polygalaceae
Synonyms: *Polygala laotica* Gagnep.
Climbing Mechanism: Scrambler-Unarmed
Distribution (Global): Assam, China South-Central, China Southeast, East Himalaya, Hainan, India, Laos, Myanmar, Thailand, Vietnam
Distribution (India): Meghalaya, Mizoram
Leaf Type: Simple
Inflorescence: Axillary or terminal, racemes, panicles, or rarely solitary
Fruit Type: Capsule
IUCN Status: Not evaluated
Reference: Mém. Soc. Phys. Genève 31(2: 2): 98 (1893)

Securidaca inappendiculata Hassk.
Polygalaceae
Synonyms: *Securidaca bracteata* A.W. Benn.
Climbing Mechanism: Scrambler-Unarmed
Distribution (Global): Assam, Bangladesh, Borneo, Cambodia, China South-Central, China Southeast, Hainan, Java, Laos, Malaya, Myanmar, Nepal, Philippines, Sumatera, Thailand, Vietnam
Distribution (India): Arunachal Pradesh, Odisha
Leaf Type: Simple, alternate, distichous
Inflorescence: Axillary, terminal, racemes, or panicles
Fruit Type: Indehiscent samara, single seeded, compressed or flattened, winged
Flowering and Fruiting: September–March
IUCN Status: Not evaluated
Notes: Found usually in moist places along streams
Reference: Pl. Jav. Rar.: 295 (1848)

Antigonon leptopus Hook. & Arn.
Polygonaceae
Synonyms: *Antigonon cordatum* M.Martens & Galeotti, *Antigonon platypus* Hook. & Arn., *Corculum leptopus* Stuntz, *Corculum leptopus var. album* Stuntz, *Polygonum cirrhosum* Moc. & Sessé ex Meisn.
Common Name: Coral vine, Honolulu creeper, Mexican creeper, Bride's tears, Chain-of-love, Hearts on a chain, Love-vine
Climbing Mechanism: Tendril Climber

Distribution (Global): Belize, El Salvador, Guatemala, Honduras, Mexico Central, Mexico Gulf, Mexico Northeast, Mexico Northwest, Mexico Southeast, Mexico Southwest, Nicaragua
Distribution (India): Telangana, Tamil Nadu, Andhra Pradesh, Puducherry, Kerala, Gujarat, Madhya Pradesh, Delhi, Rajashthan, Maharashtra, Odisha, Ahmadabad
Flowering and Fruiting: Throughout the year
IUCN Status: Not evaluated
Reference: Bot. Beechey Voy.: 308 (1838)

Fallopia convolvulus
Polygonaceae
Climbing Mechanism: Stem Twiner
IUCN Status: Not evaluated

Fallopia dumetorum
Polygonaceae
Climbing Mechanism: Stem Twiner
IUCN Status: Not evaluated

Fallopia pterocarpa
Polygonaceae
Climbing Mechanism: Stem Twiner
IUCN Status: Not evaluated

Muehlenbeckia gracillima Meisn.
Polygonaceae
Climbing Mechanism: Stem Twiner
Distribution (India): Tamil Nadu
IUCN Status: Not evaluated

Persicaria auriculata
Polygonaceae
Climbing Mechanism: Scrambler-Unarmed
IUCN Status: Not evaluated

Persicaria chinensis (L.) H.Gross
Polygonaceae
Synonyms: *Ampelygonum chinense* (L.) Lindl., *Polygonum chinense* L.
Common Name: Chinese knot weed
Climbing Mechanism: Scrambler-Unarmed
Distribution (Global): Assam, Bangladesh, Cambodia, China North-Central, China South-Central, China Southeast, East Himalaya, Hainan, India, Japan, Java, Korea, Laos, Lesser Sunda Is., Malaya, Myanmar, Nansei-shoto, Nepal, Nicobar Is., Philippines, Sri Lanka, Sumatera, Taiwan, Thailand, Tibet, Vietnam, West Himalaya

Distribution (India): Arunachal Pradesh, Mizoram, Telangana, West Bengal, Western Ghats
Flowering and Fruiting: Throughout the year
IUCN Status: Not evaluated
Notes: Distributed across a wide range of habitats: Evergreen, semi-evergreen, and moist deciduous forests
Reference: Bot. Jahrb. Syst. 49: 269 (1913)

Polygonum perfoliatum L.
Polygonaceae
Synonyms: *Persicaria perfoliata* (L.) H.Gross
Common Name: Devil's tail.
Climbing Mechanism: Stem Twiner
Leaf Type: Simple
Inflorescence: Terminal or axillary
Flowering and Fruiting: March–July
IUCN Status: Not evaluated
Reference: Syst. Nat. ed. 10, 2: 1005 (1759)

Embelia adnata Bedd. ex C.B.Clarke
Primulaceae
Synonyms: *Ribesiodes adnatum* (Bedd. ex C.B.Clarke) Kuntze
Climbing Mechanism: Scrambler-Unarmed
Distribution (Global): India
Distribution (India): Tamil Nadu
Leaf Type: Simple, alternate below and opp above
Inflorescence: Racemes/panicles, 4–6 merous
Fruit Type: Wet, drupe, globose
Flowering and Fruiting: August–December
IUCN Status: Not evaluated
Reference: J. D. Hooker, Fl. Brit. India 3: 514 (1882)

Embelia basaal (Roem. & Schult.) A.DC.
Primulaceae
Synonyms: *Embelia robusta*
Common Name: Malabar Embelia
Climbing Mechanism: Scrambler-Unarmed
Distribution (India): Eastern Ghats, Karnataka, Maharashtra, Tamil Nadu, Western Ghats
Leaf Type: Simple, alternate below and opp above
Inflorescence: Racemes/panicles, 4–6 merous
Fruit Type: Wet, drupe, globose
Flowering and Fruiting: January–April
IUCN Status: Not evaluated
Reference: Trans. Linn. Soc. London 17: 131 (1834)

Embelia canescens Jack ex Roxb.
Primulaceae
Synonyms: *Ribesiodes canescens* (Jack ex Roxb.) Kuntze
Climbing Mechanism: Scrambler-Unarmed
Distribution (Global): Bangladesh, Malaya
Leaf Type: Simple, alternate below and opp above
Inflorescence: Racemes/panicles, 4–6 merous
Fruit Type: Wet, drupe, globose
IUCN Status: Not evaluated
Reference: Fl. Ind. 2: 292 (1824)

Embelia coriacea Wall. ex A.DC.
Primulaceae
Synonyms: *Embelia macrophylla* Blume ex Scheff., *Embelia oblongata* Miq., *Ribesiodes coriaceum* (Wall. ex A.DC.) Kuntze, *Ribesiodes macrophyllum* (Blume ex Scheff.) Kuntze, *Ribesiodes oblongatum* (Miq.) Kuntze
Climbing Mechanism: Scrambler-Unarmed
Distribution (Global): Borneo, Java, Malaya, Maluku, Philippines, Sumatera, Thailand
Leaf Type: Simple, alternate below and opp above
Inflorescence: Racemes/panicles, 4–6 merous
Fruit Type: Wet, drupe, globose
IUCN Status: Not evaluated
Reference: Trans. Linn. Soc. London 17: 135 (1834)

Embelia drupacea
Primulaceae
Climbing Mechanism: Scrambler-Unarmed
IUCN Status: Not evaluated

Embelia floribunda Wall.
Primulaceae
Synonyms: *Embelia esculenta* D.Don, *Ribesiodes floribundum* (Wall.) Kuntze, *Samara esculenta* Buch.-Ham. ex D.Don.
Climbing Mechanism: Scrambler-Unarmed
Distribution (Global): Assam, Bangladesh, China South-Central, East Himalaya, Myanmar, Nepal, Tibet, West Himalaya
Distribution (India): Arunachal Pradesh, Meghalaya, Mizoram, Sikkim, West Bengal
Leaf Type: Simple, alternate below and opp above
Inflorescence: Racemes/panicles, 4–6 merous
Fruit Type: Wet, drupe, globose
IUCN Status: Not evaluated
Reference: W.Roxburgh, Fl. Ind. 2: 291 (1824)

Embelia frondosa (King ex Gamble) D.G.Long
Primulaceae
Synonyms: *Embelia gamblei* Kurz ex C.B.Clarke, *Ribesiodes gamblei* (Kurz ex C.B. Clarke) Kuntze, *Samara frondosa* King ex Gamble
Climbing Mechanism: Scrambler-Unarmed
Distribution (Global): Assam, China South-Central, East Himalaya, Myanmar, Tibet
Distribution (India): West Bengal
Leaf Type: Simple, alternate below and opp above
Inflorescence: Racemes/panicles, 4–6 merous
Fruit Type: Wet, drupe, globose
IUCN Status: Not evaluated
Reference: Edinburgh J. Bot. 56: 307 (1999)

Embelia gardneriana Wight
Primulaceae
Synonyms: *Ribesiodes gardnerianum* (Wight) Kuntze
Climbing Mechanism: Scrambler-Unarmed
Distribution (Global): India
Distribution (India): Tamil Nadu
Leaf Type: Alternate
Inflorescence: Axillary subumbellate racemes
Fruit Type: A small globose one-seeded berry
IUCN Status: Not evaluated
Reference: Icon. Pl. Ind. Orient. 4: t. 1208 (1848)

Embelia lampanii Scheff.
Primulaceae
Climbing Mechanism: Scrambler-Unarmed
Leaf Type: Simple, alternate below and opp above
Inflorescence: Racemes/panicles, 4–6 merous
Fruit Type: Wet, drupe, globose
IUCN Status: Not evaluated
Reference: Scheff. In: Comm. Myrsin. 37. (1867)

Embelia microcalyx Kurz
Primulaceae
Synonyms: *Samara microcalyx* Kurz
Climbing Mechanism: Scrambler-Unarmed
Distribution (Global): Nicobar Is.
Distribution (India): Great Nicobar Island
Leaf Type: Simple, alternate below and opp above
Inflorescence: Racemes/panicles, 4–6 merous
Fruit Type: Wet, drupe, globose
IUCN Status: Not evaluated
Reference: J. Bot. 13: 328 (1875)

Embelia myrtillus (Hook.) Kurz
Primulaceae
Synonyms: *Myrsine myrtillus* Hook., *Ribesiodes myrtillus* (Hook.) Kuntze, *Samara myrtillus* (Hook.) Kurz.
Climbing Mechanism: Scrambler-Unarmed
Distribution (Global): Java, Malaya, Myanmar
Leaf Type: Simple, alternate below and opp above
Inflorescence: Racemes/panicles, 4–6 merous
Fruit Type: Wet, drupe, globose
IUCN Status: Not evaluated
Reference: J. Asiat. Soc. Bengal, Pt. 2, Nat. Hist. 40: 67 (1871)

Embelia nutans Wall.
Primulaceae
Synonyms: *Ribesiodes nutans* (Wall.) Kuntze, *Embelia vestita* Kurz.
Climbing Mechanism: Scrambler-Unarmed
Distribution (Global): Assam, Bangladesh
Distribution (India): Meghalaya, Mizoram
Leaf Type: Simple, alternate below and opp above
Inflorescence: Racemes/panicles, 4–6 merous
Fruit Type: Wet, drupe, globose
IUCN Status: Not evaluated
Reference: W.Roxburgh, Fl. Ind. 2: 290 (1824)

Embelia parviflora Wall. ex A.DC.
Primulaceae
Synonyms: *Embelia disticha* H.R.Fletcher, *Embelia myrtifolia* Mez, *Embelia pulchella* Mez, *Ribesiodes myrtifolium* (Mez) Kuntze, *Ribesiodes parviflorum* (Wall. ex A.DC.) Kuntze, *Samara parviflora* (Wall. ex A.DC.) Kurz.
Climbing Mechanism: Scrambler-Unarmed
Distribution (Global): Assam, Bangladesh, China South-Central, China Southeast, Hainan, Malaya, Myanmar, Sumatera, Thailand, Tibet, Vietnam
Distribution (India): Arunachal Pradesh, Meghalaya
Leaf Type: Simple, alternate below and opp above
Inflorescence: Racemes/panicles, 4–6 merous
Fruit Type: Wet, drupe, globose
IUCN Status: Not evaluated
Reference: Trans. Linn. Soc. London 17: 130 (1834)

Embelia ribes Burm.f.
Primulaceae
Synonyms: *Ribesiodes ribes* (Burm.f.) Kuntze
Samara ribes (Burm.f.) Benth. & Hook.f. ex Kurz.
Climbing Mechanism: Scrambler-Unarmed

Distribution (Global): Assam, Bangladesh, Borneo, Cambodia, China South-Central, China Southeast, East Himalaya, Hainan, India, Java, Laos, Malaya, Myanmar, Nicobar Is., Philippines, Sri Lanka, Sulawesi, Sumatera, Thailand, Tibet, Vietnam
Distribution (India): Andhra Pradesh, Arunachal Pradesh, Eastern Ghats, Great Nicobar Island, Gujarat, Karnataka, Madhya Pradesh, Maharashtra, Meghalaya, Mizoram, Tamil Nadu, Uttar Pradesh, West Bengal, Western Ghats
Leaf Type: Simple, alternate below and opp above
Inflorescence: Racemes/panicles, 4–6 merous
Fruit Type: Wet, drupe, globose
Flowering and Fruiting: March–August
IUCN Status: Not evaluated
Reference: Fl. Indica: 62 (1768)

Embelia sessiliflora Kurz
Primulaceae
Synonyms: *Embelia stricta* Craib,
Ribesiodes sessiliflorum (Kurz) Kuntze
Samara sessiliflora Kurz.
Climbing Mechanism: Scrambler-Unarmed
Distribution (Global): Assam, China South-Central, Laos, Myanmar, Thailand, Vietnam
Leaf Type: Simple, alternate below and opp above
Inflorescence: Racemes/panicles, 4–6 merous
Fruit Type: Wet, drupe, globose
IUCN Status: Not evaluated
Reference: J. Asiat. Soc. Bengal, Pt. 2, Nat. Hist. 40: 66 (1871)

Embelia subcoriacea (C.B.Clarke) Mez
Primulaceae
Climbing Mechanism: Scrambler-Unarmed
Distribution (Global): Assam, Cambodia, China South-Central, China Southeast, Hainan, Laos, Myanmar, Thailand, Vietnam
Distribution (India): Arunachal Pradesh, Meghalaya
Leaf Type: Simple, alternate below and opp above
Inflorescence: Racemes/panicles, 4–6 merous
Fruit Type: Wet, drupe, globose
IUCN Status: Not evaluated
Reference: H.G.A.Engler (ed.), Pflanzenr., IV, 236: 329 (1902)

Embelia tsjeriam-cottam (Roem. & Schult.) A.DC.
Primulaceae
Synonyms: *Ardisia basaal* Roem. & Schult., *Ardisia tsjeriam-cottam* Roem. & Schult., *Dauceria acuta* Dennst., *Dauceria obtusa* Dennst., *Embelia acuta*

(Dennst.) Alston, *Embelia acutipetala* S.M.Almeida & M.R.Almeida, *Embelia basaal* (Roem. & Schult.) A.DC.
Common Name: Malabar Embelia
Climbing Mechanism: Scrambler-Unarmed
Distribution (Global): Assam, Bangladesh, Cambodia, China South-Central, East Himalaya, India, Laos, Myanmar, Nepal, Pakistan, Sri Lanka, Thailand, Vietnam, West Himalaya
Distribution (India): Andhra Pradesh, Goa, Gujarat, Himachal Pradesh, Meghalaya, Mizoram, Tamil Nadu, Uttar Pradesh, West Bengal, Western Ghats
Leaf Type: Simple, alternate below and opp above
Inflorescence: Racemes/panicles, 4–6 merous
Fruit Type: Wet, drupe, globose
Flowering and Fruiting: February–March
IUCN Status: Not evaluated
Reference: Trans. Linn. Soc. London 17: 131 (1834)

Embelia vestita Roxb.
Primulaceae
Synonyms: *Calispermum oblongifolium* (Hemsl.) Nakai, *Calispermum rude* (Hand.-Mazz.) Nakai, *Choripetalum undulatum* A.DC., *Embelia bodinieri* H.Lév., *Embelia lenticellata* Hayata, *Embelia nagushia* D.Don, *Embelia nigroviridis* C. Chen, *Embelia oblongifolia* Hemsl.
Climbing Mechanism: Scrambler-Unarmed
Distribution (Global): Assam, Cambodia, China South-Central, China Southeast, East Himalaya, Hainan, Laos, Myanmar, Nepal, Taiwan, Thailand, Tibet, Vietnam
Distribution (India): Arunachal Pradesh, Assam, Meghalaya, Mizoram, West Bengal
Leaf Type: Simple, alternate below and opp above
Inflorescence: Racemes/panicles, 4–6 merous
Fruit Type: Wet, drupe, globose
IUCN Status: Not evaluated
Reference: Fl. Ind. 2: 288 (1824)

Embelia viridiflora (A.DC.) Scheff.
Primulaceae
Synonyms: *Ardisia viridiflora* Blume ex A.DC., *Choripetalum aurantiacum* A.DC., *Choripetalum viridiflorum* A.DC., *Ribesiodes viridiflorum* (A.DC.) Kuntze, *Samara atropunctata* Arn., *Samara rheedii* Wight, *Samara viridiflora* Arn.
Climbing Mechanism: Scrambler-Unarmed
Distribution (Global): Andaman Is., Java, Nicobar Is.
Distribution (India): Great Nicobar Island
Leaf Type: Simple, alternate below and opp above
Inflorescence: Racemes/panicles, 4–6 merous
Fruit Type: Wet, drupe, globose

IUCN Status: Not evaluated
Reference: Myrsin. Arch. Ind.: 45 (1867)

Grenacheria amentacea (C.B.Clarke) Mez
Primulaceae
Synonyms: *Embelia amentacea* C.B.Clarke
Climbing Mechanism: Scrambler-Unarmed
IUCN Status: Not evaluated
Reference: H.G.A.Engler (ed.), Pflanzenr., IV, 236: 294 (1902)

Maesa andamanica Kurz
Primulaceae
Climbing Mechanism: Scrambler-Unarmed
Distribution (Global): Andaman Is., Nicobar Is.
Distribution (India): Andaman and Nicobar Islands
IUCN Status: Not evaluated
Reference: Forest Fl. Burma 2: 575 (1877)

Aconitum elwesii Stapf
Ranunculaceae
Climbing Mechanism: Scrambler-Unarmed
Distribution (Global): Assam, East Himalaya, Nepal, Tibet
Leaf Type: Middle cauline leaves long or shortly petiolate
Inflorescence: 2–4-flowered; rachis and pedicels retrorse pubescent
Fruit Type: Dry, many-seeded
Flowering and Fruiting: September–December
IUCN Status: Not evaluated
Notes: Used in folk medicine
Reference: Ann. Roy. Bot. Gard. (Calcutta) 10: 174 (1905)

Clematis acuminata DC.
Ranunculaceae
Climbing Mechanism: Scrambler-Unarmed
Distribution (Global): Assam, China South-Central, East Himalaya, India, Myanmar, Nepal, Pakistan, West Himalaya
Distribution (India): North Eastern India
Leaf Type: Leaves opposite, pinnately 3-foliolate
Inflorescence: Axillary, trichotomously branched panicles
Fruit Type: Achenes, ovate, compressed
Flowering and Fruiting: November–April
IUCN Status: Not evaluated
Notes: Usually found in evergreen forests, hedges, thickets, and scrub jungles
Reference: Syst. Nat. 1: 148 (1817)

Clematis acutangula Hook.f. & Thomson
Ranunculaceae
Climbing Mechanism: Scrambler-Unarmed
Distribution (Global): Assam, China South-Central, East Himalaya
Distribution (India): Arunachal Pradesh, Assam, Manipur, Mizoram
Leaf Type: Compound, pinnate
Inflorescence: Axillary cyme
Fruit Type: Dry, achenes
IUCN Status: Not evaluated
Reference: Fl. Ind. 1: 5 (1855)

Clematis andersonii (C.B.Clarke ex Kuntze) H.Eichler
Ranunculaceae
Climbing Mechanism: Scrambler-Unarmed
Distribution (Global): East Himalaya
Distribution (India): West Bengal
Leaf Type: Compound, pinnate
Inflorescence: Axillary cyme
Fruit Type: Dry, achenes
IUCN Status: Not evaluated
Reference: Biblioth. Bot. 31: 47 (1958)

Clematis apiculata Hook.f. & Thomson
Ranunculaceae
Climbing Mechanism: Scrambler-Unarmed
Distribution (Global): Assam
Distribution (India): Assam, Manipur
Leaf Type: Compound, pinnate
Inflorescence: Axillary cyme
Fruit Type: Dry, achenes
IUCN Status: Not evaluated
Reference: J. D. Hooker, Fl. Brit. India 1: 4 (1872)

Clematis barbellata Edgew.
Ranunculaceae
Common Name: Brown Clematis
Climbing Mechanism: Scrambler-Unarmed
Distribution (Global): Nepal, Pakistan, Tibet, West Himalaya
Distribution (India): Himachal Pradesh
Leaf Type: Ttrifoliate, with 3 ovate-lance-shaped leaflets
Inflorescence: Purple, bell-shaped flowers
Flowering and Fruiting: June–July
IUCN Status: Not evaluated
Reference: Trans. Linn. Soc. London 20: 25 (1846)

Clematis bourdillonii Dunn
Ranunculaceae
Climbing Mechanism: Scrambler-Unarmed
Distribution (Global): India
Distribution (India): Kerala, Tamil Nadu
Leaf Type: 1- or 2-ternate
Inflorescence: Axillary/Terminal panicles
Fruit Type: Achenes with feathery tails
Flowering and Fruiting: April–August
IUCN Status: Not evaluated
Notes: Usually found in evergreen forests
Reference: Bull. Misc. Inform. Kew 1914: 181 (1914)

Clematis buchananiana DC.
Ranunculaceae
Synonyms: *Clematis bucamara* Buch.-Ham. ex DC., *Clematis buchananiana var. sericea* S.K.Agarwal, *Clematis buchananii* D.Don, *Clematis loasifolia* D.Don, *Clematis wattii* Drumm. & Craib.
Common Name: Lemon Clematis, Fragrant Chinese Clematis
Climbing Mechanism: Scrambler-Unarmed
Distribution (Global): Assam, Bangladesh, China South-Central, East Himalaya, India, Myanmar, Nepal, Pakistan, Thailand, Tibet, Vietnam, West Himalaya
Distribution (India): Arunachal Pradesh, Assam, Himachal Pradesh, Manipur, Meghalaya, Mizoram, Sikkim, West Bengal
Leaf Type: Opposite, ternate or rarely pinnate
Inflorescence: Cymes axillary
Fruit Type: Achenes with long silky tails
Flowering and Fruiting: May–January
IUCN Status: Not evaluated
Notes: The paste of the roots is used as a poultice to treat swellings caused by inflammation
Reference: Syst. Nat. 1: 140 (1817)

Clematis cadmia Buch.-Ham. ex Hook.f. & Thomson
Ranunculaceae
Synonyms: *Clematis bracteata* (Roxb.) Kurz., *Clematis stronachii* Hance., *Clematis sulcata* Wall., *Thalictrum bracteatum* Roxb.
Climbing Mechanism: Scrambler-Unarmed
Distribution (Global): Assam, Bangladesh, China South-Central, China Southeast, East Himalaya, Myanmar, Vietnam
Distribution (India): Arunachal Pradesh, Assam, Manipur, West Bengal
Leaf Type: Opposite, ternately or biternately compound
Inflorescence: Solitary, axillary
Fruit Type: Achenes
Flowering and Fruiting: February–June

IUCN Status: Not evaluated
Notes: Usually found in evergreen forests
Reference: J. D. Hooker, Fl. Brit. India 1: 2 (1872)

Clematis campestris A.St.-Hil.
Ranunculaceae
Synonyms: *Clematis bangii* Rusby, *Clematis campestris var. malacophylla* W.T. Wang, *Clematis campestris var. mendocina* (Phil.) Hauman & Irigoyen, *Clematis denticulata* Vell., *Clematis dioica var. angustissima* Kuntze, *Clematis dioica* subsp. *campestris* (A.St.-Hil.)
Climbing Mechanism: Scrambler-Unarmed
Distribution (Global): Argentina Northeast, Argentina Northwest, Argentina South, Bolivia, Brazil Northeast, Brazil South, Brazil Southeast, Brazil West-Central, Chile North, Paraguay, Peru, Uruguay
Leaf Type: Compound, pinnate
Inflorescence: Axillary cyme
Fruit Type: Dry, achenes
IUCN Status: Not evaluated
Reference: Fl. Bras. Merid. 1: 4 (1824)

Clematis connata DC.
Ranunculaceae
Climbing Mechanism: Scrambler-Unarmed
Distribution (Global): China South-Central, East Himalaya, Nepal, Pakistan, Tibet, West Himalaya
Distribution (India): North Eastern India, West Bengal
Leaf Type: Compound, pinnate
Inflorescence: Axillary cyme
Fruit Type: Dry, achenes
IUCN Status: Not evaluated
Reference: Prodr. 1: 4 (1824)

Clematis flammula L.
Ranunculaceae
Synonyms: *Anemone flammula* (L.) E.H.L.Krause
Common Name: Fragrant virgin's bower, fragrant Clematis, Sweet-scented virgin's bower
Climbing Mechanism: Scrambler-Unarmed
Distribution (Global): Afghanistan, Albania, Algeria, Baleares, Bulgaria, Corse, France, Greece, Iran, Italy, Lebanon-Syria, Libya, Morocco, Pakistan, Palestine, Portugal, Sardegna, Sicilia, Spain, Transcaucasus, Tunisia, Turkey, Turkey-in-Europe, Yugoslavia
Distribution (India): Maharashtra
Leaf Type: Oppositely arranged leaves are double-compound
Fruit Type: Achenes

Flowering and Fruiting: January
IUCN Status: Not evaluated
Notes: It is popular with gardeners as a decoration along fences and trellises, or as ground cover
Reference: Sp. Pl.: 544 (1753)

Clematis florida
Ranunculaceae
Climbing Mechanism: Scrambler-Unarmed
IUCN Status: Not evaluated

Clematis glauca Willd.
Ranunculaceae
Synonyms: *Clematis daurica* Pers., *Clematis davurica* Ledeb., *Clematis gebleriana* Bong., *Clematis orientalis* L.
Climbing Mechanism: Scrambler-Unarmed
Distribution (Global): Altay, China North-Central, Kazakhstan, Kirgizstan, Krasnoyarsk, Mongolia, Qinghai, Tuva, Xinjiang
Leaf Type: Compound, pinnate
Inflorescence: Axillary cyme
Fruit Type: Dry, achenes
IUCN Status: Not evaluated
Reference: Berlin. Baumz.: 65 (1796)

Clematis gouriana Roxb. ex DC.
Ranunculaceae
Common Name: Goat's beard, Indian Clematis, Gourian Clematis, Indian travelers's joy.
Climbing Mechanism: Scrambler-Unarmed
Distribution (Global): Assam, Bangladesh, China North-Central, China South-Central, China Southeast, East Himalaya, India, Myanmar, Nepal, New Guinea, Pakistan, Philippines, Sri Lanka, Thailand, Vietnam, West Himalaya
Distribution (India): Andhra Pradesh, Arunachal Pradesh, Assam, Goa, Gujarat, Himachal Pradesh, Jharkhand, Kerala, Madhya Pradesh, Maharashtra, Manipur, Meghalaya, Mizoram, Odisha, Tamil Nadu, Tripura
Leaf Type: Oposite, pinnate, 2-pinnate, or 3-pinnate
Inflorescence: Terminal and axillary, tomentose
Fruit Type: Achenes
Flowering and Fruiting: November–April
IUCN Status: Not evaluated
Notes: Usually spreading on thickets and trees along forest borders
Reference: Syst. Nat. 1: 138 (1817)

Clematis gouriana var. mollifolia
Ranunculaceae

Climbing Mechanism: Scrambler-Unarmed
Distribution (Global): India.
Distribution (India): Tamil Nadu
Leaf Type: Compound, pinnate
Inflorescence: Axillary cyme
Fruit Type: Dry, achenes
IUCN Status: Not evaluated
Reference: Acta Phytotax. Sin. 41: 39 (2003)

Clematis grata Wall.
Ranunculaceae
Synonyms: *Clematis cordata* Royle, *Clematis taiwaniana* Hayata, *Clematis taiwaniana var. ryukiuensis* (Tamura) Tamura
Common Name: Charming Clematis.
Climbing Mechanism: Scrambler-Unarmed
Distribution (Global): Afghanistan, East Himalaya, Myanmar, Nansei-shoto, Nepal, Pakistan, Taiwan, Tibet, West Himalaya
Distribution (India): Arunachal Pradesh, Himachal Pradesh
Leaf Type: Leaves are compound with usually 5 leaflets
Flowering and Fruiting: July–September
IUCN Status: Not evaluated
Reference: Pl. Asiat. Rar. 1: 83 (1830)

Clematis graveolens Lindl.
Ranunculaceae
Synonyms: *Clematis parvifolia* Hook.f. & Thomson, *Clematis parvifolia* Edgew.
Climbing Mechanism: Scrambler-Unarmed
Distribution (Global): Afghanistan, Nepal, Pakistan, West Himalaya
Distribution (India): Himachal Pradesh
Leaf Type: Compound, pinnate
Inflorescence: Axillary cyme
Fruit Type: Dry, achenes
IUCN Status: Not evaluated
Reference: J. Hort. Soc. London 1: 307 (1846)

Clematis grewiiflora DC.
Ranunculaceae
Synonyms: *Clematis loasifolia* DC.
Climbing Mechanism: Scrambler-Unarmed
Distribution (Global): Assam, Bangladesh, East Himalaya, Myanmar, Nepal, Tibet, West Himalaya
Distribution (India): Arunachal Pradesh, Mizoram, West Bengal
Leaf Type: Opposite, pinnate; leaflets 3–5 foliolate
Inflorescence: Axillary, many flowered, trichotomously branched
Fruit Type: Achenes, Ovate

Flowering and Fruiting: October–April.
IUCN Status: Not evaluated
Notes: Usually found in evergreen forests
Reference: Syst. Nat. 1: 140 (1817)

Clematis hedysarifolia DC.
Ranunculaceae
Synonyms: *Clematis naravelioides* Kuntze
Common Name: Burman Clematis, Burman virgin's-bower, hedysarum-leaved clematis
Climbing Mechanism: Scrambler-Unarmed
Distribution (Global): India
Distribution (India): Goa, Gujarat, Maharashtra
Leaf Type: Opposite, ternately pinnate
Inflorescence: Axillary, lower ones paniculate
Fruit Type: Achenes, Ovate
Flowering and Fruiting: September–April
IUCN Status: Not evaluated
Notes: Usually found in moist deciduous and semi-Evergreen forests
Reference: Syst. Nat. 1: 148 (1817)

Clematis heynei M.A.Rau & al.
Ranunculaceae
Synonyms: *Clematis triloba* B.Heyne ex Roth
Common Name: Deccan Clematis
Climbing Mechanism: Scrambler-Unarmed
Distribution (Global): India
Distribution (India): Maharashtra
Leaf Type: Simple or trifoliate
Inflorescence: Panicles
Flowering and Fruiting: August–September
IUCN Status: Not evaluated
Notes: Rarely found in open thickets and on Fort walls
Reference: Fl. India 1: 67 (1993)

Clematis jackmanii T.Moore
Ranunculaceae
Synonyms: *Clematis contorta* Carrière, *Clematis intermedia* Carrière, *Clematis jackmanii var. rubroviolacea* Regel.
Climbing Mechanism: Scrambler-Unarmed
Leaf Type: Oppositely pinnately compound
IUCN Status: Not evaluated
Reference: Proc. Hort. Soc. London 3: 364 (1863)

Clematis javana DC.
Ranunculaceae
Synonyms: *Clematis biternata* DC., *Clematis javanica* Steud., *Clematis junghuhniana* de Vriese, *Clematis virginica* Thunb.
Climbing Mechanism: Scrambler-Unarmed
Distribution (Global): Assam, East Himalaya, Java, Lesser Sunda Is., Maluku, Myanmar, New Guinea, Philippines, Thailand, Vietnam
Leaf Type: Compound, pinnate
Inflorescence: Axillary cyme
Fruit Type: Dry, achenes
IUCN Status: Not evaluated
Reference: Syst. Nat. 1: 152 (1817)

Clematis loureiroana DC.
Ranunculaceae
Synonyms: *Clematis dioica* Lour., *Clematis filamentosa* Dunn.
Climbing Mechanism: Scrambler-Unarmed
Distribution (Global): China Southeast, Hainan, Vietnam
Distribution (India): Meghalaya
Leaf Type: Compound, pinnate
Inflorescence: Axillary cyme
Fruit Type: Dry, achenes
IUCN Status: Not evaluated
Reference: Syst. Nat. 1: 144 (1817)

Clematis meyeniana Walp.
Ranunculaceae
Climbing Mechanism: Scrambler-Unarmed
Distribution (Global): China South-Central, China Southeast, Hainan, Laos, Myanmar, Philippines, Taiwan, Thailand, Vietnam
Distribution (India): Arunachal Pradesh
Leaf Type: Compound, pinnate
Inflorescence: Axillary cyme
Fruit Type: Dry, achenes
IUCN Status: Not evaluated
Reference: Nov. Actorum Acad. Caes. Leop.-Carol. Nat. Cur. 19(Suppl. 1): 297 (1843)

Clematis munroiana Wight
Ranunculaceae
Climbing Mechanism: Scrambler-Unarmed
Distribution (Global): India
Distribution (India): Kerala, Tamil Nadu
Leaf Type: Broadly ovate
Fruit Type: Achenes

Flowering and Fruiting: June–October
IUCN Status: Not evaluated
Notes: Mostly found in evergreen and shola forests
Reference: Ill. Ind. Bot. 1: 5 (1840)

Clematis napaulensis DC.
Ranunculaceae
Synonyms: *Clematis forrestii* W.W.Sm., *Clematis montana* D.Don.
Climbing Mechanism: Scrambler-Unarmed
Distribution (Global): China South-Central, East Himalaya, India, Myanmar, Nepal, Tibet, West Himalaya
Distribution (India): West Bengal
Leaf Type: Compound, pinnate
Inflorescence: Axillary cyme
Fruit Type: Dry, achenes
IUCN Status: Not evaluated
Reference: Syst. Nat. 1: 164 (1817)

Clematis orientalis L.
Ranunculaceae
Synonyms: *Clematis aurea* A.Nelson & J.F.Macbr., *Clematis baltistanica* Qureshi & Chaudhri, *Clematis dahurica* DC., *Clematis flava* (Moench) DC.
Climbing Mechanism: Scrambler-Unarmed
Distribution (Global): Afghanistan, Altay, Buryatiya, China North-Central, East Aegean Is., Iran, Iraq, Kazakhstan, Kirgizstan, Mongolia, Nepal, North Caucasus, Oman, Pakistan, South European Russi, Tadzhikistan, Transcaucasus, Turkey, Turkmenistan, Tuva, Ukraine, Uzbekistan, West Himalaya, Xinjiang
Distribution (India): Odisha
Leaf Type: Compound, pinnate
Inflorescence: Axillary cyme
Fruit Type: Dry, achenes
IUCN Status: Not evaluated
Reference: Sp. Pl.: 543 (1753)

Clematis pettimudiensis K.M.P. Kumar, Jagadeesan & G. Prasad
Ranunculaceae
Climbing Mechanism: Scrambler-Unarmed
Distribution (Global): India
Distribution (India): Kerala
Leaf Type: Compound, pinnate
Inflorescence: Axillary cyme
Fruit Type: Dry, achenes
IUCN Status: Not evaluated
Reference: Phytotaxa 326: 289 (2017)

Clematis puberula Hook.f. & Thomson
Ranunculaceae
Climbing Mechanism: Scrambler-Unarmed
Distribution (Global): Assam, China North-Central, China South-Central, China Southeast, East Himalaya, Myanmar, Nepal, Tibet, West Himalaya
Distribution (India): Assam, Himachal Pradesh, Manipur, West Bengal
Leaf Type: Compound, pinnate
Inflorescence: Axillary cyme
Fruit Type: Dry, achenes
IUCN Status: Not evaluated
Reference: J. D. Hooker, Fl. Brit. India 1: 4 (1872)

Clematis recta L.
Ranunculaceae
Synonyms: *Anemone recta* (L.) E.H.L.Krause, *Clematis bracteosa* Banks ex Steud., *Clematis corymbosa* Poir., *Clematis hispanica* Mill., *Clematis tenuiflora* DC., *Clematis umbraticola* Schur, *Clematitis erecta* (L.) Couret-Vill.
Common Name: Spider lilly, Bridal bouquet
Climbing Mechanism: Scrambler-Unarmed
Distribution (Global): Albania, Austria, Belarus, Bulgaria, Central European Rus, Corse, Czechoslovakia, France, Germany, Hungary, Italy, Krym, North Caucasus, Poland, Romania, South European Russi, Spain, Switzerland, Transcaucasus, Turkey, Ukraine, Yugoslavia
Distribution (India): Peninsular India
Leaf Type: Compound, pinnate
Inflorescence: Axillary cyme
Fruit Type: Dry, achenes
Flowering and Fruiting: Throughout the year
IUCN Status: Not evaluated
Notes: Grown as a garden plant
Reference: Sp. Pl.: 544 (1753)

Clematis roylei Rehder
Ranunculaceae
Synonyms: *Clematis nutans* Royle
Common Name: Royle's Clematis
Climbing Mechanism: Scrambler-Unarmed
Distribution (Global): Assam, Bangladesh, India, Myanmar, Nepal, West Himalaya
Distribution (India): Andhra Pradesh, Himachal Pradesh, Madhya Pradesh, Manipur, Odisha
Leaf Type: Compound
Flowering and Fruiting: September–April
IUCN Status: Not evaluated
Notes: Found usually in shaded environments in evergreen forests
Reference: J. Arnold Arbor. 22: 575 (1941)

Clematis simplicifolia Qureshi & Chaudhri
Ranunculaceae
Synonyms: *Clematis gouriana var. gouriana.*, *Clematis cana* Wall., *Clematis indica* B.Heyne ex Roth, *Clematis martini* H.Lév., *Clematis substipulata* Kuntze
Climbing Mechanism: Scrambler-Unarmed
Distribution (Global): Assam, Bangladesh, China North-Central, China South-Central, China Southeast, East Himalaya, India, Myanmar, Nepal, New Guinea, Pakistan, Philippines, Sri Lanka, Thailand, Vietnam, West Himalaya
Distribution (India): Maharashtra
Leaf Type: Compound, pinnate
Inflorescence: Axillary cyme
Fruit Type: Dry, achenes
IUCN Status: Not evaluated

Clematis smilacifolia subsp. andamanensis
Ranunculaceae
Climbing Mechanism: Scrambler-Unarmed
Leaf Type: Compound, pinnate
Inflorescence: Axillary cyme
Fruit Type: Dry, achenes
IUCN Status: Not evaluated

Clematis smilacifolia Wall.
Ranunculaceae
Climbing Mechanism: Scrambler-Unarmed
Distribution (India): Andaman and Nicobar Islands, Andhra Pradesh, Arunachal Pradesh, Assam, Bihar, Odisha, Eastern Ghats, Great, Nicobar, Island, Karnataka, Madhya Pradesh, Mizoram, Odisha, Tamil Nadu
Leaf Type: Compound, pinnate
Inflorescence: Axillary cyme
Fruit Type: Dry, achenes
IUCN Status: Not evaluated
Reference: Asiat. Res. 13: 402 (1820)

Clematis tangutica (Maxim.) Korsh.
Ranunculaceae
Climbing Mechanism: Scrambler-Unarmed
Distribution (Global): Afghanistan, China North-Central, China South-Central, Inner Mongolia, Kazakhstan, Kirgizstan, Mongolia, Qinghai, Tadzhikistan, Tibet, West Himalaya, Xinjiang
Leaf Type: Compound, pinnate
Inflorescence: Axillary cyme
Fruit Type: Dry, achenes
IUCN Status: Not evaluated
Reference: Izv. Imp. Akad. Nauk, ser. 5, 9: 399 (1903)

Clematis theobromina Dunn
Ranunculaceae
Climbing Mechanism: Scrambler-Unarmed
Distribution (Global): India
Distribution (India): Tamil Nadu
Leaf Type: Compound, pinnate
Inflorescence: Axillary cyme
Fruit Type: Dry, achenes
IUCN Status: Not evaluated
Notes: Endemic to the Western Ghats
Reference: Bull. Misc. Inform. Kew 1914: 181 (1914)

Clematis tibetana Kuntze
Ranunculaceae
Climbing Mechanism: Scrambler-Unarmed
Distribution (Global): Nepal, Pakistan, Tibet, West Himalaya, Xinjiang
Distribution (India): Madhya Pradesh
Leaf Type: Compound, pinnate
Inflorescence: Axillary cyme
Fruit Type: Dry, achenes
IUCN Status: Not evaluated
Reference: Verh. Bot. Vereins Prov. Brandenburg 26: 172 (1885)

Clematis tortuosa
Ranunculaceae
Climbing Mechanism: Scrambler-Unarmed
IUCN Status: Not evaluated

Clematis wightiana Wall.
Ranunculaceae
Common Name: Wight's Clematis
Climbing Mechanism: Scrambler-Unarmed
Distribution (India): Andhra Pradesh, Eastern Ghats, Kerala, Maharashtra, Manipur, Mizoram, Odisha, Tamil Nadu, Telangana
Leaf Type: 3–5 foliolate
Inflorescence: Panicles
IUCN Status: Not evaluated
Reference: Wall. In: Cat. no. 4674 (1831) ex Wight & Arn., Prodr. 2. (1834).

Clematis zeylanica Poir.
Ranunculaceae
Synonyms: *Atragene lobata* Llanos, *Atragene zeylanica* L., *Naravelia lobata* (Llanos) Merr., *Naravelia zeylanica* (L.) DC.
Climbing Mechanism: Scrambler-Unarmed

Distribution (Global): Assam, Bangladesh, China South-Central, East Himalaya, India, Laos, Myanmar, Nepal, Sri Lanka, Thailand, Vietnam
Distribution (India): Tamil Nadu
Leaf Type: Leaves are 2–3-foliolate
Inflorescence: Panicles, terminal and axillary
Fruit Type: Achenes
Flowering and Fruiting: October–April
IUCN Status: Not evaluated
Notes: Ceylon Clematis is used in Ayurveda. Vine is crushed and inhaled to cure headache
Reference: J.B.A.M.de Lamarck, Encycl., Suppl. 2: 296 (1811)

Naravelia laurifolia Wall.
Ranunculaceae
Climbing Mechanism: Tendril Climber
Distribution (India): Great Nicobar Island
IUCN Status: Not evaluated

Naravelia zeylanica (L.) DC.
Ranunculaceae
Climbing Mechanism: Scrambler-Unarmed
Distribution (India): Andhra Pradesh, Arunachal Pradesh, Assam, Bihar, Odisha, Eastern Ghats, Goa, Jharkhand, Karnataka, Kerala, Madhya Pradesh, Maharashtra, Manipur, Meghalaya, Mizoram, Odisha, Tamil Nadu, Telangana, Tripura, Uttar Pradesh, West Bengal
IUCN Status: Not evaluated

Stixis scandens Lour.
Resedaceae
Synonyms: *Roydsia parviflora* Griff., *Stixis manipurensis* Deb & Rout.
Climbing Mechanism: Scrambler-Unarmed
Distribution (Global): Assam, Cambodia, China South-Central, India, Laos, Myanmar, Vietnam
Distribution (India): Assam.
Leaf Type: Simple, alternate
Inflorescence: Terminal or axillary
Fruit Type: Wet, drupe
IUCN Status: Not evaluated
Reference: Fl. Cochinch.: 295 (1790)

Stixis suaveolens (Roxburgh) Pierre
Resedaceae
Synonyms: *Roydsia suaveolens* Roxb., *Stixis suaveolens var. cochinsinensis* Pierre
Common Name: Fragrant caper vine
Climbing Mechanism: Scrambler-Unarmed

Distribution (Global): Assam, Bangladesh, Cambodia, China South-Central, China Southeast, East Himalaya, Hainan, India, Laos, Myanmar, Nepal, Thailand, Tibet, Vietnam
Distribution (India): Andaman and Nicobar Islands, Tripura, West Bengal
Leaf Type: Leaves simple, alternate
Inflorescence: Supra-axillary, racemes, or terminal panicles
Fruit Type: Drupe, ellipsoid or obovoid, pale orange when ripe
Flowering and Fruiting: February–October
IUCN Status: Not evaluated
Notes: Found mostly in mixed tropical and subtropical forests, often along streams
Reference: Bull. Mens. Soc. Linn. Paris 1: 654 (1887)

Berchemia flavescens (Wall.) Wall. ex Brongn.
Rhamnaceae
Synonyms: *Berchemia hypochrysa* C.K.Schneid., *Ziziphus flavescens* Wall.
Climbing Mechanism: Scrambler-Unarmed
Distribution (Global): China North-Central, China South-Central, East Himalaya, Myanmar, Nepal, Tibet
Distribution (India): West Bengal
Leaf Type: Simple, ovate-elliptic, papery
Inflorescence: Fasciles in narrow cymose panicles
Fruit Type: Wet, drupe
IUCN Status: Not evaluated
Reference: Mém. Fam. Rhamn.: 50 (1826)

Berchemia floribunda (Wall.) Brongn.
Rhamnaceae
Synonyms: *Berchemia fagifolia* Koidz., *Berchemia giraldiana* C.K.Schneid., *Berchemia kunitakei* Uyeki, *Berchemia laxa* Wall., *Berchemia lineata* Benth.
Climbing Mechanism: Scrambler-Unarmed
Distribution (Global): Assam, Bangladesh, China North-Central, China South-Central, China Southeast, East Himalaya, India, Japan, Korea, Laos, Myanmar, Nansei-shoto, Nepal, Thailand, Tibet, Vietnam, West Himalaya
Distribution (India): Arunachal Pradesh, Assam, Himachal Pradesh, Manipur, Meghalaya, West Bengal
Leaf Type: Simple, opposite
Fruit Type: Drupe
Flowering and Fruiting: April–July
IUCN Status: Least Concern
Reference: Mém. Fam. Rhamn.: 50 (1826)

Colubrina asiatica (L.) Brongn.
Rhamnaceae
Synonyms: *Ceanothus asiaticus* L., *Ceanothus capsularis* G.Forst.
Climbing Mechanism: Scrambler-Unarmed

Distribution (Global): Aldabra, Andaman Is., Bismarck Archipelago, Borneo, Cambodia, Caroline Is., China Southeast, Comoros, Cook Is., Fiji, Hainan, Hawaii, India, Java, Kazan-retto, Kenya, Laccadive Is., Lesser Sunda Is., Madagascar, Malaya, Maldives, Maluku, Marianas, Marquesas, Marshall Is., Mozambique, Mozambique Channel I, Myanmar, Nansei-shoto, Nauru, New Caledonia, New Guinea, Nicobar Is., Niue, Philippines, Queensland, Samoa, Seychelles, Society Is., Solomon Is., South China Sea, Sri Lanka, Sulawesi, Sumatera, Taiwan, Tanzania, Thailand, Tokelau-Manihiki, Tonga, Tuamotu, Tubuai Is., Vanuatu, Vietnam, Wallis-Futuna Is.
Distribution (India): Andaman and Nicobar Islands, Andhra Pradesh, Goa, Great Nicobar Island
Leaf Type: Alternate, ovate
Inflorescence: Cymes
Fruit Type: Capsule, globose
Flowering and Fruiting: December–May
IUCN Status: Least Concern
Notes: Grown as a hedge plant
Reference: Mém. Fam. Rhamn.: 62 (1826)

Gouania andamanica King
Rhamnaceae
Climbing Mechanism: Tendril Climber
Distribution (Global): Andaman Is.
Distribution (India): Andaman and Nicobar Islands
IUCN Status: Not evaluated
Reference: J. Asiat. Soc. Bengal, Pt. 2, Nat. Hist. 65: 382 (1896)

Gouania andamanica var. brevialata
Rhamnaceae
Climbing Mechanism: Scrambler-Unarmed
IUCN Status: Not evaluated
Reference: Bhandari & Bhansali. In: Fasc. Fl. India, 20: 37. (1990)

Gouania leptostachya DC.
Rhamnaceae
Synonyms: *Gouania leptostachya var. nainitalensis* Bhandari & Bhansali, *Naegelia dubia* Zoll. & Moritzi, *Gouania tiliifolia* Roxb.
Common Name: Slender spiked Gouania
Climbing Mechanism: Tendril Climber
Distribution (Global): Andaman Is., Assam, Bangladesh, East Himalaya, India, Java, Laos, Malaya, Myanmar, Nepal, Nicobar Is., Sumatera, Thailand, Vietnam
Distribution (India): Andaman and Nicobar Islands, Andhra Pradesh, Assam, Bihar, Odisha, Eastern Ghats, Maharashtra, Mizoram, Odisha
Leaf Type: Simple, alternate
Inflorescence: 3–8 flowered clusters or in terminal panicles

Fruit Type: Drupe
Flowering and Fruiting: August–December
IUCN Status: Not evaluated
Reference: Prodr. 2: 40 (1825)

Gouania microcarpa DC.
Rhamnaceae
Synonyms: *Gouania brandisii* Hassk., *Gouania domingensis* Blanco, *Gouania integrifolia* Kurz, *Gouania leptostachya* Villar, *Gouania tiliifolia* Rottb. ex DC.
Common Name: Tiny-fruit chawstick
Climbing Mechanism: Tendril Climber
Distribution (Global): Assam, Bismarck Archipelago, India, Myanmar, New Guinea, Sri Lanka, West Himalaya
Distribution (India): Karnataka, Maharashtra, Mizoram, Tamil Nadu
Flowering and Fruiting: October–March
IUCN Status: Not evaluated
Reference: Prodr. 2: 40 (1825)

Gouania napalensis Wall.
Rhamnaceae
Climbing Mechanism: Tendril Climber
Distribution (Global): Assam, Bangladesh, East Himalaya, India, Nepal
Distribution (India): Arunachal Pradesh, Tamil Nadu, West Bengal
Leaf Type: Simple, alternate
Inflorescence: Racemes
Fruit Type: Capsule
Flowering and Fruiting: June–November
IUCN Status: Not evaluated
Reference: W.Roxburgh, Fl. Ind. 2: 417 (1824)

Gouania tiliifolia Lam.
Rhamnaceae
Climbing Mechanism: Tendril Climber
Distribution (Global): Madagascar, Mauritius, Réunion
Distribution (India): Andaman and Nicobar Islands, Arunachal Pradesh, Assam, Manipur, Meghalaya, Tripura, West Bengal
IUCN Status: Not evaluated
Reference: Encycl. 3: 4 (1789)

Helinus lanceolatus Brandis
Rhamnaceae
Synonyms: *Mystacinus lanceolatus* (Brandis) Kuntze, *Gouania lanceolata* Wall.
Common Name: Lanceleaf Helinus
Climbing Mechanism: Tendril Climber

Distribution (Global): Bangladesh, India, Myanmar, Nepal, Pakistan, West Himalaya
Distribution (India): Andhra Pradesh, Bihar, Odisha, Eastern Ghats, Himachal Pradesh, Madhya Pradesh, Maharashtra
IUCN Status: Not evaluated
Reference: Forest Fl. N.W. India: 574 (1874)

Rhamnus napalensis (Wall.) M.A.Lawson
Rhamnaceae
Synonyms: *Celastrus tristis* H.Lév., *Rhamnus paniculiflora* C.K.Schneid., *Ceanothus napalensis* Wall.
Climbing Mechanism: Scrambler-Unarmed
Distribution (Global): Andaman Is., Assam, Bangladesh, Cambodia, China South-Central, China Southeast, Christmas I., Comoros, East Himalaya, Hainan, India, Java, Laos, Lesser Sunda Is., Madagascar, Malaya, Myanmar, Nepal, New Guinea, Nicobar Is., Philippines, Society Is., Sulawesi, Thailand, Tibet, Vietnam
Distribution (India): Andhra Pradesh, Assam, Bihar, Odisha, Mizoram, Tripura
Leaf Type: Simple, alternate, margin serrulate, recurved
Inflorescence: Axillary spikes
Fruit Type: Wet, drupe
Flowering and Fruiting: June–February
IUCN Status: Least Concern
Reference: J. D. Hooker, Fl. Brit. India 1: 640 (1875)

Rhamnus wightii Wight & Arn.
Rhamnaceae
Synonyms: *Ceanothus wightianus* Wall.
Common Name: Wight's buckthorn
Climbing Mechanism: Scrambler-Unarmed
Distribution (Global): India, Nepal, Sri Lanka
Distribution (India): Tamil Nadu, Western Ghats
Leaf Type: Simple, alternate, margin serrulate, recurved
Inflorescence: Axillary spikes
Fruit Type: Wet, drupe
Flowering and Fruiting: January–May
IUCN Status: Not evaluated
Notes: Mostly found in montane scrub jungles and shola forests
Reference: Prodr. Fl. Ind. Orient. 1: 164 (1834)

Sageretia filiformis (Roth) G.Don
Rhamnaceae
Synonyms: *Berchemia oppositifolia* Wall., *Rhamnus filiformis* Rot., *Rhamnus parviflora* J.G.Klein ex Willd., Rhamnus terminalis Buch.-Ham. ex D.Don., *Rhamnus trigyna* Buch.-Ham. ex D.Don., *Sageretia blumei* G.Don., *Sageretia oppositifolia* (Wall.) Brongn., *Sageretia par*

Climbing Mechanism: Scrambler-Armed
Distribution (Global): Andaman Is., Assam, East Himalaya, India, Myanmar, Nepal, Thailand, West Himalaya
Distribution (India): Eastern Ghats, Tamil Nadu
Leaf Type: Simple, opposite
Inflorescence: Simple/compound spikes
Fruit Type: Wet, drupes globose
Flowering and Fruiting: September–March
IUCN Status: Not evaluated
Notes: Occasionally found in evergreen forests
Reference: Gen. Hist. 2: 29 (1832)

Sageretia hamosa (Wall.) Brongn.
Rhamnaceae
Synonyms: *Ziziphus hamosa* Wall.
Climbing Mechanism: Scrambler-Armed
Distribution (Global): Assam, China South-Central, China Southeast, India, Lesser Sunda Is., Myanmar, Nepal, Philippines, Sri Lanka, Tibet, Vietnam
Distribution (India): Assam, Kerala, Tamil Nadu, Western Ghats
Leaf Type: Simple, opposite
Inflorescence: Simple/compound spikes
Fruit Type: Wet, drupes globose
Flowering and Fruiting: August–December
IUCN Status: Not evaluated
Notes: Often found in semi-evergreen and evergreen forests
Reference: Mém. Fam. Rhamn.: 53 (1826)

Sageretia parviflora G.Don
Rhamnaceae
Climbing Mechanism: Scrambler-Armed
Distribution (India): Kerala, Karnataka, Tamil Nadu
Leaf Type: Simple, opposite
Inflorescence: Simple/compound spikes
Fruit Type: Wet, drupes globose
IUCN Status: Not evaluated
Reference: Gen. Hist. 2: 29 (1832)

Scutia indica Brongn.
Rhamnaceae
Climbing Mechanism: Scrambler-Armed
IUCN Status: Not evaluated

Scutia myrtina (Burm.f.) Kurz
Rhamnaceae

Synonyms: *Catha zeylanica* (Roth) G.Don., *Adolia alba* Lam., *Adolia capensis* Kuntze
Common Name: Cat thorn, Droog-my-keel
Climbing Mechanism: Scrambler-Armed
Distribution (Global): Aldabra, Burundi, Cape Provinces, China South-Central, China Southeast, Ethiopia, India, Kenya, KwaZulu-Natal, Madagascar, Malawi, Mauritius, Mozambique, Myanmar, Northern Provinces, Rodrigues, Rwanda, Réunion, Sri Lanka, Sudan, Swaziland, Tanzania, Thailand, Uganda, Vietnam, Zambia, Zaire, Zimbabwe
Distribution (India): Andhra Pradesh, Kerala, Puducherry, Tamil Nadu, Western Ghats
Leaf Type: Oppositely arranged leaves
Inflorescence: Axillary-condensed cymes, flowers are yellow-green
Fruit Type: Spherical, smooth
Flowering and Fruiting: February-November
IUCN Status: Least Concern
Notes: Usually occur in scrub jungles
Reference: J. Asiat. Soc. Bengal, Pt. 2, Nat. Hist. 44: 168 (1875)

Scutia myrtina var. emarginata
Rhamnaceae
Climbing Mechanism: Scrambler-Armed
Distribution (Global): Andaman Is.
Flowering and Fruiting: March–June
IUCN Status: Not evaluated
Notes: Often found in dry deciduous forests
Reference: Fasc. Fl. India 20: 71 (1990)

Smythea bombaiensis (Dalzell) S.P. Banerjee & P.K. Mukh.
Rhamnaceae
Synonyms: *Ventilago bombaiensis* Dalzell
Common Name: Bombay Smythea, Bombay Ventilago
Climbing Mechanism: Scrambler-Unarmed
Distribution (Global): India.
Distribution (India): Maharashtra, Tamil Nadu
Leaf Type: Simple, glabrous
Inflorescence: Fascicles
Fruit Type: Dry, capsule
Flowering and Fruiting: February–April
IUCN Status: Not evaluated
Notes: Occurs mainly in evergreen and semi-evergreen forests
Reference: Indian Forester 96: 214 (1970)

Smythea calpicarpa Kurz
Rhamnaceae

Climbing Mechanism: Scrambler-Unarmed
Distribution (Global): Assam, Myanmar
Distribution (India): Andaman and Nicobar Islands, Assam
Leaf Type: Simple, glabrous
Inflorescence: Fascicles
Fruit Type: Dry, capsule
IUCN Status: Not evaluated
Reference: J. Asiat. Soc. Bengal, Pt. 2, Nat. Hist. 41: 301 (1872)

Smythea lanceata Summerh.
Rhamnaceae
Synonyms: *Berchemia trichantha* Miq., *Smythea dupontii* Hemsl., *Smythea novoguineensis* Scheff., *Smythea pacifica* Seem., *Smythea reticulata* King., *Smythea trichantha* (Miq.) C.K.Schneid., *Ventilago cernua* Tul., *Ventilago fasciculiflora* Merr., *Ventilago lanceata* Tul.
Climbing Mechanism: Scrambler-Unarmed
Distribution (Global): Andaman Is., Bismarck Archipelago, Borneo, Caroline Is., Fiji, Malaya, Maluku, New Guinea, Nicobar Is., Philippines, Santa Cruz Is., Seychelles, Solomon Is., Sulawesi, Sumatera, Thailand, Vanuatu
Distribution (India): Great Nicobar Island
Leaf Type: Simple, glabrous
Inflorescence: Fascicles
Fruit Type: Dry, capsule
Flowering and Fruiting: Throughout the year
IUCN Status: Not evaluated
Notes: Distributed in a wide range of habitats: Mangroves, tropical and subtropical moist broadleaf forests
Reference: Bull. Misc. Inform. Kew 1928: 389 (1928)

Smythea macrocarpa Hemsl.
Rhamnaceae
Synonyms: *Smythea lancifolia* Ridl.
Climbing Mechanism: Scrambler-Unarmed
Distribution (Global): Borneo, Malaya, Sumatera, Thailand
Distribution (India): Meghalaya, Mizoram
Leaf Type: Simple, glabrous
Inflorescence: Fascicles
Fruit Type: Dry, capsule
Flowering and Fruiting: February–December
IUCN Status: Not evaluated
Reference: Hooker's Icon. Pl. 16: t. 1558 (1886)

Ventilago bombaiensis Dalzell
Rhamnaceae
Synonyms: *Smythea bombaiensis* (Dalzell) S.P.Banerjee & P.K.Mukh.

Common Name: Bombay Smythea, Bombay Ventilago
Climbing Mechanism: Stem Twiner
Distribution (Global): India
Distribution (India): Karnataka, Kerala, Tamil Nadu, Western Ghat
Leaf Type: Simple, alternate
Inflorescence: Panicles, axillary/terminal
Fruit Type: Dry, samara
IUCN Status: Not evaluated
Notes: Usually found in evergreen forests
Reference: Indian Forester 96: 214 (1970)

Ventilago denticulata var. acuta
Rhamnaceae
Synonyms: *Berchemia laotica* Tardieu
Common Name: Toothed-leaf Red creeper
Climbing Mechanism: Stem Twiner
Distribution (Global): Andaman Is., Assam, Bangladesh, Cambodia, China South-Central, East Himalaya, India, Laos, Myanmar, Nepal, Thailand, Vietnam, West Himalaya
Leaf Type: Simple, alternate
Inflorescence: Panicles, axillary/terminal
Fruit Type: Dry, samara
IUCN Status: Not evaluated
Reference: Neue Schriften Ges. Naturf. Freunde Berlin 3: 417 (1801)

Ventilago denticulata var. bifida
Rhamnaceae
Climbing Mechanism: Stem Twiner
Leaf Type: Simple, alternate
Inflorescence: Panicles, axillary/terminal
Fruit Type: Dry, samara
IUCN Status: Not evaluated
Reference: Bhandari & Bhansali. In: Fasc. Fl. India, 20: 84. (1990)

Ventilago denticulata Willd.
Rhamnaceae
Climbing Mechanism: Stem Twiner
Distribution (India): Andaman and Nicobar Islands, Andhra Pradesh, Arunachal Pradesh, Eastern Ghats, Gujarat, Jharkhand, Kerala, Madhya Pradesh, Maharashtra, Odisha, Tamil Nadu, Telangana, West Bengal
Leaf Type: Simple, alternate
Inflorescence: Panicles, axillary/terminal
Fruit Type: Dry, samara
IUCN Status: Not evaluated
Reference: Willd. In: Ges. Naturf. Fr. Neue Schr. 3: 417. (1801).

Ventilago diffusa (G.Don) Exell
Rhamnaceae
Synonyms: Celastrus diffusus G.Don
Climbing Mechanism: Stem Twiner
Distribution (Global): Cameroon, Ethiopia, Gulf of Guinea Is., Ivory Coast, Kenya, Malawi, Nigeria, Tanzania, Uganda, Zaire
Leaf Type: Simple, alternate
Inflorescence: Panicles, axillary/terminal
Fruit Type: Dry, samara
IUCN Status: Not evaluated
Reference: Cat. Vasc. Pl. S. Tomé: 139 (1944)

Ventilago gamblei Suess.
Rhamnaceae
Synonyms: *Ventilago lanceolata* Gamble
Climbing Mechanism: Stem Twiner
Distribution (Global): India, Sri Lanka
Distribution (India): Andhra Pradesh, Eastern Ghats, Kerala, Maharashtra, Tamil Nadu
Leaf Type: Alternate
Inflorescence: Panicle-like
Fruit Type: Samaras
Flowering and Fruiting: December–July
IUCN Status: Not evaluated
Notes: Often found in semi-evergreen forests
Reference: Mitt. Bot. Staatssamml. München 1: 356 (1953)

Ventilago goughii Gamble
Rhamnaceae
Climbing Mechanism: Stem Twiner
Distribution (Global): India
Distribution (India): Tamil Nadu
Leaf Type: Simple, alternate
Inflorescence: Panicles, axillary/terminal
Fruit Type: Dry, samara
IUCN Status: Not evaluated
Notes: Distributed mostly in moist deciduous forests of the Western Ghats and the Eastern Ghats
Reference: Bull. Misc. Inform. Kew 1916: 134 (1916)

Ventilago kurzii Ridl.
Rhamnaceae
Synonyms: *Apteron lanceolatum* Kurz, *Kurzinda lanceolatum* Kuntze
Climbing Mechanism: Stem Twiner
Distribution (Global): Malaya, Myanmar

Leaf Type: Simple, alternate
Inflorescence: Panicles, axillary/terminal
Fruit Type: Dry, samara
IUCN Status: Not evaluated
Reference: Fl. Malay Penins. 5: 300 (1925)

Ventilago lanceolata Gamble
Rhamnaceae
Climbing Mechanism: Stem Twiner
Leaf Type: Simple, alternate
Inflorescence: Panicles, axillary/terminal
Fruit Type: Dry, samara
IUCN Status: Not evaluated
Reference: Gamble. In: Kew Bull. 134. (1916)

Ventilago maderaspatana Gaertn
Rhamnaceae
Common Name: Red creeper
Climbing Mechanism: Stem Twiner
Distribution (India): Andaman and Nicobar Islands, Andhra Pradesh, Arunachal Pradesh, Assam, Bihar, Odisha, Eastern Ghats, Goa, Gujarat, Karnataka, Maharashtra, Meghalaya, Odisha, Puducherry, Tamil Nadu, Telangana, Tripura, Uttar Pradesh, Western Ghats
Leaf Type: Alternate, Ovate-elliptic
Inflorescence: Panicles axillary and terminal
Fruit Type: Samara
Flowering and Fruiting: December–April
IUCN Status: Not evaluated
Notes: Distributed mostly in moist deciduous and semi-evergreen forests

Ventilago maderspatana var. fructifida
Rhamnaceae
Climbing Mechanism: Stem Twiner
Leaf Type: Simple, alternate
Inflorescence: Panicles, axillary/terminal
Fruit Type: Dry, samara
Flowering and Fruiting: December–January
IUCN Status: Not evaluated

Ventilago maingayi M.A.Lawson
Rhamnaceae
Climbing Mechanism: Stem Twiner
Distribution (Global): Java, Malaya, Myanmar, Thailand
Leaf Type: Simple, alternate
Inflorescence: Panicles, axillary/terminal

Fruit Type: Dry, samara
IUCN Status: Not evaluated
Reference: J. D. Hooker, Fl. Brit. India 1: 631 (1875)

Ziziphus apetala Hook.f.
Rhamnaceae
Climbing Mechanism: Scrambler-Armed
Distribution (Global): Assam, Bangladesh, East Himalaya, India, Nepal
Distribution (India): Arunachal Pradesh, Assam
Leaf Type: Leaves simple, alternate
Inflorescence: Cymes
Fruit Type: Drupes, yellow
IUCN Status: Not evaluated
Notes: Mostly found in open and exposed areas
Reference: Fl. Brit. India 1: 635 (1875)

Ziziphus calophylla Wall.
Rhamnaceae
Synonyms: *Berchemia calophylla* (Wall.) G.Don, *Ziziphus ornata* Miq.
Climbing Mechanism: Scrambler-Armed
Distribution (Global): Borneo, Cambodia, Laos, Malaya, Sumatera, Thailand, Vietnam
Leaf Type: Simple, alternate, 3–5 nerved
Inflorescence: Cymes in clusters
Fruit Type: Wet, drupe
IUCN Status: Not evaluated
Reference: W.Roxburgh, Fl. Ind. 2: 366 (1824)

Ziziphus funiculosa Buch.-Ham. ex Wall.
Rhamnaceae
Synonyms: *Ziziphus venulosa* Wall.
Climbing Mechanism: Scrambler-Armed
Distribution (Global): Assam, Bangladesh, East Himalaya, Laos, Myanmar, Nepal, Philippines, Thailand, Vietnam
Distribution (India): Arunachal Pradesh, Assam, Bihar, Odisha, Eastern Ghats, Tripura
Leaf Type: Simple, alternate, 3–5 nerved
Inflorescence: Cymes in clusters
Fruit Type: Wet, drupe
Flowering and Fruiting: February–May
IUCN Status: Not evaluated
Reference: J. D. Hooker, Fl. Brit. India 1: 636 (1875)

Ziziphus glaberrima (Sedgw.) Santapau
Rhamnaceae

Synonyms: *Ziziphus xylopyrus* (Retz.) Willd. *Rhamnus rotundifolia* Pers., *Rhamnus xylopyrus* Retz., *Ziziphus caracutta* Buch.-Ham. ex Roxb., *Ziziphus cuneata* Wall., *Ziziphus elliptica* Roxb., *Ziziphus glaberrima* (Sedgw.) Santapau, *Ziziphus heterogenea* Russell ex Wall.
Climbing Mechanism: Scrambler-Unarmed
Distribution (Global): Assam, Bangladesh, India, Nepal, Sri Lanka
Leaf Type: Leaves simple, alternate, stipulate
Inflorescence: Velvety, axillary Cymes
Fruit Type: Drupes, globose
Flowering and Fruiting: April–June
IUCN Status: Not evaluated
Reference: Ziziphus xylopyrus (Retz.) Willd.

Ziziphus horrida Roth
Rhamnaceae
Synonyms: *Ziziphus bhunder* Royle
Climbing Mechanism: Scrambler-Armed
Distribution (Global): India
Distribution (India): Andhra Pradesh, Tamil Nadu
Leaf Type: Simple, alternate, 3–5 nerved
Inflorescence: Cymes in clusters
Fruit Type: Wet, drupe
Flowering and Fruiting: September–February
IUCN Status: Not evaluated
Notes: Distributed mostly in dry forests and scrub jungles
Reference: J.J.Roemer & J.A.Schultes, Syst. Veg., ed. 15 bis 5: 341 (1819)

Ziziphus horsfieldii Miq.
Rhamnaceae
Synonyms: *Ziziphus palawanensis* Elmer
Climbing Mechanism: Scrambler-Armed
Distribution (Global): Andaman Is., Borneo, Java, Malaya, Nicobar Is., Philippines, Sumatera
Distribution (India): Great Nicobar Island
Leaf Type: Simple, alternate, 3–5 nerved
Inflorescence: Cymes in clusters
Fruit Type: Wet, drupe
IUCN Status: Not evaluated
Reference: Fl. Ned. Ind. 1(1): 643 (1856)

Ziziphus napeca Willd.
Rhamnaceae
Synonyms: *Ziziphus linnaei* M.A.Lawson., *Girtanneria napeca* (L.) Neck., *Rhamnus napeca* L.
Climbing Mechanism: Scrambler-Armed

Distribution (Global): Sri Lanka.
Leaf Type: Simple, alternate, 3–5 nerved
Inflorescence: Cymes in clusters
Fruit Type: Wet, drupe
IUCN Status: Not evaluated
Reference: J. D. Hooker, Fl. Brit. India 1: 635 (1875)

Ziziphus nummularia (Burm.f.) Wight & Arn.
Rhamnaceae
Synonyms: *Rhamnus microphylla* Roxb., *Rhamnus nummularia* Burm.f., *Ziziphus microphylla* Roxb., *Ziziphus nummularia var.*, *glabrescens* Bhandari & Bhansali., Ziziphus rotundifolia Lam.
Common Name: Wild jujube
Climbing Mechanism: Scrambler-Armed
Distribution (Global): Afghanistan, Gulf States, India, Iran, Iraq, Nepal, Pakistan, Palestine, Saudi Arabia, Sinai
Distribution (India): Kerala, Rajasthan, Tamil Nadu
Leaf Type: Ovate, elliptic
Inflorescence: Cymes
Fruit Type: Drupes, globose, shining, and black-red when ripe
Flowering and Fruiting: August–February
IUCN Status: Not evaluated
Notes: Often found in dry deciduous forests and scrub jungles
Reference: R.Wight, Cat. Ind. Pl.: 31 (1833)

Ziziphus oenopolia (L.) Mill.
Rhamnaceae
Synonyms: *Girtanneria oenopolia* (L.) Neck., *Rhamnus oenopolia* L., *Ziziphus albens* Roxb., *Ziziphus celtidifolia* DC.
Common Name: Jackal jujube, Indian plum
Climbing Mechanism: Scrambler-Armed
Distribution (Global): Andaman Is., Assam, Bangladesh, Cambodia, China South-Central, China Southeast, India, Laos, Lesser Sunda Is., Malaya, Myanmar, Nepal, New Guinea, Northern Territory, Philippines, Queensland, Sri Lanka, Thailand, Vietnam
Distribution (India): Andaman and Nicobar Islands, Andhra Pradesh, Arunachal Pradesh, Bihar, Odisha, Delhi, Eastern Ghats, Gujarat, Jharkhand, Karnataka, Kerala, Madhya Pradesh, Maharashtra, Mizoram, Puducherry, Rajasthan, Tamil Nadu, Telangana, Tripura, Uttar Pradesh, West Bengal, West
Leaf Type: Simple, alternate, ovate
Inflorescence: Velvety Cymes
Flowering and Fruiting: June–February
IUCN Status: Least Concern
Notes: Mostly found in moist and dry deciduous forests
Reference: Gard. Dict. ed. 8: n.° 3 (1768)

Ziziphus rugosa Lam.
Rhamnaceae
Common Name: Wild Jujube, Wrinkled jujube
Climbing Mechanism: Scrambler-Armed
Distribution (Global): Andaman Is., Assam, Bangladesh, China South-Central, East Himalaya, Hainan, India, Laos, Myanmar, Nepal, Nicobar Is., Pakistan, Sri Lanka, Thailand, Vietnam
Distribution (India): Andhra Pradesh, Arunachal Pradesh, Assam, Eastern Ghats, Goa, Great Nicobar Island, Gujarat, Karnataka, Kerala, Madhya Pradesh, Maharashtra, Odisha, Rajasthan, Tamil Nadu, Telangana, West Bengal
Leaf Type: Broadly elliptic
Inflorescence: Cymes
Fruit Type: Globose or pyriform, white when ripe
Flowering and Fruiting: November–July
IUCN Status: Not evaluated
Notes: Found often in deciduous forests, also in the plains
Reference: Encycl. 3: 319 (1789)

Ziziphus truncata Blatt. & Hallb.
Rhamnaceae
Climbing Mechanism: Scrambler-Armed
Distribution (Global): India
Leaf Type: Simple, alternate, 3–5 nerved
Inflorescence: Cymes in clusters
Fruit Type: Wet, drupe
IUCN Status: Not evaluated
Reference: J. Bombay Nat. Hist. Soc. 26: 234 (1918)

Cotoneaster ellipticus
Rosaceae
Climbing Mechanism: Scrambler-Armed
IUCN Status: Not evaluated

Prinsepia utilis Hayata
Rosaceae
Synonyms: *Cycnia spinosa* Griff.
Climbing Mechanism: Scrambler-Armed
Distribution (India): Tamil Nadu, Tripura
IUCN Status: Not evaluated
Reference: Ill. Bot. Himal. Mts. 1: 206 (1835)

Rosa odorata var. gigantea (Collett ex Cr,p.) Rehder & E.H.Wilson
Rosaceae
Climbing Mechanism: Scrambler-Armed
Distribution (India): Tamil Nadu

Leaf Type: Compound, pinnate
Inflorescence: Solitary flowers
Fruit Type: Wet, drupe
IUCN Status: Not evaluated

Rosa banksiae f. lutescens Voss
Rosaceae
Climbing Mechanism: Scrambler-Armed
IUCN Status: Not evaluated

Rosa canina L,man
Rosaceae
Synonyms: *Rosa aciphylla* A.Rau, *Rosa adenocalyx* Gren., *Rosa adscita* Déségl., *Rosa afzeliana* Fr., *Rosa agraria* Ripart ex Déségl.
Climbing Mechanism: Scrambler-Armed
Distribution (Global): Afghanistan, Albania, Algeria, Austria, Baleares, Baltic States, Belarus, Belgium, Bulgaria, Canary Is., Central European Rus, Corse, Cyprus, Czechoslovakia, Denmark, East Aegean Is., East European Russia, Finland, France, Germany, Great Britain, Greece, Hungary, Iran, Iraq, Ireland, Italy, Kazakhstan, Kirgizstan, Kriti, Krym, Lebanon-Syria, Madeira, Morocco, Netherlands, North Caucasus, Norway, Pakistan, Palestine, Poland, Portugal, Romania, Sardegna, Sicilia, South European Russi, Spain, Sweden, Switzerland, Tadzhikistan, Transcaucasus, Tunisia, Turkey, Turkey-in-Europe, Turkmenistan, Ukraine, Uzbekistan, Yugoslavia
Distribution (India): Maharashtra.
Leaf Type: Compound, pinnate
Inflorescence: Solitary flowers
Fruit Type: Wet, drupe
Flowering and Fruiting: Throughout the year
IUCN Status: Least Concern
Reference: Sp. Pl.: 491 (1753)

Rosa cathayensis (Rehder & E.H.Wilson) L.H.Bailey
Rosaceae
Synonyms: *Rosa multiflora* Bunge non Thumb
Climbing Mechanism: Scrambler-Armed
Distribution (India): Maharashtra
Leaf Type: Compound, pinnate
Inflorescence: Solitary flowers
Fruit Type: Wet, drupe
Flowering and Fruiting: Throughout the year
IUCN Status: Not evaluated
Reference: Gentes Herbarum 1: 29 (1920)

Rosa clinophylla Redout, & Thory
Rosaceae
Synonyms: *Rosa clinophylla var. glabra* (Lindl. ex Prain) Ghora & Panigrahi, *Rosa clinophylla var. parvifolia* (Lindl. ex Prain) Ghora & Panigrahi, *Rosa involucrata* Roxb. ex Lindl., *Rosa lindleyana* Tratt., *Rosa lyellii* Lindl., *Rosa palustris* Buch.-Ham. ex Lind
Climbing Mechanism: Scrambler-Armed
Distribution (Global): Assam, East Himalaya, Laos, Myanmar, Nepal, Thailand, West Himalaya
Distribution (India): Uttar Pradesh
Leaf Type: Simple, alternate
Inflorescence: Axillary, Solitary
Fruit Type: Drupe
Flowering and Fruiting: February–March
IUCN Status: Not evaluated
Reference: P.J.Redouté & C.A.Thory, Roses 1: 43 (1817)

Rosa gigantea Collett ex Cr,p.
Rosaceae
Synonyms: *Rosa macrocarpa* G.Watt ex Crép.
Common Name: Manipur wild-tea rose
Climbing Mechanism: Scrambler-Armed
Distribution (Global): Assam, China South-Central, Laos, Myanmar, Thailand, Vietnam
Distribution (India): Tamil Nadu
Leaf Type: Compound, pinnate
Inflorescence: Solitary flowers
Fruit Type: Wet, drupe
IUCN Status: Not evaluated
Reference: Bull. Soc. Roy. Bot. Belgique 27: 148 (1888)

Rosa hirsuta Ghora & Panigrahi
Rosaceae
Synonyms: *Rosa sericea var. sericea*
Climbing Mechanism: Scrambler-Armed
Leaf Type: Compound, pinnate
Inflorescence: Solitary flowers
Fruit Type: Wet, drupe
IUCN Status: Not evaluated
Reference: Bull. Bot. Surv. India 28: 177 (1986 publ. 1988)

Rosa involucrata Roxb.
Rosaceae
Synonyms: *Rosa clinophylla* Theory
Climbing Mechanism: Scrambler-Armed

Distribution (India): Bihar, Odisha, Rajasthan
Leaf Type: Compound, pinnate
Inflorescence: Solitary flowers
Fruit Type: Wet, drupe
Flowering and Fruiting: February–March
IUCN Status: Not evaluated
Reference: Ros. Monogr.: 8 (1820)

Rosa laevigata Michx.
Rosaceae
Synonyms: *Rosa amygdalifolia* Ser., *Rosa cherokeensis* Donn ex Small, *Rosa cucumerina* Tratt., *Rosa hystrix* Lindl., *Rosa laevigata var. leiocarpa* Y.Q. Wang & P.Y.Chen, *Rosa laevigata f. semiplena* T.T.Yu & T.C.Ku, *Rosa nivea* DC., *Rosa sinica* W.T.Aiton, *Rosa tern.*
Common Name: Cherokee rose
Climbing Mechanism: Scrambler-Armed
Distribution (Global): China North-Central, China South-Central, China Southeast, Hainan, Taiwan, Vietnam
Distribution (India): Meghalaya, Tamil Nadu
Leaf Type: Compound, pinnate
Inflorescence: Solitary flowers
Fruit Type: Wet, drupe
IUCN Status: Not evaluated
Reference: Fl. Bor.-Amer. 1: 295 (1803)

Rosa leschenaultiana (Redout, & Theory) Wight & Arn.
Rosaceae
Synonyms: *Rosa sempervirens var. leschenaultiana* Theory
Climbing Mechanism: Scrambler-Armed
Distribution (Global): India
Distribution (India): Arunachal Pradesh, Karnataka, Kerala, Tamil Nadu, Western Ghats
Leaf Type: Compound, pinnate
Inflorescence: Solitary flowers
Fruit Type: Wet, drupe
Flowering and Fruiting: Throughout the year
IUCN Status: Not evaluated
Notes: Found mostly in shola forests and grasslands
Reference: Prodr. Fl. Ind. Orient. 1: 301 (1834)

Rosa longicuspis Bertol.
Rosaceae
Common Name: Shillong climbing rose
Climbing Mechanism: Scrambler-Armed
Distribution (Global): Assam, China South-Central

Distribution (India): North Eastern India
Leaf Type: Compound, pinnate
Inflorescence: Solitary flowers
Fruit Type: Wet, drupe
IUCN Status: Not evaluated
Reference: Misc. Bot. 21: 15 (1861)

Rosa moschata Herrm.
Rosaceae
Synonyms: *Rosa arborea* Pers., *Rosa broteroi* Tratt., *Rosa brownii* Tratt., *Rosa glandulifera* Roxb., *Rosa manuelii* Losa, *Rosa opsostemma* Ehrh., *Rosa pissardii* Carrière, *Rosa ruscinonensis* Gren. & Déségl.
Common Name: Musk rose.
Climbing Mechanism: Scrambler-Armed
Distribution (Global): Afghanistan, Iran
Distribution (India): Meghalaya
Leaf Type: Compound, pinnate
Inflorescence: Solitary flowers
Fruit Type: Wet, drupe
IUCN Status: Not evaluated
Reference: De Rosa: 15 (1762)

Rosa multiflora Merino
Rosaceae
Synonyms: *Rosa polyanthos* Rössig
Common Name: Multiflora rose, Baby rose, Japanese rose
Climbing Mechanism: Scrambler-Armed
Distribution (Global): China North-Central, China South-Central, China Southeast, Japan, Korea
Distribution (India): Karnataka, Kerala, Madhya Pradesh, Manipur, Mizoram, Tamil Nadu
Leaf Type: Compound, pinnate
Inflorescence: Solitary flowers
Fruit Type: Wet, drupe
Flowering and Fruiting: Throughout the year
IUCN Status: Not evaluated
Reference: J.A. Murray, Syst. Veg. ed. 14: 474 (1784)

Rosa sericea Cr,p.
Rosaceae
Synonyms: *Rosa tetrapetala*, *Rosa wallichii*, *Rosa polyphylla*, *Rosa omiensis*.
Common Name: Silky rose
Climbing Mechanism: Scrambler-Armed
Distribution (Global): Assam, China South-Central, East Himalaya, Myanmar, Nepal, Tibet, West Himalaya

Leaf Type: Compound
IUCN Status: Not evaluated
Reference: Ros. Monogr.: 105 (1820)

Rosa triphylla Roxb. ex Lindl.
Rosaceae
Synonyms: *Rosa laevigata* Michx.
Climbing Mechanism: Scrambler-Armed
Distribution (India): West Bengal.
Leaf Type: Compound, pinnate
Inflorescence: Solitary flowers
Fruit Type: Wet, drupe
IUCN Status: Not evaluated
Reference: A. Beatson, Tracts St. Helena: 321 (1816)

Rubus acuminatus Sm.
Rosaceae
Climbing Mechanism: Scrambler-Armed
Distribution (Global): Assam, China South-Central, East Himalaya, Myanmar, Nepal, Vietnam.
Distribution (India): Arunachal Pradesh, Mizoram.
Leaf Type: Compound, odd-pinnate, 3–7 foliolate
Inflorescence: Corymbose panicles
Fruit Type: Wet, drupelets
IUCN Status: Not evaluated
Reference: A. Rees, Cycl. 30: no. 43 (1815)

Rubus alceifolius Poir.
Rosaceae
Synonyms: *Cumbata alcefolia* (Poir.) Raf., *Rubus alceifolius* var. *diversilobatus* (Merr. & Chun) T. T. Yu & L.T.Lu, *Rubus bullatifolius* Merr., *Rubus fimbriifer* Focke, *Rubus gilvus* Focke, *Rubus hainanensis* Focke, *Rubus monguillonii* H. Lév. & Vaniot, *Rubus roridus*
Climbing Mechanism: Scrambler-Armed
Distribution (Global): Assam, Borneo, Cambodia, China North-Central, China South-Central, China Southeast, Hainan, Java, Laos, Lesser Sunda Is., Malaya, Myanmar, Sulawesi, Sumatera, Taiwan, Thailand, Vietnam
Distribution (India): Arunachal Pradesh, Meghalaya, Mizoram
Leaf Type: Compound, odd-pinnate, 3–7 foliolate
Inflorescence: Corymbose panicles
Fruit Type: Wet, drupelets
IUCN Status: Not evaluated
Reference: J.B.A.M.de Lamarck, Encycl. 6: 247 (1804)

Rubus alpestris Blume
Rosaceae
Climbing Mechanism: Scrambler-Armed
Distribution (Global): Assam, Borneo, Java, Maluku, Sulawesi, Sumatera, Thailand, Vietnam
Leaf Type: Compound, odd-pinnate, 3–7 foliolate
Inflorescence: Corymbose panicles
Fruit Type: Wet, drupelets
IUCN Status: Not evaluated
Reference: Bijdr. Fl. Ned. Ind.: 1108 (1826)

Rubus assamensis Focke
Rosaceae
Synonyms: *Rubus bahanensis* Hand. -Mazz., *Rubus qinglongensis* Q.H.Chen & T.L. Xu.
Common Name: Assam raspberry
Climbing Mechanism: Scrambler-Armed
Distribution (Global): Assam, China South-Central, China Southeast, Tibet
Distribution (India): Arunachal Pradesh, Manipur, Meghalaya, Mizoram
Leaf Type: Compound, odd-pinnate, 3–7 foliolate
Inflorescence: Corymbose panicles
Fruit Type: Wet, drupelets
IUCN Status: Not evaluated
Reference: Abh. Naturwiss. Vereins Bremen 4: 197 (1874)

Rubus barberi H.E.Weber
Rosaceae
Synonyms: *Rubus rugosus* E.Barber
Climbing Mechanism: Scrambler-Armed
Distribution (Global): Czechoslovakia
Leaf Type: Compound, odd-pinnate, 3–7 foliolate
Inflorescence: Corymbose panicles
Fruit Type: Wet, drupelets
IUCN Status: Not evaluated
Reference: Abh. Ber. Naturkundemus. Görlitz 61(8): 35 (1987)

Rubus biflorus Bouch
Rosaceae
Synonyms: *Rubus biflorus* var. *spinocalycinus*
Common Name: Two-flower raspberry
Climbing Mechanism: Scrambler-Armed
Distribution (Global): Assam, China North-Central, China South-Central, East Himalaya, Myanmar, Nepal, Pakistan, Tibet, West Himalaya
Distribution (India): Arunachal Pradesh, Himachal Pradesh
Leaf Type: Imparipinnate, trifoliate

Fruit Type: Aggregate
IUCN Status: Not evaluated
Reference: A.Rees, Cycl. 30: no. 9 (1815)

Rubus birmanicus Hook. f.
Rosaceae
Climbing Mechanism: Scrambler-Armed
Distribution (Global): Assam, Myanmar
Distribution (India): Arunachal Pradesh, Manipur, Mizoram
Leaf Type: Compound, odd-pinnate, 3–7 foliolate
Inflorescence: Corymbose panicles
Fruit Type: Wet, drupelets
IUCN Status: Not evaluated
Reference: Fl. Brit. India 2: 331 (1878)

Rubus burkillii Rolfe
Rosaceae
Climbing Mechanism: Scrambler-Armed
Distribution (Global): Assam.
Distribution (India): Manipur.
Leaf Type: Compound, odd-pinnate, 3-7 foliolate
Inflorescence: Corymbose panicles
Fruit Type: Wet, drupelets
IUCN Status: Not evaluated
Reference: Bull. Misc. Inform. Kew 1920: 109 (1920)

Rubus dasycoccus W.C.R.Watson
Rosaceae
Synonyms: *Rubus lasiocarpus* W.C.R.Watson
Climbing Mechanism: Scrambler-Armed
Distribution (Global): Great Britain.
Leaf Type: Compound, odd-pinnate, 3–7 foliolate
Inflorescence: Corymbose panicles
Fruit Type: Wet, drupelets
IUCN Status: Not evaluated
Reference: Rep. Bot. Soc. Exch. Club Brit. Isles 10: 21 (1933)

Rubus duthieanus N.P.Balakr.
Rosaceae
Synonyms: *Rubus wallichianus* Wight & Arn.
Climbing Mechanism: Scrambler-Armed
Distribution (India): Arunachal Pradesh
Leaf Type: Compound, odd-pinnate, 3–7 foliolate
Inflorescence: Corymbose panicles
Fruit Type: Wet, drupelets

IUCN Status: Not evaluated
Reference: J. Bombay Nat. Hist. Soc. 67: 58 (1970)

Rubus ellipticus Kupcsok
Rosaceae
Synonyms: *Rubus rotundifolius*
Common Name: Yellow Himalayan raspberry
Climbing Mechanism: Scrambler-Armed
Distribution (Global): Assam, China South-Central, China Southeast, East Himalaya, India, Laos, Myanmar, Nepal, Pakistan, Philippines, Sri Lanka, Thailand, Tibet, Vietna
Distribution (India): Andhra Pradesh, Arunachal Pradesh, Bihar, Odisha, Karnataka, Kerala, Madhya Pradesh, Maharashtra, Manipur, Meghalaya, Mizoram, Tamil Nadu, Tripura, West Bengal, Western Ghats
Leaf Type: Compound
Flowering and Fruiting: November–June
IUCN Status: Least Concern
Notes: Mostly distributed in evergreen, semi-evergreen, and shola forests
Reference: A.Rees, Cycl. 30: no. 16 (1815)

Rubus fairholmianus Gardner
Rosaceae
Climbing Mechanism: Scrambler-Armed
Distribution (Global): India, Sri Lanka.
Distribution (India): Arunachal Pradesh, Kerala, Karnataka
Leaf Type: Compound, odd-pinnate, 3–7 foliolate
Inflorescence: Corymbose panicles
Fruit Type: Wet, drupelets
Flowering and Fruiting: August–December
IUCN Status: Not evaluated
Notes: Usually found in evergreen forests
Reference: Calcutta J. Nat. Hist. 8: 5 (1847)

Rubus fockei Braeucker
Rosaceae
Climbing Mechanism: Scrambler-Armed
Distribution (India): Karnataka
Leaf Type: Compound, odd-pinnate, 3–7 foliolate
Inflorescence: Corymbose panicles
Fruit Type: Wet, drupelets
Flowering and Fruiting: July–October
IUCN Status: Not evaluated
Reference: 292 Rubus-arten: 39 (1882)

Rubus formosensis Kuntze
Rosaceae
Synonyms: *Rubus nantoensis* Hayata, *Rubus randaiensis* Hayata, *Rubus rubribracteatus* F.P. Metcalf, *Rubus rugosissimus* Hayata
Climbing Mechanism: Scrambler-Armed
Distribution (Global): China Southeast, Taiwan
Leaf Type: Compound, odd-pinnate, 3–7 foliolate
Inflorescence: Corymbose panicles
Fruit Type: Wet, drupelets
IUCN Status: Not evaluated
Reference: Meth. Sp.-Beschr. Rubus: 73 (1879)

Rubus fulvus Focke
Rosaceae
Synonyms: *Rubus kasthuriae* Gandhi
Climbing Mechanism: Scrambler-Armed
Leaf Type: Compound, odd-pinnate, 3–7 foliolate
Inflorescence: Corymbose panicles
Fruit Type: Wet, drupelets
IUCN Status: Not evaluated
Reference: Sp. Rub.: 81 (1910)

Rubus gardenerianus Kuntze
Rosaceae
Climbing Mechanism: Scrambler-Armed
Distribution (India): Kerala.
Leaf Type: Compound, odd-pinnate, 3-7 foliolate
Inflorescence: Corymbose panicles
Fruit Type: Wet, drupelets
IUCN Status: Not evaluated

Rubus glomeratus Figert
Rosaceae
Climbing Mechanism: Scrambler-Armed
Distribution (India): Kerala, Tamil Nadu.
Leaf Type: Compound, odd-pinnate, 3–7 foliolate
Inflorescence: Corymbose panicles
Fruit Type: Wet, drupelets
Flowering and Fruiting: August–January
IUCN Status: Not evaluated
Notes: Often found in evergreen forests and grasslands
Reference: Deutsche Bot. Monatsschr. 22: 119 (1911)

Rubus griffithii Hook. f.
Rosaceae

Synonyms: *Rubus excurvatus* Kuntze
Climbing Mechanism: Scrambler-Armed
Distribution (Global): Assam, East Himalaya, Nepal
Distribution (India): Mizoram, West Bengal
Leaf Type: Compound, odd-pinnate, 3–7 foliolate
Inflorescence: Corymbose panicles
Fruit Type: Wet, drupelets
IUCN Status: Not evaluated
Reference: Fl. Brit. India 2: 327 (1878)

Rubus hamiltonii Hook. f.
Rosaceae
Climbing Mechanism: Scrambler-Armed
Distribution (Global): Assam, East Himalaya, Myanmar, Nepal
Distribution (India): Arunachal Pradesh, Tripura, West Bengal
Leaf Type: Compound, odd-pinnate, 3–7 foliolate
Inflorescence: Corymbose panicles
Fruit Type: Wet, drupelets
IUCN Status: Not evaluated
Reference: Fl. Brit. India 2: 328 (1878)

Rubus haridasanii Chand. Gupta & S. S. Dash
Rosaceae
Climbing Mechanism: Scrambler-Armed
Distribution (Global): East Himalaya
Distribution (India): Arunachal Pradesh
Leaf Type: Compound, odd-pinnate, 3–7 foliolate
Inflorescence: Corymbose panicles
Fruit Type: Wet, drupelets
IUCN Status: Not evaluated
Reference: Phytotaxa 289(2): 176 (2016)

Rubus hexagynus Roxb.
Rosaceae
Synonyms: *Ametron pirifolium* (Sm.) Raf., *Rubus indicus* Lesch. ex Ser.
Climbing Mechanism: Scrambler-Armed
Distribution (Global): Assam, Bangladesh, Cambodia
Distribution (India): Arunachal Pradesh, Meghalaya
Leaf Type: Compound, odd-pinnate, 3–7 foliolate
Inflorescence: Corymbose panicles
Fruit Type: Wet, drupelets
IUCN Status: Not evaluated
Reference: Fl. Ind. ed. 1832, 2: 516 (1832)

Rubus indicus Thunb.
Rosaceae
Synonyms: *Rubus wightii* Gamble
Common Name: Indian raspberry
Climbing Mechanism: Scrambler-Armed
Distribution (Global): India, Sri Lanka
Distribution (India): Kerala.
Leaf Type: Compound, odd-pinnate, 3–7 foliolate
Inflorescence: Corymbose panicles
Fruit Type: Wet, drupelets
Flowering and Fruiting: October
IUCN Status: Not evaluated
Reference: De Rubo: 5 (1813)

Rubus khasianus Cardot
Rosaceae
Climbing Mechanism: Scrambler-Armed
Distribution (Global): Assam
Distribution (India): Meghalaya
Leaf Type: Compound, odd-pinnate, 3–7 foliolate
Inflorescence: Corymbose panicles
Fruit Type: Wet, drupelets
IUCN Status: Not evaluated
Reference: Notul. Syst. (Paris) 3: 298 (1917)

Rubus kumaonensis N.P. Balakr.
Rosaceae
Synonyms: *Rubus reticulatus* Wall. ex Hook. f.
Climbing Mechanism: Scrambler-Armed
Distribution (Global): East Himalaya, Nepal, Tibet, West Himalaya
Distribution (India): Manipur
Leaf Type: Compound, odd-pinnate, 3-7 foliolate
Inflorescence: Corymbose panicles
Fruit Type: Wet, drupelets
IUCN Status: Not evaluated
Reference: J. Bombay Nat. Hist. Soc. 67: 58 (1970)

Rubus kurzii N.P. Balakr.
Rosaceae
Synonyms: *Rubus efferatus* Craib
Climbing Mechanism: Scrambler-Armed
Distribution (India): Arunachal Pradesh
Leaf Type: Compound, odd-pinnate, 3–7 foliolate
Inflorescence: Corymbose panicles
Fruit Type: Wet, drupelets

IUCN Status: Not evaluated
Reference: J. Bombay Nat. Hist. Soc. 67: 58 (1970)

Rubus leucocarpus Arn.
Rosaceae
Synonyms: *Rubus niveus* subsp. *leucocarpus* (Arn.) Focke, *Rubus niveus* var. *subglaber* (Thwaites) H.O. Saxena
Climbing Mechanism: Scrambler-Armed
Distribution (Global): Assam, India, Sri Lanka
Distribution (India): Tamil Nadu
Leaf Type: Compound, odd-pinnate, 3–7 foliolate
Inflorescence: Corymbose panicles
Fruit Type: Wet, drupelets
IUCN Status: Not evaluated
Reference: Nova Acta Phys.-Med. Acad. Caes. Leop. -Carol. Nat. Cur. 18: 334 (1836)

Rubus lineatus Reinw. ex Blume
Rosaceae
Climbing Mechanism: Scrambler-Armed
Distribution (Global): Assam, Borneo, China South-Central, East Himalaya, Java, Lesser Sunda Is., Myanmar, Nepal, Sumatera, Tibet, Vietna
Distribution (India): Manipur
Leaf Type: Compound, odd-pinnate, 3–7 foliolate
Inflorescence: Corymbose panicles
Fruit Type: Wet, drupelets
IUCN Status: Not evaluated
Reference: Bijdr. Fl. Ned. Ind.: 1108 (1826)

Rubus lucens Focke
Rosaceae
Synonyms: *Rubus laevigatus* Wall.
Climbing Mechanism: Scrambler-Armed
Distribution (Global): Assam, China South-Central
Distribution (India): Arunachal Pradesh, Manipur, Meghalaya, Mizoram
Leaf Type: Compound, odd-pinnate, 3–7 foliolate
Inflorescence: Corymbose panicles
Fruit Type: Wet, drupelets
IUCN Status: Not evaluated
Reference: Abh. Naturwiss. Vereins Bremen 4: 199 (1874)

Rubus macilentus Genev.
Rosaceae
Common Name: Lean raspberry
Climbing Mechanism: Scrambler-Armed

Leaf Type: Trifoliate, rarely simple
Fruit Type: Raspberry
IUCN Status: Not evaluated
Reference: Mém. Soc. Acad. Maine Loire 28: 42 (1873)

Rubus mesogaeus Focke
Rosaceae
Climbing Mechanism: Scrambler-Armed
Distribution (Global): Assam, China North-Central, China South-Central, China Southeast, East Himalaya, Japan, Kuril Is., Nepal, Sakhalin, Taiwan, Tibet
Distribution (India): Arunachal Pradesh
Leaf Type: Compound, odd-pinnate, 3–7 foliolate
Inflorescence: Corymbose panicles
Fruit Type: Wet, drupelets
IUCN Status: Not evaluated
Reference: Bot. Jahrb. Syst. 29: 399 (1900)

Rubus micropetalus Gardner
Rosaceae
Climbing Mechanism: Scrambler-Armed
Distribution (Global): India, Sri Lanka (Indo Malesia)
Distribution (India): Karnataka, Kerala, Tamil Nadu, Western Ghats
Leaf Type: Palmately compound 5-lobed
Inflorescence: Axillary and terminal corymbose panicles
Fruit Type: Compound berry reddish brown on ripening
Flowering and Fruiting: February–June
IUCN Status: Not evaluated
Reference: J. Nat. Hist. 8: 6 (1847)

Rubus mollucannus var. *macrocarpa* L.
Rosaceae
Climbing Mechanism: Scrambler-Armed
Leaf Type: Compound, odd-pinnate, 3–7 foliolate
Inflorescence: Corymbose panicles
Fruit Type: Wet, drupelets
IUCN Status: Not evaluated

Rubus moluccanus Matsum. & Hayata
Rosaceae
Synonyms: *Rubus moluccanus* var. *angulosus* Kalkman., *Rubus moluccanus* var. *discolor* (Blume) Kalkman., *Rubus moluccanus* var. *moluccanus*., *Rubus moluccanus* var. *neocaledonicus* Schltr., *Rubus moluccanus* var. *obtusangulus* Miq.
Common Name: Wild raspberry, Blackberry, Black cherry, Broad-leaved bramble, Ceylon blackberry, Molucca bramble, Molucca raspberry

Climbing Mechanism: Scrambler-Armed
Distribution (Global): Assam, Bismarck Archipelago, Borneo, Caroline Is., India, Java, Laos, Lesser Sunda Is., Malaya, Maluku, Myanmar, New Caledonia, New Guinea, New South Wales, Philippines, Queensland, Solomon Is., South Australia, Sri Lanka, Sulawesi, Sumatera, Thailand, Victoria, Vietnam
Distribution (India): Arunachal Pradesh, Great Nicobar Island, Maharashtra, Sikkim, Tripura, West Bengal
Leaf Type: Alternately arranged
Fruit Type: Berry, red color
Flowering and Fruiting: April–May
IUCN Status: Not evaluated
Reference: Sp. Pl.: 1197 (1753)

Rubus niveus Thunb.
Rosaceae
Common Name: Ceylon raspberry, Mysore raspberry, Mahabaleshwar raspberry, Snowpeaks raspberry
Climbing Mechanism: Scrambler-Armed
Distribution (Global): Indo-Himalaya
Distribution (India): Andhra Pradesh, Arunachal Pradesh, Himachal Pradesh, Karnataka, Kerala, Maharashtra, Tamil Nadu, Tripura, and the Western Ghats
Leaf Type: Compound, odd-pinnate, 3–7 foliolate
Inflorescence: Corymbose panicles
Fruit Type: Wet, drupelets
Flowering and Fruiting: November–July
IUCN Status: Not evaluated
Notes: Mostly found in evergreen and moist deciduous forests and also in grasslands

Rubus opulifolius Bertol.
Rosaceae
Climbing Mechanism: Scrambler-Armed
Distribution (India): Manipur
Leaf Type: Compound, odd-pinnate, 3–7 foliolate
Inflorescence: Corymbose panicles
Fruit Type: Wet, drupelets
IUCN Status: Not evaluated
Reference: Bot. 22: 16 (1862)

Rubus paniculatus C.B.Clarke
Rosaceae
Climbing Mechanism: Scrambler-Armed
Distribution (India): Arunachal Pradesh, West Bengal, Himachal Pradesh
Leaf Type: Compound, odd-pinnate, 3–7 foliolate
Inflorescence: Corymbose panicles
Fruit Type: Wet, drupelets

Flowering and Fruiting: June–August
IUCN Status: Not evaluated

Rubus pedunculosus D. Don
Rosaceae
Synonyms: *Rubus concolor* Wall., *Rubus gracilis* Roxb., *Rubus hypargyrus* Edgew., *Rubus hypargyrus* var. *concolor* (Wall. ex Hook.f.) H.Hara., *Rubus niveus* var. *concolor* Wall. ex Hook. f., *Rubus niveus* var. *hypargyrus* (Edgew.) Hook.f.
Climbing Mechanism: Scrambler-Armed
Distribution (Global): Assam, China South-Central, East Himalaya, Nepal, Tibet, West Himalaya
Leaf Type: Compound, odd-pinnate, 3–7 foliolate
Inflorescence: Corymbose panicles
Fruit Type: Wet, drupelets
Flowering and Fruiting: June–August
IUCN Status: Not evaluated
Reference: Prodr. Fl. Nepal.: 234 (1825)

Rubus pentagonus Wall. ex Focke
Rosaceae
Climbing Mechanism: Scrambler-Armed
Distribution (Global): Assam, China South-Central, East Himalaya, Myanmar, Nepal, Tibet, Vietnam, West Himalaya
Distribution (India): Arunachal Pradesh, Manipur
Leaf Type: Compound, odd-pinnate, 3–7 foliolate
Inflorescence: Corymbose panicles
Fruit Type: Wet, drupelets
Flowering and Fruiting: April–July
IUCN Status: Not evaluated
Reference: Sp. Rub.: 145 (1911)

Rubus praestans H.E.Weber
Rosaceae
Climbing Mechanism: Scrambler-Armed
Distribution (Global): Belgium, Germany
Leaf Type: Compound, odd-pinnate, 3–7 foliolate
Inflorescence: Corymbose panicles
Fruit Type: Wet, drupelets
Flowering and Fruiting: December–May
IUCN Status: Not evaluated
Reference: Mitt. Pollichia Pfälz. Vereins Naturk. 75: 203 (1988)

Rubus racemosus Roxb.
Rosaceae
Climbing Mechanism: Scrambler-Armed

Distribution (Global): India
Distribution (India): Kerala, Tamil Nadu, Western Ghats
Leaf Type: Pinnately compound imparipinnate
Inflorescence: Axillary
Fruit Type: Globose fruits purple in color
Flowering and Fruiting: December–May
IUCN Status: Not evaluated
Reference: Fl. Ind. ed. 1832, 2: 519 (1832)

Rubus ramachandrae S. S. Dash & Chand. Gupta
Rosaceae
Climbing Mechanism: Scrambler-Armed
Distribution (Global): East Himalaya
Distribution (India): Arunachal Pradesh
Leaf Type: Compound, odd-pinnate, 3–7 foliolate
Inflorescence: Corymbose panicles
Fruit Type: Wet, drupelets
IUCN Status: Not evaluated
Reference: Blumea 63: 26 (2018)

Rubus rosifolius Stokes
Rosaceae
Synonyms: *Rubus rosaefolius*
Climbing Mechanism: Scrambler-Armed
Distribution (India): Manipur, Meghalaya, Tripura
Leaf Type: Compound, odd-pinnate, 3–7 foliolate
Inflorescence: Corymbose panicles
Fruit Type: Wet, drupelets
Flowering and Fruiting: August–November
IUCN Status: Not evaluated
Notes: Occur mostly in Semi-evergreen and evergreen forests
Reference: Bot. Mat. Med. 3: 147 (1812)

Rubus sumatranus Miq.
Rosaceae
Synonyms: *Rubus asper* Wall. ex D.Don., *Rubus dolichocephalus* Hayata., *Rubus indotibetanus* Koidz., *Rubus myriadenus* H.Lév. & Vaniot., *Rubus sorbifolius* Maxim., *Rubus sumatranus* var. *myriadenus* (H.Lév. & Vaniot) W.Lee, *Rubus sumatranus* var. *suichangensis* P.L.Chiu ex
Climbing Mechanism: Scrambler-Armed
Distribution (Global): Assam, Cambodia, China North-Central, China South-Central, China Southeast, East Himalaya, Hainan, Japan, Java, Korea, Laos, Malaya, Myanmar, Nepal, Sri Lanka, Sumatera, Taiwan, Thailand, Tibet, Vietnam
Leaf Type: Compound, odd-pinnate, 3–7 foliolate

Inflorescence: Corymbose panicles
Fruit Type: Wet, drupelets
IUCN Status: Not evaluated
Reference: Fl. Ned. Ind., Eerste Bijv.: 307 (1861)

Rubus thomsonii Focke
Rosaceae
Climbing Mechanism: Scrambler-Armed
Distribution (Global): East Himalayas, Nepal, Tibet
Leaf Type: Compound, odd-pinnate, 3–7 foliolate
Inflorescence: Corymbose panicles
Fruit Type: Wet, drupelets
IUCN Status: Not evaluated
Reference: Abh. Naturwiss. Vereins Bremen 4: 198 (1874)

Rubus treutleri Hook. f.
Rosaceae
Synonyms: *Rubus arcuatus* Kuntze., *Rubus rosulans* Kuntze., *Rubus tonglooensis* Kuntze
Climbing Mechanism: Scrambler-Armed
Distribution (Global): Assam, China South-Central, East Himalaya, Nepal, Tibet
Distribution (India): Mizoram
Leaf Type: Compound, odd-pinnate, 3–7 foliolate
Inflorescence: Corymbose panicles
Fruit Type: Wet, drupelets
IUCN Status: Not evaluated
Reference: Fl. Brit. India 2: 331 (1878)

Rubus ulmifolius
Rosaceae
Climbing Mechanism: Scrambler-Armed
IUCN Status: Not evaluated

Rubus wardii Merr.
Rosaceae
Synonyms: *Rubus gigantiflorus* H.Hara., *Rubus hookeri* Focke., *Rubus macrocarpus* King ex C.B.Clarke
Climbing Mechanism: Scrambler-Armed
Distribution (Global): Assam, China South-Central, East Himalaya, Tibet
Distribution (India): West Bengal
Leaf Type: Compound, odd-pinnate, 3–7 foliolate
Inflorescence: Corymbose panicles
Fruit Type: Wet, drupelets
IUCN Status: Not evaluated
Reference: Brittonia 4: 84 (1941)

Aidia auriculata (Wall.) Ridsdale
Rubiaceae
Synonyms: *Anomanthodia auriculata* (Wall.) Hook.f., *Cupia auriculata* (Wall.) DC., *Pseudixora auriculata* (Wall.) Miq., *Randia auriculata* (Wall.) Steud., *Webera auriculata* Wall.
Climbing Mechanism: Stem Twiner
Distribution (Global): Borneo, Java, Malaya, Philippines, Sumatera, Thailand
IUCN Status: Not evaluated
Reference: Blumea 41: 161 (1996)

Aidiopsis orophila (Miq.) Ridsdale
Rubiaceae
Synonyms: *Aidia forbesii* (King & Gamble) K.M.Wong, *Aidiopsis forbesii* (King & Gamble) Tirveng., *Randia forbesii* King & Gamble, *Randia orophila* (Miq.) Hallier, *Stylocoryna orophila* Miq., *Webera orophila* (Miq.) Boerl.
Climbing Mechanism: Stem Twiner
Distribution (Global): Andaman Is., Java, Malaya, Sumatera, Thailand
IUCN Status: Not evaluated
Reference: Blumea 41: 176 (1996)

Chiococca alba (L.) Hitchc.
Rubiaceae
Synonyms: *Chiococca brachiata* var. *valida* Müll.Arg., *Chiococca capitata* Wernham, *Chiococca densifolia* Mart., *Chiococca densifolia* var. *cubensis* DC., *Chiococca floridana* (DC.) Raf., *Chiococca latifolia* Raf., *Chiococca macrocarpa* M.Martens & Galeotti., *Chiococca micr*
Climbing Mechanism: Scrambler-Unarmed
Distribution (Global): Argentina Northeast, Bahamas, Belize, Bermuda, Bolivia, Brazil North, Brazil Northeast, Brazil South, Brazil Southeast, Brazil West-Central, Cayman Is., Colombia, Costa Rica, Cuba, Dominican Republic, Ecuador, El Salvador, Florida, French Guiana, Galápagos, Guatemala, Guyana, Haiti, Honduras, Jamaica, Leeward Is., Mexican Pacific Is., Mexico Central, Mexico Gulf, Mexico Northeast, Mexico Northwest, Mexico Southeast, Mexico Southwest, Netherlands Antilles, Nicaragua, Panamá, Paraguay, Peru, Puerto Rico, Southwest Caribbean, Texas, Trinidad-Tobago, Turks-Caicos Is., Venezuela, Venezuelan Antilles, Windward Is.
IUCN Status: Least Concern
Reference: Rep. (Annual) Missouri Bot. Gard. 4: 94 (1893)

Coelospermum decipiens Baill.
Rubiaceae
Synonyms: *Morinda reticulata* Benth.
Common Name: Mapoon bush, Yellow dye
Climbing Mechanism: Stem Twiner
Distribution (Global): New Guinea, Queensland

Leaf Type: Oblanceolate to linear-lanceolate
Inflorescence: Terminal umbellate heads
Flowering and Fruiting: March–September
IUCN Status: Not evaluated
Notes: Endemic to Southern Western Ghats. Found mostly in evergreen forests
Reference: Bull. Mens. Soc. Linn. Paris 1: 218 (1879)

Dentella repens (L.) J.R.Forst. & G.Forst.
Rubiaceae
Synonyms: *Hedyotis repens* (L.) Lam., *Oldenlandia repens* L.
Common Name: Creeping Dentella, Creeping lickstoop
Climbing Mechanism: Scrambler-Unarmed
Distribution (Global): Andaman Is., Assam, Bangladesh, Cambodia, Caroline Is., China Southeast, East Himalaya, India, Java, Laccadive Is., Lesser Sunda Is., Malaya, Maluku, Marianas, Marshall Is., Myanmar, Nepal, New Caledonia, New Guinea, Northern Territory, Philippines, Queensland, South China Sea, Sri Lanka, Sulawesi, Taiwan, Thailand, Vietnam
Distribution (India): Manipur
Leaf Type: Subsessile
Inflorescence: Axillary, solitary, sessile
Fruit Type: Globular, indehiscent, tomentose
Flowering and Fruiting: Throughout the year
IUCN Status: Least Concern
Reference: Char. Gen. Pl.: 26 (1776)

Galium aparine L.
Rubiaceae
Synonyms: *Aparine vulgaris* Hill, *Asperula aparine* (L.) Besser, *Asperula aparine* var. *aparine* (L.) Nyman, *Asterophyllum aparine* (L.) K.F.Schimp. & Spenn., *Crucianella purpurea* Wulff ex Steud., *Galion aparinum* (L.) St.-Lag., *Galium aculeatissimum* Kit. ex Kanitz
Common Name: Goose grass, Common bedstraw, Coachweed, Catchweed
Climbing Mechanism: Scrambler-Armed
Distribution (Global): Afghanistan, Albania, Algeria, Altay, Austria, Azores, Baleares, Baltic States, Belarus, Belgium, Bulgaria, Buryatiya, Canary Is., Central European Rus, China North-Central, China South-Central, China Southeast, Chita, Corse, Cyprus, Czechoslovakia, Denmark, East Aegean Is., East European Russia, East Himalaya, Finland, France, Germany, Great Britain, Greece, Gulf States, Hungary, India, Iran, Iraq, Ireland, Irkutsk, Italy, Kirgizstan, Korea, Krasnoyarsk, Kriti, Krym, Lebanon-Syria, Libya, Madeira, Manchuria, Mongolia, Morocco, Nepal, Netherlands, North European Russi, Northwest European R, Norway, Oman, Pakistan, Palestine, Poland, Portugal, Romania, Sardegna, Saudi Arabia, Selvagens, Sicilia, South European Russi, Spain, Sweden, Switzerland, Tadzhikistan, Tibet, Tunisia, Turkey, Turkey-in-Europe, Tuva, Ukraine, Vietnam, West Himalaya, West Siberia, Yakutskiya, Yugoslavi

Distribution (India): Arunachal Pradesh, Himachal Pradesh
Leaf Type: Simple
IUCN Status: Not evaluated
Reference: Sp. Pl.: 108 (1753)

Galium asperifolium Wall.
Rubiaceae
Synonyms: *Galium mollugo* subsp. *asperifolium* (Wall.) Kitam.
Common Name: Rough-leaved clivers.
Climbing Mechanism: Scrambler-Unarmed
Distribution (Global): Afghanistan, Assam, Bangladesh, China South-Central, China Southeast, East Himalaya, India, Myanmar, Nepal, Pakistan, Sri Lanka, Sumatera, Thailand, Tibet, West Himalaya
Distribution (India): Andhra Pradesh, Arunachal Pradesh, Himachal Pradesh, Meghalaya, Mizoram, Odisha, Tamil Nadu
Leaf Type: Leaves whorled
Inflorescence: Axillary, Solitary
Flowering and Fruiting: March–May
IUCN Status: Not evaluated
Reference: W.Roxburgh, Fl. Ind. 1: 381 (1820)

Galium verum L.
Rubiaceae
Synonyms: *Asterophyllum galium* K.F.Schimp. & Spenn., *Galium floridum* Salisb., *Galium verum* subsp. *euverum* Hyl., *Rubia vera* (L.) Baill.
Common Name: Lady's bedstraw, Yellow bedstraw
Climbing Mechanism: Scrambler-Armed
Distribution (Global): Albania, Algeria, Altay, Austria, Baltic States, Belarus, Belgium, Bulgaria, Buryatiya, Central European Rus, China North-Central, China South-Central, China Southeast, Chita, Czechoslovakia, Denmark, East Aegean Is., East European Russia, Finland, France, Germany, Great Britain, Greece, Hungary, Iceland, Inner Mongolia, Iran, Iraq, Ireland, Irkutsk, Italy, Japan, Kamchatka, Kazakhstan, Kirgizstan, Korea, Krasnoyarsk, Krym, Kuril Is., Lebanon-Syria, Manchuria, Mongolia, Morocco, Netherlands, North Caucasus, North European Russi, Northwest European R, Norway, Pakistan, Palestine, Poland, Portugal, Primorye, Qinghai, Romania, Sakhalin, Sardegna, Sicilia, South European Russi, Spain, Sweden, Switzerland, Tasmania, Tibet, Transcaucasus, Tunisia, Turkey, Turkey-in-Europe, Turkmenistan, Tuva, Ukraine, Uzbekistan, West Himalaya, West Siberia, Xinjiang, Yakutskiya, Yugoslavia
Distribution (India): Himachal Pradesh
IUCN Status: Not evaluated
Reference: Sp. Pl.: 107 (1753)

Gynochthodes macrophylla Kurz
Rubiaceae
Climbing Mechanism: Stem Twiner
Distribution (Global): Andaman Is., Malaya, Nicobar Is., Thailand
IUCN Status: Not evaluated
Reference: J. Asiat. Soc. Bengal, Pt. 2, Nat. Hist. 41: 314 (1872)

Gynochthodes sublanceolata Miq.
Rubiaceae
Synonyms: *Gynochthodes coriacea* Blume
Climbing Mechanism: Stem Twiner
IUCN Status: Not evaluated
Reference: Fl. Ned. Ind., Eerste Bijv.: 548 (1861)

Hedyotis miqueliana Valeton
Rubiaceae
Synonyms: *Scleromitrion capitatum* Miq.
Climbing Mechanism: Scrambler-Unarmed
IUCN Status: Not evaluated
Reference: Bot. Jahrb. Syst. 44: 545 (1910)

Hedyotis scandens Roxb.
Rubiaceae
Synonyms: *Hedyotis polycarpa* R.Br. ex G.Don, *Hedyotis subscandens* Griff., nom. subnud., *Hedyotis volubilis* R.Br. ex Wall., nom. superfl., *Oldenlandia scandens* (Roxb.) Kuntze, *Petesia hita* Buch.-Ham. ex D.Don.
Common Name: Climbing diamond flowe
Climbing Mechanism: Scrambler-Unarmed
Distribution (India): Arunachal Pradesh, Assam, Bihar, Odisha, Manipur, Mizoram, Sikkim, Tripura, West Bengal
Leaf Type: Simple, opposite decussate
Inflorescence: Axillary or terminal panicles of cymes
Fruit Type: Capsule
Flowering and Fruiting: March–December
IUCN Status: Not evaluated
Reference: Fl. Ind. 1: 364 (1820)

Leucocodon reticulatum Gardner
Rubiaceae
Climbing Mechanism: Root Climber
Distribution (Global): Sri Lanka.
IUCN Status: Not evaluated
Reference: Calcutta J. Nat. Hist. 7: 5 (1847)

Manettia cordifolia Mart
Rubiaceae
Synonyms: *Guagnebina ignita* Vell., *Lygistum ignitum* (Vell.) Kuntze. +2
Common Name: Red firecracker vine
Climbing Mechanism: Scrambler-Unarmed
Distribution (Global): Argentina Northeast, Argentina Northwest, Bolivia, Brazil North, Brazil Northeast, Brazil South, Brazil Southeast, Brazil West-Central, Colombia, Paraguay, Peru, Urugua
Distribution (India): Tamil Nadu
IUCN Status: Not evaluated
Reference: Denkschr. Königl. Akad. Wiss. München 9: 95 (1824)

Morinda umbellata L.
Rubiaceae
Climbing Mechanism: Stem Twiner
Distribution (India): Andhra Pradesh, Arunachal Pradesh, Eastern Ghats, Jharkhand, Madhya Pradesh, Meghalaya, Tamil Nadu
Leaf Type: Simple, interpetiolar stipules, elliptic
Inflorescence: Umbellate heads, 5-merous
Fruit Type: wet
IUCN Status: Not evaluated

Morinda villosa Hook.f.
Rubiaceae
Climbing Mechanism: Stem Twiner
Distribution (India): Arunachal Pradesh
Leaf Type: Simple, interpetiolar stipules, elliptic
Inflorescence: Umbellate heads, 5-merous
Fruit Type: wet
IUCN Status: Not evaluated

Mussaenda erythrophylla
Rubiaceae
Climbing Mechanism: Scrambler-Unarmed
IUCN Status: Least Concern

Mussaenda frondosa L.
Rubiaceae
Climbing Mechanism: Scrambler-Unarmed
Distribution (India): Arunachal Pradesh, Maharashtra, Tamil Nadu, Tripura, Western Ghats
IUCN Status: Not evaluated

Mussaenda glabra Vahl
Rubiaceae

Climbing Mechanism: Scrambler-Unarmed
Distribution (India): Arunachal Pradesh, Manipur, Meghalaya, Mizoram, West Bengal
IUCN Status: Not evaluated

Mussaenda glabrata (Hook.f.) Hutch. ex Gamble
Rubiaceae
Climbing Mechanism: Scrambler-Unarmed
Distribution (India): Gujarat, Maharashtra, Tamil Nadu
IUCN Status: Not evaluated

Mussaenda hirsutissima (Hook.f.) Hutch. ex Gamble
Rubiaceae
Climbing Mechanism: Stem Twiner
Distribution (India): Tamil Nadu.
IUCN Status: Not evaluated

Mussaenda macrophylla Wall.
Rubiaceae
Climbing Mechanism: Scrambler-Unarmed
Distribution (India): Arunachal Pradesh, Great, Nicobar, Island, Manipur, Meghalaya, West Bengal
IUCN Status: Not evaluated

Neohymenopogon assamicus (Hook.f.) Bennet
Rubiaceae
Synonyms: *Dunnia assamica* (Hook.f.) Ridsdale
Climbing Mechanism: Scrambler-Unarmed
Distribution (Global): Assam
IUCN Status: Not evaluated
Reference: Indian Forester 107: 436 (1981)

Neurocalyx calycinus (R.Br. ex Benn.) Rob.
Rubiaceae
Synonyms: *Argostemma calycinum* R.Br. ex Benn., *Neurocalyx capitata* Benth. ex Hook.f., *Neurocalyx hookerianus* Wight, *Neurocalyx wightii* Arn.
Climbing Mechanism: Scrambler-Unarmed
Distribution (Global): India, Sri Lanka
Flowering and Fruiting: March–July
IUCN Status: Not evaluated
Reference: Proc. Amer. Acad. Arts 45: 402 (1910)

Oxyceros longiflorus (Lam.) T.Yamaz.
Rubiaceae
Synonyms: *Aidia oppositifolia* var. *floribunda* (Kurz) Alam & S.N.Uddin +22

Climbing Mechanism: Scrambler-Unarmed
Distribution (Global): Andaman Is., Bangladesh, Borneo, Cambodia, Java, Lesser Sunda Is., Malaya, Nicobar Is., Sumatera, Thailand, Vietnam
Distribution (India): Great Nicobar Island, Mizoram
Leaf Type: Simple, opposite decussate
Inflorescence: Cymes
Fruit Type: Berries
Flowering and Fruiting: April–January
IUCN Status: Not evaluated
Reference: J. Jap. Bot. 45: 339 (1970)

Oxyceros rugulosus (Thwaites) Tirveng.
Rubiaceae
Synonyms: *Aidia rugulosa* (Thwaites) Swamin., *Griffithia rugulosa* Thwaites, *Randia rugulosa* (Thwaites) Hook. f.
Climbing Mechanism: Stem Twiner
Distribution (Global): India, Sri Lanka
Distribution (India): Eastern Ghats, Maharashtra, Tamil Nadu, Western Ghats
Leaf Type: Simple
Inflorescence: Cymes lateral or axillary
Fruit Type: Berry
Flowering and Fruiting: February–September
IUCN Status: Not evaluated
Notes: Mostly found in semi-evergreen forest
Reference: Nordic J. Bot. 3: 466 (1983)

Paederia calycina Kurz
Rubiaceae
Synonyms: *Hondbesseion calycinum* (Kurz) Kuntze
Climbing Mechanism: Stem Twiner
Distribution (Global): Myanmar, Thailand
IUCN Status: Not evaluated
Reference: J. Asiat. Soc. Bengal, Pt. 2, Nat. Hist. 42: 74 (1873)

Paederia cruddasiana Prain
Rubiaceae
Synonyms: *Paederia aestivalis* Buch. -Ham. ex Wall., *Paederia autumnalis* Buch. -Ham. ex Wall., *Paederia cruddasiana* subsp. *microcarpa* (Kurz) Puff.
Climbing Mechanism: Stem Twiner
Distribution (Global): Assam, Bangladesh, China South-Central, East Himalaya, India, Myanmar, Nepal, Thailand, Vietnam
Distribution (India): West Bengal
IUCN Status: Not evaluated
Reference: J. Asiat. Soc. Bengal, Pt. 2, Nat. Hist. 67: 295 (1898)

Paederia foetida L.
Rubiaceae
Synonyms: *Apocynum foetidum* Burm.f., *Gentiana scandens* Lour.+30
Climbing Mechanism: Stem Twiner
Distribution (Global): E. Nepal to Japan and Malesia
Distribution (India): Andaman and Nicobar Islands, Andhra Pradesh, Assam, Bihar, Odisha, Jharkhand, Maharashtra, Manipur, Odisha, Telangana, Tripura, Uttar Pradesh, West Bengal
Leaf Type: Simple
Inflorescence: Axillary or terminal, paniculate, thyrsiform, corymbiform, or cymose
Fruit Type: Globose, Shiny brown capsule
Flowering and Fruiting: August–September
IUCN Status: Not evaluated
Reference: Mant. Pl. 1: 52 (1767)

Paederia lanuginosa Wall.
Rubiaceae
Synonyms: *Hondbesseion lanuginosum* (Wall.) Kuntze, *Paederia macrocarpa* Wall. ex G. Don.
Climbing Mechanism: Stem Twiner
Distribution (Global): Assam, Bangladesh, China South-Central, India, Myanmar, Thailand, Vietnam
Distribution (India): Mizoram
IUCN Status: Not evaluated
Reference: Pl. Asiat. Rar. 2: 52 (1831)

Paederia linearis Hook.f.
Rubiaceae
Synonyms: *Hondbesseion lineare* (Hook.f.) Kuntze
Climbing Mechanism: Stem Twiner
Distribution (Global): Cambodia, Laos, Myanmar, Thailand, Vietnam
IUCN Status: Not evaluated
Reference: Fl. Brit. India 3: 197 (1881)

Paederia pilifera Hook. f.
Rubiaceae
Synonyms: *Hondbesseion piliferum* (Hook.f.) Kuntze, *Paederia kerrii* Craib, *Hondbesseion wallichii* (Hook.f.) Kuntze, *Paederia wallichii* Hook.f.
Climbing Mechanism: Stem Twiner
Distribution (Global): Bangladesh, Laos, Myanmar, Thailand, Vietnam
IUCN Status: Not evaluated
Reference: Fl. Brit. India 3: 196 (1881)

Paederia verticillata Blume
Rubiaceae

Synonyms: *Hondbesseion verticillatum* (Blume) Kuntze
Climbing Mechanism: Stem Twiner
Distribution (Global): Borneo, Java, Malaya, Philippines, Sulawesi, Sumatera
IUCN Status: Not evaluated
Reference: Bijdr. Fl. Ned. Ind.: 968 (1826)

Pseudaidia speciosa (Bedd.) Tirveng
Rubiaceae
Synonyms: *Aidia rugulosa* var. *speciosa* (Bedd.) Swamin., *Griffithia speciosa* Bedd., *Randia rugulosa* var. *speciosa* Hook. f.
Climbing Mechanism: Scrambler-Unarmed
Distribution (Global): India, Sri Lanka
Distribution (India): Tamil Nadu, Western Ghats
Flowering and Fruiting: January–June
IUCN Status: Not evaluated
Notes: Usually found in evergreen forests
Reference: Bull. Mus. Natl. Hist. Nat., B, Adansonia 8: 287 (1986)

Psychotria adenophylla Wall.
Rubiaceae
Synonyms: *Chassalia hasseltiana* Miq., *Grumilea adenophylla* (Wall.) Miq., *Psychotria connata* Kurz, *Psychotria hasseltiana* (Miq.) Boerl., *Psychotria siamensis* Ridl., *Uragoga adenophylla* (Wall.) Kuntze
Climbing Mechanism: Scrambler-Unarmed
Distribution (Global): Andaman Is., Assam, Bangladesh, Cambodia, India, Java, Laos, Malaya, Myanmar, Thailand, Vietna
Distribution (India): Arunachal Pradesh
Leaf Type: Simple
Inflorescence: Axillary or terminal
Flowering and Fruiting: May–January
IUCN Status: Not evaluated
Reference: W.Roxburgh, Fl. Ind. 2: 166 (1824)

Psychotria sarmentosa Blume
Rubiaceae
Synonyms: *Mapouria sarmentosa* (Blume) K.Schum., *Uragoga sarmentosa* (Blume) Kuntze
Climbing Mechanism: Root Climber
Distribution (Global): Andaman Is., Borneo, Cambodia, India, Java, Lesser Sunda Is., Malaya, Myanmar, Nicobar Is., Philippines, Sri Lanka, Sumatera, Thailand, Vietnam
Distribution (India): Great Nicobar Island, Tamil Nadu
Flowering and Fruiting: January–March
IUCN Status: Not evaluated
Reference: Bijdr. Fl. Ned. Ind.: 964 (1826)

Rothmannia longiflora Salisb.
Rubiaceae
Synonyms: *Gardenia jasminiflora* Zipp. ex Span., *Gardenia speciosa* A.Rich., *Gardenia stanleyana* Hook. ex Lindl., *Randia longiflora* (Salisb.) T.Durand & Schinz, *Randia maculata* DC., *Randia sapinii* De Wild., *Randia spathacea* De Wild.
Climbing Mechanism: Scrambler-Unarmed
Distribution (Global): Angola, Benin, Burkina, Burundi, Cabinda, Cameroon, Central African Repu, Congo, Gabon, Gambia, Ghana, Guinea, Guinea-Bissau, Gulf of Guinea Is., Ivory Coast, Kenya, Liberia, Malawi, Nigeria, Sierra Leone, Sudan, Tanzania, Togo, Uganda, Zaire
Leaf Type: Simple, opposite
Inflorescence: Solitary flowers, trumpet shaped
Fruit Type: Wet, berry
IUCN Status: Least Concern
Reference: Parad. Lond. 1: t. 65 (1807)

Rubia angustisissima Wall. ex G.Don
Rubiaceae
Climbing Mechanism: Scrambler-Unarmed
Distribution (Global): Myanmar
Distribution (India): West Bengal
Leaf Type: Leaves whorled
Inflorescence: Cymes, dichasial
Fruit Type: Wet, drupe
IUCN Status: Not evaluated
Reference: Gen. Hist. 3: 643 (1834)

Rubia cordifolia L.
Rubiaceae
Synonyms: *Galium cordifolium* (L.) Kuntze
Common Name: Indian madder
Climbing Mechanism: Scrambler-Unarmed
Distribution (Global): Afghanistan, Amur, Angola, Assam, Bangladesh, Borneo, Burundi, Buryatiya, Cape Provinces, China North-Central, China South-Central, China Southeast, Chita, East Himalaya, Ethiopia, Free State, Greece, India, Inner Mongolia, Japan, Java, Kenya, Khabarovsk, Korea, KwaZulu-Natal, Malawi, Malaya, Mongolia, Mozambique, Myanmar, Nepal, Northern Provinces, Pakistan, Primorye, Qinghai, Rwanda, Somalia, Sri Lanka, Sudan, Sumatera, Swaziland, Taiwan, Tanzania, Tibet, Uganda, Vietnam, West Himalaya, Zambia, Zaire, Zimbabwe
Distribution (India): Andhra Pradesh, Arunachal Pradesh, Bihar, Odisha, Himachal Pradesh, Maharashtra, Manipur, Meghalaya, Mizoram, Tamil Nadu
Leaf Type: Leaves whorled
Inflorescence: Cymes, dichasial

Fruit Type: Wet, drupe
Flowering and Fruiting: August–February
IUCN Status: Not evaluated
Notes: Found mostly in semi-evergreen and moist-deciduous forests
Reference: Syst. Nat. ed. 12: 229 (1767)

Rubia hispidicaulis D.G.Long
Rubiaceae
Synonyms: *Rubia cordifolia* f. strigosa Deb & Malick
Climbing Mechanism: Scrambler-Unarmed
Distribution (Global): East Himalaya
Distribution (India): West Bengal
Leaf Type: Leaves whorled
Inflorescence: Cymes, dichasial
Fruit Type: Wet, drupe
IUCN Status: Not evaluated
Reference: Edinburgh J. Bot. 53: 109 (1996)

Rubia manjith Roxb. ex Fleming
Rubiaceae
Synonyms: *Rubia munjista* Roxb.
Climbing Mechanism: Scrambler-Unarmed
Distribution (Global): Assam, China South-Central, East Himalaya, India, Nepal, Qinghai, Tibet, West Himalaya
Distribution (India): Himachal Pradesh, Madhya Pradesh, Sikkim, West Bengal
Leaf Type: Leaves whorled
Inflorescence: Cymes, dichasial
Fruit Type: Wet, drupe
IUCN Status: Not evaluated
Reference: Asiat. Res. 11: 177 (1810)

Rubia sikkimensis Kurz
Rubiaceae
Climbing Mechanism: Scrambler-Unarmed
Distribution (Global): Assam, East Himalaya, Nepal
Distribution (India): Arunachal Pradesh, Manipur, Mizoram, West Bengal
Leaf Type: Leaves whorled
Inflorescence: Cymes, dichasial
Fruit Type: Wet, drupe
IUCN Status: Not evaluated
Reference: J. Asiat. Soc. Bengal, Pt. 2, Nat. Hist. 43: 188 (1874)

Rubia tinctorum
Rubiaceae

Climbing Mechanism: Scrambler-Unarmed
IUCN Status: Not evaluated

Rubia wallichiana Decne.
Rubiaceae
Synonyms: *Rubia asperrima* Decne.
Climbing Mechanism: Scrambler-Unarmed
Distribution (Global): China South-Central, China Southeast, East Himalaya, Hainan, Nepal, West Himalaya
Distribution (India): West Bengal
Leaf Type: Leaves whorled
Inflorescence: Cymes, dichasial
Fruit Type: Wet, drupe
IUCN Status: Not evaluated
Reference: Mém. Couronnés Acad. Roy. Sci. Bruxelles (4to) 12: 61 (1837)

Uncaria acida (Hunter) Roxb.
Rubiaceae
Synonyms: *Nauclea acida* W.Hunter., *Ourouparia acida* (W.Hunter) Baill., *Uruparia acida* (W.Hunter) Kuntze
Climbing Mechanism: Scrambler-Unarmed
Distribution (Global): Bangladesh, Borneo, Java, Lesser Sunda Is., Malaya, Myanmar, New Guinea, Philippines, Sulawesi, Sumatera, Thailand, Vietnam
Leaf Type: Simple
Inflorescence: Many in clusters
Fruit Type: Wet, drupe
IUCN Status: Not evaluated
Reference: Fl. Ind. 2: 129 (1824)

Uncaria attenuata Korth.
Rubiaceae
Synonyms: *Nauclea attenuata* (Korth.) Walp., *Uncaria athemiata* Treub., *Uncaria bulusanensis* Elmer., *Uruparia attenuata* (Korth.) Kuntze
Climbing Mechanism: Scrambler-Unarmed
Distribution (Global): Andaman Is., Borneo, Java, Malaya, Philippines, Sulawesi, Sumatera, Thailand
Leaf Type: Simple
Inflorescence: Many in clusters
Fruit Type: Wet, drupe
IUCN Status: Not evaluated
Reference: Verh. Nat. Gesch. Ned. Bezitt., Bot.: 170 (1842)

Uncaria callophylla Blume ex Korth.
Rubiaceae

Synonyms: *Uncaria avenia* Valeton., *Uncaria forbesii* Wernham., *Uncaria jasminiflora* Wall. ex Hook.f., *Uncaria luzoniensis* Merr., *Uncaria wrayi* King., *Uruparia callophylla* (Blume ex Korth.) Kuntze., *Uruparia jasminiflora* (Wall. ex Hook.f.) Kuntze
Climbing Mechanism: Scrambler-Unarmed
Distribution (Global): Borneo, Malaya, Maluku, New Guinea, Philippines, Queensland, Sumatera, Thailand
Leaf Type: Simple
Inflorescence: Many in clusters
Fruit Type: Wet, drupe
IUCN Status: Not evaluated
Reference: Verh. Nat. Gesch. Ned. Bezitt., Bot.: 170 (1842)

Uncaria canescens Korth.
Rubiaceae
Synonyms: *Nauclea canescens* (Korth.) Walp.
Climbing Mechanism: Scrambler-Unarmed
Distribution (Global): Malaya, Sumatera, Thailand
Leaf Type: Simple
Inflorescence: Many in clusters
Fruit Type: Wet, drupe
IUCN Status: Not evaluated
Reference: Verh. Nat. Gesch. Ned. Bezitt., Bot.: 172 (1842)

Uncaria cordata (Lour.) Merr.
Rubiaceae
Synonyms: *Nauclea ferruginea* Blume., *Nauclea hallii* (Korth.) Walp., *Nauclea insignis* (Bartl. ex DC.) D.Dietr., *Nauclea lanosa* Poir., *Nauclea nemorosa* (Korth.) Walp. +30
Climbing Mechanism: Scrambler-Unarmed
Distribution (Global): Andaman Is., Bangladesh, Borneo, Cambodia, Java, Laos, Lesser Sunda Is., Malaya, Maluku, Myanmar, New Guinea, Nicobar Is., Philippines, Queensland, Sulawesi, Sumatera, Thailand, Vietnam
Distribution (India): Andaman and Nicobar Islands
Leaf Type: Simple
Inflorescence: Many in clusters
Fruit Type: Wet, drupe
IUCN Status: Not evaluated
Reference: Interpr. Herb. Amboin.: 479 (1917)

Uncaria elliptica R.Br. ex G.Don
Rubiaceae
Synonyms: *Nauclea dasyoneura* (Korth.) Walp., *Nauclea elliptica* (R.Br. ex G.Don) Walp., *Uncaria brevispina* Maingay ex Hook.f., *Uncaria dasyoneura* Korth.,

Uncaria gambier Thwaites., *Uncaria rostrata* Pierre ex Pit., *Uncaria thwaitesii* (Hook.f.) Alston., *Uruparia dasy*
Climbing Mechanism: Scrambler-Unarmed
Distribution (Global): Borneo, Cambodia, Java, Malaya, Myanmar, Sri Lanka, Sumatera, Thailand
Leaf Type: Simple
Inflorescence: Many in clusters
Fruit Type: Wet, drupe
IUCN Status: Not evaluated
Reference: Gen. Hist. 3: 471 (1834)

Uncaria gambir (Hunter) Roxb.
Rubiaceae
Synonyms: *Cinchona kattukambar* J.Koenig ex Retz., *Nauclea gambir* W.Hunter., *Ourouparia gambir* (W.Hunter) Baill., *Uncaria yunnanensis* K.C.Hsia., *Uruparia gambir* (W.Hunter) Kuntze
Climbing Mechanism: Scrambler-Unarmed
Distribution (Global): Borneo, Java, Malaya, Sumatera
Leaf Type: Simple
Inflorescence: Many in clusters
Fruit Type: Wet, drupe
IUCN Status: Not evaluated
Reference: Fl. Ind. 2: 126 (1824)

Uncaria homomalla Miq.
Rubiaceae
Synonyms: *Uncaria parviflora* Ridl., *Uncaria quadrangularis* E.T.Geddes., *Uncaria tonkinensis* Havil., *Uruparia homomalla* (Miq.) Kuntze
Climbing Mechanism: Scrambler-Unarmed
Distribution (Global): Assam, Bangladesh, Cambodia, China South-Central, China Southeast, Laos, Malaya, Myanmar, Sumatera, Thailand, Vietnam
Distribution (India): Manipur
Leaf Type: Simple
Inflorescence: Many in clusters
Fruit Type: Wet, drupe
IUCN Status: Not evaluated
Reference: Fl. Ned. Ind. 2: 343 (1857)

Uncaria laevigata Wall. ex G.Don
Rubiaceae
Synonyms: *Nauclea laevigata* (Wall. ex G.Don) Walp., *Uncaria dasycarpa* Pierre ex Pit., *Uruparia laevigata* (Wall. ex G.Don) Kuntze
Common Name: Smooth Uncaria, Smooth gambier
Climbing Mechanism: Scrambler-Unarmed

Distribution (Global): Assam, Bangladesh, China South-Central, China Southeast, Laos, Myanmar, Thailand, Vietnam.
Distribution (India): Manipur
Leaf Type: Elliptic
Flowering and Fruiting: May–November
IUCN Status: Not evaluated
Reference: Gen. Hist. 3: 470 (1834)

Uncaria lanosa var. *ferrea* (Blume) Ridsdale
Rubiaceae
Synonyms: *Nauclea ferrea* Blume., *Ourouparia ferrea* (Blume) K.Schum., *Uncaria ferrea* (Blume) DC., *Uncaria horsfieldiana* Miq., *Uruparia ferrea* (Blume) Kuntze
Climbing Mechanism: Scrambler-Unarmed
Distribution (Global): Borneo, Cambodia, Java, Lesser Sunda Is., Malaya, Myanmar, New Guinea, Nicobar Is., Philippines, Sumatera, Thailand, Vietnam
Leaf Type: Simple
Inflorescence: Many in clusters
Fruit Type: Wet, drupe
IUCN Status: Not evaluated
Reference: Blumea 24: 87 (1978)

Uncaria lanosa var. *glabrata* (Blume) Ridsdale
Rubiaceae
Synonyms: *Nauclea glabrata* Blume., *Uncaria ferrea* subsp. *glabrata* (Blume) Kuntze., *Uncaria glabrata* (Blume) DC., *Uncaria lobbii* Hook.f., *Uruparia lobbii* (Hook.f.) Kuntze
Climbing Mechanism: Scrambler-Unarmed
Distribution (Global): Borneo, Java, Malaya, Sumatera, Thailand
Leaf Type: Simple
Inflorescence: Many in clusters
Fruit Type: Wet, drupe
IUCN Status: Not evaluated
Reference: Blumea 24: 86 (1978)

Uncaria lanosa Wall.
Rubiaceae
Synonyms: *Uruparia lanosa* (Wall.) Kuntze
Climbing Mechanism: Scrambler-Unarmed
Distribution (Global): Andaman Is., Bismarck Archipelago, Borneo, Cambodia, Caroline Is., Java, Lesser Sunda Is., Malaya, Maluku, Myanmar, New Guinea, Nicobar Is., Philippines, Queensland, Solomon Is., Sulawesi, Sumatera, Taiwan, Thailand, Vietnam
Leaf Type: Simple
Inflorescence: Many in clusters

Fruit Type: Wet, drupe
IUCN Status: Not evaluated
Reference: W.Roxburgh, Fl. Ind. 2: 131 (1824)

Uncaria longiflora var. *pteropoda* (Miq.) Ridsdale
Rubiaceae
Synonyms: *Uncaria longiflora* (Poir.) Merr., *Bancalus rotundifolius* Kuntze., *Bancalus rotundifolius* Kuntze. +10
Climbing Mechanism: Scrambler-Unarmed
Distribution (Global): Borneo, Malaya, Maluku, New Guinea, Philippines, Sulawesi, Sumatera, Thailand
Leaf Type: Simple
Inflorescence: Many in clusters
Fruit Type: Wet, drupe
IUCN Status: Not evaluated
Reference: Interpr. Herb. Amboin.: 480 (1917)

Uncaria macrophylla Wall.
Rubiaceae
Synonyms: *Nauclea grandifolia* Spreng., *Nauclea latifolia* F.Dietr., *Nauclea silhetana* D.Dietr., *Uncaria sessilifolia* Roxb. ex Kurz., *Uruparia macrophylla* (Wall.) Kuntze
Climbing Mechanism: Scrambler-Unarmed
Distribution (Global): Assam, Bangladesh, Cambodia, China South-Central, China Southeast, East Himalaya, Hainan, Laos, Myanmar, Thailand, Vietnam
Distribution (India): Arunachal Pradesh, Assam, Meghalaya, Mizoram, West Bengal
Leaf Type: Leaves simple, opposite decussate
Inflorescence: Axillary and terminal
Fruit Type: Capsules linear ellipsoid
Flowering and Fruiting: July–December
IUCN Status: Not evaluated
Notes: Mostly found in secondary forests.
Reference: W.Roxburgh, Fl. Ind. 2: 132 (1824)

Uncaria perrottetii (A.Rich.) Merr.
Rubiaceae
Synonyms: *Ourouparia perrottetii* (A.Rich.) Baill., *Sabicea perrottetii* A.Rich., *Uncaria ferrea* Fern.-Vill., *Uncaria hookeri* S.Vidal.
Climbing Mechanism: Scrambler-Unarmed
Distribution (Global): Borneo, Philippines
Leaf Type: Simple
Inflorescence: Many in clusters
Fruit Type: Wet, drupe
IUCN Status: Not evaluated
Reference: Philipp. J. Sci., C 8: 60 (1913)

Uncaria roxburghiana Korth.
Rubiaceae
Synonyms: *Nauclea roxburghiana* (Korth.) Walp., *Uncaria brevicarpa* Elmer., *Uncaria roxburghii* Korth., *Uruparia roxburghiana* (Korth.) Kuntze
Climbing Mechanism: Scrambler-Unarmed
Distribution (Global): Borneo, Malaya, Philippines, Sumatera
Leaf Type: Simple
Inflorescence: Many in clusters
Fruit Type: Wet, drupe
IUCN Status: Not evaluated
Reference: Verh. Nat. Gesch. Ned. Bezitt., Bot.: 172 (1842)

Uncaria scandens (Sm.) Hutch.
Rubiaceae
Synonyms: *Cephalanthus cavaleriei* H.Lév., *Nauclea pilosa* (Roxb.) Blume., *Nauclea scandens* Sm., *Uncaria pilosa* Roxb., *Uncaria wangii* F.C.How., *Uruparia pilosa* (Roxb.) Kuntze
Climbing Mechanism: Scrambler-Unarmed
Distribution (Global): Assam, Bangladesh, China South-Central, China Southeast, East Himalaya, Hainan, Laos, Myanmar, Nepal, Thailand, Tibet, Vietnam, West Himalaya
Distribution (India): Arunachal Pradesh, West Bengal
Leaf Type: Leaves simple, opposite decussate
Inflorescence: Axillary and often in terminal groups
Fruit Type: Capsule
Flowering and Fruiting: February–November
IUCN Status: Not evaluated
Reference: W.Roxburgh, Fl. Ind. 2: 130 (1824)

Uncaria sessilifructus Roxb.
Rubiaceae
Synonyms: *Nauclea ovalifolia* (Roxb.) Spreng., *Nauclea scandens* Roxb. ex Hook. f., *Nauclea sessilifructus* (Roxb.) D.Dietr., *Nauclea sessilis* Spreng., *Nauclea uncaria* D.Dietr., *Uncaria ovalifolia* Roxb., *Uruparia ovalifolia* (Roxb.) Kuntze
Climbing Mechanism: Scrambler-Unarmed
Distribution (Global): Assam, Bangladesh, China South-Central, China Southeast, East Himalaya, India, Laos, Myanmar, Nepal, Vietnam
Distribution (India): Andaman and Nicobar Islands, Andhra Pradesh, Arunachal Pradesh, Meghalaya, Mizoram, Tamil Nadu, Tripura, and West Bengal
Leaf Type: Simple
Inflorescence: Many in clusters
Fruit Type: Wet, drupe
Flowering and Fruiting: August–February
IUCN Status: Not evaluated
Reference: Fl. Ind. 2: 130 (1824)

Harrisonia bennettii A.W.Benn.
Rutaceae
Synonyms: *Harrisonia perforata* (Blanco) Merr.
Climbing Mechanism: Scrambler-Unarmed
Distribution (India): Andaman and Nicobar Islands
IUCN Status: Not evaluated
Reference: J. D. Hooker, Fl. Brit. India 1: 519 (1875)

Harrisonia brownii A.Juss.
Rutaceae
Synonyms: *Ebelingia brownii* Steud., *Harrisonia browniana* Gaudich.
Climbing Mechanism: Scrambler-Unarmed
Distribution (Global): Andaman Is., Java, Lesser Sunda Is., Maluku, Myanmar, New Guinea, Nicobar Is., Philippines, Solomon Is., Sulawesi
Distribution (India): Andaman and Nicobar Islands
IUCN Status: Least Concern
Reference: Mém. Mus. Hist. Nat. 12: 540 (1825)

Harrisonia perforata (Blanco) Merr.
Rutaceae
Synonyms: *Anisifolium pubescens* (Wall.) Kuntze, *Ebelingia paucijuga* (Benn.) Kuntze, *Fagara piperita* Blanco, *Feroniella pubescens* (Wall.) Tanaka, *Harrisonia bennetii* (Planch.) A. W. Benn., *Harrisonia citrinaecarpa* Elmer, *Harrisonia paucijuga* (Benn.) Oliv
Climbing Mechanism: Scrambler-Unarmed
Distribution (Global): Andaman Is., Bangladesh, Borneo, Cambodia, China Southeast, Hainan, Java, Laos, Lesser Sunda Is., Malaya, Myanmar, Nicobar Is., Philippines, Sulawesi, Sumatera, Thailand, Vietnam
Distribution (India): Andaman and Nicobar Islands
IUCN Status: Least Concern
Reference: Philipp. J. Sci., C 7: 236 (1912)

Luvunga eleutherandra Dalzell
Rutaceae
Synonyms: *Rhamnus oppositifolia* Noronha, *Lampetia sarmentosa* (Blume) M. Roem., *Luvunga crassifolia* Tanaka, *Luvunga eleutherandra* Dalzell, *Luvunga linearis* Tanaka, *Luvunga tavoyana* Lindl., *Triphasia sarmentosa* Blume
Climbing Mechanism: Scrambler-Armed
Distribution (Global): Borneo, Sumatera, Vietnam
Distribution (India): Peninsular India
IUCN Status: Not evaluated
Notes: Distributed in evergreen and semi-evergreen forests of the Western Ghats
Reference: J. Asiat. Soc. Bengal, Pt. 2, Nat. Hist. 39: 69 (1870)

Luvunga sarmentosa Kurz
Rutaceae
Climbing Mechanism: Scrambler-Armed
Distribution (India): Karnataka, Tamil Nadu, Western Ghats
IUCN Status: Not evaluated

Luvunga scandens (Roxb.) Buch. -Ham. ex Wight & Arn.
Rutaceae
Synonyms: *Limonia scandens* Roxb., *Luvunga motleyi* (Pierre) Oliv., *Luvunga nitida* Pierre
Common Name: Lavang lata, Indian lavang
Climbing Mechanism: Scrambler-Armed
Distribution (Global): Assam, Bangladesh, Borneo, Cambodia, China South-Central, China Southeast, Hainan, Laos, Malaya, Myanmar, Thailand, Vietnam
Distribution (India): Assam, West Benga
Leaf Type: Leaves are compound, with 3 leaflets that are lancelike and leathery
Inflorescence: Leaf stalks are channelled. Peduncles carrying 4-12 pretty large, white, fragrant flowers arise from leaf axils
Flowering and Fruiting: March–April
IUCN Status: Not evaluated
Reference: Prodr. Fl. Ind. Orient. 1: 90 (1834)

Merope angulata Swingle
Rutaceae
Synonyms: *Atalantia angulata* (Willd.) Engl., *Atalantia longispina* Kurz, *Atalantia spinosa* (Blume) Hook. ex Koord., *Citrus angulata* Willd., *Glycosmis spinosa* (Blume) D.Dietr., *Gonocitrus angulatus* (Willd.) Kurz, *Limonia angulata* (Willd.) Wight & Arn. ex Voigt
Climbing Mechanism: Scrambler-Armed
Distribution (Global): Andaman Is., Bangladesh, Borneo, India, Java, Malaya, Maluku, Myanmar, New Guinea, Nicobar Is., Philippines, Sulawesi
Leaf Type: Alternate, thickly leathery, aromatic leaves have transparent dots and measure 4.5– 16 by 2–7 cm. Leaf stalks are unwinged, leaf edges are slightly notched, and leaf tip are blunt or slightly pointed. Leaves are covered with minute glands and have a resinous
Inflorescence: Cluster/Inflorescence
Flower Location Axillary
Fruit Type: Ovoid, angular berry, with 3-4 flattened sides, aromatic when crushed, 2– 5 cm long
IUCN Status: Not evaluated
Notes: Usually restricted to landward margins of mangroves in low salinity zones
Reference: J. Washington Acad. Sci. 5: 423 (1915)

Paramignya andamanica Tanaka
Rutaceae

Climbing Mechanism: Scrambler-Armed
Distribution (Global): Andaman Is., Nicobar Is
Distribution (India): Andaman and Nicobar Islands, Tamil Nadu
IUCN Status: Not evaluated
Reference: Bull. Soc. Bot. France 75: 712 (1928)

*Paramignya ar*mata Oliv.
Rutaceae
Synonyms: *Arthromischus armatus* Thwaites
Climbing Mechanism: Scrambler-Armed
Distribution (Global): Sri Lanka, Thailand, Vietnam
Distribution (India): Andaman and Nicobar Islands, Karnataka, Tamil Nadu, Western Ghats
IUCN Status: Not evaluated
Reference: J. Proc. Linn. Soc., Bot. 5(Suppl. 2): 43 (1861)

Paramignya beddomei Tanaka
Rutaceae
Climbing Mechanism: Scrambler-Armed
Distribution (Global): India, Sri Lanka
Distribution (India): Tamil Nadu.
Flowering and Fruiting: June–January
IUCN Status: Not evaluated
Notes: Usually found in evergreen forests
Reference: J. Bot. 68: 230 (1930)

Paramignya griffithii Hook. f.
Rutaceae
Climbing Mechanism: Scrambler-Armed
Distribution (India): Arunachal Pradesh, Assam, Bihar, Odisha
IUCN Status: Not evaluated
Reference: Fl. Brit. India 1: 510 (1875)

Paramignya monophylla Wight
Rutaceae
Synonyms: *Paramignya monophylla* var. *monophylla*, *Paramignya monophylla* var. *obtusa* B.C.Stone
Common Name: Climbing Atalantia
Climbing Mechanism: Scrambler-Armed
Distribution (Global): Assam, Bangladesh, East Himalaya, India, Myanmar, Nepal, Sri Lanka, Vietna
Distribution (India): Andhra Pradesh, Arunachal Pradesh, Assam, Eastern Ghats, Goa, Karnataka, Maharashtra, Mizoram, Tamil Nadu, West Bengal, and Western Ghats
Flowering and Fruiting: October–March

IUCN Status: Not evaluated
Notes: Usually found in evergreen forests
Reference: Ill. Ind. Bot. 1: 109 (1838)

Paramignya scandens Craib
Rutaceae
Synonyms: *Atalantia scandens* (Griff.) Engl., *Citrus scandens* Griff.
Climbing Mechanism: Scrambler-Armed
Distribution (Global): Assam, Bangladesh, Cambodia, India, Laos, Malaya, Myanmar, Thailand, Vietnam
Distribution (India): Andhra Pradesh, Eastern Ghats, Odisha, Tripura
Leaf Type: Simple, alternate
Inflorescence: Axillary clusters
Fruit Type: Berries globose
Flowering and Fruiting: December–April
IUCN Status: Not evaluated
Reference: Fl. Siam. 1: 235 (1926)

Toddalia asiatica (L.) Lam.
Rutaceae
Synonyms: *Zanthoxylum asiaticum* (L.) Appelhans., *Cranzia aculeata* Oken, *Cranzia asiatica* (L.) Kuntze, *Cranzia nitida* Kuntze, *Cranzia schmidelioides* (Baker) Kuntze. +22
Common Name: Orange climber, Forest pepper, Wild orange tree
Climbing Mechanism: Scrambler-Armed
Distribution (India): Andhra Pradesh, Madhya Pradesh, Manipur, Meghalaya, Mizoram, Odisha, Sikkim, Tamil Nadu, Telangana, West Bengal, Western Ghats
Leaf Type: Leaves trifoliolate; leaflets sessile
Inflorescence: Panicles,axillary
Fruit Type: Subglobose, apiculate, glandular, orange color when ripe
Flowering and Fruiting: September–July
IUCN Status: Not evaluated
Notes: Found mostly in evergreen and shola forests, also in scrub jungles
Reference: Molec. Phylogen. Evol. 126: 43 (2018)

Triphasia trifolia (Burm.f.) P.Wilson
Rutaceae
Synonyms: *Limonia diacantha* DC., *Limonia retusa* D.Don, *Limonia trifolia* Burm. f., *Limonia trifoliata* L., *Triphasia aurantiola* Lour., *Triphasia diacantha* M. Roem., *Triphasia javanica* M.Roem., *Triphasia trifoliata* (L.) DC.
Common Name: Lime berry, Limau kiah
Climbing Mechanism: Scrambler-Armed
Distribution (Global): Andaman Is., Bangladesh, Borneo, Cambodia, Christmas I., Cocos (Keeling) Is., Java, Lesser Sunda Is., Malaya, Maluku, Myanmar, New Guinea, Nicobar Is., Philippines, Solomon Is., Vietnam

Distribution (India): Kerala, Tamil Nadu
Leaf Type: Leaves 3-foliolate
Inflorescence: Axillary, solitary cyme
Fruit Type: Berries ovoid-ellipsoid, apiculate
Flowering and Fruiting: Throughout the year
IUCN Status: Not evaluated
Reference: Torreya 9: 33 (1909)

Zanthoxylum acanthopodium DC.
Rutaceae
Synonyms: *Fagara acanthopodium* (DC.) M.Hiroe, *Zanthoxylum acanthopodium* var. *oligotrichum* Z.M.Tan, *Zanthoxylum alatum* Wall., *Zanthoxylum timbor* Wall.
Climbing Mechanism: Scrambler-Armed
Distribution (Global): Assam, Bangladesh, China South-Central, China Southeast, East Himalaya, Laos, Malaya, Myanmar, Nepal, Thailand, Tibet, Vietnam
Distribution (India): Arunachal Pradesh, Assam, Manipur, Meghalaya, West Bengal
Leaf Type: Leaves compound, imparipinnate
Inflorescence: Axillary panicles
Fruit Type: Follicles
Flowering and Fruiting: March–September
IUCN Status: Least Concern
Reference: Prodr. 1: 727 (1824)

Zanthoxylum armatum DC.
Rutaceae
Climbing Mechanism: Scrambler-Armed
Distribution (Global): Assam, Bangladesh, China North-Central, China South-Central, China Southeast, East Himalaya, India, Japan, Korea, Laos, Myanmar, Nansei-shoto, Nepal, Pakistan, Philippines, Taiwan, Thailand, Tibet, Vietnam, West Himalaya
Distribution (India): Odisha
Leaf Type: Leaves compound, imparipinnate
Inflorescence: Cymes terminal, paniculate
Fruit Type: Follicles
Flowering and Fruiting: March–October
IUCN Status: Least Concern
Notes: Often found in moist deciduous forests
Reference: Prodr. 1: 727 (1824)

Zanthoxylum burkillianum Babu
Rutaceae
Climbing Mechanism: Scrambler-Armed
Distribution (Global): East Himalaya
Distribution (India): Andhra Pradesh

Leaf Type: Compound, odd-pinnate
Inflorescence: Panicles, cymose
Fruit Type: Wet, rugose
IUCN Status: Not evaluated
Reference: Bull. Bot. Surv. India 16: 55 (1974 publ. 1977)

Zanthoxylum khasianum Hook. f.
Rutaceae
Synonyms: *Fagara khasyana* (Hook.f.) Engl., *Zanthoxylum yunnanense* C.C. Huang.
Common Name: Khasi yellow-wood
Climbing Mechanism: Scrambler-Armed
Distribution (Global): Assam, China South-Central
Distribution (India): Assam, Meghalaya
Leaf Type: Alternate or rarely nearly opposite
Inflorescence: Thyrses
IUCN Status: Not evaluated
Reference: Fl. Brit. India 1: 494 (1875)

Zanthoxylum limonella (Dennst.) Alston
Rutaceae
Synonyms: *Fagara budrunga* Roxb., *Fagara parviflora* (Benth.) Engl., *Fagara rhetsa* Roxb., *Lacuris illicioides* Buch. -Ham., *Tipalia limonella* Dennst., *Zanthoxylum budrunga* (Roxb.) DC., *Zanthoxylum crenatum* Wall., *Zanthoxylum limonella* (Dennst.) Alston, *Zanthoxy*
Common Name: Indian prickly ash
Climbing Mechanism: Scrambler-Armed
Distribution (Global): Andaman Is., Assam, Bangladesh, Borneo, Cambodia, East Himalaya, India, Java, Laos, Lesser Sunda Is., Malaya, Myanmar, New Guinea, Northern Territory, Queensland, Solomon Is., Sri Lanka, Sulawesi, Sumatera, Vietnam, Western Australia
Distribution (India): Eastern Ghats
Leaf Type: Leaves are compound, imparipinnate, alternate
Inflorescence: Cymose, Panicles
Fruit Type: Capsules
Flowering and Fruiting: March–November
IUCN Status: Least Concern
Notes: Mostly distributed in evergreen and moist deciduous forests
Reference: Prodr. 1: 728 (1824)

Zanthoxylum nitidum (Roxb.) DC.
Rutaceae
Synonyms: *Fagara nitida* Roxb.
Climbing Mechanism: Scrambler-Armed

Distribution (Global): Assam, Bangladesh, Borneo, Cambodia, China South-Central, China Southeast, East Himalaya, Hainan, Laos, Malaya, Maluku, Myanmar, Nansei-Shoto, Nepal, New Guinea, Philippines, Queensland, Sulawesi, Taiwan, Thailand, Vietnam
Distribution (India): Arunachal Pradesh, Assam, Mizoram, West Bengal
Leaf Type: Leaves compound, trifoliolate to imparipinnate
Inflorescence: Axillary panicles
Fruit Type: Follicles
Flowering and Fruiting: March–November
IUCN Status: Least Concern
Notes: Often found in evergreen forests
Reference: Prodr. 1: 727 (1824)

Zanthoxylum ovalifolium Tutcher
Rutaceae
Synonyms: *Fagara ovalifolia* (Wight) Engl., *Fagara varians* Domin, *Limonia leptostachya* Jack ex Hook.f., *Toddalia mitis* Miq. ex Hook.f., *Zanthoxylum dominianum* Merr. & L.M.Perry, *Zanthoxylum inerme* C.T.White & W.D. Francis, *Zanthoxylum lucidum* Wall., *Zanthoxylu*
Common Name: Thorny yellow-wood
Climbing Mechanism: Scrambler-Armed
Distribution (Global): Andaman Is., Assam, Bangladesh, East Himalaya, India, Lesser Sunda Is., Myanmar, Nepal, New Guinea, Queensland
Distribution (India): Andaman and Nicobar Islands, Tamil Nadu, Western Ghats
Leaf Type: Leaves are simple or compound with leaflets
Inflorescence: Panicles
Fruit Type: Capsules
Flowering and Fruiting: November–August
IUCN Status: Least Concern
Notes: Mostly found in evergreen and deciduous forests
Reference: Ill. Ind. Bot. 1: 169 (1839)

Zanthoxylum oxyphyllum Edgew.
Rutaceae
Synonyms: *Fagara oxyphylla* (Edgew.) Reeder & S.Y.Cheo, *Zanthoxylum alpinum* C.C.Huang, *Zanthoxylum taliense* C.C.Huang, *Zanthoxylum tibetanum* C.C. Huang
Climbing Mechanism: Scrambler-Armed
Distribution (Global): Assam, China South-Central, East Himalaya, Myanmar, Nepal, Tibet
Distribution (India): Arunachal Pradesh, Assam, Manipur, Meghalaya, West Bengal
Leaf Type: Leaves compound, imparipinnate
Inflorescence: Umbels or cluster of Cymes
Fruit Type: Follicles, subglobose
Flowering and Fruiting: April–September

IUCN Status: Not evaluated
Notes: Mostly found in temperate forests at higher altitudes up to 2700 m
Reference: Trans. Linn. Soc. London 20: 42 (1846)

Zanthoxylum rhetsa DC.
Rutaceae
Climbing Mechanism: Scrambler-Armed
Distribution (India): Odisha.
Leaf Type: Compound, odd-pinnate
Inflorescence: Panicles, cymose
Fruit Type: Wet, rugose
IUCN Status: Least Concern
Reference: DC. In: Prod. 1: 728. (1824)

Zanthoxylum scandens Blume
Rutaceae
Synonyms: *Fagara chinensis* Merr., *Fagara cuspidata* (Champ. ex Benth.) Engl. +10
Climbing Mechanism: Scrambler-Armed
Distribution (Global): Assam, Cambodia, China South-Central, China Southeast, Hainan, Java, Laos, Malaya, Myanmar, Nansei-shoto, Taiwan, Vietnam
Distribution (India): Mizoram
Leaf Type: Compound, odd-pinnate
Inflorescence: Panicles, cymose
Fruit Type: Wet, rugose
IUCN Status: Not evaluated
Reference: Bijdr. Fl. Ned. Ind.: 249 (1825)

Zanthoxylum tetraspermum Wight & Arn.
Rutaceae
Synonyms: *Fagara tetrasperma* (Wight & Arn.) Engl.
Climbing Mechanism: Scrambler-Armed
Distribution (Global): India, Sri Lanka
Distribution (India): Karnataka, Tamil Nadu
Leaf Type: Leaves 5–9 foliolate
Inflorescence: Panicles, axillary
Fruit Type: Capsule
Flowering and Fruiting: April–December
IUCN Status: Not evaluated
Notes: Found mostly in the transitions between evergreen and semi-evergreen forests
Reference: Prodr. Fl. Ind. Orient. 1: 148 (1834)

Sabia campanulata var. *kingiana*
Sabiaceae

Climbing Mechanism: Stem Twiner
Distribution (Global): East Himalaya
Leaf Type: Simple
Inflorescence: Solitary
Fruit Type: Wet, drupe
IUCN Status: Not evaluated
Reference: Bull. Bot. Surv. India 24: 219 (1982 publ. 1983)

Sabia campanulata Wall.
Sabiaceae
Synonyms: *Sabia campanulata* subsp. *Campanulata*, *Sabia campanulata* var. *kingiana* M.P.Nayar & S.C.Majumdar, *Sabia campanulata* subsp. *ritchieae* (Rehder & E.H.Wilson) Y.F.Wu
Common Name: Bellflower Sabia
Climbing Mechanism: Scrambler-Unarmed
Distribution (Global): Assam, China North-Central, China South-Central, China Southeast, East Himalaya, Nepal, Tibet, West Himalaya
Distribution (India): Arunachal Pradesh, Manipur, West Bengal
Leaf Type: Simple
Inflorescence: Solitary
Fruit Type: Wet, drupe
Flowering and Fruiting: March–June
IUCN Status: Not evaluated
Reference: W.Roxburgh, Fl. Ind. 2: 311 (1824)

Sabia lanceolata Colebr.
Sabiaceae
Common Name: Lanceleaf Sabia
Climbing Mechanism: Scrambler-Unarmed
Distribution (Global): Assam, Bangladesh, East Himalayas, Myanmar, Thailand, Tibet
Distribution (India): Arunachal Pradesh, Assam, Manipur, Mizoram, Tripura
Leaf Type: Simple and alternate
Inflorescence: Solitary flowers rarely 2 paired which are yellowish-green or dark purple in color
Fruit Type: Schizocarp
Flowering and Fruiting: October–April
IUCN Status: Not evaluated
Reference: Trans. Linn. Soc. London 12: 355 (1819)

Sabia limoniacea Wall. ex Hook. f. & Thomson
Sabiaceae
Synonyms: *Myrsine ardisioides* Hook. & Arn., *Myrsine hookeriana* Steud., *Androglossa reticulatum* Benth., *Sabia malabarica* Bedd., *Sabia paniculata* Seem

Climbing Mechanism: Scrambler-Unarmed
Distribution (Global): Assam, Bangladesh, Borneo, Cambodia, China South-Central, China Southeast, East Himalaya, Hainan, India, Laos, Malaya, Myanmar, Sumatera, Thailand, Vietnam
Distribution (India): Assam, Kerala, Mizoram, Tamil Nadu, Tripura, West Benga
Leaf Type: intercostae reticulate
Inflorescence: Cymes solitary and axillary to new leaves or in panicled cymes in the axils of old leaves
Fruit Type: Drupe globose
Flowering and Fruiting: April–May
IUCN Status: Not evaluated
Notes: Found mostly in evergreen forests
Reference: Fl. Ind. 1: 210 (1855)

Sabia malabarica Bedd.
Sabiaceae
Climbing Mechanism: Scrambler-Unarmed
Leaf Type: Simple
Inflorescence: Solitary
Fruit Type: Wet, drupe
IUCN Status: Not evaluated
Reference: Icon. Pl. Ind. Or.: 39 (1874)

Sabia paniculata Edgew. ex Hook. f. & Thomson
Sabiaceae
Common Name: Panicled Sabia
Climbing Mechanism: Scrambler-Unarmed
Distribution (Global): Assam, Bangladesh, China South-Central, East Himalaya, Myanmar, Nepal, Thailand
Distribution (India): Assam, Bihar, Odisha, West Bengal
Leaf Type: Simple, alternate, spirally arranged
Inflorescence: 3-6-flowered cymes which are arranged in panicles and are greenish white in color
Fruit Type: Druplets distinctly obovoid to oblong-obovoid, compressed
Flowering and Fruiting: January–April
IUCN Status: Not evaluated
Reference: Fl. Ind. 1: 211 (1855)

Sabia parviflora Wall.
Sabiaceae
Synonyms: *Sabia parviflora* subsp. *Parviflora*, *Sabia parviflora* subsp. *philippinensis* (C.B.Rob.) Water
Climbing Mechanism: Scrambler-Unarmed
Distribution (Global): Assam, Borneo, China South-Central, China Southeast, East Himalaya, Laos, Myanmar, Nepal, Philippines, Thailand, Vietnam

Distribution (India): Arunachal Pradesh, Assam, Mizoram, West Bengal
Leaf Type: Simple
Inflorescence: Solitary
Fruit Type: Wet, drupe
IUCN Status: Not evaluated
Reference: W.Roxburgh, Fl. Ind. 2: 310 (1824)

Sabia purpurea Hook.f. & Thomson
Sabiaceae
Climbing Mechanism: Scrambler-Unarmed
Distribution (Global): Assam, Bangladesh, China South-Central, East Himalaya, Myanmar, Nepal, Thailand
Distribution (India): Arunachal Pradesh, Assam, Manipur, West Bengal
Leaf Type: Leaves simple, alternate, and spirally arranged
Inflorescence: Flowers in panicles with dichotomous cymes, 12–35 flowered
Fruit Type: Schizocarp, green to red or blue, endocarp with inconspicuous midrib, with obscurely foveolate
Flowering and Fruiting: February–June
IUCN Status: Not evaluated
Notes: Found often in moist evergreen forests
Reference: Fl. Ind. 1: 209 (1855)

Salix suaveolens Andersson
Salicaceae
Synonyms: *Salix nilagirica* Miq., *Pleiarina tetrasperma* (Roxb.) N. Chao & G.T. Gong, *Salix tetrasperma* Roxb. + 19
Common Name: Indian willow
Climbing Mechanism: Scrambler-Unarmed
Distribution (Global): Indo-Malesia and South China
Leaf Type: Simple, alternate
Inflorescence: Axillary catkins (Raceme) yellowish-greenish colored flowers
Fruit Type: Capsule, 2–4 valved
Flowering and Fruiting: July—December
IUCN Status: Not evaluated
Notes: Found along riverbanks in semi-evergreen forests
Reference: Kongl. Vetensk. Acad. Handl. 1850: 491 (1851)

Azima sarmentosa (Blume) Benth. & Hook.f.
Salvadoraceae
Synonyms: *Actegeton sarmentosum* Blume, *Azima nova* Blanco, *Azima scandens* Baill., *Monetia brunoniana* Wall., *Monetia laxa* Planch., *Monetia sarmentosa* Baill., *Salvadora madurensis* Decne.
Climbing Mechanism: Scrambler-Armed
Distribution (Global): Cambodia, Hainan, Java, Laos, Lesser Sunda Is., Myanmar, New Guinea, Philippines, Sulawesi, Thailand, Vietnam

IUCN Status: Least Concern
Reference: Gen. Pl. 2: 681 (1876)

Azima tetracantha Lam.
Salvadoraceae
Synonyms: *Azima nova* J.F.Gmel., *Azima spinosissima* Engl., *Kandena spinosa* Raf., *Monetia barlerioides* L'Hér., *Monetia tetracantha* Salisb., *Fagonia montana* Miq.
Common Name: Mistletoe.
Climbing Mechanism: Scrambler-Armed
Distribution (Global): Aldabra, Angola, Bangladesh, Burundi, Cape Provinces, Comoros, India, Kenya, KwaZulu-Natal, Madagascar, Malawi, Mozambique, Myanmar, Namibia, Northern Provinces, Oman, Rwanda, Somalia, Sri Lanka, Swaziland, Tanzania, Uganda, Zambia, Zaire, Zimbabwe
Distribution (India): Andhra Pradesh, Karnataka, Odisha, Telangana
Flowering and Fruiting: August-September
IUCN Status: Least Concern
Reference: Encycl. 1: 343 (1783)

Allophylus cobbe (L.) Raeusch.
Sapindaceae
Synonyms: *Rhus cobbe* L., *Allophylus filiger* Radlk., *Allophylus serrulatus* Radlk., *Aporetica gemella* DC., *Gemella trifoliata* Lour., *Ornitrophe asiatica* Steud., *Ornitrophe cobbe* (L.) Willd., *Ornitrophe malabarica* Hiern, *Ornitrophe schmidelia* Pers.
Common Name: Indian Allophylus
Climbing Mechanism: Scrambler-Unarmed
Distribution (Global): Andaman Is., Assam, Bangladesh, Cambodia, India, Laccadive Is., Laos, Malaya, Maldives, Myanmar, Nicobar Is., Philippines, Sri Lanka, Vietnam
Leaf Type: Trifolioate
Inflorescence: Thyrses axillary
Flowering and Fruiting: May–January
IUCN Status: Not evaluated
Notes: Commonly occur in moist deciduous and semi-evergreen forests
Reference: Bot. Nomencl.: 226 (1794)

Allophylus concanicus Radlk.
Sapindaceae
Climbing Mechanism: Scrambler-Unarmed
Distribution (Global): India
Leaf Type: acuminate, cuspidate
Inflorescence: Raceme
Flowering and Fruiting: December–April
IUCN Status: Not evaluated

Reference: Sitzungsber. Math. -Phys. Cl. Königl. Bayer. Akad. Wiss. München 20: 230 (1890)

Allophylus concanicus var. *lanceolatus Radlk.*
Sapindaceae
Climbing Mechanism: Scrambler-Unarmed
IUCN Status: Not evaluated
Reference: Radlk. In: Sitzungsb. math. phys. Cl. bayer. Akad. Wiss. 20: 230. (1890).

Allophylus serratus (Hiern) Kurz
Sapindaceae
Synonyms: *Ornitrophe serrata* Roxb., *Schmidelia serrata* DC.
Common Name: Toothed-leaf Allophylus
Climbing Mechanism: Scrambler-Unarmed
Distribution (Global): Bangladesh, Cambodia, East Himalayas, India, Myanmar, Vietnam
Leaf Type: Trifolioate
Fruit Type: Drupe
Flowering and Fruiting: June–January
IUCN Status: Not evaluated
Notes: Mostly found in scrub jungles and deciduous forests
Reference: J. Asiat. Soc. Bengal, Pt. 2, Nat. Hist. 44: 185 (1876)

Cardiospermum corindum L.
Sapindaceae
Synonyms: *Cardiospermum alatum* Bremek. & Oberm., *Cardiospermum canescens* Wall., *Cardiospermum clematideum* A.Rich., *Cardiospermum corindum* f. angustisectum (Griseb.) Radlk., *Cardiospermum corindum* var. *angustisectum* (Griseb.) F.A.Barkley. +30
Common Name: Small balloon vine
Climbing Mechanism: Tendril Climber
Distribution (Global): Angola, Argentina Northeast, Argentina Northwest, Bolivia, Botswana, Brazil Northeast, Brazil South, Brazil Southeast, California, Caprivi Strip, Cayman Is., China South-Central, Colombia, Costa Rica, Cuba, Djibouti, Dominican Republic, Ecuador, El Salvador, Eritrea, Ethiopia, Galápagos, Guatemala, Guinea, the Gulf of Guinea Is., Haiti, India, Jamaica, Kenya, KwaZulu-Natal, Leeward Is., Marianas, Mexico Central, Mexico Gulf, Mexico Northeast, Mexico Northwest, Mexico Southwest, Myanmar, Namibia, Netherlands Antilles, Northern Provinces, Paraguay, Peru, Puerto Rico, Saudi Arabia, Somalia, Sri Lanka, Sudan, Swaziland, Tanzania, Trinidad-Tobago, Uganda, Venezuela, West Himalaya, Windward Is., Yemen, Zambia, Zimbabwe
Distribution (India): Andhra Pradesh, Tamil Nadu, Telangana
Leaf Type: Biternate-trifoliate
Fruit Type: Capsules
Flowering and Fruiting: Throughout the year

IUCN Status: Not evaluated
Notes: Often found in dry deciduous forests and scrub jungles
Reference: Sp. Pl. ed. 2: 526 (1762)

Cardiospermum halicacabum L.
Sapindaceae
Synonyms: *Cardiospermum acuminatum* Miq., *Cardiospermum glabrum* Schumach. & Thonn., *Cardiospermum inflatum* Salisb., *Cardiospermum luridum* Blume, *Cardiospermum moniliferum* Sw. ex Steud., *Corindum halicacabum* (L.) Medik., *Rhodiola biternata* Lour
Common Name: Balloon vine, Black winter cherry, Heart pea, Heart seed
Climbing Mechanism: Tendril Climber
Distribution (Global): Pan tropical
Distribution (India): Andaman and Nicobar Islands, Andhra Pradesh, Arunachal Pradesh, Assam, Bihar, Odisha, Delhi, Gujarat, Himachal Pradesh, Kerala, Madhya Pradesh, Manipur, Rajasthan, Tamil Nadu, Telangana, Tripura, Uttar Pradesh, West Bengal
Leaf Type: alternate, bi-ternate
Fruit Type: Capsule, peppery
Flowering and Fruiting: July–February
IUCN Status: Least Concern
Notes: Common in moist deciduous forests and in scrub jungles
Reference: Sp. Pl.: 366 (1753)

Cardiospermum halicacabum var. *luridum*
Sapindaceae
Climbing Mechanism: Tendril Climber
Distribution (India): Peninsular India
Leaf Type: Alternate-spiral, biternate
Fruit Type: Capsule
Flowering and Fruiting: July–March
IUCN Status: Not evaluated
Notes: Often found abundant in wastelands

Cardiospermum microcarpum Kunth
Sapindaceae
Synonyms: *Cardiospermum halicacabum* var. *microcarpum* (Kunth) Blume, *Cardiospermum microspermum* E.Mey., *Cardiospermum truncatum* A.Rich.
Climbing Mechanism: Tendril Climber
Distribution (Global): Angola, Argentina Northeast, Argentina Northwest, Bermuda, Bolivia, Borneo, Botswana, Brazil North, Brazil South, Brazil Southeast, Brazil West-Central, Cameroon, Cape Provinces, Cape Verde, China South-Central, China Southeast, Colombia, Cuba, Dominican Republic, East Himalaya, El Salvador, Eritrea, Ethiopia, Fiji, French Guiana, Guyana, Haiti, Hawaii, Honduras, India, Jamaica, Java, Kenya, KwaZulu-Natal, Leeward Is., Lesser

Sunda Is., Madagascar, Malaya, Maluku, Mexico Central, Mexico Southwest, Mozambique, Myanmar, Namibia, Netherlands Antilles, New Guinea, Nicaragua, Niger, Northern Provinces, Pakistan, Panamá, Paraguay, Peru, Philippines, Puerto Rico, Queensland, Saudi Arabia, Senegal, Seychelles, Society Is., Sri Lanka, Sudan, Sulawesi, Sumatera, Suriname, Swaziland, Taiwan, Tanzania, Thailand, Tonga, Trinidad-Tobago, Uruguay, Venezuela, Vietnam, West Himalaya, Windward Is., Yemen, Zambia, Zimbabw
Distribution (India): Maharashtra, Telangana
Leaf Type: Compound, even-pinnate
Inflorescence: Corymbose racemes
Fruit Type: Dry, capsule
Flowering and Fruiting: July–February
IUCN Status: Not evaluated
Notes: Occur mostly in hill slopes and deciduous forests
Reference: F.W.H.von Humboldt, A.J.A.Bonpland & C.S.Kunth, Nov. Gen. Sp. 5: 104 (1821)

Kadsura heteroclita (Roxb.) Craib
Schisandraceae
Synonyms: *Kadsura acuminata* P.Parm., *Kadsura billitonensis* A.Agostini, *Kadsura championii* C.B.Clarke, *Kadsura interior* A.C.Sm., *Kadsura parvifolia* A.Agostini, *Kadsura polysperma* Y.C.Yang, *Kadsura roxburghiana* Arn., *Kadsura wattii* C.B.Clarke, *Kadsura wighti.*
Climbing Mechanism: Scrambler-Unarmed
Distribution (Global): Assam, Bangladesh, Borneo, Cambodia, China South-Central, China Southeast, East Himalaya, Hainan, Laos, Malaya, Myanmar, Sri Lanka, Sumatera, Taiwan, Thailand, Vietnam
Distribution (India): Tamil Nadu
Leaf Type: Leaves simple, alternate, entire or distantly serrate, coriaceous, pinnately veined
Inflorescence: Flowers monoecious, solitary, axillary or subaxillary or sometimes arising from shoots below the leaves on short lateral branchlets of old wood, pedicels terete
Fruit Type: Fruits globose, indehiscent fleshy single seeded fruitlets, suspended seeds 1-2
Flowering and Fruiting: April–December
IUCN Status: Not evaluated
Notes: Found usually in evergreen forests.
Reference: Fl. Siam. 1: 28 (1925)

Schisandra elongata (Blume) Baill.
Schisandraceae
Synonyms: *Sphaerostema elongatum* Blume
Climbing Mechanism: Scrambler-Unarmed
Distribution (Global): Java

Distribution (India): Assam
Leaf Type: Simple, papyraceous
Inflorescence: Axillary to leaves
Fruit Type: Wet, berry
IUCN Status: Not evaluated
Reference: Hist. Pl. 1: 148 (1868)

Schisandra grandiflora (Wall.) Hook. f. & Thomson
Schisandraceae
Synonyms: *Kadsura grandiflora* Wall., *Sphaerostema grandiflorum* (Wall.) Blume
Common Name: Large-flowered magnolia vine
Climbing Mechanism: Scrambler-Unarmed
Distribution (Global): East Himalaya, Nepal, Tibet, West Himalaya
Distribution (India): North-eastern India
Leaf Type: Simple, papyraceous
Inflorescence: Axillary to leaves
Fruit Type: Wet, berry
Flowering and Fruiting: April–June
IUCN Status: Not evaluated
Reference: J. D. Hooker, Fl. Brit. India 1: 44 (1872)

Schisandra neglecta A.C.Sm.
Schisandraceae
Synonyms: *Schisandra lancifolia* var. *polycarpa* Z.Ho
Climbing Mechanism: Scrambler-Unarmed
Distribution (Global): Assam, Bangladesh, China South-Central, East Himalaya, Myanmar, Nepa
Distribution (India): Meghalaya, Mizoram
Leaf Type: Leaves simple, alternate, entire or distantly serrate, crowded on shoots, punctuate, thin, exstipulate
Fruit Type: Globose or obovoid, indehiscent on much-elongated torus, seeds 1-2, subglobose or reniform
Flowering and Fruiting: April–September
IUCN Status: Not evaluated
Notes: Often occurs in mixed subtropical forests (altitude of about 2,200 m)
Reference: Sargentia 7: 127 (1947)

Schisandra propinqua (Wall.) Baill.
Schisandraceae
Synonyms: *Kadsura propinqua* Wall., *Sphaerostema propinquum* (Wall.) Blum
Climbing Mechanism: Scrambler-Unarmed
Distribution (Global): Assam, China North-Central, China South-Central, China Southeast, Java, Lesser Sunda Is., Myanmar, Nepal, Thailand, Tibet, West Himalay
Distribution (India): Arunachal Pradesh, Assam, Mizoram

Leaf Type: Leaves simple, alternate, entire or distantly serrate, crowded on shoots, punctuate, thin, exstipulate
Inflorescence: Axillary, solitary, or in clusters
Fruit Type: globose or obovoid, indehiscent on a much-elongated torus, seeds 1-2, subglobose or reniform
Flowering and Fruiting: July–November
IUCN Status: Not evaluated
Reference: Hist. Pl. 1: 148 (1868)

Schisandra propinqua subsp. *axillaris* (Blume) R.M.K.Saunders
Schisandraceae
Synonyms: *Schisandra axillaris* (Blume) Hook.f. & Thomson., *Sphaerostema pyrifolium* Blume., *Sphaerostema axillare* Blume
Climbing Mechanism: Scrambler-Unarmed
Distribution (Global): Java, Lesser Sunda Is.
Distribution (India): Arunachal Pradesh, Assam, Mizoram
Leaf Type: Simple, papyraceous
Inflorescence: Axillary to leaves
Fruit Type: Wet, berry
IUCN Status: Not evaluated
Reference: Fl. Males. 13: 204 (1997)

Schisandra sphenanthera Rehder & E.H.Wilson
Schisandraceae
Synonyms: *Schisandra chinensis* var. *rubriflora* Franch., *Schisandra flaccidiramosa* C.R.Sun.
Climbing Mechanism: Scrambler-Unarmed
Distribution (Global): China North-Central, China South-Central, China Southeast
Distribution (India): Assam
Leaf Type: Simple, papyraceous
Inflorescence: Axillary to leaves
Fruit Type: Wet, berry
IUCN Status: Not evaluated
Reference: C.S.Sargent, Pl. Wilson. 1: 414 (1913)

Buddleja asiatica Lour.
Scrophulariaceae
Synonyms: *Buddleja acuminatissima* Blume, *Buddleja arfakensis* Kaneh. & Hatus., *Buddleja densiflora* Blume, *Buddleja discolor* Roth, *Buddleja neemda* Buch. -Ham. ex Roxb Buddleja amentacea Kraenzl., *Buddleja serrulata* Roth, *Buddleja subserrata* Buch. -Ham. ex D.Don
Common Name: White butterfly bush, Asian butterfly bus
Climbing Mechanism: Scrambler-Unarmed
Distribution (Global): Assam, Bangladesh, Borneo, Cambodia, China North-Central, China South-Central, China Southeast, East Himalaya, Hainan, India,

Java, Laos, Lesser Sunda Is., Malaya, Maluku, Marianas, Myanmar, Nepal, New Guinea, Pakistan, Philippines, Sulawesi, Sumatera, Taiwan, Thailand, Tibet, Vietnam, West Himalaya
Distribution (India): Madhya Pradesh, Rajasthan
Leaf Type: Opposite
Inflorescence: Terminal or axillary
Fruit Type: Capsule
Flowering and Fruiting: October–May
IUCN Status: Least Concern
Reference: Fl. Cochinch.: 72 (1790)

Buddleja madagascariensis Lam.
Scrophulariaceae
Synonyms: *Adenoplea madagascariensis* (Lam.) Eastw., *Buddleja heterophylla* Lindl., *Nicodemia madagascariensis* (Lam.) R.Parke
Common Name: Madagascar butterfly bush, Smoke bush
Climbing Mechanism: Scrambler-Unarmed
Distribution (Global): Madagascar
Distribution (India): Madhya Pradesh, Maharashtra, Rajasthan
Flowering and Fruiting: Throughout the year
IUCN Status: Least Concern
Notes: Often planted in gardens
Reference: Encycl. 1: 513 (1785)

Heterosmilax japonica Kunth
Smilacaceae
Synonyms: *Smilax bockii* Warb.
Climbing Mechanism: Tendril Climber
Distribution (India): Andhra Prades
Flowering and Fruiting: June–February
IUCN Status: Not evaluated
Reference: Enum. Pl. 5: 270 (1850)

Smilax aspera L.
Smilacaceae
Synonyms: *Smilax aspera* subsp. *balearica* (Willk. ex A.DC.) K.Richt., *Smilax mauritanica* subsp. *vespertilionis* (Boiss.) K.Richt., *Smilax aspera* subsp. *mauritanica* (Poir.) Arcang. +30
Common Name: Common Smilax, Prickly ivy, Rough bindweed, Salsaparilla
Climbing Mechanism: Tendril Climber
Distribution (Global): Albania, Algeria, Assam, Azores, Baleares, Bangladesh, Canary Is., China South-Central, Corse, Cyprus, East Aegean Is., East Himalaya, Eritrea, Ethiopia, France, Greece, India, Iran, Italy, Kenya, Kriti, Lebanon-Syria, Libya, Madeira, Morocco, Myanmar, Nepal, Pakistan, Palestine, Portugal,

Sardegna, Sicilia, Spain, Sri Lanka, Tanzania, Tibet, Tunisia, Turkey, Uganda, West Himalaya, Yugoslavia, Zambia, Zaire
Distribution (India): Andaman and Nicobar Islands, Arunachal Pradesh, Himachal Pradesh, Sikkim, Tamil Nadu, Western Ghats
Leaf Type: Alternate
Inflorescence: Inflorescence of sessile and subsessile umbels, perianth white, cream, greenish, pink, yellowish, or brownish
Fruit Type: Berry red or dry to black, 3-seeded
Flowering and Fruiting: February–June
IUCN Status: Least Concern
Notes: Found usually in evergreen and shola forests
Reference: Sp. Pl.: 1028 (1753)

Smilax aspericaulis Wall. ex A. DC.
Smilacaceae
Synonyms: *Smilax bracteata* subsp. *verruculosa* (Merr.) T.Koyama., *Smilax verruculosa* Merr., *Smilax trachyclada* Hayata
Climbing Mechanism: Tendril Climber
Distribution (Global): Andaman Is., Assam, Bangladesh, China South-Central, China Southeast, East Himalaya, Hainan, Malaya, Myanmar, Nansei-shoto, Philippines, Taiwan, Thailand, Vietnam
Distribution (India): Andaman and Nicobar Islands, Sikkim
Leaf Type: Simple, alternate, 3–5 nerved
Inflorescence: Umbels racemes in axils
Fruit Type: Wet, berry, globose
Flowering and Fruiting: October—December.
IUCN Status: Not evaluated
Reference: A.L.P.P.de Candolle & A.C.P.de Candolle, Monogr. Phan. 1: 195 (1878)

Smilax bracteata C.Presl
Smilacaceae
Synonyms: *Smilax blancoi* Kunth., *Smilax bonii* Gagnep., *Smilax divaricata* Blanco., *Smilax fistulosa* Blanco., *Smilax lyi* H.Lév., *Smilax phyllantha* Gagnep., *Smilax stenopetala* A.Gray., *Smilax trukensis* Hosok.
Climbing Mechanism: Tendril Climber
Distribution (Global): Cambodia, Caroline Is., China South-Central, China Southeast, Hainan, Japan, Laos, Malaya, Nansei-shoto, Philippines, Taiwan, Thailand, Vietnam
Distribution (India): Arunachal Pradesh, Andaman and Nicobar Islands
Leaf Type: Simple, alternate, 3–5 nerved
Inflorescence: Umbels racemes in axils
Fruit Type: Wet, berry, globose
IUCN Status: Not evaluated
Reference: Reliq. Haenk. 1: 131 (1827)

Smilax domingensis Willd.
Smilacaceae
Synonyms: *Smilax balbisiana* Kunth., *Smilax berteroi* Spreng., *Smilax canaliculata* F.W.Apt., *Smilax caudata* Lundell., *Smilax jauaensis* Steyerm. & Maguire., *Smilax microscola* (B.L.Rob.) Killip & C.V.Morton. +20
Climbing Mechanism: Tendril Climber
Distribution (Global): Belize, Bolivia, Brazil North, Brazil Northeast, Brazil South, Brazil Southeast, Brazil West-Central, Colombia, Costa Rica, Cuba, Dominican Republic, Ecuador, El Salvador, French Guiana, Guatemala, Guyana, Haiti, Honduras, Jamaica, Leeward Is., Mexico Gulf, Mexico Northeast, Mexico Southeast, Mexico Southwest, Nicaragua, Panamá, Peru, Puerto Rico, Suriname, Venezuela
Leaf Type: Simple, alternate, 3–5 nerved
Inflorescence: Umbels racemes in axils
Fruit Type: Wet, berry, globose
IUCN Status: Not evaluated
Reference: Sp. Pl., ed. 4, 4: 783 (1806)

Smilax elegans subsp. *subrecta* Noltie
Smilacaceae
Synonyms: *Smilax longibracteolata* Hook. f.
Climbing Mechanism: Tendril Climber
Distribution (Global): Assam, Bangladesh, China South-Central, East Himalaya, Myanmar, Tibet
Leaf Type: Simple, alternate, 3–5 nerved
Inflorescence: Umbels racemes in axils
Fruit Type: Wet, berry, globose
IUCN Status: Not evaluated
Reference: Edinburgh J. Bot. 51: 155 (1994)

Smilax elegans Wall. ex Kunth
Smilacaceae
Synonyms: *Smilax glaucophylla*, *Smilax parvifolia* [Illeg.]
Common Name: Elegant Smilax.
Climbing Mechanism: Tendril Climber
Distribution (Global): Assam, Bangladesh, China South-Central, East Himalaya, Myanmar, Nepal, Pakistan, Tibet, West Himalaya
Distribution (India): Arunachal Pradesh
Leaf Type: Simple, alternate, 3–5 nerved
Inflorescence: Umbels racemes in axils
Fruit Type: Wet, berry, globose
Flowering and Fruiting: April–May
IUCN Status: Not evaluated
Reference: Enum. Pl. 5: 163 (1850)

Smilax glabra Roxb.
Smilacaceae
Synonyms: *Smilax blinii* H.Lév., *Smilax calophylla* var. *concolor* C.H.Wright., *Smilax dunniana* H.Lév., *Smilax hookeri* Kunth., *Smilax lanceolata* Burm.f., *Smilax mengmaensis* R.H.Miao., *Smilax trigona* Warb.
Climbing Mechanism: Tendril Climber
Distribution (Global): Assam, Bangladesh, Cambodia, China North-Central, China South-Central, China Southeast, Hainan, Laos, Myanmar, Taiwan, Thailand, Tibet, Vietnam
Distribution (India): Andaman and Nicobar Islands
Leaf Type: Lanceolate leaves
Inflorescence: Umbel
Fruit Type: Berries blue-black, Sweet and sour in taste
Flowering and Fruiting: July–April
IUCN Status: Not evaluated
Notes: It occurs mainly in moist, evergreen, and mixed deciduous forests
Reference: Fl. Ind. ed. 1832, 3: 792 (1832)

Smilax griffithii A. DC.
Smilacaceae
Synonyms: *Smilax griffithii* var. *borii* Panigrahi & Naik., *Smilax pallescens* A. DC.
Climbing Mechanism: Tendril Climber
Distribution (Global): Assam, Bangladesh, East Himalayas, Myanmar, Thailand, Tibet
Distribution (India): Arunachal Pradesh.
Leaf Type: Simple, alternate, 3–5 nerved
Inflorescence: Umbels racemes in axils
Fruit Type: Wet, berry, globose
IUCN Status: Not evaluated
Reference: A.L.P.P.de Candolle & A.C.P.de Candolle, Monogr. Phan. 1: 198 (1878)

Smilax guianensis Vitman
Smilacaceae
Synonyms: *Smilax guianensis* var. *subarmata* O.E.Schulz., *Smilax megalophylla* Duhamel., *Smilax macrophylla* Willd., *Smilax subarmata* (O.E.Schulz) O.E. Schulz
Climbing Mechanism: Tendril Climber
Distribution (Global): Guyana, Leeward Is., Puerto Rico, Suriname, Trinidad-Tobago, Venezuela, Windward Is.
Leaf Type: Simple, alternate, 3–5 nerved
Inflorescence: Umbels racemes in axils
Fruit Type: Wet, berry, globose
IUCN Status: Not evaluated
Reference: Summa Pl. 5: 422 (1791)

Smilax hemsleyana Craib
Smilacaceae
Synonyms: *Smilax zeylanica* subsp. *hemsleyana* (Craib) T.Koyam
Climbing Mechanism: Tendril Climber
Distribution (Global): Assam, China South-Central, Myanmar, Thailan
Distribution (India): Andaman and Nicobar Islands
Leaf Type: Simple, alternate, 3–5 nerved
Inflorescence: Umbels racemes in axils
Fruit Type: Wet, berry, globose
IUCN Status: Not evaluated
Reference: Bull. Misc. Inform. Kew 1912: 409 (1912)

Smilax lanceifolia Roxb.
Smilacaceae
Common Name: Lanceleaf Smilax
Climbing Mechanism: Tendril Climber
Distribution (Global): Andaman Is., Assam, Bangladesh, Borneo, Cambodia, China South-Central, China Southeast, East Himalaya, Hainan, India, Laos, Malaya, Myanmar, Nepal, Nicobar Is., Taiwan, Thailand, Vietnam
Distribution (India): Andaman and Nicobar Islands, Arunachal Pradesh, Assam, Bihar, Odisha, Odisha, Sikkim
Leaf Type: Leaves are faintly reticulate, lanceolate
Inflorescence: Flowers are born in 1 or 2 umbels
Fruit Type: Berries are yellowish-red to black, spherical
Flowering and Fruiting: September–March
IUCN Status: Not evaluated
Reference: Fl. Ind. ed. 1832, 3: 792 (1832)

Smilax luzonensis C.Presl
Smilacaceae
Synonyms: *Smilax helferi* A.DC., *Smilax singaporensis* A. DC.
Climbing Mechanism: Tendril Climber
Distribution (Global): Cambodia, India, Java, Laos, Malaya, Myanmar, Philippines, Sumatera, Thailand, Vietnam, West Himalaya
Leaf Type: Simple, alternate, 3–5 nerved
Inflorescence: Umbels racemes in axils
Fruit Type: Wet, berry, globose
IUCN Status: Not evaluated
Reference: Reliq. Haenk. 1: 131 (1827)

Smilax menispermoidea A. DC.
Smilacaceae
Synonyms: *Smilax luteocaulis* H.Lév., *Smilax rubriflora* Rehder.
Climbing Mechanism: Tendril Climber

Distribution (Global): China North-Central, China South-Central, East Himalaya, Laos, Myanmar, Nepal, Tibet, Vietnam, West Himalaya
Leaf Type: Simple, alternate, 3–5 nerved
Inflorescence: Umbels racemes in axils
Fruit Type: Wet, berry, globose
IUCN Status: Not evaluated
Reference: A.L.P.P.de Candolle & A.C.P.de Candolle, Monogr. Phan. 1: 108 (1878)

Smilax morongii Small
Smilacaceae
Synonyms: *Smilax megacarpa* Morong.
Climbing Mechanism: Tendril Climber
Distribution (Global): Florida.
Leaf Type: Simple, alternate, 3–5 nerved
Inflorescence: Umbels racemes in axils
Fruit Type: Wet, berry, globose
IUCN Status: Not evaluated
Reference: Bull. Torrey Bot. Club 25: 605 (1898)

Smilax munita S.C.Chen
Smilacaceae
Synonyms: *Smilax hajrae* R.C.Srivast., *Smilax rigida* Wall. ex Kunth., *Smilax myrtillus* var. *rigida* Noltie
Climbing Mechanism: Tendril Climber
Distribution (Global): China South-Central, East Himalaya, Myanmar, Nepal, Tibet
Leaf Type: Simple, alternate, 3–5 nerved
Inflorescence: Umbels racemes in axils
Fruit Type: Wet, berry, globose
IUCN Status: Not evaluated
Reference: Acta Phytotax. Sin. 34: 436 (1996)

Smilax odoratissima Blume
Smilacaceae
Climbing Mechanism: Tendril Climber
Distribution (Global): Bangladesh, Borneo, Java, Lesser Sunda Is., Myanmar, Thailand
Distribution (India): Andaman and Nicobar Islands
Leaf Type: Simple, alternate, 3–5 nerved
Inflorescence: Umbels racemes in axils
Fruit Type: Wet, berry, globose
IUCN Status: Not evaluated
Reference: Enum. Pl. Javae: 19 (1827)

Smilax ovalifolia Roxb. ex D.Don
Smilacaceae

Synonyms: *Smilax columnifera* Buch. -Ham. ex D. Don., *Smilax grandifolia* Voigt., *Smilax grandis* Wall. ex Voigt., *Smilax macrophylla* Roxb., *Smilax retusa* Roxb., *Smilax roxburghii* Kunth.
Climbing Mechanism: Tendril Climber
Distribution (Global): Andaman Is., Assam, Bangladesh, Cambodia, East Himalaya, Hainan, India, Laos, Myanmar, Nepal, Nicobar Is., Thailand, Vietnam, West Himalaya
Distribution (India): Andaman and Nicobar Islands, Arunachal Pradesh, Assam, Himachal Pradesh, Sikkim, West Bengal
Leaf Type: Alternate
Inflorescence: Flowers white, in dense umbels in leaf axils
Flowering and Fruiting: September–December
IUCN Status: Not evaluated
Reference: Prodr. Fl. Nepal.: 49 (1825)

Smilax perfoliata Lou
Smilacaceae
Synonyms: *Smilax annamensis* Rendle., *Smilax laurina* Kunth
Climbing Mechanism: Tendril Climber
Distribution (Global): Assam, Bangladesh, Cambodia, East Himalaya, Hainan, India, Laos, Myanmar, Nepal, Sri Lanka, Taiwan, Thailand, Vietnam, West Himalaya
Distribution (India): Andhra Pradesh, Arunachal Pradesh, Assam, Madhya Pradesh, Odisha, Sikkim, Tamil Nadu, Telangana, Uttar Pradesh, West Bengal, and the Western Ghats
Leaf Type: Simple, alternate, 3–5 nerved
Inflorescence: Umbels racemes in axils
Fruit Type: Wet, berry, globose
Flowering and Fruiting: February–November
IUCN Status: Not evaluated
Notes: Frequently found in moist deciduous forests
Reference: Fl. Cochinch.: 622 (1790)

Smilax quadrata A.DC.
Smilacaceae
Climbing Mechanism: Tendril Climber
Distribution (Global): Assam, China South-Central, Myanmar
Distribution (India): Arunachal Pradesh
Leaf Type: Simple, alternate, 3–5 nerved
Inflorescence: Umbels racemes in axils
Fruit Type: Wet, berry, globose
IUCN Status: Not evaluated
Reference: A.L.P.P.de Candolle & A.C.P.de Candolle, Monogr. Phan. 1: 183 (1878)

Smilax saienii
Smilacaceae
Climbing Mechanism: Tendril Climber
IUCN Status: Not evaluated

Smilax scobinicaulis C.H. Wright
Smilacaceae
Synonyms: *Smilax brevipes* Warb., *Smilax martini* H.Lév. & Vaniot., *Smilax cavaleriei* H.Lév. & Vaniot., *Smilax ocreata* H.Lév. & Vaniot
Climbing Mechanism: Tendril Climber
Distribution (Global): China North-Central, China South-Central, China Southeast
Leaf Type: Simple, alternate, 3–5 nerved
Inflorescence: Umbels racemes in axils
Fruit Type: Wet, berry, globose
IUCN Status: Not evaluated
Reference: Bull. Misc. Inform. Kew 1895: 117 (1895)

Smilax vaginata Decne
Smilacaceae
Climbing Mechanism: Tendril Climber
Distribution (Global): Afghanistan, Assam, India, Pakistan, West Himalayas
Distribution (India): Himachal Pradesh
Leaf Type: Simple, alternate, 3–5 nerved
Inflorescence: Umbels racemes in axils
Fruit Type: Wet, berry, globose
IUCN Status: Not evaluated
Reference: V.Jacquemont, Voy. Inde 4: 169 (1844)

Smilax wightii A. DC.
Smilacaceae
Climbing Mechanism: Tendril Climber
Distribution (Global): East Himalaya, India, Nepal
Distribution (India): Sikkim, Tamil Nadu
Leaf Type: Simple, alternate, 3–5 nerved
Inflorescence: Umbels racemes in axils
Fruit Type: Wet, berry, globose
Flowering and Fruiting: April–October
IUCN Status: Not evaluated
Notes: Often found in moist deciduous and shola forests
Reference: A.L.P.P.de Candolle & A.C.P.de Candolle, Monogr. Phan. 1: 174 (1878)

Smilax zeylanica L.
Smilacaceae

Synonyms: *Smilax ceylanica* Oken., *Smilax collina* Kunth., *Smilax elliptica* R.Br., *Smilax elliptica* Desv., *Smilax hohenackeri* Kunth., *Smilax indica* Burm.f., *Smilax villandia* Buch. -Ham. ex Royle., *Smilax zollingeri* Kunth.
Climbing Mechanism: Tendril Climber
Distribution (Global): Assam, Bangladesh, East Himalaya, India, Java, Malaya, Myanmar, Nepal, Solomon Is., Sri Lanka
Distribution (India): Andaman and Nicobar Islands, Andhra Pradesh, Arunachal Pradesh, Assam, Goa, Gujarat, Himachal Pradesh, Jharkhand, Kerala, Madhya Pradesh, Odisha, Rajasthan, Tamil Nadu, Telangana, Tripura, Uttar Pradesh, Western Ghats
Leaf Type: Alternate
Inflorescence: Umbels, axillary
Fruit Type: Berrys, subglobose
Flowering and Fruiting: April–February
IUCN Status: Not evaluated
Notes: Mostly found in moist deciduous forests and semi-evergreen forests
Reference: Sp. Pl.: 1029 (1753)

Cestrum diurnum L.
Solanaceae
Synonyms: *Cestrum album* Ferrero ex Dunal., *Cestrum diurnum* var. *fastigiatum* (Jacq.) Stehlé., *Cestrum diurnum* var. *marcianum* Proctor., *Cestrum diurnum* var. *odontospermum* (Jacq.) O.E.Schulz., *Cestrum diurnum* var. *portoricense* O.E. Schulz., *Cestrum diurnum* var. *tinctor*
Common Name: Day jasmine, Day-blooming Cestrum, Inkberry, Tentanchinu, White chocolate jasmine
Climbing Mechanism: Scrambler-Unarmed
Distribution (Global): Cuba, Dominican Republic, Florida, Haiti, Jamaica, Leeward Is., Mexico Gulf, Mexico Northeast, Mexico Northwest, Mexico Southeast, Puerto Rico, Southwest Caribbean, Trinidad-Tobago, Windward Is.
Distribution (India): Tripura
Flowering and Fruiting: Throughout the yea
IUCN Status: Least Concern
Notes: Planted in gardens. The fruits are poisonous
Reference: Sp. Pl.: 191 (1753)

Cestrum elegans (Brongn. ex Neumann) Schltdl.
Solanaceae
Synonyms: *Cestrum elegans* var. *longiflorum* Francey., *Cestrum elegans* var. *totutla* Dunal., *Cestrum elegans* var. *truncata* Fernald. +10
Common Name: Butterfly flower, Elegant Cestrum, Purple Cestrum, Red Cestrum
Climbing Mechanism: Scrambler-Unarmed
Distribution (Global): Mexico Central, Mexico Gulf, Mexico Northeast, Mexico Southwest
Distribution (India): Maharashtra, Tamil Nad

Leaf Type: Ovate-elliptic
Inflorescence: Racemose panicles
Fruit Type: Berries are dark pink
Flowering and Fruiting: January–June
IUCN Status: Least Concern
Notes: Cultivated for ornamental purposes
Reference: Linnaea 19: 261 (1847)

Cestrum nocturnum L.
Solanaceae
Synonyms: *Cestrum graciliflorum* Dunal, *Cestrum leucocarpum* Dunal, *Cestrum multiflorum* Roem. & Schult., *Cestrum nocturnum* var. *mexicanum* O.E.Schulz, *Cestrum propinquum* M.Martens & Galeotti, *Cestrum scandens* Thibaud ex Duna
Cestrum spicatum Mill.
Common Name: Night jasmine, Queen of the night, Lady of the nigh
Climbing Mechanism: Scrambler-Unarmed
Distribution (Global): Belize, Colombia, Costa Rica, El Salvador, French Guiana, Guatemala, Honduras, Mexico Central, Mexico Gulf, Mexico Northeast, Mexico Northwest, Mexico Southeast, Mexico Southwest, Nicaragua, Panamá, Venezuela
Distribution (India): Tripura
Flowering and Fruiting: Throughout the year
IUCN Status: Least Concern
Notes: Flowers are highly fragrant. Also known as Queen- or Lady-of-the-Night
Reference: Sp. Pl.: 191 (1753)

Cestrum parqui L'Hér.
Solanaceae
Synonyms: *Cestrum adelbertii* Bakh.f., *Cestrum campestre* Griseb. +18
Common Name: Chilean flowering jasmine
Climbing Mechanism: Scrambler-Unarmed
Distribution (Global): Argentina Northeast, Argentina Northwest, Argentina South, Bolivia, Brazil South, Brazil Southeast, Brazil West-Central, Chile Central, Chile North, Chile South, Paraguay, Urugua
Distribution (India): Tamil Nadu
Leaf Type: Alternately arranged, elliptic-lanceolate
Inflorescence: Terminal compound paniculiform
Flowering and Fruiting: January–June
IUCN Status: Least Concern
Notes: Introduced as a garden plant now often found in the wild
Reference: Stirp. Nov.: 73 (1788)

Lycianthes laevis (Dunal) Bitter
Solanaceae
Synonyms: *Solanum laeve* Dunal

Climbing Mechanism: Scrambler-Unarmed
Distribution (Global): Assam, Borneo, China South-Central, China Southeast, East Himalaya, Hainan, India, Java, Laos, Lesser Sunda Is., Malaya, Nansei-shoto, Philippines, Sri Lanka, Taiwan, Thailand, Vietnam
Distribution (India): Andhra Pradesh, Madhya Pradesh, Maharashtra, Tamil Nadu, Tripura
Leaf Type: Upper leaves are often paired and unequal, elliptic-lanceolate, acuminate
Inflorescence: Flowers 2–5 in axils
Fruit Type: Berry globose, red
Flowering and Fruiting: December–May
IUCN Status: Not evaluated
Notes: Distributed in evergreen forests of the Western Ghats
Reference: Gatt. Lycianthes: 484 (1919)

Solandra grandiflora Sw.
Solanaceae
Synonyms: *Datura sarmentosa* Lam., *Datura sarmentosa* Lam., *Solandra grandiflora* var. *nitida* (Zuccagni) Voss., *Solandra hirsuta* Dunal., *Solandra minor* Griseb., *Solandra nitida* Zuccagni., *Solandra scandens* Willd., *Solandra scandens* (Vell.) Toledo
Common Name: Showy chalice vine, Cup of gold
Climbing Mechanism: Scrambler-Armed
Distribution (Global): Brazil Northeast, Brazil South, Brazil Southeast, Costa Rica, Cuba, El Salvador, Guatemala, Honduras, Jamaica, Mexico Central, Mexico Gulf, Mexico Southeast, Mexico Southwest, Nicaragua, Panamá, Puerto Rico, Southwest Caribbean, Trinidad-Tobago, Venezuela
Distribution (India): Tamil Nadu
Leaf Type: Leathery
Fruit Type: Round berries
IUCN Status: Not evaluated
Reference: Kongl. Vetensk. Acad. Handl. 8: 300 (1787)

Solanum dulcamara L.
Solanaceae
Synonyms: *Dulcamara flexuosa* Moench., *Solanum asiae-mediae* Pojark. +32
Climbing Mechanism: Scrambler-Unarmed
Distribution (Global): North Asia and Europe
Distribution (India): Peninsular India
Leaf Type: Simple
Inflorescence: Panicles, cymes/racemose
Fruit Type: Wet, berry, globose
IUCN Status: Least Concern
Reference: Sp. Pl.: 185 (1753)

Solanum flaccidum Vell.
Solanaceae
Synonyms: *Solanum convolvuliforme* St.-Lag., *Solanum convolvulus* Sendtn., *Solanum convolvulus* var. *heterophyllum* Witasek., *Solanum delilei* Dunal., *Solanum flaccidum* var. *heterophyllum* Witasek., *Solanum fultum* Schrank ex Sendtn., *Solanum luridum* Dunal
Climbing Mechanism: Scrambler-Unarmed
Distribution (Global): Brazil Northeast, Brazil South, Brazil Southeast, Paraguay
Distribution (India): Tamil Nadu
Leaf Type: Simple
Inflorescence: Panicles, cymes/racemose
Fruit Type: Wet, berry, globose
IUCN Status: Not evaluated
Reference: Fl. Flumin.: 87 (1829)

Solanum heterodoxum Dunal
Solanaceae
Synonyms: *Solanum heterodoxum* var. *heterodoxum*
Climbing Mechanism: Scrambler-Unarmed
Distribution (Global): Mexico Central, Mexico Gulf, Mexico Northeast, Mexico Northwest, Mexico Southwest, New Mexico
Leaf Type: Simple
Inflorescence: Panicles, cymes/racemose
Fruit Type: Wet, berry, globose
IUCN Status: Not evaluated
Reference: Hist. Nat. Solanum: 235 (1813)

Solanum laxum Spreng.
Solanaceae
Synonyms: *Solanum jasminoides* subvar. *glaberrimum* Kuntze +10
Common Name: Potato vine
Climbing Mechanism: Scrambler-Unarmed
Distribution (Global): Argentina Northeast, Brazil South, Brazil Southeast, Paraguay, Uruguay
Distribution (India): Tamil Nadu
Leaf Type: Simple
Inflorescence: Panicles, cymes/racemose
Fruit Type: Wet, berry, globose
IUCN Status: Not evaluated
Reference: Syst. Veg. ed. 16, 1: 682 (1824)

Solanum seaforthianum Andrews
Solanaceae
Synonyms: *Solanum botryophorum* Ridl., *Solanum cyrrhosum* Humb. & Bonpl. ex Dunal., *Solanum cyrrhosum* Humb. & Bonpl. ex Dunal., *Solanum pentadactylon*

G.Don., *Solanum prunifolium* Willd., *Solanum salignum* Willd., *Solanum seaforthianum* var. *disjunctum* O.E.Schulz., Solan
Common Name: Brazilian nightshade, Climbing nightshade, Deadly nightshade
Climbing Mechanism: Scrambler-Unarmed
Distribution (Global): Belize, Brazil Northeast, Colombia, Costa Rica, Cuba, Dominican Republic, Florida, Guatemala, Haiti, Honduras, Jamaica, Leeward Is., Mexico Gulf, Mexico Northeast, Mexico Northwest, Mexico Southeast, Mexico Southwest, Nicaragua, Panamá, Puerto Rico, Trinidad-Tobago, Venezuela, Venezuelan Antilles, Windward Is.
Distribution (India): Andhra Pradesh, Bihar, Odisha, Madhya Pradesh, Maharashtra, Odisha, Rajasthan, Tamil Nadu
Leaf Type: Leaves are mostly pinnately cut into almost leaflets
Inflorescence: star-shaped purple inflorescence
Fruit Type: Berries, red colored
Flowering and Fruiting: July–February
IUCN Status: Not evaluated
Notes: Usually found in degraded evergreen and moist deciduous forests
Reference: Bot. Repos. 8: t. 504 (1808)

Solanum spirale Roxb.
Solanaceae
Synonyms: *Solanum apoense* Elmer., *Solanum callium* C.T.White ex R.J.F.Hend., *Solanum naratida* Buch.-Ham. ex Wall., *Solanum spirale* var. *tetrasepalum* H. Zhu., *Solanum superficiens* Adelb.
Common Name: Spiral nightshade, Spiral bittersweet
Climbing Mechanism: Scrambler-Unarmed
Distribution (Global): Assam, Bangladesh, China South-Central, China Southeast, East Himalaya, Java, Laos, Myanmar, Sumatera, Thailand, Tibet, Vietnam
Distribution (India): Assam
Leaf Type: Leaves are unequally paired
Inflorescence: Short, simple spiral racemes
Flowering and Fruiting: May–July
IUCN Status: Not evaluated
Reference: Fl. Ind. 2: 247 (1824)

Solanum trilobatum L.
Solanaceae
Synonyms: *Solanum acetosifolium* Lam., *Solanum canaranum* Miq., *Solanum canaranum* Miq. ex C.B.Clarke., *Solanum fuscum* B.Heyne ex Wall., *Solanum maingayi* Kuntze., *Solanum prostratum* Raeusch., *Solanum sarmentosum* Nees
Common Name: Red pea eggplant, Thai nightshade
Climbing Mechanism: Scrambler-Armed
Distribution (Global): India, Laos, Malaya, Myanmar, Sri Lanka, Vietnam
Distribution (India): Andhra Pradesh, Bihar, Odisha, Gujarat, Maharashtra, Odisha, Puducherry, Rajasthan, Tamil Nadu

Leaf Type: Round and ovate
Inflorescence: 3-9-flowered clusters in leaf-axils
Fruit Type: Scarler, round, and red in color
Flowering and Fruiting: Throughout the year
IUCN Status: Not evaluated
Notes: Found mostly in deciduous forests
Reference: Sp. Pl.: 188 (1753)

Solanum wendlandii Hook. f.
Solanaceae
Synonyms: *Solanum mazatenangense* J.M.Coult. & Donn.Sm., *Solanum mazatenangense* J.M.Coult., *Solanum tlacotalpense* Sessé & Moc., *Solanum unguis-cati* Standl.
Common Name: Giant potato creeper, Paradise flower or Costa Rican nightshade
Climbing Mechanism: Scrambler-Armed
Distribution (Global): Colombia, Costa Rica, Honduras, Mexico Central, Mexico Gulf, Mexico Northeast, Mexico Southeast, Mexico Southwest, Nicaragua, Panamá, Peru, Venezuela, Venezuelan Antilles
Distribution (India): Maharashtra, Tamil Nadu
Leaf Type: Leaves are variable from simple to ovate, alternate
Inflorescence: 50-flowered lax scorpioid or paniculate cymes
Fruit Type: Clusters at the ends of elongated branches, dark bottle green becoming yellow to red, globose
Flowering and Fruiting: January–October
IUCN Status: Not evaluated
Notes: Found occasionally naturalized on roadsides and in montane forests
Reference: Bot. Mag. 113: t. 6914 (1887)

Stemona tuberosa Lour.
Stemonaceae
Synonyms: *Roxburghia gloriosoides* Jones ex Roxb.
Common Name: Wild Asparagus
Climbing Mechanism: Stem Twiner
Distribution (Global): Assam, Bangladesh, Cambodia, China South-Central, China Southeast, Hainan, India, Laos, Lesser Sunda Is., Maluku, Myanmar, New Guinea, Philippines, Sri Lanka, Taiwan, Thailand, Vietna
Distribution (India): Andhra Pradesh, Arunachal Pradesh, Bihar, Odisha, Odisha, Tamil Nadu, Telangana, Tripura
Leaf Type: Leaves are opposite or whorled, rarely alternate
Fruit Type: Capsules
Flowering and Fruiting: February–July
IUCN Status: Not evaluated
Notes: Often more gregarious and associated along moist places
Reference: Fl. Cochinch.: 404 (1790)

Stemonurus malaccensis (Mast.) Sleumer
Stemonuraceae
Synonyms: *Lasianthera malaccensis* Mast., *Stemonurus capitatus* Becc., *Stemonurus labuanensis* Stapf., *Urandra capitatus* (Becc.) Kuntze., *Urandra nitida* R.A.Howard
Climbing Mechanism: Stem Twiner
Distribution (Global): Borneo, Malaya, Myanmar, Sumatera, Thailand, Vietnam
Leaf Type: Simple
Inflorescence: Umbel
Fruit Type: Wet, drupe
IUCN Status: Not evaluated
Reference: Notizbl. Bot. Gart. Berlin-Dahlem 15: 243 (1940)

Enkleia malaccensis Griff.
Thymelaeaceae
Synonyms: *Enkleia andamanica* (Hutch. ex C.E.Parkinson) N.P.Balakr., *Enkleia coriacea* Hallier f., *Enkleia malayana* Griff., *Enkleia riouwensis* Hallier f., *Enkleia siamensis* (Kurz) Nevling, *Lasiosiphon scandens* Endl., *Linostoma andamanicum* Hutch. ex C.E.Parkins
Climbing Mechanism: Scrambler-Unarmed
Distribution (Global): Andaman Is., Borneo, Cambodia, Laos, Malaya, Myanmar, Sumatera, Thailand, Vietnam
IUCN Status: Not evaluated
Reference: Calcutta J. Nat. Hist. 4: 235 (1844)

Linostoma decandrum (Roxb.) Wall. ex Meisn.
Thymelaeaceae
Synonyms: *Nectandra decandra* Roxb.
Climbing Mechanism: Scrambler-Unarmed
Distribution (Global): Assam, Bangladesh, Cambodia, Laos, Myanmar, Thailand, Vietnam
Distribution (India): Mizoram, Tripura
Leaf Type: Leaves simple, opposite, elliptic to oblongelliptic or oblong-lanceolate
Inflorescence: Inflorescence
terminal, umbelliform, 2–3 flowered
Fruit Type: Fruits ovoid-ellipsoid
Flowering and Fruiting: December–March
IUCN Status: Not evaluated
Notes: It usually grows in moist and shaded environments of evergreen forests
Reference: Nomencl. Bot., ed. 2, 2: 50 (1841)

Linostoma scandens Kurz
Thymelaeaceae
Synonyms: *Enkleia malaccensis* Griff., *Enkleia andamanica* (Hutch. ex C.E. Parkinson) N.P.Balakr., *Enkleia coriacea* Hallier f., *Enkleia malayana* Griff.,

Enkleia riouwensis Hallier f., *Enkleia siamensis* (Kurz) Nevling, *Lasiosiphon scandens* Endl., *Linostoma andam*
Climbing Mechanism: Scrambler-Unarmed
Distribution (Global): Andaman Is., Borneo, Cambodia, Laos, Malaya, Myanmar, Sumatera, Thailand, Vietnam
Leaf Type: Leaves subopposite, coriaceous, golden-brown velutinous; petiole, blade ovate to elliptical
Inflorescence: Inflorescence paniculately branched, terminal, each branch with 4–14 flowers arranged like an umbel
Fruit Type: Fruit an ovoid drupe, prominently ribbed
Flowering and Fruiting: April– October
IUCN Status: Not evaluated
Notes: It usually occurs in primary forests in lowlands
Reference: Calcutta J. Nat. Hist. 4: 235 (1844)

Tropaeolum majus L.
Tropaeolaceae
Synonyms: *Cardamindum majus* (L.) Moench, *Nasturtium indicum* Garsault. +18
Common Name: Nasturtium, Garden Nasturtium, Indian cress
Climbing Mechanism: Stem Twiner
Distribution (Global): Peru
Distribution (India): Madhya Pradesh, Maharashtra, Manipur, Tamil Nadu, West Bengal
Leaf Type: Leaves simple, alternate, peltate
Inflorescence: Axillary, solitary, orange, purple, maroon colored
Fruit Type: oblate, separating into 3,1-seeded mericarp at maturity
Flowering and Fruiting: Throughout the year
IUCN Status: Not evaluated
Reference: Sp. Pl.: errata, last page (1753)

Tropaeolum peregrinum L.
Tropaeolaceae
Synonyms: *Tropaeolum aduncum* Sm., *Tropaeolum canariense* Markham, *Tropaeolum morreanum* Klatt, *Tropaeolum peregrinum* var. *weberbaueri* (Loes.) Sparre, *Tropaeolum weberbaueri* Loes., *Trophaeum aduncum* (Sm.) Kuntze
Trophaeum peregrinum (L.) Kuntze
Climbing Mechanism: Stem Twiner
Distribution (Global): Ecuador, Peru
Distribution (India): Maharashtra
Leaf Type: Simple, palm shaped
Inflorescence: Single, yellow
Fruit Type: Dry
Flowering and Fruiting: November–December

IUCN Status: Not evaluated
Reference: Sp. Pl.: 345 (1753)

Conocephalus suaveolens Blume
Urticaceae
Synonyms: *Poikilospermum suaveolens* (Blume) Merr., *Covellia composita* Miq., *Covellia grandifolia* Miq.
Balansaephytum tonkinense Drake. +20
Climbing Mechanism: Scrambler-Unarmed
Distribution (Global): Andaman Is., Assam, Bangladesh, Borneo, Cambodia, China South-Central, East Himalaya, Java, Laos, Malaya, Maluku, Myanmar, Nicobar Is., Philippines, Sulawesi, Sumatera, Thailand, Vietnam
Distribution (India): Maharashtra, Peninsular India
Leaf Type: Alternate
Fruit Type: Achenes, obovoid
Flowering and Fruiting: April–May
IUCN Status: Not evaluated
Notes: Often found in moist places, rain forests, monsoon forests, and near streams
Reference: Contr. Arnold Arbor. 8: 47 (1934)

Oreocnide frutescens (Thunb.) Miq.
Urticaceae
Synonyms: *Boehmeria frutescens* (Thunb.) Thunb., *Urtica frutescens* Thunb., *Villebrunea frutescens* (Thunb.) Blume
Climbing Mechanism: Scrambler-Unarmed
Distribution (Global): Assam to Temp. E. Asia and Peninsula Malaysia
Flowering and Fruiting: December–May
IUCN Status: Least Concern
Notes: Frequently occur in damp valleys
Reference: Ann. Mus. Bot. Lugduno-Batavi 3: 131 (1867)

Pellionia procridifolia Kurz
Urticaceae
Climbing Mechanism: Scrambler-Unarmed
Distribution (India): Great Nicobar Island
IUCN Status: Not evaluated
Reference: J. Bot. 13: 330 (1875)

Pilea microphylla (L.) Liebm.
Urticaceae
Synonyms: *Dubrueilia microphylla* (L.) Gaudich., *Parietaria microphylla* L., *Pilea muscosa* var. *microphylla* (L.) Wedd., *Pilea trianthmoides* var. *microphylla* (L.) Wedd., *Urtica microphylla* (L.) Sw.
Common Name: Artillery gun plant, Gunpowder plant, Rockweed
Climbing Mechanism: Scrambler-Unarmed

Distribution (Global): Bahamas, Belize, Bolivia, Cayman Is., Colombia, Costa Rica, Cuba, Dominican Republic, El Salvador, Florida, French Guiana, Georgia, Guatemala, Guyana, Haiti, Honduras, Jamaica, Leeward Is., Louisiana, Mexico Central, Mexico Gulf, Mexico Northeast, Mexico Northwest, Mexico Southeast, Mexico Southwest, Netherlands Antilles, Panamá, Peru, Puerto Rico, South Carolina, Southwest Caribbean, Suriname, Trinidad-Tobago, Venezuela, Venezuelan Antilles, Windward Is.
Distribution (India): Gujarat
Flowering and Fruiting: Throughout the year
IUCN Status: Not evaluated
Notes: It usually occurs as a weed in gardens
Reference: Mexic. Neldeagt. Pl.: 18 (1851)

Poikilospermum naucleiflorum (Roxburgh ex Lindl.) Chew
Urticaceae
Synonyms: *Conocephalus naucleiflorus* Lindl., *Urtica naucliflora* (Lindl.) Roxb.
Climbing Mechanism: Scrambler-Unarmed
Distribution (Global): East Himalayas, Myanmar, Nepal, Thailand, Tibet
Distribution (India): Arunachal Pradesh.
IUCN Status: Not evaluated
Reference: Gard. Bull. Singapore 20: 76 (1963)

Poikilospermum suaveolens (Blume) Merr.
Urticaceae
Synonyms: *Covellia composita* Miq., *Covellia grandifolia* Miq., *Balansaephytum tonkinense* Drake, *Conocephalus ellipticus* Trécul, *Conocephalus gratus* Miq., *Conocephalus mollis* Gagnep. +18
Common Name: Fragrant conehead
Climbing Mechanism: Scrambler-Unarmed
Distribution (Global): Andaman Is., Assam, Bangladesh, Borneo, Cambodia, China South-Central, East Himalaya, Java, Laos, Malaya, Maluku, Myanmar, Nicobar Is., Philippines, Sulawesi, Sumatera, Thailand, Vietnam
Distribution (India): Arunachal Pradesh, Assam, Great Nicobar Island, Meghalaya, Mizora
Flowering and Fruiting: January–April
IUCN Status: Not evaluated
Reference: Contr. Arnold Arbor. 8: 47 (1934)

Pouzolzia pentandra var. *wightii* (Benn. & Br.) Friis & Wilmot-Dear
Urticaceae
Synonyms: *Gonostegia pentandra* (Roxb.) Miq., *Pouzolzia wightii* Benn., *Pouzolzia wallichiana* Wight, *Memorialis concinna* Wedd., *Boehmeria melastomoidea* Griff., *Gonostegia alternifolia* Turcz. +41
Climbing Mechanism: Scrambler-Unarmed
Distribution (India): Maharashtra

Flowering and Fruiting: August–March
IUCN Status: Not evaluated
Reference: Pl. Jav. Rar.: 64 (1838)

Lantana camara L.
Verbenaceae
Synonyms: *Camara vulgaris* Benth
Common Name: Red sage
Climbing Mechanism: Scrambler-Armed
Distribution (Global): Aruba, Bahamas, Belize, Brazil North, Brazil West-Central, Cayman Is., Colombia, Costa Rica, Cuba, Dominican Republic, Ecuador, El Salvador, Guatemala, Guyana, Haiti, Honduras, Jamaica, Leeward Is., Mexico Central, Mexico Gulf, Mexico Northeast, Mexico Northwest, Mexico Southeast, Mexico Southwest, Netherlands Antilles, Nicaragua, Panamá, Puerto Rico, Southwest Caribbean, Trinidad-Tobago, Venezuela, Venezuelan Antilles, Windward Is.
Distribution (India): Bihar, Odisha, Goa, Maharashtra, Meghalaya, Odisha, Puducherry, Rajasthan, Tamil Nadu, Telangana, Tripura, and the Western Ghats
Leaf Type: Simple, opposite, decussate, attached by a short hairy petiole. The blade is small, oval, hispid, with margin toothed
Inflorescence: Flowers are in a large umbel round shape. Compact, flat-topped inflorescences supported by a peduncle, pubescent and glandular
Fruit Type: Small, fleshy, purplish black berries
Flowering and Fruiting: Almost throughout the year
IUCN Status: Not evaluated
Notes: Occurs in the bushveld, forest, forest-edge, riparian or riverine, staff villages, and roadsides
Reference: Sp. Pl.: 627 (1753)

Lantana montevidensis
Verbenaceae
Climbing Mechanism: Scrambler-Unarmed
IUCN Status: Not evaluated

Lantana salvifolia Baker & Stapf
Verbenaceae
Climbing Mechanism: Scrambler-Armed
Distribution (India): Gujarat
IUCN Status: Not evaluated
Reference: Pl. Hort. Schoenbr. 3: 18 (1798)

Petrea volubilis L.
Verbenaceae
Synonyms: *Petrea amazonica* Moldenke, *Petrea arborea* Kunth, *Petrea arborea* f. albiflora Standl., *Petrea arborea* var. *broadwayi* Moldenke, *Petrea arborea*

f. broadwayi (Moldenke) Moldenke, *Petrea arborescens* Archer ex Moldenke +28
Common Name: Purple wreath, Queen's wreath, Sandpaper vine
Climbing Mechanism: Scrambler-Unarmed
Distribution (Global): Belize, Bolivia, Brazil North, Brazil Northeast, Brazil South, Brazil Southeast, Brazil West-Central, Cayman Is., Colombia, Costa Rica, Cuba, Dominican Republic, Ecuador, El Salvador, Florida, French Guiana, Guatemala, Guyana, Haiti, Honduras, Jamaica, Leeward Is., Mexico Central, Mexico Gulf, Mexico Northeast, Mexico Southeast, Mexico Southwest, Nicaragua, Panamá, Paraguay, Peru, Puerto Rico, Suriname, Trinidad-Tobago, Venezuela, Windward Is.
Distribution (India): Andaman and Nicobar Islands, Delhi, Gujarat, Madhya Pradesh, Maharashtra, Odisha, Tamil Nadu
Leaf Type: Simple
Inflorescence: Raceme spike, axillary
Fruit Type: Drupe
Flowering and Fruiting: February–April
IUCN Status: Not evaluated
Reference: Sp. Pl.: 626 (1753)

Viburnum colebrookeanum Wall. ex DC.
Viburnaceae
Synonyms: *Viburnum lutescens* Hook. f. & Thomson
Climbing Mechanism: Scrambler-Unarmed
Distribution (Global): Assam, Bangladesh, East Himalaya, Myanmar, Nepal
Distribution (India): Manipur
Leaf Type: Simple
Inflorescence: Corymbose
Fruit Type: Wet, drupe
Flowering and Fruiting: February–June
IUCN Status: Least Concern
Reference: Prodr. 4: 325 (1830)

Ampelocissus araneosa (Dalzell) Gamble
Vitaceae
Synonyms: *Cissus araneosa* Dalzell, *Vitis araneosa* (Dalz.) M. Lawson
Common Name: Wild grape
Climbing Mechanism: Tendril Climber
Distribution (Global): India, Thailand
Distribution (India): Tamil Nadu, Andhra Pradesh, Maharashtra
Leaf Type: Compound, 3-foliolate to palmately lobed
Inflorescence: Cymes, corymbose, peduncle with tendrils
Fruit Type: Wet, berry
Flowering and Fruiting: July–March
IUCN Status: Not evaluated

Notes: Common in semi-evergreen and evergreen forests
Reference: Fl. Madras: 230 (1918)

Ampelocissus barbata (Wall.) Planch.
Vitaceae
Synonyms: *Vitis barbata* Wall., Vitis latifolia Buch. -Ham. ex Wall.
Climbing Mechanism: Tendril Climber
Distribution (Global): Andaman Is., Assam, Bangladesh, Cambodia, East Himalaya, Laos, Myanmar, Nepal, Nicobar Is., Thailand, Vietnam, West Himalaya
Distribution (India): Tripura, Mizoram, Arunachal Pradesh, West Bengal, Andaman and Nicobar Islands, North Eastern India
Leaf Type: Simple, alternate
Inflorescence: Dichotomous cymes with 5–10 flowers
Fruit Type: Drupe
Flowering and Fruiting: March–September
IUCN Status: Not evaluated
Reference: A.L.P.P.de Candolle & A.C.P.de Candolle, Monogr. Phan. 5: 372 (1887)

Ampelocissus birii P.Singh & B.V.Shetty
Vitaceae
Climbing Mechanism: Tendril Climber
Distribution (Global): India
Distribution (India): Kerala
Leaf Type: Simple, alternate
Flowering and Fruiting: May–July
IUCN Status: Not evaluated
Notes: Distributed mostly in semi-evergreen forests
Reference: Fl. Kerala 1: 738 (2005)

Ampelocissus cinnamomea (Wall. ex M.A.Lawson) Planch
Vitaceae
Synonyms: *Vitis cinnamomea* Wall., *Ampelopsis ternata* DC., *Vitis labrusca* Lour., *Vitis lanata* Wall., *Vitis obtusifolia* Buch. -Ham. ex Wall. *Vitis oppositifolia* Buch. -Ham. ex Steud., *Vitis ternata* Roth ex Roem. & Schult., *Vitis tomentosa* Heyne ex Roth, *Vitis*
Climbing Mechanism: Tendril Climber
Distribution (Global): Malaya, Sumatera
Leaf Type: Compound, 3-foliolate to palmately lobed
Inflorescence: Cymes, corymbose, peduncle with tendrils
Fruit Type: Wet, berry
IUCN Status: Not evaluated
Reference: A.L.P.P.de Candolle & A.C.P.de Candolle, Monogr. Phan. 5: 408 (1887)

Ampelocissus compositifolia (M.A.Lawson) Planch
Vitaceae

Synonyms: *Ampelocissus thyrsiflora* (Blume) Planch
Climbing Mechanism: Tendril Climber
Leaf Type: Compound, 3-foliolate to palmately lobed
Inflorescence: Cymes, corymbose, peduncle with tendrils
Fruit Type: Wet, berry
IUCN Status: Not evaluated
Reference: A.L.P.P.de Candolle & A.C.P.de Candolle, Monogr. Phan. 5: 412 (1887)

Ampelocissus divaricata (Wall. ex M.A.Lawson) Planch
Vitaceae
Synonyms: *Ampelocissus rupicola* Craib, Vitis divaricata Wall. ex M.A.Lawson
Climbing Mechanism: Tendril Climber
Distribution (Global): Assam, Cambodia, East Himalaya, India, Laos, Myanmar, Nepal, Thailand, Vietnam, West Himalaya
Distribution (India): Andhra Pradesh, Madhya Pradesh, Himachal Pradesh, Tamil Nadu, Odisha, Manipur
Leaf Type: Compound, 3-foliolate to palmately lobed
Inflorescence: Cymes, corymbose, peduncle with tendrils
Fruit Type: Wet, berry
Flowering and Fruiting: June–December
IUCN Status: Not evaluated
Notes: Distributed in evergreen forests
Reference: Vigne Amér. Vitic. Eur. 8: 375 (1884)

Ampelocissus gracilis (Wall.) Planch.
Vitaceae
Synonyms: *Vitis gracilis* Wall.
Climbing Mechanism: Tendril Climber
Distribution (Global): Borneo, Malaya, Sumatera
Leaf Type: Compound, 3-foliolate to palmately lobed
Inflorescence: Cymes, corymbose, peduncle with tendrils
Fruit Type: Wet, berry
IUCN Status: Not evaluated
Reference: A.L.P.P.de Candolle & A.C.P.de Candolle, Monogr. Phan. 5: 407 (1887)

Ampelocissus indica (L.) Planch.
Vitaceae
Synonyms: *Ampelocissus arnottiana* Planch., *Cissus indica* (L.) Walp., *Vitis indica* L.
Common Name: Red-stemmed wild grapevine
Climbing Mechanism: Tendril Climber
Distribution (Global): India, Sri Lanka
Distribution (India): Kerala, Tamil Nadu, Maharashtra
Leaf Type: Simple
Inflorescence: Racemes, Umbels

Fruit Type: Berries
Flowering and Fruiting: March–September
IUCN Status: Not evaluated
Notes: Common in forests and along hedges
Reference: Vigne Amér. Vitic. Eur. 8: 375 (1884)

Ampelocissus jacquemontii (R. Parker) Raizada
Vitaceae
Climbing Mechanism: Tendril Climber
Leaf Type: Compound, 3-foliolate to palmately lobed
Inflorescence: Cymes, corymbose, peduncle with tendrils
Fruit Type: Wet, berry
IUCN Status: Not evaluated
Reference: Suppl. Duthie's Fl. Upper Gangetic Plain 5: 44 (1976)

Ampelocissus latifolia (Roxb.) Planch
Vitaceae
Synonyms: *Cissus latifolia* (Roxb.) Peterm., *Cissus latifolia* Walp., *Cissus roxburghii* Peterm., *Vitis glabrata* Roth, *Vitis latifolia* Roxb., *Vitis montana* M. A.Lawson, *Vitis zeylanica* Russell ex Wall.
Common Name: Jungle grapevine
Climbing Mechanism: Tendril Climber
Distribution (Global): Assam, Bangladesh, East Himalaya, India, Nepal, Pakistan, Thailand, and West Himalaya
Distribution (India): Peninsular India, Gujarat, Madhya Pradesh, Assam, Uttar Pradesh, Orissa, Western Ghats, North-eastern India
Leaf Type: Simple
Inflorescence: Thyrses
Fruit Type: Berries
Flowering and Fruiting: May–December
IUCN Status: Not evaluated
Notes: Frequently found near streams in Evergreen forests
Reference: Vigne Amér. Vitic. Eur. 8: 374 (1884)

Ampelocissus rugosa
Vitaceae
Climbing Mechanism: Tendril Climber
Distribution (India): Himachal Pradesh
Leaf Type: Compound, 3-foliolate to palmately lobed
Inflorescence: Cymes, corymbose, peduncle with tendrils
Fruit Type: Wet, berry
IUCN Status: Not evaluated
Reference: Planch. In: J. Vigne Arn. 1884: 374. (1884)

Ampelocissus serrata Voss
Vitaceae
Climbing Mechanism: Tendril Climber
Leaf Type: Compound, 3-foliolate to palmately lobed
Inflorescence: Cymes, corymbose, peduncle with tendrils
Fruit Type: Wet, berry
IUCN Status: Not evaluated
Reference: Vilm. Blumengärtn. ed. 3, 1: 182, 1894

Ampelocissus sikkimensis (M.A. Lawson) Planch.
Vitaceae
Synonyms: *Vitis sikkimensis* M.A.Lawson
Climbing Mechanism: Tendril Climber
Distribution (Global): Assam, China South-Central, East Himalaya, Nepal
Distribution (India): West Bengal, Northeast India
Leaf Type: Compound, 3-foliolate to palmately lobed
Inflorescence: Cymes, corymbose, peduncle with tendrils
Fruit Type: Wet, berry
IUCN Status: Not evaluated
Reference: Vigne Amér. Vitic. Eur. 8: 375 (1884)

Ampelocissus tomentosa (B.Heyne & Roth) Planch.
Vitaceae
Synonyms: *Ampelopsis ternata* DC., *Vitis obtusifolia* Buch. -Ham. ex Wall., *Vitis oppositifolia* Buch. -Ham. ex Steud., *Vitis tomentosa* Roth, *Vitis trifida* Roth, *Vitis triloba* Roth.
Common Name: Hairy wild grape
Climbing Mechanism: Tendril Climber
Distribution (Global): Bangladesh, India, Myanmar
Distribution (India): Peninsular India, West Bengal, Gujarat, Madhya Pradesh, Orissa
Leaf Type: Compound, 3-foliolate to palmately lobed
Inflorescence: Cymes, corymbose, peduncle with tendrils
Fruit Type: Wet, berry
Flowering and Fruiting: April–November
IUCN Status: Not evaluated
Notes: Mostly found in Semi-evergreen and moist deciduous forests
Reference: Vigne Amér. Vitic. Eur. 8: 375 (1884)

Ampelocissus wightiana B.V.Shetty & Par.Singh
Vitaceae
Synonyms: *Ampelocissus erioclada* (Wight & Arn.) Planch., Vitis erioclada Wight & Arn.
Climbing Mechanism: Tendril Climber
Distribution (Global): India, Sri Lanka

Distribution (India): Tamil Nadu, Maharashtra
Leaf Type: Compound, 3-foliolate to palmately lobed
Inflorescence: Cymes, corymbose, peduncle with tendrils
Fruit Type: Wet, berry
IUCN Status: Not evaluated
Reference: Kew Bull. 44: 469 (1989)

Ampelopsis cantoniensis (Hook. & Arn.) Planch.
Vitaceae
Climbing Mechanism: Tendril Climber
Leaf Type: Simple, cordate at base
Inflorescence: Cymes, corymbose, opposite to leaves
Fruit Type: Wet, berry
IUCN Status: Not evaluated
Reference: Hort. Dendrol. 48: 11 (1853)

Ampelopsis glandulosa (Wall.) Momiy.
Vitaceae
Synonyms: *Vitis glandulosa* Wall.
Climbing Mechanism: Tendril Climber
Distribution (Global): Assam, Bangladesh, China North-Central, China South-Central, China Southeast, East Himalaya, Japan, Khabarovsk, Korea, Kuril Is., Laos, Manchuria, Myanmar, Nansei-Shoto, Nepal, Philippines, Primorye, Sakhalin, Taiwan, Vietnam
Leaf Type: Simple, cordate at base
Inflorescence: Cymes, corymbose, opposite to leaves
Fruit Type: Wet, berry
IUCN Status: Not evaluated
Reference: Bull. Univ. Mus. Univ. Tokyo 2: 78 (1971)

Ampelopsis glandulosa f. elegans (K.Koch) Momiy.
Vitaceae
Climbing Mechanism: Tendril Climber
Leaf Type: Simple, cordate at base
Inflorescence: Cymes, corymbose, opposite to leaves
Fruit Type: Wet, berry
IUCN Status: Not evaluated
Reference: Momiy. In: J. Jap. Bot., 52(1): 31. (1977)

Ampelopsis glandulosa f. elegans (K.Koch) Momiy.
Vitaceae
Climbing Mechanism: Tendril Climber
IUCN Status: Not evaluated

Ampelopsis rubifolia (Wall.) Planch.
Vitaceae
Synonyms: *Nekemias rubifolia* (Wall.) J.Wen & Z.L.Nie
Climbing Mechanism: Tendril Climber
Leaf Type: Simple, cordate at base
Inflorescence: Cymes, corymbose, opposite to leaves
Fruit Type: Wet, berry
IUCN Status: Not evaluated
Reference: A.L.P.P.de Candolle & A.C.P.de Candolle, Monogr. Phan. 5: 462 (1887)

Ampelopsis vitifolia (Boiss.) Planch.
Vitaceae
Synonyms: *Cissus vitifolia* Boiss.
Climbing Mechanism: Tendril Climber
Distribution (Global): Afghanistan, Iran, Kirgizstan, Pakistan, Tadzhikistan, Turkmenistan, Uzbekistan, West Himalayas
Leaf Type: Simple, cordate at base
Inflorescence: Cymes, corymbose, opposite to leaves
Fruit Type: Wet, berry
IUCN Status: Not evaluated
Reference: A.L.P.P.de Candolle & A.C.P.de Candolle, Monogr. Phan. 5: 454 (1887)

Cayratia anemonifolia (Zipp. ex Miq.) Suess.
Vitaceae
Climbing Mechanism: Tendril Climber
Distribution (Global): India, Lesser Sunda Is.
Distribution (India): Tamil Nadu
Leaf Type: Compound, pedately, 3–7 foliolate
Inflorescence: Axillary cyme
Fruit Type: Wet, berry
IUCN Status: Not evaluated
Notes: Mostly found in high-altitude forests
Reference: H.G.A.Engler, Nat. Pflanzenfam. ed. 2, 20d: 280, 391 (1953)

Cayratia geniculata (Blume) Gagnep.
Vitaceae
Synonyms: *Cayratia mollissima* (Wall.) Gagnep., *Cayratia rhodocarpa* (Blume) Gagnep., *Cayratia wrayi* (King) Gagnep., *Cissus alata* Blanco., *Cissus geniculata* Blume., *Cissus hirtella* Blume., *Cissus jackii* Korth. ex Miq. +10
Climbing Mechanism: Tendril Climber
Distribution (Global): Assam, Borneo, Cambodia, China South-Central, China Southeast, East Himalaya, Hainan, India, Java, Laos, Lesser Sunda Is., Malaya, Myanmar, Philippines, Sulawesi, Thailand, Tibet, Vietnam
Distribution (India): Assam, West Bengal
Leaf Type: Acuminate at the apex

Inflorescence: Extra-axillary, trichotomous cymes
Fruit Type: Berry, Obovoid
Flowering and Fruiting: March–December
IUCN Status: Not evaluated
Notes: Common in forests of Western Ghats. Fruits are used in curry
Reference: Notul. Syst. (Paris) 1: 345 (1911)

Cayratia japonica (Thunb.) Gagnep.
Vitaceae
Synonyms: *Causonis japonica* (Thunb.) Raf., *Cayratia tenuifolia* (Wight & Arn.) Gagnep. +20
Common Name: Bush killer
Climbing Mechanism: Tendril Climber
Distribution (India): Andaman and Nicobar Islands, Arunachal Pradesh, Assam, Karnataka, Kerala, Mizoram, Tamil Nadu, Tripura, and West Bengal
Leaf Type: 5-Foliolate, pedate
Fruit Type: 3-4 Seeded berries
Flowering and Fruiting: March–June
IUCN Status: Not evaluated
Notes: Often occur in moist deciduous forests
Reference: Sylva Tellur.: 87 (1838)

Cayratia mollissima (Planch.) Gagnep.
Vitaceae
Climbing Mechanism: Tendril Climber
Distribution (India): Arunachal Pradesh, Karnataka, Kerala, Tamil Nadu
Leaf Type: Compound, pedately, 3–7 foliolate
Inflorescence: Axillary cyme
Fruit Type: Wet, berry
IUCN Status: Not evaluated
Reference: Gagnep. In: Lecomte, Not. Syst. 1: 345. (1911)

Cayratia pedata (Lam.) Gagnep.
Vitaceae
Synonyms: *Melothria zeylanica* J.Koenig ex Wight & Arn., *Fusanus pedatus* (Lour.) Spreng., *Cayratia longzhouensis* W.T.Wang., *Cissus canarensis* (Dalzell) Planch., *Cissus heptaphylla* Retz. +10
Common Name: Birdfoot grapevine, Birdfoot treebine
Climbing Mechanism: Tendril Climber
Distribution (Global): Andaman Is., Assam, Bangladesh, Cambodia, China South-Central, China Southeast, Christmas I., East Himalaya, India, Java, Laos, Myanmar, Nicobar Is., Philippines, Sri Lanka, Thailand, Vietnam
Distribution (India): Andaman and Nicobar Islands, Andhra Pradesh, Arunachal Pradesh, Kerala, Madhya Pradesh, Maharashtra, Odisha, Puducherry, Tamil Nadu, Telangana, West Bengal, Western Ghats

Leaf Type: Leaves are alternate, pedate
Inflorescence: Dichasial cymes
Fruit Type: Berry, subglobose
Flowering and Fruiting: Throughout the year
IUCN Status: Vulnerable
Notes: The whole plant of Nilgiri Grape-Vine (excluding the roots) has low diuretic activity and has been a reputed remedy for cough, bronchitis, asthma, joint pain, and to check uterine reflexes
Reference: Notul. Syst. (Paris) 1: 346 (1911)

Cayratia reticulata (M.A.Lawson) Mabb.
Vitaceae
Synonyms: *Cayratia retivenia* (Planch.) Suess., *Cissus reticulata* Thwaites., *Cissus retivenia* Planch., *Columella retivenia* (Planch.) Alston., *Vitis reticulata* M.A. Lawson
Climbing Mechanism: Tendril Climber
Distribution (Global): Sri Lanka
Leaf Type: Compound, pedately, 3-7 foliolate
Inflorescence: Axillary cyme
Fruit Type: Wet, berry
Flowering and Fruiting: April–June
IUCN Status: Not evaluated
Reference: M.D.Dassanayake & al. (eds.), Revis. Handb. Fl. Ceylon 9: 456 (1995)

Cayratia roxburghii (Planch.) Gagnep.
Vitaceae
Synonyms: *Cissus roxburghii* Planch., *Vitis roxburghii* M.A.Lawson
Climbing Mechanism: Tendril Climber
Distribution (Global): India, Laos
Distribution (India): Kerala, Tamil Nadu
Leaf Type: Compound, pedately, 3–7 foliolate
Inflorescence: Axillary cyme
Fruit Type: Wet, berry
Flowering and Fruiting: January–April
IUCN Status: Not evaluated
Notes: Endemic to Southern Western Ghats
Reference: Notul. Syst. (Paris) 1: 346 (1911)

Cayratia tenuifolia (Wight & Arn.) Gagnep.
Vitaceae
Climbing Mechanism: Tendril Climber
Distribution (India): Goa, Kerala, Maharashtra, Tamil Nadu
Leaf Type: Compound, pedately, 3–7 foliolate
Inflorescence: Axillary cyme
Fruit Type: Wet, berry

IUCN Status: Not evaluated
Reference: Gagnep. In: Not. Syst. 1: 348. (1911)

Cayratia trifolia (L.) Domin
Vitaceae
Synonyms: *Causonis trifolia* (L.) Mabb. & J. Wen. + 10
Common Name: Bush grape, foxgrape, three-leaved wild vine
Climbing Mechanism: Tendril Climber
Distribution (Global): Andaman Is., Assam, Bangladesh, Bismarck Archipelago, Borneo, Cambodia, Caroline Is., China South-Central, Christmas I., East Himalaya, India, Java, Laccadive Is., Laos, Lesser Sunda Is., Malaya, Maluku, Myanmar, Nepal, New Caledonia, New Guinea, Nicobar Is., Northern Territory, Pakistan, Philippines, Queensland, Solomon Is., South China Sea, Sri Lanka, Sulawesi, Sumatera, Thailand, Vanuatu, Vietnam, West Himalaya, Western Australia
Distribution (India): Andaman and Nicobar Islands, Andhra Pradesh, Arunachal Pradesh, Goa, Great, Nicobar, Island, Kerala, Madhya Pradesh, Maharashtra, Odisha, Rajasthan, Telangana, Uttar Pradesh, West Bengal
Leaf Type: Trifoliolate
Inflorescence: Solitary Cymes
Fruit Type: Berries
Flowering and Fruiting: May–November
IUCN Status: Not evaluated
Notes: The root, ground with black pepper, is applied to boils. The root is also used as an astringent medicine.
Reference: Mabberley's Pl.-Book, ed. 4: 1101 (2017)

Cissus adnata Roxb.
Vitaceae
Synonyms: *Cissus compressa* Blume., *Cissus latifolia* Vahl., *Vitis adnata* (Roxb.) Wall., *Vitis compressa* (Blume) Backer, *Vitis monosperma* Noronha., *Vitis radiata* Zipp. ex Span., *Vitis repens* Wall., *Vitis vitiginea* var. *adnata* (Roxb.) Kuntze
Common Name: Entire-leaf wild grape
Climbing Mechanism: Tendril Climber
Distribution (Global): Assam, Bangladesh, Borneo, Cambodia, China South-Central, East Himalaya, India, Java, Laos, Lesser Sunda Is., Madagascar, Malaya, Myanmar, Nepal, New Guinea, Northern Territory, Philippines, Queensland, Sri Lanka, Thailand, Vietnam, Western Australia
Distribution (India): Andhra Pradesh, Arunachal Pradesh, Gujarat, Karnataka, Madhya Pradesh, Maharashtra, Manipur, Mizoram, Odisha, Tamil Nadu, Telangana, Tripura, Uttar Pradesh, West Bengal
Leaf Type: Compound, alternate, 3–5 Folaite
Inflorescence: Umbels
Flowering and Fruiting: August–March

IUCN Status: Not evaluated
Notes: Boiled extract of leaves is used in treating urinary tract infections and the powdered roots are antiseptic, applied to cuts and wounds
Reference: Fl. Ind. 1: 405 (1820)

Cissus aristata Blume
Vitaceae
Synonyms: *Cissus pyrrhodasys* Miq., *Cissus simplex* Blanco, *Vitis pyrrhodasys* (Miq.) Ridl., *Vitis simplex* (Blanco) Burkill.
Climbing Mechanism: Tendril Climber
Distribution (Global): Andaman Is., Assam, Bangladesh, Cambodia, China South-Central, East Himalaya, Hainan, India, Java, Malaya, Myanmar, New Guinea, Nicobar Is., Philippines, Queensland, Solomon Is., Sri Lanka, Sumatera, Thailand, Vietnam
Distribution (India): Andaman and Nicobar Islands, Kerala, Mizoram, Tamil Nadu
Leaf Type: Ovate, truncate
Inflorescence: Umbelled cymes
Fruit Type: Berries, ellipsoid
Flowering and Fruiting: May–December
IUCN Status: Not evaluated
Notes: Mostly found in deciduous forests and scrub jungles
Reference: Bijdr. Fl. Ned. Ind.: 183 (1825)

Cissus arnottiana B.V. Shetty & Par.Singh
Vitaceae
Climbing Mechanism: Tendril Climber
Distribution (Global): India, Sri Lanka
Distribution (India): Tamil Nadu
Leaf Type: Ovate, acute
Inflorescence: Cymes axillary
Fruit Type: Berries
Flowering and Fruiting: April–June
IUCN Status: Not evaluated
Reference: Kew Bull. 44: 473 (1989)

Cissus assamica (M.A. Lawson) Craib
Vitaceae
Synonyms: *Cissus ambigua* Elmer ex Merr., *Vitis assamica* M.A. Lawson
Climbing Mechanism: Tendril Climber
Distribution (Global): Assam, Bangladesh, Cambodia, China South-Central, China Southeast, East Himalaya, Hainan, India, Malaya, Maluku, Myanmar, Nepal, Philippines, Taiwan, Thailand, Tibet, Vietnam
Distribution (India): Arunachal Pradesh, Manipur, Meghalaya, Mizoram, Odisha, Tripura, West Bengal
Leaf Type: Simple, alternate

Inflorescence: Umbellate cymes
Fruit Type: Berries, black, ellipsoid
Flowering and Fruiting: July–September
IUCN Status: Not evaluated
Notes: Usually found in primary forests, in moist and cool places
Reference: Bull. Misc. Inform. Kew 1911: 31 (1911)

Cissus elongata Roxb.
Vitaceae
Synonyms: *Cayratia elongata* (Roxb.) Suess., *Vitis elongata* (Roxb.) Wall.
Climbing Mechanism: Tendril Climber
Distribution (Global): Andaman Is., Assam, Bangladesh, China South-Central, China Southeast, East Himalaya, Hainan, India, Malaya, Myanmar, Thailand, Vietnam
Distribution (India): Andaman and Nicobar Islands, Maharashtra
Leaf Type: Palmately compound
Inflorescence: Cymes, leaf opposed
Fruit Type: Wet, berry
Flowering and Fruiting: May–November
IUCN Status: Not evaluated
Reference: Fl. Ind. 1: 411 (1820)

Cissus elongata subsp. *littoralis* (Talbot) B.V. Shetty & Par.Singh
Vitaceae
Climbing Mechanism: Tendril Climber
Distribution (Global): India
Distribution (India): Peninsular India
Leaf Type: Palmately compound
Inflorescence: Cymes, leaf opposed
Fruit Type: Wet, berry
Flowering and Fruiting: July–December
IUCN Status: Not evaluated
Notes: Common in evergreen forests and along hedges
Reference: Kew Bull. 44: 475 (1989)

Cissus glyptocarpa Thwaites
Vitaceae
Synonyms: *Vitis glyptocarpa* (Thwaites) M. A. Lawson
Climbing Mechanism: Tendril Climber
Distribution (Global): India
Distribution (India): Kerala, Tamil Nadu
Leaf Type: Ovate
Inflorescence: Cymes
Fruit Type: Not seen
Flowering and Fruiting: April–October

IUCN Status: Not evaluated
Notes: Usually found in moist deciduous forests
Reference: Enum. Pl. Zeyl.: 62 (1858)

Cissus heyneana Planch.
Vitaceae
Synonyms: *Cissus rotundata* B. Heyne ex Wall., *Cissus siamica* Planch., *Cissus thwaitesii* Planch., *Vitis heyneana* var. *glabra* (Parker) H.B. Naithani & S.Biswas, *Vitis pallida* Trimen
Climbing Mechanism: Tendril Climber
Distribution (Global): Assam, India, Sri Lanka, Thailand
Distribution (India): Arunachal Pradesh, Kerala, Rajasthan, Tamil Nadu, West Bengal
Leaf Type: Palmately compound
Inflorescence: Cymes, leaf opposed
Fruit Type: Wet, berry
Flowering and Fruiting: July–October
IUCN Status: Not evaluated
Notes: Occasionally found on hills up to 800 m
Reference: Nomencl. Bot., ed. 2, 1: 372 (1841)

Cissus javana DC.
Vitaceae
Synonyms: *Cissus discolor* Blume., *Cissus marmorea* G.Don., *Cissus sessilis* Amshoff., *Cissus sicyoides* Klein ex Steud., *Vitis bracteata* Noronha., *Vitis costata* Wall., *Vitis discolor* (Blume) Dalzell., *Vitis diversifolia* Wall., *Vitis inaequalis* Wall.
Common Name: Climbing Begonia, Rex Begonia Vine, Painted Cissus
Climbing Mechanism: Tendril Climber
Distribution (Global): Andaman Is., Assam, Bangladesh, Cambodia, China South-Central, East Himalaya, India, Java, Laos, Lesser Sunda Is., Malaya, Myanmar, Nepal, Nicobar Is., Philippines, Thailand, Vietnam
Distribution (India): North-eastern India, Goa, West Bengal
Leaf Type: Ovate lance shaped
Inflorescence: Compound Umbells
Fruit Type: Berry is black to reddish-purple
Flowering and Fruiting: July–November
IUCN Status: Not evaluated
Reference: Catalogus: 39 (1823)

Cissus latifolia Lam.
Vitaceae
Synonyms: *Cissus gigantea* Bedd., *Cissus wightii* Planch., *Vitis gigantea* (Bedd.) Bedd.
Climbing Mechanism: Tendril Climber

Distribution (Global): India, Sri Lanka
Distribution (India): Kerala, Maharashtra, Tamil Nadu
Leaf Type: Ovate-Cordate
Inflorescence: Cymes leaf-opposed
Fruit Type: Berry, ovoid, purplish
Flowering and Fruiting: Throughout the year
IUCN Status: Not evaluated
Reference: Encycl. 1: 30 (1783)

Cissus pentagona Roxb.
Vitaceae
Synonyms: *Vitis pentagona* (Roxb.) Voigt
Climbing Mechanism: Tendril Climber
Distribution (Global): Andaman Is., Bangladesh, India, Myanmar
Distribution (India): Andaman and Nicobar Islands
Leaf Type: Palmately compound
Inflorescence: Cymes, leaf opposed
Fruit Type: Wet, berry
IUCN Status: Not evaluated
Reference: Fl. Ind. 1: 426 (1820)

Cissus quadrangularis L.
Vitaceae
Synonyms: *Saelanthus quadragonus* (L.) Forssk. ex J.F.Gmel., *Vitis quadrangularis* (L.) Wall. ex Wight.
Common Name: Adamant creeper, Bone setter, Edible-stemmed vine
Climbing Mechanism: Tendril Climber
Distribution (Global): Angola, Assam, Bangladesh, Benin, Burkina, Burundi, Cameroon, Central African Repu, Chad, Djibouti, Eritrea, Ethiopia, Gambia, Guinea-Bissau, India, Ivory Coast, Kenya, KwaZulu-Natal, Laccadive Is., Madagascar, Malawi, Mali, Mauritania, Mozambique, Myanmar, Niger, Nigeria, Northern Provinces, Oman, Pakistan, Rwanda, Saudi Arabia, Senegal, Somalia, Sri Lanka, Sudan, Swaziland, Tanzania, Togo, Uganda, Yemen, Zambia, Zaire, Zimbabwe
Distribution (India): Andhra Pradesh, Assam, Gujarat, Jharkhand, Karnataka, Kerala, Madhya Pradesh, Maharashtra, Puducherry, Tamil Nadu, Telangana, Tripura, West Bengal
Leaf Type: Simple, alternate distichous
Inflorescence: Umbellate cymes
Fruit Type: Berries, red when ripe
Flowering and Fruiting: October–January
IUCN Status: Not evaluated
Notes: Medicinally valuable for treating various ailments
Reference: Syst. Nat. ed. 12, 2: 124 (1767)

Cissus repanda (Wight & Arn.) Vahl
Vitaceae
Synonyms: *Rinxostylis repanda* (Wight & Arn.) Raf., *Vitis repanda* Wight & Arn., *Vitis vitiginea* var. *repanda* (Vahl) Kuntze
Common Name: Pani Bel, Wavy-leaved Cissus.
Climbing Mechanism: Tendril Climber
Distribution (Global): Andaman Is., Assam, Bangladesh, Cambodia, China South-Central, East Himalaya, Hainan, India, Laos, Myanmar, Nepal, Nicobar Is., Thailand, Vietnam, West Himalaya
Distribution (India): Andhra Pradesh, Arunachal Pradesh, Assam, Gujarat, Jharkhand, Kerala, Madhya Pradesh, Maharashtra, Manipur, Odisha, Rajasthan, Tamil Nadu, Telangana, Tripura, West Bengal
Leaf Type: Suborbicular
Inflorescence: Lax subcorymbosely branched umbelled cymes
Fruit Type: Berries, ellipsoid, pyriform
Flowering and Fruiting: February–September
IUCN Status: Not evaluated
Notes: Usually found in semi-evergreen forests
Reference: Symb. Bot. 3: 18 (1794)

Cissus repens Lam.
Vitaceae
Climbing Mechanism: Tendril Climber
Distribution (India): Andaman and Nicobar Islands, Andhra Pradesh, Arunachal Pradesh, Assam, Goa, Kerala, Maharashtra, Manipur, Meghalaya, Mizoram, Odisha, Tamil Nadu, Tripura, West Bengal
Leaf Type: Palmately compound
Inflorescence: Cymes, leaf opposed
Fruit Type: Wet, berry
IUCN Status: Not evaluated
Reference: Lam. In: Encycl. 1: 31. (1789)

Cissus reticulata Blume ex Planch
Vitaceae
Synonyms: *Vitis reticulata* M.A. Lawson
Climbing Mechanism: Tendril Climber
Distribution (Global): Java
Distribution (India): Maharashtra
Leaf Type: Palmately compound
Inflorescence: Cymes, leaf opposed
Fruit Type: Wet, berry
Flowering and Fruiting: April–June
IUCN Status: Not evaluated
Reference: A.L.P.P.de Candolle & A.C.P.de Candolle, Monogr. Phan. 5: 443 (1887)

Cissus rotundifolia Vahl
Vitaceae
Synonyms: *Vitis forskahlii* Blatt., *Vitis rotundifolia* (Vahl) Deflers.
Common Name: Venezuelan treebine, Arabian wax Cissus, Peruvian grape ivy.
Climbing Mechanism: Tendril Climber
Distribution (Global): Burundi, Djibouti, Eritrea, Ethiopia, Kenya, KwaZulu-Natal, Malawi, Mauritius, Mozambique, Northern Provinces, Oman, Réunion, Saudi Arabia, Somalia, Sudan, Swaziland, Tanzania, Uganda, Yemen, Zaire, Zimbabwe
Distribution (India): Tamil Nadu, Telangana
Leaf Type: Simple or occasionally lobed
Inflorescence: Lax Cymes/umbel-like cymes
Fruit Type: Berry, red when ripe
IUCN Status: Not evaluated
Reference: Symb. Bot. 3: 19 (1794)

Cissus spectabilis Hochst. ex Planch.
Vitaceae
Climbing Mechanism: Tendril Climber
Distribution (India): North-eastern India, West Bengal
Leaf Type: Palmately compound
Inflorescence: Cymes, leaf opposed
Fruit Type: Wet, berry
IUCN Status: Not evaluated
Reference: DC. Monog. Phan. 5: 609 in obs. (1887)

Cissus subramanyamii B.V.Shetty & Par.Singh
Vitaceae
Climbing Mechanism: Tendril Climber
Distribution (Global): India
Distribution (India): Tamil Nadu
Leaf Type: Palmately compound
Inflorescence: Cymes, leaf opposed
Fruit Type: Wet, berry
IUCN Status: Not evaluated
Reference: Fl. India 5: 293 (2000)

Cissus vitiginea L.
Vitaceae
Synonyms: *Vitis linnaei* Wall. ex Wight & Arn.
Common Name: South Indian treebine
Climbing Mechanism: Tendril Climber
Distribution (India): Andhra Pradesh, Kerala, Madhya Pradesh, Maharashtra, Odisha, Puducherry, Tamil Nadu, Telangana, West Bengal
Leaf Type: Simple

Inflorescence: Umbellate cymes
Fruit Type: Berry, Ovoid
Flowering and Fruiting: May–December
IUCN Status: Not evaluated
Notes: Occasional in plains and spreading over thickets in scrub jungles up to 900 m

Cissus woodrowii (Stapf ex Cooke) Santapau
Vitaceae
Synonyms: *Vitis woodrowii* T.Cooke
Common Name: Woodrow's grape tree
Climbing Mechanism: Tendril Climber
Distribution (Global): India
Distribution (India): Maharashtra
Leaf Type: Palmately compound
Inflorescence: Cymes, leaf opposed
Fruit Type: Wet, berry
Flowering and Fruiting: June–October
IUCN Status: Not evaluated
Reference: Kew Bull. 3: 276 (1948)

Cyphostemma auriculatum (Roxb.) P. Singh & B.V.Shetty
Vitaceae
Synonyms: *Cayratia auriculata* (Roxb.) Gamble, *Cissus auriculata* Roxb., *Cissus helleborifolia* Zipp. ex Span., *Vitis auriculata* (Roxb.) Wall., *Vitis erythroclada* (Roxb.) Kurz.
Common Name: Eared Cyphostemma
Climbing Mechanism: Tendril Climber
Distribution (Global): Taxon 35: 596 (1986)
Distribution (India): Assam, Kerala, Maharashtra, Odisha, Tamil Nadu, Telangana
Leaf Type: Digitate, 5-foliolate
Inflorescence: Divaricate Cymes
Fruit Type: 1-Seeded red round berry
Flowering and Fruiting: May–October
IUCN Status: Not evaluated
Notes: Common on evergreen forests
Reference: Taxon 35: 596 (1986)

Cyphostemma niveum (Hochst. ex Schweinf.) Desc.
Vitaceae
Synonyms: *Cissus nivea* Hochst. ex Schweinf., *Vitis pannosa* Baker
Climbing Mechanism: Tendril Climber
Distribution (Global): Eritrea, Ethiopia
IUCN Status: Not evaluated
Reference: Naturalia Monspel., Sér. Bot. 18: 225 (1967)

Cyphostemma oxyphyllum (A.Rich.) Vollesen
Vitaceae
Climbing Mechanism: Tendril Climber
IUCN Status: Not evaluated
Reference: Vollesen. In: Nordic J. Bot., 4(1): 37. (1984)

Cyphostemma setosum (Roxb.) Alston
Vitaceae
Synonyms: *Cissus psammophila* Gilg & M.Brandt, *Cissus setosa* Roxb., *Cyphostemma psammophilum* (Gilg & M.Brandt) Desc., *Vitis setosa* (Roxb.) Wall. ex Wight.
Common Name: Bristly wild grape
Climbing Mechanism: Tendril Climber
Distribution (Global): India, Malawi, Mozambique, Sri Lanka, Tanzania, Zambia
Distribution (India): Andhra Pradesh, Madhya Pradesh, Tamil Nadu, Telangana
Leaf Type: Trifoliate rarely lower ones simple
Fruit Type: Berry, bright red when ripe
Flowering and Fruiting: September–January
IUCN Status: Not evaluated
Notes: The leaves are applied to ulcers to promote suppuration and to assist in the extraction of Guinea worms
Reference: H.Trimen, Handb. Fl. Ceylon 6(Suppl.): 53 (1931)

Cyphostemma trilobata (Lam.) M.R.Almeida
Vitaceae
Synonyms: *Cissus trilobata* Lam., *Puria trilobata* (Lam.) N.C.Nair, *Vitis heterophylla* Wall.
Climbing Mechanism: Tendril Climber
Distribution (Global): India, Laccadive Is., Sri Lanka
Distribution (India): Peninsular India, Maharashtra
Flowering and Fruiting: October–December
IUCN Status: Not evaluated
Notes: Common in the slope of hills
Reference: Fl. Maharashtra 1: 273 (1996)

Leea macrophylla Roxb. ex Hornem.
Vitaceae
Synonyms: *Leea angustifolia* M.A.Lawson, *Leea cinerea* M.A.Lawson, *Leea coriacea* M.A.Lawson. +10
Common Name: Large-Leaf Leea
Climbing Mechanism: Tendril Climber
Distribution (Global): Andaman Is., Assam, Bangladesh, Cambodia, China South-Central, East Himalaya, India, Laos, Myanmar, Nepal, Thailand
Distribution (India): Maharashtra

Leaf Type: Leaves Simple, alternate, 1- or 3-foliolate or 1–3 pinnate, ovate-cordate, conspicuously large-lower leaves
Inflorescence: Tiny greenish-white flowers in corymbose cymes
Fruit Type: Berries are depressed-globose, black
Flowering and Fruiting: July–September
IUCN Status: Not evaluated
Notes: Distributed in evergreen forests of the Western Ghats
Reference: Hort. Bot. Hafn. 1: 251 (1813)

Leea sambucina M. A. Lawson
Vitaceae
Synonyms: *Leea indica* (Burm.f.) Merr., *Staphylea indica* Burm.f., *Aquilicia ottilis* Gaertn., *Aquilicia sambucina* L., *Leea arborea* Sieber ex Miq., *Leea biserrata* Miq. +25
Common Name: Bandicoot berry
Climbing Mechanism: Tendril Climber
Distribution (Global): Andaman Is., Assam, Bangladesh, Bismarck Archipelago, Borneo, Cambodia, Caroline Is., China South-Central, China Southeast, East Himalaya, Fiji, Hainan, India, Java, Laccadive Is., Laos, Lesser Sunda Is., Malaya, Maluku, Myanmar, Nepal, New Guinea, Nicobar Is., Philippines, Santa Cruz Is., Solomon Is., Sri Lanka, Sulawesi, Sumatera, Thailand, Vanuatu, Vietnam
Distribution (India): Bihar, Odisha
Leaf Type: The leaves are double compound or triple compound (2 or 3-pinnate leaflets) dentate
Inflorescence: Cymes, peduncle, paired, axillary
Fruit Type: Berry purple
Flowering and Fruiting: March–August
IUCN Status: Not evaluated
Reference: Philipp. J. Sci. 14: 245 (1919)

Parthenocissus neilgherriensis Planch.
Vitaceae
Synonyms: *Vitis himalayana* Brandis, *Vitis semicordata* var. *roylei* King ex.R Parker +6
Climbing Mechanism: Tendril Climber
Distribution (India): Tamil Nadu
IUCN Status: Not evaluated

Parthenocissus quinquefolia (L.) Planch.
Vitaceae
Synonyms: *Ampelopsis engelmannii* (Koehne & Graebn.) Rehder, *Ampelopsis graebneri* Bolle, *Ampelopsis hederacea* DC., *Ampelopsis hederacea* var. *dumetorum* Focke +62
Common Name: Virginia creeper

Climbing Mechanism: Tendril Climber
Distribution (Global): Alabama, Arkansas, Bahamas, Bermuda, Connecticut, Cuba, Delaware, District of Columbia, El Salvador, Florida, Georgia, Guatemala, Illinois, Indiana, Iowa, Kansas, Kentucky, Louisiana, Maine, Maryland, Massachusetts, Mexico Central, Mexico Gulf, Mexico Northeast, Mexico Northwest, Mexico Southeast, Mexico Southwest, Michigan, Minnesota, Mississippi, Missouri, Nebraska, New Hampshire, New Jersey, New York, North Carolina, Ohio, Oklahoma, Ontario, Pennsylvania, Québec, Rhode I., Saskatchewan, South Carolina, South Dakota, Tennessee, Texas, Vermont, Virginia, West Virginia, Wisconsin
Leaf Type: Palmately compound
Fruit Type: Berries
IUCN Status: Not evaluated
Reference: A.L.P.P.de Candolle & A.C.P.de Candolle, Monogr. Phan. 5: 448 (1887)

Parthenocissus semicordata (Wall.) Planch.
Vitaceae
Synonyms: *Vitis semicordata* Wall.
Common Name: Himalayan woodbine
Climbing Mechanism: Tendril Climber
Distribution (Global): Assam, Bangladesh, China North-Central, China South-Central, China Southeast, East Himalaya, India, Java, Laos, Malaya, Myanmar, Nansei-shoto, Nepal, Pakistan, Taiwan, Thailand, Tibet, Vietnam, West Himalaya
Distribution (India): Himachal Pradesh, Kerala, Mizoram, Tamil Nadu, West Bengal
Leaf Type: Trifoliate
Inflorescence: Umbellate cymes
Fruit Type: Blackberry
Flowering and Fruiting: May–October
IUCN Status: Not evaluated
Reference: A.L.P.P.de Candolle & A.C.P.de Candolle, Monogr. Phan. 5: 451 (1887)

Pterisanthes heterantha M.A. Lawson
Vitaceae
Synonyms: *Embamma heteranthum* Griff., *Pterisanthes heterotricha* Merr.
Climbing Mechanism: Tendril Climber
Distribution (Global): Malaya
IUCN Status: Not evaluated
Reference: J. D. Hooker, Fl. Brit. India 1: 664 (1875)

Pterisanthes pedata M.A. Lawson
Vitaceae
Climbing Mechanism: Tendril Climber
Distribution (Global): Malaya

IUCN Status: Not evaluated
Reference: J. D. Hooker, Fl. Brit. India 1: 664 (1875)

Pterisanthes polita Planch.
Vitaceae
Synonyms: *Ampelocissus sinuosa* (Merr.) J. Wen & Boggan, *Pterisanthes coriacea* Korth. ex-King, *Pterisanthes gladiata* Steenis, *Pterisanthes parvifolia* Merr., *Pterisanthes sinuosa* Merr., *Vitis polita* Miq.
Climbing Mechanism: Tendril Climber
Distribution (Global): Borneo, Malaya, Myanmar, Philippines, Sumatera, Thailand
IUCN Status: Not evaluated
Reference: J. D. Hooker, Fl. Brit. India 1: 663 (1875)

Tetrastigma andamanicum (King) Suess.
Vitaceae
Synonyms: *Vitis andamanica* King
Climbing Mechanism: Tendril Climber
Distribution (Global): Andaman Is., Nicobar Is.
Distribution (India): Andaman and Nicobar Islands
Leaf Type: Compound, 3–7 foliolate
Inflorescence: Cymes, panicled
Fruit Type: Wet, berry
IUCN Status: Not evaluated
Reference: Mitt. Bot. Staatssamml. München 1: 356 (1953)

Tetrastigma angustifolium (Roxb.) Planch.
Vitaceae
Synonyms: *Cissus angustifolia* Roxb., *Vitis angustifolia* (Roxb.) Wall.
Climbing Mechanism: Tendril Climber
Distribution (Global): Assam, Bangladesh, Myanmar, Sumatera
Distribution (India): Mizoram, West Bengal
Leaf Type: Compound, 3–7 foliolate
Inflorescence: Cymes, panicled
Fruit Type: Wet, berry
IUCN Status: Not evaluated
Reference: A.L.P.P.de Candolle & A.C.P.de Candolle, Monogr. Phan. 5: 439 (1887)

Tetrastigma bracteolatum (Wall.) Planch.
Vitaceae
Synonyms: *Vitis bracteolata* Wall.
Climbing Mechanism: Tendril Climber
Distribution (Global): Assam, Bangladesh, East Himalaya, Myanmar, Nepal
Distribution (India): Andhra Pradesh, Arunachal Pradesh, Assam, Manipur, Mizoram, Tripura, West Bengal
Leaf Type: Leaves digitately compound, ternate

Inflorescence: Divaricate cymes
Fruit Type: Berries, black when ripe, globose
Flowering and Fruiting: March–October
IUCN Status: Not evaluated
Reference: A.L.P.P.de Candolle & A.C.P.de Candolle, Monogr. Phan. 5: 428 (1887)

Tetrastigma campylocarpum (Kurz) Planch.
Vitaceae
Synonyms: *Tetrastigma alcicorne* Haines., *Vitis alcicorois* (Haines) Haines., *Vitis campylocarpa* Kurz
Vitis parkeri Gagnep. ex Osmaston.
Climbing Mechanism: Tendril Climber
Distribution (Global): Assam, Cambodia, China South-Central, East Himalaya, Myanmar, Nicobar Is., Thailand
Distribution (India): Great, Nicobar, Island, Mizoram
Leaf Type: Palmately compound, Penta foliate
Fruit Type: Berries, globose
IUCN Status: Not evaluated
Notes: Usually found in evergreen forests
Reference: A.L.P.P.de Candolle & A.C.P.de Candolle, Monogr. Phan. 5: 437 (1887)

Tetrastigma canarense (Dalziel) Gamble
Vitaceae
Synonyms: *Melothria zeylanica* J. Koenig ex Wight & Arn., *Fusanus pedatus* (Lour.) Spreng., *Cayratia longzhouensis* W.T. Wang. + 12
Climbing Mechanism: Tendril Climber
Distribution (Global): Andaman Is., Assam, Bangladesh, Cambodia, China South-Central, China Southeast, Christmas I., East Himalaya, India, Java, Laos, Myanmar, Nicobar Is., Philippines, Sri Lanka, Thailand, Vietnam
Distribution (India): Andhra Pradesh, Tamil Nadu
Leaf Type: Compound, imparipinnate
Fruit Type: Berries
IUCN Status: Not evaluated
Reference: Notul. Syst. (Paris) 1: 346 (1911)

Tetrastigma corymbosum D.G. Long
Vitaceae
Synonyms: *Tetrastigma rumicispermum* var. *rumicispermum*., *Cayratia kiujiangensis* C.Y.Wu., *Vitis tuberculata* Wall.
Common Name: Warty chestnut vine
Climbing Mechanism: Tendril Climber
Distribution (Global): Assam, Bangladesh, China South-Central, East Himalaya, Laos, Malaya, Myanmar, Nepal, Thailand, Tibet, Vietnam
Distribution (India): West Bengal
Leaf Type: Leaves are pedately 5-foliolate

Inflorescence: Umbels
Fruit Type: Berries, spherical
Flowering and Fruiting: April–May
IUCN Status: Not evaluated

Tetrastigma dubium (Lawson) Planch.
Vitaceae
Synonyms: *Tetrastigma garrettii* Gagnep., *Tetrastigma henryi* Gagnep., *Tetrastigma henryi* var. *mollifolium* W.T. Wang., *Tetrastigma lunglingense* C. Y.Wu & W.T.Wang., *Tetrastigma tenue* Craib., *Vitis dubia* M.A.Lawson., *Vitis oxyphylla* Wall.
Climbing Mechanism: Tendril Climber
Distribution (Global): Assam, Bangladesh, Borneo, Cambodia, China South-Central, East Himalaya, Laos, Malaya, Myanmar, Nepal, Thailand, Tibet, Vietnam
Distribution (India): Arunachal Pradesh, Manipur, Mizoram, Tripura, West Benga
Leaf Type: Leaves digitately compound, ternate
Inflorescence: Divaricately branched, corymbose cymes
Fruit Type: Berries, globose
Flowering and Fruiting: September–February
IUCN Status: Not evaluated
Reference: A.L.P.P.de Candolle & A.C.P.de Candolle, Monogr. Phan. 5: 437 (1887)

Tetrastigma gamblei B.V. Shetty & P. Singh
Vitaceae
Climbing Mechanism: Tendril Climber
Distribution (Global): India
Distribution (India): Kerala, Maharashtra, Tamil Nadu
Leaf Type: Leaves 3-foliolate, glabrous
Inflorescence: Umbelled cymes
Fruit Type: Berries, globose
Flowering and Fruiting: November–March
IUCN Status: Not evaluated
Notes: Frequently found in moist deciduous forests of the Western Ghats
Reference: Kew Bull. 42: 933 (1987)

Tetrastigma hookeri (M.A. Lawson) Planch.
Vitaceae
Synonyms: *Vitis hookeri* M.A. Lawson
Climbing Mechanism: Tendril Climber
Distribution (Global): Assam, India, Java, Malaya, Vietnam
Leaf Type: Leaves pedately compound
Inflorescence: Compact, corymbose cymes
Fruit Type: Berries, Ovoid
Flowering and Fruiting: September–May

IUCN Status: Not evaluated
Reference: A.L.P.P.de Candolle & A.C.P.de Candolle, Monogr. Phan. 5: 428 (1887)

Tetrastigma lanceolarium (Roxb.) Planch.
Vitaceae
Synonyms: *Tetrastigma leucostaphylum* (Dennst.) Alston., *Cissus assimilis* Kurz ex M. A. Lawson. +15
Climbing Mechanism: Tendril Climber
Distribution (Global): Assam, Bangladesh, Cambodia, East Himalaya, India, Java, Laos, Lesser Sunda Is., Malaya, Myanmar, Nepal, Sumatera, Thailand, Vietnam
Distribution (India): Andaman and Nicobar Islands, Andhra Pradesh, Arunachal Pradesh, Assam, Odisha, Tamil Nadu, Tripura, Uttar Pradesh
Leaf Type: Pinnately compound 3–5 foliate
Fruit Type: Berry
Flowering and Fruiting: March–August
IUCN Status: Not evaluated
Notes: Mostly found in slopes of hills at lower elevations
Reference: Taxon 26: 539 (1977)

Tetrastigma leucostaphylum (Dennst.) Alston
Vitaceae
Synonyms: *Cissus feminea* Roxb., *Tetrastigma kunstleri* (King) Craib +15
Climbing Mechanism: Tendril Climber
Distribution (Global): Assam, Bangladesh, Cambodia, East Himalaya, India, Java, Laos, Lesser Sunda Is., Malaya, Myanmar, Nepal, Sumatera, Thailand, Vietnam
Distribution (India): Andaman and Nicobar Islands, Andhra Pradesh, Arunachal Pradesh, Assam, Karnataka, Kerala, Madhya Pradesh, Maharashtra, Manipur, Meghalaya, Mizoram, Tamil Nadu, West Bengal, Western Ghats
Leaf Type: 3–5 foliate simple
Inflorescence: Axillary/subcorymbose cymes
Fruit Type: Berries, globose, fleshy
Flowering and Fruiting: January–August
IUCN Status: Not evaluated
Notes: Distributed across a wide range of habitats: Evergreen, semi-evergreen, moist deciduous, and shola forests
Reference: Taxon 26: 539 (1977)

Tetrastigma muricatum (Wall.) Gamble
Vitaceae
Synonyms: *Tetrastigma leucostaphylum* (Dennst.) Alston., *Cissus assimilis* Kurz ex M. A. Lawson. +15
Climbing Mechanism: Tendril Climber
Distribution (India): Western Ghats
Leaf Type: Compound, 3–7 foliolate
Inflorescence: Cymes, panicled

Fruit Type: Wet, berry
IUCN Status: Not evaluated

Tetrastigma nilagiricum (Miq.) B.V. Shetty
Vitaceae
Synonyms: *Tetrastigma nilagiricum.*, *Tetrastigma glycosmoides* Planch.
Climbing Mechanism: Tendril Climber
Distribution (Global): India, Sri Lanka
Distribution (India): Kerala
Leaf Type: Leaves 3–5 foliolate, pedate
Inflorescence: Umbelled cymes
Fruit Type: Berries, globose
Flowering and Fruiting: December–March
IUCN Status: Not evaluated
Notes: Occurs mostly in shola forests
Reference: Kew Bull. 44: 477 (1989)

Tetrastigma obovatum Gagnep.
Vitaceae
Synonyms: *Vitis obovata* M. A. Lawson
Climbing Mechanism: Tendril Climber
Distribution (Global): Assam, Bangladesh, China South-Central, Laos, Myanmar, Thailand, Vietnam
Distribution (India): Arunachal Pradesh, Meghalaya, Mizoram, Tripura
Leaf Type: Leaves digitately compound 5–7 foliate
Inflorescence: Subcorymbose cymes
Fruit Type: Berries, globose
Flowering and Fruiting: January–November
IUCN Status: Not evaluated
Notes: Often found in evergreen forests
Reference: Notul. Syst. (Paris) 1: 266 (1910)

Tetrastigma obtectum (Wall. ex M.A. Lawson) Planch. ex Franch.
Vitaceae
Climbing Mechanism: Tendril Climber
Distribution (Global): Assam, China North-Central, China South-Central, China Southeast, East Himalaya, Nepal, Taiwan, Vietnam, West Himalaya
Distribution (India): Arunachal Pradesh, Mizoram
Leaf Type: Compound, 3–7 foliolate
Inflorescence: Cymes, panicled
Fruit Type: Wet, berry
IUCN Status: Not evaluated
Reference: Bull. Soc. Bot. France 33: 458 (1886)

Tetrastigma pedunculare (Wallich ex Lawson) Planch.
Vitaceae
Climbing Mechanism: Tendril Climber
Distribution (Global): Borneo, Lesser Sunda Is., Malaya, Sumatera, Thailand
Leaf Type: Compound, 3–7 foliolate
Inflorescence: Cymes, panicled
Fruit Type: Wet, berry
IUCN Status: Not evaluated
Reference: A.L.P.P.de Candolle & A.C.P.de Candolle, Monogr. Phan. 5: 438 (1887)

Tetrastigma planicaule (Hook.f.) Gagnep.
Vitaceae
Synonyms: *Cayratia neurosa* (Kurz) H.N. Naithani., *Vitis neurosa* Kurz., *Tetrastigma laoticum* Gagnep., *Vitis planicaulis* Hook.f.
Climbing Mechanism: Tendril Climber
Distribution (Global): Assam, Bangladesh, China South-Central, China Southeast, East Himalaya, Laos, Sri Lanka, Thailand, Tibet, Vietnam
Distribution (India): Andaman and Nicobar Islands, Arunachal Pradesh, Assam, West Bengal
Leaf Type: Leaves digitately compound, 3–5 foliolate
Inflorescence: Corymbose cyme
Fruit Type: Berries, black
Flowering and Fruiting: April–September
IUCN Status: Not evaluated
Notes: Frequently occur in primary forests growing near a water course
Reference: Notul. Syst. (Paris) 1: 319 (1910)

Tetrastigma rumicispermum (M.A. Lawson) Planch.
Vitaceae
Synonyms: *Vitis rumicisperma* M. A. Lawson
Climbing Mechanism: Tendril Climber
Distribution (Global): Assam, Bangladesh, China South-Central, East Himalaya, Laos, Malaya, Myanmar, Nepal, Thailand, Tibet, Vietnam
Distribution (India): Mizoram
Leaf Type: Leaves are pedately 5-foliolate
Flowering and Fruiting: January–May
IUCN Status: Not evaluated
Notes: Rarely found in the Eastern Ghats
Reference: A.L.P.P.de Candolle & A.C.P.de Candolle, Monogr. Phan. 5: 429 (1887)

Tetrastigma serrulatum (Roxb.) Planch.
Vitaceae
Synonyms: *Cissus napaulensis* DC. +20
Common Name: Toothed-leaf chestnut vine
Climbing Mechanism: Tendril Climber

Distribution (India): Andaman and Nicobar Islands, Arunachal Pradesh, Himachal Pradesh, Manipur, Meghalaya, Mizoram, Sikkim, Tripura, West Bengal
Leaf Type: Digitately compound, 5-foliate
Inflorescence: Corymb-like cymes
Fruit Type: Berries
Flowering and Fruiting: March–June
IUCN Status: Not evaluated
Reference: A.L.P.P.de Candolle & A.C.P.de Candolle, Monogr. Phan. 5: 432 (1887)

Tetrastigma sulcatum (P. Lawson) Gamble
Vitaceae
Synonyms: *Vitis sulcata* M. A. Lawson
Climbing Mechanism: Tendril Climber
Distribution (Global): India
Distribution (India): Andhra Pradesh, Goa, Karnataka, Kerala, Tamil Nadu, Western Ghats
Leaf Type: Leaves 3–5-foliolate
Fruit Type: Berry
Flowering and Fruiting: November–August
IUCN Status: Not evaluated
Notes: Often found in evergreen and moist deciduous forests
Reference: Fl. Madras: 229 (1918)

Tetrastigma thomsonianum Planch.
Vitaceae
Climbing Mechanism: Tendril Climber
Distribution (Global): Assam, Bangladesh, East Himalaya, Sumatera
Distribution (India): Arunachal Pradesh, Manipur, Meghalaya
Leaf Type: Leaves compound, palmately 5-foliolate
Inflorescence: Umbellate cymes
Fruit Type: Berry, reddish-brown
Flowering and Fruiting: October–December
IUCN Status: Not evaluated
Notes: Usually found in evergreen forests
Reference: A.L.P.P.de Candolle & A.C.P.de Candolle, Monogr. Phan. 5: 439 (1887)

Vitis alcicornis Haines
Vitaceae
Climbing Mechanism: Tendril Climber
Distribution (India): Bihar, Odisha
Leaf Type: Simple, lobed
Inflorescence: Panicles, leaf opposed
Fruit Type: Wet, berry
IUCN Status: Not evaluated
Reference: Bedd. In: Trans. L. Soc. 25: 213. (1865)

Vitis anamalayana Bedd.
Vitaceae
Synonyms: *Parthenocissus semicordata* var. *semicordata*., *Ampelopsis heterophylla* Blume., *Ampelopsis himalayana* Royle., *Ampelopsis neilgherrensis* Wight., *Cissus heterophylla* Walp.
Climbing Mechanism: Tendril Climber
Distribution (Global): Assam, Bangladesh, China North-Central, China South-Central, China Southeast, East Himalaya, India, Java, Laos, Malaya, Myanmar, Nansei-shoto, Nepal, Pakistan, Taiwan, Thailand, Tibet, Vietnam, West Himalaya
Leaf Type: Simple, lobed
Inflorescence: Panicles, leaf opposed
Fruit Type: Wet, berry
IUCN Status: Not evaluated
Notes: Usually found in evergreen forests at high ranges

Vitis araneosa Miq.
Vitaceae
Synonyms: *Pterisanthes miquelii* Planch., *Cissus araneosa* Miq.
Climbing Mechanism: Tendril Climber
Distribution (Global): Malaya, Sumatera
Leaf Type: Simple, lobed
Inflorescence: Panicles, leaf opposed
Fruit Type: Wet, berry
IUCN Status: Not evaluated
Reference: A.L.P.P.de Candolle & A.C.P.de Candolle, Monogr. Phan. 5: 418 (1887)

Vitis cissoides (Blume) Backer
Vitaceae
Climbing Mechanism: Tendril Climber
Leaf Type: Simple, lobed
Inflorescence: Panicles, leaf opposed
Fruit Type: Wet, berry
IUCN Status: Not evaluated

Vitis flexuosa Thunb.
Vitaceae
Climbing Mechanism: Tendril Climber
Leaf Type: Simple, lobed
Inflorescence: Panicles, leaf opposed
Fruit Type: Wet, berry
IUCN Status: Not evaluated
Reference: Thunb. In: Trans. L. Soc. 2: 332. (1794)

Vitis furcata M. A. Lawson
Vitaceae
Synonyms: *Cissus furcata* (M.A. Lawson) Gagnep., *Cissus purpurascens* Zipp. ex Miq., *Vitis rostrata* Miq.
Climbing Mechanism: Tendril Climber
Distribution (Global): Borneo, Malaya, New Guinea, Thailand
Leaf Type: Simple, lobed
Inflorescence: Panicles, leaf opposed
Fruit Type: Wet, berry
IUCN Status: Not evaluated
Reference: A.L.P.P.de Candolle & A.C.P.de Candolle, Monogr. Phan. 5: 500 (1887)

Vitis gardneri M. A. Lawson
Vitaceae
Synonyms: *Cissus gardneri* Thwaites
Climbing Mechanism: Tendril Climber
Distribution (Global): Sri Lanka
Leaf Type: Simple, lobed
Inflorescence: Panicles, leaf opposed
Fruit Type: Wet, berry
IUCN Status: Not evaluated
Reference: Enum. Pl. Zeyl.: 63 (1858)

Vitis glaberrima Wall.
Vitaceae
Synonyms: *Cissus nodosa* Blume., *Cissus glaberrima* (Wall.) Steud., *Cissus glaberrima* Planch., *Cissus subtetragona* Planch., *Vitis nodosa* (Blume) Miq., *Vitis verrucaria* Noronha
Climbing Mechanism: Tendril Climber
Distribution (Global): Andaman Is., Borneo, Cambodia, China South-Central, China Southeast, Hainan, Java, Laos, Malaya, Myanmar, Philippines, Thailand, Vietnam
Leaf Type: Simple, lobed
Inflorescence: Panicles, leaf opposed
Fruit Type: Wet, berry
IUCN Status: Not evaluated
Reference: Catalogus: 39 (1823)

Vitis glauca Wall.
Vitaceae
Synonyms: *Cissus repens* Lam., *Cissus blumeana* Steud., *Cissus cerifera* Teijsm. & Binn., *Cissus cordata* Roxb., *Cissus diffusa* (Miq.) Amsh., *Cissus glaucoramea* Planch. +10
Common Name: Creeping treebine
Climbing Mechanism: Tendril Climber

Distribution (Global): Andaman Is., Assam, Bangladesh, Borneo, Cambodia, China South-Central, China Southeast, Christmas I., East Himalaya, India, Java, Laos, Lesser Sunda Is., Malaya, Maluku, Myanmar, Nepal, New Caledonia, Nicobar Is., Philippines, Queensland, Taiwan, Thailand, Vietnam
Leaf Type: Simple, ovate
Inflorescence: Umbells
Fruit Type: Berry
Flowering and Fruiting: July–November
IUCN Status: Not evaluated
Notes: Found in both deciduous and evergreen forests
Reference: Encycl. 1: 31 (1783)

Vitis glyptocarpa M. A. Lawson
Vitaceae
Synonyms: *Cissus glyptocarpa* Thwaites
Climbing Mechanism: Tendril Climber
Distribution (Global): India
Leaf Type: Simple, lobed
Inflorescence: Panicles, leaf opposed
Fruit Type: Wet, berry
Flowering and Fruiting: April–October
IUCN Status: Not evaluated
Notes: Found mostly in moist deciduous, and evergreen forests
Reference: Enum. Pl. Zeyl.: 62 (1858)

Vitis helferi M.A. Lawson
Vitaceae
Synonyms: *Ampelocissus helferi* (M. A. Lawson) Planch.
Climbing Mechanism: Tendril Climber
Distribution (Global): Andaman Is., Myanmar
Leaf Type: Simple, lobed
Inflorescence: Panicles, leaf opposed
Fruit Type: Wet, berry
IUCN Status: Not evaluated
Reference: A.L.P.P.de Candolle & A.C.P.de Candolle, Monogr. Phan. 5: 411 (1887)

Vitis heyneana Roem. & Schult.
Vitaceae
Synonyms: *Vitis jacquemontii* R. Parker
Climbing Mechanism: Tendril Climber
Distribution (India): Andhra Pradesh, Odisha
Leaf Type: Simple, lobed
Inflorescence: Panicles, leaf opposed
Fruit Type: Wet, berry
Flowering and Fruiting: February–August

IUCN Status: Not evaluated
Reference: Roem. & Schult. In: Syst. 5: 318. (1819)

Vitis jacquemontii R. Parker
Vitaceae
Climbing Mechanism: Tendril Climber
Leaf Type: Simple, lobed
Inflorescence: Panicles, leaf opposed
Fruit Type: Wet, berry
IUCN Status: Not evaluated
Reference: R. Parker. In: For. Fl. Punjab ed. 2: 559. (1924)

Vitis labruscana L.H. Bailey
Vitaceae
Synonyms: *Vitis* × *alexanderi* Prince ex Jacques., *Vitis* × *alexandrina* Fisch., *Vitis* × *isabellae* Otto & A. Dietr., *Vitis* × *prolifera* Raf.
Climbing Mechanism: Tendril Climber
Leaf Type: Simple, lobed
Inflorescence: Panicles, leaf opposed
Fruit Type: Wet, berry
IUCN Status: Not evaluated
Reference: J. Soc. Agron. Prat. 1829: 409 (1829)

Vitis linnaei Wall.
Vitaceae
Synonyms: *Cissus vitiginea* L., *Cissus angulata* Lam., *Cissus indica* J.Koenig ex Steud., *Cissus vitiginea* var. *cochinchinensis* DC., *Vitis vitiginea* (L.) W.Theob., *Vitis vitiginea* var. *cochinchinensis* (DC.) Kuntze, *Vitis vitiginea* var. *typica* Kuntze
Common Name: South Indian treebine
Climbing Mechanism: Tendril Climber
Distribution (Global): Bangladesh, India, Myanmar, Sri Lanka
Leaf Type: Simple
Inflorescence: Umbellate cymes
Fruit Type: Berry, Ovoid
Flowering and Fruiting: May–December
IUCN Status: Not evaluated
Notes: Frequently found in dry and moist deciduous forests
Reference: Sp. Pl.: 117 (1753)

Vitis lonchiphylla M. A. Lawson
Vitaceae
Synonyms: *Cissus lonchiphylla* Thwaites
Climbing Mechanism: Tendril Climber
Distribution (Global): Sri Lanka

Leaf Type: Simple, lobed
Inflorescence: Panicles, leaf opposed
Fruit Type: Wet, berry
IUCN Status: Not evaluated
Reference: Enum. Pl. Zeyl.: 62 (1858)

Vitis macrostachya Miq.
Vitaceae
Synonyms: *Ampelocissus spicifer* (Griff.) Planch., *Cissus spicifera* Griff., *Nothocissus spicifera* (Griff.) Latiff., *Vitis spicifera* (Griff.) Kuntze
Climbing Mechanism: Tendril Climber
Distribution (Global): Borneo, Malaya, Sumatera, Thailand
Leaf Type: Simple, lobed
Inflorescence: Panicles, leaf opposed
Fruit Type: Wet, berry
IUCN Status: Not evaluated
Reference: A.L.P.P.de Candolle & A.C.P.de Candolle, Monogr. Phan. 5: 406 (1887)

Vitis mollissima Wall.
Vitaceae
Synonyms: *Cayratia geniculata* (Blume) Gagnep., *Cayratia rhodocarpa* (Blume) Gagnep., *Cayratia wrayi* (King) *Gagnep., Cissus* alata Blanco., *Cissus geniculata* Blume., *Cissus hirtella* Blume., *Cissus jackii* Korth. ex Miq. +10
Climbing Mechanism: Tendril Climber
Distribution (Global): Assam, Borneo, Cambodia, China South-Central, China Southeast, East Himalaya, Hainan, India, Java, Laos, Lesser Sunda Is., Malaya, Myanmar, Philippines, Sulawesi, Thailand, Tibet, Vietnam
Leaf Type: Ovate
Inflorescence: Extra-axillary, trichotomous cymes
Fruit Type: Berry, ovoid
Flowering and Fruiting: March–December
IUCN Status: Not evaluated
Notes: Distributed across a wide range of habitats: Evergreen, semi-evergreen, and moist deciduous forests
Reference: Notul. Syst. (Paris) 1: 345 (1911)

Vitis montana Raf.
Vitaceae
Synonyms: *Vitis riparia* Michx. *Vitis columbina* Raf., *Vitis concolor* Raf., *Vitis cordifolia* var. *riparia* (Michx.) A.Gray, *Vitis dimidiata* Raf., *Vitis illinoensis* W.R.Prince, *Vitis incisa* Jacq. +10
Climbing Mechanism: Tendril Climber
Distribution (Global): Alabama, Arkansas, Colorado, Connecticut, District of Columbia, Illinois, Indiana, Iowa, Kansas, Kentucky, Louisiana, Maine, Manitoba, Maryland, Massachusetts, Mexico Northeast, Michigan, Minnesota,

Mississippi, Missouri, Montana, Nebraska, New Brunswick, New Hampshire, New Jersey, New Mexico, New York, North Carolina, North Dakota, Nova Scotia, Ohio, Oklahoma, Ontario, Oregon, Pennsylvania, Québec, Rhode I., South Dakota, Tennessee, Texas, Vermont, Virginia, West Virginia, Wisconsin, Wyoming
Leaf Type: Simple
Inflorescence: Pyramidal-thyrsoid
Fruit Type: Berry, globose, black when ripe
Flowering and Fruiting: May–June
IUCN Status: Not evaluated
Notes: Usually found near streams in evergreen forests
Reference: Fl. Bor.-Amer. 2: 231 (1803)

Vitis nitida M. A. Lawson
Vitaceae
Synonyms: *Ampelocissus nitida* (M. A. Lawson) Planch.
Climbing Mechanism: Tendril Climber
Distribution (Global): Malaya
Leaf Type: Simple, lobed
Inflorescence: Panicles, leaf opposed
Fruit Type: Wet, berry
IUCN Status: Not evaluated
Reference: A.L.P.P.de Candolle & A.C.P.de Candolle, Monogr. Phan. 5: 411 (1887)

Vitis novemfolia Wall.
Vitaceae
Climbing Mechanism: Tendril Climber
Leaf Type: Simple, lobed
Inflorescence: Panicles, leaf opposed
Fruit Type: Wet, berry
IUCN Status: Not evaluated

Vitis obovata Raf.
Vitaceae
Climbing Mechanism: Tendril Climber
Distribution (India): Assam
Leaf Type: Simple, lobed
Inflorescence: Panicles, leaf opposed
Fruit Type: Wet, berry
IUCN Status: Not evaluated

Vitis pedicellata M. A. Lawson
Vitaceae
Climbing Mechanism: Tendril Climber
Leaf Type: Simple, lobed

Inflorescence: Panicles, leaf opposed
Fruit Type: Wet, berry
IUCN Status: Not evaluated
Reference: J. D. Hooker, Fl. Brit. India 1: 650 (1875)

Vitis polystachya Wall.
Vitaceae
Synonyms: *Ampelocissus polystachya* (Wall. ex M. A. Lawson) Planch.
Climbing Mechanism: Tendril Climber
Distribution (Global): Andaman Is., Malaya, Myanmar
Leaf Type: Simple, lobed
Inflorescence: Panicles, leaf opposed
Fruit Type: Wet, berry
IUCN Status: Not evaluated
Reference: A.L.P.P.de Candolle & A.C.P.de Candolle, Monogr. Phan. 5: 411 (1887)

Vitis rheedei Wight & Arn.
Vitaceae
Synonyms: *Cyphostemma trilobata* (Lam.) M.R. Almeida
Climbing Mechanism: Tendril Climber
Leaf Type: Simple, lobed
Inflorescence: Panicles, leaf opposed
Fruit Type: Wet, berry
Flowering and Fruiting: October–December
IUCN Status: Not evaluated
Notes: Mostly found in semi-evergreen and moist deciduous forests

Vitis roxburghii Wight & Arn.
Vitaceae
Synonyms: *Cayratia roxburghii* (Planch.) Gagnep., *Cissus roxburghii* Planch.
Climbing Mechanism: Tendril Climber
Distribution (Global): India, Laos
Leaf Type: Simple, lobed
Inflorescence: Panicles, leaf opposed
Fruit Type: Wet, berry
Flowering and Fruiting: January-April.
IUCN Status: Not evaluated
Notes: Usually found in evergreen forests
Reference: Notul. Syst. (Paris) 1: 346 (1911)

Vitis sagittifolia M. A. Lawson
Vitaceae
Synonyms: *Cissus hastata* Miq., *Vitis hastata* (Miq.) Miq.
Climbing Mechanism: Tendril Climber

Distribution (Global): Bangladesh, Borneo, Java, Laos, Malaya, New Guinea, Philippines, Queensland, Sumatera, Thailand, Vietnam
Leaf Type: Simple, lobed
Inflorescence: Panicles, leaf opposed
Fruit Type: Wet, berry
IUCN Status: Not evaluated
Reference: Fl. Ned. Ind., Eerste Bijv.: 517 (1861)

Vitis setosa (Roxb.) Wall. ex Wight
Vitaceae
Synonyms: *Cissus psammophila* Gilg & M. Brandt., *Cissus setosa* Roxb., *Cyphostemma psammophilum* (Gilg & M.Brandt) Desc., *Cyphostemma setosum* (Roxb.) Alston.
Common Name: Bristly wild grape
Climbing Mechanism: Tendril Climber
Distribution (Global): India, Malawi, Mozambique, Sri Lanka, Tanzania, Zambia
Leaf Type: 3(5)-foliolate, simple
Inflorescence: Cymes
Fruit Type: Berry, bright red
Flowering and Fruiting: September–January
IUCN Status: Not evaluated
Notes: Commonly occur in dry deciduous forests and scrub jungles
Reference: H. Trimen, Handb. Fl. Ceylon 6(Suppl.): 53 (1931)

Vitis tuberculata Wall.
Vitaceae
Synonyms: *Tetrastigma rumicispermum* var. *rumicispermum*., *Cayratia kiujiangensis* C.Y. Wu, *Tetrastigma corymbosum* D.G.Long.
Common Name: Warty chestnut vine
Climbing Mechanism: Tendril Climber
Distribution (Global): Assam, Bangladesh, China South-Central, East Himalaya, Laos, Malaya, Myanmar, Nepal, Thailand, Tibet, Vietnam
Distribution (India): Assam, Mizoram, Tripura
Leaf Type: Leaves compound, imparipinnate
Inflorescence: Dichotomous, umbellate-cymose
Fruit Type: Berries
Flowering and Fruiting: April–May
IUCN Status: Not evaluated
Reference: Wall. In: Cat. n. 6014. (1831)

Vitis vinifera L.
Vitaceae
Synonyms: *Cissus vinifera* (L.) Kuntze, *Maerklinia viridis* Bronner, *Noachia macrophylla* Bronner, *Palatina dichotoma* Bronner. +42
Common Name: Common grape vine, Wine grape, European grape, Grape

Climbing Mechanism: Tendril Climber
Distribution (Global): Pan tropical
Distribution (India): Delhi, Madhya Pradesh, Maharashtra, Odisha, Rajasthan, Tamil Nadu, Telangana, Tripura, West Bengal
Leaf Type: Leaves are alternate, palmately lobed
Fruit Type: Berries
Flowering and Fruiting: Throughout the year
IUCN Status: Least Concern
Reference: Sp. Pl.: 202 (1753)

Vitis vulpina L.
Vitaceae
Synonyms: *Vitis amara* Raf., *Vitis cordifolia* Michx., *Vitis cordifolia* var. *sempervirens* Munson., *Vitis floridana* Raf., *Vitis hyemalis* Dum.Cours., *Vitis illex* L.H. Bailey., *Vitis longifolia* Raf., *Vitis missouriensis* W.R.Prince., *Vitis sylvestris* W.Bartram
Climbing Mechanism: Tendril Climber
Distribution (Global): Alabama, Arkansas, Delaware, District of Columbia, Florida, Georgia, Illinois, Indiana, Iowa, Kansas, Kentucky, Louisiana, Maryland, Michigan, Mississippi, Missouri, Nebraska, New Jersey, New York, North Carolina, Ohio, Oklahoma, Ontario, Pennsylvania, South Carolina, South Dakota, Tennessee, Texas, Virginia, West Virginia
Leaf Type: Simple, lobed
Inflorescence: Panicles, leaf opposed
Fruit Type: Wet, berry
IUCN Status: Not evaluated
Reference: Sp. Pl.: 203 (1753)

Yua thomsonii (M.A. Lawson) C.L.Li
Vitaceae
Synonyms: *Cayratia thomsonii* (M. A. Lawson) E. Reid & M. Chandler., *Cissus thomsonii* (M. A. Lawson) Planch., *Parthenocissus thomsonii* (M. A. Lawson) Planch., *Psedera thomsonii* (M. A. Lawson) Stuntz., *Vitis thomsonii* M.A. Lawson., *Yua chinensis* C.L.Li.
Climbing Mechanism: Tendril Climber
Distribution (Global): Assam, China South-Central, China Southeast, Nepal, Taiwan
IUCN Status: Not evaluated
Reference: Acta Bot. Yunnan. 12: 5 (1990)

5.4 Learnings and Recommendations

Handling a large dataset is a time-consuming task for countries like India. Here are some of the most common challenges encountered during dataset preparation:

1. *Data collection and geographical boundaries*: Some of the pioneer floristic studies, such as Flora of British India and Flora of the Presidency of Bombay, were published long before India's bifurcation and thus include Flora from outside of present-day India. We removed the unique species from those Flora, which have not been reported elsewhere in India.
2. *Assigning a precise growth form*: Many scramblers are classified as climbers in some studies but as free-standing shrubs in others. For example, in various studies, *Ziziphus oenoplia*, a common scrambler in India, has been classified as a shrub, small tree, or climber, adding to the difficulty in determining the growth form. Similarly, scramblers, such as members of the genus *Capparis* (Capparaceae), are challenging to classify neatly into a single growth form due to their ability to grow as a bush or as a trailing liana, depending on the environment. We considered the habit described in the Flora with broader geographical coverage for those species designated with more than one growth form, as it would provide a better description of the species based on multiple occurrences across different habitats.
3. *Confounding terminologies and assigning climbing mechanisms*: Multiple terminologies for the same species were frequently observed across the Flora. While previous research focused primarily on twiners and tendril-bearers, more recent research has included up to nine different climbing strategies. As a result, when multiple growth/climbing strategies were recorded for a single species, we assigned the climbing mechanisms by consulting the most recent manuals, including Floras and relevant research articles, as well as field observations for some species.
4. *Data duplication*: Data duplication was yet another challenge frequently encountered during the data compilation, mainly because of the synonyms. The nomenclature and family names often change over time. Therefore, we conducted a thorough systematic revision following the most recent APG IV classification. As a result, we removed over 20,000 entries that were either synonyms or duplicates and could have over-predicted the species richness.
5. *New species discovery*: There are many new climber species reported from India every year, to the extent that the list of discoveries over a decade accounted for roughly 10% of India's total climbing flora. Therefore, leaving discoveries off the checklist could significantly impact diversity.
6. *Taxonomy vs. ecological studies*: Our understanding of climber diversity is primarily based on floristic surveys and ecological inventories. The exclusive checklist based on ecological inventories underrepresented climbers by 40%. This is primarily because most inventories are biased toward woody climbers, leaving a significant proportion of herbaceous climbers. Herbaceous climbers make up 37% of India's total climber flora. Furthermore, the higher size-class

threshold set for liana inventories may result in less coverage of the diversity of climbers. As a result, estimating climber diversity based on ecological data may arbitrarily reduce climber diversity. Similarly, Floras records may easily overlook scramblers and new species records. Therefore, we integrated taxonomic and ecological studies to estimate climber diversity precisely.

Despite their potential, climbers are likely one of the most underexplored plant groups. One major impediment is a lack of precise data on their diversity, such as how many species exist and their contribution to the diversity of the plant kingdom. Thus, we attempted to estimate the climbing flora of one of the world's most biodiverse countries, India. In this chapter, I presented one of the most precise and comprehensive methods for estimating the richness of climber species within a defined geographical boundary. Despite its focus on Indian climbing flora, it is adaptable and replicable elsewhere. The baseline data generated by this research will benefit several ecological, taxonomic, phylogenetic, and evolutionary studies of climbers in the future. When climbers continue to increase in number and diversity globally, this work will serve as the basis for our efforts to create a global climber database (Plates 5.21, 5.22, 5.23, 5.24, 5.25, 5.26, 5.27, 5.28, 5.29, 5.30, 5.31, 5.32, 5.33, 5.34, 5.35, 5.36, 5.37, 5.38, 5.39, 5.40, 5.41, 5.42, 5.43, 5.44, 5.45, 5.46, 5.47, 5.48, and 5.49).

Plate 5.21 *Parsonsia alboflavescens* (**a**); *Grewia flavescens* (**b**). Picture credit: Dhatchanamoorthy

Plate 5.22 *Capparis spinosa* (**a**); *C. brevispina* (**b**). Picture credit: Subbaiah Karuppusamy (**a**); Vivek Pandi (**b**)

Plate 5.23 *Gardneria ovata* (**a**); *Eleagnus kologa* (**b**). Picture credit: Subbaiah Karuppusamy

Plate 5.24 *Derris scandens* (**a**); *Tetracera akara* (**b**). Picture credit: Subbaiah Karuppusamy

Plate 5.25 *Thunbergia mysorensis* (**a**); *Maerua oblongifolia* (**b**); *Strychnos lenticellata* (**c**). Picture credit: Subbaiah Karuppusamy (**a**); Dhatchanamoorthy (**b**); Vivek Pandi (**c**)

Plate 5.26 Flowers of *Capparis zeylanica*. Picture credit: Vivek Pandi

Plate 5.27 Flowers of *Capparis sepiaria (a) and C. rotundifolia (b)*. Picture credit: Vivek Pandi

Plate 5.28 Flowers of *Elaeagnus conferta (**a**) and Solanum seaforthianum (**b**).* Picture credit: Vivek Pandi

Plate 5.29 General habit of *Smilax zeylanica*, a monocot climber (**a**–**c**). Picture credit: Vivek Pandi

Plate 5.30 Flowers and the fruits (follicles) of *Tylophora indica* (**a** & **b**). Picture credit: Vivek Pandi

Plate 5.31 General habit and fruits of *Diploclisia glaucescens* (**a** & **b**). Picture credit: Vivek Pandi

Plate 5.32 General habit and fruits of *Uvaria naraum* (**a** & **b**). Picture credit: Vivek Pandi

Plate 5.33 Flowers of *Sarcostemma acidum* (**a**); *Canavalia gladiata* (**b**); *Macroptilium atropurpureum (c)*. Picture credit: Vivek Pandi (**a**); Dhatchanamoorthy (**b** & **c**)

Plate 5.34 *Clitoria ternatea (**a**); Pergularia daemia (**b**).* Picture credit: Vivek Pandi (**a**); Dhatchanamoorthy (**b**)

Plate 5.35 Flowers of *Getonia floribunda* (***a***); *Ampelocissus sp.* (**b**). Picture credit: Vivek Pandi

Plate 5.36 *Ichnocarpus frutescens* (**a**); *Coccinia grandis* (**b**); *Secamone emetica* (**c**). Picture credit: Vivek Pandi

Plate 5.37 *Reissantia indica* (**a**); *Celastrus paniculatus* (**b**). Picture credit: Vivek Pandi (**a**); Dhatchanamoorthy (**b**)

Plate 5.38 Fruits of *Plecospermum spinosum* (**a**); *Hugonia mystax* (**b**). Picture credit: Vivek Pandi

Plate 5.39 Flowers and fruits of *Premna corymbosa*. Picture credit: Vivek Pandi

Plate 5.40 *Carissa carandas*. Picture credit: Vivek Pandi

Plate 5.41 *Argyreia elliptica.* Picture credit: Dhatchanamoorthy

Plate 5.42 *Cardiospermum corindum* (***a***); *Decalepis hamiltonii* (***b***). Picture credit: Dhatchanamoorthy

Plate 5.43 *Mucuna pruriens* (***a***); *Mukia maderaspatana* (***b***). Picture credit: Dhatchanamoorthy

Plate 5.44 *Morinda umbellata (**a**); Salacia chinensis (**b**).* Picture credit: Dhatchanamoorthy

Plate 5.45 *Tetrastigma serrulatum (**a**); Gnetum ula (**b**).* Picture credit: Dhatchanamoorthy

Plate 5.46 *Adenia khondala (**a**); Miquelia dentata (**b**); Olax scandens (**c**).* Picture credit: Subbaiah Karuppusamy (**a**); Vivek Pandi (**b** & **c**)

Plate 5.47 *Tiliacora acuminata* (***a***); *Pachygone ovata* (***b***). Picture credit: Vivek Pandi

Plate 5.48 *Cayratia pedata* (***a***); *Pyrenacantha volubilis* (***b***); *Cansjera rheedi* (***c***). Picture credit: Vivek Pandi

Plate 5.49 *Rivea hypocrateriformis (**a**); Jasminum angustifolium (**b**); Trichosanthes tricuspidata (**c**); Adenia wightiana (**d**); Strychnos lenticellata (**e**).* Picture credit: Vivek Pandi

5.5 Data Availability

This data set is publicly available and accessible via:

DRYAD: Climbing flora of India.

Pandi, Vivek; Babu, Kanda Naveen (2022), Climbing Flora of India, Dryad, Dataset, https://doi.org/10.5061/dryad.d7wm37q45

This dataset contains the following data:

Climbing flora of India_v1. xlxs (Contains data in a Microsoft Excel spreadsheet of compiled list of climbing plants from India with up-to-date information on nomenclature, family, woodiness, climbing mechanism, and IUCN status).

Data are available under the terms of the Creative Commons Zero "No rights reserved" data waiver (CC0 1.0 Public domain dedication).

References

Acevedo-Rodríguez, P., Ackerman, J. D., Anderson, C., Austin, D., Biral, L., Calvo, J., Delprete, P., Dressler, S., Feuillet, C., & Liede Schumann S, et al. 2015. Lianas and climbing plants of the Neotropics. National Museum of Natural History; Accessed 2022 Mar 6. https://naturalhistory.si.edu/research/botany/research/lianas-and-climbingplants-neotropics

Anbarashan, M., & Parthasarathy, N. (2013). Diversity and ecology of lianas in tropical dry evergreen forests on the Coromandel Coast of India under various disturbance regimes. *Flora-Morphology, Distribution, Functional Ecology of Plants, 208*(1), 22–32.

Cai, Z. Q., Schnitzer, S. A., Wen, B., Chen, Y. J., & Bongers, F. (2009). Liana communities in three tropical forest types in Xishuangbanna, South-West China. *Journal of Tropical Forest Science*, 252–264.

Pandi, V., Babu, N. K., Anbarashan, M., Reddy, S. C., Borgohain, J., Shynyan, K., Mathew, A. A., Rakshith, H., Joseph, J., Kennedy, V. N., & Parthasarathy, N. (2022). Taxonomic estimates of climbing plants in India: How many species are out there? *Écoscience*, 1–19.

POWO. (2022). *Plants of the World Online*. Facilitated by the Royal Botanic Gardens.

Putz, F. E., & Chai, P. (1987). Ecological studies of lianas in Lambir national park, Sarawak, Malaysia. *The Journal of Ecology*, 523–531.

Vivek, P., & Babu, K. N. (2022). Climbing flora of India. *Dryad [Dataset]*. https://doi.org/10.5061/dryad.d7wm37q45

Climber Radiations in Angiosperms: How Many Families Have Climbers?

6

Abstract

In this chapter, I present the most recent and comprehensive list of global climber plant families. The purpose of this chapter was to look into climber representations across flowering plant families. I also intended to test Gentry's hypothesis of specialized climbing mechanisms promoting species diversification across climber families. The findings show that climbers are present in nearly half (46%) of the flowering plant families, which is significantly higher than previous estimates. The 194 plant families studied in this study had a wide range of climbing mechanisms. Gentry's hypothesis was supported by the Indian climbing flora, which included tendril climbers in four of the top five families, but not at the genus level. Using the Indian climber dataset ($n = 2624$ species; 102 families), a weak positive correlation was found between the diversity of climbing mechanisms and the species richness of climbing families ($r^2 = 0.22$; $p < 0.001$) and genus ($r^2 = 0.29$; $p < 0.001$). Therefore, I propose an alternate hypothesis linking the diversity of climbing mechanisms and species richness among the climber-representing genus and families. The data of climbing plant families presented in this chapter added significantly to the existing dataset (~20%) and will find applications as a fundamental element in climber ecology and evolution.

Keywords

Phylogenetic diversity · Climbing mechanisms · Gymnosperms · Eudicots · Monocots · Magnoliids · Diversification

V. Pandi, *Taxonomy and Ecology of Climbers: Climbing Plants of India*,
https://doi.org/10.1007/978-981-19-8645-1_6

6.1 Overview

Although climbing habits have been found in fossil Pteridosperms (385–398 Mya) (Burnham, 2009, 2015), extant Pteridophytes (Gianoli, 2015), and Gymnosperms (Gianoli, 2015; Pandi et al., 2022), their phylogenetic breadth peaked with the evolution of flowering plants (Pandi et al., 2022). Climbers and climbing mechanisms have evolved several times within the Angiosperms (Sousa-Baena et al., 2018). In fact, there are plant families that are only climbers (e.g., Convolvulaceae, Menispermaceae, and Cucurbitaceae). Gentry (1991) estimated that climbers exist in at least 131 plant families. Gianoli (2015) later updated the list with 171 Tracheophytes families (160 Angiosperms; 38.4 percent) having climbers. Climbers, on the other hand, are one of the most visible elements of tropical forest ecosystems (Pérez-Salicrup, Schnitzer, & Putz, 2004; Schnitzer & Bongers, 2002; Vivek & Parthasarathy, 2018), and they may be distributed in at least half of the 416 Angiosperm families (Chase et al., 2016). Furthermore, data on the diversity of climbing strategies across different climber families is insufficient to understand their origin and evolution. Some families, such as Fabaceae, have evolved multiple climbing strategies within closely related species, whereas Convolvulaceae have only one climbing strategy. The macro-evolutionary pattern of association between climbing mode and diversification, according to Gianoli (2015), has not received the quantitative attention it deserves. Therefore, I attempt to present an updated list of global climbing plant families and their diverse climbing mechanisms to fill these fundamental gaps in climber ecology and evolution. I also examine Gentry's (1991) hypothesis, which states that more specialized climbing mechanisms may promote species diversification among climbing clades. This study will shed light on the patterns of climber radiation and the evolution of various climbing mechanisms in Angiosperms.

6.2 Methodology

We conducted a thorough survey of research articles, Floras (printed and online), databases, books, book chapters, and research reports to compile a list of global climbing plant families and the variety of climbing mechanisms reported under each family. In search engines such as Web of Science, SCOPUS, Research Gate, Google Scholar, and web sources, we used keywords such as lianas, climbers, scramblers, tendrils, lianes, vines, climbing plants, and herbaceous climbers. We filtered over 300 references that met the above search criteria. However, the sources that provided at least the species and family details were only considered for data compilation. In addition, for the estimation of climber representations in flowering plant families, a dataset by Gallagher (2015) was used for the climbing flora of Australia; Hu, Li, and Li (2010) for China; Hu and Li (2015) for Eurasia and North Africa, Pandi et al. (2022) for India, and Acevedo et al. (Acevedo-Rodríguez et al., 2015 onwards) for the Neotropics. The initial pooled dataset consisted of a list of species and families from various sources. Following APG IV classification, we then applied the nomenclatural changes for species to avoid synonyms or duplicates, and the families were updated following Chase et al. (2016). The number of climbing mechanisms used in the studies ranged from two to nine. The number of climbing mechanisms used in the studies ranged from two to nine. All species, however, were assigned to one of six major climbing categories: twining, tendril climbing, armed scrambling, unarmed scrambling, root climbing, and hook climbing. More than one-third of the 25,000+ raw data points lacked information on climbing mechanisms. As a result, there were multiple representations for estimating the diversity of climbing mechanisms for many families but none for others. After removing duplicates and categorizing climbing strategies, I arrived at the final list of global climbing plant families with a wide range of climbing mechanisms. For data analysis and pictorial representations, I used WorldFlora (Kindt, 2020), the ggplot2 packages in R Version 4.1.2 (R Core Team, 2021), and MEGA X 11.0.10.

6.3 Results and Discussion

The extensive survey of global climbing plant families revealed that climbers are phylogenetically widespread across the Angiosperm phylogeny, from the ANA grade Austrobaileyales to the Eudicots' Asterales and Dipsacales (Table 6.1, Fig. 6.1). As predicted, nearly half of all flowering plant families (46.6%) had at least one climber species: represented in more than two-thirds of all orders, spread across 194 families (Table 6.1). This is 20% and 32% higher than Gianoli's (2015) and Gentry's (1991) previous estimates of global climber families, respectively. The number of orders and families classified as Angiosperms has increased significantly between APG I (1998) and APG IV (Chase et al., 2016), causing confounding comparisons. Our comparison with Gianoli (2015), on the other hand, is not skewed for the following reasons: (1) after the revision of families following APG IV (Chase et al., 2016), only two new climber families, Peraceae and Petiveriaceae, were added to the current list; (2) The number of climber families from Gianoli's list remained intact, with no families lost due to the merging after APG IV.

Climbers were found in more than 80% of Monocot orders and 75% of Eudicot orders of Angiosperms (Table 6.1). Climbers were found in more than half of the families classified as Malpighiales and Caryophyllales and in 90% of the families classified as Lamiales. In contrast, climbers are absent from more than 75 percent of Monocot families. Whether the loss of secondary growth during evolution in Monocots (Ragni & Greb, 2018; Spicer & Groover, 2010) has not facilitated the development of complex growth forms, such as large-bodied lianas, leaving limited options for novelties in bauplan and flexibility required for climbing? The role of novel secondary thickening cambium (Rudall, 1991) in promoting monocot climber diversification is also unknown. More research is needed to determine whether bifacial cambium and woodiness can promote species diversification among the extra monocot families. We make this claim for the following reasons: First, approximately two-thirds of the Indian (Pandi et al., 2022; also see Chapter 5) and tropical Australian climbing floras are woody (Gallagher, 2015). Second, even in the

Table 6.1 Global representation of climbers in flowering plant families along with their diversity of climbing mechanisms (TW—twiners, TC—tendril climbers, SC-A—armed-scramblers, SC-UA—unarmed-scramblers, HC—hook climbers, RC—root climbers)

Orders and family		Diversity of climbing mechanisms
Angiosperms Orders: 46/64 Families: 194/416		TW, TC, SC-A, SC-UA, HC, RC
ANA GRADE Orders: 1/3 Families: 3/7		TW, SC-UA
Austrobaileyales	Austrobaileyaceae	NA
	Schisandraceae	TW, SC-UA
	Trimeniaceae	NA
MAGNOLIIDS Orders: 3/4 Families: 8/18		TW, TC, SC-A, SC-UA, HC
Piperales	Aristolochiaceae	TW, TC
	Piperaceae	TW, RC
Laurales	Hernandiaceae	TW, SC-UA
	Lauraceae	TW, SC-A
	Monimiaceae	SC-UA
	Siparunaceae	SC-UA
Magnoliales	Annonaceae	TW, SC-UA, HC
	Myristicaceae	SC-UA
MONOCOTS Orders: 9/11 Families: 24/103		TW, TC, SC-A, SC-UA, HC, RC
Alismatales	Araceae	TW, SC-UA, RC
Dioscoreales	Dioscoreaceae	TW, TC, SC-A
Pandanales	Cyclanthaceae	SC-UA
	Pandanaceae	RC
	Stemonaceae	TW
Liliales	Alstroemeriaceae	TW, SC-UA
	Colchicaceae	TC
	Liliaceae	NA
	Petermanniaceae	NA
	Philesiaceae	NA
	Smilacaceae	TW, TC, SC-A, SC-UA, HC
	Ripogonaceae	NA
Asparagales	Amaryllidaceae	NA
	Asparagaceae	TW, SC-A, SC-UA
	Asphodelaceae	NA
	Orchidaceae	TW, SC-UA
Arecales	Arecaceae	TW, SC-A, SC-UA
Poales	Bromeliaceae	NA
	Cyperaceae	SC-UA

(continued)

Table 6.1 (continued)

Orders and family		Diversity of climbing mechanisms
	Flagellariaceae	SC-UA
	Poaceae	TW, SC-UA
	Restionaceae	NA
Commelinales	Commelinaceae	TW
Zingiberales	Marantaceae	SC-UA
EUDICOTS Orders: 33/44 Families: 159/312		TW, TC, SC-A, SC-UA, HC, RC
Ranunculales	Berberidaceae	SC-UA
	Lardizabalaceae	TW, SC-UA
	Menispermaceae	TW, TC, SC-UA
	Papaveraceae	SC-UA
	Ranunculaceae	TW, TC, SC-UA, HC
Proteales	Sabiaceae	TW, SC-UA
Dilleniales	Dilleniaceae	TW, SC-UA
Saxifragales	Aphanopetalaceae	NA
	Crassulaceae	NA
	Grossulariaceae	NA
	Hamamelidaceae	NA
	Iteaceae	NA
	Peridiscaceae	NA
	Saxifragaceae	NA
Vitales	Vitaceae	TW, TC, SC-A, RC
Zygophyllales	Zygophyllaceae	NA
Celastrales	Celastraceae	TW, TC, SC-A, SC-UA, HC, RC
Oxalidales	Connaraceae	TW, SC-UA
	Cunoniaceae	NA
	Elaeocarpaceae	NA
	Huaceae	NA
	Oxalidaceae	TW, SC-UA
Malphighiales	Achariaceae	NA
	Calophyllaceae	SC-UA
	Chrysobalanaceae	NA
	Clusiaceae	SC-UA
	Dichapetalaceae	TW, SC-UA
	Erythroxylaceae	NA
	Euphorbiaceae	TW, SC-UA
	Hypericaceae	NA
	Ixonanthaceae	NA
	Linaceae	HC
	Lophopyxidaceae	SC-UA
	Malpighiaceae	TW, SC-UA
	Ochnaceae	TW

(continued)

Table 6.1 (continued)

Orders and family		Diversity of climbing mechanisms
	Pandaceae	NA
	Passifloraceae	TW, TC
	Peraceae	TW
	Phyllanthaceae	SC-UA
	Rhizophoraceae	NA
	Salicaceae	SC-UA
	Trigoniaceae	TW
	Violaceae	TW, SC-UA
Fabales	Fabaceae	TW, TC, SC-A, SC-UA, HC
	Polygalaceae	TW, TC, SC-A, SC-UA
Rosales	Cannabaceae	TC, SC-A, SC-UA
	Elaeagnaceae	SC-A, SC-UA, HC
	Moraceae	SC-A, SC-UA, RC
	Rhamnaceae	TW, TC, SC-A, SC-UA
	Rosaceae	TC, SC-A, SC-UA, HC
	Ulmaceae	NA
	Urticaceae	TC, SC-UA
Cucurbitales	Begoniaceae	SC-UA
	Coriariaceae	SC-UA
	Cucurbitaceae	TW, TC
Geraniales	Geraniaceae	SC-UA
Myrtales	Combretaceae	TW, TC, SC-A, SC-UA, RC
	Lythraceae	SC-UA
	Melastomataceae	TW, SC-UA
	Myrtaceae	NA
	Onagraceae	SC-UA
	Vochysiaceae	NA
Crossosomatales	Stachyuraceae	SC-UA
Sapindales	Anacardiaceae	SC-UA
	Burseraceae	NA
	Meliaceae	NA
	Rutaceae	TW, SC-A, SC-UA, HC
	Sapindaceae	TW, TC, SC-A, SC-UA, HC
	Simaroubaceae	NA
Huerteales	Gerrardinaceae	NA
Malvales	Malvaceae	TW, SC-A, SC-UA
	Thymelaeaceae	TW, SC-UA
Brassicales	Brassicaceae	TW
	Capparaceae	SC-A, SC-UA, HC
	Caricaceae	SC-UA
	Pentadiplandraceae	NA
	Resedaceae	SC-UA
	Salvadoraceae	SC-A

(continued)

Table 6.1 (continued)

Orders and family		Diversity of climbing mechanisms
	Tovariaceae	NA
	Tropaeolaceae	TW, SC-UA
Beriberidopsidales	Berberidopsidaceae	SC-UA
Santalales	Loranthaceae	SC-A
	Olacaceae	TC, SC-UA
	Opiliaceae	TW, SC-A
	Santalaceae	NA
Caryophyllales	Aizoaceae	NA
	Amaranthaceae	TW, SC-UA
	Ancistrocladaceae	SC-UA, HC
	Asteropeiaceae	NA
	Barbeuiaceae	NA
	Basellaceae	TW
	Cactaceae	TW, SC-A, SC-UA
	Caryophyllaceae	TW, HC
	Dioncophyllaceae	NA
	Droseraceae	NA
	Drosophyllaceae	NA
	Nepenthaceae	TC
	Nyctaginaceae	TW, SC-A, SC-UA, HC
	Petiveriaceae	TW, SC-UA
	Phytolaccaceae	TW, SC-A
	Plumbaginaceae	TW, SC-UA
	Polygonaceae	TW, TC, SC-UA
	Portulacaceae	NA
	Stegnospermataceae	SC-UA
	Talinaceae	NA
Cornales	Cornaceae	HC
	Hydrangeaceae	TW, SC-UA, RC
	Loasaceae	TW
Ericales	Actinidiaceae	TW, SC-UA
	Balsaminaceae	NA
	Cyrillaceae	NA
	Ebenaceae	NA
	Ericaceae	SC-UA
	Lecythidaceae	NA
	Marcgraviaceae	SC-UA
	Pentaphylacaceae	NA
	Polemoniaceae	TW, TC
	Primulaceae	TW, SC-UA
	Sapotaceae	NA
	Theaceae	NA
Icacinales	Icacinaceae	TW, TC, SC-UA

(continued)

Table 6.1 (continued)

Orders and family		Diversity of climbing mechanisms
Solanales	Convolvulaceae	TW, SC-A
	Montiniaceae	NA
	Solanaceae	TW, SC-A, SC-UA
Gentianales	Apocynaceae	TW, SC-A, SC-UA, TC
	Gelsemiaceae	TW, SC-UA
	Gentianaceae	TW
	Loganiaceae	TW, TC, SC-A, SC-UA, HC
	Rubiaceae	TW, TC, SC-A, SC-UA, HC
Lamiales	Acanthaceae	TW, SC-A, SC-UA, TC
	Bignoniaceae	TW, TC, SC-A, SC-UA, RC
	Calceolariaceae	TW
	Gesneriaceae	SC-UA
	Lamiaceae	TW, SC-UA
	Lentibulariaceae	NA
	Oleaceae	TW, SC-UA
	Orobanchaceae	TW, SC-UA
	Paulowniaceae	NA
	Pedaliaceae	NA
	Plantaginaceae	TW, SC-UA
	Schlegeliaceae	SC-UA
	Scrophulariaceae	SC-UA
	Stilbaceae	NA
	Verbenaceae	TW, TC, SC-A, SC-UA
Boraginales	Boraginaceae	TW, TC, SC-A, SC-UA
Aquifoliales	Aquifoliaceae	NA
	Cardiopteridaceae	TW
	Stemonuraceae	TW
Asterales	Asteraceae	TW, SC-UA
	Campanulaceae	TW, SC-UA
	Goodeniaceae	NA
	Rousseaceae	NA
	Stylidiaceae	NA
Apiales	Apiaceae	NA
	Araliaceae	TW, SC-A, SC-UA, RC
	Griseliniaceae	SC-UA
	Pittosporaceae	NA
Dipsacales	Caprifoliaceae	TW, SC-UA
	Viburnaceae	SC-UA

absence of aggressive climber radiations, Monocots are one of the most species-rich groups (Tang et al., 2017); thus, climbing innovation is irrelevant to species diversification. Nonetheless, more mechanistic explanations are needed to understand why

Fig. 6.1 Hypothesized phylogenetic tree depicting the relationship among the global climbing plant families (194 Angiosperm families). The tree is based on the species-level phylogenetic information of 194 climber species randomly chosen from their representative families (see Table 6.1 for more information)

approximately three-fourths of the Monocot families are non-climbers when the monocots account for about 25% of all known flowering plants (Anderson & Janssen, 2009; Tang et al., 2017).

Across the angiosperm families, the diversity of climbing mechanisms varied significantly (range: 1–6). (Table 6.1). Except for those facilitated by roots, Fabaceae had evolved all climbing mechanisms (Table 6.1). On the other hand, some species-poor families displayed multiple climbing mechanisms across their species (Table 6.1). Vitaceae and Menispermaceae had a single climbing mechanism. The reasons why some families evolved different climbing strategies within closely related species while others evolved specific climbing strategies remain unknown. Climbers with varying modes of climbing show differential distribution along the

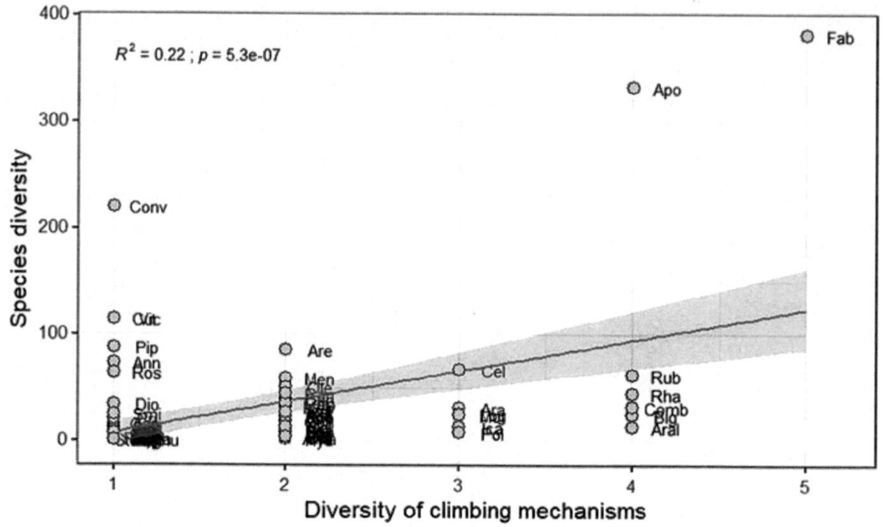

Fig. 6.2 Relationship between the diversity of climbing mechanisms and species diversity among the climbing flora of India ($n = 104$ families and 2624 species)

forest succession, allowing them to explore a variety of habitats (Gianoli, 2015; Parthasarathy et al., 2015).

Using the climbing flora of India, I tested Gentry's hypothesis that more speciose climber families have more specialized climbing mechanisms. Gentry's prediction was confirmed by the Indian climbing flora, as four of the top five climber families in India had tendril climbers, arguably the most specialized climbing mechanism (Vaughn & Bowling, 2010). Still, none of the top five genera had tendrils.

Gianoli (2015) reported a similar finding while investigating the climbing flora of Panama. I recommend more research using global datasets to confirm or refute Gentry's hypothesis because some of the most common families in the Neotropics, such as the Bignoniaceae, are underrepresented in India and vice versa. Using the same dataset, we also investigated whether the diversity of climbing mechanisms was related to species richness among climber families, but I focused on the diversity of climbing mechanisms rather than the presence/absence of tendrils. Interestingly, I found a weak positive correlation between the diversity of climbing mechanisms and species richness among the Indian climbing genera ($n = 585$; $r^2 = 0.29$; $p = 0.001$) and families ($n = 104$; $r^2 = 0.22$; $p = 0.001$) (Figs. 6.2 and 6.3).

6.4 Conclusion

Climbers are phylogenetically more diverse and represented in more families than previously estimated. More research is needed to understand better the role of bifacial cambium and the degree of woodiness in promoting species diversification

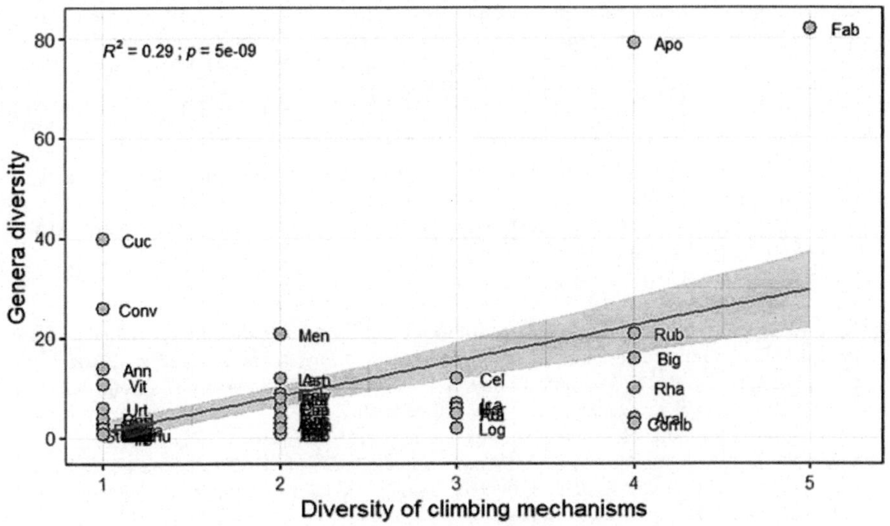

Fig. 6.3 Relationship between the diversity of climbing mechanisms and genera diversity among the climbing flora of India ($n = 102$ families and 585 genera)

among climbers. The comparisons within India's climbing flora were limited by a lack of species-level information on climbing mechanisms on a global scale. Despite this, I found a weak positive correlation between the diversity of climbing mechanisms and the richness of climber families. Gentry's claim was supported by Indian climbing flora at the family level but not at the genus level. Therefore, I propose an alternative hypothesis: *diversity in climbing mechanisms may promote species richness within climber families*. However, more data at the species level is required to strengthen the hypothesis.

References

Acevedo-Rodríguez, P., Ackerman, J. D., Anderson, C., Austin, D., Biral, L., Calvo, J., et al. (2015). Lianas and climbing plants of the Neotropics: National Museum of Natural History. Retrieved March 6, 2022, from https://naturalhistory.si.edu/research/botany/research/lianas-and-climbingplants-neotropics

Anderson, C. L., & Janssen, T. (2009). Monocots. In S. B. Hedges & S. Kumar (Eds.), *Timetree of life* (pp. 203–212). Oxford University Press.

Angiosperm Phylogeny Group. (1998). An ordinal classification for the families of flowering plants. *Annals of the Missouri Botanical Garden, 85*, 531–553.

Burnham, R. J. (2009). An overview of the fossil record of climbers: Bejucos, sogas, trepadoras, lianas, cipós, and vines. *Revista Brasileira de Paleontologia, 12*(2), 149–160.

Burnham, R. J. (2015). Climbing plants in the fossil record: Paleozoic to present. In *Ecology of lianas* (pp. 203–220). Wiley.

Chase, M. W., Christenhusz, M. J. M., Fay, M. F., Byng, J. W., Judd, W. S., Soltis, D. E., Mabberley, D. J., Sennikov, A. N., Soltis, P. S., & Stevens, P. F. (2016). An update of the

Angiosperm Phylogeny Group classification for the orders and families of flowering plants: APG IV. *Botanical Journal of the Linnean Society, 181*(1), 1–20.

Gallagher, R. V. (2015). Climbing plant diversity in Australia: Taxonomy, biogeography and functional traits. In *Ecology of lianas* (pp. 104–115). Wiley.

Gentry, A. H. (1991). The distribution and evolution of climbing plants. In *The biology of vines* (Vol. 351). Cambridge University Press.

Gianoli, E. (2015). Evolutionary implications of the climbing habit in plants. In *Ecology of lianas* (pp. 239–250). Wiley.

Hu, L., & Li, M. (2015). Diversity and distribution of climbing plants in Eurasia and North Africa. In *Biodiversity of lianas* (pp. 57–79). Springer.

Hu, L., Li, M., & Li, Z. (2010). The diversity of climbing plants in the spermatophyte flora of China. *Biodiversity Science, 18*(2), 198.

Kindt, R. (2020). WorldFlora: An R package for exact and fuzzy matching of plant names against the World Flora Online taxonomic backbone data. *Applications in Plant Sciences, 8*(9), e11388.

Pandi, V., Babu, N. K., Anbarashan, M., Reddy, S. C., Borgohain, J., Shynyan, K., Mathew, A. A., Rakshith, H., Joseph, J., Kennedy, V. N., & Parthasarathy, N. (2022). Taxonomic estimates of climbing plants in India: How many species are out there? *Ecoscience, 29*, 1–19.

Parthasarathy, N., Muthuramkumar, S., Muthumperumal, C., Vivek, P., Ayyappan, N., & Reddy, M. S. (2015). Liana composition and diversity among tropical forest types of peninsular India. In *Ecology of lianas* (pp. 36–49). Wiley.

Pérez-Salicrup, D. R., Schnitzer, S. A., & Putz, F. E. (2004). Community ecology and management of lianas. *Forest Ecology and Management, 190*, 1.

R Core Team. (2021). R: A language and environment for statistical computing. R Foundation for Statistical 752 Computing, Vienna, Austria. Retrieved from https://www.R-project.org/

Ragni, L., & Greb, T. (2018, July). Secondary growth as a determinant of plant shape and form. In *Seminars in cell & developmental biology* (Vol. 79, pp. 58–67). Academic Press.

Rudall, P. (1991). Lateral meristems and stem thickening growth in monocotyledons. *The Botanical Review, 57*(2), 150–163.

Schnitzer, S. A., & Bongers, F. (2002). The ecology of lianas and their role in forests. *Trends in Ecology & Evolution, 17*(5), 223–230.

Sousa-Baena, M. S., Sinha, N. R., Hernandes-Lopes, J., & Lohmann, L. G. (2018). Convergent evolution and the diverse ontogenetic origins of tendrils in angiosperms. *Frontiers in Plant Science, 9*, 403.

Spicer, R., & Groover, A. (2010). Evolution of development of vascular cambia and secondary growth. *New Phytologist, 186*(3), 577–592.

Tang, C. Q., Orme, C. D. L., Bunnefeld, L., Jones, F. A., Powell, S., Chase, M. W., et al. (2017). Global monocot diversification: Geography explains variation in species richness better than environment or biology. *Botanical Journal of the Linnean Society, 183*(1), 1–15.

Vaughn, K. C., & Bowling, A. J. (2010). Biology and physiology of vines. In J. Janick (Ed.), *Horticultural reviews* (Vol. 38). Wiley. https://doi.org/10.1002/9780470872376.ch1

Vivek, P., & Parthasarathy, N. (2018). Contrasting leaf-trait strategies in dominant liana and tree species of Indian tropical dry evergreen forest. *Flora, 249*, 143–149.